Geophysical Monograph 138

Inside the Subduction Factory

John Eiler
Editor

American Geophysical Union
Washington, DC

Geophysical Monograph Series

Geophysical Monograph 138

The Subduction Factory

John Eiler
Editor

American Geophysical Union
Washington, DC

Library of Congress Cataloging-in-Publication Data
Inside the subduction factory / John Eiler, editor.
p. cm. – (Geophysical monograph ; 138)
Includes bibliographical references.
ISBN 0-87590-997-3
1. Subduction zones. I. Eiler, John, 1967 – II. Series.

QE511.46.S82 2003
551.1'36—dc22

ISSN 0065-8448
ISBN 0-87590-997-3

Cover:
Low-oblique aerial photograph of erupting Fukutoku-oka-no-ba and extinct Minami Iwo Jima volcanoes in the central Izu-Bonin-Mariana arc (24°16'N, 141°29'E). The picture was taken from a airplane of the Hydrographic and Oceanographic Department of Japan on the morning of Jan. 21, 1986, looking south. Courtesy of the Hydrographic and Oceanographic Department of Japan.

CONTENTS

CONTENTS

Section IV: Synthesis

PREFACE

Subduction zones helped nucleate and grow the continents, they fertilize and lubricate the earth's interior, they are the site of most subaerial volcanism and many major earthquakes, and they yield a large fraction of the earth's precious metals. They are obvious targets for study—almost anything you learn is likely to impact important problems—yet arriving at a general understanding is notoriously difficult: Each subduction zone is distinct, differing in some important aspect from other subduction zones; fundamental aspects of their mechanics and igneous processes differ from those in other, relatively well-understood parts of the earth; and there are few direct samples of some of their most important metamorphic and metasomatic processes. As a result, even first-order features of subduction zones have generated conflict and apparent paradox. A central question about convergent margins, for instance—how vigorous magmatism can occur where plates sink and the mantle cools—has a host of mutually inconsistent answers: Early suggestions that magmatism resulted from melting subducted crust have been emphatically disproved and recently just as emphatically revived; the idea that melting is fluxed by fluid released from subducted crust is widely held but cannot explain the temperatures and volatile contents of many arc magmas; generations of kinematic and dynamic models have told us the mantle sinks at convergent margins, yet strong evidence suggests that melting there is often driven by upwelling. In contrast, our understanding of why volcanoes appear at ocean ridges and "hotspots"—although still presenting their own chestnuts—are fundamentally solved problems.

Over the last several years, the "subduction problem" has changed in response to advances in several sub-disciplines that collectively redefine, and are likely soon to solve, some of our most deeply entrenched debates. Data on the mineral physics and melting properties of hydrous silicate rocks have improved to the point that one can use them as detailed constraints on models of convection and magmatism; geophysical data have directly imaged details of subduction zones that were previously inferred; numerical models are beginning to capture key, in some cases counter-intuitive, kinematic behaviors and their dynamical forcings; geochemical data have proliferated and their general features abstracted in ways analogous to our long-standing, relatively deep understanding of volcanism at mid-ocean ridges. The emerging picture is different from what we held just a few years ago. The present volume articulates this new picture and points the way toward the next set of challenges faced by scientists interested in subduction zones.

It is common for volumes of this type to divide themselves into two types of papers: those that broadly summarize a field, much as a text book or review paper might, and those that present recent scientific results, much as an article contributed to a journal. We have done something different. The papers in this volume can be sub-divided by the region or discipline they consider (more on this below) and contain pedagogical "coverage" of the background to their subjects. At the same time, contributing authors advance new ideas that consider recent discoveries or advances in observational or experimental method while focusing on fundamental questions in subduction zone research: How does the earth's mantle move near subducting plates; what does this movement mean for the thermal evolution of subducting slabs; what, in turn, does this thermal evolution imply about the metamorphism of, and phases escaping from, the slab; and what, in detail, are the consequences (rheological, magmatic, and chemical) of delivering those escaping phases to the overlying mantle?

We have organized the volume so that the papers will build upon one another in a logical way. At the same time, the authors have not slavishly conformed to such a structure, and you will find several major themes developed in detail throughout the book. The introduction briefly summarizes the last century of scientific thought about what we now recognize as subduction zones, providing a context for readers with little background in the field. Section I initiates the discussion with a set of three papers on the subducted slab: Chapter 1 discusses the thermal structure of slabs, in particular their upper surface; Chapter 2 analyzes the geochemistry of slabs, in particular the compositions of fluids and melts that might be released from them due to heating; and Chapter 3 reviews geophysical methods for interrogating shallow subduction-zone structure, including the electromagnetic methods.

In section II, we focus on the mantle wedge—that part of the mantle between the down-going slab and the over-riding plate. The dynamics and melting behavior of the mantle wedge is arguably the centerpiece of current debates about subduction zones, so this section contains the largest number and most mechanistically detailed papers in the volume. We start, in Chapter 4, with geophysical constraints on the structure and kinematics of the mantle wedge, followed by a discussion of experimental and observational constraints on the rheology of upper mantle rocks in Chapter 5—a specialized problem but one that underlies some of the most important issues facing subduction zone research. Then, in Chapter 6, we discuss experimental constraints on the way in which the mantle wedge melts in response to the introduction of slab-derived hydrous phases and end this section with Chapter 7—a particularly forward-looking proposition for a means of geophysically mapping hydrogen concentration (arguably the controlling parameter for rheology and melting properties) in the upper mantle, including the mantle wedge.

Section III brings together a set of papers on another side of subduction zone research: the need to come to grips with historical and geographic idiosyncrasies of individual subduction zones. These papers explore the geology, geophysics, and geochemistry of three subduction zones that are the focus of much recent research, including two that are 'focus areas' identified by the Margins program of the National Science Foundation. These are the Central American arc, the Izu-Bonin-Mariana subduction zone system in the western Pacific, and the Aleutian island arc—Chapters 8, 9, and 10, respectively. It is important to note that these papers develop arguments that are certain to prompt future research on subduction zones generally and are not simply reviews of prior work.

The book concludes with two papers in section IV that advance focused arguments with broad implications for the behavior of subduction zones, based on syntheses of much of the material discussed in more detail throughout the rest of the volume. These include a proposal for the processes that control the location of subduction zone volcanoes—a problem worth re-visiting whenever our understanding of convergent margins advances—and a radical suggestion, backed by evidence from igneous and metamorphic petrology and thermal models, that subducted slabs are often, if not always, heated above their melting points. This is an idea that was deeply held thirty years ago, discarded almost entirely for most of the last twenty years, and is apparently ready to be revived for a new and more sophisticated look.

This volume derives from a conference, "The Subduction Factory," held in Eugene, Oregon in late summer of 2000. The conference was an inspiration to many of those who attended because it brought together key players in the study of subduction zones at a special time of discovery and conceptual advances in their respective fields. A majority of speakers and a few particularly active attendees were asked to contribute chapters to this volume, reporting on the key results they presented, summarizing the contextual background of their relevant sub-disciplines and, perhaps most importantly, reflecting on relationships between their advances and what they learned from others at the meeting. In the two years that have passed since this meeting, the ideas it generated have retained much of their immediacy and matured as authors reflected on one another's work.

We thank, first and foremost, the authors who contributed to this volume, both for the thought and work they put into their own contributions and for their patience with the collective effort. We are also grateful to the many reviewers who helped improve these chapters—particularly those asked to deal with the longer and more technically detailed contributions and those who had to work under a demanding deadline. Similarly, we thank Allan Graubard, our acquisitions editor, Bethany Matsko, our production editor, and the rest of the AGU book production staff, who provided much needed guidance and repeatedly demonstrated the patience of saints. Finally, we thank the organizers of the "Subduction Factory" conference and the administrators and staff in NSF's Margins program that supported both that meeting and this volume.

John M. Eiler
California Institute of Technology

Introduction

John M. Eiler

Division of Geological and Planetary Sciences, California Institute of Technology, Pasadena, California

Geological models of subduction zones impact thinking about many of the central problems in the structure, dynamics, chemistry, and history of the solid earth. Should those models change, the effects will reach across the earth sciences. We are currently in the midst of such a change, brought on by several causes. First, the earth science community recently began an organized, multi-disciplinary study of subduction zones by way of the National Science Foundation's "Margins initiative." This has improved the quality of our descriptions of key focus areas and should continue to do so for the next several years. Less purposefully but of equal importance, several of the debates in subduction zone research have evolved in recent years, making us revisit both recent and old observations in a new light. This book consists of descriptions of these advances, written by the some of the leading participants. I describe the origin and organization of the book in the preface; here I review the context of previous thought regarding convergent margins, with particular focus on the link between tectonics and magmatism, and point out questions raised in the following chapters.

ANCIENT HISTORY

For much of geology's modern era, models of subduction zones have been inseparable from "the andesite problem"; that is, the question of the origin of andesitic rocks that are abundant in many magmatic arcs. This specialized petrologic problem is disproportionately important both because andesite makes up a large fraction of subduction zone products, and because andesites compositionally resemble average continental crust (e.g., *Rudnick and Fountain,* 1995). Moreover, the literature on the "andesite problem" has been a clearing-house for ideas about the geology of convergent margins generally.

It was recognized near the beginning of the 20th century that andesitic rocks are highly concentrated into young, geologically active belts surrounding the Pacific ocean basin. These belts were collectively referred to as the "the andesite line" and interpreted as evidence for a specialized kind of magmatism that occurs only at the boundary between continental land masses and ocean basins (*Marshall,* 1911). The earliest widely remembered explanation of andesite genesis was Bowen's hypothesis that they, like many other silicic magmas, are liquids residual to crystallization-differentiation of basalt (*Bowen,* 1928). By this interpretation, andesites are ultimately products of basaltic magmatism and, to project his ideas into the plate tectonic world-view, therefore are close relatives of volcanic rocks from mid-ocean-ridges and intraplate volcanic centers. More detailed and geological interpretations of andesites and associated rocks at this time called on geosynclinal theory and seem peculiar and non-physical when looked at with the benefit of hindsight. Figure 1a is the author's attempt to synthesize several of these ideas. A noteworthy weakness of such models is that they offered no clear explanation of why andesites are common in Pacific-rim magmatic arcs while rare elsewhere.

Inside the Subduction Factory
Geophysical Monograph 138

10.1029/138GM01

PRE-PLATE-TECTONIC HUNCHES

Several discoveries over the course of the 1950s and 1960s pointed out the special character of andesites from magmatic arcs and the link between tectonics, deep seismicity, and magmatism. First, it was recognized that seismicity associated with volcanic arcs is organized into Benioff zones, suggesting that oceanic crust founders into the mantle at what we now recognize as convergent margins (*Benioff,* 1954). Moreover, experimental and applied petrology provided controversial but growing evidence that mafic magmas are derived from partial melting of peridotite, that peridotite is the major constituent of the upper mantle, that water lowers the melting points of silicates, and that crystallization-differentiation of anhydrous basalt, as envisioned by Bowen's earliest work, produces typical andesites only under special circumstances (see *Ringwood,* 1975, for a review of these early experiments). The unique compositional character (i.e., high silica content and proportion of andesite) of lavas from magmatic arcs and similarities between andesite and the continents as a whole were recognized (*Polvedaart,* 1955). Finally, there were growing suspicions that partial melting of peridotites to low degrees and in the presence of water produces unusually siliceous melts. *Polvedaart* (1955) made among the earliest such suggestion, although he did so based on the outlandish contemporary belief that the primitive crusts of the moon and mars consist almost entirely of rhyolite pumice. *O'Hara* (1965) made a similar but more well-founded suggestion based on the phase equilibria of idealized systems.

The relationships among these discoveries and their overall significance were not widely recognized until the plate tectonic revolution was underway (for instance, a standard petrologic text of this era, *Turner and Verhoogen,* 1960, does not link Benioff zones and andesitic volcanism). However, speculative inklings appeared surprisingly early. For example, *Polvedaart* (1955) suggested that andesites are partial melts of rocks in the hanging walls of Benioff-zone faults (i.e., the mantle wedge), formed in the presence of water released from foundered (i.e., subducted) ocean crust, and silicic compared to basalts because of the effect of water on the melting behavior of peridotite. A charitable interpretation of his hypothesis is illustrated in Figure 1b; similar models with sounder petrologic basis and more explicit ties to the concept of subduction would crop up as plate tectonic theory took form (*McBirney,* 1969; *Wyllie,* 1971). Perhaps the greatest long-term impact of this hypothesis was to inspire experimental studies of peridotite-water systems (e.g., *Kushiro et al.,* 1968), which eventually became a centerpiece of modern thinking about arc magmatism. However, the idea that andesites are primary partial melts of hydrous mantle, while containing elements of the modern understanding of subduction zones, conflicted with evidence that andesites are complexly evolved rocks that interact extensively with the crust through which they erupt (see *Turner and Verhoogen,* 1960, and *Ringwood,* 1975, for contemporary summaries of these criticisms). Hence the "andesite problem": the secondary origin suggested by Bowen is unable to explain why andesites are common in magmatic arcs and rare elsewhere (and gets details of the petrology and chemistry wrong); a primary origin explains the association of andesites with Benioff zones, but gets the chemistry and petrology substantially wrong.

THE LAST BIG THING

Responses to this dilemma played out during the formulation of plate tectonic theory, and therefore both built on the preceding arguments and integrated them with new insights about earth's plates. The following geophysical, geochemical, and experimental observations were particularly important: Magmatic arcs were shown to be clearly associated with convergent margins and subduction of ocean lithosphere; active convergent-margin volcanism was found to consistently occur ca. 80 to 150 km above the top of the down-going slab, suggesting control by a single pressure and/or temperature-dependent reaction in the slab; partial melting of high-pressure mafic rocks (e.g., eclogites) were shown to have compositions broadly similar to those of convergent-margin andesites; and the new tools of isotope and trace-element geochemistry demonstrated that convergent-margin magmas sample the top of the subducting plate (see *Gill,* 1981, for review of these developments).

These new constraints (and a variety of abstruse petrologic arguments) were used to suggest that magmatism at convergent margins is ultimately driven by melting of basalt in the down-going plate. Simple versions of this hypothesis equated andesites with slab melts (*Hamilton,* 1969; *Oxburgh and Turcotte,* 1970); more sophisticated versions called upon reaction between slab-derived melts and the overlying mantle wedge, upwelling and decompression melting of that hybrid mantle to produce basaltic and andesitic melts, and complex processes of differentiation, mixing and crustal contamination prior to eruption (Figure 1c; *Ringwood,* 1975). These suggestions obviously require that the top of subducted lithosphere melts, placing lower limits on its conductive and frictional heating (e.g., *Oxburgh and Turcotte,* 1970), and makes predictions about the compositions of residues of melting in deeply subducted slabs, which stir

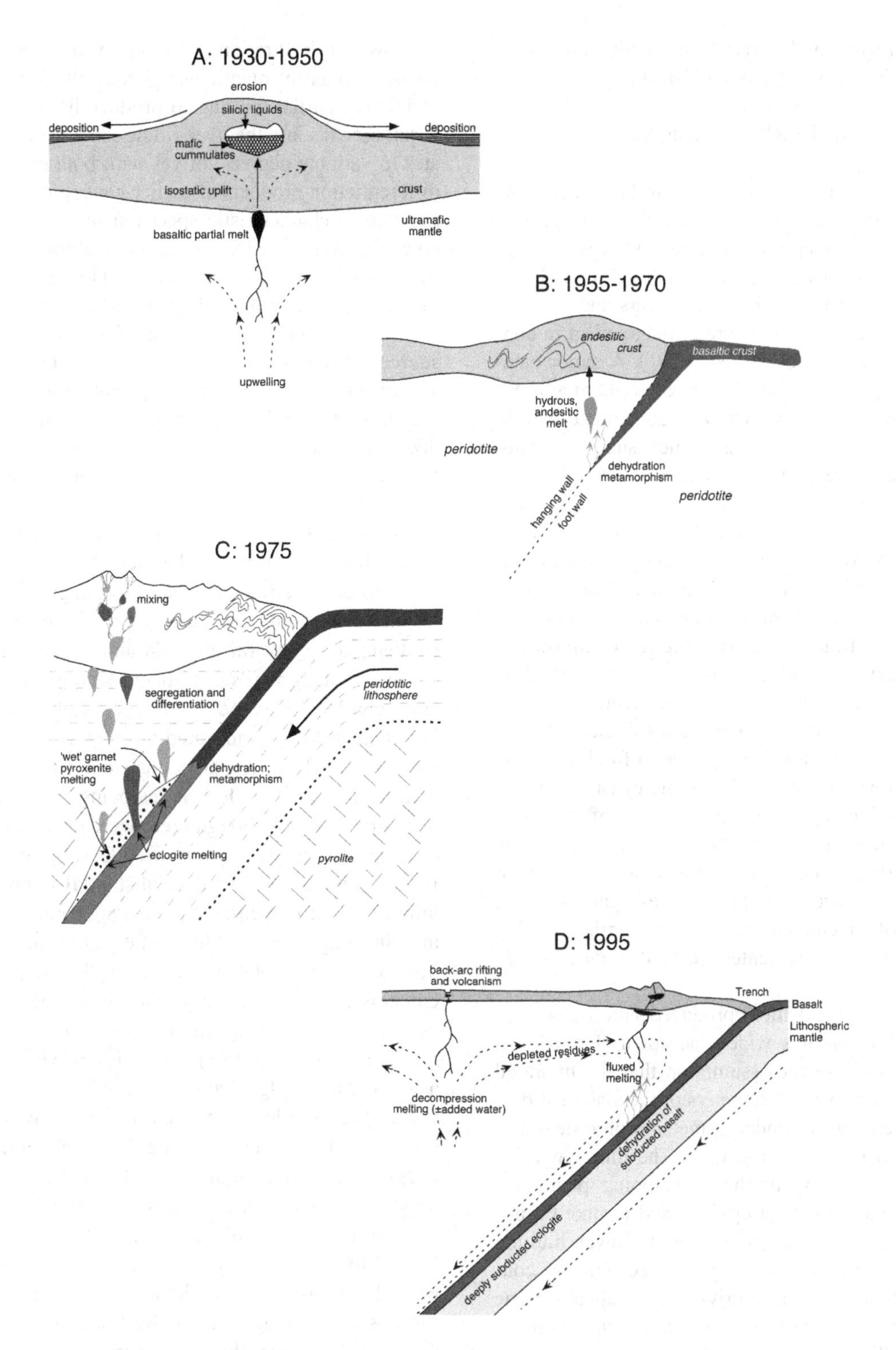

Figure 1. Schematic illustrations of conceptual models for the origin of igneous rocks at convergent margins. Panels A and B are the author's attempts to summarize ideas expressed in *Bowen* (1928), *Polvedaart* (1955), and their contemporaries. Panel C is after *Ringwood*, 1975. Panel D is the author's synthesis of ideas expressed in numerous recent studies of subduction zone magmatism and kinematics.

into the deeper mantle and provide chemically and mineralogically exotic sources of future volcanism.

THE MODERN PARADIGM

Ringwood's model (Figure 1c) and similar suggestions involving slab melting can still be found in recent papers, and broadly similar concepts are suggested in several chapters of this book. However, the "slab melting" model was widely discarded approximately twenty years ago in favor of what remains the standard picture of magmatism in convergent plate boundaries (described below). A major catalyst for this change came from kinematic and thermal models, supported by new geophysical and geological data on plate ages, positions and motions, which suggested subducted lithosphere rarely, if ever, becomes hot enough to melt (*Peacock,* 1991; *Davies and Stevenson,* 1992). A second key was the growing volume and quality of geochemical data pointing to water as the agent driving sub-arc melting (*Garcia et al.,* 1979; *Stolper and Newman,* 1994), and to the importance of geochemical fractionations between solids and aqueous fluids in determining the compositions of components extracted from subducting slabs (*Brenan et al.,* 1995). Uranium-series isotope geochemistry re-enforced this evidence, showing that fluid-soluble U is preferentially removed from the slab relative to fluid-insoluble Th, and further demonstrated a time-scale of ca. 10^4 years for transport of incompatible elements out of the slab, through the mantle wedge, and to the surface as components of lavas (e.g., *Elliott et al.,* 1997). Finally, subtle details of the geochemistry of arc lavas provided evidence that the mantle wedge often contains residues of partial melting beneath back-arc spreading centers (e.g., *Woodhead et al.,* 1993).

These and related observations produced a model for subduction zones that remains widely accepted (Figure 1d). This model begins with the assumption that the hydrated basalts in the upper several kilometers of the slab heat during subduction enough to undergo metamorphic dehydration reactions but not enough to melt. The thin veneer of pelagic sediments on top of the down-going plate also undergoes dehydration metamorphism, and (unlike basalt) might also partially melt by virtue of its lower melting point. The base of the mantle wedge is mechanically coupled to the subducting plate, driving an asthenospheric motion referred to as "corner flow" in which mantle is drawn horizontally into the wedge, dragged down along the top of the descending slab, compressed and cooled. At no point does the mantle heat or upwell beneath the arc, and thus we should expect convergent margins to be a-magmatic. However, infiltration of slab-derived aqueous fluids into the hot core of the mantle wedge radically lowers its solidus and drives partial melting to produce hydrous basalt. This basalt ascends into the over-riding plate, where it differentiates to variable degrees, mixes with both crustal melts and differentiation products of earlier generations of basalt, and produces a characteristic spectrum of intrusive and extrusive igneous rocks including abundant andesites. Subduction and corner flow can lead to diffuse upwelling in the back-arc (several hundred km behind the magmatic arc), driving seafloor spreading and basaltic volcanism. The sources of this back-arc magmatism are enriched in water, suggesting the slab-derived fluids not only pass through the mantle wedge, but also are released from the slab after it is deeply subducted or are carried out of the mantle wedge by convection of metasomatized asthenosphere. Figure 1d illustrates the essential elements of this model.

The "standard" model summarized above remains widely held because it is richly detailed and accounts for a wide range of observations. It is hard to imagine at present that its major elements are far wrong. However, it has been challenged on several grounds and appears as close to a major overhaul as it's ever been. One of the first significant challenges came from the study of *Plank and Langmuir* (1988), who noted that the major element compositions of arc lavas are consistent with the extents of melting of their mantle sources being controlled by adiabatic upwelling. Although there were previous suggestions of diapirism in the mantle wedge (Figure 1c; see also *Wyllie,* 1971), this study substantially broadened the importance of upwelling as a driving force for convergent margin magmatism. Along the same lines, there are instances of arc-related magmas with low primary water contents (i.e., they are dry, and not simply because they degassed) and/or very high liquidus temperatures (e.g., *Sisson and Bronto,* 1998), inconsistent with the picture of a cool mantle wedge that melts only because it is fluxed by water. The petrologists tell us of exotic lavas (high-magnesium andesites and "Adkites"; see Kelemen, this volume) that match many of the expected properties of slab melts and are surprisingly widely distributed—apparently heedless of common models of slab temperature. Moreover, the kinematics and mantle rheology behind the "standard" model are grossly inconsistent with the observed topography of back-arc basins (*Billen and Gurnis,* 2001). Finally, the occurrence of earthquakes at great depth and the distribution of seismic activity into "double" seismic zones have opened new possibilities for where and when slabs devolatilize (e.g., *Peacock,* 2001).

QUESTIONS

The points listed above raise new questions, the answers to which will certainly enrich our understanding of convergent margins and might make us overhaul key elements of the "standard" model. Below is a list of five questions that I've kept in mind while editing this volume. Although not all are discussed in every chapter, they crop up in most and I suggest them as themes to consider as you read:

How does the asthenosphere deform in response to subduction? This is a technical question because only specialists in dynamical modeling have the tools to answer it properly. However, its solution will deeply impact everyone working on convergent margins because asthenospheric motions control where and how the mantle melts, where fluids go once they are released from the slab, the velocities and topographies of the plates, and how hot the slab gets as it descends into the mantle. There is enough evidence to tell us mantle motions in Figure 1d are often, if not usually wrong in detail; what will the right picture look like?

Why does the mantle melt at convergent margins? It seems obvious that fluids or volatile-rich melts released from subducted slabs are involved in melting beneath arcs, yet evidence for decompression melting is widespread and in some cases it seems to sample dry sources. No model of magmatism in the "subduction factory" can be complete without explaining the causes, relative importance, and interaction of these two driving forces for mantle melting.

Do "slabs" often melt? The answer to this question was a firm and confident "no" just a few years ago, but the issue has been raised again and a forceful "yes" is given in one chapter of this book (*Kelemen et al.,* this volume). The correct answer to this question will place first-order constraints on thermal models of subducted slabs, dynamic models of convection deep in the mantle wedge, and interpretations of the origins of arc lavas.

Where and how much does the slab dehydrate? It has generally been assumed that most water released from slabs comes from hydrated basalts, that it is evolved below the arc volcanoes, and that it leaves relatively dry residual rocks. In contrast, metamorphic and experimental petrology tell us that slabs dehydrate throughout their first ca. 200 km of subduction (*Schmidt and Poli,* 1998), seismicity suggests dehydration reactions might occur deep within the ultramafic portions of the slab (*Peacock,* 2001), dynamic models permit fluids to move large horizontal distances before reaching a zone of partial melting in the mantle wedge (*Davies and Stevenson,* 1992; M. Spiegelman, pers. com.), and experimentalists have produced exotic phases that might host water retained in deeply subducted slabs (*Schmidt and Poli,* 1998 and references therein). These new arguments must somehow be integrated with our model for subduction zone processes; in particular, it must be shown how they lead to relatively simple distributions of volcanoes in magmatic arcs.

Are ocean arcs the seeds of the continents? It is often noted that andesites from magmatic arcs resemble the average composition of the continents. However, oceanic arcs generally contain abundant basalts in addition to these andesites and have bulk compositions that are not appropriate continental building blocks. Several resolutions to this dilemma have been suggested; a final answer is likely to come from more complete and detailed understanding of the crustal structure and geological evolution of a few representative magmatic arcs and their derived sediments. This volume's chapters on focus regions provide examples of the cutting edge of that understanding.

REFERENCES

Benioff, H. (1954) Orogenesis and deep crustal structure: additional evidence from seismology. *Geol. Soc. Am. Bull., 65,* 385-400.

Billen, M.I. and M. Gurnis (2001) A low viscosity wedge in subduction zones. *Earth Planet. Sci. Lett., 193,* 227-236.

Bowen, N.L. (1928) The evolution of igneous rocks. Princeton University Press, Princeton, NJ, 332 pp.

Brenan, J.M., H.F. Shaw, and F.J. Ryerson (1995) Experimental evidence for the origin of lead enrichment in convergent-margin magmas. *Nature, 378,* 54-56.

Davies, J.H. and D.J. Stevenson (1992) Physical model of source region of subduction zone volcanics. *Jour. Geophys. Res., 97,* 2037-2070.

Elliott, T., T. Plank, A. Zindler, W. White, and B. Bourdon (1997) Element transport from slab to volcanic front at the Mariana arc. *Jour. Geophys. Res., 102,* 14991-15019.

Garcia, M.O., N.W.K. Liu, and D.W. Muenow (1979) Volatiles in submarine volcanic rocks from the Mariana island arc and trough. *Geochim. Cosmochim. Acta, 43,* 305-312.

Gill, J. (1981) Orogenic andesites and plate tectonics. Springer-Verlag, New York. 390 pp.

Hamilton, W. (1969) Mesozoic California and the underflow of Pacific mantle. *Geol. Soc. Amer. Bull., 80,* 2409.

Kushiro, I., H.S. Yoder, and M. Nishikawa (1968) Effect of water on the melting of enstatite. *Bull. Geol. Soc. Am., 79,* 1685-1692.

Marshall, P. (1911) Oceania: Handbook of Regional Geology, v. 7, no. 2, 36 pp.

McBirney, A.R. (1969) Compositional variations in Cenozoic calc-alkaline suites of Central America. *Oreg. Dep. Geol. Minera. Ind. Bull., 65,* 185.

O'Hara (1965) Primary magmas and the origin of basalts. Scottish *J. Geol., 1,* 19-40.

Oxburgh, E.R. and D.L. Turcotte (1970) Thermal structure of island arcs. *Geol. Soc. Amer. Bull., 81,* 1665.

Peacock, S.M. (1991) Numerical simulation of subduction zone pressure temperature time paths—constraints on fluid production and arc magmatism. *Phil. Trans. Roy. Soc. Lond. Series A, 335,* 341-353.

Peacock, S.M. (2001) Are the lower planes of double seismic zones caused by serpentine dehydration in subducting oceanic mantle? *Geology, 29,* 299-302.

Plank, T. and C.H. Langmuir (1988) An evaluation of the global variations in the major element chemistry of arc basalts. *Earth Planet. Sci. Lett., 90,* 349-370.

Polvedaart, A. (1955) Chemistry of the earth's crust. *Geol. Soc. Am. Special Paper 62,* p. 119-144.

Ringwood, A.E. (1975) Composition and Petrology of the Earth's Mantle. McGraw-Hill, USA, 617 pp.

Rudnick, R.L. and D.M. Fountain (1995) Nature and composition of the continental crust—a lower crustal perspective. *Rev. Geophys., 33,* 267-309.

Schmidt, M.W. and S. Poli (1998) Experimentally based water budgets for dehydrating slabs and consequences for arc magma generation. *Earth Planet. Sci. Lett., 163,* 361-379.

Sisson, T.W. and S. Bronto (1998) Evidence for pressure-release melting beneath magmatic arcs from basalt at Galunggung, Indonesia. *Nature, 391,* 883-886.

Stolper, E.M. and S. Newman (1994) The role of water in the petrogenesis of Mariana trough magmas. *Earth Planet. Sci. Lett., 121,* 293-325.

Turner, F.J. and J. Verhoogen (1960) Igneous and metamorphic petrology. McGraw-Hill, New York.

Woodhead, J., S. Eggins, and J. Gamble (1993) High-field-strength and transition element systematics in island-arc and back-arc basin basalts—evidence for multiphase melt extraction and a depleted mantle wedge. *Earth Planet. Sci. Lett., 114,* 491-504.

Wyllie, P.J. (1971) Role of water in magma generation and initiation of diapiric uprise in the mantle. *Jour. Geophys. Res., 76,* p. 1328-1338.

John M. Eiler, California Institute of Technology, Geological Planetary Sciences, 1200 East California Boulevard, MC 170-25, Pasadena, CA 91125

Thermal Structure and Metamorphic Evolution of Subducting Slabs

Simon M. Peacock

Department of Geological Sciences, Arizona State University, Tempe, Arizona

Variations in subduction-zone seismicity, seismic velocity, and arc magmatism reflect differences in the thermal structure and metamorphic reactions occurring in subducting oceanic lithosphere. Current kinematic and dynamical models of subduction zones predict cool slab-mantle interface temperatures less than one-half of the initial mantle temperature. Weak rocks along the slab-mantle interface likely limit the rate of shear heating; surface heat flux measurements and other observations suggest interface shear stresses are 0 - 40 MPa, consistent with this expectation. Thermal models of the NE Japan, Izu-Bonin, and Aleutian subduction zones predict slab-mantle interface temperatures of ~500 °C beneath the volcanic front. In such cool subduction zones, subducting oceanic crust transforms to eclogite at depths > 100 km and temperatures are too low to permit partial melting of subducted sediments or crust. In the Nankai subduction zone, where the incoming Philippine Sea Plate is unusually warm, predicted interface temperatures beneath sparse Holocene volcanoes are ~800 °C and eclogite transformation, slab dehydration reactions, and intermediate-depth seismicity occur at < 60 km depth. The geometry and vigor of mantle-wedge convection remains considerably uncertain; models incorporating strongly temperature-dependent mantle viscosity predict significantly higher slab-mantle interface temperatures.

INTRODUCTION

Subducting lithospheric plates are the cool, downwelling limbs of mantle convection and the negative buoyancy of subducting slabs (slab pull) drives plate tectonics [*Forsyth and Uyeda,* 1975]. Subduction zones are regions of intense earthquake activity, explosive volcanism, and complex mass transfer between the crust, mantle, hydrosphere, and atmosphere. In this contribution, I present subduction-zone thermal models that provide a framework for discussing the petrological and seismological processes that occur in subducting slabs (defined herein as the subducting sediments, oceanic crust, and oceanic mantle). Specific issues to be discussed include uncertainties regarding mantle-wedge convection, metamorphic reactions in the subducting plate, the origin of arc magmas, and subduction-zone earthquakes.

Inside the Subduction Factory
Geophysical Monograph 138

10.1029/138GM02

GENERAL OBSERVATIONS REGARDING SUBDUCTION ZONES

Subducting slabs are cool because oceanic lithosphere, the cold upper boundary layer of Earth's internal convection, descends into the mantle more rapidly than heat conduction warms the slab. The chilling effect of subduction is recorded by surface heat flux measurements < 0.03 W m^{-2} in subduction-zone forearcs (one-half of the average global surface heat flux). In well-studied subduction zones like Cascadia, forearc heat flux systematically decreases from the trench to the volcanic front [*Hyndman and Wang,* 1995].

Cold subducting slabs are well resolved as high-velocity regions in seismic tomography studies [e.g., *Zhao et al.,* 1994]. Low-temperature, high-pressure metamorphic rocks (blueschists, eclogites) provide an important record of the unusually cool temperatures at depth in subduction zones [e.g., *Carswell,* 1990; *Peacock,* 1992; *Hacker,* 1996].

Despite subducting slabs being cool compared to the surrounding mantle, almost all subduction zones are distinguished by active arc volcanism, which requires that rocks melt somewhere in the subduction zone system. Early thermal models of subduction zones assumed a priori that arc magmas were derived from direct melts of the subducting slab and these models incorporated high rates of shear heating along the slab-mantle interface in order to supply the required heat [e.g., *Oxburgh and Turcotte,* 1970; *Turcotte and Schubert,* 1973]. Over time, this view has evolved and most arc magmas are now thought to represent partial melts of the mantle wedge induced by infiltration of aqueous fluids derived from the subducting slab [e.g., *Gill,* 1981; *Hawkesworth et al.,* 1993]. Current thermal models of subduction zones call upon lower rates of shear heating and predict that slab melting only occurs in unusually warm subduction zones characterized by young incoming lithosphere and slow convergence [e.g., *Peacock et al.,* 1994]. The complex origin of arc magmas, however, remains an area of active research and debate.

THERMAL STRUCTURE OF SUBDUCTION ZONES

The thermal structure of subduction zones has been investigated using analytical [e.g., *Molnar and England,* 1990] and numerical techniques [e.g., *Toksöz et al.,* 1971; *Peacock,* 1990a; *Davies and Stevenson,* 1992; *Peacock et al.,* 1994; *Kincaid and Sacks,* 1997]. These studies have identified a number of important parameters that control the thermal structure of a subduction zone (Figure 1) including: (1) convergence rate, (2) thermal structure of the incoming lithosphere, which is primarily a function of lithospheric age but is also affected by hydrothermal cooling and the thickness of insulating sediments, (3) geometry of the subducting slab, (4) rate of shear heating (= shear stress x convergence rate), and (5) vigor and geometry of flow in the mantle wedge [see review by *Peacock,* 1996]. The first three parameters are relatively well constrained whereas the rate of shear heating and mantle wedge flow are considerably uncertain.

Calculated slab temperatures decrease with increasing convergence rate and increasing age of the incoming lithosphere. Most western Pacific subduction zones, such as the Kamchatka-Kurile-Honshu and Izu-Bonin-Mariana systems, are characterized by rapid convergence of old, cool lithosphere; subducted slabs in these subduction zones are relatively cool. In contrast, the young incoming lithosphere and modest convergence rates of subduction zones such as Nankai and Cascadia lead to relatively warm subducted slabs.

At shallow depths (<50 km), temperatures along the slab-mantle interface during the earliest stages of underthrusting are predicted to equal the average of the surface temperature (T_s) and the initial (pre-subduction) mantle temperature at the depth of interest (T_i) [*Molnar and England,* 1990]. Continued underthrusting removes heat from the upper plate and interface temperatures decrease to less than $0.5\ (T_s + T_i)$. High rates of shear heating increase interface temperatures, but the low surface heat flux observed in forearcs requires that advective cooling, and not shear heating, controls the shallow thermal structure of subduction zones. In general, there is good agreement among the different thermal models presented in the literature and much of the apparent variation in published thermal structures results from different rates of shear heating. Recent studies, based on surface heat flow measurements and other data, suggest shear stresses in subduction zones are of order 10 MPa and range from 0 to 40 MPa (Table 1). Shear heating, while an important heat source, is not the primary control on temperatures in the subducting slab.

At depths greater than ~50 km, convection in the overlying mantle wedge strongly influences slab temperatures. Induced mantle-wedge convection warms the subducting slab and a cool boundary layer forms in the mantle wedge adjacent to the slab [e.g., *McKenzie,* 1969]. Mantle-wedge

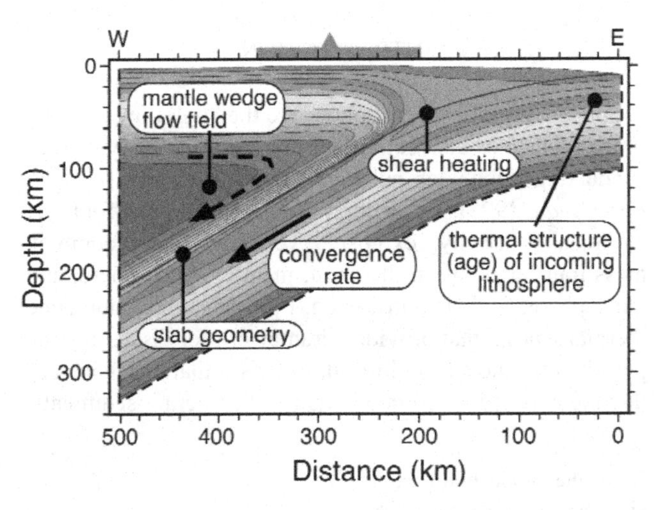

Figure 1. Important parameters which govern the thermal structure of a subduction zone.

convection increases slab-mantle interface temperatures by ~200-250 °C based on comparing *Molnar and England's* [1990] analytical expressions with the results of numerical models incorporating wedge convection (Figure 2). However, models with mantle-wedge convection still predict slab-mantle interface temperatures less than 0.5 (T_s + T_i) (Figure 2) [e.g., *Davies and Stevenson,* 1992; *Furukawa,* 1993; *Peacock et al.,* 1994; *Peacock,* 1996; *Kincaid and Sacks,* 1997].

THERMAL-PETROLOGIC MODELS OF COOL AND WARM SUBDUCTION ZONES

Recently, we constructed a set of two-dimensional, finite-element, thermal models of four subduction zones—NE Japan, Izu-Bonin, the Aleutians, and Nankai—in order to test thermal models against seismological and magmatic observations (Plate 1) [*Peacock and Wang,* 1999; *Peacock and Hyndman,* 1999; this study]. These models solve the steady-state heat transfer equation including terms for heat conduction, advection, and heat sources; in the case of Nankai we used a transient solution in order to account for the subduction of the young Shikoku basin. For each subduction zone, the geometry of the subducting slab is defined using seismic reflection and refraction studies at shallow levels and Wadati-Benioff zone seismicity at deeper levels. Our models include two heat sources: radiogenic heat production in the upper-plate crust and shear heating along the plate boundary from the trench to 70 km depth. We neglect

Table 1. Recent estimates of subduction zone shear stresses.

Subduction zone	Shear stress (MPa)	Reference
(1) Match of thermal models to surface heat flow		
Continental	14 - 27	*Tichelaar and Ruff* [1993]
Cascadia	~0	*Hyndman and Wang* 1995]
Nankai	~0	*Peacock and Wang* [1999]
NE Japan	10	*Peacock and Wang* [1999]
Kermadec	40 ± 17	*von Herzen et al.* [2001]
(2) Blueschists (high P – low T conditions)		
Franciscan	< 20 - 30	*Peacock* [1992]
Mariana	18 ± 8	*Peacock* [1996]
(3) Dynamical modeling of trench topography		
Oceanic	15 - 30	*Zhong and Gurnis* [1994]
(4) Upper plate stress field		
Cascadia	< 10	*Wang et al.* [1995]

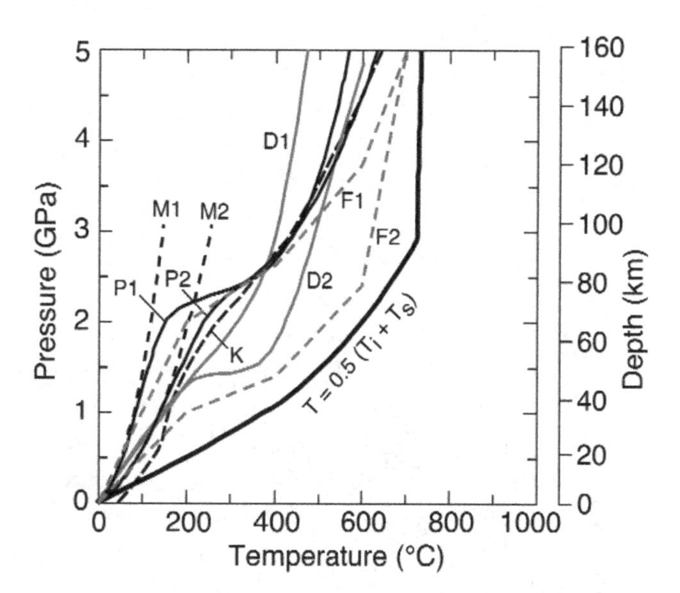

Figure 2. Predicted pressure-temperature conditions along the slab-mantle interface based on published kinematic and dynamical models. In each case predicted interface temperatures are less than one-half the sum of the surface temperature (T_s) and the initial, pre-subduction mantle temperature at depth (T_i). M1, M2; *Molnar and England's* [1990] analytical expressions for convergence rate, V = 100 and 30 mm yr^{-1}, respectively. D1, D2; *Davies and Stevenson's* [1992] numerical solutions for V = 72 mm yr^{-1} and 60° and 30°, respectively. F1, F2; *Furukawa's* [1993] numerical solutions for V = 100 mm yr^{-1} and a slab-wedge coupling depth of 100 and 40 km, respectively. P1, P2; *Peacock et al.'s* [1994] numerical solutions for V = 100 and 30 mm yr^{-1}, respectively. K, *Kincaid and Sack's* [1997] numerical solution for V = 100 mm yr^{-1}.

heat transported by fluids and heat consumed (released) by endothermic (exothermic) reactions; both fluid advection and metamorphic reaction enthalpies are proportional to the amount of H_2O involved, which in subduction zones is very limited at depths > 10 km [*Peacock,* 1987; *Peacock,* 1990b]. The thermal structure of the incoming plate is fixed using an oceanic geotherm of appropriate age [*Stein and Stein,* 1992]. The arc-side boundary is defined by either a continental geotherm (surface heat flux = 0.065 W m^{-2}) or a 20 Ma oceanic geotherm in the case of the Izu-Bonin model. The surface temperature is fixed at 0 °C and the temperature at the base of the 95-km-thick subducting plate is fixed at 1450 °C [*Stein and Stein,* 1992]. Where material flows out of the model grid, no horizontal conductive heat flow is permitted.

These thermal models are "kinematic" in the sense that the slab geometry and convergence rate are model inputs, but we use a dynamical model for flow in the mantle wedge. In a pure "dynamical" model, the slab geometry and subduction

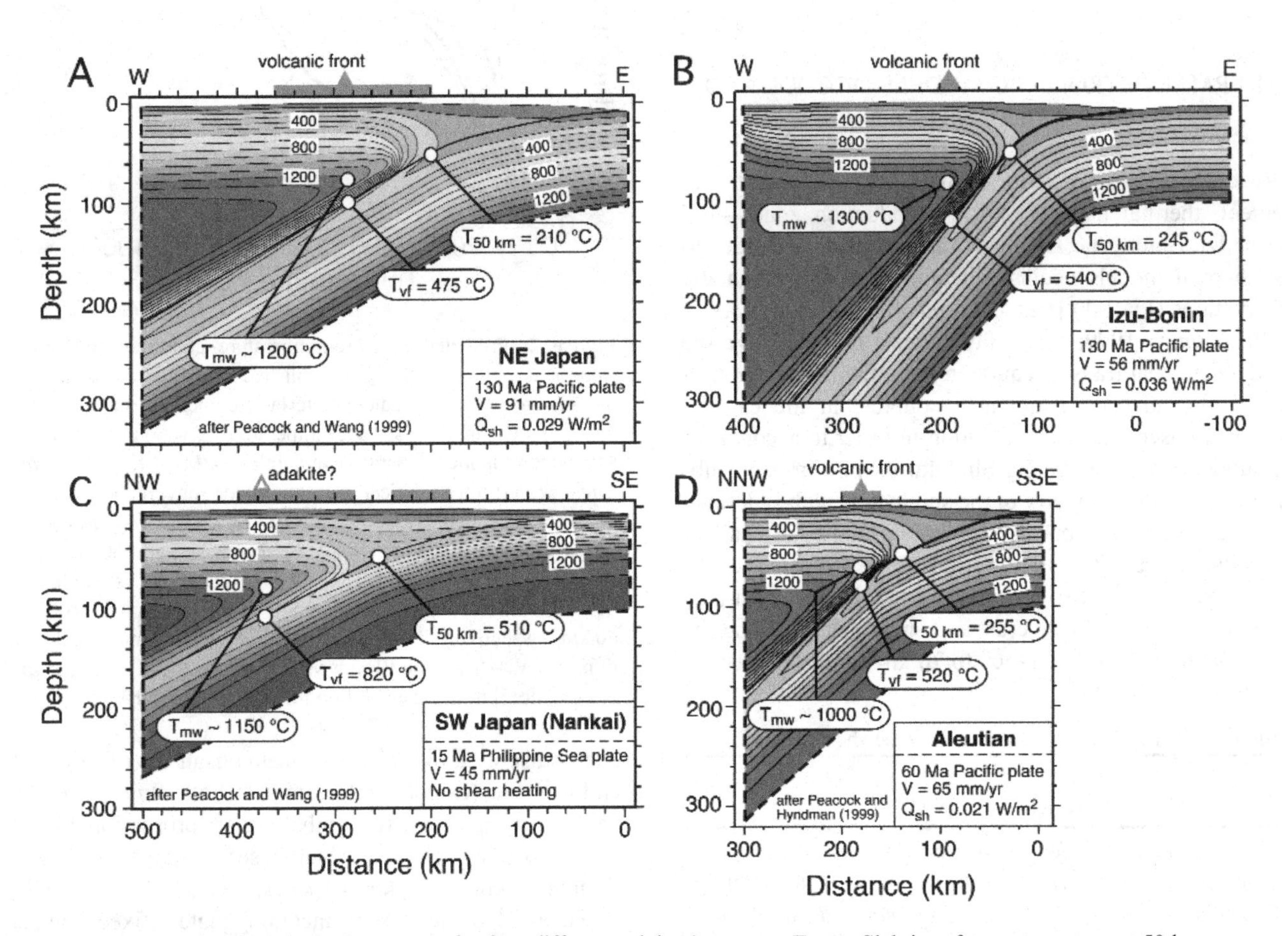

Plate 1. Calculated thermal structures for four different subduction zones. T_{50km}, Slab interface temperature at 50 km depth; T_{vf}, slab interface temperature directly beneath volcanic front; T_{mw}, maximum temperature in the mantle wedge directly beneath volcanic front. (A) NE Japan (Honshu) after *Peacock and Wang* [1999]. (B) SW Japan (Nankai) after *Peacock and Wang* [1999]. (C) Izu-Bonin (32 °N). (D) Aleutian Islands (Umnak) after *Peacock and Hyndman* [1999].

rate would be calculated based on internal and external forces. Velocities in the subducting slab are set equal to the orthogonal convergence rates determined using the NUVEL-1A plate motion model [*DeMets et al.,* 1994]. For the mantle wedge, we use a simple dynamical flow model of Newtonian viscous corner flow [*Batchelor,* 1967] driven by a no-slip boundary condition along the top of the subducting plate. Mantle-wedge flow occurs beneath a 50-km-thick rigid lithosphere and is truncated in the tip of the mantle wedge to satisfy surface heat flux data [*Peacock and Wang,* 1999].

Thermal models constructed for three cool subduction zones (NE Japan, Izu-Bonin, the Aleutians) and one relatively warm subduction zone (Nankai) are depicted in Plate 1. Calculated slab-mantle interface temperatures in the Nankai subduction zone are ~300 °C warmer than interface temperatures in the NE Japan, Izu-Bonin, and Aleutian subduction zones (Figure 3B). The high slab temperatures in Nankai reflect the hot incoming Philippine Sea Plate [*Wang et al.,* 1995; *Peacock and Wang,* 1999]. At 50 km depth, the calculated temperatures along the slab-mantle interface range from 210 to 255 °C for the three cool subduction zones as compared to 510 °C for the warmer Nankai subduction zone. Beneath the volcanic front the slab-mantle interface temperatures for the three cool subduction zones are similar, ranging from 475 to 540 °C. In contrast, beneath the sparse Holocene volcanoes in the Nankai subduction zone, the calculated slab-mantle interface temperature is 820 °C. Note the apparent paradox—warmer slab interface temperatures correlate with less productive volcanic arcs.

The key to resolving this paradox lies in the recognition that most arc magmas are generated not by slab melting, but by slab-derived aqueous fluids that infiltrate the hot core of the overlying mantle wedge [e.g., *Gill,* 1981]. In subduction zones with cool slabs, hydrous minerals are stable to depths > 100 km and H_2O is released beneath hot (> 1300 °C) mantle wedge capable of undergoing H_2O-flux melting. In subduction zones with warm slabs, hydrous minerals breakdown at depths < 100 km and H_2O infiltrates forearc mantle that is too cool to undergo partial melting. In the extreme, several subduction zones that lack volcanic arcs are characterized by very warm slabs as a result of flat slab geometry or subducting ridges.

Qualitatively, the calculated thermal structure of the mantle wedge is similar in all four subduction zones which is not surprising given the simple viscous corner-flow model. Heat transfer in the mantle wedge occurs primarily by advection such that isotherms in the mantle wedge closely follow material flow lines. Basaltic lavas present in many arcs require mantle temperatures greater than 1300 °C [e.g., *Tatsumi et al.,* 1983]. In all models the nose of the 1300 °C isotherm occurs at 80-95 km depth where the top of the underlying slab lies at 115-130 km depth. There is considerable variation among the four subduction zones in the depth to the subducting slab beneath the volcanic front. In the Izu-Bonin subduction zone, the depth to the slab beneath the volcanic front is 120 km as compared to only 80 km in the Aleutian subduction zone. This variation results in maximum mantle-wedge temperatures directly beneath the volcanic front ranging from ~1300 °C in the Izu-Bonin subduction zone to ~1000 °C in the Aleutian subduction zone. The uncertainties in mantle-wedge flow, discussed below, lead to considerable uncertainties in the calculated thermal structure of the mantle wedge beneath the volcanic arc. For example, higher wedge temperatures beneath the Aleutian arc would be generated by models assuming a temperature-dependent mantle viscosity and shallower viscous coupling depth.

CONVECTION IN THE MANTLE WEDGE

The geometry and vigor of mantle-wedge convection are poorly known and represent a major uncertainty in thermal models of subduction zones. Subducting slabs induce convection in the mantle wedge through mechanical and thermal coupling. Mechanical coupling (viscous traction) between the subducting slab and the base of the mantle wedge drives corner flow [*McKenzie,* 1969], a type of forced convection. The subducting slab also cools the base of the mantle wedge driving thermal convection (free convection); the cooler, denser base of the mantle wedge tends to sink together with the subducting slab [e.g., *Rabinowicz et al.,* 1980]. Thermal convection becomes increasingly important with decreasing viscosity. Hydration and partial melting may also be significant sources of buoyancy in parts of the mantle wedge.

In forced convection models, such as the models presented above, flow in the mantle wedge is driven solely by mechanical coupling to the subducting slab. For the case of constant wedge viscosity, the resulting corner-flow velocity field may be calculated analytically [*Batchelor,* 1967]. In constant viscosity corner flow, material flows into the mantle wedge along subhorizontal flow lines parallel to the base of the overriding lithosphere. Toward the wedge corner, the flow lines bend downward and become subparallel to the top of the subducting slab. In these models, mantle wedge flow is driven by the boundary conditions. The extent to which hot mantle flows into the forearc region depends on the thickness of the overlying "rigid" lithosphere [*Rowland and Davies,* 1999] and the depth at which slab-mantle coupling begins [*Furukawa,* 1993].

Forced convection mantle-wedge flow fields calculated using a *T*-dependent viscosity are qualitatively similar to constant viscosity flow fields because the velocity boundary condition along the top of the slab controls the overall flow structure [*Davies and Stevenson,* 1992]. Depending primarily on the model boundary conditions, temperature-dependent viscosity flow fields may closely resemble constant-viscosity corner flow fields [e.g., *Davies and Stevenson,* 1992] or may exhibit a significant upward component of motion beneath the overriding lithosphere [e.g., *Furukawa,* 1993]. The component of upward motion depends, in part, on the depth of the high-temperature isotherms on the arc-side boundary relative to the depth at which slab-wedge coupling begins [*Furukawa,* 1993; *Rowland and Davies,* 1999]. In *Furukawa's* [1993] models the upward component of wedge flow increases as the depth at which slab coupling begins decreases. Decreasing the depth of slab-wedge coupling from 100 km to 40 km increases slab surface temperatures by as much as 300 °C at 65 km depth (Figure 2) [*Furukawa,* 1993]. At present, the depth at which full slab-wedge coupling begins is poorly constrained, but shallow coupling depths < 70 km result in high-temperature mantle flow beneath the forearc, and therefore can be ruled out by the observed low heat flow in forearcs. In forced convection models, the subducting slab drags the base of the mantle wedge downward. In models with strong temperature-dependent viscosity, the down-dragged wedge material is replaced by hot mantle and calculated interface temperatures are several hundred degrees warmer than constant viscosity models [*Kiefer et al.,* 2001].

More realistic models of the mantle wedge flow require dynamical calculations incorporating buoyancy forces (free convection) [e.g., *Davies and Stevenson,* 1992; *Kincaid and Sacks,* 1997]. Local buoyancy forces generated by partial melting can modify the slab-induced flow field in the mantle wedge [Davies and Stevenson, 1992]. If the viscosity of the mantle wedge is low ($<6 \times 10^{18}$ Pa s), then the buoyancy generated by partial melting could lead to appreciable upward flow or flow reversal [*Davies and Stevenson,* 1992]. Calculations by *Iwamori* [1997], using a mantle viscosity of 10^{21} Pa s, generated thermal buoyancy driven flow approximately $1/3$ as vigorous as mechanically driven flow. This result depends strongly on the mantle viscosity structure; temperature-dependent mantle viscosities will result in more vigorous flow that should elevate temperatures in the core of the mantle wedge. *Kincaid and Sack's* [1997] dynamical models showed that the material in the mantle-wedge corner cools and stagnates with time, which is consistent with the low surface heat flow observed in forearcs.

Similar slab-mantle interface temperatures are predicted by thermal models incorporating forced convection (viscous traction) and combined forced and free convection (viscous traction and thermal buoyancy) (Figure 2). In all models a cool boundary layer forms along the base of the mantle wedge and heat must be conducted across this layer in order to reach the slab. For a wide range of subduction parameters, calculated slab-mantle interface temperatures are less than $0.5\,(T_s + T_i)$ (Figure 2). However, numerical modeling efforts have yet to explore the possibility of rapid mantle-wedge convection where wedge velocities exceed slab velocities.

Rapid mantle-wedge convection could occur if mantle viscosities are lower than previously considered and a growing body of evidence suggests this may be the case. Laboratory measurements indicate that olivine is dramatically weaker in the presence of water [*Karato and Wu,* 1993]; such conditions would likely exist in the mantle wedge above the dehydrating slab. In the northern Cascadia subduction zone, post-seismic uplift data are consistent with viscoelastic deformation models with a mantle-wedge viscosity of 10^{18} to 10^{19} Pa s [*Wang et al.,* 1994] and post-glacial rebound data indicate a mantle wedge viscosity of 5×10^{18} to 5×10^{19} Pa s [*James et al.,* 2000]. Dynamical models of the Tonga-Kermadec subduction zone by *Billen and Gurnis* [in press] require a low viscosity mantle wedge (a factor of 10 less viscous than the surrounding asthenosphere) in order to decouple the slab from the overriding plate and provide a better match to topographic, gravity, and geoid observations.

The complex rheological structure of the mantle wedge remains a formidable challenge to determining the geometry and vigor of mantle-wedge flow. The viscosity of mantle-wedge peridotite can vary over many orders of magnitude due to variations in temperature (T), pressure (P), strain rate, bulk composition, the amount of hydration, and the local presence of aqueous fluids and silicate melts. Relatively weak rock types (metasediments, metamorphosed oceanic crust, serpentinite) present along the slab-mantle interface will likely control the degree of viscous coupling between the subducting slab and overlying mantle wedge [*Yuen et al.,* 1978]. Current modeling efforts by several different groups are systematically examining these complexities, but the key to constraining the geometry and vigor of mantle-wedge flow may well lie in the seismological and arc geochemical observations.

METAMORPHIC EVOLUTION OF SUBDUCTING SLABS

During subduction, sediments, oceanic crust, and oceanic mantle undergo metamorphic transformations that increase the density of the subducting slab. Many of these metamorphic reactions involve the breakdown of hydrous minerals and release substantial amounts of H_2O [e.g., *Poli and Schmidt,* 1995; *Schmidt and Poli,* 1998]. Most of the H_2O liberated from subducting slabs at depths greater than 10 km is derived from variably hydrated basalts and gabbros in the

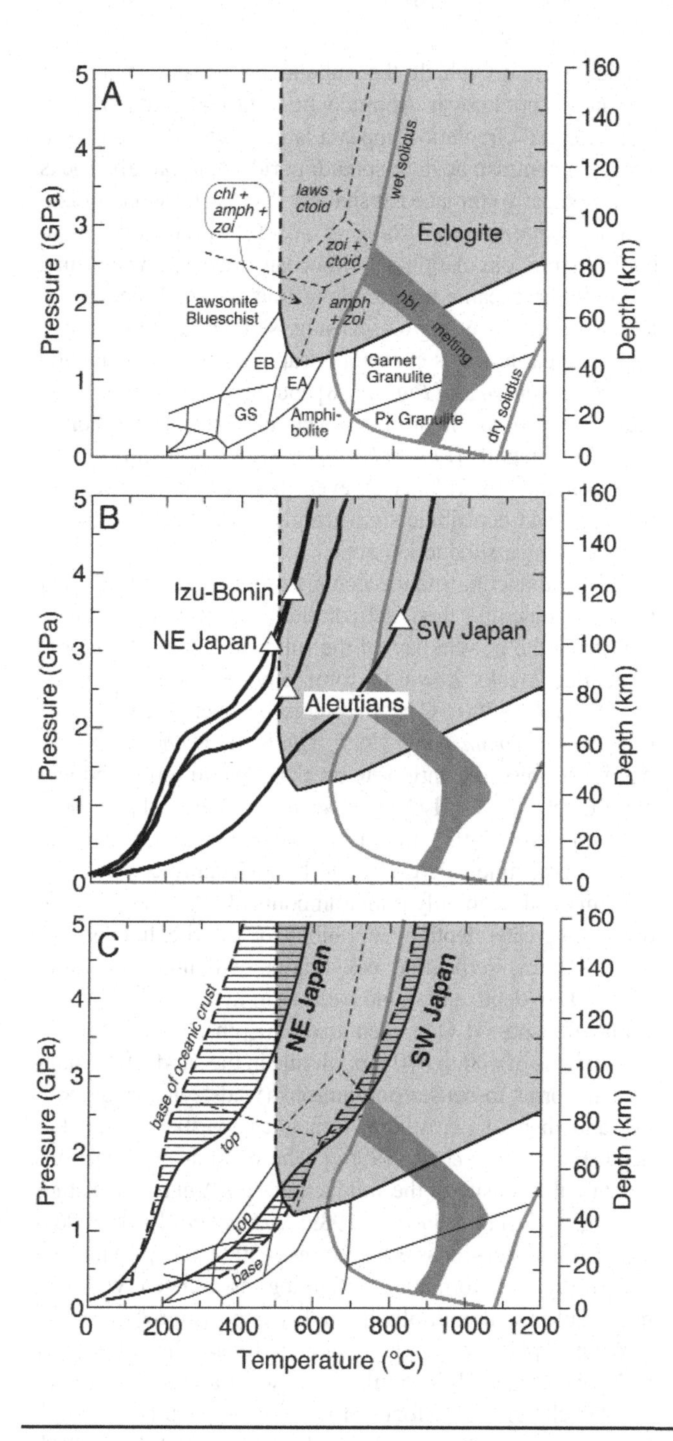

subducting oceanic crust [e.g., *Peacock,* 1990a]. Globally, 1 to 2 x 10^{12} kg of bound H_2O is subducted each year with hydrous minerals in the altered oceanic crust accounting for ~90-95% of this flux [*Ito et al.,* 1983; *Peacock,* 1990a; *Bebout,* 1996].

The volume and composition of pelagic and terrigenous sediments subducted in different subduction zones varies considerably due to variable input, offscraping, and underplating [e.g., *von Huene and Scholl,* 1991]. At shallow depths (<10 km), large amounts of pore waters are expelled by sediment compaction [*Moore and Vrolijk,* 1992]. Structurally bound H_2O is released from sediments during the transformation of opal to quartz (~80 °C), the dehydration of clay mineral to form mica (100-180 °C), and chlorite breakdown (400-600 °C) [*Moore and Vrolijk,* 1992]. In warm subduction zones, mica will dehydrate and/or partially melt at $T \sim 800$ °C.

H_2O subducted as part of the oceanic crust dominates the H_2O flux into subduction zones. Drill holes and hydrogeologic data show that the uppermost kilometer of the oceanic crust has high porosities of ~10% [e.g., *Becker et al.,* 1989; *Fisher,* 1998]. Collapse of this porosity at temperatures of perhaps 300-500 °C will expel substantial amounts of pore water. Alternatively, interstitial pore water may react to form low-temperature minerals such as zeolites that subsequently dehydrate as the crust subducts. The most important reactions in subducting oceanic crust involve the transformation to eclogite, a relatively dense, anhydrous rock consisting primarily of garnet and omphacite (Na-Ca clinopyroxene) (Figure 3). In a given subduction zone, the depth and nature of eclogite formation and slab dehydration reactions depends on the *P-T* conditions encountered by the subducting oceanic crust. In the relatively warm Nankai subduction zone, subducted oceanic crust passes through the greenschist facies and the transformation to eclogite may occur at ~50 km depth. Calculated *P-T* paths for Nankai intersect mafic partial melting reactions at ~100 km depth and the uppermost oceanic crust may possibly melt (see discussion of adakites below). In contrast, calculated *P-T* paths for relatively cool subduction zones like NE Japan pass through the blueschist facies and eclogite may not form until depths > 100 km (Figure 3C, 4).

Figure 3. Calculated pressure-temperature (*P-T*) paths and metamorphic conditions encountered by subducting oceanic crust. (A) *P-T* diagram constructed for metabasaltic compositions showing metamorphic facies (solid lines), hydrous minerals stable in the eclogite facies (italics), and partial melting reactions (dark gray lines) [see references in *Peacock et al.,* 1994; *Poli and Schmidt,* 1995; *Peacock and Wang,* 1999]. EA, epidote-amphibolite facies; EB, epidote-blueschist facies; GS, greenschist facies; Px Granulite, pyroxene granulite facies; amph, amphibole, chl, chlorite, ctoid, chlorotoid, laws, lawsonite; hbl, hornblende; zoi, zoisite. (B) Calculated *P-T* paths for top of the subducting oceanic crust in four subduction zones. Triangles represent *P-T* conditions directly below the volcanic front. (C) Calculated *P-T* conditions for top and base of oceanic crust subducted beneath NE and SW Japan [after *Peacock and Wang,* 1999].

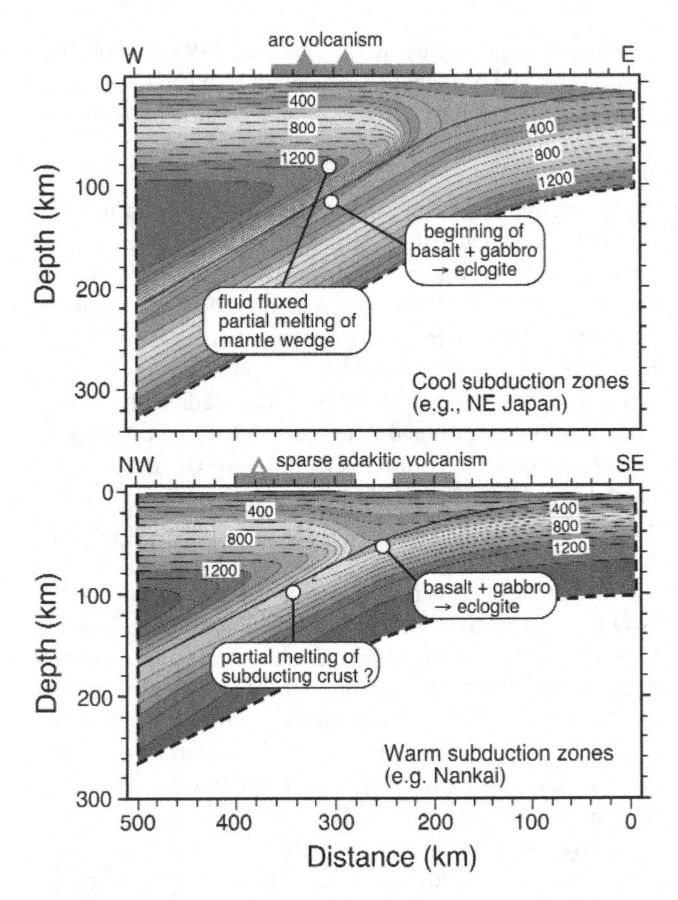

Figure 4. Location of important petrologic processes in (A) cool and (B) warm subduction zones.

In cool subducting slabs, hydrous minerals stable in the blueschist and low-temperature eclogite facies are capable of transporting H_2O to depths > 200 km. Recent experiments do not support earlier models linking the volcanic front to specific dehydration reactions in the subducting slab. On the contrary, the progressive metamorphism of metabasalts involves complex, continuous reactions that occur over a range in *P-T* space [e.g., *Spear,* 1993; *Schmidt and Poli,* 1998] and temperature-dependent dehydration reactions will be smeared out over a considerable depth range. Experiments conducted on metabasaltic compositions demonstrate that important hosts for H_2O in subducting oceanic crust include amphibole, lawsonite, phengite (mica), chlorite, talc, chlorotoid, and zoisite [e.g., *Pawley and Holloway,* 1993; *Poli and Schmidt,* 1995]. Amphibole (2 wt % H_2O) and lawsonite (11 wt % H_2O) are particularly important hosts for H_2O in subducting mafic rocks. In cool subducting slabs, amphibole dehydrates at ~75 km depth whereas lawsonite remains stable to depths > 200 km [*Pawley and Holloway,* 1993; *Poli and Schmidt,* 1995].

The extent to which the subducting oceanic mantle is hydrated is not known. At moderate to fast spreading ridges, hydrothermal circulation appears largely restricted to crustal levels; in contrast, at slow-spreading ridges, ultramafic rocks are tectonically emplaced at shallow depths and serpentinization is common [e.g., *Buck et al.,* 1998]. In addition to hydrothermal circulation at mid-ocean ridges, seawater may infiltrate the oceanic mantle along fracture zones and in the trench—outer rise region causing serpentinization. Large outer-rise earthquakes commonly rupture the oceanic mantle [e.g., *Christensen and Ruff,* 1988] and these highly permeable faults may promote hydrothermal circulation and alteration of the oceanic mantle [*Peacock,* 2001]. Because serpentine minerals contain ~13 wt % H_2O, even small amounts of serpentinization can contribute significantly to the total amount of H_2O entering a subduction zone.

If the subducting mantle contains serpentine and/or other hydrous minerals, then dehydration reactions will release H_2O from the mantle part of the subducting slab. Antigorite serpentine breaks down to form olivine + orthopyroxene + H_2O at 600-700 °C at pressures between 2 and 5 GPa [*Ulmer and Trommsdorff,* 1995; *Wunder and Schreyer,* 1997]. At higher pressures antigorite breaks down to form olivine + hydrous phase A + H_2O [*Wunder and Schreyer,* 1997; *Bose and Navrotsky,* 1998]. In warm subducting slabs, we would expect serpentine dehydration to be essentially complete by 100 km depth with only minor amounts of H_2O possibly subducted to greater depth by amphibole or chlorite. In cool subducting slabs, serpentine dehydration will not occur until 150-250 km depth and stable hydrous minerals such as phase-A may transport H_2O to even greater depths.

At depths of 400 to 670 km, olivine in the subducting mantle transforms to denser polymorphs (wadsleyite, ringwoodite) and then to perovskite + magnesiowustite at 660 km depth [e.g., *Kirby et al.,* 1996a]. These solid-solid reactions increase the density of the subducting slab. A number of studies [e.g., *Green and Burnley,* 1989; *Kirby et al.,* 1991; 1996a] suggest that deep-focus earthquakes (>300 km depth) may be caused by transformational faulting asso-ciated with the metastable reaction of olivine to denser polymorphs.

What happens to the H_2O liberated from the subducting slab? We expect H_2O to migrate upward and updip, primarily by channelized (focused) flow along faults and high permeability horizons [*Peacock,* 1990a; *Bebout,* 1991]. Most water may ultimately reach the seafloor or land surface. The widespread occurrence of low salinity fluids emanating from accretionary prisms provides direct evidence for deep dehydration fluids reaching the seafloor [*Moore and Vrolijk,* 1992]. The H_2O generated by compaction and low-temperature reactions should elevate fluid pressures and promote faulting along the subduction thrust. As discussed below, H_2O released at depths > 40 km may trigger intraslab earthquakes through dehydration embrittlement [*Kirby et al.,* 1996b].

A portion of the water released from the subducting slab infiltrates the overlying mantle wedge forming hydrous minerals, such as serpentine, brucite, and talc. Hydration will dramatically alter the rheological properties of the wedge, particularly along the slab-mantle interface. In the Mariana forearc, active serpentine mud volcanoes provide dramatic evidence for hydration of the mantle wedge [*Fryer et al.,* 1999]. Serpentine and other hydrous minerals in the forearc mantle may control the downdip limit of subduction thrust earthquakes [*Hyndman et al.,* 1997; *Peacock and Hyndman,* 1999]. At depths >100 km, water released from the subducting slab can trigger partial melting in the overlying mantle wedge.

Paleosubduction zones contain blueschists and low-temperature eclogites that provide insight into the thermal and petrologic structure of subducting slabs. Rigorously inverted metamorphic P-T data is difficult because, in general, we do not know the convergence rate at the time these rocks were subducted. Blueschist-facies metabasaltic clasts recovered from an active serpentine mud volcano in the Mariana forearc record T = 150-250 °C at P = 0.5-0.6 GPa [*Maekawa et al.,* 1993] and suggest shear stresses of ~20 MPa along the subduction thrust [*Peacock,* 1996]. Blueschist-facies metasedimentary rocks record subsolidus conditions along the slab interface and are consistent with thermal models with modest to no shear heating [*Peacock,* 1992].

Not all paleosubduction zones record high P/T conditions. For example, amphibolite-facies rocks in the Santa Catalina schist complex (southern California) record peak metamorphic conditions of T = 640-750 °C at P = 0.8-1.1 GPa [*Sorensen and Barton,* 1987]. Thermal models predict that slab interface temperatures are relatively high during the initial stages of subduction and decrease over time [*Peacock,* 1990a; *Peacock et al.,* 1994]. Such cooling is well recorded by the successive underplating of greenschist-facies (~500-600 °C) and blueschist-facies (~300-400 °C) units on Santa Catalina Island [*Platt,* 1975; *Bebout,* 1991; *Grove and Bebout,* 1995]. Similarly, eclogite blocks (T ~ 500 °C, P ~ 1-1.5 GPa) in the Franciscan Complex yield the oldest radiometric ages and formed during the early stages of subduction [*Cloos,* 1985].

ARC VOLCANISM

Arc lavas provide important information about subduction zones and petrological and geochemical data may be inverted to gain insight into the thermal and petrologic structure at depth. Basalts are common in many arcs which strongly suggests partial melting of the ultramafic mantle wedge as opposed to the mafic oceanic crust. Glass inclusions in mafic arc lavas exhibit a wide range in H_2O contents from 0.2 to 6 wt % H_2O [e.g., *Sisson and Grove,* 1993; *Roggensack et al.,* 1997; *Newman et al.,* 2000]. Relatively wet magmas reflect the importance of H_2O flux melting in the mantle wedge [e.g., *Gill,* 1981] whereas relatively dry arc magmas may result from adiabatic decompression of the mantle [e.g., *Sisson and Bronto,* 1998]. In the Cascades volcanic arc, dry basaltic magmas last equilibrated with the mantle at T = 1300-1450 °C and P = 1.2-2.2 GPa corresponding to depths of 36-66 km [*Elkins Tanton et al.,* 2001]. Similar mantle-magma equilibration conditions have been inferred by *Tatsumi et al.* [1983] and *Sisson and Bronto* [1998]. Mantle-wedge temperatures in the models depicted in Plate 1 exceed 1300 °C, but at substantially greater depth (>80 km) and generally behind the volcanic front. Most likely this discrepancy results from the simple constant-viscosity wedge-flow model; models employing a temperature-dependent viscosity [e.g., *Furukawa, 1993; Kiefer et al.,* 2001] yield higher mantle-wedge temperatures at shallower depths. In addition, mantle-magma equilibration temperatures may record locally hot mantle conditions beneath arc volcanoes and may not reflect mantle wedge temperatures beneath the entire arc.

Specific minor and trace elements of arc lavas (e.g., K, other large-ion lithophile elements, B, Be, Th, and Pb) appear to be derived from the subducting slab [e.g., *Gill,* 1981; *Hawkesworth et al.,* 1993; *Plank and Langmuir,* 1993; *Davidson,* 1996; *Elliot et al.,* 1997]. Geochemical studies of high-pressure metamorphic rocks have shown that white mica (specifically, a Si-rich muscovite called phengite) is the dominant host for many of these "slab" elements [*Domanik et al.,* 1993]. Most of these slab elements are readily transported in aqueous fluids, but recent experimental mineral-fluid partitioning data suggest that the efficient transport of Be and Th from slab sediments into arc magmas may require sediment melting [*Johnson and Plank,* 1999].

Melting of pelagic sediment at 2-4 GPa requires T > 650-800 °C [*Nichols et al.,* 1996; *Johnson and Plank,* 1999]. Thermal models suggest that such high temperatures are achieved along the slab interface only in unusually warm subduction zones, such as Nankai and Cascadia, where young, hot lithosphere enters the trench. Thermal models of most subduction zones suggest slab interface temperatures are 150-300 °C cooler than required for sediment melting. There are several possible ways to reconcile the high temperatures required for sediment melting with the lower temperatures predicted by thermal models. Rapid convection in the mantle wedge might lead to higher slab temperatures than predicted by existing models in which wedge flow velocities are less than slab velocities. Subducted sediments may melt, but not at the slab interface; for example, sediments might be emplaced into the warmer overlying wedge by diapirism or tectonic intrusion. Alternatively, aqueous fluids may effectively transfer Th and Be into the mantle wedge if fluid fluxes are sufficiently high or if the fluid contains different complexing anions than the experiments. Finally, Th and Be mineral-fluid partition coefficients may be substantially lower (i.e., more strongly partitioned into the fluid) at 500 °C than at 700 °C (the experimental conditions) because the stable mineralogy is different.

Adakites are relatively rare high-Mg andesites with distinctive geochemical characteristics (e.g., light REE enrichment, heavy REE depletion, high Sr) that suggest they formed from partial melts of subducted, eclogite-facies oceanic crust [*Kay,* 1978; *Defant and Drummond,* 1990]. Thermal models predict that partial melting of oceanic crust should only occur where very young oceanic lithosphere is subducted [*Peacock et al.,* 1994]. Recent adakitic lavas present in southern Chile, southwest Japan (Daisen and Sambe volcanoes), Cascadia (Mt. St. Helens), and Panama are underlain by unusually warm subducting crust [e.g., *Peacock et al.,* 1994; Plate 1C]. The type locality for adakite, Adak Island in the Aleutians, represents a striking exception to the general correlation between adakites and warm subducting oceanic crust [*Kay,* 1978; *Yogodzinski and Kelemen,* 1998]. Because the convergence rate and plate age are similar, the thermal structure of the Aleutian arc at Adak Island should be similar to cool thermal structure at Umnak Island, located 600 km to the east (Plate 1D). I do not have a good explanation for the Adak Island adakite locality – perhaps there is another eclogite source, such as a remnant slab, in the Adak mantle wedge? If adakites were common in volcanic arcs, then one might argue that the thermal models are in serious error, but adakites are relatively rare and, in general, occur where young, warm lithosphere is being subducted.

SEISMOLOGICAL OBSERVATIONS

In global and regional seismic tomographic studies, subducting slabs are readily imaged as high velocity, low attenuation regions which reflect the overall cool nature of the slab [e.g., *Zhao et al.,* 1994]. More detailed seismological investigations, using converted phases and waveform dispersion, reveal a thin (<10 km thick), dipping low-velocity layer coinciding with the zone of thrust and intermediate-depth earthquakes [e.g., *Hasegawa et al.,* 1994; *Helffrich,* 1996; *Abers,* 2000]. Thin dipping low-velocity layers have been observed in the Alaska, central Aleutian, Cascadia, northern Kurile, NE Japan, and Nankai subduction zones [e.g., *Fukao et al.,* 1983; *Matsuzawa et al.,* 1986; *Cassidy and Ellis,* 1993; *Abers and Sarker,* 1996; *Helffrich,* 1996; *Abers,* 2000]. In contrast, a high-velocity layer is observed in the Tonga-Kermadec subduction zone [*Ansell and Gubbins,* 1986]. The seismic velocity of eclogite is comparable to mantle peridotite, thus the dipping low seismic-velocity layer is generally interpreted as subducted oceanic crust that has not transformed to eclogite. The low-velocity layer extends to 60 km depth beneath SW Japan [*Fukao et al.,* 1983] and to 150 km depth beneath NE Japan [*Hasegawa et al.,* 1994], in good agreement with the predicted depth of eclogite transformation (Figure 3C) [*Peacock and Wang,* 1999]. Alternatively, the deeper extent of the low-velocity layer in the cooler NE Japan subduction zone may reflect the sluggish kinetics of the anhydrous gabbro to eclogite reaction [*Kirby et al.,* 1996b].

Subduction zones are regions of intense earthquake activity reflecting complex stresses generated by the interaction between the forces that drive and resist subduction, slab deformation (bending, unbending, flexure), thermal expansion, and metamorphic densification reactions [e.g., *Isacks and Barazangi,* 1977; *Spence,* 1987]. At depths > 40 km, high pressure and temperature should inhibit brittle behavior, but earthquakes in subduction zones occur as deep as 670 km. *Kirby et al.* [1996b] proposed that intermediate-depth earthquakes (50-300 km depth) are triggered by dehydration embrittlement associated with the transformation of metabasalt and metagabbro to eclogite within subducting oceanic crust. *Davies* [1999] proposed a related hypothesis linking intermediate-depth earthquakes to hydrofracturing of the subducting oceanic crust. Earthquakes that define the lower plane of double seismic zones, observed in a number of cool subducting slabs, may be triggered by serpentine dehydration reactions [*Peacock,* 2001].

The depth extent of intraslab earthquakes in NE and SW Japan agrees well with the predicted depth of dehydration reactions in the subducting oceanic crust [*Peacock and Wang,* 1999; *Hacker et al.,* 2000]. Beneath NE Japan, intraslab earthquake activity peaks at 125 km depth and extends to >200 km depth [*Hasegawa et al.,* 1994; *Kirby et al.,* 1996b]. Calculated *P-T* paths for NE Japan predict garnet-forming dehydration reactions will begin in the subducted oceanic crust at ~110 km depth and hydrous minerals remain stable to >160 km depth (Figure 3C). Beneath Shikoku (SW Japan), intraslab seismicity ceases at 50-65 km depth [*Nakamura et al.,* 1997]. Calculated *P-T* paths for SW Japan suggest major dehydration of the subducted oceanic crust should occur at ~50 km depth associated with the formation of eclogite; hydrous minerals may persist to 90 km depth (Figure 3C). The lack of intraslab earthquakes at depths > 65 km beneath SW Japan may reflect the onset of ductile slab behavior at $T > 600$ °C [*Peacock and Wang,* 1999] or the relatively small amount of dehydration expected after eclogite formation.

RECENT ADVANCES

Significant progress has been made on many fronts since the Subduction Factory Theoretical and Experimental Institute meeting in Eugene, Oregon, in August, 2000. In this section, I highlight recent advances in the thermal modeling of subduction zones and the connection between intermediate-depth earthquakes and metamorphic dehydration reactions in the subducting slab.

New numerical models have explored the effect of mantle-wedge rheology on the thermal and melting structure of subduction zones. *Van Keken et al.* [2002] constructed a new set of finite element models with high spatial resolution (400 m) and temperature- and stress-dependent olivine rheology for the mantle wedge. Compared to *Peacock and Wang's* [1999] isoviscous model (Plate 1A), *van Keken et al.'s* [2002] models predict higher temperatures within the mantle wedge and along the subduction interface (Figures 5 and 6). Predicted slab interface temperatures beneath the NE Japan volcanic front of 475 °C for the isoviscous case [*Peacock and Wang,* 1999] increase to 610 °C with higher spatial resolution and to 810 °C with more realistic olivine rheology [*van Keken et al.,* 2002] (Figure 6). The higher predicted temperatures suggests that partial melting may occur along the slab interface. Similarly, *Kelemen et al.* [this volume] found that using a T-dependent viscosity for the mantle wedge leads to possible melting of subducted crust over a wider range of convergence rates and plate ages than previously suggested (e.g., subduction of 50 Ma lithosphere at 60 mm/yr). Predicted temperatures within the subducting slab are less affected by the assumed mantle-wedge rheology and slab P-T paths remain subsolidus (Figure 6). *Van Keken et al.* [2002] suggest that the high thermal gradients perpendicular to slab interface could explain the conflicting geochemical evidence for sediment melting and basalt dehydration.

Compared to isoviscous mantle-wedge models, models with a T-dependent mantle-wedge viscosity yield substantial decompression melting in the wedge and high temperatures at shallow depth beneath the arc that are in better agreement with petrological observations [*Kelemen et al.,* this volume; *van Keken et al.,* 2002; *Conder et al.,* 2002]. In isoviscous mantle-wedge models, the 1300 °C wedge isotherm is not reached until 80-90 km depth (e.g., Plate 1). In T-dependent mantle-wedge viscosity models, the 1300 °C isotherm is reached at shallower depths of 45 to 60 km beneath the arc [*Conder et al.,* 2002; *Kelemen et al.,* this volume; van *Keken et al.,* 2002]. In *Conder et al.'s* [2002] model, flow induced by the subducting slab erodes the lower part of the overriding plate leading to enhanced decompression melting and flow into the corner of the mantle wedge. Similar thinning of the upper thermal boundary layer occurs in the models presented by *Kelemen et al.* [this volume] and *van Keken et al.* [2002].

Significant differences among the thermal models remain, most notably with respect to the viscous coupling along the

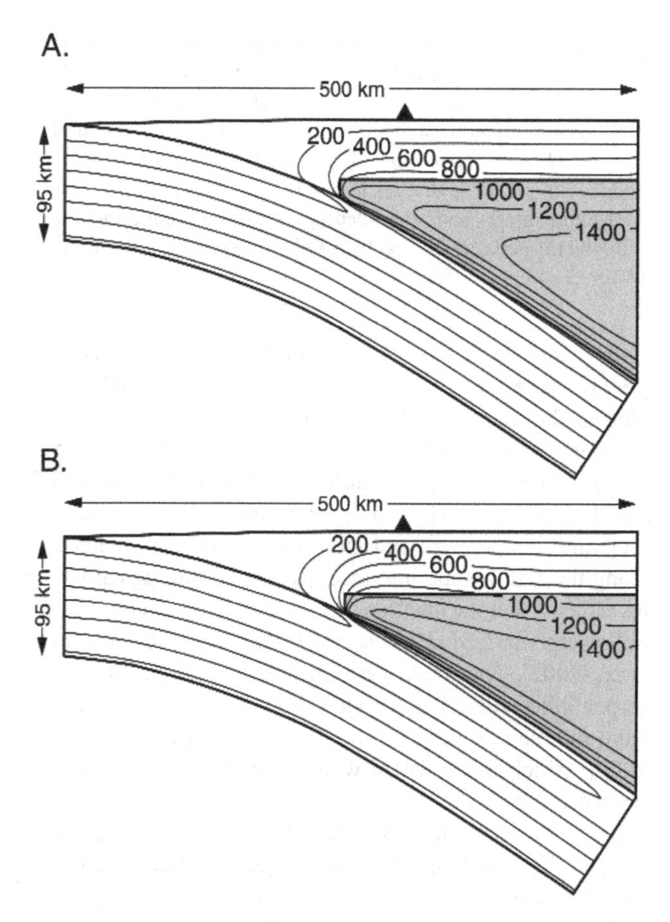

Figure 5. Calculated thermal structure of the NE Japan subduction zone assuming (A) isoviscous rheology and (B) temperature- and stress-dependent olivine rheology for the mantle wedge [*van Keken et al.,* 2002]. Models use the same geometry, heat sources, and boundary conditions as isoviscous model for NE Japan presented by *Peacock and Wang* [1999]. Gray shading = mantle wedge, black triangle = volcanic front. Contour interval = 200 °C.

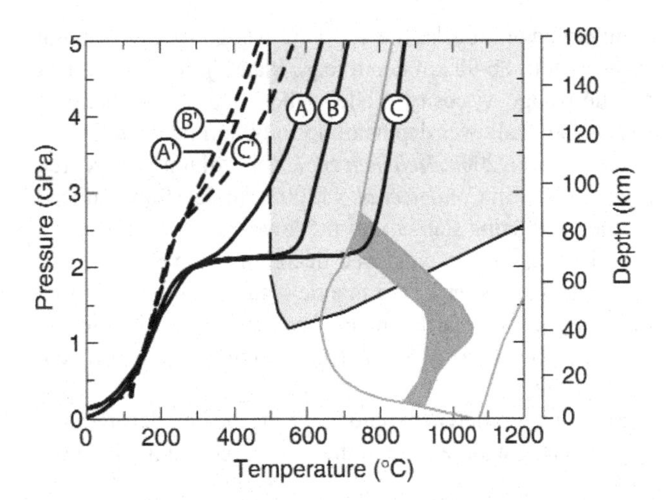

Figure 6. Calculated pressure-temperature conditions for top (solid lines) and base (dashed lines) of oceanic crust subducted beneath NE Japan for three different finite element models. A, A' = isoviscous mantle wedge with spatial resolution on the order of 5 km [*Peacock and Wang,* 1999]. B, B' = isoviscous mantle wedge with 400 m spatial resolution [*van Keken et al.,* 2002]. C, C' = temperature- and stress-dependent mantle wedge rheology with 400 m spatial resolution [*van Keken et al.,* 2002]. See Figure 3A for key to phase diagram.

slab-wedge interface. As demonstrated by *Furukawa* [1993], the depth at which the slab becomes viscously coupled to the mantle wedge strongly influences the thermal structure of the mantle wedge and the degree of decompression melting. As discussed above, the mechanical nature of the slab-wedge interface likely depends on many variables including temperature and the presence of weak materials along the slab-mantle interface. Recent seismological studies indicate that the mantle wedge beneath the forearc is partially serpentinized [*Kamiya and Kobayashi,* 2000; *Bostock et al.,* 2002]. The weak rheology and positive buoyancy of serpentinite will act to isolate the hydrated forearc wedge from the mantle-wedge convection system [*Bostock et al.,* 2002], which is consistent with the low surface heat flow observed in forearcs.

In most thermal models of subduction zones, the velocity field of the subducting plate is defined kinematically [e.g., *Peacock,* 1990; *Davies and Stevenson,* 1992]. *King* [2001] reviewed evidence indicating slabs deform (thicken) significantly as they descend through the upper mantle. Compared to kinematic models, dynamical models that permit slab thickening should yield cooler slab interiors because heat must be conducted a greater distance into the slab interior from the upper and lower slab surfaces [*King,* 2001].

Several groups, using very different approaches, have constructed phase diagrams for metamorphosed basalts in order to gain insight into the dehydration of subducted oceanic crust. Phase diagrams of metabasalts have been constructed based on experimental studies [*Poli and Schmidt,* 1995; *Schmidt and Poli,* 1998], thermodynamic data and a free energy minimization strategy [*Kerrick and Connolly,* 2001], and petrological field observations combined with thermodynamic calculations of key reactions [*Hacker et al.,* 2002a]. At low temperatures and high pressure, where experiments are difficult to conduct and natural samples are rare, there are considerable differences among the proposed phase diagrams, particularly with respect to the stability fields of lawsonite, amphibole, and chlorotoid. At 500 °C and 3 GPa, estimates of the H_2O content of fully hydrated metabasalt range from 2.6 wt% H_2O [*Kerrick and Connolly,* 2001] to 1.0 wt% H_2O [*Schmidt and Poli,* 1998] to 0.3 wt% H_2O [*Hacker et al.,* 2002a]. An accurate understanding of the extent and location of dehydration reactions in cool subducting crust requires resolving these differences.

Considerable theoretical and observational advances have been made in our understanding of subduction-zone earthquakes and their connection to metamorphic dehydration reactions in the subducting plate. In an integrated petrological-seismological study of four subduction zones, *Hacker et al.* [2002b] found a strong correlation between intermediate-depth seismicity patterns and the predicted location of dehydration reactions in the subducting oceanic crust and uppermost mantle; this correlation supports *Kirby et al.'s* [1996] dehydration embrittlement hypothesis for intraslab earthquakes. *Seno and Yamanaka's* [1996] proposal that double seismic zones are linked to the dehydration of serpentine in subducting mantle has received additional petrological and seismological support [*Seno et al.,* 2001; Peacock, 2001; *Omori et al.,* 2002]. Based on a statistical analysis of 360 fault plane solutions in the Tonga subduction zone, *Jiao et al.* [2000] demonstrated that earthquakes down to 450 km depth occurred along preexisting asymmetric fault systems that formed prior to subduction. Hydrous minerals present along the preexisting fault zones may explain how these faults zones remain weak to great depth. *Tibi et al.* [2002] found that rupture areas of six large intermediate-depth earthquakes extended 25-50 km parallel to the strike of the slab, but only 4-13 km perpendicular to the slab surface. The orientation of the rupture areas is consistent with the reactivation of trench-parallel faults, with at least the larger earthquakes rupturing the slab mantle. In the SW Japan subduction zone, unusual deep (~30 km) long-period (1-20 Hz) tremors, which persist for several days to weeks, appear to be related to fluids released by slab dehydration reactions [*Obara,* 2002]. Fluids have also been proposed to explain the upward migration of a high *Vp/Vs* region from the hypocenter of the M_W = 8.0 Antofagasta, Chile 1995 thrust earthquake [*Husen and Kissling,* 2001]

CONCLUDING THOUGHTS

The thermal-petrologic models presented in this paper are consistent with a broad range of seismological and arc geochemical observations. For most subduction zones, these models predict that subducting sediments, crust, and mantle will undergo subsolidus metamorphic reactions. Partial melting of subducting materials is predicted to occur only in subduction zones like Nankai where young incoming lithosphere subducts relatively slowly. Tantalizing evidence, such as the Adak Island adakite and mineral-fluid partitioning data on Be and Th, suggests that calculated temperatures along the slab-mantle interface may be too low, at least locally. The incorporation of temperature-dependent viscosity models for the mantle wedge increase calculated slab-mantle interface temperatures, but this is just one of a number of steps we need to take in order to accurately simulate flow in the mantle wedge.

There are a number of important subduction-zone processes that are poorly understood at present and limit our understanding of subduction zones. In my opinion, some of the most important questions to be investigated are (1) the vigor and geometry of mantle-wedge flow induced by the subducting slab and upper-plate extension; (2) variations in thermal parameters and rheology as a function of *P, T,* and composition; (3) metabasaltic phase equilibria and reaction kinetics at low temperatures and high pressure (*e.g.,* 500 °C and 3 GPa); and (4) the amount and distribution of H_2O in the oceanic crust and mantle prior to subduction.

Acknowledgments. I thank the organizers of the NSF Margins Subduction Factory Theoretical and Experimental Institute for arranging a very stimulating and productive conference in Eugene, Oregon. I also thank John Eiler and two anonymous reviewers for providing detailed, constructive reviews. This research was supported by NSF grants EAR 97-25406 and 98-09602

REFERENCES

Abers, G. A., Hydrated subducted crust at 100-250 km depth, *Earth Planet. Sci. Lett., 176,* 323-330, 2000.

Abers, G. A., and G. Sarker, Dispersion of regional body waves at 100- to 150-km depth beneath Alaska: In situ constraints on metamorphism of subducted crust, *Geophys. Res. Lett., 23,* 1171-1174, 1996.

Ansell, J. H., and D. Gubbins, Anomalous high-frequency wave propagation from the Tonga-Kermadec seismic zone to New Zealand, *Geophys. J. R. Astro. Soc., 85,* 93-106, 1986.

Batchelor, G. K., *An introduction to fluid dynamics,* 615 pp., Cambridge University Press, Cambridge, 1967.

Bebout, G. E., Geometry and mechanisms of fluid flow at 15 to 45 kilometer depths in an early Cretaceous accretionary complex, *Geophys. Res. Lett., 18,* 923-926, 1991

Bebout, G. E., Volatile transfer and recycling at convergent margins: Mass balance and insights from high-P/T metamorphic rocks, in *Subduction: Top to Bottom,* edited by Bebout, G. E., *et al.,* 179-193, AGU Geophysical Monograph *96,* Washington, D.C., 1996.

Becker, K., and 24 others, Drilling deep into young oceanic crust, hole 514B, Costa Rica rift, *Rev. Geophys., 27,* 79-102, 1989.

Billen, M. I., and M. Gurnis, A low viscosity wedge in subduction zones, *Earth Planet. Sci. Lett.,* in press, 2001.

Bose, K., and A. Navrotsky, Thermochemistry and phase equilibria of hydrous phases in the system $MgO-SiO_2-H_2O$: Implications for volatile transport to the mantle, J. *Geophys. Res., 103,* 9713-9719, 1998.

Bostock, M.G., R.D. Hyndman, S. Rondenay, and S.M. Peacock, An inverted continental Moho and serpentinization of the forearc mantle, Nature, 417, 536-538, 2002.

Buck, W. R., P. T. Delaney, J. A. Karson, and Y. Lagabrielle (eds.), *Faulting and Magmatism at Mid-Ocean Ridges,* AGU Geophysical Monograph 106, Washington, D.C., 1998.

Carswell, D. A. (ed.), *Eclogite Facies Rocks,* 396 pp., Blackie, London, 1990.

Cassidy, J. F., and R. M. Ellis, S wave velocity structure of the northern Cascadia subduction zone, *J. Geophys. Res., 98* , 4407-4421, 1993.

Christensen, D. H., and L. J. Ruff, Seismic coupling and outer rise earthquakes, *J. Geophys. Res., 93,* 13,421-13,444, 1988.

Cloos, M., Thermal evolution of convergent plate margins: Thermal modeling and reevaluation of isotopic Ar-ages for blueschists in the Franciscan complex of California, *Tectonics, 4* , 421-433, 1985.

Conder, J.A., D.A. Weins, and J. Morris, On the decompression melting structure at volcanic arcs and back-arc spreading centers, Geophys. Res. Lett., in press, 2002.

Davidson, J. P., Deciphering mantle and crustal signatures in subduction zone magmatism, in *Subduction: Top to Bottom,* edited by Bebout, G. E., et al., 251-262, AGU Geophysical Monograph 96, Washington, D.C., 1996.

Davies, J. H., The role of hydraulic fractures and intermediate-depth earthquakes in generating subduction-zone magmatism, *Nature, 398,* 142-145, 1999.

Davies, J. H. and D. J. Stevenson, Physical model of source region of subduction zone volcanics, *J. Geophys. Res., 97,* 2037-2070, 1992.

Defant, M. J., and M. S. Drummond, Derivation of some modern arc magmas by melting of young subducted lithosphere, *Nature, 347,* 662-665, 1990.

DeMets, C., R. G. Gordon, D. F. Argus, and S. Stein, Effect of recent revisions to the geomagnetic reversal timescale on estimates of current plate motions, *Geophys. Res. Lett., 21,* 2191-2194, 1994.

Domanik, K. J., R. L. Hervig, and S. M. Peacock, Beryllium and boron in subduction zone minerals: an ion microprobe study, *Geochim. Cosmochim. Acta, 57,* 1993.

Elkins Tanton, L. T., T. L. Grove, and J. Donnelly-Nolan, Hot, shallow mantle melting under the Cascades volcanic arc, *Geology, 29,* 631-634, 2001.

Elliott, T., T. Plank, A. Zindler, W. White, and B. Bourdon, Element transport from slab to volcanic front at the Mariana arc, *J. Geophys. Res., 102,* 14991-15019, 1997.

Fisher, A. T., Permeability within the basaltic oceanic crust, *Rev. Geophys., 36,* 143-182, 1998.

Forsyth, D., and S. Uyeda, Relative importance of driving forces of plate motion, Geophys. *J. Roy. Astro. Soc., 43,* 163-200, 1975.

Fryer, P., C. G. Wheat, and M. J. Mottl, Mariana blueschist mud volcanism: Implications for conditions within the subduction zone, *Geology, 27,* 103-106, 1999.

Fukao, Y., S. Hori, and M. Ukawa, A seismological constraint on the depth of basalt-eclogite transition in a subducting oceanic crust, *Nature, 303,* 413-415, 1983.

Furukawa, Y., Depth of the decoupling plate interface and thermal structure under arcs, *J. Geophys. Res., 98,* 20,005-20,013, 1993.

Gill, J., *Orogenic Andesites and Plate Tectonics,* 390 pp., Springer-Verlag, New York, 1981.

Green, H. W. I., and P. C. Burnley, A new self-organizing mechanism for deep-focus earthquakes, *Nature, 341,* 733-737, 1989.

Grove, M., and G. E. Bebout, Cretaceous tectonic evolution of coastal southern California: Insights from the Catalina Schist, *Tectonics, 14,* 1290-1308, 1995.

Hacker, B. R., Eclogite formation and the rheology, buoyancy, seismicity, and H_2O content of oceanic crust, in *Subduction: Top to Bottom,* edited by Bebout, G. E., *et al.,* 337-346, AGU Geophysical Monograph *96,* Washington, D.C., 1996.

Hacker, B. R., G. A. Abers, and S. M. Peacock, Phase transformations and the buoyancy, seismicity, and H_2O contents of subduction zones (abs.), *Eos Trans. AGU, 81,* F1372, 2000.

Hacker, B.R., G.A. Abers, and S.M. Peacock, Subduction Factory 1. Theoretical mineralogy, density, seismic wave speeds, and H_2O content, J. Geophys. Res., in press, 2002a.

Hacker, B.R., S.M. Peacock, G.A. Abers, and S.D. Holloway, Subduction Factory 2. Intermediate-depth earthquakes in subducting slabs are linked to metamorphic dehydration reactions, J. Geophys. Res., in press, 2002b.

Hasegawa, A., S. Horiuchi, and N. Umino, Seismic structure of the northeastern Japan convergent plate margin: A synthesis, *J. Geophys. Res., 99,* 22,295-22,311, 1994.

Hawkesworth, C. J., K. Gallagher, J. M. Hergt, and F. McDermott, Mantle and slab contributions in arc magmas, *Ann. Rev. Earth Planet. Sci., 21,* 175-204, 1993.

Helffrich, G., Subducted lithospheric slab velocity structure: Observations and mineralogical inferences, in *Subduction: Top to Bottom,* edited by Bebout, G. E., *et al.,* 215-222, AGU Geophysical Monograph *96,* Washington, D.C., 1996.

Husen, S., and E. Kissling, Postseismic fluid flow after the large subduction earthquake of Antofagasta, Chile, Geology, 29, 847-850, 2001.

Hyndman, R. D., and K. Wang, The rupture zone of Cascadia great earthquakes from current deformation and the thermal regime, *J. Geophys. Res., 100,* 22,133-22,154, 1995.

Hyndman, R. D., M. Yamano, and D. A. Oleskevich, 1997, The seismogenic zone of subduction thrust faults, *Island Arc, 6,* 244-260, 1997.

Isacks, B. L., and M. Barazangi, Geometry of Benioff zones: Lateral segmentation and downwards bending of the subducted lithosphere, in *Island Arcs, Deep Sea Trenches, and Back Arc Basins,* edited by M. Talwani and W. C. Pitman, 99-114, AGU, Washington, D.C., 1977.

Ito, E., D. M. Harris, and A. T. J. Anderson, Alteration of oceanic crust and geologic cycling of chlorine and water, *Geochim. Cosmochim. Acta, 47,* 1613-1624, 1983.

Iwamori, H., Heat sources and melting in subduction zones, *J. Geophys. Res., 102,* 14,803-14,820, 1997.

James, T. S., J. J. Clague, K. Wang, and I. Hutchinson, Postglacial rebound at the northern Cascadia subduction zone, *Quaternary Sci. Rev., 19,* 1527-1541, 2000.

Jiao, W., P.G. Silver, Y. Fei, and C.T. Prewitt, Do intermediate- and deep-focus earthquakes occur on preexisting weak zones? An examination of the Tonga subduction zone. J. Geophys. Res., 105, 28,125-28,138, 2000.

Johnson, M. C., and T. Plank, Dehydration and melting experiments constrain the fate of subducted sediments, *Geochem. Geophys. Geosyst., 1,* 1999.

Kamiya, S., and Y. Kobayashi, Seismological evidence for the existence of serpentinized wedge mantle, Geophys. Res. Lett., 27, 819-822, 2000.

Karato, S., and P. Wu, Rheology of the upper mantle: A synthesis, *Science, 260,* 771-778, 1993.

Kay, R. W., Aleutian magnesian andesites: melts from subducted pacific ocean crust, *J. Volc. Geotherm. Res., 4,* 117-132, 1978.

Kelemen, P.B., J.L. Rilling, E.M. Parmentier, L. Mehl, and B.R. Hacker, Flow in the mantle wedge beneath subduction-related magmatic arcs, in Eiler, J. (ed.), The Subduction Factory, AGU Geophysical Monograph, in press, 2002.

Kerrick, D.M., and J.A.D. Connolly, Metamorphic devolatilization of subducted oceanic metabasalts: implications for seismicity, arc magmatism and volatile recycling, Earth Planet. Sci. Lett., 189, 19-29, 2001.

Kiefer, B., P. van Keken, and S. M. Peacock, The thermal structure of subduction zones (abs.), *Eos Trans. AGU, 82,* S411, 2001

Kincaid, C., and I. S. Sacks, Thermal and dynamical evolution of the upper mantle in subduction zones, *J. Geophys. Res., 102,* 12,295-12,315, 1997.

King, S.D., Subduction zones: Observations and geodynamic models, Phys. Earth Planet. Interiors, 127, 9-24, 2001.

Kirby, S. H., W. B. Durham, and L. A. Stern, Mantle phase changes and deep-earthquake faulting in subducting lithosphere, *Science, 252,* 216-225, 1991.

Kirby, S. H., S. Stein, E. A. Okal, and D. C. Rubie, Metastable mantle phase transformations and deep earthquakes in subducting oceanic lithosphere, *Rev. Geophys., 34,* 261-306, 1996a.

Kirby, S. H., E. R. Engdahl, and R. Denlinger, Intraslab earthquakes and arc volcanism: dual physical expressions of crustal and uppermost mantle metamorphism in subducting slabs, in *Subduction: Top to Bottom,* edited by Bebout, G. E., *et al.,* 195-214, AGU Geophysical Monograph *96,* Washington, D.C., 1996b.

Maekawa, H., M. Shozui, T. Ishii, P. Fryer, and J. A. Pearce, Blueschist metamorphism in an active subduction zone, *Nature, 364,* 520-523, 1993.

Matsuzawa, T., N. Umino, A. Hasegawa, and A. Takagi, Upper mantle velocity structure estimated from PS - converted wave beneath the north-eastern Japan Arc, *Geophys. J. Roy. Astro. Soc., 86,* 767-787, 1986.

McKenzie, D. P., Speculations on the consequences and causes of plate motions, *Geophys. J. Roy. Astro. Soc., 18,* 1-32, 1969.

Molnar, P., and P. C. England, Temperatures, heat flux, and frictional stress near major thrust faults, *J. Geophys. Res., 95,* 4833-3856, 1990.

Moore, J. C., and P. Vrolijk, Fluids in accretionary prisms, *Rev. Geophys.,* 30, 113-135, 1992.

Nakamura, M., H. Watanabe, T. Konomi, and K. Miura, Characteristic activities of subcrustal earthquakes along the outer zone of southwestern Japan, *Annals Disaster Prevention Res. Inst., Kyoto Univ., 40 B-1,* 1997.

Newman, S., E. Stolper, and R. Stern, H2O and CO2 in magmas from the Mariana arc and back arc systems, *Geochem. Geophys. Geosyst., 1,* 2000.

Nichols, G. T., P. J. Wyllie, and C. R. Stern, Experimental melting of pelagic sediment, constraints relevant to subduction, , in Subduction: Top to Bottom, edited by Bebout, G. E., *et al.,* 293-298, AGU Geophysical Monograph *96,* Washington, D.C., 1996.

Obara, K., Nonvolcanic deep tremor associated with subduction in southwest Japan, Science, 296, 1679-1681, 2002.

Omori, S., S. Kamiya, H. Maruyama, and D. Zhao, Morphology of the intraslab seismic zone and devolatilization phase equilibria of the subducting slab peridotite, Bull. Earthq. Res. Inst. Univ. Tokyo, 76, 455-478, 2002.

Oxburgh, E. R., and D. L. Turcotte, Thermal structure of island arcs, *Geol. Soc. Amer. Bull., 81,* 1665-1688, 1970.

Pawley, A. R., and J. R. Holloway, Water sources for subduction zone volcanism: new experimental constraints, *Science, 260,* 664-667, 1993.

Peacock, S. M., Thermal effects of metamorphic fluids in subduction zones, *Geology, 15* , 1057-1060, 1987.

Peacock, S. M., Fluid processes in subduction zones, *Science, 248,* 329-337, 1990a.

Peacock, S. M., Numerical simulation of metamorphic pressure-temperature-time paths and fluid production in subducting slabs, *Tectonics, 9,* 1197-1211, 1990b.

Peacock, S. M., Blueschist-facies metamorphism, shear heating, and P-T-t paths in subduction shear zones, *J. Geophys. Res., 97,* 17,693-17,707, 1992.

Peacock, S. M., Thermal and petrologic structure of subduction zones, in *Subduction: Top to Bottom,* edited by Bebout, G. E., *et al.,* 119-133, AGU Geophysical Monograph *96,* Washington, D.C., 1996.

Peacock, S. M., Are the lower planes of double seismic zones caused by serpentine dehydration in subducting oceanic mantle?, *Geology, 29,* 299-302, 2001.

Peacock, S. M., and R. D. Hyndman, Hydrous minerals in the mantle wedge and the maximum depth of subduction thrust earthquakes, *Geophys. Res. Lett., 26,* 2517-2520, 1999.

Peacock, S. M., T. Rushmer, and A. B. Thompson, Partial melting of subducting oceanic crust, *Earth Planet. Sci. Lett., 121,* 227-244, 1994.

Peacock, S. M., and K. Wang, Seismic consequences of warm versus cool subduction zone metamorphism: Examples from northeast and southwest Japan, *Science, 286,* 937-939, 1999.

Plank, T. A., and C. Langmuir, Tracing trace elements from sediment input to volcanic output at subduction zones, *Nature, 362,* 739-742, 1993.

Platt, J. P., Metamorphic and deformational processes in the Franciscan Complex, California: Some insights from the Catalina Schist terrane, *Geol. Soc. Amer. Bull., 86,* 1337-1347, 1975.

Poli, S., and M. W. Schmidt, H2O transport and release in subduction zones: Experimental constraints on basaltic and andesitic systems, *J. Geophys. Res., 100,* 22,299-22,314, 1995.

Rabinowicz, M., B. Lago, and C. Froidevaux, Thermal transfer between the continetal asthenosphere and the oceanic subducting lithosphere: Its effect on subcontinental convection, *J. Geophys. Res., 85,* 1839-1853, 1980.

Roggensack, K., R. L. Hervig, S. B. McKnight, and S. N. Williams, Explosive basaltic volcanism from Cerro Negro volcano: Influence of volatiles on eruptive style, *Science, 277,* 1639-1642, 1997.

Rowland, A., and J. H. Davies, Buoyancy rather than rheology controls the thickness of overriding mechanical lithosphere at subduction zones, *Geophys. Res. Lett., 26,* 3037-3040, 1999.

Schmidt, M. W., and S. Poli, Experimentally based water budgets for dehydrating slabs and consequences for arc magma generation, *Earth Planet. Sci. Lett., 163* , 361-379, 1998.

Seno, T., D. Zhao, Y. Kobayashi, and M. Nakamura, Dehydration of serpentinized slab mantle: Seismic evidence from southwest Japan, Earth Planets Space, 53, 861-871, 2001.

Sisson, T. W., and T. L. Grove, Temperatures and H2O contents of low-MgO high-alumina basalts, *Contrib. Mineral. Petrol., 113,* 167-184, 1993.

Sisson, T. W., and S. Bronto, Evidence for pressure-release melting beneath magmatic arcs from basalt at Galunggung, Indonesia, *Nature, 391,* 883-886, 1998.

Sorensen, S. S., and M. D. Barton, Metasomatism and partial melting in a subduction complex: Catalina Schist, southern California, *Geology, 15,* 115-118, 1987.

Spear, F. S., *Metamorphic Phase Equilibria and Pressure-Temperature-Time Paths,* 799 pp., Mineralogical Society of America, Washington D.C., 1993.

Spence, W., Slab pull and the seismotectonics of subducting lithosphere, *Rev. Geophys., 25,* 55-69, 1987.

Stein, C. A., and S. Stein, A model for the global variation in oceanic depth and heat flow with lithospheric age, *Nature, 359,* 123-129, 1992.

Tatsumi, Y., M. Sakuyama, H. Fukuyama, and I. Kushiro, Generation of arc basalt magmas and thermal structure of the mantle wedge in subduction zones, *J. Geophys. Res., 88,* 5815-5825, 1983.

Tibi, R., G. Bock, and C.H. Estabrook, Seismic body wave constraint on mechanisms of intermediate-depth earthquakes, J. Geophys. Res., 107, 10.1029/2001JB000361, 2002.

Tichelaar, B. W., and L. J. Ruff, Depth of seismic coupling along subduction zones, *J. Geophys. Res., 98,* 2017-2037, 1993.

Toksöz, M. N., J. W. Minear, and B. R. Julian, Temperature field and geophysical effects of a downgoing slab, *J. Geophys. Res., 76,* 1113-1138, 1971.

Turcotte, D. L., and G. Schubert, Frictional heating of the descending lithosphere, *J. Geophys. Res., 78,* 5876-5886, 1973.

Ulmer, P., and V. Trommsdorff, Serpentine stability to mantle depths and subduction-related magmatism, *Science, 268,* 858-861, 1995.

van Keken, P.E., B. Kiefer, and S.M. Peacock, High resolution models of subduction zones: Implications for mineral dehydration reactions and the transport of water into the deep mantle, Geochem., Geophys., Geosys., in press, 2002.

von Huene, R., and D. W. Scholl, Observations at convergent margins concerning sediment subduction, subduction erosion, and the growth of continental crust, *Rev. Geophys., 29,* 279-316, 1991.

von Herzen, R., C. Ruppel, P. Molnar, M. Nettles, S. Nagihara, and G. Ekström, A constraint on the shear stress at the Pacific-Australian plate boundary from heat flow and seismicity at the Kermadec forearc, *J. Geophys. Res., 106,* 6817-6833, 2001.

Wang, K., H. Dragert, and H. J. Melosh, Finite element study of uplift and strain across Vancouver Island, *Can. J. Earth Sci., 31,* 1510-1522, 1994.

Wang, K., T. Mulder, G. C. Rogers, and R. D. Hyndman, Case for low coupling stress on the Cascadia subduction fault, *J. Geophys. Res., 100,* 12,907-12,918, 1995.

Wunder, B., and W. Schreyer, Antigorite: High-pressure stability in the system MgO-SiO_2-H_2O (MSH), *Lithos, 41,* 213-227, 1997.

Yogodzinski, G. M., and P. B. Kelemen, Slab melting in the Aleutians: implications of an ion probe study of clinopyroxene in primitive adakite and basalt, *Earth Planet. Sci. Lett., 158,* 53-65, 1998.

Yuen, D. A., L. Fleitout, G. Schubert, and C. Froidevaux, Shear deformation zones along major transform faults and subducting slabs, *Geophys. J. Roy. Astro. Soc., 54,* 93-119, 1978.

Zhao, D., A. Hasegawa, and H. Kanamori, Deep structure of Japan subduction zone as derived from local, regional, and teleseismic events, *J. Geophys. Res., 99,* 22,313-22,329, 1994.

Zhong, S. J., and M. Gurnis, Controls on trench topography from dynamic-models of subducted slabs, *J. Geophys. Res., 99,* 15,683-15,695, 1994.

S. M. Peacock, Department of Geological Sciences, Arizona State University, Box 871404, Tempe, Arizona, 85287-1404. (e-mail: peacock@asu.edu)

Tracers of the Slab

Tim Elliott

Department of Earth Sciences, University of Bristol, Bristol, UK

A new global compilation of high precision trace element and isotopic analyses from mafic island arc lavas is explored to highlight the key geochemical features of volcanic front lavas. Two distinct components from the slab can be identified in island arc lavas. One component dominates the budgets of most incompatible trace elements and another affects a smaller range of elements, most notably Ba, Pb and Sr. These contrasting slab components require different ultimate sources and mechanisms of transport to the mantle wedge. The first component is argued to be a melt of the down-going sediment, while the second is likely an aqueous fluid derived from the altered mafic oceanic crust. U-series nuclides constrain the timing of slab component additions. The sediment component is close to ^{238}U-^{230}Th equilibrium implying >350ky since the last major fractionation of U and Th, plausibly the time since sediment melting. Lavas dominated by the 'fluid' component can have extreme ^{226}Ra-^{230}Th disequilibrium together with elevated ratios of the stable Ba/Th analogue. In its simplest interpretation, this observation implies only a few thousand years elapse between 'fluid' release from the slab, and eruption of this component in arc lavas. In most cases the two subduction components appear to be added to a recently melt-depleted mantle. This suggests the mantle wedge is fed with material processed through a back-arc melting regime. These first order geochemical observations and inferences place important constraints on physical models of the subduction zone.

1. INTRODUCTION

In deciphering the complex geochemical signatures of arc lavas there is general agreement about the involvement of a slab component. Indeed the very production of abundant arc lavas from a slab-cooled mantle implicates a key subducted contribution to the mantle wedge, namely water. Beyond acknowledging its general presence, the nature, magnitude and associated transport processes of slab derived material have been topics of much debate. A detailed understanding of the role of slab derived components in arc lavas places major constraints on the physical workings of the subduction zone and is required to assess the composition of a deep recycled residue after subduction zone processing.

Much has been written on slab tracers [e.g. *Gill,* 1981; *Pearce,* 1982; *Hawkesworth et al.,* 1993; *Tatsumi and Eggins,* 1995; and *Davidson,* 1996], but there are significant recent developments that are worth taking stock. The rapid development of high quality inductively coupled plasma mass-spectrometry (ICP-MS) has provided precise analyses for a wide range of elements of interest across the large span of concentrations experienced at arcs [e.g. *Eggins et al.,* 1997; and *Elliott et al.,* 1997]. In arc lavas there is much less redundancy in the behaviour of incompatible elements compared to mid-ocean ridge basalt (MORB) and ocean island basalt (OIB) magmatism. Thus the use of different elemen-

Inside the Subduction Factory
Geophysical Monograph 138

10.1029/138GM03

tal tracers can strongly affect the conclusions derived [e.g. *Morris and Hart,* 1983]. The discussion here is based on recent ICP-MS data sets, which allow a full range of tracers to be examined. Additionally, ICP-MS considerably improves accuracy of measurement of certain key elements, notably Nb and Ta, which occur at very low abundances in some arc lavas.

ICP-MS trace element analyses provide a framework to understand more fully the information yielded by an increasing panoply of isotopic tools. Analytical advances have introduced the new stable light isotope systems $^{7}Li/^{6}Li$ [e.g. *Moriguti and Nakamura,* 1998; *Chan et al.,* 2002; and *Tomascak et al.,* 2002] and $^{11}B/^{10}B$ [*Palmer,* 1991; *Ishikawa and Nakamura,* 1994; *Smith et al.,* 1997; and *Ishikawa and Tera,* 1999], resulted in higher precision oxygen isotope data [e.g. *Macpherson et al.,* 1998; *Eiler et al.,* 2000; and *Dorendorf et al.,* 2000] and allowed increasingly extensive data sets of ^{10}Be and Hf isotope measurements [e.g. *Morris et al.,* in press; and *Pearce et al.,* 1999]. Additionally, mass-spectrometric measurements of ^{238}U-^{230}Th [*Reagan et al.,* 1994; *Turner et al.,* 1996; *Elliott et al.,* 1997; *Regelous et al.,* 1997; *Turner et al.,* 1997; *Turner et al.,* 1998; and *Turner et al.,* 1999] ^{230}Th-^{226}Ra [*Chabaux et al.,* 1999; *Turner et al.,* 2000; and *Turner et al.,* 2001] and ^{235}U-^{231}Pa systems [*Pickett and Murrell,* 1997; and *Bourdon et al.,* 1999] provide novel information on the rates of transport of slab derived material.

In this contribution I re-examine global chemical variations of island arc lavas. To this end I have constructed a database that uses only samples for which sufficiently extensive elemental and isotopic data are reported (see Appendix A for details). Additionally I have removed from the database all samples with SiO_2>56% to minimize any effects of differentiation and crustal contamination on elemental and isotopic ratios. Finally, I focus only on compositional trends in lavas from the arc front, the major site of subduction magmatism.

2. A WORKING MODEL OF ARC LAVA GEOCHEMISTRY

The principal inputs of incompatible elements into subduction zone are sediment and the underlying, mafic oceanic crust (Figure 1). At least the upper 500m of the latter are altered by interaction with seawater and this process significantly elevates the concentration of some incompatible elements, e.g. U and the alkalis [e.g. *Staudigel et al.,* 1996]. Both sediment and mafic crust have much higher incompat-

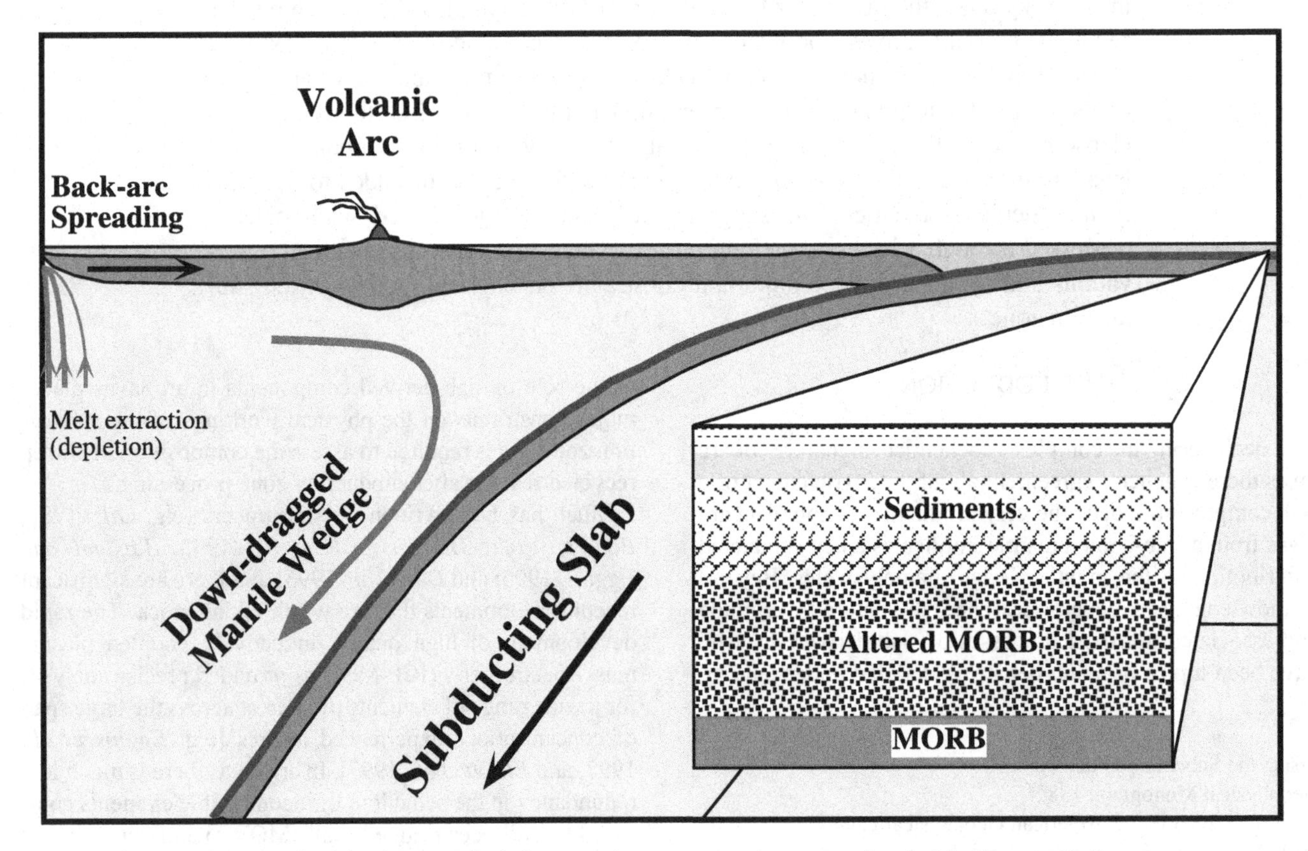

Figure 1. Cartoon of subduction zone depicting some key sources and processes invoked in this contribution.

ible element concentrations than typical mantle rocks and so have the potential to impart a strong chemical signature if added to the mantle wedge. The mantle wedge itself is most plausibly of a composition similar to that of ubiquitous upper mantle, as sampled by MORB. Additionally, where back-arc spreading is active, the mantle fed into the subduction zone may have been depleted by prior partial melting in the back arc [e.g. *McCulloch and Gamble,* 1991]. From a basic conceptual model of the subduction zone (Figure 1), it might thus be expected that arc lavas represent a mix of three main components:

1. depleted mantle wedge
2. subducted sediment
3. (altered) mafic oceanic crust.

3. OVERVIEW OF INCOMPATIBLE ELEMENT SYSTEMATICS

The elements and associated isotope systems discussed in this contribution are generally termed incompatible, implying that they preferentially enter a melt phase relative to a solid residue. This is somewhat loose terminology since most elements are compatible in some, albeit relatively uncommon, phases. In the case of sediment melting discussed below, accessory minerals can be stable in which many traditionally incompatible elements are compatible, e.g. Nb in rutile. Thus 'incompatible' in this contribution is taken to mean incompatible during normal mantle melting, i.e. with an olivine, orthopyroxene, clinopyroxene and aluminous phase assemblage. The term 'depleted' perhaps also requires some clarification. Depleted signatures are those that show low incompatible element abundances and low ratios of more to less incompatible elements, which in time evolve into low $^{87}Sr/^{86}Sr$, high $^{143}Nd/^{144}Nd$, for example. Any, or all, of these characteristics may be used to infer depletion. Lava sources are depleted by removal of melt, but this process can be masked by subsequent addition of 'enriched' material, such as subducted sediments.

It is common practice when discussing subduction zone magmatism to classify incompatible elements as 'fluid mobile' or 'fluid immobile'. These groups closely correspond to other monikers such as large ion-lithophile elements (LILE) and high-field strength elements (HFSE) respectively. The rationale is that large ions (e.g. K^+, Ba^{2+}) more readily partition into an aqueous fluid phase than small highly charged ions (e.g. Zr^{4+}, Nb^{5+}). The issues become more blurred for elements such as the light rare earth elements (LREE, e.g. La^{3+}) and actinides (e.g. Th^{4+}) that have quite large ionic radii but are also quite highly charged. These elements are treated by some as 'fluid mobile' LILE and by others as 'fluid-immobile' elements similar to the HFSE. At low temperatures LREE and Th are not readily moved by aqueous fluids, but at higher temperatures there is conflicting experimental evidence about their fluid mobility. In this contribution I try to place observational constraints on elemental behaviour, rather than make prior assumptions.

Although a crude tool, the extended Masuda-Coryell diagram [*Masuda,* 1962; and *Coryell et al.,* 1963] or 'spidergram', provides both a useful overview of the behaviour of incompatible elements and illustrates the key features that need to be explained in subduction zone magmas (Figure 2). The ordering of elements along the abscissa has traditionally been empirical, with the aim of generating a regular depletion toward the left (more incompatible) end of the diagram for MORB [e.g. *Hofmann,* 1988]. Reliable determinations of partition coefficients appropriate for mantle melting, recently made more accessible by ion-microprobe analysis, has added an experimental basis to this ordering [e.g. *Hart and Dunn,* 1993; *Hauri et al.,* 1994; *Blundy et al.,* 1998; and *Salters and Longhi,* 1999]. Given the underlying aim of producing a smooth pattern for MORB, concentration spikes or troughs on the spidergram point to processes at work other than those involved in normal upper mantle melting.

Plots of average island arc lava compositions are shown in Figure 2a, and it is clear that their spider-traces are far from smooth. With its a priori smooth pattern, a MORB spider-gram can be reasonably well described with rather few elements. This is clearly not the case in arc lavas, where a fuller range in trace elements is required to characterise a sample properly. There are notable positive spikes at Sr and Pb and a large negative spike at Nb (and Ta), relative to adjacent elements. The negative niobium anomaly has long been stressed as a key signature of arc volcanics [e.g. *Gill,* 1981]. More subtly, it can be noted that Th is less abundant relative to its highly incompatible compatriots (e.g. Ba and U) in the most depleted lavas. This negative Th anomaly becomes less apparent while the negative Nb anomaly tends to become more pronounced in the more enriched lavas, i.e. those with generally elevated incompatible element abundances (Figure 2a). Any model of arc lava formation needs address these major features.

In a general sense, the unevenness of the arc lava spider-traces mimics the nature of oceanic sediments (Figure 2b), although the compositions of subducted sediment packages can vary considerably from arc to arc [*Plank and Langmuir,* 1998]. It is also clear that the highly incompatible elements, to the left of the spidergram, are relatively the least abundant in depleted mantle, to which slab components are

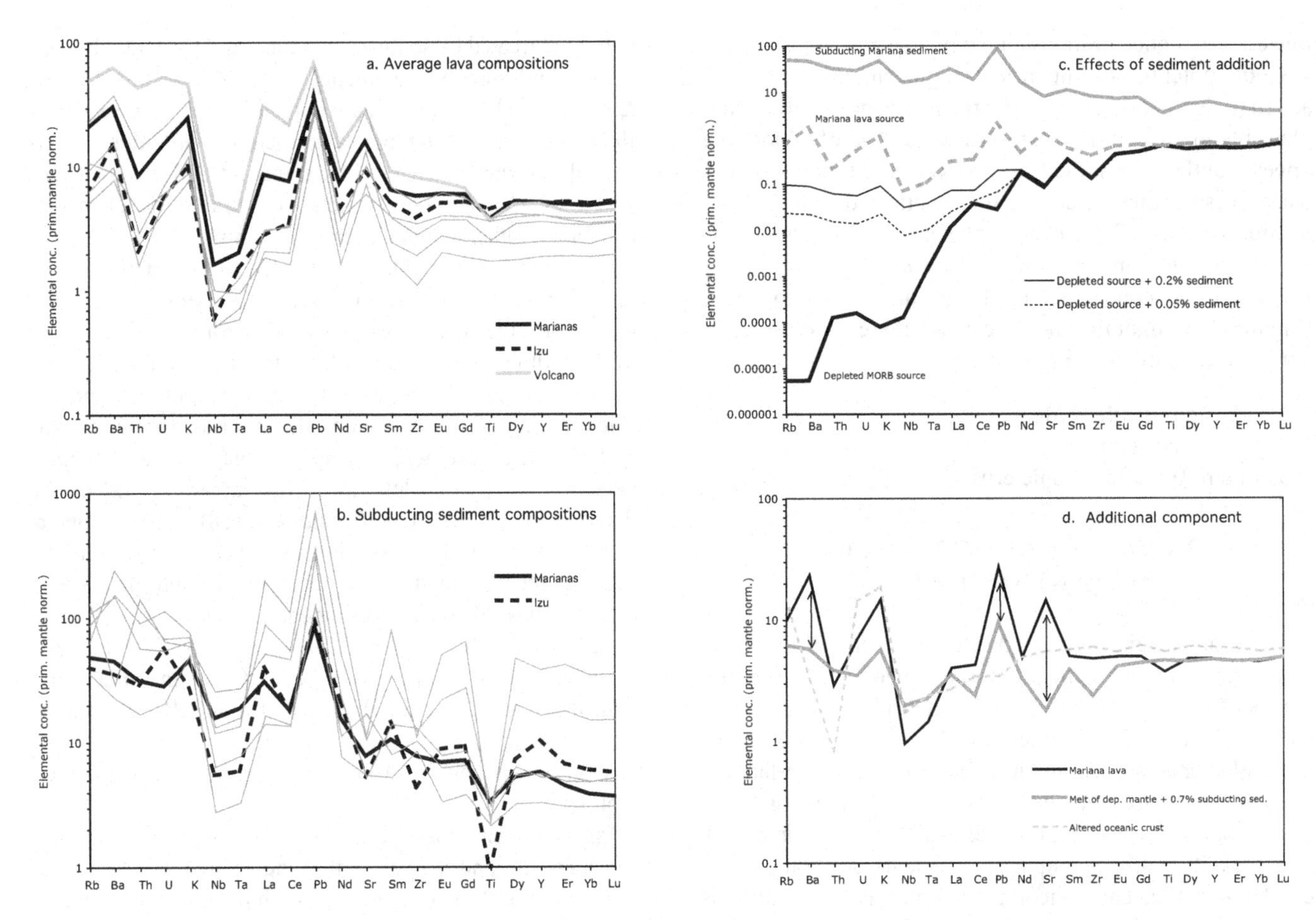

Figure 2. Primitive mantle normalised [*Sun & McDonough,* 1989] spidergrams. (a) average lava compositions for individual oceanic island arcs (from the compilation of this study). Three Western Pacific arcs are picked out to highlight global range, Volcano arc (solid grey line), Marianas central island province (solid black line) and Izu arc (dashed black line) (b) average subducting sediment compositions [*Plank and Langmuir,* 1998] for arcs shown in (a) - note no average composition available for Volcano arc sediment (c) compositions of a calculated depleted MORB source, average subducting Mariana sediment and 0.05% and 0.2% (by mass) mixes of the sediment with depleted mantle. A calculated source of a typical Mariana lava is also shown for reference. A normal MORB source is calculated assuming the average MORB composition of *Sun and McDonough* [1989] represents a 10% accumulated fractional melt. The melt residue after 5% fractional melting of this MORB source is plotted as the depleted MORB source. The Mariana lava source is calculated by assuming GUG13 [*Elliott et al.,* 1997] represents a 5% accumulated fractional melt. Melting calculations assume 0.5% porosity, modes and melting reactions after *Robinson et al.* [1998], and clinopyroxene/melt partition coefficients, which dominate the bulk partition coefficients, from *Hart and Dunn* [1992] taken as appropriate for these relatively large degree melts. For elements used here, not measured by *Hart and Dunn* [1992], interpolations have been made using the procedure of *Wood and Blundy* [1997] (d) typical Mariana lava (GUG 13) compared to a 5% accumulated fractional melt of the depleted mantle source from (c) to which has been added 0.7% of the average subducting Mariana sediment- melting calculations as in (c). Altered oceanic crust composite of *Staudigel et al.* [1996] plotted as reference. Double headed arrows highlight the most significant mismatches between calculated and observed lava elemental abundances, that further do not correspond to obvious enrichments in the altered mafic oceanic crust.

added (Figure 2c). Indeed the contrast between depleted mantle and subducted sediment is up to seven orders of magnitude. Thus the highly incompatible element composition of an arc lava source is strongly influenced by even very small fractions of sediment addition (Figure 2c). The difference in elemental concentrations between mantle and sediment diminishes to the right of the spider-diagram, and thus the relative influence of the sediment on the budget of these elements in the arc lava source dramatically decreases (Figure 2c). Judicious selection of elements can thus inform about different aspects of the arc lava petrogenesis. Moderately incompatible element abundances at the right of the spidergram will dominantly reflect melting and fractionation processes, whereas the highly incompatible elements will chiefly document the effects of slab additions.

As a rough illustration of the effect of sediment addition to the mantle wedge, Figure 2d shows a 0.7% (by weight) mix of average subducted Mariana sediment with depleted mantle, which has then been melted. The result is compared with a typical Mariana arc lava [*Elliott et al.,* 1997]. This tiny amount of sediment addition to the mantle wedge can elevate many highly incompatible element concentrations to values comparable to an arc lava source. However, it is worth re-expressing this modelled mass-fraction of sediment added to the mantle in the more tangible terms of the fraction of total sediment subducted. In a very simple scenario, the thickness of the sediment veneer descending the Mariana trench (500m, density ~1800kgm^{-3}) can be compared to a mantle melting column from the Benioff zone to the bottom of the arc front crust (~100km, density ~3300kgm^{-3}). This represents a mass-ratio of sediment-mantle of ~0.25%. While an unrealistic physical mixing model, this calculation illustrates that apparently very modest sediment-mantle ratios can readily account for the fate of all subducted sediment.

Although a large fraction of the incompatible element characteristics of the average Mariana lava can be modelled by the crude mix discussed above, there are some notable deficiencies (Figure 2d). Firstly, the negative niobium anomaly produced is not large enough. Secondly, the model fails to fully replicate the magnitude of the positive Ba, U, K, Pb and Sr spikes. Additional contributions from the altered oceanic crust can help account for the spikes and the effects of sediment melting can explain the relatively low Nb abundances [*Elliott et al.,* 1997]. The compelling evidence for these arguments revolves around carefully chosen trace element and isotope ratios (see below), which illustrate specific issues in greater detail than can be gleaned from logarithmic plots of a swathe of element abundances. Thus having seen some useful general features we bid the spidergrams adieu.

3.1. Depleted Mantle

It is a difficult task to see through possible additions from altered oceanic crust and sediment to attempt to characterise the composition of the mantle wedge prior to subduction contributions. An approach championed by Woodhead and coworkers [e.g. *Woodhead et al.,* 1993] has been to look at the abundances and ratios of high field strength elements (HFSE). As evident in Figure 2b, these elements are less abundant in sediments relative to other elements of comparable incompatibility and so are least enriched by sediment addition (Figure 2d). Furthermore, high field strength elements are not readily transported by fluids [e.g. *Tatsumi et al.,* 1986] and therefore they are little affected by a possible slab derived 'fluid' component. Although less dominated by subduction additions, high field strength element abundances in arc lavas (especially Nb and Ta for example) are still influenced by contributions from the subducted sediment (e.g. Figure 2d), and so only provide a *minimum* constraint on the depletion of the mantle wedge.

Woodhead et al. [1993] convincingly discussed this issue using the less incompatible HFSE Ti and Zr together with Sc and V. With the advent of reliable ICP-MS data, the more highly incompatible Nb and Ta can be used to provide yet more compelling evidence for a depleted mantle protolith in the source of many arc lavas. The advantage of using the more highly incompatible elements (i.e. those to the left of the spidergram) is two-fold:

a) Prior mantle depletion should more markedly affect the most highly incompatible element abundances (see Figure 2c)

b) Fractional crystallisation, especially of titano-magnetite, and variations in melting affect ratios of the less incompatible HSFE in the erupted lavas, but little influence highly incompatible element ratios.

It is striking that unlike other elements of comparable incompatibility (e.g. Th), Nb concentrations in some arc lavas reach values *lower* than in MORB (Figure 3a). Given that the MORB mantle reservoir is normally understood to be a depleted reservoir, it appears that some arc lava sources are even more depleted.

Although Figure 3a provides a striking first order observation, a plot of simple elemental abundances is a limited tool. Liquid lines of descent in arc lavas suites can be confused by mixing and crystal accumulation [e.g. *Gill,* 1981] and so quantitative fractionation correction is not always easy. Moreover, the elemental abundances in melts depend not only on the composition of the mantle source but also the degree of melting, which remains a matter of debate for the poorly understood melting regime beneath arcs.

A rather more robust measure of source depletion can be obtained via the Nb/Ta ratio. As two highly incompatible elements with the same valency and apparently identical ionic radius [Shannon, *1976*], the difficulty in fractionating Nb and Ta during melting has long been recognised. Thus melts should faithfully preserve the Ta/Nb of their source. In residues of a near fractional melting process, however, even highly incompatible elements are fractionated from one another [e.g. *Johnson et al.,* 1989]. ICP-MS measurements provide sufficiently high quality data to assess small

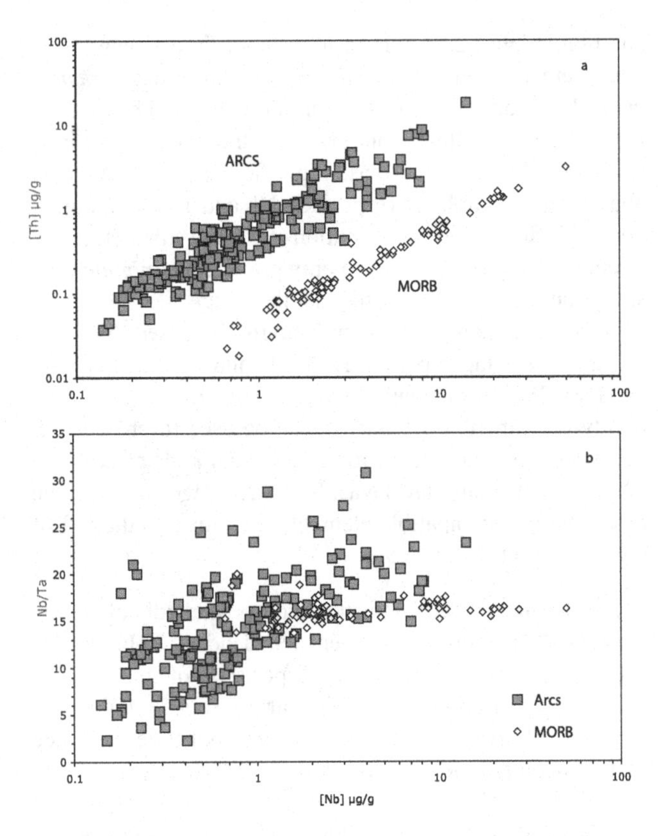

Figure 3. Plots of (a) [Nb]versus[Th], both µg/g b) [Nb], µg/g, vs. weight ratio Nb/Ta Filled light grey squares represent island arc lavas from the compilation of this study, and open diamonds represent MORB, see Appendix A for data sources.

variations in Nb/Ta ratios, which can therefore be used as a tracer of prior melt depletion in the arc source [*Plank and White,* 1995; *Eggins et al.,* 1997; and *Elliott et al.,* 1997]. Experimentally determined distribution coefficients of Ta and Nb in common mantle minerals show that Nb is more incompatible than Ta [*Green et al.,* 1989], and so melt depletion should result in a mantle residue with Nb/Ta ratios less than chondrite (~17) at low Nb abundances. This characteristic is observed in arc lavas, Figure 3b, and provides strong evidence for the involvement of a mantle source in arcs more depleted than the ambient upper mantle.

In the simple cartoon of the subduction zone (Figure 1), this depletion is readily accomplished by feeding the mantle wedge with material that has first passed though the back-arc melting regime. The depletion evident in high field strength element abundances and ratios should thus be a recent phenomenon. This conjecture can be tested using the Hf isotope system, in which long-lived REE/HFSE variations should translate into variable $^{176}Hf/^{177}Hf$. Salters and Hart [1991] investigated Fijian lavas which show a wide range of arc-related HFSE depletions and found no correlation with their Hf isotope ratios. They therefore concluded that the HFSE depletion was recent.

3.2. Two Subduction Components

Many influential, general models of arc magmatism have attempted to account for composition of arc volcanics by adding a single component from the subducting slab to their source [e.g. *Tatsumi et al.,* 1986; *Morris et al.,* 1990; and *McCulloch and Gamble,* 1991]. Yet, an inescapable conclusion from detailed studies on individual arcs is that in many cases there is evidence for (at least) two discrete subduction components [*Kay,* 1980; *Ellam and Hawkesworth,* 1988; *Plank and Langmuir,* 1992; *McDermott et al.,* 1993; *Miller et al.,* 1994; *Reagan et al.,* 1994; *Turner et al.,* 1996; *Elliott et al.,* 1997; *Regelous et al.,* 1997; *Turner et al.,* 1997; and *Class et al.,* 2000]. By looking only at certain elements, a single component may dominate and so mute the contribution of other additions. For example, in Figure 3b it might appear that there is a unique subduction component lying at the top right hand side of the plot, with high Nb/Ta and high [Nb]. If a larger suite of trace elements is examined, however, it generally becomes apparent that this cannot be the case.

To exemplify this point, Figure 4 shows a plot using the chondrite normalised ratio of La/Sm, as a more familiar index of enrichment than [Nb] or Nb/Ta. Sediment addition to the mantle wedge, for example, should both increase incompatible element contents (e.g. Nb or La) and also cause light rare earth enrichment (i.e. increasing La/Sm) in the arc lava source (e.g. Figure 2c). Although La/Sm is also possibly affected by different degrees of melting, for the large degrees of melting proposed for arc front lavas [*Plank and Langmuir,* 1988] this should not be significant. Thus La/Sm should act as a useful tracer of the source composition of mafic, arc front lavas. From Figure 4 it is clear that the most enriched arc lavas, potentially containing the greatest sedimentary component in their source, are the least different from MORB in terms of their Ba/Th. Conversely, the light rare earth depleted lavas have extreme Ba/Th ratios, orders of magnitude higher than values found in other oceanic lavas.

Thus two components need to be added to depleted upper mantle to generate the range of arc lava sources responsible variability in arc lava compositions shown in Figure 4. Component 1 causes light rare earth enrichment and component 2 generates large relative Ba abundances. Below I examine the other characteristic chemical signatures associated with each end-member and conclude that component 1 is of sedimentary origin and component 2 is derived from the mafic altered oceanic crust.

4. COMPONENT 1 (SEDIMENT)

4.1. Trace Element Systematics

Increasing La/Sm in the global dataset of arc lavas is also associated with dramatically increasing incompatible element contents, such as Th (Figure 5a), La (Figure 5b) and Nb (Figure 3a). It is worth noting that although the actual abundances of Nb in arc lavas are low compared to the other elements (i.e. the negative Nb anomaly, Figure 2a), there is still a two orders of magnitude increase in Nb contents in arc lavas as they become more LREE enriched. As was illustrated in Figure 2c, increasing incompatible element contents and LREE enrichment in the arc lava source is consistent with larger proportions of sediment added to the mantle wedge. Such behaviour is not, however, diagnostic of this process.

Further geochemical systematics are shown in Figure 6. There is a general decrease in $^{143}Nd/^{144}Nd$ (Figure 6a) and a scattered increase in Th/Nb (Figure 6b) with light rare earth enrichment. When individual arcs are identified within the general array in Figure 6b, it becomes apparent that on the arc-scale there are good correlations of Th/Nb and La/Sm. The importance of the indices used in Figure 6 is that $^{143}Nd/^{144}Nd$ is a robust tracer of source composition, unaffected by differences in melting process or fractionation. The general negative correlation of $^{143}Nd/^{144}Nd$ with La/Sm implies that the more commonly available trace element ratio index of enrichment is indeed a reasonable tracer of source. The Th/Nb ratio is a simple expression of the classic 'negative niobium anomaly' evident in spidergrams such as Figure 2a. Notably, the Th/Nb of arc lavas are considerably higher than values observed in MORB or indeed OIB. Oceanic sediments characteristically have negative

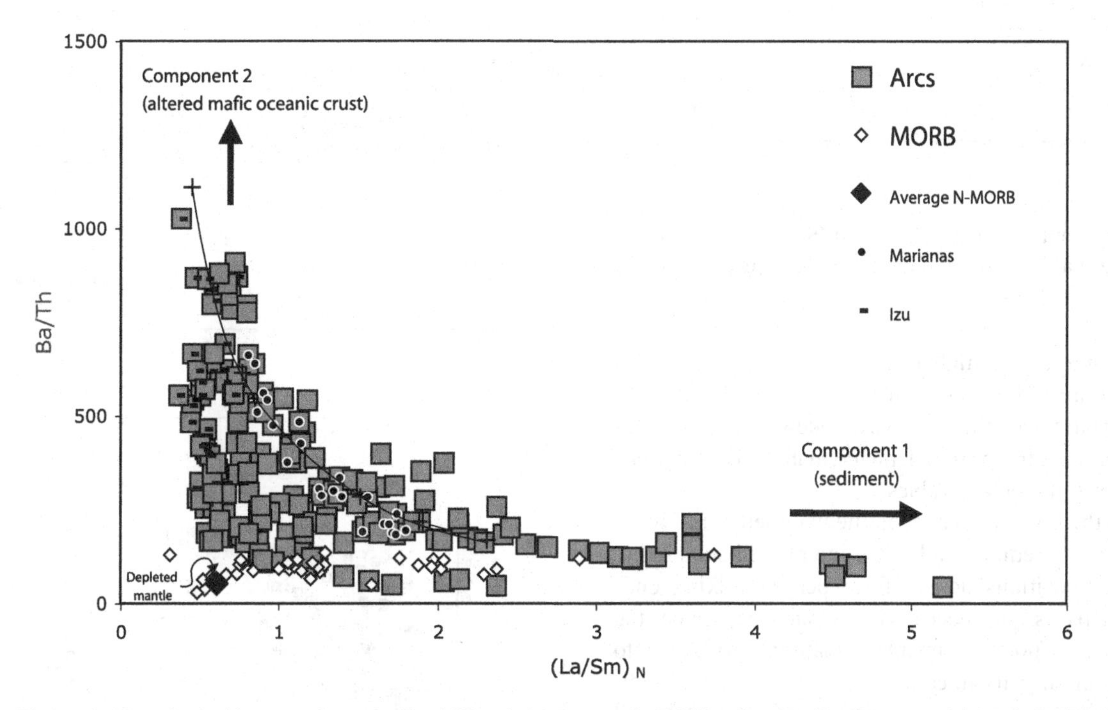

Figure 4. Plot of primitive mantle normalised [*Sun & McDonough,* 1989] ratio of La/Sm versus weight ratio Ba/Th for island arc (filled grey squares) and MORB (open diamonds) from the compilation of this study (see Appendix A for data sources). The MORB samples cover a full range of composition found, but are weighted to rather enirched compositions (see Appendix A). Thus for further reference, an average depleted mantle composition is inferred from the N-MORB of *Sun and McDonough* [1989]. Additionally samples from the Mariana central island province and the Izu arc are highlighted with black filled circles and black dashes respectively. A mixing calculation, marked with vertical crosses joined with a solid line, has been plotted to show the effect of variable the additional of sediment with a constant fluid flux. Average subducting Mariana sediment [*Plank and Langmuir,* 1998] is mixed with a depleted MORB source (as calculated for Fig 2) in proportions of 0.005, 0.01,0.015, 0.03 and 0.1%, and to this mix is added constant fraction of Ba (4% of the amount that would be fluxed by 100% sediment addition) to simulate the fluid (which is assumed to carry no REE nor Th).

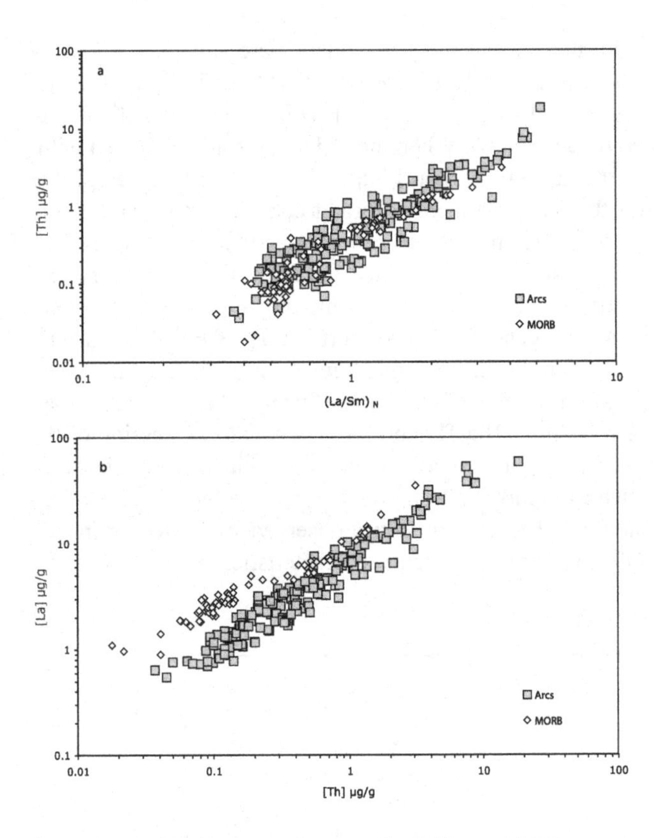

Figure 5. (a) Primitive mantle normalised [Sun and McDonough 1989] ratios of La/Sm vs. [Th] in μg/g (b) [Th] in μg/g vs. [La] in μg/g, symbols as Fig 4.

niobium anomalies, i.e. high Th/Nb, and so the systematics in Figure 6a are in keeping with (variable) sediment addition to the sources of the arc lavas. Likewise $^{143}Nd/^{144}Nd$ values of arc lavas trend from depleted mantle compositions toward lower sedimentary values.

Although the sense of change in the parameters in Figure 6 is indicative of sediment addition, the plotted subducting sediment compositions do not form perfect mixing end-members. This is not necessarily a surprise, since the process that transports sedimentary material from slab to wedge may modify its trace element ratios. Thus a sedimentary slab component need not have the same La/Sm and Th/Nb as the bulk subducting sediment. If sediment addition can be otherwise clearly implicated, the fractionation of these key elemental ratios can then provide important evidence about the sediment transport process.

4.2. Key Sediment Signatures

a) Perhaps most simply, Plank and Langmuir [1993] have elegantly demonstrated that the outputs of key incompatible elements at various arcs world-wide correspond closely to the sediment fluxes of these elements into the trench.

b) Once possible surficial contamination can be ruled out the presence of the short-lived cosmogenic nuclide ^{10}Be in arc lavas provides unambiguous evidence for sediment involvement [*Morris and Tera,* 1989]. The majority of arcs appear to have significant ^{10}Be abundances [*Morris et al.,* in press]. Since the ^{10}Be ($t_{1/2}$ ~1.5My) in incoming sediment has decayed in all but the surface veneer (<10 My), not only does the presence of ^{10}Be indicate a sediment component, but also that even the very top of the sediment column is subducted to depth.

c) Pb isotopes provide another highly sensitive tracer of sediment involvement, having three related isotope ratios to finger-print incoming sediment. Woodhead [1989] summarized long established observations [e.g. *Meijer,* 1976] of arc lava arrays trending from the MORB array toward aver-

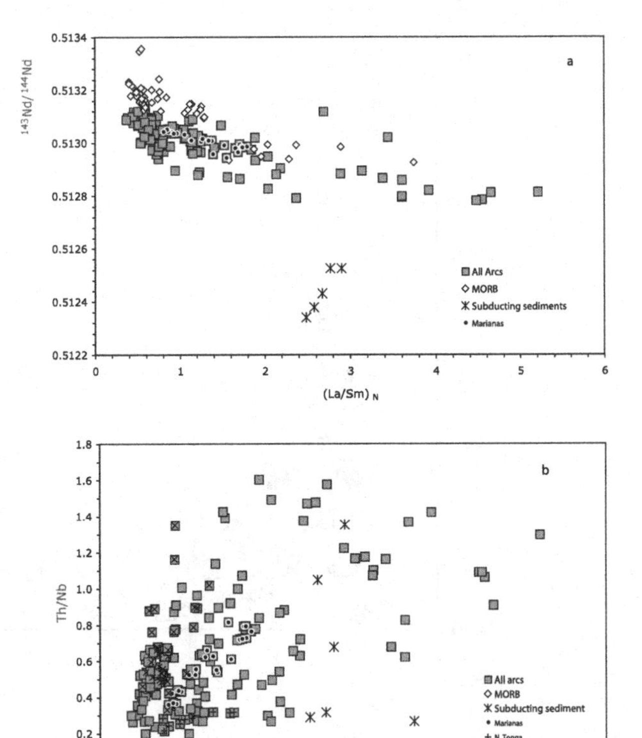

Figure 6. Primitive mantle normalised [*Sun and McDonough,* 1989] ratios of La/Sm vs. (a)$^{143}Nd/^{144}Nd$ and (b) weight ratio of Th/Nb. Symbols as Fig 3 and 4, with additional symbols to highlight N. Tonga (Tafahi and Niuatoputapu) and S. Tonga/Kermadec arc samples, black vertical and black diagonal crosses respectively. Bulk subducting sediment compositions plotted as stars [*Plank and Langmuir,* 1998].

age local sediment compositions. Further case studies show along arc variations in the Pb isotope ratios of erupted lavas matching along strike variability in incoming sedimentary Pb isotope compositions, e.g. the Lesser Antilles [*White and Dupré,* 1986], Banda [*Vroon et al.,* 1995] and Philippines [*McDermott et al.,* 1993]. Furthermore, two contrasting Pb isotopic signatures in the heterogeneous subducting assemblage in the Tongan arc are strikingly replicated in erupted volcanics [*Turner and Hawkesworth,* 1997; and Wendt et al., 1997]. Two arrays in Tongan lavas can be observed (Figure 7); one that trends from the MORB array toward the local pelagic sediment composition and another that heads toward the composition of the subducting Louisville ridge and associated volcaniclastics.

d) The rare earth elements generally behave as a coherent group of 3+ cations with gradually decreasing ionic radii. The electronic configurations of Eu and Ce result in their anomalous behaviour relative to systematic variations within the group as a whole, such that under reducing conditions Eu forms Eu^{2+} and in oxidising conditions Ce forms Ce^{4+} ions. As a consequence, some deep sea sediments show anomalously low Ce concentrations relative to the other REE [e.g. *Ben Othman et al.,* 1989; *Toyoda et al.,* 1990; and *Plank and Langmuir,* 1998], so-called negative Ce anomalies. The sediment subducting at the Marianas has a notable Ce anomaly [*Hole et al.,* 1984; *Lin,* 1992; and *Elliott et al.,* 1997] and it is striking that Mariana arc lavas show increasing negative Ce anomaly with many of the other indices of sediment addition [Woodhead, 1989; and *Elliott et al.,* 1997]. Since at mantle redox conditions Ce does not form 4+ ions, these negative Ce anomalies are linked to the subducting slab. Not all subducting sediment assemblages have Ce anomalies however and so this is not a universal tracer.

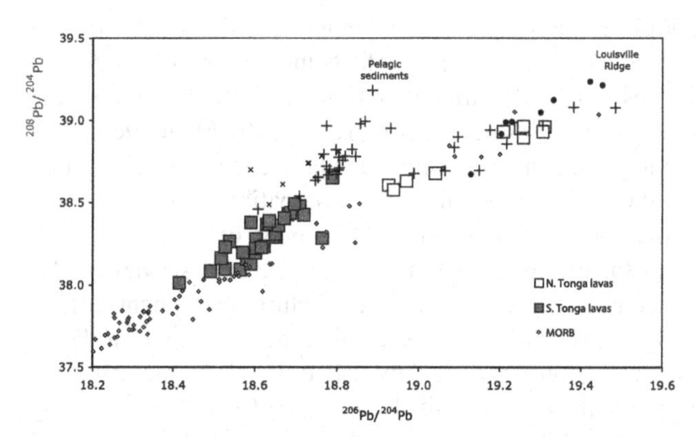

Figure 7. $^{206}Pb/^{204}Pb$ versus $^{208}Pb/^{204}Pb$ for N. Tongan (Tafahi and Niuatoputapu) and S. Tongan/Kermadec arc, open squares and filled grey squares respectively (data from this compilation), Pacific MORB, small diamonds, pelagic and volcaniclastic sediments off-board the Tonga-Kermadec arc, vertical crosses [*Turner et al.,* 1998; *Ewart et al.,* 1998; and *Gamble et al.,* 1998], pelagic sediments from the Pacific as a whole, diagonal crosses [*Ben Othman et al.,* 1989] and samples from the Louisville ridge, filled circles [*Park,* 1990]. The N. Tongan lavas clearly trend toward a Pb isotopic composition defined by the Louisville ridge and associated volcaniclastics measured in the lower section of ODP Hole 204 [*Turner et al.,* 1998; and *Ewart et al.,* 1998], which is quite distinct from the rest of the Tonga-Kermadec arc that forms an array toward appropriate pelagic clay compositions.

Datasets that include the more specific tracers of sediment outlined above do not always contain sufficient trace element measurements to relate directly with the trace element trends discussed for component 1. Yet it is highly unlikely that sediment addition occurs without generating the trace element features observed in Figure 3, 5 and 6. Moreover, in the cases where negative Ce anomalies are observed, these do clearly correlate with light rare earth enrichment [*Woodhead,* 1989; and *Elliott et al.,* 1997].

4.3. Nature of Sediment Transport

An important issue is the means by which sediment is transported into the mantle wedge. As discussed above, there is evidence in the chemistry of arc lavas that it is not a bulk sediment component, but a fractionated sedimentary component that is added to the arc lava source. Most plausibly, this fractionation is the related to sediment melting.

Figure 8 shows again the data from Figure 6b, but additionally highlights the composition of subducting sedimentary components for three specific arcs. Mixing of the local sediment compositions with depleted upper mantle clearly cannot generate the array of compositions defined by the associated arc lavas. To generate the source of the most enriched arc lavas requires addition of a component with even higher Th/Nb than the subducting sediment. Melting of the sediment with residual rutile would result in higher Th/Nb of a melt relative to bulk composition. Both Nb and Ta are strongly compatible in rutile [e.g. *Green and Pearson,* 1987], but otherwise this accessory phase little influences the bulk partitioning of other highly incompatible trace elements [e.g. *Foley et al.,* 2000]. Residual rutile was noted in the pelagic sediment melting experiments of *Nichols et al.* [1994], and it is readily saturated in siliceous sediment melts [*Ryerson and Watson,* 1987]. The sediment melting experiments of *Johnson and Plank* [1999], however, did not find residual rutile. Interestingly, the sediment composition used by *Johnson and Plank* [1999] was similar to the bulk composition represented in Figure 8 which has the highest starting Th/Nb. For this composition, the relative Th/Nb fractionation required to generate a suitable slab

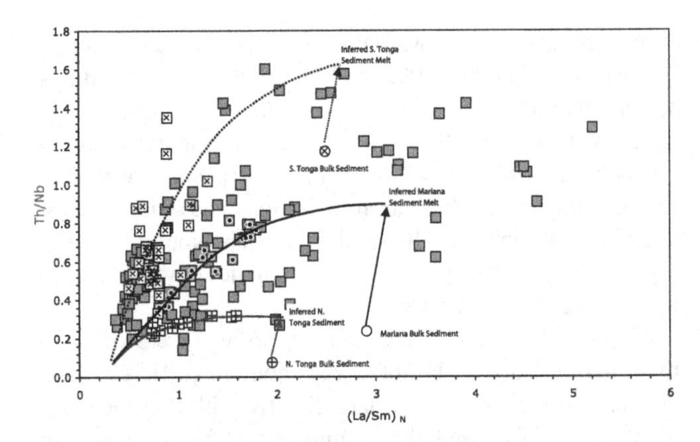

Figure 8. As Figure 6b, but additionally showing compositions of bulk subducting sediment assemblages for the three arc segments highlighted [*Plank and Langmuir,* 1998]. Arrows qualitatively indicate the direction of fractionation required of sediments to allow them to represent a reasonable end-members for the qualitatively sketched mixing arrays.

end-member is the smallest and does not necessarily require the presence of residual rutile.

Further detailed arguments for the need to invoke a fractionated sediment component can be found in individual arc case studies [e.g. *Elliott et al.,* 1997, *Turner et al.,* 1997, *Class et al.,* 2000]. Although a fractionated sediment composition appears to be required, there is some debate as to whether it necessarily needs to be a melt. The significance of inferring sediment melting is that the sediment solidii are rather too high for commonly accepted models of slab surface temperatures [*Nichols et al.,* 1994; and *Johnson and Plank,* 1999]. In trying to reconcile this mismatch, the point has been made that a high pressure and temperature fluid may have rather similar properties to a melt, and so the two may be difficult to distinguish geochemically [e.g. *Stolper and Newman,* 1994].

Johnson and Plank [1999], however, make strong arguments that the geochemical characteristics of the sediment component do require it to be a melt. A key observation is the strong enrichment of Th in the sediment component (e.g Figure 5a). The experiments of Johnson and Plank [1999] suggest that Th is not readily carried in the high pressure aqueous fluid phase released from sediment. It is also important to stress that the element transport properties of the phase that does carry the sediment component is clearly very different than that of other slab contribution, component 2, discussed below. For example, most incompatible element budgets are significantly elevated by addition of component 1 (Figure 5). Component 2, in contrast, selectively enriches rather few elements. To my mind, the simplest rationalisation of this observation is that components 1 and 2 are carried by different phases with different element transport properties, namely a melt and aqueous fluid respectively.

5. COMPONENT 2 (ALTERED MAFIC OCEANIC CRUST)

5.1 Trace Element Systematics

In the discussion of the sediment component above, the second slab component was rather ignored. This was possible by examining elements that can be shown empirically to be little effected by component 2. Thus the role of sediment—mantle mixing could be investigated without additional complication. The rationale behind identifying elements influenced by component 2 is illustrated in Figure 9. In certain plots, it is evident that lavas at the LREE depleted end of the array have incompatible element ratios more divergent from MORB values than the sediment rich end (e.g. Figure 4, Figure 9a). In other cases, the depleted end of the arc lava array appears to trend back toward a depleted mantle field (Figure 6). Intermediate scenarios are also apparent (Figure 9b). These systematics indicate a highly variable influence of a second component on the budgets of different elements. By examining further incompatible element ratios in this manner, it can be shown that component 2 significantly affects the budgets of Ba, Sr and Pb, less markedly influences those of K, Rb, U and does not perceptibly contribute LREE, Th, Nb, Zr (Figure 9c).

Simple incompatible element abundance plots, also indictae the role of component 2. Incompatible elemental abundances increase with inferred amount of sediment input, even for elements such as Ba that are also strongly influenced by component 2. The key feature for elements affected by the second component is that the overall *variability* in elemental abundance is diminished. For example, while Th varies in the arc lava suite by a factor of 100, Ba varies only by factor of 30 (Figure 9d). Thus addition of component 2 moderates the range in certain element abundances by boosting their concentration in the sediment poor lavas.

Another important constraint on the addition of component 2 can be gleaned from the curved inverse correlation evident for some arc lava suites in Figure 4. As already discussed at length, variable LREE enrichment can be related to variable addition of a sedimentary component. The addition of a second component need not be linked, and so the inverse correlation in Figure 4 is not necessarily expected. Nevertheless, if the sedimentary component is allowed to vary but the fluid component is near constant, then the form observed in Figure 4 can be reasonably reproduced. In other words, a source with a large sedimentary addition is so

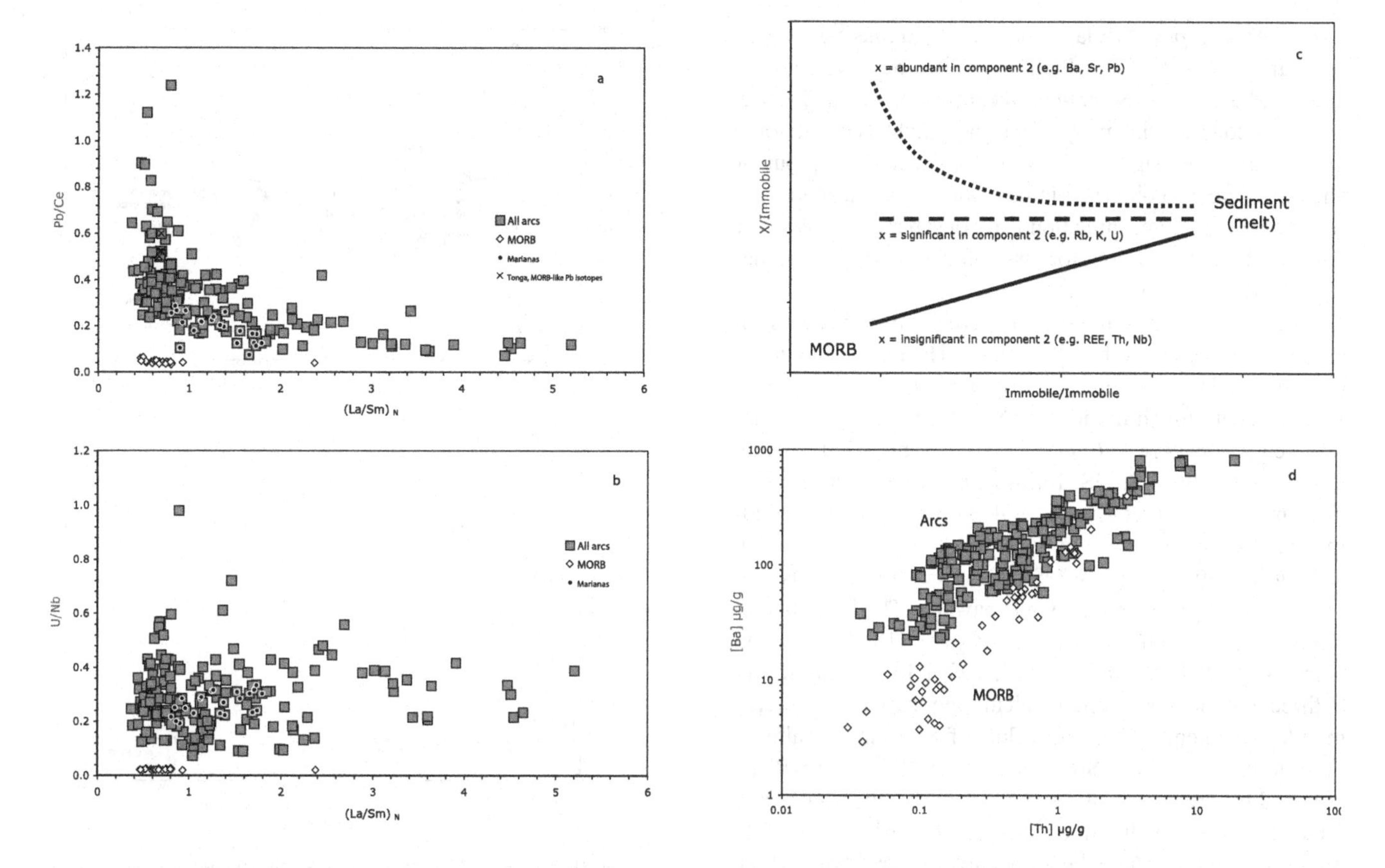

Figure 9. Primitive mantle normalised La/Sm [*Sun and McDonough,* 1989] vs. (a) Pb/Ce weight ratios and (b) U/Nb weight ratios. Symbols as Figure 4, but additionally samples from Tofua, the Tongan lavas with the most MORB-like Pb isotope signatures (see Fig 7) are shown as crosses (c) generalised cartoon depicting the systematics of arc lava arrays using different elements in plots such as Figure 4, Figure 9a and b, see text for details (d) [Th] vs. [Ba], both in µg/g (symbols as Figure 4).

dominated by the sediment contribution to all element budgets that the fluid component has little further effect, whereas in sources where relatively little sediment has been added the effect of fluid addition on certain elements is marked.

Some within-arc trends, however, do not mimic this general pattern. For example, highly depleted arcs (where minor sediment addition does not mask the effects of prior melt depletion) such as Izu, New Britain and Tonga show variable Ba/Th at similar degrees of LREE enrichment (e.g. Figure 4) which clearly cannot be accounted for by variable sediment and near constant addition of a second component [*Taylor and Nesbitt*, 1998; and *Woodhead et al.,* 1998]. It appears that variability in the composition and flux of component 2 is further control that is of importance in shaping the compositions of some arcs where the sediment influence is minor, and role of component 2 dominant.

5.2. Key Signatures of Altered Mafic Oceanic Crust

The key chemical finger-print of component 2, identified above, is a large contribution of 2+ cations and a somewhat less important addition of 1+ cations. Altered mafic oceanic crust has enhanced alkali contents as a result of low temperature alteration (Figure 2d). The striking enrichment of 2+ cations evident in component 2, however, is not a signature of altered mafic oceanic crust, and indeed Ba is highly depleted [e.g. Staudigel et al 1996]. Nevertheless, as with the sedimentary component, there exist powerful isotopic arguments that strongly implicate the role of the altered oceanic crust as the ultimate source of component 2. Once this is established, then the mismatch between the relative elemental abundances inferred for component 2 and the altered oceanic crust protolith can be addressed.

5.2.1 Sr isotopes. Arc lavas have a ubiquitous positive Sr anomalies (Figure 2), which can also be conveniently expressed as high Sr/Nd ratios. Accumulation of plagioclase, which is common in many arc lavas, might account for a minor part of the signature [e.g. *Vukadinovic,* 1993], but Sr anomalies are also observed in near aphyric lavas. Moreover, while plagioclase accumulation can contribute to 'excess' Sr contents, it cannot account for associated excesses of the other 2+ cations Ba and Pb.

$^{87}Sr/^{86}Sr$ provides a tracer of the source of the excess Sr, and by inference the other 2+ cations. The excess Sr cannot be dominantly sedimentary in origin as this would result in arc lavas with much higher $^{87}Sr/^{86}Sr$ than observed, Figure 10a [e.g. *Hawkesworth et al.,* 1993; and Figure 10a]. Nevertheless, the $^{87}Sr/^{86}Sr$ ratios of most arc lavas are significantly elevated relative to MORB. Having ruled out sediments, this implicates the role of altered oceanic crust. During alteration, the $^{87}Sr/^{86}Sr$ of the oceanic crust is elevated by isotopic exchange with seawater. The best characterised site of altered oceanic crust (417/418) has a mean $^{87}Sr/^{86}Sr$ ~0.7046 [*Staudigel et al.,* 1996]. Figure 10a shows a three component mixing calculation between depleted mantle, sediment and a large flux of Sr from the altered oceanic crust. At high Sr/Nd values, the $^{87}Sr/^{86}Sr$ will be buffered by the composition of Sr from the altered oceanic crust. The average altered oceanic crust $^{87}Sr/^{86}Sr$ of 0.7046 does not well reproduce the global array of arc lavas, which would be more consistent with a slightly less radiogenic value. Either site 417/418 is not representative of global mean altered oceanic crust or perhaps there is some interaction of component 2 with the mantle wedge before mixing to form the arc lava source. Nevertheless, from Figure 10a it is clear that a Sr rich component with $^{87}Sr/^{86}Sr$ ~0.7035-0.7045 is required and this is most plausibly derived from the altered mafic oceanic crust.

5.2.2 Pb isotopes. Similar arguments apply to the lead isotopic systems as those for Sr. Subducting sediments generally have markedly elevated Pb concentrations (Figure 2b) and so in dealing with the Pb budget of the sediment contribution is more significant than for Sr. As emphasised earlier, the involvement of sediment has frequently been inferred from the trend of many arc lava suites toward the lead isotopic composition of the local sediment. It is thus the other end of the array that is of importance here, and the arguments have previously been cogently discussed by *Miller et al.* [1994]. In brief, sediment poor lavas frequently have significant positive Pb concentration anomalies even though their Pb isotopic compositions are close to MORB. For example, most arc lavas have Pb/Ce greater than the near constant value in MORB or

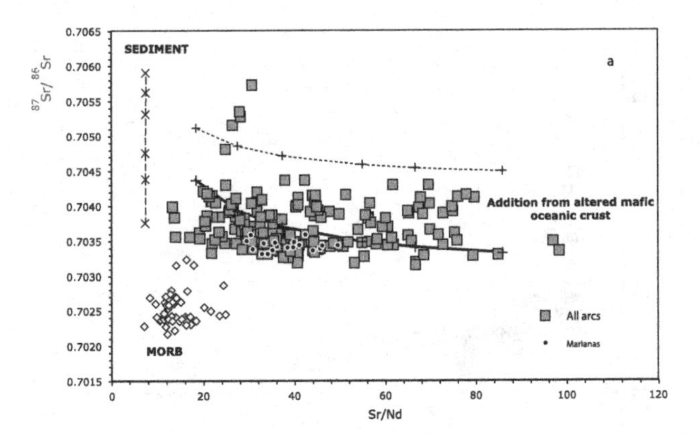

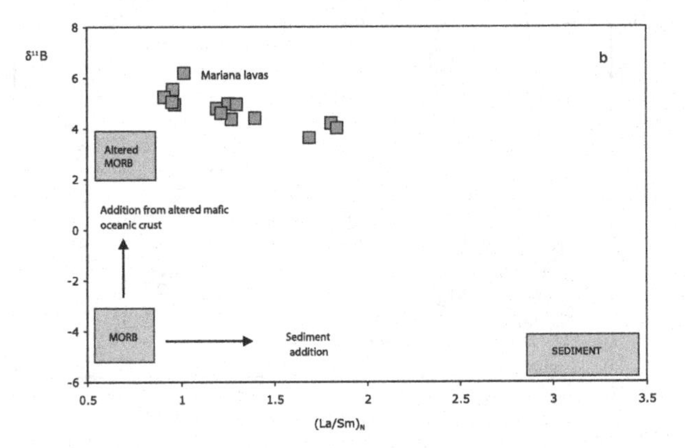

Figure 10. (a) Sr/Nd versus $^{87}Sr/^{86}Sr$ for island arc lavas and MORB (symbols as Figure 4). Using the Marianas as an example, several calculations are shown for mixing between (i) depleted MORB source (see Figure 2) and average Mariana sediment [*Elliott et al.,* 1997], at sediment weight fractions of 0.005, 0.01, 0.15, 0.03, 0.05 and 0.1, marked by diagonal crosses joined by dashed line (ii) as (i) but with a fluid component also simulated by adding extra Sr (constant in all cases and equal to 15.5% the amount of Sr that would be fluxed by 100% sediment addition) with an isotopic composition of average altered MORB [*Staudigel et al.,* 1996], marked with vertical crosses joined by dashed line (iii) as in (ii) but using an ad hoc lower $^{87}Sr/^{86}Sr$ of the fluid component (namely 0.7033) to better fit the data array, vertical crosses joined with solid line. Although the value of $^{87}Sr/^{86}Sr$ was picked to fit the data, it should be noted that the average altered oceanic is at present determined by a single DSDP site. (b) primitive mantle normalised [*Sun and McDonough,* 1989] La/Sm against $\delta^{11}B$ for Mariana lavas [*Ishikawa and Tera,* 1999; and *Woodhead,* 1989] with approximate fields for MORB, altered MORB and pelagic sediment [*Chaussidon and Jambon,* 1994; *Smith et al.,* 1995; and *Ishikawa and Nakamura,* 1994]. The least sediment rich lavas show the highest $\delta^{11}B$, qualitatively consistent with a larger altered oceanic crust contribution. All samples, however, have higher $\delta^{11}B$ than the estimate for altered oceanic crust, which may indicate a component of isotopic fractionation during transport of slab B to the sub-arc mantle.

OIB. Notably the Tongan samples in Figure 7 that plot on the MORB Pb isotopic array have very high Pb/Ce, much greater than MORB (Figure 9a).

The most likely source for the additional Pb is again the subducting mafic oceanic crust. Unlike Sr, the Pb isotope ratio is little affected by seafloor alteration processes, and so retains MORB values. Alternatively it has also been proposed that the additional lead may be simply derived by enhanced fluid partitioning of lead from the mantle wedge [e.g. *Hawkesworth et al.,* 1993]. Given the highly depleted mantle protolith composition inferred for many arcs (e.g. Figure 2c), it is unlikely that large enough volumes of mantle are involved to contribute sufficient Pb. Moreover, for the Izu and Mariana arcs there is a useful contrast between the Pb isotope compositions of the slab and mantle wedge to resolve these possibilities. The mantle wedge beneath the Marianas and Izu arcs likely has a characteristic Indian-ocean type Pb isotope composition [*Hickey-Vargas,* 1991], whereas the subducting mafic oceanic crust has a Pacific Pb signature. A significant Pacific Pb isotope component is inferred for both Izu and Mariana arc lavas [*Elliott et al.,* 1997; *Taylor and Nesbitt,* 1998], which thus emphasises the importance of a Pb flux from the altered mafic oceanic crust.

5.2.3 Stable isotopes. The constraints from Sr and Pb isotope systematics on the origin of the fluid component are strong. It would be useful, however, to have specific tracers of contributions from the altered oceanic crust, just as ^{10}Be provides unequivocal evidence of sediment involvement. Light stable isotope systems potentially yield such tracers, as stable isotope ratios are only significantly fractionated by low temperature processes near the Earth's surface.

A robust stable isotope tracer should be able to distinguish between material from sediment and altered mafic oceanic crust. This is not the case for oxygen and lithium systems, where both sediment and altered mafic oceanic crust generally have heavy ratios (high δ^{18}O and high δ^{7}Li) compared to mantle values. These systems thus provide potential tracers of generic slab derived material, but not of specific components. Oxygen and lithium isotope signatures in arc lavas are both strongly influenced by mantle contributions [*Eiler et al.,* 2000; and *Tomascak et al.,* 2002], and so yield valuable information on this frequently masked component. They are not, however, sensitive tracers of slab derived material.

The B isotope system shows some potential as a distinctive tracer of altered mafic oceanic crust. Boron is a highly incompatible element and so slab contributions should overwhelm the depleted mantle. Furthermore, the limited pelagic sediments so far analysed have B isotopic signature lighter (i.e. lower δ^{11}B) than mantle values [e.g. *Ishikawa and Nakamura,* 1994], whereas the altered oceanic crust is heavier [*Smith et al.,* 1995]. An uncertainty in interpretation of B isotope data arises from potential B isotopic fractionations during incremental B loss from the slab during subduction. Both experimental [*You et al.,* 1996] and observational studies [e.g. *Moran et al.,* 1992] suggest a large proportion of B is lost during prograde metamorphic dehydration reactions in the descending slab. Fractionation of the B isotope ratio appears to occur during this process [*Peacock and Hervig,* 1999; and *Rose et al.,* 2001. Thus the B isotopic signature of arc lavas likely reflects an integrated history of isotope fractionation both during protolith dehydration and transport to the lava source. Thus heavy δ^{11}B is not an unequivocal tracer of altered oceanic crust, but still holds some promise.

The number of B isotopic studies of arcs remain rather small, and few of the samples analysed are uncontaminated lavas from oceanic arc fronts. Within the criteria set for this review, discussion is thus restricted to the work of *Ishikawa and Tera* [1999] on the Marianas Islands. The data from the latter study are replotted in Figure 10b, to show that the most LREE depleted lavas have the heaviest B isotope ratios. Taken at face value, this is in keeping with the derivation of component 2 from the altered mafic oceanic crust. Samples with higher La/Sm have lower δ^{11}B, in keeping with a larger sedimentary contribution to their B signature. All lavas have rather heavy B isotope ratios relative MORB, sediment and even altered MORB. This may imply the importance of altered oceanic crust in the boron budget of arc lavas, but may also indicate enrichment of δ^{11}B in a slab derived fluid relative to the residual solid [e.g. *Peacock and Hervig,* 1999]

5.3. Nature of Transport of Altered Mafic Crustal Component

Given the isotopic evidence for the involvement of altered oceanic crust, we can assess what fractionation occurs during transfer of elements from this part of the slab to the mantle wedge. In detail, the are many complicating factors that makes this calculation highly under-constrained. To highlight the first order issues, however, I simply use the example discussed in Figure 2. It is assumed that the elemental budget not accounted for by 0.7% sediment addition (for simplicity the fractionations associated with sediment melting are ignored) is contributed by the altered oceanic crust. In Figure 11a I plot the elemental abundances required in component 2 relative to their abundances in the altered mafic oceanic crust.

The elements most clearly enriched in component 2 relative to the mafic altered oceanic crust are all 2+ ions, frequently assumed to be mobile in aqueous fluids derived from the slab [e.g. *Gill,* 1981]. The empirical elemental fractionations illustrated in Figure 11a can be compared with an expanding high pressure experimental dataset of solid-fluid partitioning (Figure 11b). It should be noted that there is rather poor agreement between the different experimental datasets. To illustrate contrasting trends within the mass of data, the fluid-clinopyroxene partitioning data from *Brenan et al.* [1995a and b] and *Stalder et al.* [1998] are highlighted (Figure 11b). Notably the *Brenan et al.* [1995a and b] data show marked differences in partitioning behaviour for a range of 'incompatible' elements, whereas the *Stalder et al.* [1998] data show rather smooth variations in their fluid-clinopyroxene partitioning. Some of these differences presumably reflect both a range in experimental approaches and conditions. Nevertheless a better convergence of experimental results appears desirable.

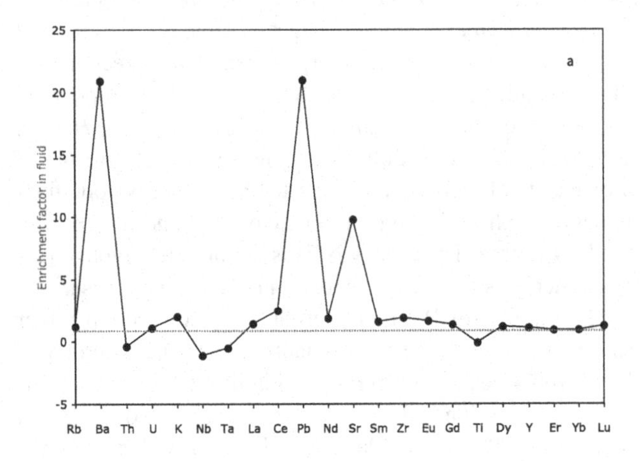

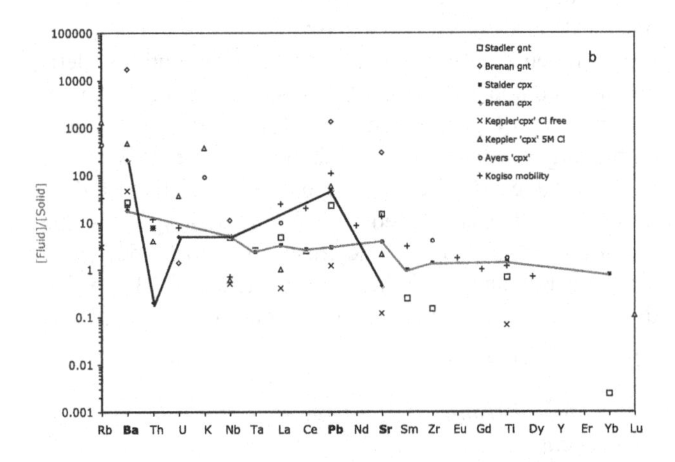

Although some features inferred from arc lavas can be identified in the experimental data, e.g. the high relative mobility of Ba and Pb, no dataset reproduces all the features deduced from variations in lava compositions (Figure 11a). For example, although the experiments of *Brenan et al.* [1995a and b] and *Keppler* [1996] show a number of important features in keeping with empirical deductions, in both studies Sr is determined to be as immobile in the fluid as Th, which is not in keeping with the inferences made in this contribution.

In summary, transport of material from altered mafic oceanic crust to the arc lava source needs to selectively fractionate elements that are highly incompatible during normal mantle melting (e.g. Ba from Th). This is in contrast to the sediment component in which a full range of incompatible elements are moved from the slab to arc source. While an aqueous fluid derived from the altered oceanic crust probably best accounts for the signature of component 2, this is not yet clearly established experimentally.

The apparent 'fluid' mobility of elements such as Th determined in some experiments has been welcomed by workers trying to model arc lavas using a *single* subduction flux. In

Figure 11. (a) Estimate of relative enrichment of elements in component 2 compared to its altered mafic oceanic crustal source. The calculation adds 0.7% Mariana sediment to a depleted MORB source (as in Fig 2), and then fractionated material from the altered oceanic crust [*Staudigel et al.,* 1996] is added. A given weight fraction of altered oceanic crust is added (2%), but for each element an additional multiplier is calculated which allows this addition to fully account for the Mariana source chemistry. These factors are shown in (a) and represent the required fractionation of elements during transport of material from altered oceanic crust to the mantle wedge as component 2. This is a simple calculation that does not consider element fractionations during sediment melt transport, for example, but the clear fractionations required for Ba, Sr and Pb are likely to be robust. (b) experimental data on aqueous/solid fluid partition coefficients to compare with the empirically derived values in (a). Experimental methodologies include mineral-fluid partitioning [*Brenan et al.,* 1995a and b; *Stalder et al.,* 1998], silicate-aqueous fluid partitioning, converted to mineral aqueous fluid partitioning using given $D^{mineral}_{silicate}$ melt [*Keppler,* 1996; *Ayers et al.,* 1997] and element mobility resulting from serpentine dehydration [*Kogiso et al.,* 1997], which is converted to a partition coefficient assuming 5% fluid release). Only single respresentative mineral analyses from many of the studies are shown. Two clinopyroxene samples, one from the experiments of *Brenan et al.* [1995a and b], black line, and one from *Stalder et al.* [1998], grey line, have been picked out to highlight the very different general element behaviours observed in different experiments. Although the experiments of *Brenan et al.* [1995a and b] show some of the 'bumpiness' observed in (a), Sr mobility is not reproduced.

such models a 'fluid' needs to transport all elements. Given the observational constraints presented above, a single slab component cannot explain the geochemical systematics of arc lavas, and so it is not *necessary* to invoke fluid mobility for elements such Th. If aqueous-fluid partitioning does efficiently transport the full range of elements discussed, then an alternative mechanism needs to be invoked to supply the highly specific suite of elements associated with component 2. Acknowledging and reconciling these experimental and geochemical observations represents an important goal in the near future, that will hopefully provide important constraints on the conditions under which fluid partitioning occurs.

5.4. Additional Components?

The treatment above identified variable sediment addition to the mantle wedge as the dominant control on arc lava composition. It was then noted that this single slab flux was not sufficient to account for all the geochemical systematics, and so a second slab component was invoked. It is possible that the latter, residual chemical signature in fact comprises several different components. This issue is difficult to resolve, but some further information is provided courtesy of the so-called fluid mobile elements (FME).

The aim of this contribution is to identify the controls on the chemistry of the volumetrically dominant arc front lavas. Yet there are valuable constraints to be gleaned from compositional variations in the low volume, 'cross-arc' volcanics. Cross-arc volcanics form chains of volcanoes that run perpendicular to the trench, from the arc front out into the back arc. Significantly, they may monitor progressive changes along a 2-D transect of the arc.

Key features in cross-arc lavas are the general increase in many highly incompatible elements away from the arc front [*Ryan et al.,* 1996]. This has been attributed to smaller degrees of melting related to a decreasing water supply at increasing distances from the trench. Not all highly incompatible elements show this behaviour. Elements such as B, As, Sb and Cs are notable for their *decrease* in concentrations in cross-arc volcanoes away from the arc front [*Ryan et al.,* 1996]. The behaviour of these elements is thus comparable to that deduced for water. Moreover, progressive depletions of these elements with increasing metamorphic grade and devolatilisation have been documented in exhumed subduction terrains [e.g. *Moran et al.,* 1992]. It has therefore been proposed that B, As, Sb and Cs effectively track water, and so have been dubbed fluid mobile elements.

The FME are not commonly measured, and the dataset remains small. For the samples considered in this study, FME systematics in arc front lavas closely follow that of elements linked to the altered mafic crust [see *Noll et al.,* 1996]. Yet, the behaviour of many of the key elements (e.g. Ba) that comprise the altered mafic crust component are different in the cross-arc chains relative to the FME [see *Ryan et al.,* 1996]. This most likely reflects different efficiencies of element transport during slab dehydration, which will in turn relate to the location of their addition to the mantle wedge. As a speculative alternative, different processes may be responsible for the transport of FME and other elements in component 2. For example, it is possible that while the FME are transported by an aqueous fluid, the other component 2 elements are carried by a melt from the altered mafic oceanic crust. The solidus of hydrated basalt is not greatly different from some hydrated sediments [e.g. *Johnson and Plank,* 1999]. Given the powerful arguments for sediment melting, it is possible that under comparable conditions, the altered mafic crust also melts. Intriguingly, recent partitioning experiments suggest that even anhydrous melts, if sufficiently sodic, can suitably fractionate key elements to produce compositions that resemble those of component 2 [Bennett et al, submitted].

6. RATES OF PROCESSES AT ARCS

The U-series nuclides provide a set of chronometers, sensitive to different processes and time-scales. Very briefly, the decay of U to Pb (^{238}U-^{206}Pb and ^{235}U-^{207}Pb) occurs in as a chain of short-lived intermediate nuclides. Since the head of this cascade is fed by a parent significantly longer lived than any of the daughters, the system naturally establishes a steady state, termed secular equilibrium. At secular equilibrium the rate of decay (i.e. measurable activity) of all the nuclides is equal, and set by the abundance of the long-lived parent. Geological process often cause fractionation between the different daughter elements which results in disequilibrium between some or all of the nuclides in the chain. The time-scale of return to equilibrium of any adjacent nuclide pair is controlled by the half-life of the shortest lived nuclide, and as a rule of thumb discernible disequilibrium persists for some five times this half-life. Most of the U-series nuclides have half-lives rather too short to be of use in igneous systems. However, ^{230}Th, ^{226}Ra and ^{231}Pa have half-lives of 75380, 1600 and 32760y respectively. Thus looking at departures of the activity ratios, denoted by parentheses, of (^{238}U/^{230}Th), (^{226}Ra/^{230}Th) and (^{231}Pa/^{235}U) from equilibrium (i.e. 1) provides invaluable information about the fractionation of these elements in the last 1000-300000y.

Recent developments of mass-spectrometric techniques allow precise measurement of U-Th-Ra and U-Pa in arc

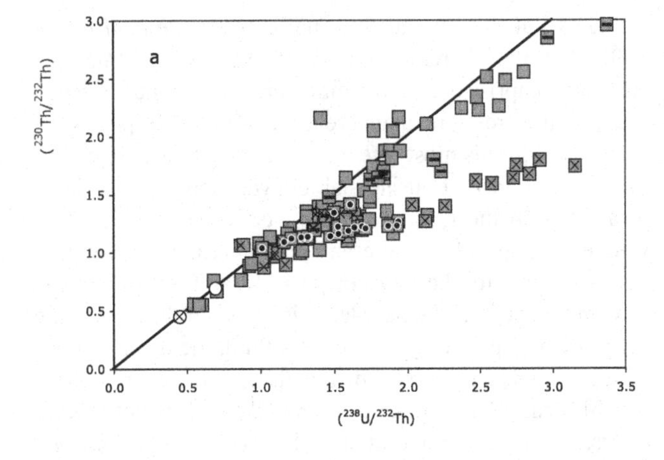

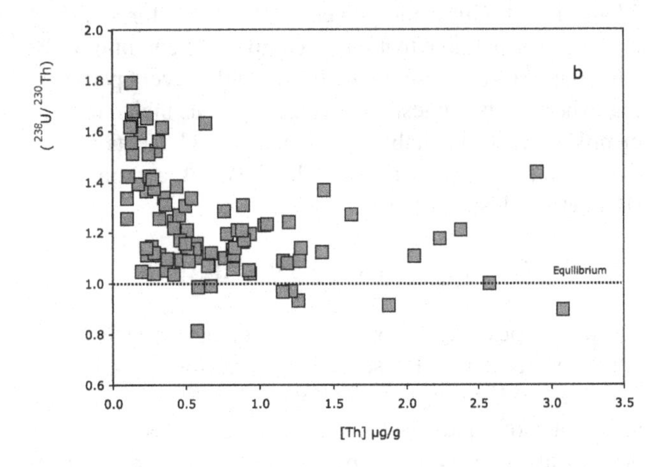

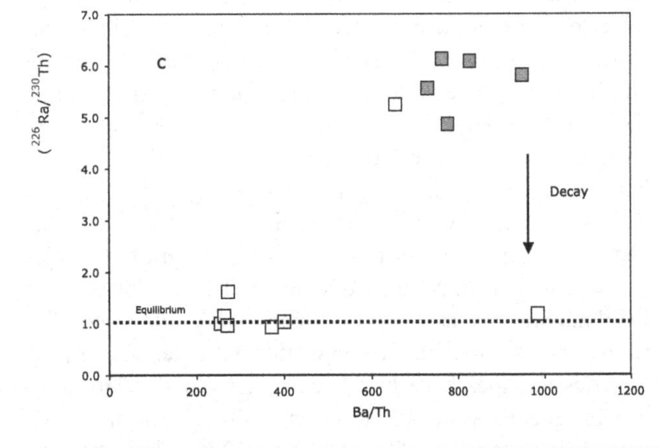

lavas, and combination of all three systems promises tight constraints on the timing of element transfer and melting processes. Dealing with the necessary back-ground to understand fully the detailed systematics of U-series tracers is beyond the scope of this contribution. Here I will briefly explore some of the more straight forward inferences that can be made from U-Th analyses and the surprisingly large Ra excesses observed in some arc basalts [e.g. *Capaldi et al.*, 1983; *Rubin et al.*, 1989; *Gill and Williams,* 1990; *Gill et al.*, 1993; *Reagan et al.*, 1994; *Hoogewerff et al.*, 1997; *Clark et al.*, 1998; *Chabaux et al.*, 1999, *Turner et al.*, 2000; and *Turner et al.*, 2001].

The U-Th system has been the most thoroughly studied of the U-series nuclides and there is currently a fairly extensive database of high precision mass-spectrometric measurements (Figure 12a). U-Th data is conventionally shown on a so-called equiline diagram [*Allègre,* 1968], in which $(^{238}U/^{232}Th)$ is plotted against $(^{230}Th/^{232}Th)$, such that secular equilibrium is represented by the 'equiline', a slope with a gradient of unity. Samples to the left of this line (which include most MORB and OIB) have been recently enriched in ^{230}Th, or depleted in ^{238}U, while samples to the right have experienced recent ^{238}U addition, or ^{230}Th depletion [*Allègre and Condomines,* 1982]. A striking feature of many island arcs lavas is that they plot to the right of the equiline, i.e. have ^{238}U-excesses [*Newman et al.,* 1984]. An additional important observation is that large ^{238}U-excesses are restricted to the depleted, and by inference, sediment poor samples, Figure 12b [see also *McDermott and Hawkesworth,* 1991; and *Condomines and Sigmarsson,* 1993]. Only where little sediment has been added to the arc lava source, does the U from the altered oceanic crust dominant the U budget sufficiently to cause major disequilibrium. This is in accord with the two component model described above. There is the important caveat, however, that the sediment component was either in secular equilibrium or had decayed back to equilibrium since its addition to the sub-arc mantle.

These simple observations have significant implications on the timing of addition of the slab components [*Elliott et al.,* 1997]. For the sediment component to be in secular

Figure 12. (a) Equiline plot of island arc lava compilation (see Appendix A). Solid diagonal line shows secular equilibrium. Several individual arc lava suites with arrays of different slope are picked out: Marianas (filled black circles), S. Tonga/Kermadec (diagonal crosses), New Bismark (long black dashes), together with bulk subducting sediment composition in the Mariana (open circle) and Tonga (circle containing black cross)- no sediment estimate available for New Britain (b) Degree of disequilibrium, $(^{238}U/^{230}Th)$, vs. [Th] µg/g for island arc lavas samples. Note extreme ^{238}U excesses at low [Th] (c) Initial $(^{226}Ra/^{230}Th)$ vs. Ba/Th from Tongan suite of *Turner et al.* [2000]. Filled symbols show samples <56% SiO_2, open symbols show samples >56% SiO_2, which may have experienced some near surface ^{226}Ra decay during differentiation (vector of radioactive decay indicated).

equilibrium requires that at least 350ky have elapsed since the last fractionation of U and Th. The U/Th systematics of arc lavas suggest many sediment packages experienced U-Th fractionation during subduction, which must therefore have occurred >350ky beforearc basalt generation. Most subducting sediments have low U/Th (Figure 12a). Given their high concentrations of these highly incompatible elements, even a small amount of sediment addition to the depleted mantle will rapidly decrease its U/Th ratio. The high U/Th (and $^{230}Th/^{232}Th$) of most arc lavas (Figure 12a) suggests that melting likely increased the U/Th of the added sediment component relative to subducting sediment. Although difficult to fractionate by melting in the presence of common mantle phases, sediment melting is likely to produce residual accessory phases that can significantly fractionation U from Th [e.g. *Johnson and Plank,* 1999].

In gross terms, the mere observation that many depleted lavas show U excesses imply that the time between U addition and eruption is less than 350ky. In detail, however, the equiline plot is an isochron diagram and so straight line arrays potentially have age significance, from >350ky for a slope of 1 to 0y for a perfectly horizontal array. Many individual arcs appear to define sub-horizontal arrays (Figure 12a), implying ages of 30-100ky, although New Britain, [*Gill et al.,* 1993; and Figure 12a] and Kamchatka, [*Turner et al.,* 1998] are notable exceptions with arrays lying close to the equiline (≥350ky). At face value the sub-horizontal arrays imply a period of 30-100ky between U enrichment, presumably by dehydration of the altered oceanic crust, and magma eruption. Thus the combined processes of fluid migration, melting and melt movement to the surface can be argued to occur within this time window. However, arrays on an isochron plot can also reflect other processes [see Elliott et al., 1997], and tighter constraints on the timing of addition of material from the altered oceanic crust can be obtained from Ra disequilibrium.

As already discussed, perhaps the most prominent signature of addition of material from the altered oceanic crust is elevated Ba/Th, as illustrated in Figure 4. Ra is an alkaline earth like Ba, just one period removed, and so should have very similar geochemical behaviour. Thus the ratio $^{226}Ra/^{230}Th$ is a tailor-made timepiece of component 2 addition, to complement the stable Ba/Th ratio. Both early alpha counted data [*Hoogewerff et al.,* 1997] and recent mass-spectrometric data [*Turner et al.,* 2001] document some extreme Ra enrichments, $(^{226}Ra/^{230}Th)>6$, that in many cases are correlated with Ba/Th (Figure 12c). Thus, not only is Ba/Th increased by component 2 addition, but this process occurs on a timescale rapid even compared to the 1600y half-life of ^{226}Ra.

In more detail, the increase in Ba/Th of arc lavas relative to the depleted mantle, can be gauged from Figure 4 to be up to a factor of 20. This is a factor of three higher than the largest increase of ^{226}Ra over ^{230}Th relative to the better defined baseline of secular equilibrium, i.e. 1. Using the Ba enrichment as a proxy for initial $(^{226}Ra/^{230}Th)$, this gives a timescale of slightly less than two half lives of ^{226}Ra decay or some 3000y. This clearly places very stringent constraints on element transport rates associated with component 2 addition.

A key point is the presence not only of extreme $(^{226}Ra/^{230}Th)$, but that these frequently correlate with the stable element analogue Ba/Th [*Reagan et al.,* 1994; *Hoogewerff et al.,* 1997; *Turner et al.,* 2000; and *Turner et al.,* 2001]. An initial puzzle of disequilibria in MORB was that for these large degrees of melting no significant fractionation of highly incompatible elements was expected. Development of simple time dependent melting models [*McKenzie,* 1985] readily illustrated that disequilibria in short-lived nuclides can be generated without any net fractionation in stable element pairs in the erupted lavas. Accounting for Ra excesses as extreme as 6 would be problematic even in the context of simple time dependent melting models. More fundamentally, the coupling of $(^{226}Ra/^{230}Th)$ and Ba/Th ratios implies that the Ra disequilibrium is not a simple consequence of a time dependent melting process that generates arc lavas.

Thus ^{226}Ra excesses correlated to Ba/Th in arc lavas seem clearly related to element fractionation as a result of slab dehydration. In the simplest model this then suggests ~3000y between slab dehydration and eruption of lavas containing this component. If the process of fluid migration through the mantle wedge can preserve or produce $(^{226}Ra/^{230}Th)$ ratios as great as 6, then the interpretation is somewhat changed. In this case, the timescale of 3000y relates to the time between mixing of sediment and fluid components and eruption of the lava. After mixing, any ^{226}Ra in-growth processes should influence both enriched and depleted end-members comparably, thus destroying correlations with Ba/Th. If mixing of the two slab derived components occurs shallow in the mantle wedge, this places much less severe constraints on melt movement rates than the simplest scenario. It remains to be demonstrated, however, that melt percolation in the mantle wedge can generate large ^{226}Ra excesses.

Finally, the possible age significance of the arrays in Figure 12a, needs to be discussed in relation to the shorter time-scales derived from $(^{226}Ra/^{230}Th)$. In this context recent $(^{231}Pa/^{235}U)$ data is of significance. From expected similarities in the geochemical behaviour of Pa and Th it might be anticipated that the ^{238}U-^{230}Th and ^{235}U-^{231}Pa systems also behave rather similarly. In MORB both show excesses of the daughter [e.g. *Goldstein et al.,* 1993], although the Pa excesses are much more extreme (~200% ^{231}Pa excesses, compared ~20% ^{230}Th excesses), presumably due to the greater incompatibility of Pa relative to Th. In an initial survey of arcs, it was notable that most lavas show ^{231}Pa excesses, even those that have (^{230}Th/^{238}U)<1 [*Pickett and Murrell,* 1997]. The Tongan lavas, with some of the largest (^{238}U/^{230}Th), have only modest (^{231}Pa/^{235}U) deficits [*Pickett and Murrell,* 1997; and *Bourdon et al.,* 1999]. A reasonable explanation is that initial ^{231}Pa deficits, caused by U addition from the altered mafic crust, are counter-acted by subsequent melting effects. Analogy to the MORB case indeed suggests that melting should most significantly affect the ^{235}U-^{231}Pa pair. If the time-scale of melting is such that the ^{235}U-^{231}Pa is affected, then so too will the ^{238}U-^{230}Th, albeit it less strongly. Thus it is unlikely that the arrays in Figure 12a represent true ages, and *Thomas et al.* [in press] show for two arc lava suites that combined U-Th-Pa systematics cannot reflect decay alone. The melting times reflected by U-Th-Pa and the very short transit of Ra through the melting system may nevertheless appear contradictory. However, U, Th, Pa and Ra have different residence times in the melting regime [see *Spiegelman and Elliott,* 1993], and while Ra may pass rapidly through, Th and U can, and presumably do, spend longer in the residual solid phases. This is a natural consequence of the melting process. Other workers, however, have proposed alternative explanations for the issue [see *Turner et al.,* 2000].

7. CONCLUSIONS

This review provides a summary of recent observations that strongly supports some long contended notions of three dominant contributions to most mafic island arc lavas: depleted mantle, sediment melt and a dehydration fluid from the altered oceanic crust. It is evident that (at least) two components are required from the subducted slab, and future models of subduction zones cannot ignore the clear chemical evidence for two very different chemical additions to the mantle wedge. A light rare earth element enriched component carries a full range of incompatible elements and dominates the budgets of most of these elements. Variations in the amount and composition of this light rare earth element enriched component both between arcs and within arcs creates the predominant geochemical variations observed in most island arc lavas locally and globally. The second component has an important influence only on the budgets of certain incompatible elements, dominantly 2+ cations (Ba, Ra, Sr and Pb) and to a lesser extent other so-called large ion lithophiles (Rb, K, U). The radically different compositions of these two slab components implies very different sources and/or transport mechanisms for these slab additions. A wide range of evidence from isotopic and incompatible element ratios implicates the components as a sediment melt and an aqueous fluid derived from the altered oceanic crust. U-series data provides information on the timing of addition of these slab components. At least 350ky elapses between sediment melting and eruption of lavas. The period between the last major ^{226}Ra-^{230}Th fractionation, possibly slab dehydration, and eruption of arcs lavas is only a few thousand years. Geochemical studies of island arcs are now sufficiently developed that these observations need to be considered in new models of subduction zone behaviour.

APPENDIX A

The compilation comprises recent data sets which, in all but one case, report ICP-MS trace element data together with a range of isotopic data. Specifically the data are from, Lesser Antilles [*Turner et al.*, 1996], South Sandwich Islands [*Pearce et al.*, 1995], Vanuatu [*Peate et al.*, 1997], New Britain [*Woodhead et al.*, 1998], Tonga-Kermadec [*Regelous et al.*, 1997; *Wendt et al.*, 1997; *Ewart et al.*, 1998; and *Turner et al.*, 1997, see below], Marianas-Central Island Province [*Elliott et al.*, 1997], N. Marianas and Volcano arc [*Bloomer et al.*, 1989; *Lin et al.*, 1990; and *Peate et al.*, 1998], Izu [*Taylor and Nesbitt*, 1998]. The compilation is restricted to:

(a) *Island arcs* The effect of continental crust and lithosphere (albeit it young, in cases such as the Cascades and Central America) can serve to complicate issues.

(b) *Basalts and basaltic andesites* (all samples used have <56 wt% SiO_2) Crustal assimilation and/or crystallisation of accessory phases, that may fractionate key trace element ratios, are likely to have a major influence only in more evolved rocks.

(c) *Volcanic Front* Samples from 'cross-arc' trends were excluded as they have been shown [e.g. *Ryan et al.*, 1996] to show distinct geochemical variations related to their position, that do not necessarily pertain to the processes responsible for the dominant volcanism at the arc front.

The only data set without ICP-MS trace element data is the Lesser Antilles [*Turner et al.*, 1996] where element concentrations should be high enough to allow reliable measurement by the techniques used (X-ray fluorescence and instrumental neutron activation analysis). Furthermore, one data set from Tonga [*Turner et al.*, 1998] has been largely excluded due to unusual and non-systematic Nb abundances in some low concentration samples, which greatly differ from comparable samples reported by *Ewart et al.* [1998] . The *Turner et al.* [1998] dataset has only been used for samples reanalysed for ICP-MS tracers by T. Plank (pers comm), which are now compatible with the other dataset.

δ^{11}B analyses are not available for many of the samples in the compilation, and so Fig 10b uses samples other than those used in the other plots. Much of the arc lava boron isotope work does not meet the criteria of this study being from cross-arc samples [e.g. *Ishikawa and Nakamura*, 1994; and *Ishikawa and Tera*, 1997], evolved and contaminated samples [*Smith et al.*, 1997] or samples with no additional chemical data [*Palmer*, 1991]. Fig 10b uses only the data of *Ishikawa and Tera* [1999] who analysed Mariana (Central Island Province) samples of *Woodhead* [1989]. Nevertheless, these samples show very similar characteristics to those of *Elliott et al.* [1997] used in other plots.

The data plotted in Fig 12 is mass-spectrometric data from (a sub-set) of the samples of this compilation but sometimes reported in further publications to those cited above e.g. Vanuatu data [*Turner et al.*, 1999]. Additionally, a single older alpha counted data set for New Britain [*Gill et al.*, 1993] is included an unusual island arc that defines a steep array in the equiline diagram at high ($^{230}Th/^{232}Th$). These New Britain samples are similar, but not the same as those used elsewhere in this contribution [*Woodhead et al.*, 1998]

As reference for the island arc lava variations, MORB data are plotted, but it is difficult to find suitable datasets with the full range of trace elements discussed in this contribution analysed to good precision. As a small, but hopefully representative dataset, samples from the 'Mainz' MORB suite, [*Jochum et al.*, 1983; *Jochum et al.*, 1986; *Ito et al.*, 1987; and *Jochum et al.*, 1993] with trace elements determined by spark source mass-spectrometry, and a N. Atlantic suite from *Dosso et al.* [1993], analysed by refined INAA and long counting XRF techniques. This sample set includes a disproportionate number of 'enriched' samples, and so covers the full range observed in MORB but does not clearly weight the more common LREE depleted samples.

Acknowledgments. Many thanks to Marc Hirschmann and Terry Plank for instigating the "Inside the Subduction Factory" workshop in Eugene, Oregon, which was the catalyst for much thought and this contribution. NSF funding to attend this meeting is gratefully acknowledged. The ideas in this contribution have been shaped by discussions with many people, but collaboration with Terry Plank is particularly noted. Reviews from Jeff Ryan, Terry Plank and John Eiler helped pull the manuscript into shape.

REFERENCES

Allègre, C.J., ^{230}Th dating of volcanic rocks. A comment, *Earth Planet. Sci. Lett.*, 3, 338.

Allègre, C.J., and M. Condomines, Basalt genesis and mantle structure studied through Th-isotopic geochemistry, *Nature*, 299, 21-24, 1982.

Ayers, J.C., S.K. Dittmer, and G.D. Layne, Partitioning of elements between peridotite and H_2O at 2.0-3.0GPa and 900-1100°C and application to models of subduction zone processes, *Earth Planet. Sci. Lett.*, 150, 381-398, 1997.

Bennett, S., J. Blundy, and T. Elliott, The effect of sodium and titanium on crystal-melt partitioning of trace elements, *Geochim. Cosmochim. Acta*, submitted.

Ben Othman, D., W.M. White, and J. Patchett, The geochemistry of marine sediments, island arc magma genesis, and crust-mantle recycling, *Earth Planet. Sci. Lett.*, 94, 1-21, 1989.

Bloomer, S.H., R.J. Stern, E. Fisk, and C.H. Geschwind, Shoshonitic volcanism in the Northern Mariana arc 1. Mineralogic and major and trace element characteristics *J. Geophys. Res.*, 94, 4469-4496, 1989.

Blundy, J.D., J.A.C. Robinson, and B.J. Wood, Heavy REE are compatible in clinopyroxene on the spinel lherzolite solidus, *Earth Planet. Sci. Lett.*, 160, 493-504, 1998.

Bourdon, B., S.P. Turner, and C.J. Allègre, Melting dynamics beneath the Tonga-Kermadec island arc inferred from ^{231}Pa-^{235}U systematics, *Science*, 286, 2491-2493, 1999.

Brenan, J., H.F. Shaw, F.J. Ryerson, and D.L. Phinney, Mineral-aqueous fluid partitioning of trace elements at 900°C and 2.0GPa: constraints on the trace element chemistry of mantle and deep crustal fluids, *Geochim. Cosmochim. Acta, 59*, 3331-3350, 1995a.

Brenan, J.M., H.F. Shaw, and F.J. Ryerson, Experimental evidence for the origin of lead enrichment in convergent-margin magmas, *Nature, 378*, 54-56, 1995b.

Capaldi, G., M. Cortini, and R. Pece, U and Th decay-series disequilibria in historical lavas from the Eolian Islands, Tyrrhenian Sea *Isot. Geosci.*, 1, 39-55, 1983.

Chabaux, F., C. Hémond, and C.J. Allègre, ^{238}U-^{230}Th-^{226}Ra disequilibria in the Lesser Antilles arc: implications for mantle metasomatism, *Chem. Geol.*, 153, 171-185, 1999.

Chan, L.H., W.P. Leeman, and C.F. You, Lithium isotopic composition of Central American Volcanic Arc lavas: implications for modification of subarc mantle by slab-derived fluids: Correction, *Chem Geol.*, 160, 255-280, 2002.

Chaussidon, M. and, A. Jambon, Boron content and isotopic composition of oceanic basalts: geochemical and cosmochemical implications, *Earth Planet. Sci. Lett.*, 121, 277-291, 1994.

Class, C., D. M. Miller, S. L. Goldstein, and C. H. Langmuir, Distinguishing melt and fluid subduction components in Umnak Volcanics, Aleutian Arc, *Geochem. Geophys. Geosyst.*, 1, 1999GC000010, 2000.

Clark, S.K., M.K. Reagan, and T. Plank, Trace element and U-series systematics for 1963-1965 tephras from Irazu Volcano, Costa Rica: implications for magma generation processes and transit times, *Geochim. Cosmochim. Acta*, 62, 2689-2699, 1998.

Condomines, M., and O. Sigmarsson, Why are so many arc magmas close to ^{238}U-^{230}Th radioactive equilibrium?, *Geochim. Cosmochim. Acta*, 57, 4491-4497, 1993.

Coryell, C.D., J.W. Chase, and J.W. Winchester, A procedure for geochemical interpretation of terrestrial rare earth abundance patterns, *J. Geophys. Res.*, 68, 559-566, 1963.

Davidson, J.D., Deciphering mantle and crustal signatures in subduction zone magmatism, *AGU monograph 96*, edited by G.E. Bebout et al., pp251-262, 1996.

Dorendorf, F., U. Wiechert, and G. Wörner, Hydrated sub-arc mantle: a source for the Kluchevskoy volcano, Kamchatka/Russia, *Earth Planet. Sci. Lett.*, 175, 69-86, 2000.

Dosso, L., H. Bougault, and J.-L. Joron, Geochemical morphology of the North Mid-Atlantic ridge, 10-24°N: trace element-isotope complementarity, *Earth Planet. Sci. Lett.*, 120, 443-462, 1993.

Eggins, S.M., J.D. Woodhead, L.P.J. Kinsley, G.E. Mortimer, P. Sylvester, M.T. McCulloch, J.M. Hergt, and M.R. Handler, A simple method for the precise determination of >=40 trace elements in geological samples by ICPMS using enriched isotope internal standardisation, *Chem. Geol.*, 134, 311-326, 1997.

Eiler, J.M, A. Crawford, T. Elliott, K.A. Farley, J.W. Valley, and E.M. Stolper, Oxygen isotope geochemistry of oceanic-arc lavas, *J. Petrol.*, 41, 229-256, 2000.

Ellam, R.M., and C.J. Hawkesworth, Elemental and isotopic variations in subduction related basalts: evidence for a three component model, *Contrib. Mineral. Petrol.*, 98, 72-80, 1988.

Elliott, T., T. Plank, A. Zindler, W. White, and B. Bourdon, Element transport from slab to volcanic front at the Mariana arc, *J. Geophys. Res.*,102, 14991-15019, 1997.

Ewart, A., K.D. Collerson, M. Regelous, J.I. Wendt, and Y. Niu, Geochemical evolution within the Tonga-Kermadec-Lau arc-back-arc systems: the role of varying mantle wedge composition in space and time *J. Petrol*, 39, 331-368, 1998.

Foley, S.F., M.G. Barth, and G.A. Jenner, Rutile/melt partition coefficients for trace elements and an assessment of the influence of rutile on the trace element characteristics of subduction zone magmas, *Geochim. Cosmochim. Acta.*, 64, 933-938, 2000.

Gill, J.B., *Orogenic Andesites and Plate Tectonics*, pp. 330, Springer-Verlag, Berlin, 1981.

Gill, J.B., and R.W. Williams, Th isotope and U-series studies of subduction-related volcanic rocks, *Geochim. Cosmochim. Acta*, 54, 1427-1442, 1990.

Gill, J.B., J.D. Morris, and R.W. Johnson, Timescales for producing the geochemical signature of island arc magmas: U-Th-Po and Be-B systematics in recent Papua New Guinea lavas, *Geochim. Cosmochim. Acta*, 57, 4269-4283, 1993.

Goldstein, S.J., M.T. Murrell, and R.W. Williams, ^{231}Pa and ^{230}Th chronology of mid-ocean ridge basalts. *Earth Planet. Sci. Lett.*, 115, 151-159, 1993.

Green, T.H., and N.J. Pearson, An experimental study of Nb and Ta partitioning between Ti-rich minerals and silicate liquids at high temperature and pressure, *Geochim. Cosmochim. Acta*, 51, 55-62, 1987.

Green, T.H., S.H. Sie, C.G. Ryan, and D.R. Cousens, Proton microprobe-determined partitioning of Nb, Ta, Zr, Sr and Y between garnet, clinoproxene and basaltic magma at high pressure and temperature, *Chem. Geol.*, 74, 201-216, 1989.

Hart, S.R., and T. Dunn, Experimental cpx/melt partitioning of 24 trace elements, *Contrib. Mineral. Petrol.*, 113, 1-8, 1993.

Hauri, E.H., T.P. Wagner, and T.L. Grove, Experimental and natural partitioning of Th, U, Pb and other trace elements between garnet, clinopyroxene and basaltic melts, *Chem. Geol.*, 117, 149-166, 1994.

Hawkesworth, C.J., K. Gallagher, J.M. Hergt, and F. McDermott, Mantle and slab contributions in arc magmas, *Ann. Rev. Earth Planet. Sci.*, 21, 175-204, 1993.

Hickey-Vargas, R.L., Isotope characteristics of submarine lavas from the Philippine Sea: implications for the origin of arc and basin magmas of the Philippine tectonic plate, *Earth Planet. Sci. Lett.*, 107, 290-304, 1991.

Hofmann, A.W., Chemical differentiation of the Earth: the relationship between mantle, continental crust and oceanic crust, *Earth Planet Sci. Lett.*, 90, 297-314, 1988.

Hole, M.J., A.D. Saunders, G.F. Marriner, and J. Tarney, Subduction of pelagic sediments: implications for the origin of Ce-anomalous basalts from the Mariana Islands, *J. Geol. Soc. London*, 141, 453-472, 1984.

Hoogewerff, J.A., M.J. van Bergen, P.Z. Vroon, J. Hertogen, R. Wordel, A. Sneyers, A. Nausution, J.C. Varekamp, H.L.E. Moens, and D. Mouchel, U-series , Sr-Nd-Pb isotope and trace element systematics across an active island arc-continent collision zone: implications for element transfer at the slab-wedge interface *Geochim. Cosmochim. Acta*, 61, 1057-1072, 1997.

Ishikawa, T., and E. Nakamura E., Origin of the slab component in arc lavas from across-arc variation of B and Pb isotopes, *Nature*, 370, 205-208, 1994.

Ishikawa, T., and F. Tera, Source, composition and distribution of the fluid in the Kurile mantle wedge: Constraints from across-arc variations of B/Nb and B isotopes *Earth Planet. Sci. Lett.*, 152, 123-138, 1997.

Ishikawa, T., and F. Tera, Two isotopically distinct fluid components involved in the Mariana arc: Evidence from Nb/B ratios and B, Sr, Nd, and Pb isotope systematics *Geology*, 27, 83-86, 1999.

Ito, E., W.M. White, and C. Göpel, The O, Sr, Nd and Pb isotope geochemistry of MORB, *Chem Geol.*, 62, 157-176, 1987.

Jochum, J.P., A.W. Hofmann, E. Ito, H.M. Seufert, and W.M. White, K, U and Th in mid-oceanic ridge basalt glasses and heat production, K/U and K/Rb in the mantle, *Nature, 306*, 431-436, 1983.

Jochum, K.P., H.M. Seufert, B. Spettel, and H. Palme, The solar-system abundances of Nb, Ta and Y, and the relative abundances of refractory lithophile elements in differentiated planetary bodies, *Geochim. Cosmochim. Acta, 50*, 1173-1183, 1986.

Jochum, K.P., A.W. Hofmann, and H.M. Seufert, Tin in mantle-derived rocks: constraints on Earth evolution, *Geochim. Cosmochim. Acta.*, 57, 3585-3595, 1993.

Johnson, K.T.M., H.J.B. Dick, and N. Shimizu, Melting in the oceanic upper mantle: an ion microprobe study of diopsides in abyssal peridotites, *J. Geophys. Res.*, 95, 2661-2678, 1990.

Johnson, M.C., and T. Plank, Dehydration and melting experiments constrain the fate of subducted sediments, *Geochem. Geophys. Geosys.*, 1, 1999GC000014, 1999.

Kay, R.W., Volcanic arc magmas: implications of a melting-mixing model for element recycling in the crust-upper mantle system, *J. Geol.*, 88, 497-522, 1980.

Keppler, H., Constraints from partitioning experiments on the composition of subduction-zone fluids, *Nature*, 380, 237-240, 1996.

Kogiso, T., Y. Tatsumi, and S. Nakano, Trace element transport during dehydration processes in the subducted oceanic crust: 1. Experiments and implications for the origin of ocean island basalts *Earth Planet. Sci. Lett.*, 148, 193-205, 1997.

Lin, P.-N., R.J. Stern, J. Morris, and S.H. Bloomer, Nd- and Sr-isotopic compositions of lavas from the northern Mariana ad southern Volcano arcs: implications for the origin of island arc melts *Contrib. Mineral. Petrol.,* 105, 381-392, 1990.

Lin, P.-N., Trace element and isotopic characteristics of western Pacific pelagic sediments: implications for the petrogenesis of Mariana Arc magmas, *Geochim. Cosmochim. Acta,* 56, 1641-1654, 1992.

McCulloch, M.T., and J.A. Gamble, Geochemical and geodynamical constraints on subduction zone magmatism, *Earth Planet. Sci. Lett.,* 102, 358-374, 1991.

McDermott, F.M., and C. Hawkesworth, Th, Pb, and Sr isotope variations in young island arc volcanics and oceanic sediments, *Earth Planet. Sci . Lett.,* 104, 1-15, 1991.

McDermott, F., M.J. Defant, C.J. Hawkesworth, R.C. Maury, and J.L. Joron, Isotopic and trace element evidence for three component mixing in the genesis of the North Luzon arc lavas, *Contrib. Mineral. Petrol.,* 113, 9-23, 1993.

McKenzie, D., ^{230}Th-^{238}U disequilibrium and the melting process beneath ridge axes, *Earth Planet. Sci. Lett.,* 72, 149-157, 1985.

Macpherson, C.G., J.A. Gamble, and D.P. Mattey, Oxygen isotope geochemistry of lavas from an oceanic to continental arc transition, Kermadec-Hikurangi margin, SW Pacific, *Earth Planet. Sci. Lett.,* 160, 609-621, 1998.

Masuda, A., Regularities in variation of relative abundances of lanthanide elements and an attempt to analyse separation-index patterns of some minerals, *J. Earth Sci. Nagoya Univ.,* 10, 173-187, 1962.

Meijer, A., Pb and Sr isotopic data bearing on the origin of volcanic rocks from the Mariana island arc system, *Geol. Soc. Am. Bull.,* 87, 1358-1369, 1976.

Miller, D.M., S.L. Goldstein, and C.H. Langmuir, Cerium/lead and lead isotope ratios in arc magmas and the enrichment of lead in the continents, *Nature,* 368, 514-520, 1994.

Moran, A.E., V.B. Sisson, and W.P. Leeman, Boron depletion during progressive metamorphism: implications for subduction processes, *Earth Planet. Sci. Lett.,* 111, 331-349, 1992.

Moriguti, T., and E. Nakamura, Across-arc variation of Li isotopes in lavas and implications for crust/mantle recycling at subduction zones, *Earth Planet. Sci. Lett.,* 163, 167-174, 1998.

Morris, J.D., and S.R. Hart, Isotopic and incompatible element constraints on the genesis of island arc volcanics from Cold Bay and Amak Island, Aleutians, and implications for mantle structure, *Geochim. Cosmochim. Acta,* 47, 2015-2030, 1983.

Morris, J., and F. Tera, ^{10}Be/^{9}Be in mineral separates and whole rocks from volcanic arcs: implications for sediment subduction, *Geochim. Cosmochim. Acta,* 53, 3197-3206, 1989.

Morris, J., W.P. Leeman, and F. Tera, The subducted component in island arc lavas: constraints from Be isotopes and B-Be systematics, *Nature,* 344, 31-36, 1990.

Morris, J., J. Gosse, and S. Brachfeld, Cosmogenic 10Be and the solid earth: studies in geomagnetism, subduction zone processes and active tectonics, *Reviews in Mineralogy,* in press.

Newman, S., J.D. MacDougall, and F.C. Finkel, ^{230}Th-^{238}U disequilibrium in island arcs: evidence from the Aleutians and the Marianas, *Nature,* 308, 268-270, 1984.

Nichols, G.T., P.J. Wyllie, and C.R. Stern, Subduction zone melting of pelagic sediments constrained by melting experiments, *Nature,* 371, 785-788, 1994.

Noll, P.D., H.E. Newsom, W.P. Leeman, and J.G. Ryan. The role of hydrothermal fluids in the production of subduction zone magmas: evidence from siderophile and chalcophile trace elements and boron, *Geochim. Cosmochim. Acta,* 60, 587-611, 1996.

Park, K.-H., Sr, Nd and Pb isotopic studies of ocean island basalts: constraints on their origin and evolution *PhD thesis (unpub.),* Columia University, New York, pp252, 1990.

Palmer, M.R., Boron-isotope systematics of Halmahera arc (Indonesia) lavas: evidence for the involvement of the subducted slab, *Geology,* 19, 215-217, 1991.

Peacock, SM., and R.L. Hervig,, Boron isotopic composition of subduction-zone metamorphic rocks, *Chem. Geol.,* 160, 281-290, 1999.

Pearce, J.A., Trace element characteristics of lavas from destructive plate boundaries, in *Andesites,* edited by R.S. Thorpe, pp. 525-548, Wiley, New York, N.Y. 1982.

Pearce, J.A., P.E. Baker, P.K. Harvey, and I.W. Luff, Geochemical evidence for subduction fluxes, mantle melting and fractional crystallisation beneath the South Sandwich Island arc *J. Petrol.,* 36, 1073-1109, 1995.

Pearce, J.A., P.D. Kempton, G.M. Nowell, and S.R. Noble, Hf-Nd element and isotope perspective on the nature and provenance of mantle and subduction components in Western Pacific arc-basin systems, *J. Petrol.,* 40, 1579-1611, 1999.

Peate, D.W., J.A. Pearce, C.J. Hawkesworth, H. Colley, C.M.H. Edwards, and K. Hirose, Geochemical variations in Vanuatu arc lavas: the role of subducted material and a variable mantle wedge composition *J. Petrol.,* 38, 1331-1358, 1997.

Peate, D.W. and J.A. Pearce, Causes of spatial compositional variations in Mariana arc lavas: trace element evidence *Isl. Arc,* 7, 479-495, 1998.

Pickett, D.A., and M.T. Murell, Observations of ^{231}Pa-^{235}U disequilibrium in volcanic rocks, *Earth Planet. Sci. Lett.,*148, 259-271, 1997.

Plank, T., and C.H. Langmuir, An evaluation of the global variations in the major element chemistry of arc basalts, *Earth Planet. Sci. Lett., 90,* 349-370, 1988.

Plank, T., and C.H. Langmuir, Sediments melt and basaltic crust dehydrates at subduction zones, *EOS, Trans. AGU,* 73, 637, 1992.

Plank, T., and C.H. Langmuir, Tracing trace elements from sediment input to volcanic output at subduction zones, *Nature,* 362, 799-742, 1993.

Plank, T., and W. White, Nb and Ta in arc and mid-ocean ridge basalts, *EOS, Trans. AGU,* 76, 655, 1995.

Plank, T., and C.H. Langmuir, The chemical composition of subducting sediment and its consequences for the crust and mantle, *Chem. Geol.,* 145, 325-394, 1998.

Reagan, M.K., J.D. Morris, E.A. Herrstrom, and M.T. Murrell, Uranium series and beryllium isotope evidence for an extended history of subduction modification of the mantle below Nicaragua, *Geochim. Cosmochim. Acta,* 58, 4199-4212, 1994.

Regelous, M., K.D. Collerson, A. Ewart, and J.I. Wendt, Trace element transport rates in subduction zones: evidence from Th, Sr and Pb isotope data for Tonga-Kermadec arc lavas, *Earth Planet. Sci. Lett.,* 150, 291-302, 1997.

Robinson, J.A.C., B.J. Wood, and J.D. Blundy, The beginning of melting of fertile and depleted peridotite at 1.5GPa, *Earth Planet. Sci. Lett.,* 155, 97-111, 1998.

Rose, E.F., N. Shimizu, G.D. Layne, and T.L. Grove, Melt production beneath Mt. Shasta from boron data in primitive melt inclusions, *Science,* 293, 281-283, 2001.

Rubin, K.H., G.E. Wheller, M.O. Tanzer, J.D. MacDougall, R. Varne, and R. Finkel, ^{238}U decay series systematics of young lavas from Batur volcano, Sunda Arc *J. Volcanol. Geotherm. Res.,* 38, 215-226, 1989.

Ryan, J., J. Morris, G. Bebout , and W.P. Leeman, Describing chemical fluxes in subduction zones: insights from 'depth profiling' studies of arc and fore-Arc rocks, *AGU monograph 96,* edited by G.E. Bebout et al., pp263-268, 1996.

Ryerson, F.J., and E.B. Watson, Rutile saturation in magmas: implications for Ti-Nb-Ta depletion in island arc basalts, *Earth Planet. Sci. Lett.,* 86, 225-239, 1987.

Salters, V.J.M., and S.R. Hart, The mantle sources of ocean ridges, islands and arcs: the Hf isotope connection, *Earth Planet. Sci. Lett.,* 104, 364-380, 1991.

Salters, V.J.M., and J. Longhi, Trace element partitioning during the initial stages of melting beneath mid-ocean ridges, *Earth Planet. Sci. Lett.,* 166,15-30, 1999.

Shannon, R.D., Revised effective ionic radii and systematic studies of interatomic distances in halides and chalcogenides, *Acta Crystallogr.,* 32, 751-767, 1976.

Smith, H.J., A.J. Spivack, H. Staudigel, and S.R. Hart, The boron isotopic composition of altered oceanic crust, *Chem. Geol.,* 126, 119-135, 1995.

Smith, H.J., W.P. Leeman, J. Davidson, and A.J. Spivack, The B isotopic composition of arc lavas from Martinique, Lesser Antilles, *Earth Planet. Sci. Lett.,* 146, 303-314, 1997.

Spiegelman, M., and T. Elliott, Consequences of melt transport for uranium series disequilibrium in young lavas. *Earth Planet. Sci. Lett.,* 118, 1-20, 1993.

Stalder, R., S.F. Foley, G.P. Brey, and I. Horn, Mineral aqueous fluid partitioning of trace elements at 900-1200 degrees C and 3.0-5.7 GPa: New experimental data for garnet, clinopyroxene, and rutile, and implications for mantle metasomatism, Geochim. Cosmochim. Acta, 62, 1781-1801, 1998.

Staudigel, H., T. Plank, W. White, and H.-U. Schmincke, Geochemical fluxes during seafloor alteration of the basaltic upper oceanic crust: DSDP sites 417 and 418, *AGU monograph 96,* edited by G.E. Bebout et al., pp19-38, 1996.

Stolper, E., and S. Newman, The role of water in the petrogenesis of the Mariana Trough magmas, *Earth Planet. Sci. Lett.,* 121, 293-325, 1994.

Sun, S.-s., and W.F. McDonough, Chemical and isotopic systematics of oceanic basalts: implications for mantle composition and processes, in *Magmatism in the Ocean Basins,* editted by A.D. Saunders, and M.J. Norry, *Geol. Soc. Spec. Publ.,* 42, pp 313-345, 1989.

Tatsumi, Y., and S. Eggins, *Subduction Zone Magmatism,* pp211, Blackwell, Oxford, 1995.

Tatsumi, Y., D.L. Hamilton, and R.W. Nesbitt, Chemical characteristics of fluid phase released from a subducted lithosphere and origin of arc magmas: evidence from high pressure experiments and natural rocks, *J. Volcanol. Geotherm. Res.,* 29, 293-309, 1986.

Taylor, R.N., and R.W. Nesbitt, Isotopic characteristics of subduction fluids in an intraoceanic setting, Izu-Bonin Arc, Japan, *Earth Planet. Sci. Lett.,* 164, 79-98, 1998.

Thomas, R.B., M.M. Hirschmann, H. Cheng, M.K. Ragan, and R.L. Edwards, (^{231}Pa/^{235}U)-(^{230}Th/^{238}U) of young mafic volcanic rocks from Nicaragua and Costa Rica and the role of flux melting on U-series systematics of arc lavas, *Geochim. Cosmochim. Acta,* in press.

Tomascak, P.B., E. Widom, L.D. Benton, S.L. Goldstein, and J.G. Ryan, The control of lithium budgets in island arcs, *Earth Planet. Sci. Lett.,* 196, 227-238, 2002.

Toyoda, K., Y. Nakamura, and A. Masuda, Rare earth elements of Pacific pelagic sediments, *Geochim. Cosmochim. Acta,* 54, 1093-1103, 1990.

Turner, S.P., C.J. Hawkesworth, P. van Calsteren, E. Heath, R. Macdonald, and S. Black, U-series isotopes and destructive plate margin magma genesis in the Lesser Antilles, *Earth Planet. Sci. Lett.,* 142, 191-207, 1996.

Turner, S.P., and C.J. Hawkesworth, Constraints on flux rates and mantle dynamics beneath island arcs from Tonga-Kermadec lava geochemistry, *Nature,* 389, 568-573, 1997.

Turner, S.P., C.J. Hawkesworth, N.W. Rogers, J. Bartlett, T. Worthington, J. Hergt, J. Pearce, and I. Smith, ^{238}U-^{230}Th disequilibria, magma petrogenesis, and flux rates beneath the depleted Tonga-Kermadec island arc, *Geochim. Cosmochim. Acta,* 61, 4855-4884, 1997.

Turner, S.P., F. McDermott, C.J. Hawkesworth, and P. Kepezhinskas, A U-series study of lavas from Kamchatka and the Aleutians: constraints on source composition and melting processes, *Contrib. Mineral. Petrol.,* 133, p217-234, 1998.

Turner, S.P., D.W. Peate, C.J. Hawkesworth, S.M. Eggins, and A.J. Crawford, Two mantle domains and the timescales of fluid transfer beneath the Vanuatu arc, *Geology,* 27, 963-966, 1999.

Turner, S.P., B. Bourdon, C.J. Hawkesworth, and P. Evans, ^{226}Ra-^{230}Th evidence for multiple dehydration events, rapid melt ascent and the timescales of differentiation beneath the Tonga-Kermadec island arc, *Earth Planet. Sci. Lett.,* 179, 581-593, 2000.

Turner, S.P., P. Evans, and C.J. Hawkesworth, Ultrafast source-to-surface movement of melt at island arcs from ^{226}Ra-^{230}Th systematics, *Science,* 292, 1363-1366, 2001.

Vroon, P.Z., M.J. van Bergen, G.J. Klaver, and W.M. White, Strontium, neodymium and lead isotopic and trace element signatures of the East Indonesian sediments: provenance and implications for Banda Arc magma genesis, *Geochim. Cosmochim. Acta,* 59, 2573-2598, 1995.

Vukadinovic, D., Are Sr enrichments in arc basalts due to plagioclase accumulation? *Geology*, 21, 611-614, 1993.

Wendt, J.I., M. Regelous, K.D. Collerson, and A. Ewart, Evidence for a contribution from two mantle plumes to island arc lavas from northern Tonga, *Geology*, 25, 611-614, 1997.

White, W.M., and B. Dupré, Sediment subduction and magma genesis in the Lesser Antilles: isotopic and trace element constraints, *J. Geophys. Res.,* 91, 5927-5941, 1986.

Wood, B.J., and J.D. Blundy, A predictive model for rare earth element partitioning between clinopyroxene and anhydrous silicate melt, *Contrib. Mineral. Petrol.,* 129, 166-181, 1997.

Woodhead, J.D., Geochemistry of the Mariana arc (western Pacific): source composition and processes, *Chem. Geol.,* 76, 1-24, 1989.

Woodhead, J., S. Eggins, and J. Gamble, High field strength and transition element systematics in island arc and back-arc basin basalts: evidence for multiphase melt extraction and a depleted mantle wedge, *Earth Planet. Sci. Lett.,* 114, 491-504, 1993.

Woodhead, J.D., S.M. Eggins, and R.W. Johnson, Magma genesis in the New Britain island arc: further insights into melting and mass transfer processes, *J. Petrol.,* 39, 1641-1668, 1998.

You, C.-F., P.R. Castillo, J.M. Gieskes, L.H. Chan, and A.J. Spivak, Trace element behaviour in hydrothermal experiments: implications for fluid processes at shallow depths in subduction zones, *Earth Planet. Sci. Lett.,* 140, 41-52, 1996.

T. Elliott, Department of Earth Sciences, Wills Memorial Building, Queens Road, University of Bristol, Bristol, BS8 1RJ, U.K. (tim.elliott@bristol.ac.uk)

Basic Principles of Electromagnetic and Seismological Investigation of Shallow Subduction Zone Structure

George Helffrich

Earth Sciences, University of Bristol, Bristol, United Kingdom

This chapter reviews the basic concepts involving waves that electromagnetic and seismic investigations rely upon, for intiutive understanding by non-specialists in either field. Following sections dealing with the general properties of waves, further specialized to each method, the review summarizes investigative results from key subduction zone studies that illustrate how the methods were used.

1. INTRODUCTION

The aim of this chapter is to introduce the principles of electromagnetic and seismological investigation techniques to inform the non-specialist of the ways in which the methods can be used to infer the shallow structure of subduction zones. This background material is provided so that future researchers may plan investigations of subduction margins in an informed way. The emphasis is on conveying a set of simple, useful concepts that provide intuition on the strengths and weaknesses of each method. Other references in this spirit that provided inspiration for the emphases in this contribution are *Eberhart-Phillips et al.* (1995), which focuses on electrical and seismic investigation of fault-related fluid flow and processes, *Wannamaker et al.* (1989), and *Jiracek* (1995), both of which give overviews of electromagnetic methods, and *Jones* (1993, 1998) who focuses on joint use of the methods to attack subduction-related problems.

2. PROPERTIES OF WAVES

2.1. Fundamentals

Interrogating the earth with either electromagnetic or seismic methods involves the use of a wave, so a useful starting point is a review of the nomenclature and basic concepts applicable to all types of waves. Waves arise from the solution to a mathematical relationship between the time and the space variation in some property. Simplifying space as position along a single line x and time as t, a wave w is any function with the following form

$$w(t,x) = g(at - bx) \quad , \tag{1}$$

where a and b are constant scale factors and g is any function whatsoever. Because it is possible to make any periodic function g out a sum of sine and cosine functions (by Fourier series), one can just as well write equation (1) as

$$\begin{aligned} w(t,x) &= g(at-bx) \\ &= \sum_{n=0}^{\infty} c_n \cos[n(at-bx)] \\ &= \sum_{n=0}^{\infty} c_n \cos\left[an\left(t-\frac{b}{a}x\right)\right] \end{aligned} \tag{2}$$

c_n is some factor (possibly zero) that makes g out of a sum of particular cosine functions, and the product an is a multiplicative factor of time t that you can interpret as a frequency of wave oscillation. While possibly seeming complex, this leads to a very simple way to think about any wave. Because it is possible to express any periodic wave as a series of cosine functions, all we need to consider when characterizing a wave is only the properties of a sinusoidal (or cosinusoidal) oscillation.

Inside the Subduction Factory
Geophysical Monograph 138

10.1029/138GM04

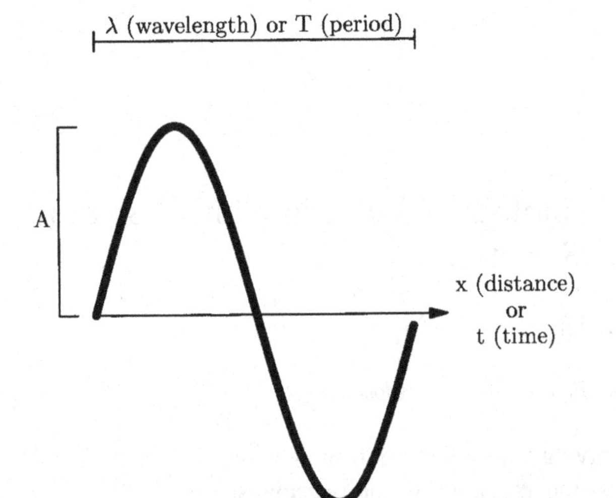

Figure 1. Key characteristics of waves. The horizontal axis may be either time or space, depending on whether you consider the spatial variation in amplitude at a fixed time, or the temporal change in amplitude at a fixed point in space.

Figure 1 shows the fundamental properties resulting from this abstraction. A wave may be viewed as oscillating in time or in space. Its amplitude is A (measured from zero). The wave speed v of the wave is related to the oscillation frequency f and the wavelength in space λ via

$$v = f\lambda = \frac{\lambda}{T} \,. \tag{3}$$

The second expression uses the period of oscillation T rather than the frequency (period is $1/f$). For electromagnetic waves, the wave speed is close to the speed of light $c = 3 \times 10^8$ m/sec, and the variations in wave speed in earth materials are small, less than a percent. Seismic wave speeds are vastly lower. P-wave speeds are 6–9 km/sec in the upper mantle, and S-wave speeds 3–6 km/sec. In contrast, their variations are much greater: variations up to 100% occur in the upper mantle, and are still larger if the rest of the earth is included.

2.2. Polarization

As waves propagate in space, they oscillate. The direction of oscillation is the polarization direction of the wave. Electromagnetic waves propagate in pairs polarized perpendicular to each other and travel at the same speed. In contrast, seismic waves travel in triplets, each orthogonal to the others, but at different speeds. On this account, one thinks of paired electromagnetic waves, but as independently-propagating seismic waves because the seismic waves travel at different speeds and soon separate from one another in space. One usually distinguishes only two types of seismic waves, P and S waves, because the two S polarizations travel at the same speed in isotropic material (whose wave speed doesn't depend on the direction the wave travels). Both S-type polarizations are simply called the S wave.

3. ELECTROMAGNETIC WAVES

3.1. Polarization and Rock Properties

One of the electromagnetic wave pairs is the oscillation of the electric field (E) while the other is the magnetic field

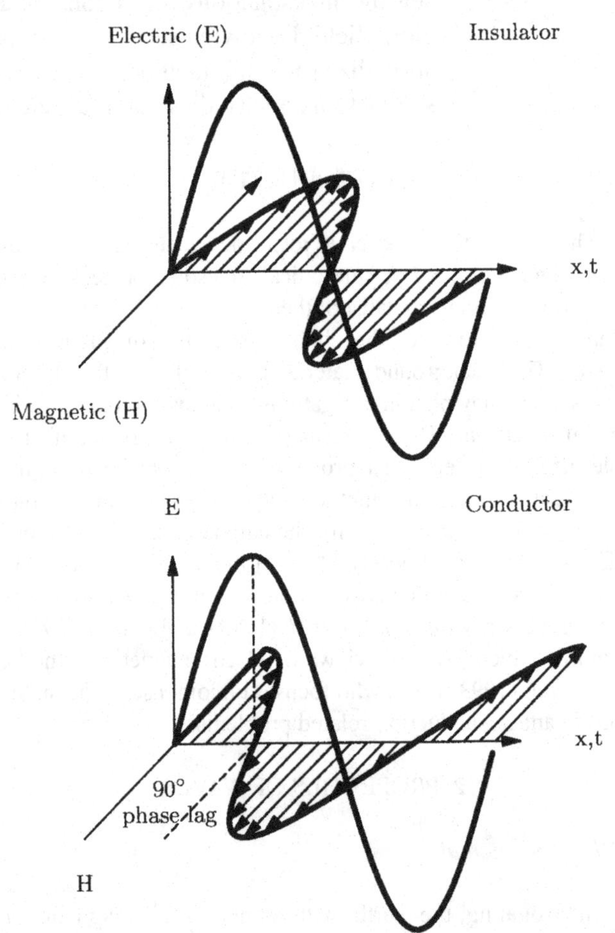

Figure 2. Oscillation of paired electromagnetic waves in an insulator and a conductor. The horizontal axis is time or space. Vertical axes are electric and magnetic field variations, polarized at right angles to each other. Lag between corresponding points in a cycle of oscillation is the phase lag, which depends on the conductivity of the material.

oscillation (H). Though they oscillate perpendicular to one another in space, their oscillation is not necessarily synchronized in time. The lag between the peak of the E oscillation and the peak of the H oscillation is called the phase lag. The difference in angle between contemporaneous points in their oscillation cycle is the way the phase lag is measured. The most useful property of electromagnetic waves is the behavior of the phase lag in different materials (Figure 2). In electrically insulating material, they are in synchrony (or in phase, meaning that the phase lag is zero), while in perfectly conducting material they are out of synchrony by a quarter cycle, or 90°. Thus by measuring the phase lag of the electromagnetic wave pair, one can infer the material's ability to conduct electricity. The conductivity varies substantially in rocks, making phase lag measurements potentially an extremely valuable way to differentiate rock types.

3.2. Resistivity and Resistance

Conductivity is not the conventional way to describe a rock's electrical properties. Curiously, it is the inverse of conductivity or resistivity ρ, and not the more familiar resistance. The reason is that resistivity is a size-independent characteristic of the medium, whereas resistance varies with size. As a concrete example, suppose some current I is flowing through a cylindrical conductor down an electric field gradient (Figure 3). If one considers some area section S through the conductor, it is easy to see that more current will flow through S if S is larger. Divide the current by S to eliminate this dependence, and call it the current density j

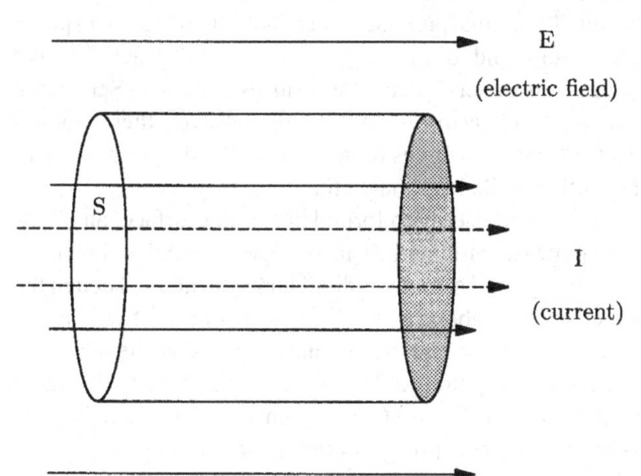

Figure 3. Current I (dashed lines) flowing down a cylindrical conductor of area S due to an applied electric field E (solid lines). In order to eliminate the dependence of the current flow on the conductor area, one uses the current density $j = I/S$ rather than I in characterizing the electrical properties of materials.

$$j = \frac{I}{S} \tag{4}$$

Next assume a linear relation between electric field strength E and the current flow, now that it is independent of the size of the current carrier

$$j = \sigma E \ . \tag{5}$$

σ is the conductivity; $1/\sigma = \rho$ is the resistivity.

To get these relations into a standard form, use the definition of E, which is the voltage gradient, or voltage change with distance change Δx

$$E = \frac{\Delta V}{\Delta x} \tag{6}$$

and combine it with (5), which gives the relation between E and j.

$$\frac{\Delta V}{\Delta x} = E = \frac{1}{\sigma} j = \rho j = \rho \frac{I}{A} \ . \tag{7}$$

This yields a relation between voltage ΔV and the rest of the quantities I, Δx, ρ and S. Compare this with Ohm's Law, by analogy

$$\begin{aligned} V &= IR \quad \text{Ohms law} \\ \Delta V &= I \frac{\Delta x p}{S} \ . \end{aligned} \tag{8}$$

Looking at the role in Ohm's Law played by the analogous quantity in equation (7), deduce that

$$R = \frac{\Delta x \rho}{S} \ , \quad \text{or that} \quad \rho = \frac{RS}{\Delta x} \tag{9}$$

which reveals that the resistance R and the resistivity ρ are scaled versions of one another, and that the units of ρ are ohm-meters.

Rock resistivities vary quite dramatically depending on the rock type and whether it is suffused with fluid (Figure 4). Crystalline rocks have the highest resistivities, briny fluids lower, and graphite films even lower still (recall that graphite has commercial electrical uses in arc-lighting and electrical motors). Thus regions with crystalline rocks (the upper crust and oceanic upper mantle) tend to be more resistive, and rocks with fluids are less so. Silicate melt, curiously, is fairly conductive, possibly due to its ionic nature.

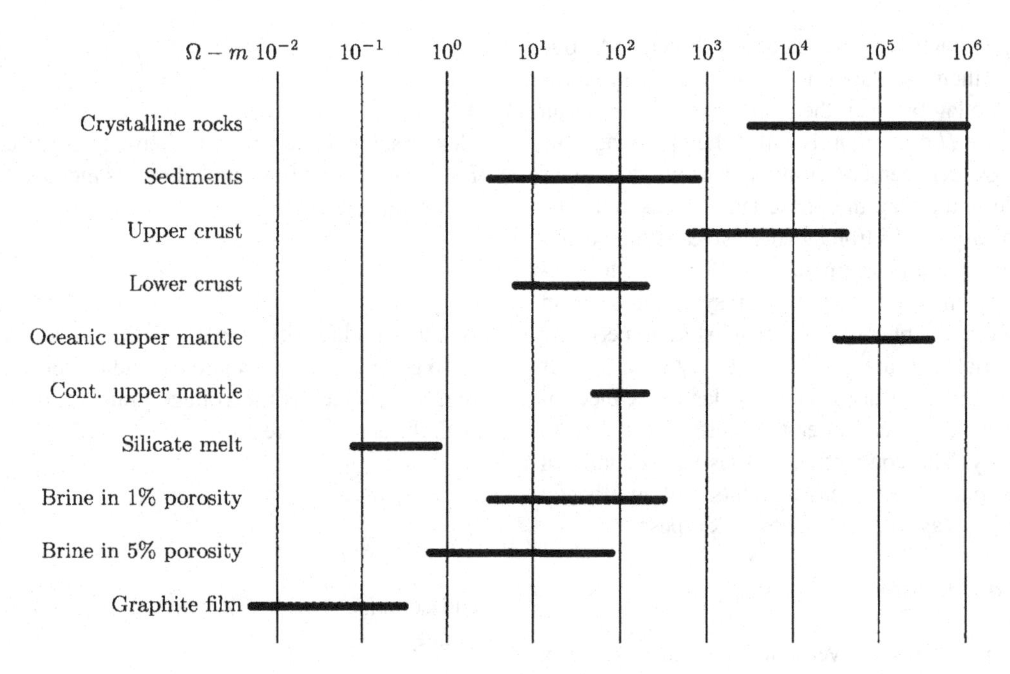

Figure 4. Geologically-relevant resistivities (redrawn from *Haak and Hutton*, 1986).

3.3. Penetration

Electromagnetic waves travel through free space, the atmosphere and rock with different degrees of ease. The more conductive the material, the harder it is for the wave to penetrate the substance. To characterize the attenuation of the wave, a useful concept is the skin depth δ, or the depth at which the amplitude decays to $1/e$ of what it had when it entered the substance (Figure 5). The skin depth is (*Parker*, 1980)

$$\delta = \sqrt{\frac{\rho \varepsilon_0 c^2}{\pi f}} \, . \tag{10}$$

The only unfamiliar quantity in this formula is ε_0 or the free-space electrical permittivity. Fortunately, this is virtually constant for nearly all materials, leading to the rule-of-thumb

$$\delta \text{ (in meters)} = 500 \sqrt{\frac{\rho \text{ (in ohm} - \text{m)}}{f \text{ (in Hz)}}} \, . \tag{11}$$

The penetration depth increases with higher resistivity (increasing ρ) and with lower frequency (decreasing f). Fairly low frequencies – 1–0.0001 Hz or 1–10000 seconds period – are required to penetrate the earth, given a rock resistivity in the mid-range of upper mantle materials (Figure 4).

3.4. The Magnetotelluric Method

For investigating the lower crust and upper mantle, the skin depth must be on the order of 10^5–10^6 meters. Taking an approximate ρ range of 10^1–10^3 ohm-m, by use of (11) one deduces that *very* long-period energy sources are called for. Stable artificial sources of energy at these frequencies are difficult to obtain, but ambient natural sources may be recruited for this purpose. They include solar-ionosphere interactions and distant lightning and sprite activity that induce atmospheric electrical field oscillations (Schumann resonance) (*Parkinson*, 1983). This use of these natural electromagnetic sources to investigate the deeper reaches of the earth is called the magnetotelluric method.

The method is easy to formulate. At the surface, an investigator measures E and H in two perpendicular directions along the ground at many individual frequencies. Though the wave source is above ground, the waves interact with materials far below it, making the net response at the surface a mixture of the source and the wave-rock interaction (Figure 6). Relating the E and H components in the x direction and its orthogonal direction y through a matrix operator Z

$$\begin{matrix} \mathbf{E} & = & \mathbf{Z} & \mathbf{H} \\ \begin{bmatrix} E_x \\ E_y \end{bmatrix} & = & \begin{bmatrix} Z_{xx} & Z_{xy} \\ Z_{yx} & Z_{yy} \end{bmatrix} & \begin{bmatrix} H_x \\ H_y \end{bmatrix} \end{matrix} \tag{12}$$

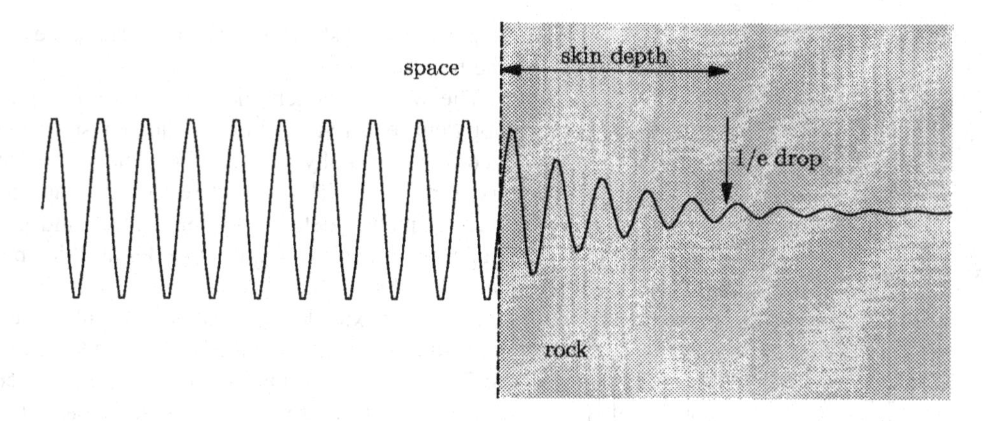

Figure 5. Decay of electromagnetic wave as it propagates into a material from free space is characterized by the skin depth. This is the distance traveled before the wave's amplitude attenuates to $1/e$ of its free-space amplitude.

one finds the apparent resistivity ρ_a, which depends on frequency f, by

$$\rho_{a,xy}(f) = \frac{|Z_{xy}(f)|^2}{2\pi f \varepsilon_0} \tag{13a}$$

and its associated phase lag ϕ_{xy} through

$$\phi_{xy}(f) = \tan^{-1}\left(\frac{\Im[Z_{xy}(f)]}{\Re[Z_{xy}(f)]}\right) \quad . \tag{13b}$$

The remaining three components of ρ and $\varnothing$ are found analogously.

3.5. Modelling Magnetotelluric Data

There is an unfortunate ambiguity in drawing inferences about the earth's electrical structure from magnetotelluric data. While it is straightforward to calculate ρ_a and $\varnothing$ profiles given a profile of ρ with depth, inverting a set of observations of ρ_a and $\varnothing$ at individual frequencies is fundamentally indeterminate (*Parker*, 1980). Surmounting this forces one to adopt a view of the resistivity structure before inverting the data. Different *a-priori* assumptions about the structure lead to quite different results. *Parker* (1980) and *Parker and Whaler* (1981) investigated this problem thoroughly; see these sources for detailed results. One example suffices for the purposes of this tutorial, which is how one actual ρ_a and $\varnothing$ profile may be fit by three different models (Figure 7): as a series of infinitely thin conductive layers, as a suite of layers of constant conductivity, or with the most gently-varying changes in space. While all show conductive layers near the surface, near 350 km depth, and near 600 km, any further inferences about the structure would be unwarranted. Note that while the total conductance increases with depth, whether one inferred an increase or a decrease in conductivity with depth would depend on the *a-priori* assumptions. This is because electromagnetic response is governed by the skin-depth-integrated resistivity (the total resistance, in other words), meaning there is a tradeoff between layer thickness and layer resistivity.

Stable methodological approaches and computational power restricts modelling to two-dimensional linear transects, where depth variation in ρ_a is found as a function of distance. The two presently popular methods are the rapid relaxation inverse (*Smith and Booker*, 1991) and Occam's inversion (*de Groot-Hedlin and Constable*, 1990). The magnetotelluric data permit some assessment of whether linear variation of the resistivity is justified because in horizontally

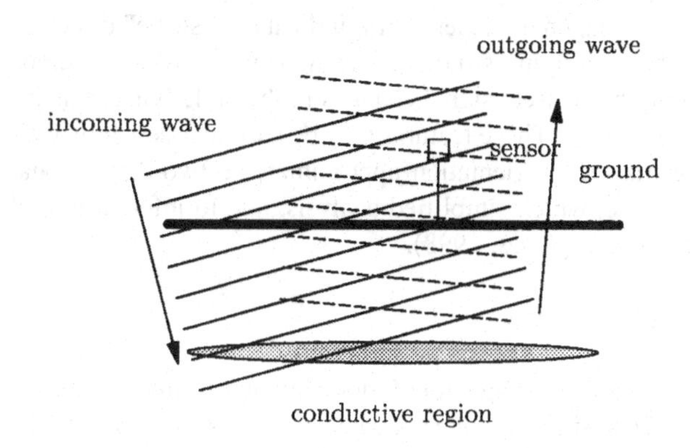

Figure 6. An electromagnetic wave from an atmospheric source interacts with electrically conductive structures in the ground. The response, measured by a sensor at the surface, is a combination of the source and induced waves in the ground. Lines shows wavefronts of incoming wave from source, and dashed lines show the wavefronts of the induced waves emanating from the conductive heterogeneity.

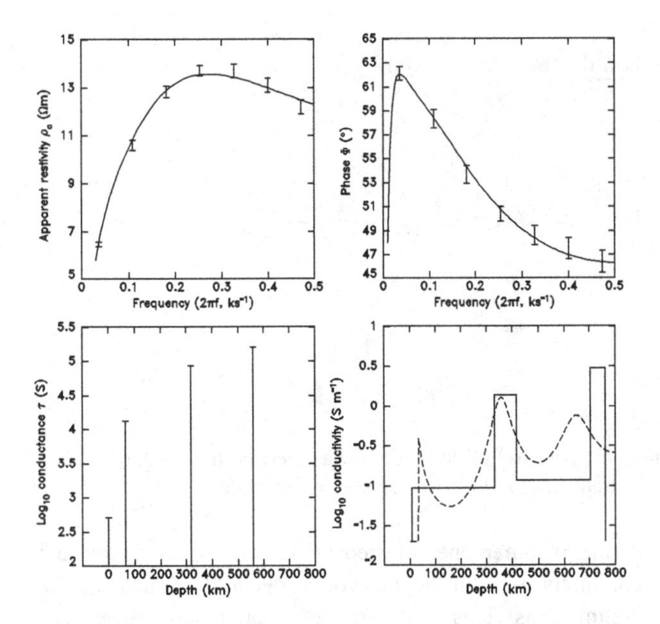

Figure 7. Figures after *Parker and Whaler* (1981) illustrating how a set of apparent resistivity and phase lag measurements at seven frequencies (top) may be fit with three different conductivity profiles: infinitely thin layers of finite conductivity (left), layers of constant conductivity (right, solid) and the smoothest profile (right, dashed). All yield conductive layers near the surface and near 400 km depth, but vary in the detailed conductivity profile with depth.

layered structures, the off-diagonal elements of **Z**, Z_{xy} and Z_{yx} (eq. 13b) should be zero. Nonzero off-diagonal elements indicate that the structures are not horizontally layered, and rotation of the x-y coordinate frame to one that minimizes the off-diagonal elements of Z indicates a "strike" direction of the underlying structure (*Jones*, 1998). Strike variations along the transect signal changes in the underlying resistivity geometry. These features may be used to assess how well the transect's orientation approximates a two-dimensional structure, which simplified analysis, and to infer structural orientation (*Jones*, 1998).

3.6. Case Studies

The earliest magnetotelluric campaign to investigate the electrical signature of a subduction margin appears to be the EMSLAB experiment, conducted across the northern Oregon coast in North America (*Booker and Chave*, 1989). An element of this deployment was a linear array of 54 sensors along a 200 km on-land transect perpendicular to the coast and an offshore leg (*Wannamaker and Hohmann*, 1991). Consideration of geological and physiographic trends in the area dictated the array's orientation. Variation in geoelectrical strike was minor except at the eastern end of the transect.

The workers undertook trial-and-error modelling of the apparent resistivity and phase data, constrained by external geological and physiographic information: ocean bathymetry, sediment thickness and resistivities, and a deep, conductive mantle below 400 km. They found their observations fit a model with a resistive 40-km thick plate, dipping at ~ 25°. Above the plate lay a conductive layer a few km thick which extended to about 60 km depth. They also found a conductive layer under the High Cascades, extending down to 70 km, but not connected to the dipping layer. By examining misfit caused by removing some elements from their models, they determined which features were robust enough to be required by the data in the context of the rest of their model. On this basis, the dipping conductive layer, the sub-Cascades conductor, and the absence of a linkage between the two are all robust features of the model. In conjunction with a seismic reflection profile near the transect, it appears that the conductor coincides with the seismically-defined plate dip, within observational uncertainty. Unlike the Vancouver Island part of the Cascade margin (*Hyndman*, 1988), fluid release, migration of dissolved species in the fluid and consequent precipitation in cooler environs is not required by the data here. The low resistivities may be explained by in-situ fluid evolution, however.

An example of a study using contemporary inversion methods is *Arzate et al.* (1995). The investigators studied the electrical properties of the Costa Rican margin, with the aim of defining the dip of the subducted Cocos plate. While an investigation of the Cascadian margin using both magnetotelluric and seismic data found that the conductive layer did not coincide with the seismically-defined plate interface (*Hyndman*, 1988), sediment subduction dominates at the Costa Rican trench and heat flow is low (*Arzate et al.*, 1995), justifying the assumption that the conductivity structure maps the plate interface. The investigators deployed 14 magnetotelluric stations in two converging linear transects in the form of a V narrowing inward from the coast. They recorded the horizontal components of E and H at a range of periods between 0.001–2000 sec. At short periods (high frequencies), the electrical structure is directionally invariant, suggesting that conductive structures in the crust are horizontally layered. At lower frequencies electrical anisotropy emerges, with a reasonably constant N-S trending electrical strike in one section but a more variable one along the other. This suggests greater subsurface heterogeneity. At lower frequencies still the structure reverts to directional invariance. The picture this information portrays is of lateral homogeneity in electrical structure both shallow and deep, and a directional dependence at intermediate levels.

Two-dimensional inversion of the data collected along the profiles revealed a conductive layer dipping inland from the coast, with a significant change in resistivity between 13 and 30 km depth. Resolution of a structure's thickness is difficult in any magnetotelluric study (*Jones*, 1993), and though the boundary of the conductive structure dips, the body's geometry is difficult to discern. In combination with regional heat flow data, a nearby seismic refraction study, regional seismicity patterns, and gravity surveying, the change in resistivity roughly agrees with dips inferred by the other methods, justifying the assumption that its cause is a sedimentrich layer atop the subducted slab (*Arzate et al.*, 1995). More summaries of combined magnetotelluric studies may be found in *Jones* (1998).

4. SEISMIC WAVES

In addition to the general properties summarized earlier, seismic waves have two further concepts worth introducing in this tutorial. The first, and simplest one, is that the different types of seismic waves—P and S—convert from one to the other when they cross an interface between two elastically different materials. Provided they can be detected, these wave conversions afford a means of telling when material properties change.

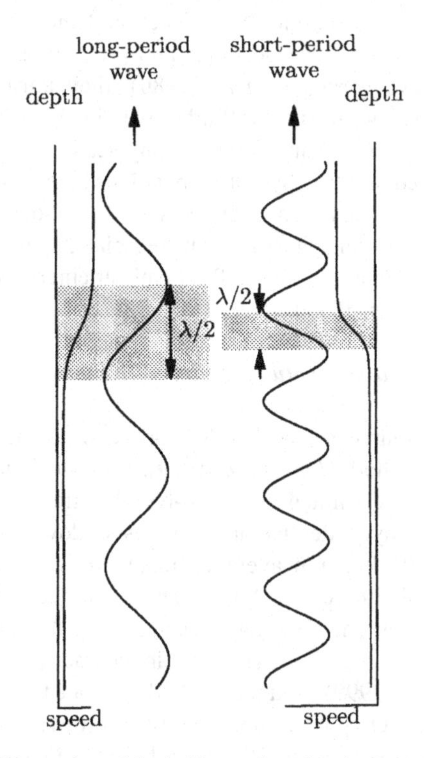

Figure 8. Illustration of the half-wavelength rule. In order to get a conversion from a change in wave material, the change must occur within about ½ of a wavelength. To detect a gradual change in wave speed (left), a longer period wave must be used. Conversely, to locate an abrupt change in wave speed (right), a shorter period wave must be used. A wave reflecting from the top or bottom crosses through the layer twice (on the way in and the way out), so the rule for them is ¼ wavelength.

The second, related, concept is how abrupt the material property change must be to cause a conversion between P and S waves (Figure 8). The distance through which a change in material properties must extend to be detectable to a seismic wave is about one-quarter of the wavelength of the incoming wave for a reflection (*Richards*, 1972), and one-half of a wavelength for a transmission (*Bostock*, 1999). If longer than this, the change is too gradual to cause a wave conversion. Locating more gradual changes in material properties requires scrutiny with longer wavelengths (and periods by equation (3)). Conversely, short wavelengths must be used to locate thin, sharp boundaries.

4.1. Refraction and Reflection Surveying

The quarter-wavelength rule leads to the need to use short-period, high-frequency energy sources to explore thinly-layered materials like the sub-seafloor and subducted oceanic crust. Typically, the sources are bubbles released from compressed-air chambers towed behind ships in water, and explosive charges or earthquakes on land. Following the energy release that probes the structure, one records a time interval containing reflections or transmissions of the seismic energy through the structure—typically a few tens of seconds for shallow investigations. Each pulse in the record corresponds to a discrete wave arrival from somewhere in the structure. What you know is the pulse's arrival time, but what you want is the position in space where the pulse originated, and the wave speed in the material it travelled through. Given distance x and speed v, time is

$$t = \frac{x}{v} , \tag{14}$$

leading to many combinations of x and v that yield the same t. The principal analysis problem is figuring the proper x and v in the material being studied.

This typically is achieved by changing the positions of the energy sources and the recorders to illuminate progressively deeper parts of the structure with seismic energy. Modelling the shallower material's thickness and wave speeds permits the deeper material's contribution to the total travel time to be determined. Reflection surveying tries to gain information about layer positions by locating recorders

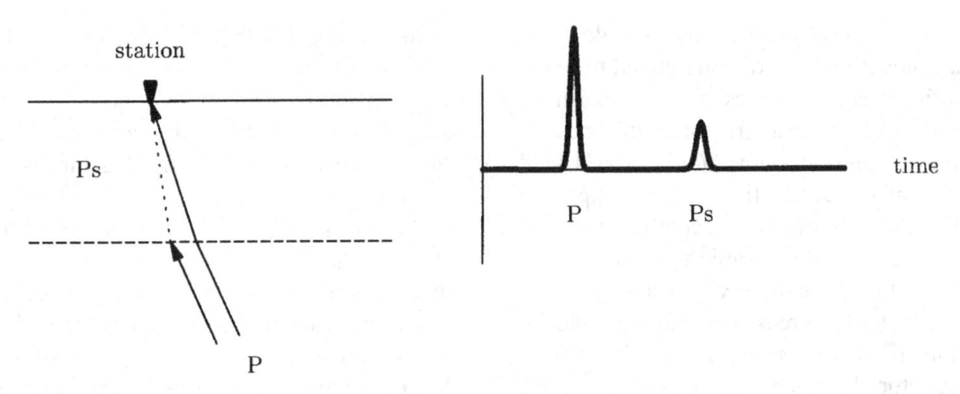

Figure 9. Illustration of the receiver function method. (*left*) P-wave emitted by distant earthquake approaches station on surface from below, and yields P-to-S conversion at layer boundaries below the station. The later-arriving S wave (Ps, dotted line), yields information about the layering through its time lag after P and its amplitude relative to P. (*right*) Seismogram showing simulated P-wave arrival and later Ps conversion. If the velocity structure is known, the time lag between P and Ps may be converted to the layer's depth.

close to the seismic energy source, so the seismogram contains near-vertical reflections from successively deeper layers. These give good constraints on elastic property changes at a specific point midway between the source and recorder. When the sources and receivers are separated more widely, the waves travel horizontally along the layers rather than vertically through them, yielding much better constraints on the average wave speed in the layer. In combination, they alleviate the ambiguity in the layer thickness – speed tradeoff mentioned earlier.

One contemporary example of this procedure is *Christeson et al.'s* [1999] study of the Costa Rican coast. They deployed sensors on the sea-bottom and onshore in a line perpendicular to the coast and towed a seismic source behind a ship, which allowed a variety of ray paths to be taken to the recorders. They solved for the shallow velocity structure on the trench slope and on land, then progressively deeper structures down to the subducting plate's surface and into the subducted plate itself. The observational program revealed that there was a half-kilometer thick low-velocity layer on top of the slab and that oceanic layer 2 (extrusive igneous rock) was slow, with wave speeds of 3.5–4 km/sec. This latter feature suggested that it represented a permeable, hydrothermally cooled zone (*Christeson et al.*, 1999) The study also showed that layering in the subducted Cocos plate persisted after the plate was subducted to 20 km depth.

The ANCORP group conducted another study of a subduction margin, this one in northern Chile (*ANCORP Working Group*, 1999). They deployed seismometers on land and used explosions to generate seismic waves, which travelled essentially vertically, reflecting off contrasting lithologies. By using high frequency sources, they were able to locate thin layers in the forearc basement and the subducted plate. They found two highly reflective patches in the forearc: one at about 70 km depth, and another between 20–30 km depth. Because fluids and rock porosity reduce seismic wave speeds (*Mavko*, 1980), they attributed the reflective patches to sites of fluid production by dehydration reactions (at 70 km) and fluid porosity sealing by low-temperature metamorphic hydration reactions at shallow levels. The deep reflective zone appears to be associated with increased seismicity in the subducted plate, suggesting the seismicity increase is due to fluid embrittlement (*ANCORP Working Group*, 1999).

4.2. Converted Wave Studies—Receiver Functions

Natural source seismology's workhorse is the receiver function method (*Langston*, 1979). This technique uses recordings of earthquakes at land-based stations to infer the subsurface layer depths and contrasts, down to depths reaching 700 km. On average, about one earthquake per month is suitable for this type of study. At layer boundaries below the station, upgoing P waves from the earthquake convert to S waves that arrive later in the trace (Figure 9). *Li and Nabelek* (1999) used the method to locate the subducted plate under Oregon in the Pacific Northwest of North America. *Cassidy et al.* (1998) similarly used the method to locate the edge of the subducted Juan de Fuca plate under the northern part of Vancouver Island. Both studies exploit the sizeable velocity contrasts that arise at or near the top of subducted oceanic lithosphere. This could either be the oceanic Moho (the abrupt change in P wavespeeds to around 8 km/sec) or sediment or extrusive basaltic materials at the plate's upper surface, which lead to contrasts of 5-8% there

(*Helffrich and Abers*, 1997), and generates S waves by the P-to-S conversion from the vicinity of the plate surface. *Rondenay et al.* (2001) also located the subducting Juan de Fuca plate boundary under Oregon with the method and found the dip of the plate increases to the east from 12° to 27° and is visible deeper than 100 km. The steepening dip coincides with thickening of the seismic signature of the oceanic crust, suggesting a locus of ongoing metamorphic reactions exists at this depth.

4.3. Converted Wave Studies—Local Ps Conversions

Provided that they are well-located, local earthquakes may be used in an analogous way to distant earthquakes to find the subducted plate's boundary. P-to-S (and S-to-P) conversions at layer interfaces produce additional arrivals whose analysis can yield spatial maps of the plate interface (Figure 10). *Matsuzawa et al.* (1986, 1987) first used the method to locate the subducted plate boundary in northern Honshu, Japan. *Zhao et al.* (1992) pioneered using these additional arrivals in tomographic studies to locate layer interfaces better. More recently, *Ohmi and Hori* (2000) conducted a similar analysis in central Japan, north of Tokyo. They found earthquake clusters at the subducted plate interface and were able to separate the earthquakes in the subducted plate from those in the overriding plate.

These converted phase studies reveal layering 5-15 km thick at the top of the subducted plate. The velocity contrasts between the material outside the plate and the layer ranges from 6-14% slower, and the layers extend down to at least 150 km in some cases. The localities displaying in-slab layering are Japan (*Matsuzawa et al.*, 1986, 1987), Alaska (*Helffrich and Abers*, 1997), and Taiwan (*Lin et al.*, 1999).

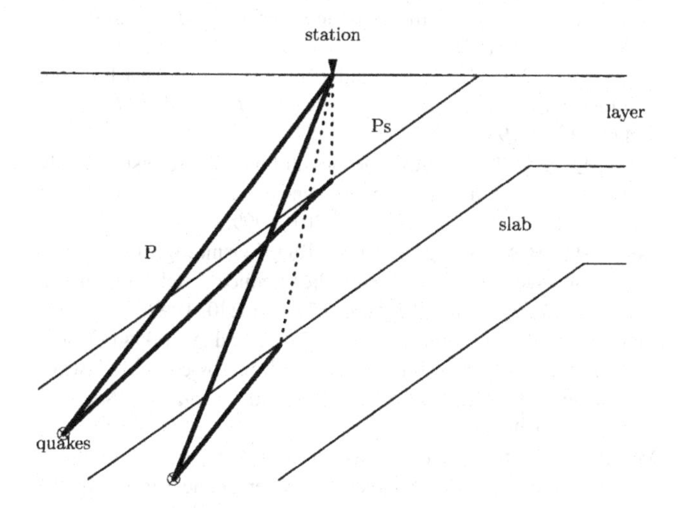

Figure 10. Local earthquakes in a subduction zone can generate P-to-S wave conversions from slab layering in a fashion analogous to the mechanism shown in Figure 9. P-wave legs shown with solid lines and S-wave legs shown with dotted lines. The time lags and Ps amplitudes depend on the location of the interface and its properties.

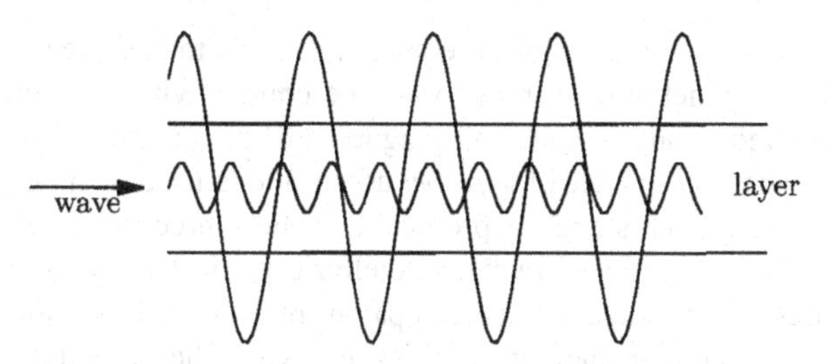

Figure 11. Sketch of how waves with different wavelengths interact with a layer that they propagate along. Oscillations of low frequency waves involve more of the surrounding material. Increasingly higher frequencies eventually confine their oscillation to the layer itself. Depending on whether the layer is slow or fast relative to its surroundings, the high frequencies might arrive late (if slow) or early (if fast) relative to the low frequency components of a seismic wave packet.

4.4. Frequency-dependent Travel Times and Layering

The typical idealization of a seismic wave is that it travels as a thin ray. The higher the wave's frequency, the better this approximation is. Oscillations involving lower frequency components of a seismic wave involve more of the surrounding medium. Thus, if a wave travels along a layer of finite thickness, the high-frequency, short-wavelength portions involve elastic deformations in the layer itself, whereas elastic deformation associated with the low frequency components incorporates the surrounding material (Figure 11). This can lead to frequency-dependent travel times, documented in Tonga/Kermadec (*Gubbins and Snieder*, 1991; *Ansell and Gubbins*, 1986; *van der Hilst and Snieder*, 1996), Alaska, the Aleutians, Japan and the Marianas (*Abers and Sarker*, 1996; *Abers*, 2000). If the layer is faster than its surrounding material, the high frequencies arrive early. Conversely, if the layer is slow, low frequencies arrive early. *Abers* (2000) showed that the low velocity layers are 1-5.5 km thick (with a factor of two uncertainty), are 5-7% slower than surrounding mantle, and extend to at least 250 km depth.

5. DISCUSSION AND CONCLUSIONS

It should be clear from the previous study summaries that seismic methods provide more detailed views of the subsurface than do electromagnetic ones. This is principally because the wavelengths are longer, in order to achieve sufficient penetration depths. Also, the conductivity structure inferred from a suite of magnetotelluric measurements depends on the set of *a-priori* assumptions used to model the data. Both share the problem that the source of the signal is indeterminate. In magnetotelluric studies, many structures can produce the same apparent resistivity; in seismological studies, many geometries may yield the same travel times. Proper design of source and receiver placement ameliorates the problems in the seismological case, but there does not seem to be similar flexibility in the magnetotelluric case due to lack of low-frequency source control. Thus magnetotelluric studies are typically interpreted along with other data to draw structural inferences (*Hyndman*, 1988; *Cassidy and Ellis*, 1991; *Arzate et al.*, 1995; *Jones*, 1998). The attractive aspect of electromagnetic methods is that they are sensitive to a completely different set of rock properties than those that govern seismic phenomena. Thus they provide a complementary investigative techniques with which properly constructed hypotheses may be tested (*Cassidy and Ellis*, 1991; *Jones*, 1998).

Acknowledgments. I thank the anonymous reviewer and the volume editor, John Eiler, for the grammatical and scientific clarifications they suggested. Prose sometimes gets away from you. I also thank the volume editors for their patience.

REFERENCES

Abers, G. A., Hydrated subducted crust at 100-250 km depth, *Earth Planet. Sci. Lett., 176*, 323-330, 2000.

Abers, G. A., and G. Sarker, Dispersion of regional body waves at 100-150 km depth beneath alaska: In situ constraints on metamorphism of subducted crust., *Geophys. Res. Lett., 23*, 1171-1174, 1996.

ANCORP Working Group, Seismic reflection image revealing offset andean subduction-zone earthquake locations into oceanic mantle, *Nature, 397*, 341-344, 1999.

Ansell, J., and D. Gubbins, Anomalous high-frequency wave propagation from the tonga-kermadec seismic zone to new zealand., *Geophys. J. R. astron. Soc., 85*, 93-106, 1986.

Arzate, J. A., M. Mareschal, and D. Livelybrooks, Electrical image of the subducting cocos plate from magnetotelluric observations, *Geology, 23*, 703-706, 1995.

Booker, J. R., and A. D. Chave, Introduction to the special section on the emslab-juan de fuca experiment, *J. Geophys. Res., 94*, 14,093-14,098, 1989.

Bostock, M. G., Seismic waves converted from velocity gradient anomalies in the earth's upper mantle, *Geophys. J. Int., 138*, 747-756, 1999.

Cassidy, J. F., and R. M. Ellis, Shear wave constraints on a deep crustal reflective zone beneath vancouver island, *J. Geophys. Res., 96*, 19,843-19,851, 1991.

Cassidy, J. F., R. M. Ellis, C. Karavas, and G. C. Rogers, The northern limit of subducted juan de fuca plate system, *J. Geophys. Res., 103*, 26,949-26,961, 1998.

Christeson, G. L., K. D. McIntosh, T. H. Shipley, E. R. Flueh, and H. Goedde, Structure of the costa rica convergent margin, offshore nicoya peninsula, *J. Geophys. Res., 104*, 25,443-25,468, 1999.

de Groot-Hedlin, C., and S. Constable, Occam's inversion to generate smooth, two-dimensional models from magnetotelluric data, *Geophysics, 55*, 1613-1624, 1990.

Eberhart-Phillips, D., W. D. Stanley, B. D. Rodriguez, and W. J. Lutter, Surface seismic ad electrical methods to detect fluids, *J. Geophys. Res., 100*, 12,919-12,936, 1995.

Gubbins, D., and R. Snieder, Dispersion of p waves in subducted litohsphere: Evidence for an eclogite layer, *J. Geophys. Res., 96*, 6321-6333, 1991.

Haak, V., and V. R. S. Hutton, Electrical resistivity in continental lower crust., in *The nature of the lower continental crust*, edited by J. B. Dawson, D. A. Carswell, J. Hall, and K. H. Wedepohl, vol. 24 of *Geol. Soc. London Spec. Pub.*, pp. 35-49, Geological Society, London, 1986.

Helffrich, G., and G. A. Abers, Slab low-velocity layer in the eastern aleutian subduction zone, *Geophys. J. Int., 130*, 640-648, 1997.

Hyndman, R. D., Dipping seismic reflectors, electrically conductive zones, and trapped water in the crust over a subducting plate, *J. Geophys. Res., 93*, 13,391-13,405, 1988.

Jiracek, G. R., Geoelectromagnetics charges on, *Rev. Geophys., 33*, 169-176, 1995.

Jones, A. G., Electromagnetic images of modern and ancient subduction zones, *Tectonophysics, 219*, 29-45, 1993.

Jones, A. G., Waves of the future: Superior influences from collocated seismic and electromagnetic experiments, *Tectonophysics, 286*, 273-298, 1998.

Langston, C. A., Structure under mount rainier, washington, inferred from teleseismic body waves, *J. Geophys. Res., 84*, 4749-4762, 1979.

Li, X.-Q., and J. Nabelek, Deconvolution of teleseismic body waves for enhancing structure beneath a seismometer array, *Bull. Seismol. Soc. Am., 89*, 190-201, 1999.

Lin, C.-H., B.-S. Huang, and R.-J. Rau, Seismological evidence for a low-velocity layer within the subducted slab of southern taiwan, *Earth Planet. Sci. Lett., 174*, 231-240, 1999.

Matsuzawa, T., N. Umino, A. Hasegawa, and A. Takagi, Upper mantle velocity structure estimated from ps-converted wave beneath north-eastern japan arc, *Geophys. J. R. astron. Soc., 86*, 767-787, 1986.

Matsuzawa, T., N. Umino, A. Hasegawa, and A. Takagi, Estimation of the thickness of a low-velocity layer at the surface of the descending oceanic plate beneath the north-eastern japan arc by using syhthesized ps-wave, *Tohoku geophys. J., 31*, 19-29, 1987.

Mavko, G., Velocity and attenuation in partially molten rocks., *J. Geophys. Res., 85*, 5173-5189, 1980.

Ohmi, S., and S. Hori, Seismic wave conversion near the upper boundary of the pacific plate beneath the kanto district, japan, *Geophys. J. Int., 141*, 136-148, 2000.

Parker, R. L., The inverse problem of electromagnetic induction: Existence and construction of solutions based upon incomplete data, *J. Geophys. Res., 85*, 4421-4425, 1980.

Parker, R. L., and K. A. Whaler, Numerical methods for establishing solutions to the inverse problem of electromagnetic induction, *J. Geophys. Res., 86*, 9574-9584, 1981.

Parkinson, W. D., *Introduction to geomagnetism*, Scottish Academic Press, Edinburgh, 1983.

Richards, P. G., Seismic waves reflected from velocity gradient anomalies within the earth's upper mantle, *Zeit. fur Geophysik, 38*, 517-527, 1972.

Rondenay, S., M. G. Bostock, and J. Schragge, Multiparameter two-dimensional inversion of scattered teleseismic body waves 3. application to the cascadia 1993 data set, *J. Geophys. Res., 106*, 30,795-30,807, 2001.

Smith, J. T., and J. R. Booker, Rapid inversion of two and three-dimensional magnetotelluric data, *J. Geophys. Res., 96*, 3905-3922, 1991.

van der Hilst, R. D., and R. Snieder, High-frequency precusors to p wave arrivals in new zealand: Implications for slab structure, *J. Geophys. Res., 101*, 8473-8488, 1996.

Wannamaker, P. E., and G. W. Hohmann, Electromagnetic induction studies, *Rev. Geophys., 29*, 405-415, 1991.

Wannamaker, P. E., J. R. Booker, A. G. Jones, A. D. Chave, J. H. Filloux, H. S. Waff, and L. K. Law, Resistivity cross section through the juan de fuca subduction system and its tectonic implications, *J. Geophys. Res., 94*, 14,127-14,144, 1989.

Zhao, D., A. Hasagewa, and S. Horiuchi, Tomographic imaging of p and s wave velocity structure beneath northeastern japan, *J. Geophys. Res., 97*, 19,909-19,928, 1992.

Seismological Constraints on Structure and Flow Patterns Within the Mantle Wedge

Douglas A. Wiens and Gideon P. Smith

Department of Earth and Planetary Sciences, Washington University, St. Louis, Missouri

The mantle wedge of a subduction zone is characterized by low seismic velocities and high attenuation, indicative of temperatures approaching the solidus and the possible presence of melt and volatiles. Tomographic images show a low velocity region above the slab extending from 150 km depth up to the volcanic front. The low velocities result at least partially from volatiles fluxed off the slab, which lower the solidus and thus raise the homologous temperature of the wedge material. Subduction zones with active back-arc spreading centers also show large low velocity and high attenuation seismic anomalies beneath the backarc basin, indicating that a broad zone of magma production feeds the backarc spreading center. The magnitude of the velocity anomaly is consistent with the presence of approximately 1% partial melt at depths of 30–90 km. The best-imaged arc-backarc system, the Tonga-Lau region, suggests that the zone of backarc magma production is separated from the island arc source region within the depth range of primary magma production. However, the anomalies merge at depths greater than 100 km, suggesting that small slab components of backarc magmas may originate through interactions at these depths. Slow velocities extend to 400 km depth beneath backarc basins, and these deep anomalies may result from the release of volatiles transported to the top of the transition zone by hydrous minerals in the slab. Observations of seismic anisotropy can provide direct evidence for the flow pattern in the mantle wedge. Slab parallel fast directions suggesting along-arc flow within the mantle wedge are found in most, but not all, subduction zones. The Tonga-Lau region shows a complex pattern of fast directions, with along-strike fast directions beneath the Tonga island arc and convergence-parallel fast directions to the west of the Lau backarc spreading center. The pattern of flow in the Lau backarc is consistent with southward mantle flow inferred from geochemical data. Geodynamic modeling suggests several possible mechanisms of flow within the mantle wedge, which may help explain the diverse observations. Viscous coupling between the backarc flow and the downgoing plate should produce induced flow within the backarc, with flow directions parallel to the convergence direction. In contrast, subduction zone roll-

Inside the Subduction Factory
Geophysical Monograph 138

10.1029/138GM05

back may produce along-arc flow. This latter model matches the observations in most subduction zones suggesting that viscous coupling does not exert a strong control on the flow pattern in the mantle wedge.

1. INTRODUCTION

The mantle wedge is the locus of the most important processes in the subduction factory, including island arc magmatism and back-arc spreading. Seismological studies can place observational constraints on the source regions and transport paths of the magmas and fluids that represent the output of the subduction factory. Seismic velocity and attenuation anomalies are functions of mantle composition (including volatiles), temperature, and melt content. Therefore, accurate mapping of seismological anomalies can help constrain the spatial distribution of temperature anomalies and melt transport paths within the wedge. In addition, the seismic anisotropy of mantle materials is a function of the strain field, which results from the flow pattern. Therefore seismic anisotropy measurements also place constraints on the mantle flow pattern within the wedge. Thus seismological observations will provide key observational constraints necessary for construction of physically realistic geodynamic models of processes in the mantle wedge.

In this paper, we first review the seismological observables and their relationship to physical parameters such as temperature and melt content. We will then present a status report on progress that has been made in constraining subduction processes in the mantle wedge using seismology.

2. RELATIONSHIP OF SEISMIC OBSERVABLES TO PHYSICAL PROPERTIES

2.1. P and S Wave Velocities

2.1.1. Effect of temperature. The relationship between *P* and *S* wave velocities and temperature can be decomposed into anharmonic and anelastic effects [*Karato*, 1993], where anelastic effects result from velocity dispersion due to attenuation [*Aki and Richards*, 1980]. The anharmonic variation of *P* (V_p) and *S* (V_s) velocity with temperature (T) for mantle materials is relatively well known from laboratory experiments at ultrasonic frequencies. Single crystal laboratory experiments suggest values of about -0.62×10^{-4} °K^{-1} for $\partial \ln V_p/\partial T$ and -0.76×10^{-4} °K^{-1} for $\partial \ln V_s/\partial T$ for olivine at upper mantle pressures and temperatures [*Anderson et al.*, 1992; *Isaak*, 1992].

The anelastic effects on the seismic velocity are sometimes neglected, but are actually important for the mantle wedge, where attenuation may be large. In an anelastic medium with frequency-independent attenuation (Q^{-1}), the seismic velocity shows dispersion as given by *Kanamori and Anderson* [1977] and *Karato* [1993]

$$V(\omega,T) = V_0(T)\,[1 + (Q^{-1}/\pi)\ln \omega\tau(T)] \qquad (1)$$

where ω is frequency, V_0 (T) is the anharmonic velocity, which is frequency independent, and τ(T) is the relaxation time. Table 1, from *Karato* [1993], shows the effect of attenuation on the effective temperature derivatives of the P and S velocity.

Studies of the attenuation structure of the mantle wedge suggest *P* wave Q (Q_P) values of 90–200 and *S* wave Q (Q_S) values of 35–100, with some variation caused by the frequency dependence of attenuation [*Sekiguchi*, 1991; *Flanagan and Wiens*, 1994; *Flanagan and Wiens*, 1998; *Roth et al.*, 1999; *Tsumura et al.*, 2000]. Therefore, the anelastic effect is about as large as that solely due to the anharmonic derivatives (table 1). Assuming a Q_P of 130 and a Q_S of 75, with a Q_P/Q_S ratio of 1.75 as observed [*Roth et al.*, 1999] and typical upper mantle seismic velocities, results in velocity derivatives of 1.0 ms^{-1}/°K for $\partial Vp/\partial T$ and 0.8 ms^{-1}/°K for $\partial Vs/\partial T$. So a change of 100°C would produce changes in *P* and *S* velocity of 0.10 and 0.08 km/s respectively, or about 1.2% for *P* waves and 1.8% for *S* waves.

The ratio of *S* to *P* wave anomalies can be diagnostic of the physical process causing the velocity anomalies [*Hales and Doyle*, 1967; *Koper et al.*, 1999]. The parameter $\nu = \partial \ln Vs/\partial \ln Vp$, giving the fractional change in V_S relative to the fractional change in Vp, is often used to describe the *S* to *P* anomaly ratio. For temperature changes, ν is generally

Table 1. Temperature derivatives of elastic wave velocities for olivine in the upper mantle (from Karato, [1993]).

Q	$\partial \ln V_p / \partial T$ (10^{-4} K^{-1})	$\partial \ln V_S / \partial T$ (10^{-4} K^{-1})
50	–2.18	–2.32
100	–1.40	–1.54
200	–1.01	–1.15
300	–0.88	–1.02
400	–0.82	–0.96
∞	–0.62	–0.76

Note: Temperature derivatives for the anharmonic case (Q = ∞) are from *Isaak* [1992].

about 1.2 for rocks of mantle composition [*Kern and Richter*, 1981]. Including velocity dispersion caused by attenuation gives a ν of about 1.4 for anomalies caused by temperature effects [*Koper et al.,* 1999].

2.1.2. Effect of melt fraction. It is widely recognized that partial melt can have a major effect on seismic velocities. Partial melt forms in pockets that assume a shape that minimizes the total free energy of the grain boundary and melt interfaces. The effect of melt on seismic velocities depends on the geometry of the melt pockets as well as the melt fraction. Our knowledge of the effect of partial melt comes from both laboratory experiments and analytical calculations of the mechanical response of material with melt-filled inclusions. Unfortunately, experiments to determine the velocity at seismic frequencies in the presence of partial melt and at the pressure and temperature of the upper mantle are difficult and have not been carried out.

Ultrasonic experiments on peridotite at high pressure and temperature suggest that there is no discontinuous change in either *P* or *S* velocity at the solidus [*Sato et al.*, 1989a]. *P* and *S* waves show similar behavior and a rapid drop in velocity is observed only if the melt fraction becomes larger than about 2%, after which the velocity decreased by about 2% for every 1% of partial melt. The effect of pressure can be accounted for by using the homologous temperature (normalizing the temperature by the solidus temperature). Since these results were obtained at ultrasonic frequencies they are unlikely to adequately represent the anelastic components of the velocity variations.

Analytical studies have generally calculated the mechanical response of melt inclusions assuming idealized shapes, including cracks [*O'Connell and Budianski*, 1974], tubes [*Mavko*, 1980], and ellipsoids [*Schmeling*, 1985]. Melt distributed along thin films or cracks has a large effect on the shear modulus, thus reducing the *S*-wave velocity with relatively small partial melt, whereas melt distributed in spheroidal inclusions shows smaller effects. An alternative methodology for calculating the mechanical response based on grain-boundary contiguity [*Takei*, 1998] has also recently been developed and tested using analog materials [*Takei*, 2000].

Recently, *Hammond and Humphreys* [2000b] used a finite element method to calculate the effect of melt on the shear and bulk modulus. They based their calculation on images of the melt geometry from laboratory experiments, which suggest that at small melt fractions (<1%) melt is distributed in triple junction tubules, but begins to form films at higher melt fraction [*Waff and Faul*, 1992; *Faul et al.*, 1994]. The *Hammond and Humphreys* [2000b] model (Figure 1) shows only small reductions in velocity for less than 1% partial melt, as melt is contained in tubules, but the velocity reduction becomes large for larger melt fractions where melt forms films. The total velocity reduction for a melt fraction of 2% relative to a reference temperature of 1000°C is 22% for *S* waves and 11% for *P* waves. The effect of larger melt fractions is much greater for *S* waves than for *P* waves, such that $\nu = \partial \ln Vs/\partial \ln Vp = 2.2$ for anomalies caused by melt. These results suggest that seismic anomalies caused by melt fractions smaller than 1% may be difficult to distinguish seismologically from the effects of temperature, but larger melt fractions should be easily detectable from the large amplitude seismic anomalies and the large *S* to *P* anomaly ratio (ν).

2.1.3. Effect of composition and volatile content. The effects of mineral composition on seismic observables are generally thought to be smaller than the effects of temperature heterogeneity and partial melt in the upper mantle [*Sobolev et al.*, 1996; *Goes et al.*, 2000]. For example, observed P wave velocities for a variety of peridotites from the continental upper mantle show relatively small variations (< 2%) despite significant variations in modal composition [*Sobolev et al.*, 1996]. Strong variations in the Mg/(Mg+Fe) ratio, however, will have a significant effect on the seismic velocity [*Akimoto*, 1972; *Jordan*, 1979]. In particular, highly depleted harzburgites may be characterized by seismic velocities that are higher by ~2% when compared to undepleted peridotites [*Jordan*, 1979].

There is some evidence that differences in volatile content can produce major variations in the seismic velocity. The presence of relatively small concentrations of dissolved water has been shown to lower the solidus of olivine and weaken olivine aggregates by several orders of magnitude [*Karato et al.*, 1986; *Hirth and Kohlstedt*, 1995]. *Hirth and Kohlstedt* [1996], *Karato and Jung* [1998], and *Karato* [2003] (this volume) suggest that the weakening effects of water will lower the seismic velocity by increasing seismic attenuation, thus reducing the observed seismic velocity as in equation 1 above. Simple calculations suggest that the shear wave velocity may be lowered as much as 6% by modest amounts of water, and *Gaherty et al.* [1996] suggest that the low velocity zone may result from the water content of the asthenosphere beneath the water-depleted lithospheric lid. The effects of water, high temperatures, and partial melt all can substantially reduce seismic velocity and are difficult to distinguish when using seismic velocity measurements alone.

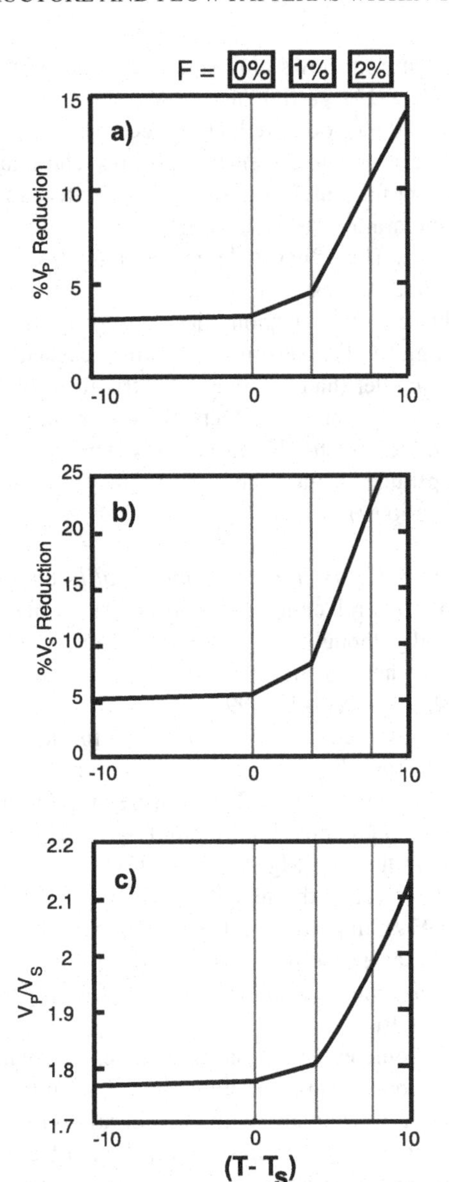

Figure 1. (a) Vp and (b) Vs velocity reduction and (c) the Vp/Vs ratio as a function of temperature and partial melt content (after *Hammond and Humphreys,* [2000b]). The horizontal axis plots temperature (in °C) relative to the solidus. Thin vertical lines indicate melt fractions of F= 0%, 1%, and 2%. The velocity reductions were calculated using a finite element method incorporating experimentally observed melt topologies [*Faul et al.,* 1994]. Melt percent as a function of temperature is calculated for water-free mantle at 30 km depth. This calculation assumes that the melt is contained in tubules at melt fractions below 1%. Attenuation is assumed to vary exponentially with temperature as described by *Jackson et al.* [1992], with Q values of $Q_\alpha = 90$ and $Q_\beta = 40$ at the solidus (T_S = 1240°C). Velocity reduction is calculated relative to a reference temperature T_0 = 1000°C; lower reference temperatures would give somewhat larger velocity reductions.

2.2. Seismic Attenuation

2.2.1. Effect of temperature. Seismic attenuation (Q^{-1}) is a measure of the decay of the seismic wave amplitude due to absorption of energy by an anelastic medium. A variety of experiments show that seismic attenuation is a strong function of temperature, particularly for temperatures near the solidus [*Berckhemer et al.*, 1982; *Sato et al.*, 1989b; *Karato and Spetzler*, 1990; *Jackson et al.*, 1992; *Gribb and Cooper,* 1998; *Jackson,* 2000]. Experiments on dunite at high temperatures suggest an *S* wave attenuation relationship at 1 Hz frequency [*Jackson et al.*, 1992]

$$Q_\beta^{-1}(T) = 5 \times 10^5 \exp [(-201 \text{ kJ/mol})/RT] \quad (2)$$

Using this relationship, the *S* wave attenuation (Q_β^{-1}) increases by more than an order of magnitude between temperatures of 1000°C and 1200°C. The pressure sensitivity of this relationship is not known, but there is evidence that the effect of pressure can be accounted for by using the homologous temperature (i.e. normalizing the temperature by the solidus temperature) [*Sato et al.*, 1989b; *Getting et al.*, 1997].

If all the attenuation were associated with the shear modulus, the ratio of Q_α/Q_β would be 2.25 for a Poisson solid, and about 2.45 for observed upper mantle elasticities [*Anderson et al,* 1965]. In contrast, if equal amounts of attenuation were associated with shear and compression, the ratio would be 1. The ratio of Q_α/Q_β is generally assumed to be in the range of 1.5–2.4 based on both experiments and seismic observations, with one recent observational study suggesting 1.75 for the Tonga mantle wedge [*Roth et al.*, 1999].

2.2.2. Effect of melt fraction. The effect of melt on seismic attenuation has been approached with both experimental and analytical methods. Experiments suggest that there is little intrinsic increase in attenuation due to small quantities (less than about 5%) of partial melt [*Berckhemer et al.*, 1982; *Gribb and Cooper*, 2000]. This is in agreement with analytical calculations that have investigated possible melt-related attenuation mechanisms, including viscous shear relaxation of the melt, melt squirt, and melt-enhanced grain boundary sliding. In each case, the major relaxation effects of these processes occur well outside the seismic frequency band, suggesting that small amounts of partial melting have little effect on seismic attenuation [*Schmeling*, 1985; *Hammond and Humphreys*, 2000a].

2.2.3. Effect of volatile content, composition, and grain size. The presence of even small amounts of water enhances anelasticity [*Karato and Jung,* 1998; *Karato,* this volume]. Thus suggests that attenuation is a strong function of volatile content. It is usually assumed that attenuation is such a strong function of temperature and volatile content that other compositional effects are clearly secondary. Several studies have suggested that attenuation is a function of homologous temperature [*Sato et al.,* 1989b; *Getting et al.*, 1997]. Materials of different solidus temperatures must show different attenuation at a particular temperature, and normalization by the solidus temperature may allow approximate correction for different mineral composition [*Getting et al.,* 1997]. There is also evidence that attenuation is a function of grain size, with smaller grains showing greater attenuation [*Gribb and Cooper*, 1998; *Jackson*, 2000]. However, graingrain size variation may play a smaller role in mantle attenuation variation than in the laboratory due to the large graingrain size of mantle ultramafic rocks [*Jackson,* 2000].

2.2.4. Identifying partial melt. Table 2 provides a qualitative analysis of the relative effect of various factors on the seismic observables in the upper mantle. Because temperature anomalies, volatiles, and small amounts of partial melt can produce similar seismic observations, it is often difficult to definitively associate observed anomalies with a particular cause. However, the simultaneous mapping of several different seismic observables offers a greater opportunity to distinguish between the various factors. The different responses of attenuation and seismic velocities to partial melt suggests that combined observations of seismic velocity and attenuation may be particularly valuable in distinguishing partial melt from large sub-solidus temperature anomalies [*Forsyth*, 1993]. Melt has the greatest effect on *S*-wave velocities, a smaller effect on *P-wave* velocities, and relatively little effect on attenuation. Conversely, high sub-solidus temperatures have an exponential effect on attenuation but a smaller effect on seismic velocities. Therefore, regions of partial melt should be distinguishable by their exceptionally slow *S*-wave velocity anomalies and large *S* to *P* anomaly ratios in the absence of a correspondingly large attenuation anomaly. Distinguishing between high temperature anomalies and the presence of volatiles will remain difficult until better laboratory data are available.

2.3. Velocity Anisotropy

Observations of anisotropy, where the seismic velocity depends on the propagation or polarization direction, can have a variety of causes. These include layering of different isotropic layers [*Backus*, 1965], cracks whether dry, fluid or melt filled [*Crampin and Booth*, 1985; *Kendall*, 1994], or systematic orientation of mantle minerals, such as olivine, whose crystallographic structure is anisotropic [*Hess*, 1964]. Where anisotropy can be associated with an increased ordering and alignment of mineral crystal orientations the ordering can in turn be related to the mechanical deformation of the rock [*Nicolas and Christensen*, 1987]. Detailed study of anisotropic structures therefore provides an opportunity to relate seismic observations to deformation processes in the Earth, including mantle flow. The mantle flow in arcs probably influences or controls many processes, such as the path of melt from the source region as well as the distribution of geochemical anomalies and is therefore of broad importance.

Systematic orientation of anisotropic minerals appears to be the dominant mechanism causing anisotropy in the upper-most mantle. This preferred orientation of minerals appears to consist of alignment of one of the crystallographic axes (either a, b, or c), with the other two axes forming a girdle. *Hess* [1964] noted that in the oceans the fast azimuths of P_n are sub-parallel to the spreading directions. Hess interpreted this as preferential orientation of the olivine b-axes, but a later laboratory study by *Raleigh* [1968] led to the proposal that the anisotropy is due to alignment of the a-axes.

To interpret these seismological observations one must be able to relate the seismic velocities observed to the velocities of realistic mantle materials. Laboratory studies have been performed on single crystals. However, these produce only estimates of anisotropy in 'pure' crystals. In the real Earth, minerals are rarely 100% aligned and so velocities have also been measured in naturally occurring rocks, such as overthrust pieces of mantle or xenoliths, and hot-pressed

Table 2. Qualitative effect of various parameters on seismic observables.

	Temperature	Partial Melt	Composition	Water Content
V_P anomaly	Large	Large	Modest	Modest-Large?
V_s anomaly	Large	Very Large	Modest	Modest-Large?
$\nu = \partial \ln V_s / \partial \ln V_p$	1.2–1.5	2.0–2.3	variable	~ 1.5 ?
Q^{-1} anomaly	Large	Small	Small	Large ?

polycrystalline aggregates, where such factors as composition and water content can be varied.

In olivine, the a-axes are the fastest direction and the b-axes the slowest [*Verma*, 1960]. Although examination of rock fabrics has produced examples of b-axes alignment [*Christensen and Crosson*, 1968; *Ave'Lallement and Carter*, 1970], a-axes alignment is more common [*Christensen*, 1984; *Mainprice and Silver*, 1993; *Ji et al.*, 1994]. In the latter case it is believed that the orientation is controlled by the finite strain [*Williams*, 1977]; the a-axes align in the direction of shear and the b-axes normal to it. This observation gives rise to the conventional assumption that the mantle flow direction corresponds to the direction of preferential orientation of olivine a-axes and the anisotropic fast direction.

Laboratory experiments that determine induced anisotropy under an applied deformation also favor alignment of the a-axes. In early experiments examination of the lattice preferred orientation (LPO) was usually performed on materials under uniaxial compression. However, theoretical models show that the type of strain is important in predicting the resulting LPO [*Wenk et al.*, 1991] and in the upper mantle simple shear is a much more common deformation mechanism than uniaxial compression. Subsequent experiments on simple shear deformation of olivine aggregates at high temperatures and pressures (~1,500°K and 300 MPa) yielded a-axis alignment in the direction of flow at large strain [*Zhang and Karato*, 1995]. Small strains produced only slight rotation from the flow direction. Later studies performed on olivine aggregates under greater deformation conditions showed that stress-controlled orientations can also develop but that it is associated with grain boundary migration processes during recrystallization and growth [*Zhang et al.*, 2000].

Water content has been shown to affect the deformation of olivine and enhance grain-boundary migration suggesting it may be important in determining the LPO. *Jung and Karato* [2001] demonstrated that wet and dry aggregates of olivine exhibited distinctly different LPO when they undergo simple shear deformation under conditions of high stress, temperature, and pressure. They showed that the olivine c-axis, for example, is nearly perpendicular to the shear direction under dry conditions whereas under wet conditions it is sub-parallel to the shear direction. It is unclear at this time what factors govern the deformation slip system for water rich mantle rocks and thus the relationship between the mantle flow direction and the fast direction of seismic anisotropy. Since the mantle wedge near the slab is likely to have significant water content due to slab dehydration and fluid migration, if the laboratory simulations of *Jung and Karato* [2001] reflect the true behavior of olivine in the Earth, then they have strong implications for the interpretation of anisotropy within a backarc basin. However, this slip system is seldom observed in natural rocks. A recent study of mantle peridotites from the relect Talkeetna arc shows yet another slip system, which differs from both the generally assumed one and that observed in the laboratory [*Mehl et al.,* 2002]. In this slip system the olivine fast direction is sub-parallel to the mantle flow direction, as conventionally assumed. In view of this uncertainty, for the purposes of this review we continue to make the conventional assumption that the fast anisotropic direction delineates the mantle flow direction.

Other factors that can influence the strength of anisotropy include the effect of temperature and pressure, which can affect both the strength of anisotropy [*Kern*, 1993] and the rate of reorientation [*Goetze and Kohlstedt*, 1973; *Mainprice and Nicolas*, 1989]. At high temperatures, diffusion and grain boundary mobility enhance orientation [*Mainprice and Nicolas*, 1989]. Indeed, if conditions are such that melt is formed, anisotropy may be further increased by increased grain size [*Hirn et al.*, 1995] or reduced by increasing the degree of diffusion creep [*Cooper and Kohlstedt*, 1986] or embrittlement [*Davidson et al.*, 1994]. A review of many of these effects is given by *Savage* [1999].

Geodynamical flow models can be linked to anisotropy observations in several ways. One approach assumes that the maximum P wave velocity aligns with the local direction of maximum finite extension, and the minimum P wave velocity aligns with minimum finite extension [*Fischer et al.,* 2000; *Hall et al.*, 2000]. The magnitude of the anisotropy is scaled by the axial ratios of the finite strain ellipsoid [*Ribe*, 1992]. More sophisticated approaches such as *Blackman et al.* [1996] and *Tommasi* [1998] predict LPO and the resulting anisotropic signature from a particular flow pattern by explicitly linking thermo-mechanical finite element models to simulations of LPO development.

3. THE SEISMIC VELOCITY AND ATTENUATION STRUCTURE OF THE MANTLE WEDGE

The most detailed information on the 2-D and 3-D variation of seismic velocity beneath island arcs comes from body wave tomography. High-resolution tomography requires either a dense array of recording stations or a dense array of earthquakes, and therefore has been carried out in only a few locations. Here we discuss both a well-studied island arc (Japan) and an arc with an accompanying backarc spreading center (the Tonga-Lau system). Northeast Japan is the best-imaged island arc due to the exceptionally large seismological dataset built up by years of recording with a dense array of permanent land seismic stations [*Zhao and Hasegawa*, 1993; *Nakajima et al.*, 2001]. The Tonga-Lau system is the best imaged island arc with an active back-arc spreading system due to the high seismicity rate of the Tonga slab and the temporary deployment of 12 land seismic stations (the SPASE experiment) and 30 ocean bottom seismographs (the LABATTS experiment) during 1993-1995 [*Wiens et al.*, 1995; *Koper et al.*, 1998].

3.1. Seismic Evidence on the Distribution of Volatiles and Magma Beneath the Volcanic Front

Body wave travel time tomography is the highest resolution seismological method for imaging the upper mantle structure beneath volcanic arcs. This method uses observed *P* or *S* wave travel times along many crossing raypaths to reconstruct the 2-D or 3-D velocity structure within the Earth. Several studies have computed 3-D tomographic models to reconstruct the *P wave* velocity variations of NE Japan with a resolution of 25–50 km [*Hasagawa et al.*, 1991; *Zhao et al.*, 1992, *Nakajima et al.*, 2001] (Plate 1). These images show a dipping upper mantle low velocity zone with *P wave* anomalies of up to 5% located about 50–75 km above the slab and connecting to shallow crustal low velocity anomalies immediately beneath the volcanic front. The most recent studies have also obtained high-resolution S-wave tomographic images and images of the P-to-S ratio [*Nakajima et al.*, 2001]. These images show S-wave anomalies with an amplitude of 6% and high P-to-S velocity ratios of 1.85 in the same positions as the low P-velocities. Later work suggests that the dipping low velocity feature in the mantle wedge is widespread throughout Japan, and that it extends to depths of about 150–200 km [*Zhao and Hasegawa*, 1993; *Zhao et al.*, 1994]. Relatively high attenuation ($Q_\alpha \sim 150$) has also been found in the same regions from lower resolution seismic attenuation tomography [*Sekiguchi*, 1991; *Tsumura et al.*, 1996; *Tsumura et al.*, 2000].

A dipping, low velocity zone beneath the volcanic arc and above the subduction zone is also observed in the Tonga-Lau region [*Zhao et al.*, 1997] (Plate 2). This image resolves *P* velocity variations along a 1400 km long region with a resolution of 50–75 km. A low velocity anomaly of up to 6% occurs above the slab and extends from the volcanic arc to a depth of about 150 km.

Similar low velocity zones have been observed at other arcs, including Alaska [*Zhao et al.*, 1995], Kyushu [*Zhao et al.*, 2000], and Kamchatka [*Gorbatov et al.*, 1999]. *Graeber and Asch* [1999] find a dipping zone of high P/S velocity ratios (low S velocity) above the subducting slab in the Andes. These observations suggest that a dipping, low velocity zone above the slab may be a ubiquitous feature of the subduction zone velocity structure. The slow velocity anomalies seem to be located at shallower depth and closer to the forearc in warmer subduction zones, where slabs may dehydrate at shallower depths [*Zhao et al.*, 2000].

There is evidence that this low velocity anomaly below the volcanic front does not represent direct seismic observations of substantial partial melt concentrations. The magnitude of the anomaly, 3–5% for P waves and 4–7% for S-waves, is less than would be expected for melt concentrations of greater than 1%. Anomalies of this magnitude can be reasonably explained by sub-solidus temperature variations of 200°C–400°C or, more likely, by hydration of the mantle wedge.

The relatively modest S wave anomalies and Vp/Vs ratio (or $\partial \ln Vs/\partial \ln Vp$) changes also argues against a widespread region of significant in-situ melt. Vp/Vs anomalies are found in both Japan [*Nakajima et al.*, 2001] and the Andes [*Myers et al*, 1998; *Graeber and Asch*, 1999] but in both cases the maximum Vp/Vs ratio is about 1.85, constraining partial melt concentrations to be less than about 1%. The Japan results are compatible with a η ($\partial \ln Vs/\partial \ln Vp$) value of about 1.8, which is elevated with respect to the expected values from purely thermal effects (1.4) but less than 2.2 expected for anomalies caused by melt [*Hammond and Humphreys*, 2000a]. Obviously there is melt in localized regions beneath the volcanic front but the evidence is against large scale (> 25 km) regions with significant (> 1%) melt concentration.

Instead, the dipping low velocity anomaly probably represents a hydrated region of the mantle wedge, and the source region of the melt. Most petrologic and geodynamic models suggest that island arc magmas are formed through the hydration of the mantle wedge by fluids from dehydration reactions in the subducting crust [*Tatsumi*, 1989; *Peacock*, 1990; *Davies and Stevenson*, 1992; *Iwamori*, 1998; *Schmidt and Poli*, 1998; *Mibe et al.*, 1999]. The presence of water will significantly reduce the seismic velocities of the hydrated region of the mantle wedge. Small amounts of water in olivine reduce the melting temperature, resulting in mantle temperatures that are near the solidus and produce slow velocity anomalies from anelastic effects [*Hirth and Kohlstedt*, 1996; *Karato*, 1993; *Karato and Jung*, 1998]. In addition, larger amounts of fluid will result in hydrous phases, which generally show much lower seismic velocities than mantle peridotite [*Christensen and Mooney*, 1995]. The low velocity regions may also denote warmer upwelling regions in the

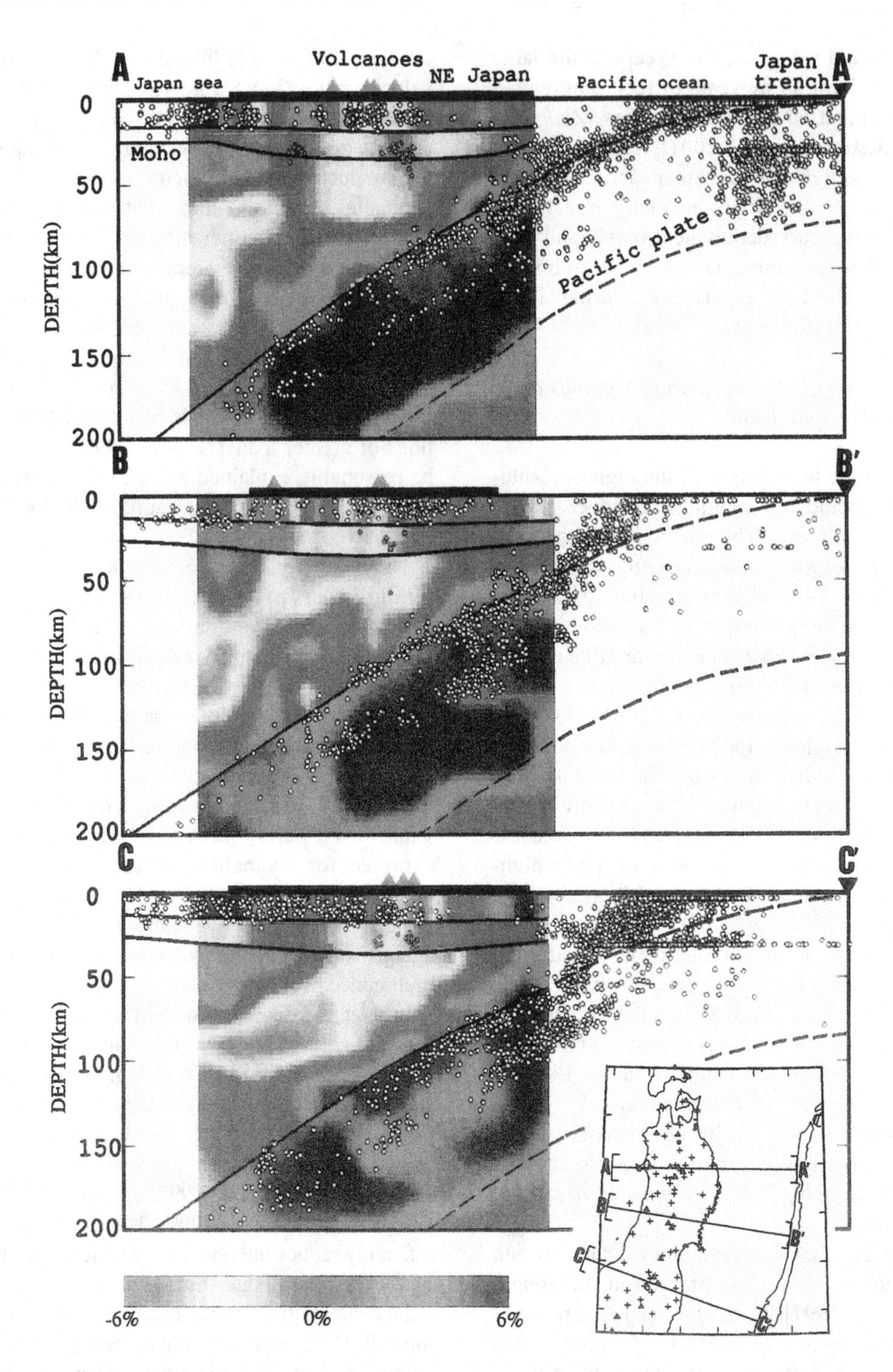

Plate 1. Vertical cross sections of P wave velocity perturbations (in percent) relative to an average Japan arc model in northern Japan along profiles AA', BB', and CC' (after *Zhao et al.,* [1992]). The slices are taken from a 3-D tomographic model computed from ~14,000 *P* wave arrivals and additional converted phases. Red and blue colors denote low and high velocity respectively, and microearthquakes within a 60 km width along the profiles are shown as open circles. The land area and active volcanoes are shown at the top of each figure by bold horizontal lines and red triangles.

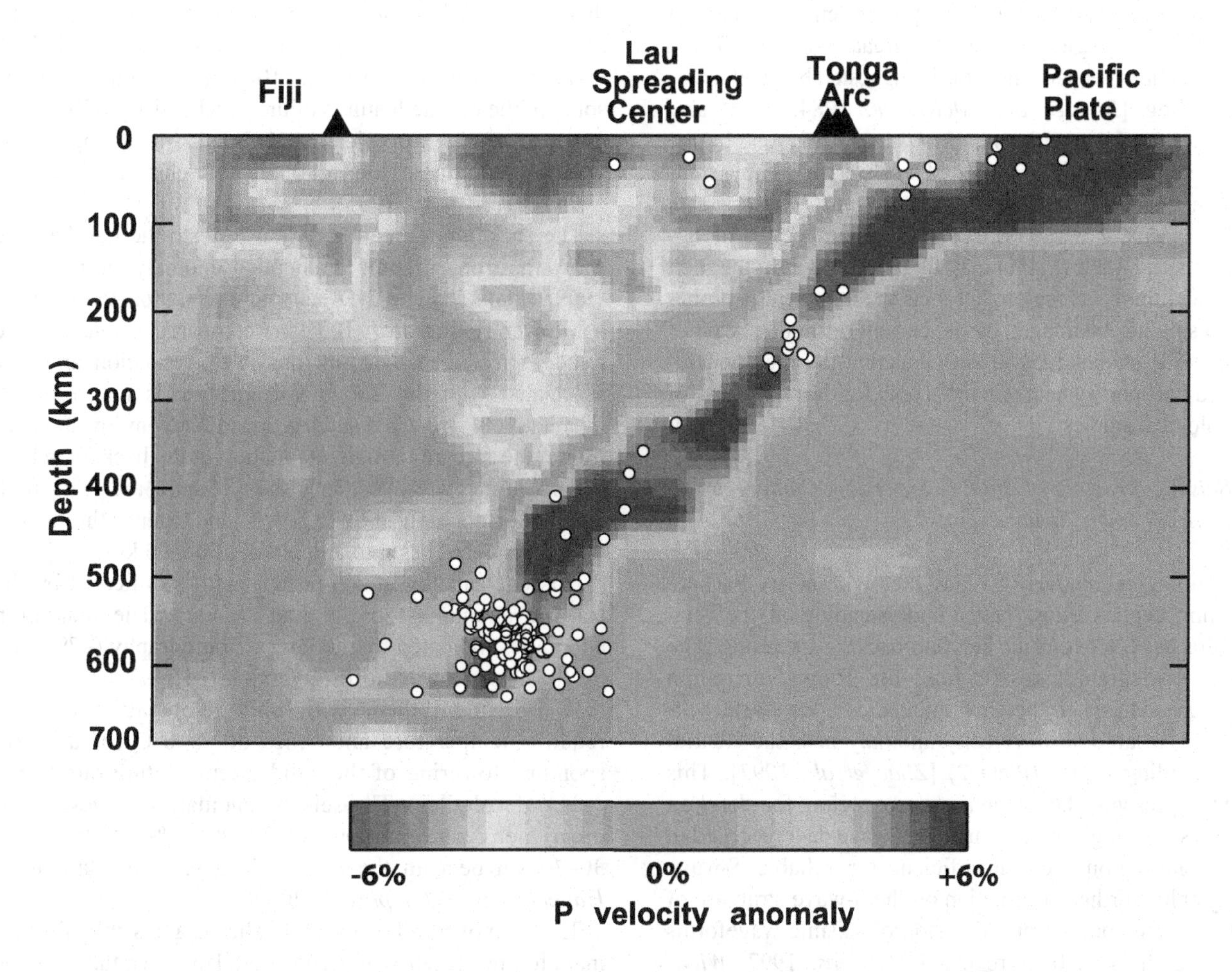

Plate 2. Vertical cross section of P wave velocity anomalies perpendicular to the strike of the Tonga subduction zone from a tomographic inversion (after *Zhao et al.,* [1997]). This image was computed from ~41,000 local and teleseismic *P wave* arrivals. Solid triangles denote active volcanoes. Earthquakes within 40 km of the cross section are shown as circles. The velocity perturbation scale relative to the IASPEI91 velocity model is shown at the bottom.

mantle wedge where small amounts of decompression melting may occur, as predicted by geodynamic models [*Andrews and Sleep,* 1974; *Conder et al.,* 2002].

Thus the low velocities above the slab in the tomographic images represent imaging of a hydrated region in the mantle wedge. This region represents the source zone for the island arc magmas erupted at the volcanic front. Significant quantities of melt (> 1%) are not imaged at present and may not occur over large regions (25 km and greater) due to efficient melt extraction mechanisms. This is suggested by geodynamic modeling [*Richter and McKenzie*, 1984; *Scott and Stevenson*, 1986], experiments [*Kohlstedt*, 1992], and by the rapid melt transport times indicated by U-series isotopic studies [*Hawkesworth et al.*, 1997; *Turner and Hawkesworth*, 1997; *Bourdon et al.*, 1999]. Naturally, some pockets of significant partial melt (> 1%) may likely exist within the melt source and transport region, but their spatial extent is probably too small to be imaged by the current resolution scale of tomographic studies (25–50 km). Imaging the melt generation and transport system remains a goal for the next phase of seismological studies.

3.2. Seismic Structure of Backarc Spreading Centers and Relationship to Arc Structure

Seismological studies of island arcs with active backarc spreading centers allow better understanding of the relationships between volcanic arcs and backarc spreading. The *P wave* tomographic results from the Tonga-Lau region shows, in addition to the slow anomalies beneath the volcanic arc, a large slow velocity anomaly near the Central Lau spreading center (Plate 2) [*Zhao et al.*, 1997]. This region represents the magma source region for the Lau backarc spreading center. Although *S*-wave data recorded in the Tonga region were insufficient for reliable *S*-wave tomography, further information on the *S*-wave structure of the Lau basin comes from inversion of seismic waveforms traversing the Lau backarc [*Xu and Wiens*, 1997; *Wiens et al.,* 2002]. Although the spatial resolution of this method is inferior to *P wave* tomography, the depth variation of seismic velocity averaged along the ray path is accurately recovered. These results (Figure 2) show exceptionally low S-wave velocities beneath the Lau backarc at depths of 40-100 km, with a maximum S-wave heterogeneity of about 17% relative to the old Pacific Plate. These depths correspond to the depths of the low velocities in the P wave tomography (Plate 2) and also represent the expected depths of primary MORB magma production [*Shen and Forsyth*, 1995]. Comparison of the various tomographic models and the waveform inversion results suggest that the maximum velocity anomalies are greater beneath the backarc spreading center than beneath the arc volcanic front.

The slow velocity regions beneath the Tonga arc and the Lau back arc are geographically separated at shallow levels but merge at depths of about 100–200 km depth. This suggests that although the arc and backarc systems are separated at shallow levels, where most of the magma is generated, there may be some interchange between the magma systems at depths great than 100 km. Interaction with slab-derived volatiles at depths greater than 100 km may help to explain some of the unique features of the geochemistry of back-arc magmas relative to typical MOR basalts, including excess volatiles and large ion lithophile enrichment [*Hawkins*, 1995; *Pearce et al.*, 1995].

The entire upper mantle region beneath the Lau backarc shows a strong seismic attenuation anomaly [*Roth et al.*, 1999] (Plate 3). The 2-D seismic attenuation tomography has lower resolution (~100 km) compared to the P wave tomography, such that any possible attenuation anomaly associated with the Tonga volcanic arc is not resolved. Qualitatively, there is a sudden transition from low attenuation in the forearc to high attenuation in the backarc, which occurs immediately beneath the volcanic arc. The high attenuation anomaly shows $Q_\alpha \sim 95$ and extends throughout the Lau backarc basin to a depth of 150–200 km.

It is unclear how much partial melt is indicated by the seismic anomalies in the Lau backarc. The maximum anomalies indicated by the P wave tomography (–7%) and by the S-wave waveform inversion (–15 %) are so large that to explain them with only temperature changes requires temperature anomalies of ~500°C (or a corresponding lowering of the solidus temperature due to the effect of volatiles). The seismic anomalies suggest maximum melt concentrations of about 1.5% at depths of 30–70 km beneath the Lau basin using the relations of *Hammond and Humphreys* [2000b].

The ratio of S to P wave anomalies is a possible discriminant for the presence of partial melt. However, there are too few good S-wave arrivals to perform S-wave tomography of the Tonga-Lau region. *Koper et al.* [1999] evaluated the parameter $\nu = \partial \ln Vs / \partial \ln Vp$ for raw arrival times with raypaths traversing the upper mantle beneath the Lau backarc and found $\nu = 1.2$–1.3, in agreement with anomalies generated by temperature changes but much less than the $n \sim 2.2$ expected for anomalies caused by partial melt.

One other promising method for evaluating the presence of melt is the observed Vp-Qp^{-1} relationship. This has not been used previously since attenuation and velocity tomographic analyses are seldom available for the same location. *Roth et al.* [2000] compared the relationships between Vp and Qp^{-1}

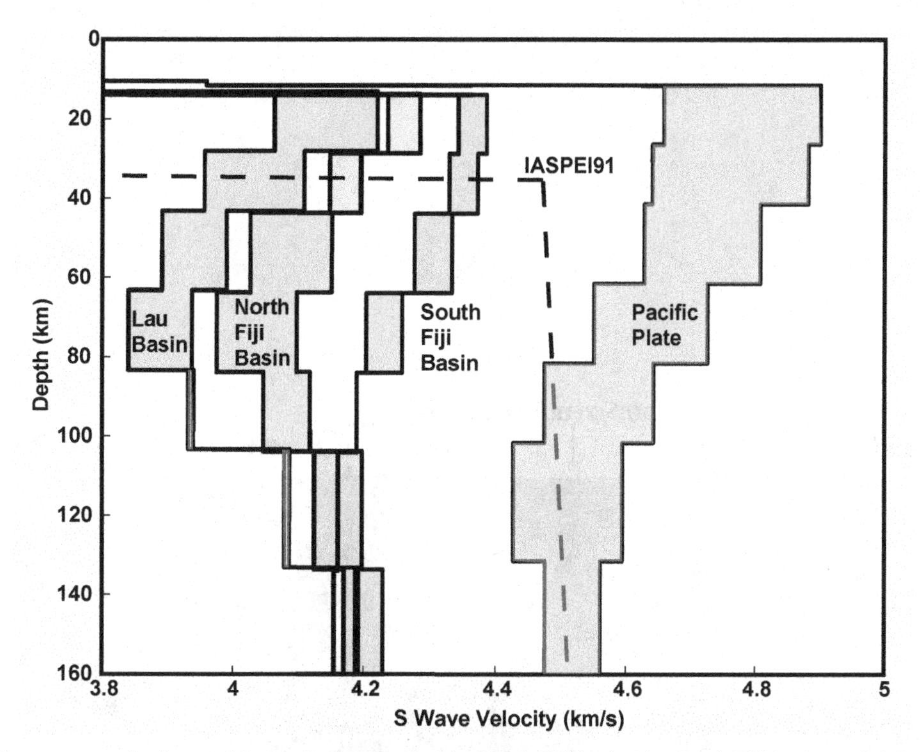

Figure 2. Shear wave velocity models for the Lau Basin, the North Fiji basin, the South Fiji basin, and the old Pacific Plate (from *Wiens et al.* [2002]). Models were obtained by inversion of the entire vertical and transverse wavetrains at regional distance stations (distance 500 to 1600 km) between 5 and 60 mHz frequencies. Radial anisotropy is shown as separate SH and SV structures for each region. Back arcs with active spreading centers (the Lau Basin and the North Fiji Basin) show exceptionally slow seismic velocities in the uppermost mantle, whereas the inactive South Fiji Basin shows an intermediate structure and the Pacific Plate is exceptionally fast. The velocity structures of the backarc basins are indistinguishable beneath 100 km, but remain slower than the Pacific Plate thoughout the depth range studied. Radial anisotropy is largely confined to the upper 80 km and is smaller in magnitude than in the old Pacific Plate.

obtained from the Tonga-Fiji results with the predictions of attenuation-corrected velocity derivatives [*Karato*, 1993] and the experimental results of *Jackson et al.*, [1992], assuming the anomalies resulted from temperature changes alone (Figure 3). The experimental predictions provide a reasonable match to the observations at depths greater than 100 km. The results for depths < 100 km are scattered, and some points have large velocity anomalies relative to the attenuation anomalies. These points also correspond to the largest backarc velocity anomaly locations in the *Zhao et al.*, [1997] results. These results suggest that the anomalies in the Tonga-Fiji back-arc region can be explained in terms of temperature anomalies, with the exception of the largest magnitude shallow velocity anomalies, which may be due to partial melt. In summary, the tomographic results for the Lau backarc probably imply some small component (~ 1%) of partial melt in the depth range of 30-90 km beneath the Lau basin, although it is difficult to understand the lack of a larger Vs/Vp anomaly in the raw *P* and *S* traveltimes.

3.3. The Deep Structure Beneath Back-arc Basins–Evidence for Deep Volatile Release?

Seismological results can also provide clues about deeper processes in the mantle wedge. Both plate 2 and figure 2 show substantial seismic anomalies in the Lau backarc at depths greater than the inferred maximum depth of primary melting, which is thought to be about 70 km [*Shen and Forsyth*, 1995]. The *S*-wave heterogeneity between the Pacific plate and the Lau basin is about 14% at a depth of 20 km, 17% at 70 km, and 9% at 120 km. The slow velocities beneath the backarc basins are resolvable by the data down to a depth of about 200 km. The large amplitude of the *S*-wave anomaly at depths of 100–150 km suggests the possibility of small fractions of water-induced melting at depths significantly greater than that of primary melting (~70 km) [*Hirth and Kohlstedt*, 1996; *Hirschmann et al.*, 1999]. Smaller slow anomalies at depths of 150–200 km may result from the proximity of the geotherm to the wet solidus with-

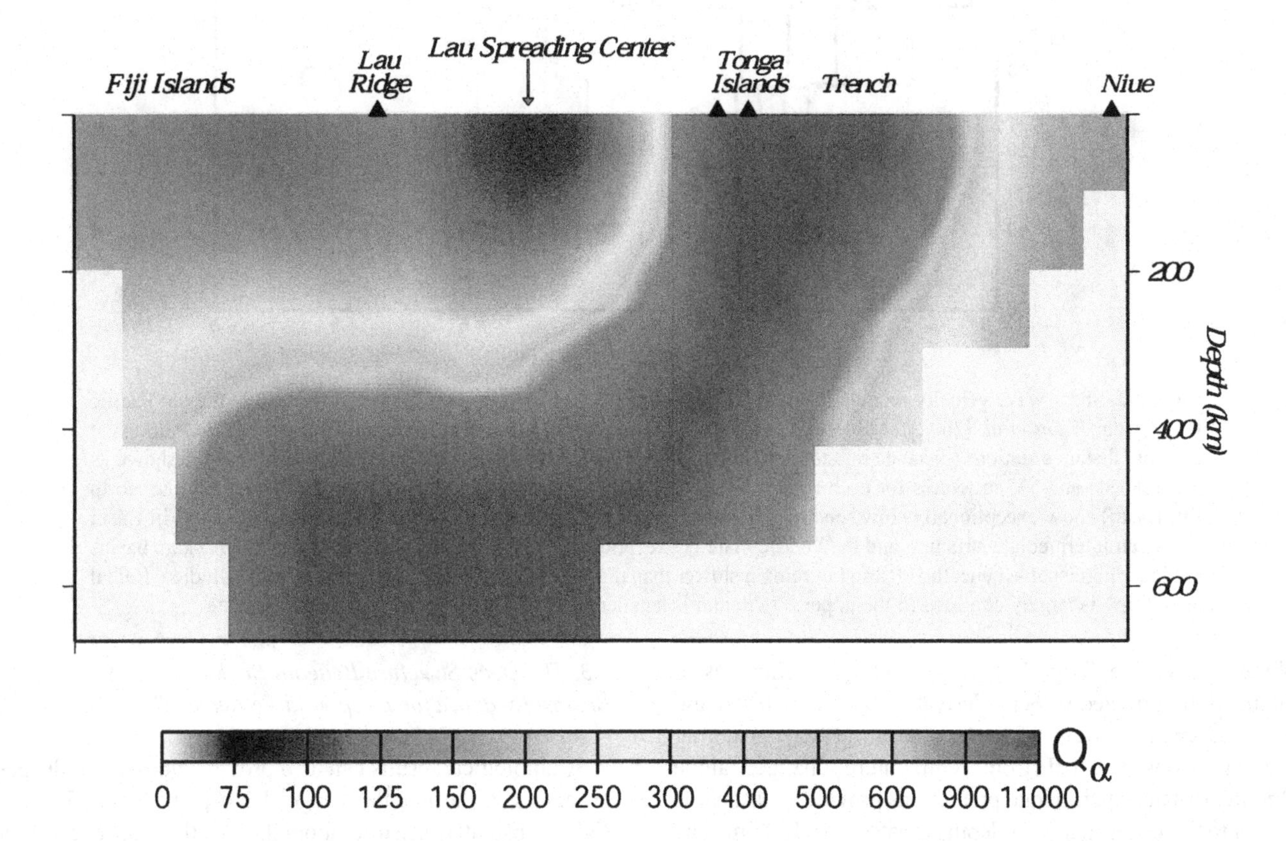

Plate 3. Cross section perpendicular to the Tonga subduction zone showing variations in P wave attenuation (Q_α) (from *Roth et al.*, [1999]). Results were derived from a 2-D tomographic inversion of differential body wave travel times recorded by land stations and ocean bottom seismographs. Red color indicates high attenuation (low Q) and blue indicates low attenuation.

in the upwelling material beneath the ridge, or from subsolidus effects of volatiles in the mantle wedge.

The P wave tomography results show smaller anomalies (2–3%) extending to depths of ~400 km beneath the Lau backarc basin (Plate 2). Slow velocity anomalies to depths of 400 km beneath backarc basins have also been detected by lower resolution tomographic inversions in the Kurile and Izu-Bonin arcs [*Van der Hilst et al.*, 1991; *Nolet*, 1995]. Such deep slow velocity anomalies have not been detected beneath mid-ocean ridges [*Webb and Forsyth*, 1998], and thus are unlikely to represent deep structure that is endemic to spreading centers, but rather processes unique to backarc basins. The most likely explanation is that seismic velocities in the deeper parts of the mantle wedge are reduced by hydration reactions as water and other volatiles are released from the slab [*Nolet*, 1995; *Zhao et al.*, 1997].

Low temperature slabs such as Tonga may be cold enough that water can be transported, either as hydrous minerals or as ice-VII [*Bina and Navrotsky*, 2000], to the stability depths of dense hydrous magnesian silicate phases [*Thompson*, 1992; *Schmidt and Poli*, 1998]. The dehydration of these phases in

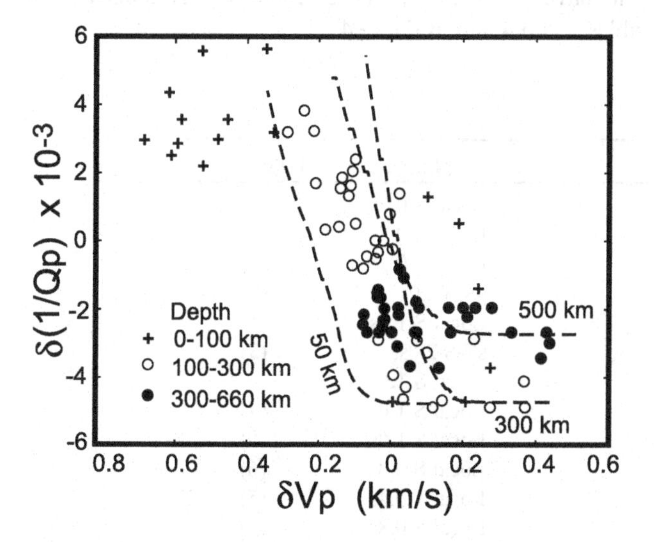

Figure 3. The relationship between velocity (δVp) and attenuation (δQ^{-1}) anomalies in the Tonga-Lau region (from *Roth et al.*, [2000]). Points represent the velocity and attenuation anomalies at a particular node in the attenuation (Plate 3) and smoothed P wave velocity models. Higher δQ^{-1} indicates low Q and higher δVp indicates fast velocity anomalies in the tomographic models. The dashed are the temperature and attenuation anomalies expected from experimental relations and the assumption that all anomalies are due to temperature heterogeneity. Theoretical velocity perturbations are generated using the relations in *Karato* [1993] incorporating the effects of anelasticity, and the attenuation perturbations are generated from *Jackson et al.,* [1992], extrapolated to various depths by scaling Q with homologous temperature.

the slab at depths of about 300–400 km may produce widespread hydration in the adjacent mantle wedge near these depths. The fact that the low velocity anomalies seem to stop at depths near 400 km in several arcs may be significant; as the slab a-olivine is transformed to wadsleyite (β-Mg_2SiO_4) at these depths, which can readily incorporate large amounts of water [*Smyth*, 1994]. Therefore, slab dehydration may cease at the depth of the α to β transformation. The flux of volatiles from the slab into the mantle wedge beneath the backarc basin to depths of 300–400 km may lower the viscosity of the upper mantle [*Karato et al.*, 1986; *Hirth and Kohlstedt*, 1996] and be an important factor in localizing backarc extension in backarc basins. The rapid and episodic character of backarc spreading [*Jurdy and Stefanick*, 1983] may thus be related to the hydration of the mantle wedge at depth by the subducting slab.

4. SEISMIC ANISOTROPY RESULTS: CONSTRAINTS ON MANTLE FLOW PATTERNS

4.1. A review of Anisotropy Observations in Arcs

In an anisotropic medium vertically arriving shear waves can split into two orthogonally polarized S-waves, quasi-SH and quasi-SV, that travel at different speeds. This 'splitting' is related to the strength and path length of anisotropy traversed, and the polarization of the first arriving S-wave gives the anisotropic fast-azimuth. Estimates of these parameters can be obtained by a variety of methods (e.g. *Crampin et al.*, 1980; *Bowman and Ando*, 1987; *Silver and Chan*, 1991]). These analysis methods produce much the same result but use differing functions to compare the two S-waves over a range of rotation angles and time-offsets to find the splitting-time and fast-azimuth [*Silver and Chan*, 1991]. Comparison of different phases (e.g. *S, ScS, SKS, SKKS, PKS*) can help localize the depth distribution of the anisotropic signature. We discuss splitting measurements here as they are also geographically well constrained and have been performed in a wide range of subduction zone environments (e.g. *Ando*, 1984; *Bowman and Ando*, 1987; *Shih et al.*, 1991; *Kaneshima and Silver*, 1992; *Fischer and Yang*, 1994; *Russo and Silver*, 1994; *Yang et al.*, 1995; *Fouch and Fischer*, 1996; *Gledhill and Gubbins*, 1996; *Fouch and Fischer*, 1998; *Fischer et al.*, 1998; *Smith et al.*, 2001] making comparison possible.

For South America, *Russo and Silver* [1994] report small local *S* splitting (~0.1s) and large (3s) *SKS* splitting times and infer that there is thus little wedge anisotropy. They instead propose strong anisotropy beneath the slab due to horizontal trench-parallel flow consistent with *Alvarez* [1982] to explain trench-parallel orientations. However, contradictory fast-azimuths are observed in a variety of locations [*Russo and*

Silver, 1994; *Helffrich et al.*, 2002]. In addition, in the mantle wedge beneath Columbia, *Shih et al.*, [1991] found splitting up to 0.4s with fast directions sub-parallel to the plate boundary. Similar results are also found in Peru, Brazil, and central Chile [*Polet et al.*, 2000] but in northern Chile fast azimuths are parallel to the absolute plate motion (APM) of the Nazca plate [*Bock*, 1998; *Polet et al.*, 2000]. These local S-wave measurements strongly indicate anisotropy in the mantle wedge with varying fast-azimuths.

In the eastern Aleutians local *S*-wave splitting is again small (up to 0.35s) indicating little (~1%) anisotropic organization in the wedge. However, the orientation is highly consistentat 60 degrees, roughly parallel to the volcanic arc. A similar arc-parallel orientation is observed using *sS-S* phase pairs and direct *S* phases [*Fischer and Yang*, 1994] and *SKS* phases [*Peyton et al.*, 2001] in the northern Kamchatka peninsula and in *S* and *SKS* in the southern Kurils [*Fouch and Fischer*, 1996]. However, in the region of the Kurils between these two observations arc-perpendicular fast azimuths are observed [*Fischer and Yang*, 1994].

Trench parallel orientations are also observed in western Honshu [*Fouch and Fischer*, 1996; *Hiramatsu et al.*, 1998] and the Japan Sea whereas in northern Japan [*Sandvol and Ni*, 1997] and Izu-Bonin [*Fouch and Fischer*, 1996; *Hiramatsu et al.*, 1998] orientations are parallel to the absolute plate motion (APM) of the Pacific plate . In the Marianas *S*, *ScS* and *SKS* measurements indicate small splitting (0.1–0.4s) roughly 20° from the APM of the downgoing plate [*Fouch and Fischer*, 1998].

In the Hikurangi (New Zealand) subduction zone trench parallel fast directions are observed in *SKS* phases [*Gledhill and Gubbins*, 1996; *Marson-Pidgeon et al.*, 1999]. In Tonga, observations at back-arc islands show striking correlation between the anisotropic fast direction and the plate convergence direction [*Fischer and Wiens*, 1996] and exhibit unusually large local S splitting times (>1s). However, closer to the Tonga trench the fast direction is trench parallel [*Smith et al.*, 2001].

Many other studies have reported splitting in subduction zone environments from a variety of shear-phases. However, in the fore-going discussion we have not included observations where the location of the anisotropy remains unconstrained. Only those observations that are able to distinguish between anisotropy above and below the slab have been noted. These results are summarized in tables 3 and 4 and in figure 4.

Table 3: Splitting occurring within the mantle wedge.

Subduction Zone	Fast Direction	Splitting	References
Northern Chile	Arc parallel close to trench	Local S 0.5s	(1)
	Parallel to Nazca APM in east	Local S 0.5s	(1)
Aleutians	Arc Parallel	Local S 0.35s	(2)
Mid-Kurils	Parallel convergence direction	sS-S 2.0s	(3)
		Local S 1.0s	
Kamchatka	Arc Parallel	sS-S 2.0s,	(3)
		Local S 1.0s	
Northern Japan	Parallel convergence direction	Local S 1.0s	(4)
Izu-Bonin	Parallel convergence direction	Local S 1.0s	(5)
	10° from convergence direction	Local S 0.8s	(6)
Japan	Arc Parallel	Local S 0.4s	(5)
Western Honshu	Arc Parallel	Local S 0.8s	(5)
	10° from Arc Parallel	Local S 0.8s	(6)
Marianas	20 from APM	S, ScS, SKS 0.4s	(7)
Tonga	Parallel convergence direction (Backarc)	SKS, Local S >1s	(8)
	Arc Parallel (Volcanic arc)	SKS, Local S >1s	(9)
Hikurangi	No shallow wedge splitting	Local S	(10)
(New Zealand)	Arc parallel	SKS, 1.5s	(11)

(1) Polet et al. 2000, (2) Yang et al. 1995, (3) Fischer and Yang, 1994; (4) Sandvol and Ni, 1997; (5) Fouch and Fischer, 1996; (6) Hiramatsu et al. 1998; (7) Fouch and Fischer, 1998; (8) Fischer and Wiens, 1996; (9) Smith et al., 2001; (10) Gledhill and Stuart, 1996; (11) Marson-Pidgeon et al., 2001.

Absolute plate motion (APM).

Table 4. Splitting observations attributed to anisotropy beneath the slab.

Subduction Zone	Fast Direction	Splitting	References
South America	Arc parallel	SKS 3.0s	(1) (2)
	Arc perpendicular at 16S, 24S and 33S	SKS 3.0s	(1)
Northern Chile	Nazca APM	SKS 1s	(3) (4)
Brazil, Peru	Arc Parallel	SKS <1s	(4)
Kamchatka	Arc Parallel	SKS 0.6s	(5)
Sakhalin Island	40° from convergence direction	SKS 1.5s	(6) (7)
N. Japan (43N)	Sub-parallel convergence	SKS 1.0s	(6)
N. Japan (41N)	Arc Parallel	SKS 1.0s	(7)
Western Honshu	Arc Parallel	SKS 1.5s	(7)
Hikurangi (New Zealand)	Arc Parallel	SKS 1.2s	(8)

(1) Russo and Silver, 1994; (2) Shih et al., 1991 (3) Polet et al. 2000, (4) Bock, Kind, Rudloff, Asch, 1998; (5) Peyton et al., 2001; (6) Sandvol and Ni, 1997; (7) Fouch and Fischer, 1996; (8) Gledhill and Gubbins, 1996.

Absolute plate motion (APM).

4.2. The Tonga-Lau System: A Complex Pattern of Anisotropic Orientations

In the Tonga-Lau region splitting observations are geographically distributed across the arc and backarc. This region displays a very complex pattern of strong seismic anisotropy. Figure 5 shows the averaged splitting results at both land and ocean bottom seismograph (OBS) stations in the region [*Smith et al.*, 2001]. Results at the land based stations on the Fiji platform to the west are roughly parallel with the convergence direction and can be explained by large-scale counterflow induced by the subduction of the Pacific plate.

However, this simple pattern of fast azimuths paralleling the absolute plate motion direction is not apparent all the way across the basin. Indeed, many of the fast vectors are almost perpendicular to the plate motion and there is a strong variation in the magnitude of the splitting with greatly reduced splitting times towards the center of the basin. 2D flow modeling instead predicts an almost uniform splitting time across the basin [*Fischer et al.*, 2000]. In order to explain the trends within this data we need to appeal to more complex modeling that takes into account factors such as the backarc spreading center and changes in the slab dip through time.

Splitting times in the center of the basin are both greatly reduced and almost orthogonal to the plate motion direction; this phenomenon occurs geographically close to the spreading center and may be related to it. Near the Tonga volcanic arc we observe strong along-strike fast directions. Many of the raypaths of the split S-phases observed at the Tonga arc stations are from intermediate depth events (150–300 km depth), with the entire path length within the mantle wedge. This indicates that the strong along-strike fast directions must be explained by processes within the mantle wedge and cannot be attributed to anisotropy within the subducted plate.

Possible mechanisms that could explain trench-parallel flow include the effect of slab rollback on mineral alignment as modeled in the laboratory by *Buttles and Olson* [1998], which should be significant in Tonga, which has experienced rapid roll back during the past 4 Ma [*Hamburger and Isaaks,* 1987]. This physical modeling experiment investigated the effect of dip angle and rollback on material flow, and produced results consistent with the observations of anisotropic fast directions obtained in the eastern Lau basin and Tonga arc.

There is also strong geochemical evidence to suggest that along trench mantle flow is occurring in the Tonga-Lau region. *Gill and Whelan* [1989] attribute a change in Fiji magmatism from arc-like to OIB to influx of the Samoan Plume around 3 Ma. A similar explanation for high Nb relative to other high field strength elements in lavas at the Islands Tafehi and Niuatoputapu at the northern end of the Tonga-Kermadec subduction zone is also proposed [*Wendt et al.*, 1997]. *Poreda and Craig* [1992] use helium isotope data to infer channeling of the Samoan Plume towards the Peggy Ridge at the northern end of the Lau Basin. A later study by *Turner and Hawkesworth* [1998] mapped the presence of the high He^3:He^4 ratios further south into the Lau backarc. Such isotope signatures, which are characteristic of the Samoan Plume, led Turner and Hawkesworth to propose that this was evidence of flow of shallow mantle, from the Samoan Plume into the Lau Basin parallel to the trench, through a gap in the subducting Pacific plate (see Figure 6). The anisotropy observations are thus remarkably consistent with the flow model proposed on the basis of geochemical data.

4.3. Models for Arc Flow Patterns and Relationship to observations

Explaining the diversity of anisotropy patterns in subduction zone systems is difficult. Clearly a simple model relating anisotropic fast directions to global models of mantle flow will not predict all of the observations. The simplest

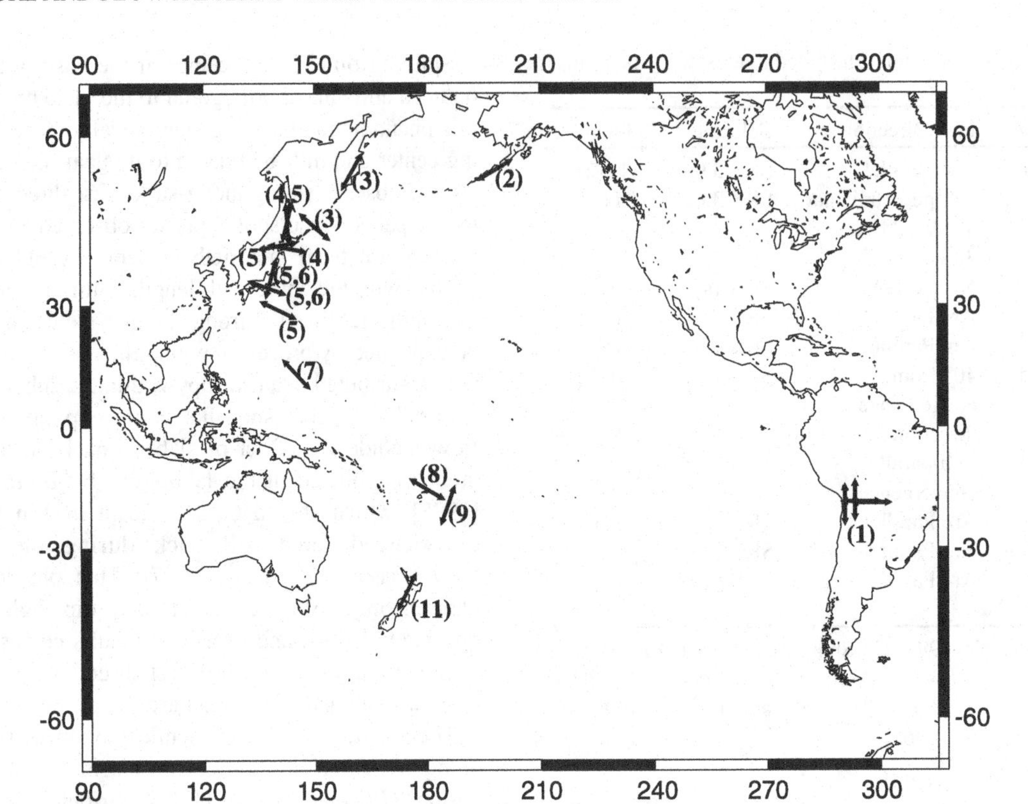

Figure 4. Schematic map of subduction zone fast anisotropy directions from splitting studies. All splitting measurements shown are those in Table 2 that have been determined to occur within the mantle wedge. Numbers correspond to the references in Table 2. The azimuth of each vector is the fast splitting direction, however lengths are kept constant due to the different phases used in the various studies.

approach for modeling anisotropy in subduction zones incorporates the strain resulting from flow coupled to the downgoing plate (Figure 7). This predicts a fairly uniform pattern of anisotropy paralleling the absolute plate motion [*McKenzie*, 1979; *Ribe*, 1989] which would match observations in Izu-Bonin and the western part of Tonga-Lau. Indeed, *Fischer et al.* [2000] demonstrate that given the right assumptions relating the strength of LPO to strain the magnitude of these splitting times could also be predicted using a simple 2D mantle flow model (Figure 7). However, such an approach will obviously not fit the along-strike fast direction observations from the Aleutians, Japan, and the Tonga arc itself. The majority of the arcs studied actually show a predominance of along-strike fast directions for mantle wedge anisotropy, suggesting that viscous coupling to the downgoing plate may be relatively unimportant for driving backarc flow.

Mechanisms which could produce trench-parallel fast-directions include the effect of melt [*Kendall*, 1994]. If melt forms in pockets or films then the geometry of these can significantly affect the shear-wave velocity. It is reasonable to assume that such melt-filled cracks will preferentially open parallel to the direction of minimum deviatoric compressive stress [*Fischer et al.,* 2000]. In the case of 2D flow in the mantle wedge one would expect the maximum compression to occur in the direction of convergence. This results in the prediction of trench-parallel melt sheets and an associated along arc fast direction. However, it seems reasonable to assume that if such a mechanism were responsible for the along-strike fast axes the mechanism should apply equally to the Izu-Bonin, where instead convergence parallel fast directions are observed.

One possible explanation for the differences between subduction zones may lie in the variations in subduction angle, convergence angle, and rollback between regions. *Buttles and Olson* [1998] showed that when there is a large component of slab rollback, the mantle flow pattern was dominated by mantle inflow to the backarc region around the ends of the slab. Their results indicated that this should occur almost independent of plate dip. Instead the amount of rollback a slab has experienced is the determining factor in trench-parallel a-axis alignment.

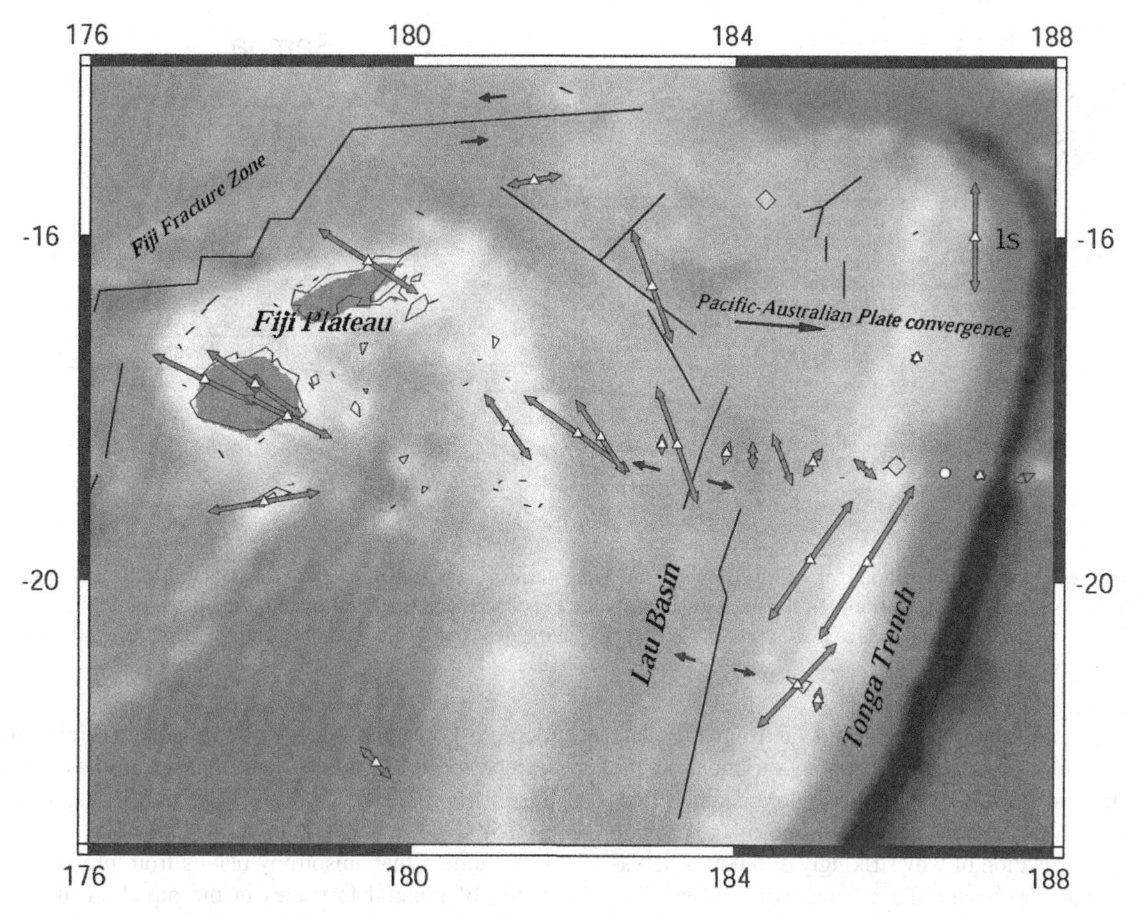

Figure 5. Shear wave splitting across the Tonga-Lau region [after *Smith et al.,* 2001]. Stations are shown as triangles. Splitting vectors are the result of stacking and averaging all measurements made at each station. Combined splitting observations are plotted at the stations as a vector. The azimuth of each vector is the fast splitting direction and its length is proportional to the splitting time. A 1s splitting vector is shown for scale.

We must also take into account the presence of other structures, such as a back-arc spreading center, in accurately accounting for splitting patterns. *Blackman et al.* [1996] used a finite element flow model to predict deformation in the vicinity of a spreading ridge. They used different starting parameters to obtain both buoyant and passive models. Regardless of which model they used they obtained two significant results. First, close to the spreading center they saw a reduction in the amplitude of the splitting. Second, they observed a reversal in the polarization of the first arriving shear-wave. Such local deviations in flow will also affect the pattern of observed splitting and were noted in the observations for Tonga [*Smith et al.*, 2001].

It seems clear that flow patterns, and their relationship to anisotropic observations in the mantle wedge, are complicated. We appear to have reached a stage in the study of seismic anisotropy where the resolution of our observations currently out-strips the resolution of our models. Although global flow patterns show good correlation with the long period anisotropic signature of the mantle [*Tanimoto and Anderson*, 1984] these predictions break down at the shorter length scales in the back-arc due to the complex flow patterns induced by the plate boundary. In addition the continent scale models such as proposed by *Russo and Silver* [1994] also appear unable to account for the small scale variations present in the data without resorting to additional variables such as changes in slab dip affecting local flow.

Although fast-directions varied from one subduction zone to the next, until recently observations suggested that at least within a single back-arc the fast-azimuth was constant, usually either paralleling the APM or the trench-strike. While this remained the case, 2D models of mantle flow driven by coupling to the downgoing plate or melt anisotropy remained a viable means of explaining the flow. While such

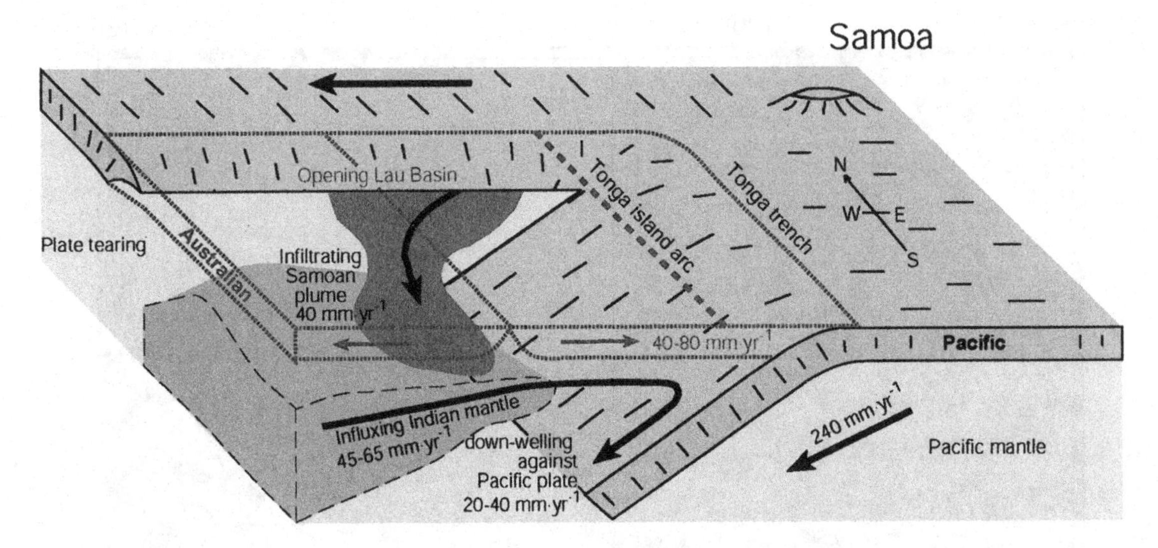

Figure 6. Schematic diagram of mantle flow in the Tonga-Lau region from geochemical data (after *Turner and Hawkesworth* [1998]). Seismological data also suggests the existence of north-south oriented mantle flow within the Tonga Arc and Lau backarc [*Smith et al.,* 2001], in agreement with the Helium isotope data [*Turner and Hawkesworth,*1998].

simple models still appear to be valid far from the plate boundary the new observations mapping variations in the fast azimuth across the backarc require more detailed flow models. In addition, the new models must be able to provide a consistent explanation of why strikingly different observations are obtained between different subduction zones.

5. CONCLUSIONS

Seismology can now provide important constraints on the locations of the melt source regions and the pattern of flow in the mantle wedge. The magma source regions for island arc volcanics are imaged as dipping low velocity zones overlying the subducting slab at depths of 50–150 km. However, the imaged regions may not correspond to actual partial melt concentrations, and may instead represent hydrated regions of the mantle. The largest seismic anomalies occur at depths of 30–90 km beneath backarc spreading centers and probably correspond to no more than 1% partial melt concentrations. The generally small partial melt concentrations in both the arc and backarc regions are consistent with models for rapid and efficient melt extraction, as suggested by geodynamic modeling, experimental work, and isotope studies. In the Tonga-Lau system, anomalies related to the volcanic arc merge with anomalies beneath the backarc spreading center at depths greater than 100 km, suggesting that small slab components of backarc magmas may originate through interactions at greater depths. Low amplitude slow velocity seismic anomalies extend to 400 km depth beneath backarc basins, and may result from the release of volatiles transported to depths of 300–400 km by hydrous minerals in the slab.

Assuming that anisotropy results from preferential orientation of mineral fast axes in the direction of flow, shear wave splitting observations suggest that most, but not all, subduction zones are characterized by along-strike mantle flow within the mantle wedge. This suggests that counterflow in the mantle wedge induced by viscous coupling with the downgoing plate is generally weak. The flow patterns of several subduction zones appear to be dominated by inflow due to slab rollback and other local factors.

Clearly major questions remain regarding the pathways of volatiles and melt through the mantle, the concentration of volatiles and melt at various locations, and how these pathways are related to mantle flow. Progress within the next ten years is likely to come from a number of different sources. Many of the issues are related to problems of spatial resolution. Large-scale experiments consisting of up to 100 stations and the accumulation of even larger datasets in areas of permanent networks will help improve resolution beyond the 25–75 km limits that are currently possible. This may enable further resolution of melt pockets within hydrated regions overlying the subducting slab beneath the volcanic arc, or delineation of the shape of the melt producing region beneath the backarc spreading center. Greater density of high quality stations will also permit much better resolution

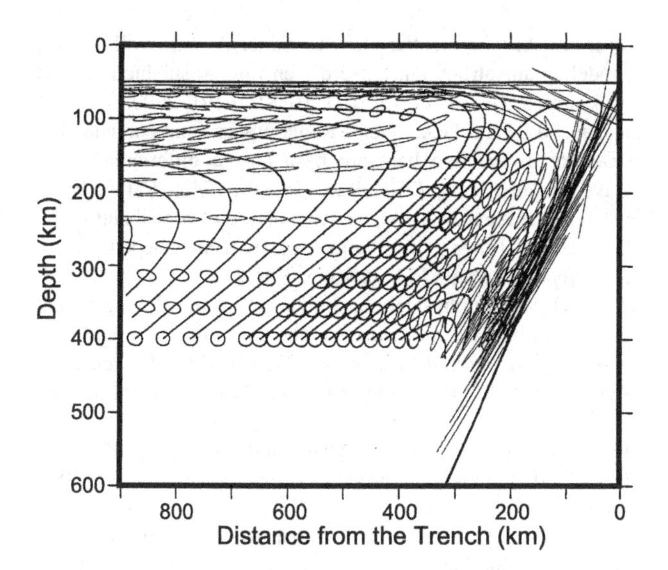

Figure 7. Streamlines and finite strain ellipses for a 2-D model of corner flow induced by viscous coupling with the downgoing slab [after *Fischer et al.*, 2000]. Strain is assumed to accumulate as material rises past a depth of 410 km, and this model assumes a constant viscosity. Models such as this allow prediction of observed seismic anisotropy using finite strain scaling or polycrystalline plasticity models [*Blackman et al.*, 1996; *Tommasi*, 1998; *Hall et al*, 2000].

of the spatial and depth variability of seismic anisotropy, thus placing better constraints on the mantle flow pattern.

Progress will also arise from studies combining disparate datasets. Whereas many previous seismological studies of the mantle wedge structure have emphasized *P* wave travel time tomography, future work will integrate *P, S,* and attenuation structure to place better constraints on material properties. Anisotropy studies will also benefit from more investigations going beyond the local *S* and teleseismic *SKS* splitting work reported here. It is likely that various phase combinations, such as the *sS-S* differential method developed by *Fischer and Yang* [1994], will be important additional sources of information. Future work will also attempt to synthesize the diverse range of other measures of anisotropy (e.g. propagation anisotropy in both P_n and surface waves as well as polarization anisotropy of surface waves) that are available.

Finally, much of the progress must come from experimental and modeling work necessary to understand the seismological measurements. Further experimental work is needed to help clarify the relationship between *P* and *S* velocity and attenuation anomalies and the presence of volatiles and melt. Experimental work is also needed to help better understand the relationship of seismic anisotropy to strain and stress associated with mantle flow, and the possible complicating effects of composition, volatile content, grain size, temperature, and pressure. More detailed 3-D geodynamical modeling applied to specific arc geometries is also necessary, since the available anisotropy studies suggest that the specific pattern of mantle flow is variable and dependent on the specific geometry and tectonic history of a particular arc. The results of these studies will be a more detailed mapping of the mantle flow pattern and the pathways of volatiles and melt through the subduction factory.

Acknowledgments. We thank James Conder for helpful comments, and Dapeng Zhao, Bill Hammond, Karen Fischer, and Simon Turner for use of their figures. We also thank John Eiler and an anonymous reviewer for helpful suggestions. Much of the research described here used equipment and data management resources provided by the Incorporated Research Institutions in Seismology (IRIS), and ocean bottom seismograph equipment used in collaboration with Leroy Dorman, John Hildebrand, and Spahr Webb. This research was supported by the National Science Foundation under grants EAR9219675, OCE9314446, EAR9614199, EAR9615402, EAR9712311, and OCE0002527.

REFERENCES

Aki, K., and P.G. Richards, *Quantitative Seismology, Theory and Methods*, W. H. Freeman and Company, San Francisco, 1980.

Akimoto, S.-I., The system $MgO\text{-}FeO\text{-}SiO_2$ at high pressures and temperatures—Phase equilibria and elastic properties, *Tectonophysics*, *13*, 161-187, 1972.

Alvarez, W., Geological evidence for the geographical pattern of mantle return flow and the driving mechanism of plate tectonics, *J. Geophys. Res*, *87*, 6697-6710, 1982.

Anderson, O.L., D. Isaak, and H. Oda, High-temperature elastic constant data on minerals relevant to geophysics, *Rev. Geophys.*, *30*, 57-90, 1992.

Anderson, D. L., A. Ben-Menahem, and C. B. Archambeau, Attenuation of seismic energy in the upper mantle, *J. Geophys. Res.*, *70*, 1441-1448, 1965.

Andrews, D. H., and N. H. Sleep, Numerical modeling of tectonic flow behind island arcs, *Geophys. J. Roy. Astr. Soc.*, 38, 237-251, 1974.

Ando, M., ScS polarization anisotropy around the Pacific Ocean, *J. Phys. Earth*, *32*, 179-196, 1984.

Ave'Lallement, H.G., and N.L. Carter, Syntectonic recrystallization of olivine and models of flow in the upper mantle, *Bull. Geol. Soc. Am.*, *81*, 2203, 1970.

Backus, G.E., Possible forms of seismic anisotropy of the uppermost mantle under oceans, *J. Geophys. Res.*, *70*, 3429-3429, 1965.

Berckhemer, H., W. Kampfmann, E. Aulbach, and H. Schmeling, Shear modulus and Q of forsterite and dunite near partial melting from forced-oscillation experiments, *Physics of the Earth and Planetary Interiors*, *29*, 30-41, 1982.

Bina, C.R., and A. Navrotsky, Possible presence of high-pressure ice in cold subducting slabs, *Nature*, *408*, 844-847, 2000.

Blackman, D.K., J.-M. Kendall, P.R. Dawson, H.R. Wenk, D. Boyce, and J.P. Morgan, Teleseismic imaging of subaxial flow at mid-ocean ridges: traveltime effects of anisotropic mineral texture in the mantle, *Geophys. J. Int.*, *127*, 415-426, 1996.

Bock, G., R. Kind, A. Rudloff, G. Asch, Shear wave anisotropy in the upper mantle beneath the Nazca plate in northern Chile, *J. Geophys. Res.*, *103*, 24333-24345, 1998.

Bourdon, B., S. Turner, and C. Allegre, Melting dynamics beneath the Tonga-Kermadec Island Arc inferred from 231Pa-235U systematics, *Science*, *286*, 2491-2493, 1999.

Bowman, J.R., and M. Ando, Shear-wave splitting in the upper-mantle wedge above the Tonga subduction zone, *Geophys. J. R. Astr. Soc.*, *88*, 25-41, 1987.

Buttles, J., and P. Olson, A laboratory model of subduction zone anisotropy, *Earth and Plan. Sci. Lett.*, *164*, 245-262, 1998.

Christensen, N.I., The magnitude, symmetry and origin of upper mantle anisotropy based on fabric analyses of ultramafic tectonites, *Geophys. J. R. Astr. Soc.*, *76*, 89-111, 1984.

Christensen, N.I., and R.S. Crosson, Seismic anisotropy in the upper mantle, *Tectonophysics*, *6*, 93-107, 1968.

Christensen, N.I., and W.D. Mooney, Seismic velocity structure and composition of the continental crust: A global view, *J. Geophys. Res.*, *100*, 9761-9788, 1995.

Conder, J. A., D. A. Wiens, and J. Morris, On the deompression melting structure at volcanic arcs and back-arc spreading centers, *Geophys. Res. Lett.*, *29*, 10.1029/2002GL015390, 2002.

Cooper, R.F., and D.L. Kohlstedt, Rheology and structure of olivine-basalt partial melts, *J. Geophys. Res.*, *91*, 9315-9323, 1986.

Crampin, S., and D.C. Booth, Shear-wave polarizations near the North Anatolian fault, II, Interpretation in terms of crack-induced anisotropy, *Geophys. J. R. Astron. Soc.*, *83*, 75-92, 1985.

Crampin, S., R. Evans, B. Ucer, M. Doyle, J.P. Davis, G.V. Yegorkina, and A. Miller, Observations of dilatancy induced polarization anomalies and earthquake prediction, *Nature*, *286*, 874-877, 1980.

Davidson, C., S.M. Schmid, and L.S. Hollister, Role of melt during deformation in the deep crust, *Terra Nova*, *6*, 133-142, 1994.

Davies, J.W., and D.J. Stevenson, Physical model of source region of subduction zone volcanics, *J. Geophys. Res.*, *97*, 2037-2070, 1992.

Farra, V., and L.P. Vinnik, Shear-wave splitting in the mantle of the Pacific, *Geophys. J. Int.*, *119*, 195-218, 1994.

Faul, U.H., D.R. Toomey, and H.S. Waff, Intergranular basaltic melt is distributed in thin, elongated inclusions, *Geophys. Res. Lett.*, *21*, 29-32, 1994.

Fischer, K.M., and D.A. Wiens, The depth distribution of mantle anisotropy beneath the Tonga subduction zone, *Earth Planet. Sci. Lett.*, *142*, 253-260, 1996.

Fischer, K.M., and X. Yang, Anisotropy in Kuril-Kamchatka subduction zone structure, *Geophys. Res. Lett.*, *21* (1), 5-8, 1994.

Fischer, K.M., M.J. Fouch, D.A. Wiens, and M.S. Boettcher, Anisotropy and flow in Pacific subduction zone back-arcs, *Pure Appl. Geophys.*, *151*, 463-475, 1998.

Fischer, K. M., Parmentier, E. M., Stine, A. R., Wolf, E. R., Modeling anisotropy and plate driven flow in subduction zone back-arcs, *J. Geophys. Res.*, *105*, 16181-16191, 2000.

Flanagan, M.P., and D.A. Wiens, Radial upper mantle attenuation structure of inactive back-arc basins from differential shear wave measurements, *J. Geophys. Res.*, *99*, 15469-15485, 1994.

Flanagan, M.P., and D.A. Wiens, Attenuation of broadband P and S waves in Tonga: Observations of frequency dependent Q, *Pure appl. geophys.*, *153*, 345-375, 1998.

Forsyth, D.W., Geophysical constraints on mantle flow and melt generation beneath mid-ocean ridges, in *Mantle flow and melt generation at mid-ocean ridges*, edited by J.P. Morgan, D.K. Blackman, and J.M. Sinton, pp. 1-65, AGU, Washington DC, 1993.

Fouch, M.J., and K.M. Fischer, Mantle anisotropy beneath northwest Pacific subduction zones, *J. Geophys. Res.*, *101*, 15,987-16,002, 1996.

Fouch, M.J., and K.M. Fischer, Shear wave anisotropy in the Mariana subduction zone, *Geophys. Res. Lett.*, *25*, 1221-1224, 1998.

Gaherty, J.B., T.H. Jordan, and L.S. Gee, Seismic structure of the upper mantle in a central Pacific corridor, *J. Geophys. Res.*, *10*, 22,291-22,310, 1996.

Getting, I.C., S.J. Dutton, P.C. Burnley, S. Karato, and H.A. Spetzler, Shear attenuation and dispersion in MgO, *Phys. Earth Planet. Int.*, *99*, 249-257, 1997.

Gill, and Whelan, Postsubduction ocean island alkali basalts in Fiji, *J. Geophys. Res.*, *94*, 4579-4588, 1989.

Gledhill, K., and D. Gubbins, SKS splitting and the seismic anisotropy of the mantle beneath the Hikurangi subduction Zone, New Zealand, *Phys. Earth and Planet. Int.*, *95* (3-4), 227-236, 1996.

Goes, S., R. Govers, and P. Vacher, Shallow mantle temperatures under Europe from P and S wave tomography, *J. Geophys. Res.*, *105*, 11,153-11,170, 2000.

Goetze, C., and D.L. Kohlstedt, Laboratory study of dislocation climb and diffusion in olivine, *J. Geophys. Res.*, *78*, 5961-5971, 1973.

Gorbatov, A., J. Dominguez, G. Suarez, V. Kostoglodov, D. Zhao, and E. Gordeev, Tomographic imaging of the P wave velocity structure beneath the Kamchatka peninsula, *Geophys. J. Int.*, *137*, 269-279, 1999.

Graeber, F.M., and G. Asch, Three-dimensional models of P wave velocity and P-to-S velocity ratio in the southern central Andes by simultaneous inversion of local earthquake data, *J. Geophys. Res.*, *104*, 20,237-20,256, 1999.

Gribb, T. T., and R. F. Cooper, Low-frequency shear attenuation on polycrystalline olivine: Grain boundary diffusion and the phyhsical significance of the Andrade model for viscoelastic rheology, *J. Geophys. Res.*, *103*, 27267-27279, 1998.

Gribb, T.T., and R.F. Cooper, The effect of an equilibrated melt phase on the shear creep and attenuation behavior of polycrystalline olivine, *Geophys. Res. Lett.*, *27*, 2341-2344, 2000.

Hales, A. L., and H. A. Doyle, P and S travel time anomalies and their interpretation, *Geophys. J. Roy. Astron. Soc.*, *13*, 403-415, 1967.

Hall, C.E., K.M. Fischer, and E.M. Parmentier, The influence of plate motions on three-dimensional back-arc mantle flow and shear wave splitting, *J. Geophys. Res.*, *105*, 28009-28033, 2000.

Hamburger, M. W., and Isacks, B. L., Deep earthquakes in the Southwest Pacific: A tectonic interpretation, *J. Geophys. Res.*, *92,* 13841-13854, 1987.

Hammond, W.C., and E.D. Humphreys, Upper mantle seismic wave attenuation: Effects of realistic partial melt distribution, *J. Geophys. Res.*, *105*, 10,987-11,000, 2000a.

Hammond, W.C., and E.D. Humphreys, Upper mantle seismic wave velocity: The effect of realistic partial melt geometries, *J. Geophys. Res.*, 105, 10975-10986, 2000b.

Hasegawa, A., D. Zhao, S. Hori, A. Yamamoto, and S. Horiuchi, Deep structure of the northeastern Japan arc and its relationship to seismic and volcanic activity, *Nature*, *352*, 683-689, 1991.

Hawkesworth, C.J., S.P. Turner, F. McDermott, D.W. Peate, and P. van Calsteren, U-Th isotopes in arc magmas: Implications for element transfer from the subducted crust, *Science*, *276*, 551-555, 1997.

Hawkins, J.W., Evolution of the Lau basin - Insights from ODP leg 135, in *Active margins and marginal basins of the western Pacific*, edited by B. Taylor, and J. Natland, pp. 125-173, AGU, Washington DC, 1995.

Helffrich, G., D. A. Wiens, E. Vera, S. Barrientos, P. Shore, S. Robertson, R. Adaros, A teleseismic shear-wave splitting study to investigate mantle flow around South America and implications for plate-driving forces, *Geophy. J. Int.*, *149,* F1-F7, 2002.

Hess, H.H., Seismic anisotropy of the uppermost mantle under oceans, *Nature*, *203*, 629-631, 1964.

Hiramatsu, Y., M. Ando, T. Tsukuda, and T Ooida, Three-dimensional image of the anisotropic bodies beneath central Honshu, Japan, *Geophys. J. Int.*, *135*, 801-816, 1998.

Hirn, A., M. Jiang, M. Sapin, J. Diaz, A. Nercessian, Q.T. Lu, J.C. Lepine, D.N. Shi, M. Sachpazi, M.R. Pandy, K. Ma, and J. Gallart, Seismic anisotropy as an indicator of mantle flow beneath the Himlayas and Tibet, *Nature*, *375*, 571-574, 1995.

Hirschmann, M.M., P.D. Asimow, M.S. Ghiorso, and E.M. Stolper, Calculation of peridotite partial melting from thermodynamic models of minerals and melts. III. Controls on isobaric melt production and the effect of water on melt production, *J. Petrology*, *40*, 831-851, 1999.

Hirth, G., and D. Kohlstedt, Water in the oceanic upper mantle: Implications for rheology, melt extraction and the evolution of the lithosphere, *Earth Planet Sci Lett*, *144*, 93-108, 1996.

Hirth, G., and D.L. Kohlstedt, Experimental constraints on the dynamics of the partially molten upper mantle, 2, Deformation in the dislocation creep regime, *J. Geophys. Res.*, *100*, 15,441-15,450, 1995.

Isaak, D.G., High-temperature elasticity of iron-bearing olivines, *J. Geophys. Res.*, *97* (B2), 1871-1885, 1992.

Iwamori, H., Transportation of H_2O and melting in subduction zones, *Earth Planet. Sci. Lett.*, *160*, 65-80, 1998.

Jackson, I., Laboratory measurement of seismic wave dispersion and attenuation: recent progress, in *The earth's deep interior: Mineral physics and tomography from the atomic to the global scale,* edited by S.-I Karato et al., Geophysical Monograph, 117, American Geophysical Union, Washington D.C., 2000.

Jackson, I., M.S. Paterson, and J.D. FitzGerald, Seismic wave dispersion and attenuation in Aheim Dunite: an experimental study, *Geophys. J. Int.*, *108*, 517-534, 1992.

Ji, S., X. Zhao, and D. Francis, Calibration of shear-wave splitting in the subcontinental upper mantle beneath active orogenic belts using ultramafic xenoliths from the Canadian Cordillera and Alaska, *Tectonophysics*, *239*, 1-27, 1994.

Jordan, T.H., Mineralogies, densities, and seismic velocities of garnet lherzolites and their geophysical significance, in *The mantle sample: Inclusions in kimberlites and other volcanics*, edited by F.R. Boyd, and H.O.A. Meyer, pp. 1-14, American Geophysical Union, Washington D.C., 1979.

Jung, H., and S.-i. Karato, Water-induced fabric transitions in olivine, *Science, 293,* 1460-1463, 2001.

Jurdy, D.M., and M. Stefanick, Flow models for back-arc spreading, *Tectonophysics*, *99*, 191-206, 1983.

Kanamori, H., and D.L. Anderson, Importance of physical dispersion in surface-wave and free-oscillation problems, *Rev. Geophys. Space Phys.*, *15*, 105-112, 1977.

Kaneshima, S., and P.G. Silver, A search for source side mantle anisotropy, *Geophys. Res. Lett.*, *19* (10), 1049-1052, 1992.

Karato, S., Importance of anelasticity in the interpretation of seismic tomography, *Geophys. Res. Lett.*, *20*, 1623-1626, 1993.

Karato, S, Mapping water content in the upper mantle, in *The Subduction Factory,* edited by J. Eiler and G. Abers, in press, 2002.

Karato, S., and H. Jung, Water, partial melting and the origin of the seismic low velocity and high attenuation zone in the upper mantle, *Tectonophysics*, *157*, 193-207, 1998.

Karato, S., M.S. Paterson, and J.D. FitzGerald, Rheology of synthetic olivine aggregates: Influence of grain size and water, *J. Geophys. Res.*, *91*, 8151-8176, 1986.

Karato, S., and H.A. Spetzler, Defect microdynamics in minerals and solid-state mechanisms of seismic wave attenuation and velocity dispersion in the mantle, *Rev. Geophys.*, *28*, 399-421, 1990.

Kendall, J.-M., Teleseismic arrivals at a mid-ocean ridge: Effects of mantle melt and anisotropy, *Geophys. Res. Lett.*, *21*, 301-304, 1994.

Kern, H., P- and S-wave anisotropy and shear wave splitting at presure and temperature in possible mantle rocks and their relations to the rock fabric, *Phys. Earth and Planet. Int.*, *78*, 245-256, 1993.

Kern, H., and A. Richter, Temperature derivatives of compressional and shear wave velocities in crustal and mantle rocks at 6 kbar confining pressure, *J. Geophys.*, *49*, 47-56, 1981.

Kohlstedt, D.L., Structure, rheology and permeabililty of partially molten rocks at low melt fraction, in *Mantle flow and melt generation at mid-ocean ridges*, edited by J.P. Morgan, B.D. K., and J.M. Sinton, pp. 103-122, AGU, Washington D. C., 1992.

Koper, K.D., D.A. Wiens, L.M. Dorman, J.A. Hildebrand, and S.C. Webb, Modeling the Tonga slab: Can travel time data resolve a metastable olivine wedge?, *J. Geophys. Res., 103*, 30079-30100, 1998.

Koper, K.D., D.A. Wiens, L.M. Dorman, J.A. Hildebrand, and S.C. Webb, Constraints on the origin of slab and mantle wedge anomalies in Tonga from the ratio of S to P anomalies, *J. Geophys. Res., 104*, 15089-15104, 1999.

Mainprice, D., and A. Nicolas, Development of shape and lattice preferred orientations: Applications to the seismic anisotropy of the lower crust, *J. Struct. Geol., 11*, 175-189, 1989.

Mainprice, D., and P.G. Silver, Interpretation of SKS-waves using samples from the subcontinental mantle, *Phys. Earth Planet. Int., 78*, 257-280, 1993.

Marson-Pidgeon, M. K. Savage, K. Gledhill, G. Stuart, Seismic anisotropy beneath the lower half of the North Island, New Zealand, *J. Geophys. Res., 104*, 20,277-20,286, 1999.

Mavko, G.M., Velocity and attenuation in partially molten rocks, *J. Geophys. Res., 85*, 5412-5426, 1980.

McKenzie, D., Finite deformation during fluid flow, *Geophys. J., 58*, 689-715, 1979.

Mehl, L., Hacker, B. R., and G. Hirth, Arc-parallel flow within the mantle wedge: Evidence from the accreted Talkeetna arc, South Central Alaska, *J. Geophys. Res.,* in press, 2002.

Mibe, K., T. Fuji, and A. Yasuda, Control of the location of the volcanic front in island arcs by aqueous fluid connectivity in the mantle wedge, *Nature, 401*, 259-262, 1999.

Myers, S.C., S. Beck, G. Zandt, and T. Wallace, Lithospheric-scale structure across the Bolivian Andes from tomographic images of velocity and attenuation for P and S waves, *J. Geophys. Res., 103*, 21,233-21,252, 1998.

Nakajima, J., Matsuzawa, T., Hasegawa, A., and D. Zhao, Three dimensional structure of Vp, Vs, and Vp/Vs beneath northeastern Japan: Imlications for arc magmatism and fluids, *J. Geophys. Res., 106,* 21843-21857, 2001.

Nicolas, A., and N.I. Christensen, Formation of anisotropy in upper mantle peridotites—a review, *Rev. Geophysics, 25*, 111-123, 1987.

Nolet, G., Seismic evidence for the occurrence of volatiles below 200 km depth in the Earth, in *Processes of Deep Earth and Planetary Volatiles*, edited by K. Farley, pp. 22-32, Am. Inst. Phys., New York, 1995.

O'Connell, R.J., and B. Budianski, Seismic velocities in dry and saturated cracked solids, *J. Geophys. Res., 79*, 5412-5426, 1974.

Peacock, S.M., Fluid processes in subduction zones, *Science, 248*, 329-337, 1990.

Pearce, J.A., M. Ernewein, S.H. Bloomer, L.M. Parson, B.J. Murton, and L.E. Johnson, Geochemistry of Lau Basin volcanic rocks: influence of ridge segmentation and arc proximity, in *Volcanism associated with extension at consuming plate boundaries*, edited by J.L. Smellie, pp. 53-75, Geological Society, London, 1995.

Peyton, V., V. Levin, J. Park, M. Brandon, J. Lees, E. Gordeev, A. Ozerov, Mantle flow at a slab edge: Seismic anisotropy in the Kamchatka region, *Geophys. Res. Lett., 28*, 379-382, 2001.

Polet, J., P.G. Silver, S. Beck, T. Wallace, G. Zandt, S. Ruppert, R. Kind, and A. Rudloff, Shear wave anisotropy beneath the Andes from the BANJO, SEDA, and PISCO experiments, *J. Geophys. Res., 105*, 6287-6304, 2000.

Poreda, R.J., and H. Craig, He and Sr isotopes in the Lau Basin mantle: depleted and primitive mantle components, *Earth Planet. Sci. Lett., 113*, 487-493, 1992.

Raleigh, C.B., Mechanisms of plastic deformation of olivine, *J. Geophys. Res., 73*, 4213-4223, 1968.

Ribe, N.M., Seismic anisotropy and mantle flow, *J. Geophys. Res., 94*, 4123-4223, 1989.

Ribe, N.M., On the relation between seismic anisotropy and finite strain, *J. Geophys. Res., 97* (B6), 8737-8747, 1992.

Richter, F.M., and D. McKenzie, Dynamical models for melt segregation from a deformable matrix, *J. Geol., 92*, 729-740, 1984.

Roth, E.G., D.A. Wiens, L.M. Dorman, J. Hildebrand, and S.C. Webb, Seismic attenuation tomography of the Tonga-Fiji region using phase pair methods, *J. Geophys. Res., 104*, 4795-4810, 1999.

Roth, E.G., D.A. Wiens, and D. Zhao, An empirical relationship between seismic attenuation and velocity anomalies in the upper mantle, *Geophys. Res. Lett., 27*, 601-604, 2000.

Russo, R.M., and P.G. Silver, Trench-parallel flow beneath the Nazca Plate from seismic anisotropy, *Science, 263*, 1105-1111, 1994.

Sandvol, E., and J. Ni, Deep azimuthal seismic anisotropy in the southern Kurile and Japan subduction zones, *J. Geophys. Res., 102,* 9911-9922, 1997.

Sato, H., S. Sacks, and T. Murase, The use of laboratory velocity data for estimating temperature and partial melt fraction in the low velocity zone: Comparison with heat flow data and electrical conductivity studies, *J. Geophys. Res., 94*, 5689-5704, 1989a.

Sato, H., S. Sacks, T. Murase, G. Muncill, and H. Fukuyama, Q_P-melting temperature relation in peridotite at high pressure and temperature: Attenuation mechanism and implications for the mechanical properties of the upper mantle, *J. Geophys. Res., 94*, 10,647-10,661, 1989b.

Savage, M.K., Seismic anisotropy and mantle deformation: What have we learned from shear wave splitting?, *Rev. Geophys., 37* (1), 65-106, 1999.

Schmeling, H., Numerical models on the influence of partial melt on elastic, anelastic and electric properties of rocks. Part I: elasticity and anelasticity, *Phys. Earth Planet. Int., 41*, 34-57, 1985.

Schmidt, M.W., and S. Poli, Experimentally based water budgets for dehydrating slabs and consequences for arc magma generation, *Earth Planet Sci Lett, 163*, 361-379, 1998.

Scott, D.R., and D.J. Stevenson, Magma ascent by porous flow, *J. Geophys. Res., 91*, 9283-9296, 1986.

Sekiguchi, S., Three-dimensional Q structure beneath the Kanto-Tokai district, Japan, *Tectonophysics, 195*, 83-104, 1991.

Shen, Y., and D.W. Forsyth, Geochemical constraints on initial and final depths of melting beneath mid-ocean ridges, *J. Geophys. Res., 100* (B2), 2211-2238, 1995.

Shih, X.R., J.F. Schneider, and R.P. Meyer, Polarities of P and S waves, and shear wave splitting observed from the Bucaramanga nest, Colombia, *J. Geophys. Res., 96* (B7), 12,069-12,082, 1991.

Silver, P.G., and W.W. Chan, Shear wave splitting and subcontinental mantle deformation, *J. Geophys. Res.*, *96*, 16429-16454, 1991.

Smith, G.P., D.A. Wiens, K.M. Fisher, L.M. Dorman, S.C. Webb, and J.A. Hildebrand, A complex pattern of mantle flow in the Lau backarc, *Science*, 292, 713-716, 2001.

Smyth, J.R., A crystallographic model for hydrous wadsleyite: An ocean in the earth's interior?, *Am. Mineral.*, *79*, 1021-1024, 1994.

Sobolev, S.V., H. Zeyen, G. Stoll, F. Werling, R. Altherr, and K. Fuchs, Upper mantle temperatures from teleseismic tomography of French Massif Central including effects of composition, mineral reactions, anharmonicity, anelasticity, and partial melt, *Earth Planet. Sci. Lett.*, *139*, 147-163, 1996.

Takei, Y., Constitutive mechanical relations of solid-liquid composites in terms of grain-boundary contiguity, *J. Geophys. Res.*, *103*, 18,183-18,204, 1998.

Takei, Y., Acoustic properties of partially molten media studied on a simple binary system with a controllable dihedral angle, *J. Geophys. Res.*, *105*, 16665-16682, 2000.

Tanimoto, T., and D.L. Anderson, Mapping convection in the mantle, *Geophys. Res. Lett.*, *11*, 287-290, 1984.

Tatsumi, Y., Migration of fluid phases and genesis of basalt magmas in subduction zones, *J. Geophys. Res.*, *94*, 4697-4707, 1989.

Thompson, A.B., Water in the Earth's upper mantle, *Nature*, *358*, 295-302, 1992.

Tommasi, A., Forward modeling of the development of seismic anisotropy in the upper mantle, *Earth Planet. Sci. Lett.*, *160*, 1-13, 1998.

Tsumura, N., A. Hasegawa, and S. Horiuchi, Simultaneous estimation of attenuation structure, source parameters and site response spectra—application to the northeastern part of Honshu, Japan, *Phys. Earth Planet. Int.*, *93*, 105-121, 1996.

Tsumura, N., S. Matsumoto, S. Horiuchi, and A. Hasagawa, Three-dimensional attenuation structure beneath the northeastern Japan arc estimated from spectra of small earthquakes, *Tectonophysics, 319,* 241-260, 2000.

Turner, S., and C. Hawkesworth, Constraints on flux rates and mantle dynamics beneath island arcs from Tonga-Kermadec lava geochemistry, *Nature*, *389*, 568-573, 1997.

Turner, S., and C. Hawkesworth, Using geochemistry to map mantle flow beneath the Lau basin, *Geology*, *26*, 1019-1022, 1998.

Van der Hilst, R., R. Engdahl, W. Spakman, and G. Nolet, Tomographic imaging of subducted lithosphere below northwest Pacific island arcs, *Nature*, *353*, 37-42, 1991.

Verma, R.K., Elasticity of some high-density crystals, *J. Geophys. Res.*, *65*, 757-766, 1960.

Waff, H.S., and U.H. Faul, Effects of crystalline anisotropy on fluid distribution in ultramafic partial melts, *J. Geophys. Res.*, *97*, 9003-9014, 1992.

Webb, S.C., and D.W. Forsyth, Structure of the upper mantle under the EPR from waveform inversion of regional events, *Science*, *280*, 1227-1229, 1998.

Wendt, J.I., M. Regelous, K.D. Collerson, and A. Ewart, Evidence for a contribution from two mantle plumes to island arc lavas from northern Tonga, *Geology*, *25*, 611-614, 1997.

Wenk, H.-R., K. Bennett, G.R. Canova, and A. Molinari, Modelling plastic deformation of peridotite with the self-consistent theory, *J. Geophys. Res.*, *96*, 8337-8349, 1991.

Wiens, D. A., G. P. Smith, and S. D. Robertson, Seismological structure and mantle flow patterns in the Lau backarc basin, *EOS Trans Am. Geophys. Un.*, *83*, 346, 2002.

Wiens, D.A., P.J. Shore, J.J. McGuire, E. Roth, M.G. Bevis, and K. Draunidalo, The Southwest Pacific Seismic Experiment, *IRIS Newsletter*, *14*, 1-4, 1995.

Williams, P.F., Foliation: a review and discussion, *Tectonophysics*, *39*, 305, 1977.

Xu, Y., and D.A. Wiens, Upper mantle structure of the Southwest Pacific from regional waveform inversion, *J. Geophys. Res.*, 102, 27439-27451, 1997.

Yang, X., K.M. Fischer, and G.A. Abers, Seismic anisotropy beneath the Shumagin Islands segment of the Aleutian-Alaska subduction zone, *J. Geophys. Res.*, *100*, 18,165-18,178, 1995.

Zhang, S., and S. Karato, Lattice preferred orientation of olivine agregates deformed in simple shear, *Nature*, *375*, 774-777, 1995.

Zhang, S., S. Karato, J. FitzGerald, U. H. Faul, and Y. Zhou, Simple shear deformation of olivine aggregates, *Tectonophysics, 316,* 133-152, 2000.

Zhao, D., K. Asamori, and H. Iwamori, Seismic structure and magmatism of the young Kyushu subduction zone, *Geophys. Res. Lett.*, *27*, 2057-2060, 2000.

Zhao, D., D. Christensen, and H. Pulpan, Tomographic imaging of the Alaska subduction zone, *J. Geophys. Res*, *100*, 6487-6504, 1995.

Zhao, D., and A. Hasegawa, P wave tomographic imaging of the crust and upper mantle beneath the Japan Islands, *J. Geophys. Res.*, *98*, 4333-4353, 1993.

Zhao, D., A. Hasegawa, and S. Horiuchi, Tomographic imaging of P and S wave velocity structure beneath northeastern Japan, *J. Geophys. Res.*, *97*, 19,909-19928, 1992.

Zhao, D., A. Hasegawa, and H. Kanamori, Deep structure of the Japan subduction zone as derived from local, regional, and teleseismic events, *J. Geophys. Res.*, *99*, 22313-22329, 1994.

Zhao, D., Y. Xu, D.A. Wiens, L. Dorman, J. Hildebrand, and S. Webb, Depth extent of the Lau back-arc spreading center and its relationship to the subduction process, *Science*, *278*, 254-257, 1997.

Gideon P. Smith and Douglas A. Wiens Department of Earth and Planetary Sciences, Washington University, 1 Brookings Dr., St. Louis, MO 63130, doug@seismo.wustl.edu

Rheology of the Upper Mantle and the Mantle Wedge: A View from the Experimentalists

Greg Hirth

Department of Geology and Geophysics, Woods Hole Oceanographic Institution, Woods Hole, Massachusetts

David Kohlstedt

Department of Geology and Geophysics, University of Minnesota, Minneapolis, Minnesota

In this manuscript we review experimental constraints for the viscosity of the upper mantle. We first analyze experimental data to provide a critical review of flow law parameters for olivine aggregates and single crystals deformed in the diffusion creep and dislocation creep regimes under both wet and dry conditions. Using reasonable values for the physical state of the upper mantle, the viscosities predicted by extrapolation of the experimental flow laws compare well with independent estimates for the viscosity of the oceanic mantle, which is approximately 10^{19} Pa s at a depth of ~100 km. The viscosity of the mantle wedge of subduction zones could be even lower if the flux of water through it can result in olivine water contents greater than those estimated for the oceanic asthenosphere and promote the onset of melting. Calculations of the partitioning of water between hydrous melt and mantle peridotite suggest that the water content of the residue of arc melting is similar to that estimated for the asthenosphere. Thus, transport of water from the slab into the mantle wedge can continually replenish the water content of the upper mantle and facilitate the existence of a low viscosity asthenosphere.

INTRODUCTION

From the alteration of crust at oceanic spreading centers to the migration of melt through the lithosphere at convergent margins, the rheology of rocks plays a key role in the dynamics and chemical fluxes associated with the "Subduction Factory". Consider the following examples: (1) Altered oceanic lithosphere is a source of fluid for hydrous arc magmatism. The amount of water available for melt generation depends on the depth and distribution of hydrothermal alteration of the oceanic lithosphere, which in turn depends on the depth to the brittle-plastic transition. (2) Subducted pelagic sediments also supply fluids to the source of arc magmatism. In this case, the amount of water available depends on the rheology and permeability of the compacting sediments. (3) The kinetics and spatial distribution of dehydration and melting reactions depend on the thermal structure of the slab-mantle interface, which is partly controlled by rheology through the effects of shear heating. (4) The temperature at the slab-mantle interface and in the mantle wedge also depend on the kinematics and dynamics of convection, which in turn are controlled by the viscosity of the mantle wedge. (5) The rheology, fluid

Inside the Subduction Factory
Geophysical Monograph 138

10.1029/138GM06

distribution and permeability of the slab determine whether fluids migrate into the mantle wedge by porous flow, transport in hydrofractures or advection in diapirs. (6) Finally, melt migration to the base of the crust beneath the arc front and transport of material to back arc spreading centers are governed by mantle flow in the wedge.

It is beyond the scope of this article to address each of these topics in detail. We therefore concentrate on experimental constraints on the viscosity of the mantle wedge. We first review experimental studies used to determine flow law parameters for both dislocation creep and diffusion creep under dry and wet conditions. These data provide insights into the Subduction Factory, as well as other upper mantle tectonic settings, and input for geodynamic models. We then emphasize topics for which progress is needed from experimentalists. Finally, we compare the experimental results with independent geophysical constraints on mantle viscosity.

Viscosity of the Mantle Wedge

The thermal structure of the mantle wedge and the processes that accommodate melt migration to arc volcanoes are in large part controlled by the viscosity of the mantle. Some portions of the mantle wedge may have a lower viscosity than any other region of the upper mantle. Estimates for the viscosity of the oceanic asthenosphere, based on geophysical observations and experimental data on the rheology of peridotite, range from 10^{18} to 10^{19} Pas [e.g., *Melosh,* 1976, *Craig and McKenzie,* 1986; *Hager,* 1991; *Karato and Wu,* 1993; *Hirth and Kohlstedt,* 1996]. There are several reasons why the viscosity of the mantle wedge could be even lower than that in the oceanic asthenosphere. First, the water content in parts of the mantle wedge could be higher than that in the asthenosphere due to the influx of fluids from the subducted slab. Experimental observations indicate that the viscosity of olivine aggregates decreases with increasing water content [e.g., *Mei and Kohlstedt,* 2000b; *Karato,* this issue]. Olivine in the oceanic asthenosphere contains ~1000 H/10^6Si, which is approximately 20% of the solubility at a depth of 120 km [*Hirth and Kohlstedt,* 1996]. At similar depths the mantle wedge may contain as much as 5 times more water (i.e., it may be water saturated). Second, because of the higher water content, drying out of the nominally anhydrous minerals due to partial melting of the wedge will not result in as large of an increase in viscosity as proposed for the melting region of mid-ocean ridges [e.g., *Hirth and Kohlstedt,* 1996; *Phipps Morgan,* 1997]. In fact, depending on the degree of melting, the flux of volatiles from the subducting slab, and the amount of melt retained in the source, partial melting may decrease the viscosity of the wedge. For a constant olivine water content, experimental data indicate that the viscosity of olivine aggregates decreases approximately exponentially with increasing melt content [*Kelemen et al.,* 1997; *Kohlstedt et al.,* 2000; *Mei et al.,* 2002]. Finally, based on the petrological and geophysical data discussed by *Kelemen et al.* [this issue], the temperature in the mantle wedge may be approximately the same as that in the oceanic asthenosphere. Taken together, these observations suggest that in some regions of the mantle wedge the viscosity may be up to an order of magnitude lower than that in the oceanic asthenosphere. In the following sections, we provide a detailed review of the rheological data used to obtain these constraints on mantle viscosity.

Experimental Background

Experimental studies constrain the magnitude of mantle viscosity. However, due to the required extrapolation from laboratory to geologic conditions, the accuracy of these constraints is not as high as their precision. Thus, the laboratory data provide stronger constraints on the change in viscosity as a function of pressure, temperature, grain size, water content and melt fraction. Laboratory experiments are usually conducted near upper mantle temperatures. Therefore, the primary limitation on accuracy comes from the relatively large extrapolation in stress. For example, by extrapolating two orders of magnitude in stress (e.g., from 100 to 1 MPa), an uncertainty in the stress exponent of ± 0.5 results in $\pm$ one order of magnitude uncertainty in viscosity. Another potential problem is that the dominant deformation mechanism may change from dislocation creep to diffusion creep with decreasing stress, provided that grain growth is kinetically inhibited.

While there is uncertainty associated with extrapolation of laboratory results, several observations justify applying experimental flow laws to the study of deformation at geologic conditions. First, a comparison of microstructures in experimentally and naturally deformed peridotites indicates that the same slip systems are active [e.g., *Nicolas,* 1986]. The analysis of lattice preferred orientations in both naturally and experimentally deformed rocks demonstrates that strain is mostly accommodated by slip on (010)[100] [e.g., *Tomassi et al.,* 1999; *Zhang et al.,* 2000], the easiest slip system in olivine [e.g., *Durham and Goetze,* 1977; *Bai et al.,* 1991]. Second, the conditions under which changes in deformation mechanism occur can be constrained using a combination of experimental data and theoretical arguments. For example, diffusion creep processes can be studied using experimentally engineered fine-grained aggregates. Viscosity in the diffusion creep regime increases non-linearly with increasing grain size. Thus, rocks with a grain size of 10 μm (which is two to three orders of magnitude smaller than the grain size in most of the mantle) will deform six to nine orders of magnitude faster by diffusion creep in the laboratory than larger grained rocks in the mantle. Likewise, because viscosity in the dislocation creep regime decreases non-linearly with increasing stress, experiments can be carried out to high strain in both deformation regimes at laboratory timescales.

The strategy of using fine-grained aggregates synthesized from natural rocks and minerals has lead to several breakthroughs in our understanding of the rheology of peridotite. The primary benefit of these types of experiments is that a large number of extensive parameters can be controlled independently, including melt fraction, grain size and water content. In addition, sample-to-sample variability is significantly decreased by the application of standard protocols for sample preparation. Many of the flow law parameters in which we have the highest confidence are determined this way. However, one drawback of this procedure is that many experiments are conducted near the transition between diffusion and dislocation creep. To obtain the highest resolution in flow law parameters for individual creep mechanisms, the component of strain rate from competing deformation processes to the total strain rate must be taken into account. This problem is outlined below in our review of flow law parameters for diffusion and dislocation creep.

CONSTRAINTS ON FLOW LAW PARAMETERS

Theoretical treatments and experimental observations demonstrate that the rheological behavior of rocks, metals and ceramics is well described by a power law dependence of strain rate ($\dot{\varepsilon}$) on differential stress (σ). For olivine aggregates, we use a power law of the form

$$\dot{\varepsilon} = A\sigma^n d^{-p} fH_2O^r \exp(\alpha\phi)\exp\left(-\frac{E^*+PV^*}{RT}\right) \quad (1)$$

where A is a constant, n is the stress exponent, d is grain size, p is the grain size exponent, fH_2O is water fugacity, r is the water fugacity exponent, ϕ is melt fraction, α is a constant, E^* is the activation energy, V^* is the activation volume, R is the gas constant, and T is absolute temperature. In the sections below, we review experimental constraints for these flow law parameters for deformation in the diffusion creep and dislocation creep regimes, as well as provide some constraints on flow law parameters for deformation accommodated by grain boundary sliding. We first review the stress dependence of deformation, followed by analyses of the influence of temperature, pressure, water content, and melt fraction. We do not review the flow law parameters for individual deformation mechanisms seperately because our analysis includes resolving the components of strain rate from competing deformation processes. For those readers wishing to skip the details of how these parameters are constrained, the values of the flow law parameters are summarized in Table 1.

In addition to the parameters in equation (1), the rheology of olivine aggregates also depends on oxygen fugacity (fO_2^q) and silica activity. However, for practical application of flow laws to geodynamic modeling, we have incorporated the influence of oxygen fugacity into A and E^*. Experimental observations indicate that $\dot{\varepsilon} \propto fO_2^q$, where the exponent q is $\leq 1/6$ for a wide range of deformation conditions [*Ricoult and Kohlstedt,* 1985; *Bai et al.,* 1991]. Thus, because the range of oxygen fugacities expected in the mantle is small, the effect of changes in oxygen fugacity on viscosity is relatively minor compared to changes in creep rate that occur due to variations in temperature, water content or pressure. Also, since olivine and pyroxene are both present in most mantle rocks, the silica activity is fixed through out most of the mantle.

Stress Dependence and Grain Size Dependence in the Diffusion Creep Regime

The flow law for diffusion creep under dry conditions is well constrained by experiments conducted on fine-grained rocks [*Karato et al.,* 1986; *Hirth and Kohlstedt,* 1995a, *Gribb and Cooper,* 2000]. These studies built on the pioneering uniaxial hot-pressing experiments of *Schwenn and Goetze* [1978] and *Cooper and Kohlstedt* [1984]. Under dry conditions, the stress exponent ($n = 1.0 \pm 0.1$) and grain size exponent ($p = 3.0 \pm 0.5$) reported by *Hirth and Kohlstedt* [1995a] are in agreement with theoretical predictions for creep dominated by grain boundary diffusion [e.g., *Coble,* 1963]. Several other studies have also yielded values for n and p which are within error of those predicted by the Coble creep equation [e.g., *Cooper and Kohlstedt,* 1984]. As discussed by *Hirth and Kohlstedt* [1995a], the slightly lower value of p and higher value of n reported by Karato et al. [1986] likely reflect a component of dislocation creep to the overall strain rate of the samples. Values of $n \approx 1$ and $p \approx 3$ are also observed under water saturated conditions [*Karato et al.,* 1986; *Mei et al.,* 2000a]. Based on the agreement between these data and the theoretical predictions, we use values of $n = 1$ and $p = 3$ in evaluation of deformation data obtained under conditions for which more than one creep mechanism contributed significantly to flow.

Table 1: Rheological Parameters for Equation (1).

	A^a	n	p	r^b	α	E^* (kJ/mol)	V^* (10^{-6} m^3/mol)
dry diffusion	1.5×10^9	1	3	-	30	375 ± 50	2-10
wet diffusion	$2.5\times10^{7\,d}$	1	3	0.7–1.0	30	375 ± 75	0-20
wet diffusion (constant C_{OH})^c	1.0×10^6	1	3	1	30	335 ± 75	4
dry dislocation	1.1×10^5	3.5 ± 0.3	0	-	30-45	530 ± 4	(see Table 2)
wet dislocation	1600	3.5 ± 0.3	0	1.2 ± 0.4	30-45	520 ± 40	22 ± 11
wet dislocation (constant C_{OH})^e	90	3.5 ± 0.3	0	1.2	30-45	480 ± 40	11
dry GBS, T>1250°C	4.7×10^{10}	3.5	2	-	30-45	600[f]	(see Table 2)[g]
dry GBS, T<1250°C	6500	3.5	2	-	30-45	400[f]	(see Table 2)[g]

[a]For stress in MPa, f_{H_2O} in MPa (or C_{OH} in H/10^{6}Si) and grain size in μm.
[b]Uncertainty in r is correlated with uncertainty in V^*
[c]Example calculation for C_{OH} = 1000 H/10^{6}Si, d = 10 mm, T = 1400°C, P = 1 GPa, σ = 0.3 MPa:
$\dot{\varepsilon} = (1.0\times10^6)*(0.3)^1*(10,000)^{-3}*(1000)^1*\exp[-(335000+10^9*4\times10^{-6})/(8.314*1673)] = 7.8\times10^{-15}$/s
[d]Value for A is given for r = 1.
[e]Example calculation for C_{OH} = 1000 H/10^{6}Si, T = 1400°C, P = 1 GPa, σ = 0.3 MPa:
$\dot{\varepsilon} = (90)*(0.3)^{3.5}*(1000)^{1.2}*\exp[-(480000+10^9*11\times10^{-6})/(8.314*1673)] = 2.5\times10^{-12}$/s
[f]The activation energy for GBS is assumed to be that for slip on (010)[100], which changes with increasing temperature [*Bai et al.,* 1991]. The values given here include the effect of temperature on oxygen fugacity.
[g]The value for V^* is assumed to be the same as that for dislocation creep.

Diffusion creep data can be used to estimate the grain boundary diffusivity of the slowest diffusing species in olivine. Diffusion creep is limited by the slowest diffusing species moving along its fastest path. Therefore, creep rate is governed by an effective diffusivity, $D_{eff} \approx [D_{gm} + \tau\delta D_{gb}/d]$, where D_{gm} is the grain matrix diffusivity, D_{gb} is the grain boundary diffusivity, δ is the grain boundary width, d is the grain size, and τ is a "tortuosity" constant; we use a value of τ = 1.5. In Figure 1, we compare the rates of grain matrix diffusion under dry conditions of Mg (D^{Mg}_{gm}) [*Gaetani and Watson,* 2000] and Si (D^{Si}_{gm}) [*Houlier et al.,* 1990] to rates for bulk grain boundary diffusion of Mg (D^{Mg}_{bulk}) [from *Watson,* 1991 (see *Hirth and Kohlstedt,* 1995a)] and Si (D^{Si}_{bulk}) [*Farver and Yund,* 2000] at a temperature of 1300°C, where $D_{bulk} = \tau\delta D_{gb}/d$. In addition, we plot the effective diffusivity determined from the rate of diffusion creep. In this case, we plot $\tau\delta D_{gb}/d$, where the product δD_{gb} is calculated from observed creep rates using the theoretical Coble creep flow law [e.g., *Karato et al.,* 1986; *Hirth and Kohlstedt,* 1995a]. These relationships indicate that diffusion creep is limited by grain boundary diffusion of silicon [e.g., *Hirth and Kohlstedt,* 1995a, *Farver and Yund,* 2000]. Based on the observation that both the grain boundary and grain matrix diffusivities for oxygen lie between the respective values for Mg and Si [e.g., *Fisler and Mackwell,* 1994; *Ryerson et al.,* 1989], we have omitted them from this figure. The diffusion data for Mg are included to check whether conditions exist where $D^{Si}_{bulk} > D^{Mg}_{gm}$, as suggested by *Karato et al.* [1986]. A transition to diffusion creep limited by grain matrix diffusion (i.e., Nabarro-Herring creep) is not predicted unless that grain size is greater than approximately 1m. This calculation, which demonstrates that Nabarro-Herring creep is unlikely to

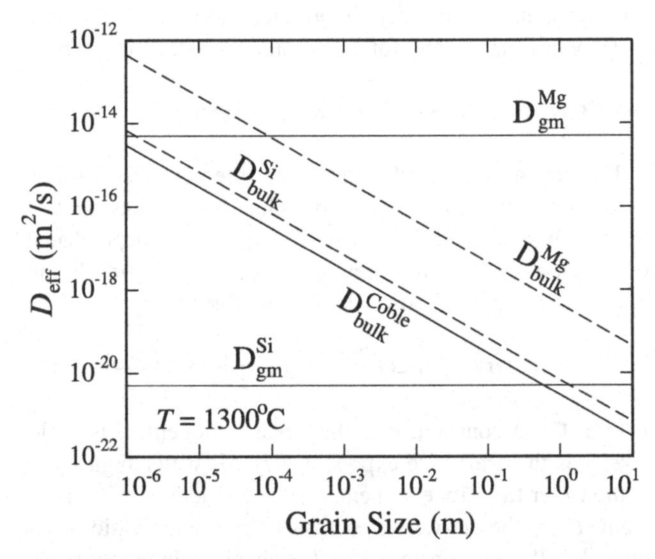

Figure 1. Plot of effective diffusivity at 1300°C versus grain size, comparing bulk diffusivities of silicon and magnesium via grain boundary transport to grain matrix (gm) diffusivities. The bulk diffusivity calculated using diffusion creep deformation data for olivine aggregates is also shown (D^{Coble}_{bulk}). These relationships indicate that diffusion creep is limited by D^{Si}_{bulk} unless the grain size is greater than ~1 m.

occur in experiments or at any conditions in the upper mantle, further justifies using $p = 3$ to normalize diffusion creep data to a common grain size. A similar analysis can not be carried out for hydrous conditions because of a lack of diffusion data.

Stress Dependence in the Dislocation Creep Regime: Dry Conditions

Much of the highest resolution mechanical data, such as that shown in Figure 2a, are acquired on fine-grained olivine aggregates deformed at the transition between diffusion creep and dislocation creep. To determine the stress exponent in the dislocation creep regime, we have accounted for the diffusion creep component of the data in two ways. In the first approach, we use a non-linear least squares fit of the data at constant temperature and pressure to the equation

$$\dot{\varepsilon} = \dot{\varepsilon}_{\text{disl}} + \dot{\varepsilon}_{\text{diff}} = A_{disl}\sigma^n + A_{diff}\sigma d^{-3} \qquad (2)$$

For the experimental results presented in Figure 2a, the non-linear fit gives $n = 3.3 \pm 0.6$. In the second approach, we subtract the strain rate of the corresponding diffusion creep flow law from the total strain rate, which leaves only the dislocation creep component, and linearly fit only the highest stress data. In this case, we obtain a value of $n = 3.6 \pm 0.5$.

These values for n are comparable to those determined for coarser grained natural rocks deformed at the same conditions [e.g., *Chopra and Paterson,* 1984]. From a global inversion of their data to determine n, E^*, and A, *Chopra and Paterson* obtained $n = 3.6 \pm 0.2$. Their analysis included some data at differential stresses significantly greater than the confining pressure and, therefore, may be influenced by a change in deformation mechanism to semi-brittle flow [e.g., *Evans et al.,* 1990] or "power law breakdown" [e.g., p. 81 *Poirier,* 1985]. By refitting their data excluding all points at differential stresses greater than 300 MPa (i.e., the confining pressure), we obtain $n = 3.4 \pm 0.2$ (using $E^* =$ 530 kJ/mol to normalize data to a constant temperature). By fitting only data at a constant temperature of 1573 K and stresses below 300 MPa (this sample is the largest satisfying these criteria, with only 4 data points), we obtain $n = 3.7 \pm 0.4$.

The values of n determined for the polycrystalline samples agree well with those determined for olivine single crystals [e.g., *Kohlstedt and Goetze,* 1974; *Durham and Goetze,* 1977; *Bai et al.,* 1991]. Values of n for the three major slip systems in olivine have been determined over a large range in temperature and oxygen fugacity. Under all of these conditions, n is observed to be in the range of 3.5. Because single crystal experiments can be conducted at 1 atm, the data have high precision at low strain rates and relatively high temperatures. Values of n of ~3.5 are observed at stresses as low as 15 MPa [*Bai et al.,* 1991, *Jin et al.,* 1994]. These data are extremely valuable as they

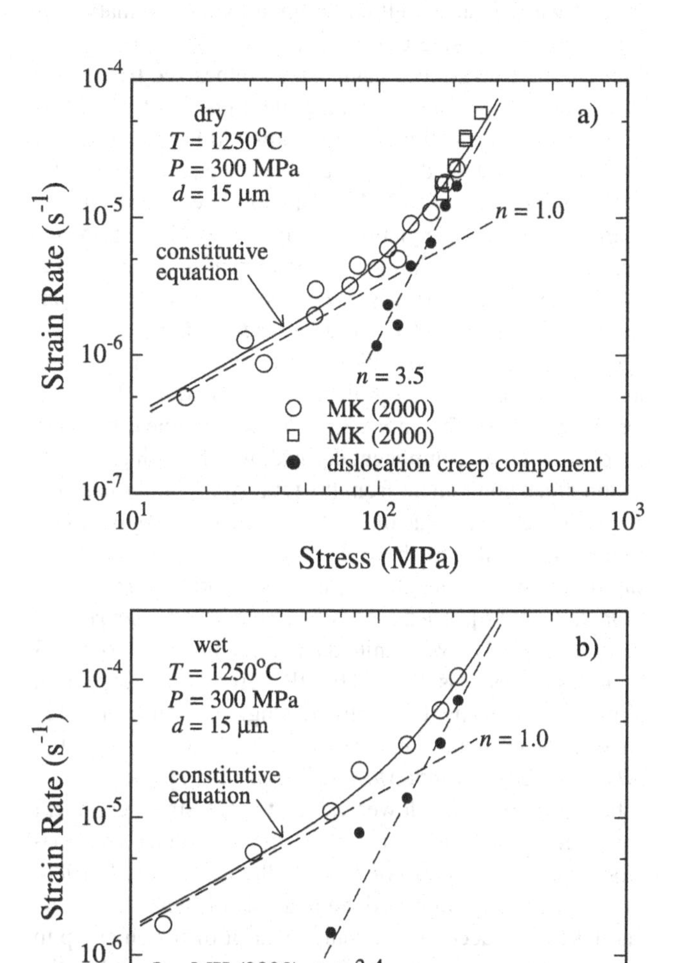

Figure 2. Plots of strain rate versus differential stress for fine-grained olivine aggregates deformed under dry (a) and wet (b) conditions. For both data sets, a transition from diffusion creep to dislocation creep occurs with an increase in differential stress. Non-linear fits to the data using equation 2 (labeled constitutive equation) and linear fits to the high stress dislocation creep component of the total strain rate are shown in both figures. These data are from the studies of *Mei and Kohlstedt* [2000a, 2000b].

significantly decrease the uncertainty in extrapolation to the lower stresses appropriate for the asthenosphere.

Stress Dependence in Dislocation Creep Regime: Wet Conditions

The stress dependence of dislocation creep in the presence of water is also well constrained by a combination of experiments on single crystals, coarse-grained natural dunites, and fine-grained aggregates synthesized from natural crystals. A data set illustrating the transition from diffusion creep to dislocation creep is shown in Figure 2b for an experiment conducted under water-saturated conditions at a confining pressure of 300 MPa. The non-linear least squares fit of the data using equation (2) gives $n = 3.4 \pm 0.5$. A linear fit of the dislocation creep component gives $n = 3.6 \pm 0.6$. The average value of n determined in this manner from the experiments of *Mei and Kohlstedt* [2000b] (7 experiments) is 3.8 ± 0.6. We re-analyzed similar data from the study of *Karato et al.* [1986] and determined $n = 3.6 \pm 0.3$ for the non-linear fit and $n = 3.5 \pm 0.3$ for the linear fit to the dislocation creep component derived by subtracting the diffusion creep component from the total strain rate. *Karato et al.* [1986] report a value of n in the range 3-3.5. Again, these values are similar to those determined for coarse-grained natural dunites and single crystals. As reported in *Hirth and Kohlstedt* [1996], a least squares linear fit to *Chopra and Paterson's* [1981] wet dunite data at a temperature of 1473 K and stresses less than 300 MPa gives $n = 3.4 \pm 0.3$. Similarly, experiments on olivine single crystals deformed under hydrous conditions yields $n = 3.0 \pm 0.1$ to 3.8 ± 0.2 [*Mackwell et al.*, 1985; *Mackwell et al.*, 1997].

In summary, while lower values for the stress exponent have been reported for olivine aggregates under wet conditions [e.g., *Mei and Kohlstedt,* 2000b; *Karato et al.*, 1986; *Karato and Jung,* in press], an analysis of these same data that takes into account the component of diffusion creep to the total strain rate indicates that the stress exponent for dislocation creep is 3.5 ± 0.3 for both wet and dry conditions.

Activation Energy for Diffusion Creep: Dry Conditions

The activation energy for creep and the pre-exponential constant are determined by normalizing diffusion creep data to constant values of stress and grain size using $n = 1$ and $p = 3$. The temperature dependence of diffusion creep determined from several suites of compression experiments is illustrated in Figure 3a. The contribution of dislocation creep has been removed from these data using the fitting procedure described above. The magnitude of the strain rates observed in these four studies agree well. However, relatively few experiments were specifically designed to determine the activation energy for diffusion creep. Indeed, for two of these studies all experiments were conducted at one temperature. Activation energies determined from two experiments on fine-grained dunite range from 310 ± 40 kJ/mol to 440 ± 80 kJ/mol [*Hirth and Kohlstedt,* 1995a]. These values are similar to the value of 430 ± 70 kJ/mol determined by a linear fit to all of the dunite data shown in Figure 3a. Figure 3a also includes data for fine-grained lherzolite samples [*Kohlstedt and Zimmerman,* 1996], which indicate that pyroxene has little effect on deformation in the diffusion creep regime. This observation suggests that diffusion rates along olivine-olivine, pyroxene-pyroxene, and olivine-pyroxene interfaces are similar. If the lherzolite data are included in the fit, the activation energy is lowered to 320 ± 30 kJ/mol.

The activation energy for diffusion creep under dry conditions has also been determined from compression experiments [*Kohlstedt et al.,* 2000], densification experiments [*Schwenn and Goetze,* 1978; *Cooper and Kohlstedt,* 1984], and low strain tortional creep experiments [*Gribb and Cooper,* 1998, 2000] conducted at room pressure. Activation energies determined from densification experiments of $E^* = 380 \pm 105$ kJ/mol [*Cooper and Kohlstedt,* 1984] and $E^* = 360 \pm 120$ kJ/mol [*Schwenn and Goetze,* 1978] agree well with those determined from the high-pressure compression experiments shown in Figure 3a. By contrast, the activation energy determined from the low-pressure compression experiments ($E^* = 530$ kJ/mol, no uncertainty given [*Kohlstedt et al.,* 2000]) is significantly greater. This value may be influenced by cavitation during creep. Likewise, the activation energy determined from the torsion tests of $E^* = 700 \pm 30$ kJ/mol, *Gribb and Cooper* [1998, 2000] is significantly higher. *Gribb and Cooper* [2000] suggest that the higher values for the activation energy result from grain boundary segregation of incompatible components. Based on the agreement between the high-pressure creep tests and the densification tests we conclude that the activation energy is 375 ± 50 kJ/mol.

Activation Energy for Diffusion Creep: Wet Conditions

Few experiments have been conducted to constrain the activation energy for diffusion creep under wet conditions. *Mei and Kohlstedt* [2000a] conducted the only experiments specifically designed to determine the activation energy. Least squares fits to data from two experiments illustrated in Figure 3b show remarkable agreement with $E^* = 380 \pm 20$ kJ/mol and 410 ± 40 kJ/mol. In Figure 3b, the dislocation creep component for each individual experiment is subtracted from the total strain rate using the fitting

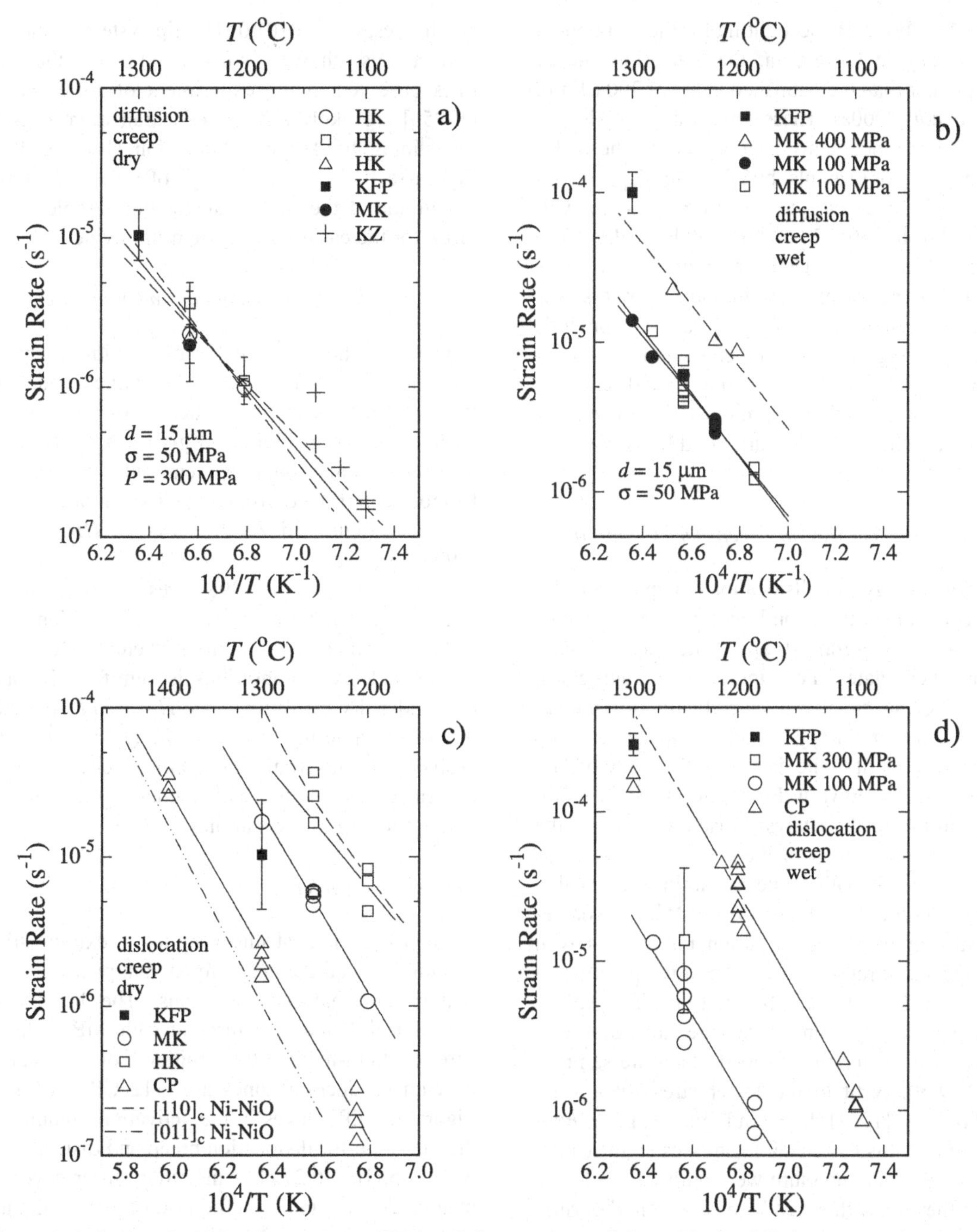

Figure 3. Arrhenius plots of strain rate versus 1/T for olivine aggregates deformed in the diffusion creep regime (a and b) and dislocation creep regime (c and d) under both dry and wet conditions. (a) Dry diffusion creep data: The solid line and dashed lines show the Arrhenius relationships for $E^* = 375 \pm 50$ kJ/mol. (b) Wet diffusion creep data: Least squares fits shown for the two 100 MPa data sets give $E^* = 380 \pm 20$ kJ/mol and 410 ± 40 kJ/mol, respectively. (c) Dry dislocation creep data: Least squares fits to the data of *Chopra and Paterson* [1984] (CP) and *Mei and Kohlstedt* [2000] (MK) give $E^* = 530 \pm 40$ and $E^* = 550 \pm 20$ kJ/mol, respectively. Strain rates for olivine single crystals [*Bai et al.*, 1991] are also shown. (d) Wet dislocation creep data: A least squares fit to the 100 MPa data of *Mei and Kohlstedt* (MK) gives $E^* = 510 \pm 70$ kJ/mol, which is in good agreement with the value of 530 ± 30 kJ/mol determined for the data of Chopra and Paterson (CP). HK = *Hirth and Kohlstedt* [1995a,b]; KFP = *Karato et al.* [1986]; MK = *Mei and Kohlstedt* [2000a,b]; KZ = *Kohlstedt and Zimmerman* [1996].

procedure detailed above. By accounting for the component of dislocation creep in these data, the activation energy appears to be greater than the published value of 300 kJ/mol [*Mei and Kohlstedt,* 2000a]. These values are not significantly influenced by uncertainty in n and E^* in the dislocation creep regime. For example, by allowing E^*_{disl} to vary from 450 to 550 kJ/mol and n_{disl} to vary from 3 to 4, the values of E^*_{diff} vary from 350 ± 30 to 440 ± 40 kJ/mol. A fit to the more limited data set at 400 MPa in Figure 3b gives 300 ± 60 kJ/mol. This experiment is not influenced by dislocation creep since the stress is low (~25 MPa). We conclude that the activation energy for diffusion creep under wet conditions is 375 ± 75 kJ/mol. Thus, for the limited data sets available at this time, the difference in the activation energy for diffusion creep under wet and dry conditions does not appear to be significant.

Activation Energy for Dislocation Creep: Dry Conditions

The activation energy for dislocation creep under dry conditions is constrained by a combination of experiments on single crystals, fine-grained aggregates and coarse-grained natural aggregates. For fine-grained aggregates, only a handful of experiments were specifically designed to measure the activation energy in the dislocation creep regime. A least squares fit to the data from the study of *Mei and Kohlstedt* [2000b] shown in Figure 3c gives E^* = 530 ± 40 kJ/mol. Similar to the analyses discussed above, the component of diffusion creep to the total strain rate was removed prior to the fit. This value is within error of that determined by *Chopra and Paterson* [1984] for coarse-grained natural dunites. Also, the strain rate agrees well with the average strain rate determined at a temperature of 1300°C by *Karato et al.* [1986]. In addition, these values for activation energy are within error of the value of 450 ± 100 kJ/mol determined for fine-grained lherzolite samples [*Zimmerman,* 1999]. A fit to the lower stress data from *Chopra and Paterson* [1984] shown in Figure 3c gives E^* = 550 ± 20 kJ/mol. In this case, data at stresses significantly greater than the confining pressure were omitted due to the possibility that they are influenced by a change in deformation mechanism. For comparison, the flow laws for olivine single crystals deformed on their easiest (i.e., (010)[100]) and hardest (i.e., (010)[001]) slip systems under dry conditions are also shown. As discussed below, the higher strain rates observed in the experiment of *Hirth and Kohlstedt* [1995b] are likely a result of a transition to grain boundary sliding limited by dislocation slip. As illustrated in Figure 3c, the activation energy of 520 ± 110 kJ/mol and the magnitude of the strain rate for this sample are similar to those for the easiest slip system in olivine.

Activation Energy for Dislocation Creep: Wet Conditions

The activation energy for dislocation creep under wet conditions is similar to that determined under dry conditions. A least squares fit to the data of *Mei and Kohlstedt* [2000b] shown in Figure 3d gives E^* = 510 ± 70 kJ/mol. This value is in good agreement with the value of 530 ± 30 kJ/mol[1] determined from data for Anita Bay dunite also plotted in Figure 3d. As discussed by *Hirth and Kohlstedt* [1996], the lower value reported for these samples by *Chopra and Paterson* [1981] results from including high stress data in the fit and an underestimation of the stress exponent. The activation energy determined for the Aheim dunite under wet conditions is significantly smaller than that under dry conditions [see *Hirth and Kohlstedt,* 1996]; the lower value for Aheim dunite (E^* = 310 ± 90 kJ/mol) probably results from a concomitant decrease in the water content of the olivine with increasing temperature due to partial melting of the samples.

Pressure Dependence

One of the largest uncertainties in extrapolating experimental data is the large difference in pressure between laboratory and natural conditions. The highest resolution mechanical data are acquired at ~300 MPa, while the pressure in high-temperature regions of the upper mantle beneath subduction zones are 2-12 GPa. As illustrated in Figure 4a, differences in the activation volume reported in the literature for dislocation creep and dislocation recovery of $V^* = 5\times10^{-6}$ to 27×10^{-6} m^3/mol result in several orders of magnitude uncertainty in the viscosity at the greatest depths in the upper mantle. Over the full depth range in the mantle wedge, the effect of pressure on viscosity can be as large or even larger than the temperature dependence on viscosity. For example, with $V^* = 15\times10^{-6}$ m^3/mol, the change in viscosity from 100 to 400 km resulting from the increase in pressure is more than 4 orders of magnitude, similar to the effect of a 400K change in temperature. As discussed below, the extrapolation of laboratory data for the activation volume for creep under wet conditions is further complicat-

[1] In *Hirth and Kohlstedt* [1996] we reported a value of Q = 515 ± 25 kJ/mol based on an analysis of a different subset of the same data from *Chopra and Paterson* [1981]. The value reported here is lower because we removed several data points determined at stresses ranging from 350 to 400 MPa for consistency with the other analyses reported in this paper.

ed by competing effects of the role of water and pressure, specifically the effect of pressure on the concentration of water in minerals and, hence, on viscosity.

Pressure Dependence Under dry Conditions

The activation volume for dislocation creep under dry conditions has been directly determined by deformation experiments and inferred from dislocation recovery experiments. The resulting values for V^* are listed in Table 2 and plotted in Figure 4b as a function of the maximum pressure tested in each study. The values for V^* determined from deformation experiments are somewhat higher than those determined in recovery studies. In both cases, the data indicate that V^* decreases with increasing pressure.

Similar to the experimental observations in Figure 4b, theoretical treatments predict that the activation volume decreases with increasing pressure. Two general types of relationships have been explored. The first, based on the elastic strain required for creation and migration of point defects, is formulated in terms of the pressure and temperature dependence of elastic moduli. The analysis of *Sammis et al.* [1981] indicates that calculations based on dilatational strain provide a better fit to experimental data than those determined using shear strain. In this case [see *Sammis et al.* [1981] after *Zener* [1942] and *Keyes* [1963]],

$$V^* = E^*[\partial \ln K_T/\partial P)_T - 1/K_T]\times [1 - (\partial \ln K_T/\partial \ln T)_P - \alpha_T T]^{-1} \quad (3)$$

where K_T is the isothermal bulk modulus and α_T is the coefficient of thermal expansion. This relationship, plotted in Figure 4b using elastic constants and derivatives for single crystal olivine [*Kumazawa and Anderson,* 1969], provides a reasonable representation of the experimental deformation data.

Founded on theoretical analyses relating melting temperature and vacancy concentration, and the observation that activation energy increases with increasing melting temperature (T_m), V^* for diffusion has been estimated using the pressure dependence of the melting temperature [e.g., *Weertman,* 1970]. In this case

$$V^* = E^*(\partial T_m/\partial P)/T_m. \quad (4)$$

A modification of this description for V^* has been suggested by Sammis et al. [1981], however, this change only modestly influences V^* for upper mantle conditions. The change in V^* with increasing pressure (depth) determined using this homologous temperature approach is illustrated

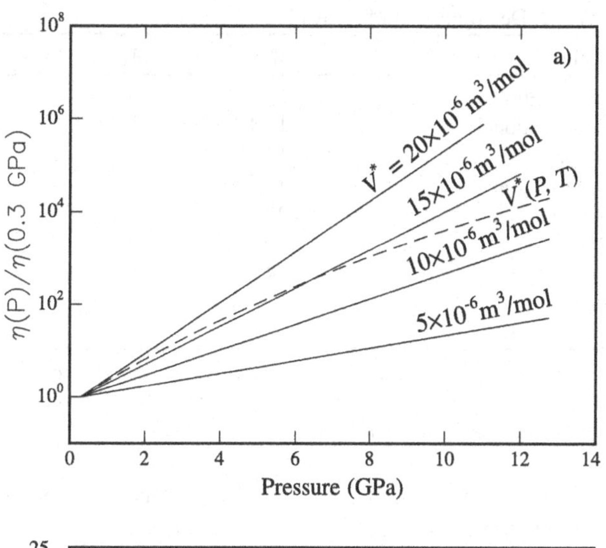

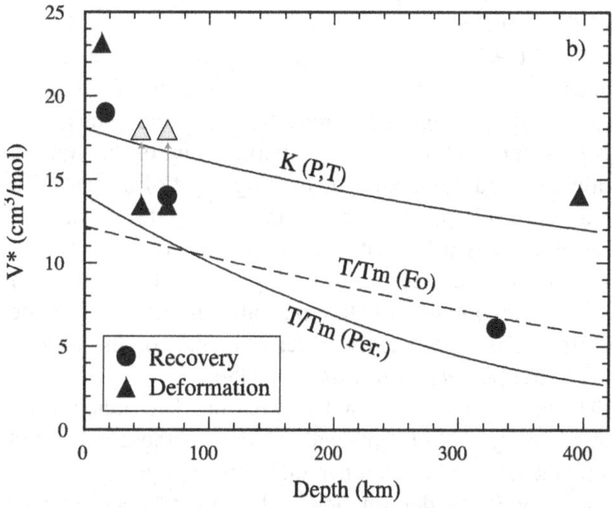

Figure 4. Plots showing the pressure dependence of deformation. (a) Plot of normalized viscosity versus pressure for a temperature of 1350°C and E^* = 530 kJ/mol. The viscosities calculated for constant values of V^* are normalized by the value at 0.3 GPa to emphasize the large effect of pressure on viscosity. The normalized viscosity for a V^* that decreases with increasing pressure ($V^*(P,T)$, calculated using equation 3) is also shown. (b) Comparison of experimental determinations of V^* to theoretical relationships calculated using the pressure and temperature dependence of the bulk modulus ($K(P,T)$ from equation 3) and the effect of pressure on the melting temperature of forsterite (T/Tm (Fo), calculated using equation 4) and the liquidus of peridotite (T/Tm (Per.), calculated using equation 4). The data sources for V^* determined from deformation experiments and recovery experiments are listed in Table 2. As discussed by *Kohlstedt et al.* [1980] an additional uncertainty in V^* arises due to the pressure dependence of thermocouple EMF. The gray symbols and arrows on the deformation data points at 45 and 60 km reflect this uncertainty.

Table 2. Determination of Activation Volumes

Technique	V* (10^{-6} m^3/mol)	P range (GPa)	Reference
Deformation	23	0.2-0.4	*Kohlstedt and Wang* [2001][b]
Deformation	13.4 (18)[a]	0.5-1.5	*Ross et al.* [1979]
Deformation	14 (18)[a]	0.3-2	*Karato and Jung* [2002]
Deformation	14	0.3-15	*Karato and Rubie* [1997]
Deformation	27	0.6-2.0	*Borch and Green* [1989]
Recovery	19[c]	10^{-4}-0.5	*Kohlstedt et al.* [1980]
Recovery	14	10^{-4}-2.0	*Karato and Ogawa* [1982]
Recovery	6	10^{-4}-10	*Karato et al.* [1993]
Diffusion (Si)	-2	5-10	*Bejina et al.* [1997]

[a]Higher value is corrected for pressure effect on thermocouple emf.
[b]Also, *Wang et al.*, Activation volume for dislocation creep in olivine (manuscript in preparation).
[c]Corrected value from *Karato* [1981].

in Figure 4b by using the pressure dependence on the melting temperature of forsterite [*Ohtaini and Kumazawa,* 1981] and the liquidus of lherzolite [*Zhang and Herzberg,* 1994]. We used the liquidus temperature of lherzolite to approximate the solidus of Fo_{88} olivine at high pressure. We emphasize that the appropriate scaling for the homologous temperature approach is to use the melting point of the mineral grains, instead of the eutectic melting point of the rock. To underscore this point, note that the creep rate of partially molten dunite with a melt fraction of <1% is approximately the same as the creep rate of single crystal olivine even though the solidus of the partially molten aggregate (~1100°C) is significantly lower than that of the olivine (~1800°C) [*Hirth and Kohlstedt,* 1995b].

The lower values of V^* at high pressure determined using the recovery experiments are similar to those predicted using the homologous temperature approach (Figure 4b). In general both the deformation and recovery data are reasonably bracketed using the theoretical treatments. While the agreement between the homologous temperature approach and the recovery data is encouraging, we note that microstructural observations suggest that climb is not the limiting process for creep on (010)[001], the hardest slip system in olivine [*Bai and Kohlstedt,* 1992]. Based on the von Mises criterion, the hardest slip system is expected to limit deformation during dislocation creep [*Paterson,* 1969].

Surprisingly, the activation volume determined for Si self-diffusion in olivine, $V^* = 1.9\times10^{6} \pm 2.4\times10^{-6}$ m^3/mol [*Bejina et al.,* 1997], is considerably lower than that observed for recovery. The essentially negligible effect of pressure is consistent with values of D_{Si} determined at ambient pressure by *Houlier et al.,* [1990] and *Dohman et al.* [2002]. Since silicon is the slowest diffusing species in olivine, it is expected to limit the rate of dislocation climb. However, the rate of climb determined from recovery experiments is several orders of magnitude greater than that expected based on the experimental data for D_{Si}. Climb requires diffusion on distances comparable to the dislocation spacing (i.e., between 50 to 500 nm for experimental samples). Given $D_{Si} = 5\times10^{-21}$ m^2/s at 1300°C [*Houlier et al.,* 1990], diffusion distances for Si over times of 0.01-1 hour are on the order of 0.4-4 nm. By contrast, experimental observations from both deformation experiments [e.g., *Mackwell et al.,* 1985] and recovery experiments [e.g., *Goetze and Kohlstedt,* 1973] demonstrate considerable dislocation climb over the same time periods. Interestingly, the rates of climb do correspond well with the rate of oxygen diffusivity [e.g., *Ryerson et al.,* 1989; *Dohman,* 2002], which are approximately two orders of magnitude greater than D_{Si} at experimental conditions. However, the activation energy for oxygen diffusion is considerably smaller than that observed for creep.

The large difference between the activation volume for D_{Si} and that for recovery, together with the discrepancy between D_{Si} and the rate of recovery, suggests that dislocation climb in olivine is not limited by matrix diffusion of silicon. As suggested above, one explanation for this apparent discrepancy is that climb is limited by oxygen matrix diffusion. In this case, the effective diffusion rate for silicon must be greater than that for oxygen matrix diffusion, indicating that silicon may diffuse primarily along dislocation cores (i.e., pipe diffusion). Alternatively, climb could be limited by silicon pipe diffusion. Interestingly, if deformation is limited by pipe diffusion, the Orowan equation predicts that the $\dot{\varepsilon} \propto v\rho^2$ [e.g., *Frost and Ashby,* 1982], where v, the dislocation velocity (i.e., climb velocity), is proportional to σ and ρ is the dislocation density. In general, $\rho \propto \sigma^q$, where q is often assumed to be 2 based on simple dislocation geometries. However, for olivine $\rho \propto \sigma^{1.37}$ [*Bai and Kohlstedt*, 1992], thus giving , $\dot{\varepsilon} \propto \rho^{3.7}$ which is within error of that determined from experimental studies for olivine (see above).

In practical determinations of the activation volume it is important to account for the potential change in V^* with increasing pressure. Specifically, because

$$V^*(P) = -(\partial \ln\dot{\varepsilon}/\partial P)RT + P\partial V/\partial P, \quad (5)$$

then $V^*(P) = -(\partial \ln\dot{\varepsilon}/\partial P)RT$ only if $\partial V/\partial P = 0$. However, since V^* apparently decreases with increasing pressure, determining $(\partial \ln\dot{\varepsilon}/\partial P)RT$ at high pressure underestimates V^*, unless the data are compared to similar data at low pressure. This problem can be seen in Figure 4a, where the pressure dependence of viscosity predicted by equation (3) is plotted versus pressure. Note that the slope of the relation at high pressure is significantly smaller than that predicted for the constant values of V^*. In practice, V^* in equation 1 is a "chord V^*", representing $RT[(\ln\dot{\varepsilon})_P - (\ln\dot{\varepsilon})_{P=0}]/P$. The values for V^* shown in Table 1 for which the pressure range extends to 1 atm (i.e., 1×10^{-4} GPa) provide a good approximation for the chord V^*.

There are limited constraints on the activation volume for diffusion creep under dry conditions. The comparison of diffusion creep data on samples deformed at 1 atm and 300 MPa gives a value in the range of 2×10^{-6} to 10×10^{-6} m^3/mol [*Kohlstedt et al.*, 2000]. The range given here reflects uncertainties associated with the correction for cavitation during creep at ambient pressures.

Combined Water and Pressure Dependence Under wet Conditions

Under water-saturated conditions, an increase in pressure influences creep rates in two ways. First, there is a direct effect: creep rates decrease due to an activation volume effect. Second, there is an indirect effect: creep rate increases with increasing pressure because (a) the solubility of water in nominally anhydrous minerals increases with increasing water fugacity [*Kohlstedt et al.*, 1996], (b) water fugacity increases with increasing pressure for a fixed water activity, and (c) creep rates increase with increasing water content [e.g., *Mei and Kohlstedt,* 2000a,b]. Following *Mei and Kohlstedt* [2000a,b] and *Karato and Jung* [in press], we show a least squares fit to experimental data for samples deformed under water-saturated conditions at various pressures (Figure 5a). The fit to the global data set (excluding the results of *Borch and Green,* 1989[2]) gives values of the water fugacity exponent of $r = 1.2 \pm 0.4$ and $V^* = 22 \pm 11$.

[2]The results of *Borch and Green* [1989] are not included in the fit shown in Figure 5a because the water content in their samples was not controlled. As discussed in *Hirth and Kohlstedt* [1996], although *Borch and Green* [1989] reported that their samples were dry, subsequent FTIR analyses [*Young et al.*, 1993] showed that they may have contained enough water to saturate olivine at 1.0 GPa.

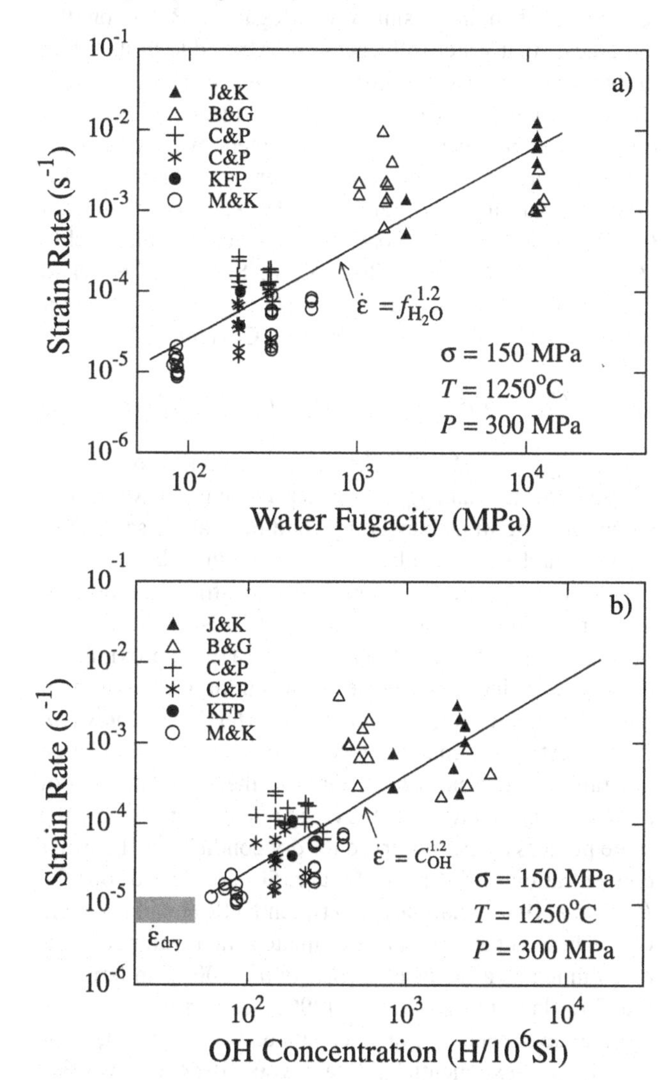

Figure 5. Plots of strain rate versus water fugacity (a) and water content (b) for olivine aggregates deformed in the dislocation creep regime. The data are normalized to a pressure of 300 MPa using the value of V^* determined from a non-linear fit to the global data set after the data were normalized to a constant stress and temperature using $n = 3.5$ and $E^* = 520$ kJ/mol. The strain rate determined under dry conditions at the same temperature and stress is shown by the shaded box in 5b. J&K = *Jung and Karato* [2001]; B&G = *Borch and Green* [1989]; C&P = *Chopra and Paterson* [1984]; KFP = *Karato et al.* [1986]; M&K = *Mei and Kohlstedt.*

As suggested by Karato and Jung [in press], the higher-pressure data can improve the resolution of the competing effects of water fugacity and activation volume.

Following *Hirth and Kohlstedt* [1996] and *Karato* [this volume], for geodynamic applications it is more practical to consider how strain rate changes with increasing water content rather than increasing water fugacity. Based on the hypothesis that water influences creep rate due to its effect on point defect concentrations [e.g., *Mackwell et al.,* 1985; *Kohlstedt and Mackwell,* 1998], the influence of water on creep rate should depend on water content even when $a_{H_2O} < 1$. In Figure 5b we plot strain rate versus water content for the same data used in Figure 5a. The water concentration (C_{OH}) is calculated for Fo_{90} olivine using the relationship [*Kohlstedt et al.,* 1996; Zhao, Y.-H., S.B. Ginsberg, and D.L. Kohlstedt, Solubility of Hydrogen in Olivine: Dependence on Temperature and Iron Content, in prep.]

$$C_{OH} = A_{H2O} \exp[-(E_{H2O} + PV_{H2O})/RT] f_{H2O} \quad (6)$$

where $A_{H2O} = 26$ H/10^6 Si/MPa, $E_{H2O} = 40$ kJ/mol, $V_{H2O} = 10\times10^{-6}$ m^3/mol and f_{H2O} is water fugacity in MPa. The strain rate determined under dry conditions at the same temperature and stress is also shown in Figure 5b. The least squares fit in Figure 5b intersects the dry strain rate value at a water content of approximately 50 H/10^{6}Si. This observation, and the fact that strain rate increases approximately linearly with increasing water concentration, is consistent with the point defect hypothesis [*Mei and Kohlstedt,* 2000b]. At water contents below ~50 H/10^{6}Si, the charge neutrality condition, which controls the concentration of point defects, changes and the creep rate is controlled by the same process as that observed for dry conditions. The water concentrations shown in Figure 5b are based on the *Paterson* [1982] relationship between FTIR absorbance and C_{OH}. These values may underestimate actual water contents by as much as a factor of 3 [*Bell et al.,* 2002]. In terms of the flow law, the *Bell et al.* [2002] calibration does not change the dependence of creep rate on water content, however, the pre-exponential constant would decrease by a factor of three.

Experimental studies also demonstrate that creep rates in the diffusion creep regime increase with increasing water content [e.g., *Mei and Kohlstedt,* 2001a]. The effect of water fugacity or water content on deformation in the diffusion creep regime has been determined by conducting experiments under water-saturated conditions as a function of pressure [*Mei and Kohlstedt,* 2000a]. The results of these experiments are consistent with the power law formulation for the effect of strain rate on water fugacity illustrated by equation 1. The value of the water fugacity exponent (r) in the diffusion creep regime is 1.0±0.3. As discussed by Mei and Kohlstedt (2001a), the uncertainty in r is again related to the concomitant effect of the PV^* term in equation 1 on strain rate. *Mei and Kohlstedt* [2000a] concluded that the effect of water on creep rate in the diffusion creep regime arises due to the influence of water fugacity on the concentration of point defects in the grain boundaries. This conclusion is consistent with the observation that the values for n, p and E^* in the diffusion creep regime are similar under wet and dry conditions.

Grain Boundary Sliding

Dislocation creep is generally assumed to be independent of grain size. However, at conditions near the transition from diffusion creep to dislocation creep, strain rates for dry olivine aggregates in the dislocation creep regime increase with decreasing grain size. This grain size effect is illustrated in Figure 6a, together with flow laws for olivine single crystals. As shown in Figure 6b, there is no discernable grain size dependence for dislocation creep under hydrous conditions.

Based on the observation that creep rates for fine-grained olivine aggregates are similar to the strain rates of olivine single crystals deformed on the easiest slip system, *Hirth and Kohlstedt* [1995b] concluded that these samples deformed by grain boundary sliding (GBS) accommodated by a dislocation creep process. This hypothesis is further supported by the relatively large influence of melt on creep rate, the lack of grain flattening observed in the largest grains, and the observation that the GBS process is observed near the transition between dislocation creep and diffusion creep. Grain boundary sliding is implicit at these conditions because of the necessity for GBS in the diffusion creep regime [*Raj and Ashby,* 1971]. Constitutive laws for dislocation accommodated GBS are of the form $\dot{\varepsilon} \propto \sigma^n / d^p$, where $n_{gbs} \approx 2$ to 3 and $p_{gbs} \approx 2$ to 1 [e.g., *Langdon,* 1994]. To explore the grain size sensitivity in the GBS regime, we plot strain rate versus grain size in Figure 7a. The contribution of diffusion creep to the total strain rate of these samples has been removed using the fitting procedures described in the section on stress dependence. The stress exponent for these samples is $n = 3.5 \pm 0.3$ (Figure 2c). For grain sizes between 10 and 60 μm, the data indicate a grain size exponent between 1 and 2. Recent experiments in which creep rate is observed to decrease during grain growth of individual samples indicate a grain size exponent in the GBS regime of $p = 1.8 \pm 0.2$ [*Kohlstedt and Wang,* 2001]. No difference in creep rate is observed between natural dunite samples with a grain sizes of 100 μm and those with a grain size of 900 μm.

The identification of GBS for olivine is complicated since $n_{gbs} \approx n_{disl}$. In this case, a transition from one mechanism to the other will occur with a change in grain size, but not with a change in stress. This scenario is illustrated schematically

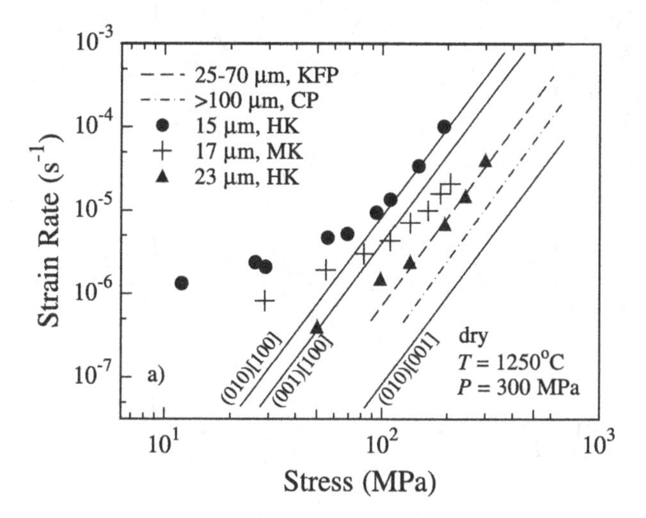

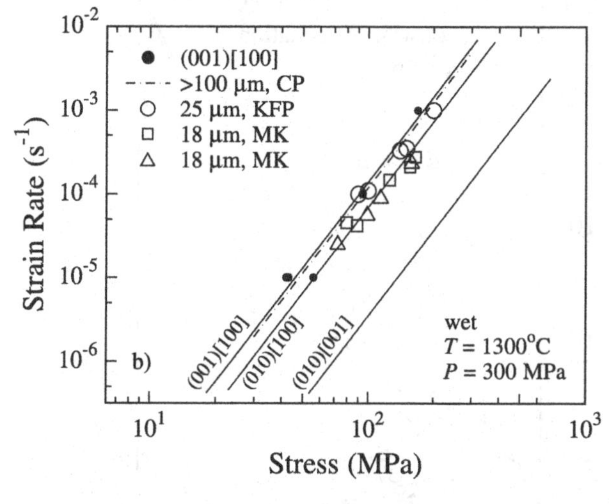

Figure 6. Plots of strain rate versus differential stress for olivine aggregates with different grain sizes deformed under dry and wet conditions. (a) Under dry conditions, at a constant stress, the strain rate in the dislocation creep regime increases with decreasing grain size. This observation suggests that creep of the fine-grained olivine aggregates occurs by dislocation-accommodated grain boundary sliding. For the finest-grained samples, a transition to diffusion creep is illustrated with decreasing differential stress. (b) Under wet conditions, there is no apparent effect of grain size on strain rate. Creep rates for olivine single crystals deformed in orientations to promote slip on different slip systems under dry [*Bai et al.*, 1991] and wet [*Mackwell et al.*, 1985] conditions are also shown. KFP = *Karato et al.* [1986]; C&P = *Chopra and Paterson* [1984]; H&K = *Hirth and Kohlstedt*, 1995; M&K = *Mei and Kohlstedt.*

in plots of strain rate versus differential stress for different grain sizes in Figures 7b-7d using the constitutive law

$$\dot{\varepsilon} = \dot{\varepsilon}_{disl} + \dot{\varepsilon}_{diff} + [1/\dot{\varepsilon}_{gbs} + 1/\dot{\varepsilon}_{easy}]^{-1} \qquad (7)$$

with $n_{gbs} = 3.5$ and $p_{gbs} = 2$. A key feature of this constitutive law, which was motivated by similar data for deformation of fine-grained ice [*Goldsby and Kohlstedt,* 2001; *Durham et al.,* 2001], is that GBS and slip on the easiest slip system ($\dot{\varepsilon}_{easy}$) are both required to accommodate deformation. Based on these observations, we conclude that the importance of GBS under geologic conditions can be explored by extrapolating the strain rates for diffusion creep, dislocation creep of coarse-grained aggregates and dislocation creep of olivine on its easiest slip system. As summarized in Table 1, we assume that E^*_{gbs} is the same as that for dislocation creep of olivine on its easiest slip system.

The lack of grain size dependence for dislocation creep processes under wet conditions and the observation that creep rates even in the coarse-grained natural samples are similar to the easiest slip systems suggests that von Mises criterion is satisfied without slip on the hardest slip system. One explanation for these observations is that, to meet compatibility requirements, a significant amount of strain occurs by dislocation climb under hydrous conditions. Climb accommodates as much as 10% strain at high temperatures under dry condition [*Durham and Goetze,* 1977]. Microstructural observations suggest that even more strain is accommodated by climb under hydrous conditions [e.g., *Mackwell et al.,* 1985]. Hence, under hydrous conditions, dislocation climb apparently accommodates the strain that is accommodated by either GBS or slip on the hardest slip system under dry conditions.

Influence of melt content

The role of melt on the creep properties of mantle aggregates has been studied extensively during the last 15 years. We will not review the results of these studies in detail here; readers are referred to recent review articles on the subject [*Kohlstedt et al.,* 2000; *Xu et al.,* 2002]. For the purposes of this paper, we summarize the experimental data for the influence of melt fraction (ϕ) on strain rate in Figure 8. For both wet and dry conditions, the data at $\phi \lesssim 0.12$ are well described by an exponential relationship $\dot{\varepsilon} \propto \exp(\alpha\phi)$, where α is a constant between 25-30 for the diffusion creep regime and between 30-45 for the dislocation creep regime. For a given stress, this relationship can also be written $\eta \propto \exp(-\alpha\phi)$, where η is effective viscosity. Analyses of data for partially molten lherzolites with pyroxene contents as high as 35% indicate a more modest effect of melt on strain

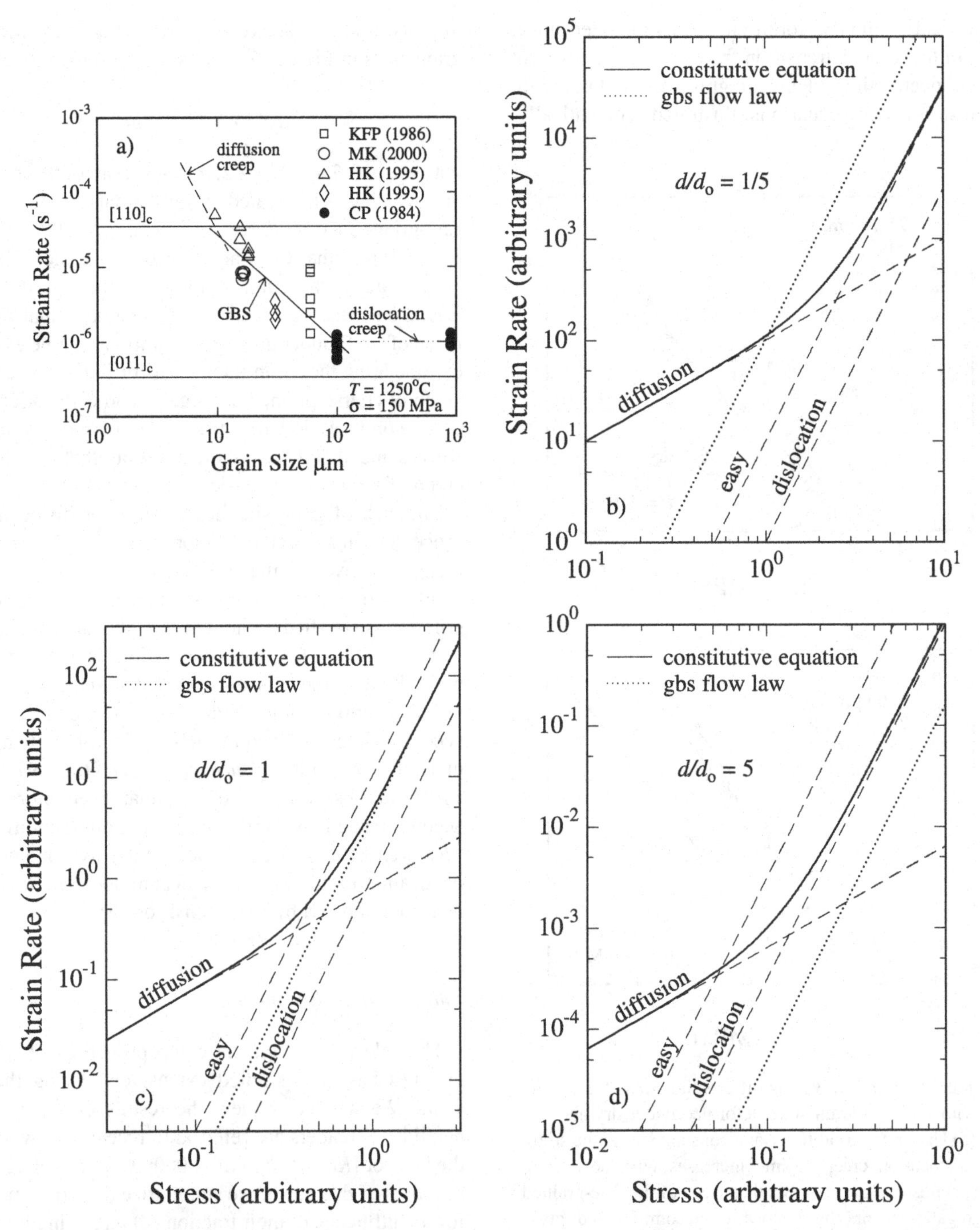

Figure 7. Plots of strain rate versus grain size. (a) Plot showing the effect of grain size on strain rate for fine-grained olivine aggregates deformed in the dislocation creep regime. The data are normalized to a constant stress using $n = 3.5$ after the component of diffusion creep to the total strain rate is subtracted (e.g., see Figure 2). A linear fit to the data over the grain size interval of 10-60 μm (labeled GBS) gives a grain size exponent of $p \approx 2$, consistent with models for dislocation-accommodated GBS. For comparison, flow laws for diffusion creep, dislocation creep of coarse-grained olivine and olivine single crystals are also shown. (b-d) Schematic plots of strain rate versus grain size calculated using equation 7 illustrating how transitions between the deformation mechanisms occur with changes in grain size. The units of stress and strain rate in Figures 7b-7d are arbitrary.

rate [*Zimmerman,* 1999]. The exponential relationship still provides a good fit to the data, but the constant α is ~20 in the diffusion creep regime and ~25 in the dislocation creep regime. One explanation for this difference between dunite and lherzolite samples is that the presence of pyroxene reduces the amount of grain boundary area wetted by melt. Thus, because the melt does not support shear stress, the effective stress for partially molten lherzolite is smaller than that for partially molten dunite at the same melt fraction. Experimental observations indicate that basaltic melt does not wet pyroxene-pyroxene grain boundaries as effectively as olivine-olivine grain boundaries [*Toramaru and Fujii,* 1986; *Daines and Kohlstedt,* 1993].

The exponential relationship between strain rate and melt fraction is empirical. For the diffusion creep regime, the effect of melt on creep rate is significantly greater than that predicted by theoretical treatments in which melt topology is assumed to be controlled by isotropic interfacial energies [*Cooper et al.,* 1989]. The enhanced strain rates are explained by deviations from ideal melt topologies that arise due to anisotropic interfacial energies [e.g., *Hirth and Kohlstedt,* 1995a; *Mei et al.,* 2002] and changes in grain coordination with increasing melt fraction [*Renner et al.,* 2002]. The reader is referred to these studies for more detailed analyses.

The dependence of melt fraction on strain rate in the dislocation creep regime is also much greater than predicted by theoretical models. The effect of melt fraction on strain rate for partially molten dunite suggests that $\dot{\varepsilon} \propto (\sigma_{\text{eff}})^n$, where $\sigma_{\text{eff}} \propto \sigma/(1\text{-}x)$ and x is the fraction of grain boundary area replaced by melt [*Hirth and Kohlstedt,* 1995b; *Mei et al.,* 2002]. This relationship emphasizes the importance of grain boundary stresses during creep and suggests the potential importance of GBS in the dislocation creep regime. An intriguing possibility is that the presence of melt may promote dislocation accommodated GBS, resulting in a relaxation of the von Mises criterion. If so, a significant decrease in viscosity could arise due to the presence of melt fractions as small as ~1%. A key factor for developing these ideas is the determination of the influence of melt on strain rate for coarse-grained olivine aggregates. This sort of study is somewhat tricky, due to the problem of maintaining "textural equilibrium" at the relatively high strain rates necessary in laboratory experiments.

In conclusion, our understanding of the effects of melt on the rheology of partially molten peridotite is continuing to improve. For example, an important problem that will receive continued attention over the next few years is understanding the relationships between grain scale melt redistribution and viscosity [e.g., *Renner et al.,* 2000; *Holtzman et al.,* 2002]. At present, while more sophisticated analyses have been presented by us and others, for practical reasons we conclude that the exponential relationships between melt fraction and strain rate/viscosity shown in Figure 8 for $\phi \lesssim 0.12$ are appropriate first-order approximations for application in geodynamic models.

EXTRAPOLATION OF EXPERIMENTAL DATA TO THE OCEANIC MANTLE AND THE MANTLE WEDGE

In this section we examine predictions for the viscosity of the oceanic mantle and the mantle wedge based on extrapolation of laboratory data. In addition, we compare these experimental constraints with independent geophysical estimates for upper mantle viscosity. Because viscosity is likely to be stress-dependent in at least some parts of the upper mantle (i.e., dislocation creep is the dominant deformation mechanism), we first illustrate the importance of assumptions about stress and strain rate conditions on the effects of temperature and pressure on viscosity, and then compare these results to approximations sometimes used in numerical studies of mantle convection.

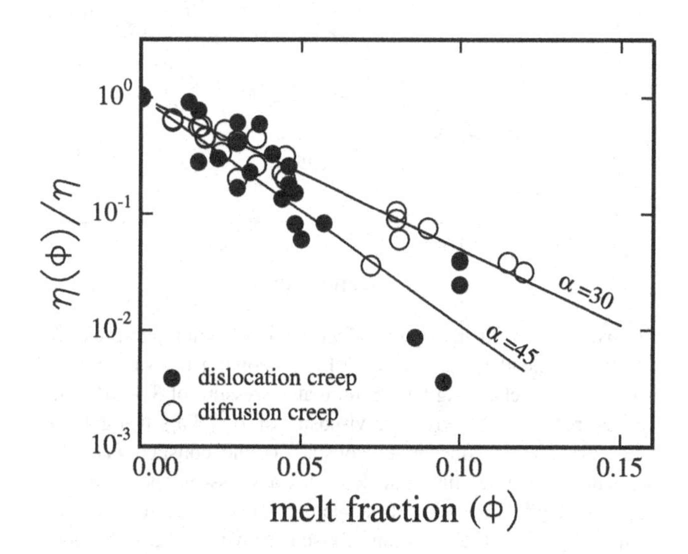

Figure 8. Plot of normalized viscosity versus melt fraction for olivine aggregates deformed in the diffusion creep and dislocation creep regimes. For each data point, the viscosity is normalized by the viscosity of a melt-free aggregate deformed at the same conditions. For both deformation regimes, the exponential relationship $\eta \propto \exp(-\alpha\phi)$ provides a good first-order approximation for the decrease in viscosity with increasing melt fraction with values of α ranging from 30-45. Data are from the studies of *Cooper and Kohlstedt* [1984], *Beeman and Kohlstedt* [1993], *Hirth and Kohlstedt* [1995a,b], *Kohlstedt and Zimmerman* [1996], *Bai et al.* [1997], *Gribb and Cooper* [2000] and *Mei et al.* [2002].

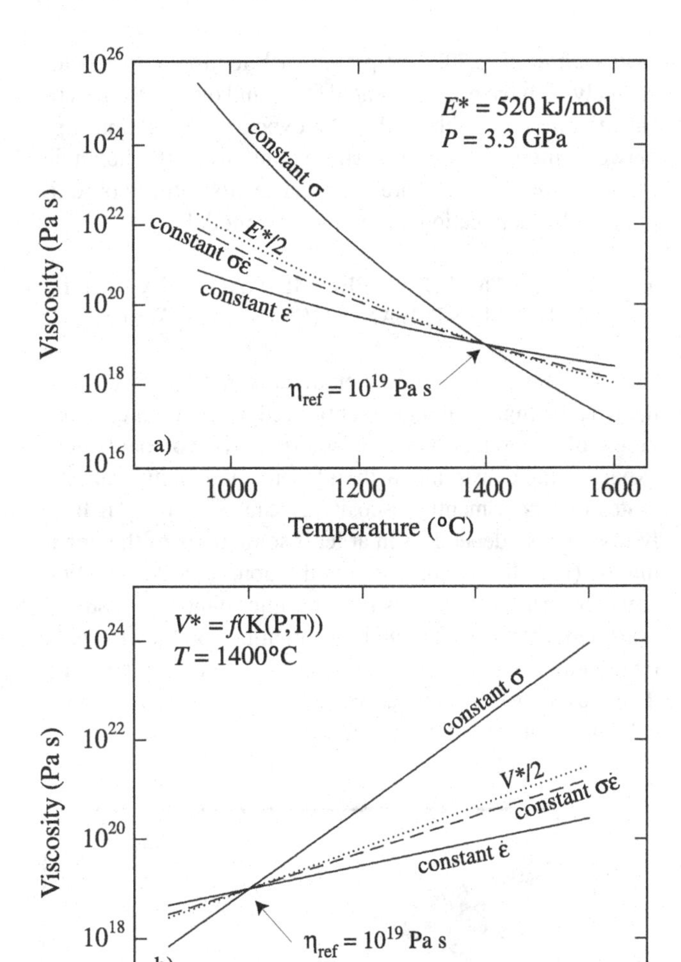

Figure 9. Plots illustrating the effects of temperature and pressure on stress-dependent viscosity. (a) Plots showing the variation in viscosity with changing temperature at a pressure of 3.3 GPa calculated relative to a reference viscosity of 10^{19} Pa s using E^* = 520 kJ/mol for a constant $\dot{\varepsilon}$, constant σ and constant $\dot{\varepsilon}\sigma$. The influence of temperature calculated for a stress-independent viscosity with E^* = 260 kJ/mol (labeled $E^*/2$) is shown for comparison. (b) Plots of the variation in viscosity with increasing depth (pressure) calculated using equation 3 for V^* at a constant temperature of 1400°C.

Temperature and Pressure Dependence of Non-linear Viscosity

The relatively large values of E^* and V^* for dislocation creep indicate that the viscosity of the upper mantle is strongly dependent on temperature and pressure. For example, at upper mantle temperatures under constant stress conditions, a change in temperature of only 30 K results in a factor of two change in viscosity for E^* = 520 kJ/mol. Similarly, viscosity decreases an order of magnitude with a 100 K increase in temperature. However, when applying dislocation creep flow laws to mantle conditions, it is important to consider whether deformation occurs closer to constant stress or to constant strain rate conditions [e.g., *Christensen,* 1989]. As illustrated in Figure 9, the pressure and temperature dependences of viscosity are much greater at a constant stress than at a constant strain rate. Theoretical treatments suggest that the viscosity of convecting systems may evolve such that deformation occurs at a constant rate of viscous dissipation (constant $\dot{\varepsilon}\sigma$) [e.g., *Christensen,* 1989], which results in temperature and pressure effects intermediate between the end-member conditions (Figure 9). The temperature and pressure effects on viscosity at constant $\dot{\varepsilon}\sigma$ are also well fit by assuming that viscosity is independent of stress (i.e., a linear relationship between stress and strain rate) and decreasing both E^* and V^* by a factor of ~2 [*Christensen,* 1989]. This procedure has been employed in modeling studies [e.g., *Braun et al.,* 2000; *Toomey et al.,* 2002] due to numerical instabilities that arise for fully non-linear viscosity formulations in problems with large variations in temperature and pressure.

Viscosity Profiles for the Oceanic Mantle and Mantle Wedge

In Figure 10, we plot the viscosity of the upper mantle as a function of depth predicted by extrapolation of experimental flow laws. Viscosity profiles for adiabatic oceanic mantle are shown in Figure 10a for dislocation creep of olivine with a water content of 1000 H/10^6Si, stresses of 0.1 and 1.0 MPa and a potential temperature of 1350°C. To calculate the viscosity profiles for a constant water content, we decreased the activation parameters to account for the temperature and pressure dependence on water content in olivine (i.e., $E^*_{eff} = E^* - E_{H2O}$ and $V^*_{eff} = V^* - V_{H2O}$), where E_{H2O} and V_{H2O} are the values from Equation (6) [*Mei and Kohlstedt,* 2000b, *Karato and Jung,* in press]. We use a water content of 1000 H/10^6Si based on the study of *Hirth and Kohlstedt* [1996]; flow law parameters calculated for a constant water content are listed in Table 1.

As discussed above, uncertainty in the value of V^* causes a large uncertainty in the absolute value of mantle viscosity. The profiles in Figure 10a are calculated using V^* determined from Equation 3 to account for the increase in V^* with increasing pressure; the value for V^* calculated from Equation (3) is within error of that determined from the fit in Figure 5. For practical reasons, it is easier to run numerical experiments using a constant value of V^*. As shown in Figure 10a, a value of V^* = 11×10^{-6} m^3/mol

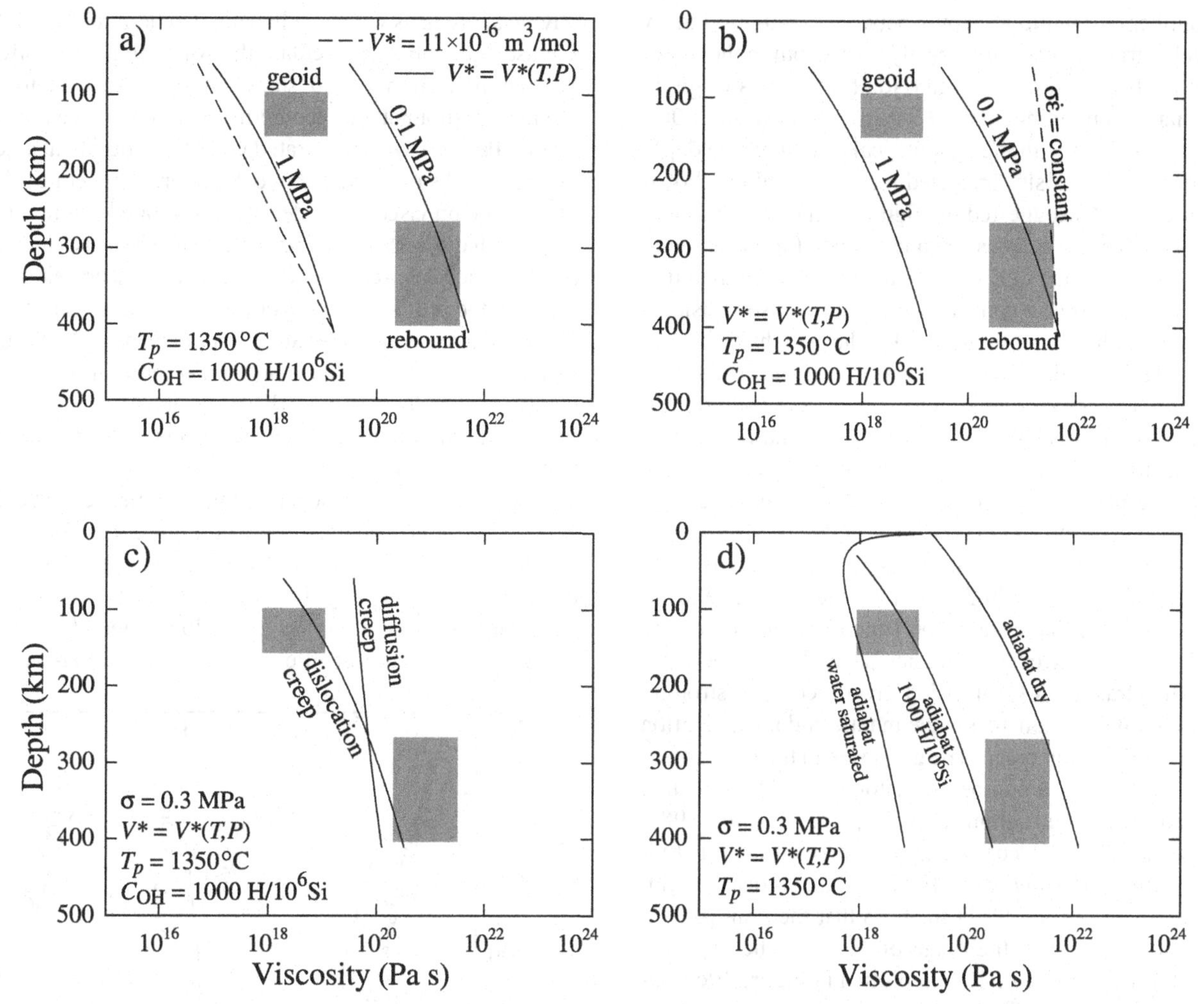

Figure 10. Plots of viscosity versus depth for the upper mantle. (a) Viscosity profiles for adiabatic mantle with a potential temperature of 1350°C. The solid lines show viscosities predicted for dislocation creep of olivine at stresses of 0.1 and 1 MPa with a water content of 1000 H/10^{6}Si and V^* from equation 3. The dashed line shows the viscosity for a constant $V^* = 11\times10^{-6}$m^3/mol. Geophysical estimates for mantle viscosity determined from analyses of the geoid and post-glacial rebound are also shown. (b) The same plot as Figure 10a with the addition of the viscosity predicted for constant $\dot{\varepsilon}\sigma$. (c) Viscosity profiles for dislocation creep calculated for a constant stress of 0.3 MPa and diffusion creep assuming a grain size of 10 mm. (d) Viscosity profiles calculated for water-saturated conditions compared with profiles calculated for dry conditions and for a constant water content of 1000 H/10^{6}Si.

provides a relatively good fit to the profile calculated assuming a pressure-dependent V^* and also agrees with the fit in Figure 5. The large difference in the magnitude of viscosity calculated for stresses of 0.1 and 1.0 MPa emphasizes the uncertainty in accuracy (as opposed to precision) associated with extrapolation of the dislocation creep flow laws to asthenospheric conditions.

The change in viscosity with depth predicted for the constant stress viscosity profiles is similar to that suggested by independent geophysical estimates (Figure 10). Based on analysis of the geoid, the viscosity of the oceanic mantle asthenosphere is $\leq 2\times10^{19}$ Pa s [*Craig and McKenzie,* 1986; *Hager,* 1991]. The numerical models presented by *Craig and McKenzie* also indicate that the relatively low viscosity asthenosphere must be restricted to depths less than ~200 km. At the base of the olivine stability field (i.e., at depths above the α–β transition) a viscosity of ~10^{21} Pa s is estimated by the analysis of post-glacial rebound [e.g., *Peltier,* 1998]. The applicability of the rebound viscosity to the oceanic mantle is suggested by the similarity of seismic velocity [e.g., *Gaherty and Jordan,* 1995] and electrical conductivity [e.g., *Hirth et al.,* 2000] for oceanic mantle and sub-continental mantle at depths between ~300-400 km. A relatively large increase in viscosity with depth in the upper mantle is also suggested by some joint inversions of geoid and dynamic topography data [*Panasyuk and Hager,* 2000].

For comparison with the geophysical estimates, the depth dependence of viscosity predicted for constant $\dot{\varepsilon}\sigma$ is shown in Figure 10b. As suggested by Figure 9, the change in viscosity with depth (i.e., pressure and temperature) calculated with a constant value of $\dot{\varepsilon}\sigma$ is considerably smaller than that calculated assuming a constant stress and is also considerably smaller than that suggested by the geophysical constraints. At face value, this observation is inconsistent with the notion that convecting systems evolve to a condition of constant viscous dissipation. A possible explanation for this apparent discrepancy is that small-scale convection at the base of the plate results in an increase in strain rate (and therefore stress) and a concomitant decrease in viscosity.

Based on the observation that seismic anisotropy is often limited to the upper 200-250 km of the mantle, *Karato* [1992] suggested that a transition from dislocation creep to diffusion creep may occur with increasing depth. The depth at which a transition to diffusion creep occurs is strongly dependent on the grain size in the mantle and the activation volume for diffusion creep. The grain size in the upper mantle is influenced by a number of factors, including dynamic recrystallization and grain growth [e.g., *Karato,* 1984; *Evans et al.,* 2001]. A combination of microstructural studies [e.g., *Ave Lallement et al.,* 1980] and theoretical considerations [*Evans et al.,* 2001] suggests that the grain size in the upper mantle is in the range of 10 mm. The viscosity profile for a grain size of 10 mm plotted in Figure 10c indicates a transition to diffusion creep occurs at a depth of ~250 km if V^* for diffusion creep is ~4×10^{-6} m^3/mol, which is within the bound of 0-20×10^{-6} m^3/mol determined experimentally [*Mei and Kohlstedt,* 2000a]. Recalling the discussion above, the large uncertainty in V^* for diffusion creep at a fixed water activity is due to the competing effects of pressure on creep rate under wet conditions. For all values of V^*, the grain size must be considerably smaller the 10 μm for diffusion creep to be the dominant deformation mechanism at depths of ~100 km.

An important caveat to the hypothesis that a transition to diffusion creep occurs with increasing depth is that grain size tends to evolve toward a value where both deformation mechanisms accommodate the same strain rate [e.g., *de Bresser et al.,* 2001]. Thus, unless grain growth is kinetically inhibited (e.g., by second phase pinning), a complete transition to diffusion creep will not occur.

To illustrate the maxium effect of water on the viscosity of the mantle, we show a viscosity profile for water-saturated olivine aggregates with an adiabatic geotherm in Figure 10d. We use equation 6 to calculate the water content of olivine. For comparison, we also show viscosity profiles for dry olivine on an adiabatic geotherm. A minimum viscosity of 5×10^{17} Pa s for water-saturated conditions occurs at a depth of ~50 km. As discussed above, there are competing effects of pressure on viscosity under water-saturated conditions. At depths shallower than ~50 km, the viscosity increases because the water-content decreases with decreasing pressure and the product PV^* becomes small relative to E^*, such that the decrease in temperature with decreasing depth has a greater effect on viscosity than the decrease in pressure. At depths greater than 50 km, the viscosity increases because the PV^* term dominates over the effects of both increasing water content and increasing temperature.

A possible range of viscosity in the mantle wedge beneath subduction zones is illustrated in Figure 11. One bound on the viscosity is shown by a profile calculated with a geotherm defined by the vapor-saturated solidus at a constant strain rate of 10^{-15} s^{-1}. The lower bound on viscosity is shown by the profile calculated for water-saturated

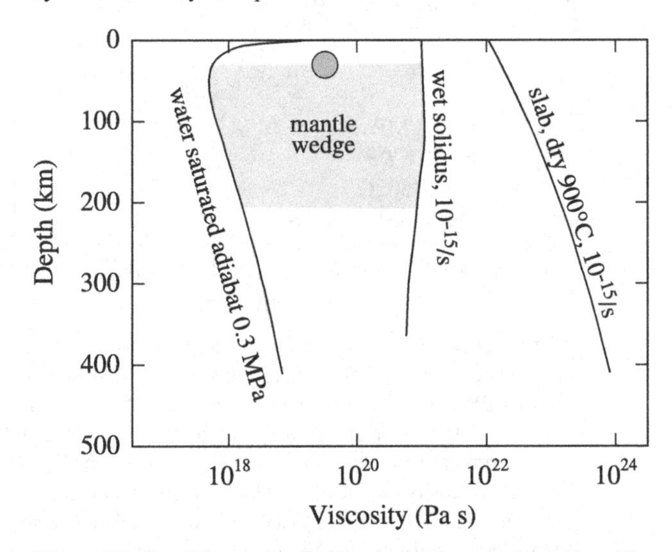

Figure 11. Possible range of viscosity in the mantle wedge. The low viscosity predicted for a depth of ~50 km, calculated for an adiabatic geotherm, water-saturated olivine and a constant stress of 0.3 MPa, provides a minimum estimate for the viscosity in the mantle wedge of subduction zones. The upper bound on viscosity of the mantle wedge is calculated using a geotherm defined by the vapor-saturated solidus and a constant strain rate of 10^{-15}/s. The filled circle illustrates the viscosity of olivine aggregates calculated using estimates for pressure, temperature and water content derived from the composition of primitive arc magmas. The viscosity predicted for dry conditions at a constant temperature of 900°C illustrates the possible contrast in viscosity between the mantle wedge and the subducted slab.

conditions with an adiabatic geotherm. Because the temperature in the mantle wedge is likely always above the solidus, the viscosities could be somewhat lower if there is a significant amount of melt present. For example, based on the relationships in Figure 8, with a melt fraction of 2% the viscosity would be reduced by an additional factor of two. However, it is important to realize that the low viscosities shown in Figure 11 require both high temperatures and high water contents. Such a situation would necessitate a large flux of water into an adiabatically convecting mantle.

The temperature and water content of olivine in the mantle wedge can be constrained using petrologic data on arc lavas. For example, analyses of primitive andesites with ~6 wt% H_2O indicate equilibration with mantle peridotite at ~1200°C and 1.0 GPa [*Baker et al.,* 1994; *Gaetani and Grove,* this volume]. In Figure 12 we show the water content of olivine in equilibrium with water-saturated melt and basaltic melts with water contents of 2-6 wt%, using the distribution coefficients of *Hirth and Kohlstedt* [1996]. We previously hypothesized that the distribution coefficient for water between olivine and basaltic melt increases with increasing depth (pressure) because of differences in the solubility mechanism of water between the two phases (indicated by differences in the effect of water fugacity on water content). Recent measurements of these distribution coefficients are similar to those that we estimated using the solubility data [*Koga et al.,* 2002]. These relationships indicate that at a depth of ~30 km (pressure of 1 GPa), olivine in equilibrium with basalt with 6 wt% H_2O will contain 400-500 $H/10^6Si$. Using our dislocation creep flow law, we calculate a viscosity of 3×10^{19} Pas for olivine aggregates with a water content of 500 $H/10^6Si$, at a temperature of 1200°C, a pressure of 1.0 GPa and a differential stress of 0.3 MPa. This value is shown by the circle in Figure 11.

Numerical modeling studies also suggest that the viscosity of the mantle wedge is relatively low. *Billen and Gurnis* [2001] demonstrate that constraints on the geoid, gravity and topography above subuction zones are satisified if the viscosity of the mantle wedge is at least an order of magnitude less than that of the surrounding asthenosphere. Similar models suggest that the viscosity in the wedge may be as low as 4×10^{18} Pa s [Han, L., and Gurnis, M., Subduction in semi-dynamic models and application to the Tonga-Kermadec subduction zone, unpublished manuscript]. Based on the extrapolations of laboratory data shown in Figure 11, such a low viscosity requires both relatively high water contents and temperatures.

The contrast in viscosity between a subducted slab and the mantle wedge is difficult to quantify due to the non-linear nature of rheology. For illustrative purposes, in Figure 11 we show the viscosity of a dry slab with a constant temperature of 900°C and a constant strain rate of $10^{-15}s^{-1}$. We use a dry rheology for the slab based on the hypothesis that water is removed from the oceanic lithosphere during melting at oceanic spreading ridges [e.g., *Hirth and Kohlstedt,* 1996]. Due to the strong temperature-dependence on viscosity, the colder interior regions of the slab would have an even greater viscosity than that predicted in Figure 11.

We reiterate that the magnitudes of viscosity shown in Figure 10 and Figure 11 are strongly dependent on the differential stress. The stress state in the wedge will be partly controlled by buoyancy resulting from temperature variations and the presence of melt. Ultimately, quantification of these effects will require analyses of dynamic models that couple the details of mantle rheology discussed in this paper with the dynamics of fluid migration into the wedge and the melting processes within the wedge.

Could the Subduction Factory Produce the oil of Plate Tectonics?

One of the requirements for Earth-like plate tectonics is that the boundaries of the plates must be considerably weak-

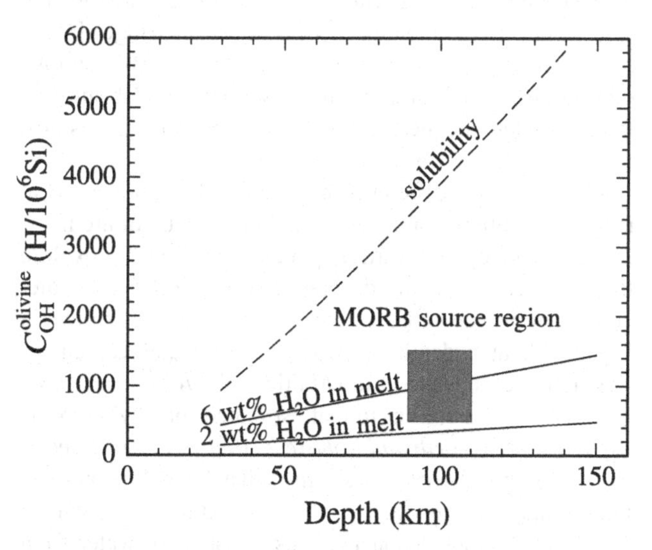

Figure 12. Plot of the water content of olivine in equilibrium with hydrous melt as a function of depth. The dashed line labeled solubility shows the water content of olivine under water-saturated conditions. The solid lines show the water content of olivine in equilibrium with basaltic melts with constant water contents of 2 wt.% and 6 wt.%. The shaded box shows the water content estimated for olivine in the oceanic asthenosphere. This plot suggests that the water content of olivine in the residue of arc magmatism is similar to that of olivine in the asthenosphere.

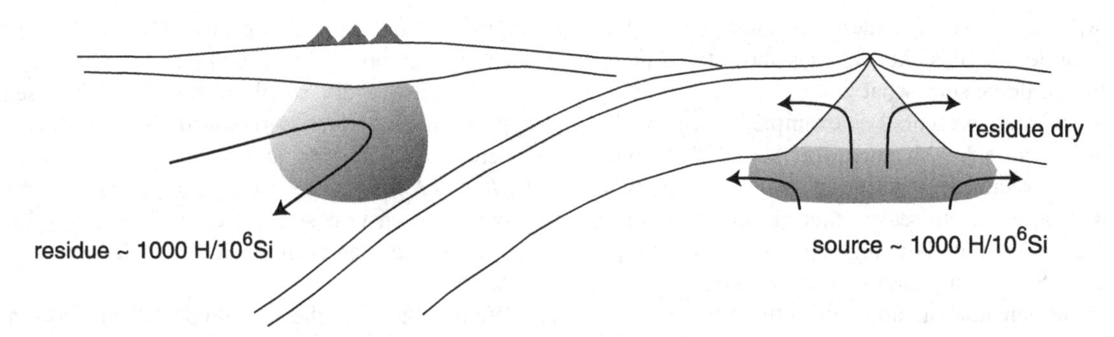

Figure 13. Cartoon illustrating fundamental differences in the role of water in the melting regions of arcs and ridges. At mid-ocean ridges, the melting process produces a dry residue. By contrast, the melting process at arcs produces a hydrous residue with a water content similar to that estimated for the oceanic asthenosphere, i.e., the MORB source region. Mantle flow lines are indicated by curved arrows. Shading indicates water content in the melting regions, with darker shades for higher water contents.

er than the plates. In addition to the plate boundaries defined by the subduction zones, the melting processes at arcs can influence tectonics by producing mantle with a water content similar to that of the asthenosphere. The relatively high water content of nominally anhydrous minerals in the oceanic mantle can create a rheological contrast between the asthenosphere and the overlying lithosphere. The viscosity of the lithosphere is increased relative to that of the asthenosphere partly because water is extracted from the residue during the melting process at oceanic ridges [*Hirth and Kohlstedt,* 1996; *Phipps Morgan,* 1997]. By contrast, due to the flux of fluids from the subducting slab into the mantle wedge, the melting process at subduction zones produces a relatively water-rich residue.

Another major difference between melting processes at ridges and subduction zones is that the residue of arc melting is advected downward by plate-driven flow. To explore the significance of this difference, we estimate the water content of olivine in the residue of arc melting. As shown in Figure 12, at a depth of ~100 km the water content of olivine in equilibrium with basaltic melt with a water content of 6 wt% is similar to that estimated for olivine in the MORB source [*Hirth and Kohlstedt,* 1996]. As illustrated in Figure 13, this observation suggests that the melting process in arcs may set the water content of nominally anhydrous phases in the asthenosphere. Thus transport of water from the slab into the mantle wedge can continually replenish the water content of the upper mantle and therefore facilitate the existence of a low viscosity asthenosphere.

Acknowledgments. We are grateful to the many colleagues with whom we have discussed the issues presented in this manuscript. In particular, we want to thank Shenghua Mei, Steve Mackwell, Mark Zimmerman, Peter Kelemen, Glenn Gaetani, Brian Evans, Magali Billen, Mike Braun, Laurent Montesi and Mathew Jull. We are also grateful for helpful reviews from Jed Mosenfelder and John Eiler. In addition, we want to thank Karen Hanghoj for assistance in preparing the manuscript. Finally, we would like to thank Sally Gregory Kohlstedt and Ann Mulligan for patience during several week long get-togethers where the authors hashed over endless details on olivine rheology. This work was supported by OCE-0099316 (GH), EAR-9910899 (GH), EAR-9906986 (DLK), OCE-0002463 (DLK), EAR-0126277 (DLK) and INT-0123224 (DLK).

REFERENCES

Ave Lallemant, H.G., J.-C., Mercier, N.L. Carter, and J.V. Ross, Rheology of the upper mantle: inferences from peridotite xenoliths, *Tectonophys.,* 70, 85-113, 1980.

Bai, Q., Mackwell, S.J., and Kohlstedt, D.L., High-temperature creep of olivine single crystals 1. Mechanical results for buffered samples. *J. Geophys. Res.,* 96, 2441-2463, 1991.

Bai, Q., and D.L. Kohlstedt, High-temperature creep of olivine single crystals; 2, Dislocation structures, *Tectonophysics,* 206, 1-29, 1992.

Bai, Q., Z. Jin, H.W. Green, Experimental investigation of partially molten peridote at upper mantle pressure and temperatue, in M. Holness (ed) *Deformation enhanced Fluid Transport in the Earth's Crust and Mantle,* Chapmann & Hall, 1997.

Baker, M.B., T.L. Grove, and R. Price, Primitive basalts and andesites from the Mt. Shasta region, N. California: products of varying melt fraction and water content, *Contrib. Mineral Petrol.,* 118, 111-129, 1994.

Beeman, M.L., and D.L. Kohlstedt, Deformation of olivine-melt aggregates at high temperatures and confining pressures, *J. Geophys. Res.,* 98, 6443-6452, 1993.

Bejina, F., P. Raterron, J. Zhang, O. Jaoul, and R. C. Lierbermann, Activation volume of silicon diffusion in San Carlos olivine, *Geophys. Res. Lett.,* 24, 25972600, 1997.

Bell, D.R., G.R. Rossman, A. Maldene, D. Endisch, and F. Rauch, Hydroxide in olivine: a quantitative determination of the absolute amount and calibration of the IR spectrum, *J. Geophys. Res.*, in press, 2002.

Billen, M.I., and M. Gurnis, A low viscosity wedge in subduction zones, *Earth Planet. Sci. Lett.*, 193, 227-236, 2001.

Borch, R.S., and H.W. Green II, Deformation of peridotite at high pressure in a new molten salt cell: Comparison of traditional and homologous temperature treatments, *Phys. Earth Planet. Inter.*, 55, 269-276, 1989.

Braun, M. G. Hirth, and E.M. Parmentier, The effects of deep, damp melting on mantle flow and melt generation beneath mid-ocean ridges, *Earth Planet. Sci. Lett.*, 176, 339-356, 2000.

Chopra, P.N., and M.S. Paterson, The experimental deformation of dunite, *Tectonophys.*, 78, 453-473, 1981.

Chopra, P.N. and M.S. Paterson, The role of water in the deformation of dunite, *J. Geophys. Res.*, 89, 7861-7876, 1984.

Christensen, U.R., *Mantle rheology, constitution, and convection, in Mantle convection; plate tectonics and global dynamics*, Peltier, W. Richard (editor), 595-655, 1989.

Coble, R.L., A model for boundary diffusion controlled creep in polycrysatlline materials, *J. Appl. Phys.*, 34, 1679-1682, 1963.

Cooper, R.F., and D.L. Kohlstedt, Solution-precipitation enhanced diffusional creep of partially molten olivine basalt aggregates during hot-pressing, *Tectonophysics*, 107, 207-233, 1984.

Cooper, R.F., D.L. Kohlstedt, and K. Chyung, Solution-precipitation enhanced creep in solid-liquid aggregates which display a non-zero dihedral angle, *Acta Metall.*, 37, 1759-1771, 1989.

Craig, C.H., and D. McKenzie, The existence of a thin low-viscosity layer beneath the lithosphere, *Earth Planet. Sci. Lett.*, 78, 420-426, 1986.

Daines, M. J., and D. L. Kohlstedt, A laboratory study of melt migration, *Phil. Trans. R. Soc. Lond. A*, 342, 43-52, 1993.

de Bresser, J.H.P., J.H. ter Heege, and C.J. Spiers, Grain size reduction by dynamic recrystallization: can it result in major rheological weakening? *Int. J. Earth Sci. (Geologische Rundschau)*, 90, 28-45, 2001.

Dohmen, R., S. Chakraborty, and H.-W. Becker, Si and O diffusion in olivine and implications for characterizing plastic flow in the mantle, submitted, 2002.

Durham, W.B., and C. Goetze, Plastic flow of oriented single crystals of olivine, 1, Mechanical data, *J. Geophys. Res.*, 82, 5737-5753, 1977.

Durham, W.B., L.A. Stern, and S.H. Kirby, Rheology of ice I at low stress and elevated confining pressure, *J. Geophys. Res.*, 106, 11,031-11,-42, 2001.

Evans, B., J. Fredrich, and T.-f Wong, The brittle-ductile transition in rocks; recent experimental and theoretical progress, in *The brittle-ductile transition in rocks,* (editors) Duba, A. G., W.B. Durham, J.W. Handin, and H.F. Wang, Geophysical Monograph, 56, 1-20, 1990.

Evans, B., J. Renner, and G. Hirth, A few remarks on the kinetics of grain growth in rocks, *Int. J. Earth Sci., (Geologishe Rundschau)*, 90, 88-103, 2001.

Farver, J.R., and R.A. Yund, Silicon diffusion in forsterite aggregates; implications for diffusion accommodated creep, *Geophys. Res. Lett.*, 27, 2337-2340, 2000.

Fisler, D. K., and S.J. Mackwell, Kinetics of diffusion-controlled growth of fayalite, *Phys. Chem. Min.*, 21, 156-165, 1994.

Frost, H.J., and M.F. Ashby, *Deformation Mechanism Maps*, Pergamon Press, 1982.

Gaetani, G.A., and E.B. Watson, Open system behavior of olivine-hosted melt inclusions, *Earth Planet. Sci. Lett.*, 183, 27-41, 2000.

Gaetani, G.A., and T.L. Grove, Experimental constraints on melt generation in the mantle wedge, this issue.

Gaherty, J.B., and T.H. Jordan, Lehmann discontinuity as the base of an anisotropic layer beneath continents, Science, 268, 1468-1471, 1995.

Goetze, C., and D.L. Kohlstedt, Laboratory Study of dislocation climb and diffusion in olivine, *J. Geophys. Res.*, 78, 5961-5971, 1973.

Goldsby, D.L., and D.L. Kohlstedt, Superplastic deformation of ice: Experimental observations, J. Geophys. Res., 106, 11,017-11,030, 2001.

Gribb, T.T., R.F. Cooper, Low-frequency shear attenuation in polycrystalline olivine; grain boundary diffusion and the physical significance of the Andrade model for viscoelastic rheology, *J. Geophys. Res.*, 103, 27,267-27,279, 1998.

Gribb, T.T., R.F. Cooper, The effect of an equilibrated melt phase on the shear creep and attenuation behavior of polycrystalline olivine, *Geophys. Res. Lett.*, 27, 2341-2344, 2000.

Hager, B.H., Mantle viscosity: A comparison of models from post-glacial rebound and from the geoid, plate driving forces, and advected heat flux, in: *Glacial Isostasy, Sea level and Mantle Rheology*, R. Sabadini et al., eds., Kluwer Academic Publishers, Netherlands, pp. 493-513, 1991.

Hirth, G., and D.L. Kohlstedt, Experimental constraints on the dynamics of the partially molten upper mantle: Deformation in the diffusion creep regime, *J. Geophys. Res*, 100, 1981-2001, 1995a.

Hirth, G., and D.L. Kohlstedt, Experimental constraints on the dynamics of the partially molten upper mantle 2. Deformation in the dislocation creep regime, *J. Geophys. Res.*, 100, 15,441-15,449, 1995b.

Hirth, G., and D.L. Kohlstedt, Water in the oceanic upper mantle: implications for rheology, melt extraction and the evolution of the lithosphere, *Earth Planet. Sci. Lett.*, 144, 93-108, 1996.

Hirth, G., Evans, R.L., and A.D. Chave, Comparison of continental and oceanic mantle electrical conductivity: Is the Archean lithosphere dry?, *Geochemistry, Geophysics, and Geosystems* *(G^3)*, 1, Paper Number 2000CG000048, 2000.

Holtzman, B., M.E. Zimmerman, D.L. Kohlstedt, and J. Phipps Morgan, Interactions of deformation and fluid migration I: Melt segregation in the viscous regime, *Eos Trans. AGU*, 82(47), 2001

Houlier, B., M. Cheraghmakani, and O. Jaoul, Silicon diffusion in San Carlos Olivine, *Phys. Earth Planet. Int.*, 62, 329-340, 1990.

Jin, Z. M., Q. Bai, and D.L. Kohlstedt, High-temperature creep of olivine crystals from four localities, *Phys. Earth Planet. Int.*, 82, 55-64, 1994.

Karato, S.-I., Comment on 'The effect of pressure on the rate of dislocation recovery in olivine' by D.L. Kohlstedt, H.P.K. Nichols, and Paul Hornack, *J. Geophys. Res.*, 86, 9319, 1981.

Karato, S., and M. Ogawa, High-pressure recovery of olivine; implications for creep mechanisms and creep activation volume, *Phys. Earth Planet. Int.*, 28, 102-117, 1982.

Karato, S., Grain-size distribution and rheology of the upper mantle, *Tectonophys.*, 104, 155-176, 1984.

Karato, S., M.S. Paterson and J.D. FitzGerald, Rheology of synthetic olivine aggregates: influence of grain size and water, J. Geophys. Res. 91, 8151-8176, 1986.

Karato, S., On the Lehmann Discontinuity, *Geophys. Res. Lett.*, 19, 2255-2258, 1992.

Karato, S., Wu, P., Rheology of the upper mantle; a synthesis, *Science*, 260, 771-778, 1993.

Karato, S., D.C., Rubie, and H. Yan, Dislocation recovery in olivine under deep upper mantle conditions; implications for creep and diffusion, *J. Geophys. Res.*, 98, 9761-9768, 1993.

Karato, S., and D.C. Rubie, Toward an experimental study of deep mantle rheology; a new multianvil sample assembly for deformation studies under high pressures and temperatures, *J. Geophys. Res.*, 102, 20,111-20, 1997.

Karato, S., Mapping water content in the upper mantle, this issue.

Karato, S., and H. Jung, Effects of pressure on high-temperature dislocation creep in olivine, in press, *Philosophical Magazine A.*, 2002.

Kelemen, P.B., G. Hirth, N. Shimizu, M. Spiegelman and H.J.B. Dick, A review of melt migration processes in the adiabatically upwelling mantle beneath spreading ridges, *Philos. Trans. R. Soc. London A*, 355, 283-318, 1997.

Kelemen, P.B., J.L. Rilling, E.M. Parmentier, L. Mehl, and B.R. Hacker, Thermal structure due to solid-state flow in the mantle wedge beneath arcs, [this issue]

Keyes, R.W., Continuum models of the effect of pressure on activated processes, *in Solids Under Pressure,* edited by W. Paul and D. M. Warschauer, 71-99, McGraw-Hill, New York, 1963.

Koga, K, E. Hauri, M. Hirshmann, and D. Bell, Hydrogen concentration analyses using SIMS and FTIR: Comparsion and calibration for nominally anhydrous minerals, *G-cubed,* in press, 2002.

Kohlstedt, D.L., and C. Goetze, Low-stress high-temperature creep in olivine single crystals, *J. Geophys. Res.*, 79, 2045-2051, 1974.

Kohlstedt, D. L., H.P.K. Nichols, and P. Hornack, P., The effect of pressure on the rate of dislocation recovery in olivine, *J. Geophys. Res.*, 85, 3122-3130, 1980.

Kohlstedt, D.L., H. Keppler and D.C. Rubie, Solubility of water in a, b, and g phases of $(Mg,Fe)_2SiO_4$, *Contrib. Min. Pet.*, 123, 345-357, 1996.

Kohlstedt, D. L., and M. E. Zimmerman, Rheology of partially molten mantle rocks, *Ann. Rev. Earth Planet. Sci.*, 24, 41-62, 1996.

Kohlstedt, D.L., and S.J. Mackwell, Diffusion of hydrogen and itrinsic point defects in olivine, *Z. Phys. Chem.*, 207, 147-162, 1998.

Kohlstedt, D.L., Q. Bai, Z.-C. Wang, and S. Mei, Rheology of partially molten rocks, in *Physics and Chemistry of Partially Molten Rocks,* ed. N.Bagdassarov, D. Laporte, A.B.Thompson, Kluwer Academic Publishers, 3-28, 2000.

Kohlstedt, D.L. and Wang, Z., Grain-Boundary Sliding Accommodated Dislocation Creep in Dunite, *Eos Trans. AGU,* 82(47), 2001.

Kumazawa, M., and O. Anderson, Elastic moduli, pressure derivatives, and temperature derivatives of single-crystal olivine and single-crystal forsterite, *J. Geophys. Res.*, 74, 5961-5972, 1994.

Langdon, T.G., A unified approach to grain boundary sliding in creep and superplasticity, *Acta Metall.*, 42, 2437-2443, 1994.

Mackwell, S.J., K.L. Kohlstedt, M.S. Paterson, The role of water in the deformation of olivine single crystals, *J. Geophys. Res.*, 90, 11319-11333, 1985.

Mackwell, S.J., G. Hirth, and D.L. Kohlstedt, Water weakening of olivine single crystals, *Deformation mechanisms in nature and experiment, Conference,* Basel, Switzerland, 1997.

Mei, S., and D.L. Kohlstedt, Influence of water on deformation of olivine aggregates; 1, Diffusion creep regime, *J. Geophys. Res.*, 105, 21,457-21,469, 2000a

Mei, S., and D.L. Kohlstedt, Influence of water on plastic deformation of olivine aggregates 2. Dislocation creep regime, J. Geophys. Res., 105, 21471-21481, 2000b.

Mei, S., W. Bai, T. Hiraga, and D.L. Kohlstedt, Ifluence of melt on plastic deformation of olivine-basalt aggregates under hydrous conditions, *Earth Planet. Sci. Lett.*, in press, 2002.

Melosh, H.J., Nonlinear stress propagation in the Earth's upper mantle, J. Geophys. Res., 81, 5621-5632, 1976

Nicolas, A., Structure and petrology of peridotites: clues to their geodynamic environment, *Rev. Geophys.*, 24, 875-895, 1986.

Ohtani, E. , and M. Kumazawa, Melting of forsterite Mg2SiO4 up to 15 GPa, Phys. *Earth Planet. Int.* 27, 32-38, 1981.

Panasyuk, S.V., and B.H. Hager, Inversion for mantle viscosity profiles constrained by dynamic topography and the geoid, and their estimated errors, *Geophys. J. Int.*, 143, 821-836, 2000.

Paterson, M.S., The ductility of rocks, in *Physics of Strength and Plasticity,* edited by A.S. Argon, 377-392, MIT Press, Cambridge, Mass., 1969.

Paterson, M.S., The determination of hydroxyl by infrared absorption in quartz silicate glasses and similar materials, *Bull. Mineral.*, 105, 20-29, 1982.

Peltier, W.R., Global glacial isostasy and relative sea level: Implications for solid earth geophysics and climate system dynamics, in *Dynamics of the Ice Age Earth,* P. Wu (editor), 17-54, 1998.

Phipps Morgan, J. The generation of a compositional lithosphere by mid-ocean ridge melting and its effect on subsequent off-axis hot spot upwelling and melting, Earth Planet Sci. Lett., 146, 213-232, 1997.

Poirier, J.-P., *Creep of Crystals,* Cambridge University Press, 1985.

Raj, R., and M.F. Ashby, On grain boundary sliding and diffusional creep, *Metall. Trans.*, 2, 1113-1127, 1971.

Renner, J., B. Evans, and G. Hirth, On the rheologically critical melt percentage, *Earth Planet. Sci. Lett.*, 181, 585-594, 2000.

Renner, J., K.Viskupic, G. Hirth, and B. Evans, Melt extraction from partially molten peridotites, submitted, *Geochemistry, Geophysics, and Geosystems (G^3),* in press, 2002.

Ricoult, D. L., and D.L. Kohlstedt, Experimental evidence for the effect of chemical environment upon the creep rate of olivine, in *Point defects in minerals,* Schock, R.N. (editor) Geophysical Monograph, 31, 171-184, 1985.

Ross, J. V., H.G. Ave Lallemant, and N.L. Carter, Activation volume for creep in the upper mantle, *Science,* 203, 261-263, 1979.

Ryerson, F. J., W.B. Durham, D.J., Cherniak, and W.A. Lanford, Oxygen diffusion in olivine; effect of oxygen fugacity and implications for creep, *J. Geophys. Res.,* 94, 4105-4118, 1989.

Sammis, C.G., J.C. Smith and G. Shubert, A critical assessment of estimation methods for activation volume, *J. Geophys. Res.,* 86, 10,707-10,718, 1981.

Schwenn, M. B., and C. Goetze, Creep of olivine during hot-pressing, *Tectonophysics,* 48, 41-60, 1978.

Tommasi, A., D. Mainprice, G. Canova, and Y. Chastel, Viscoplastic self-consistent and equilibrium-based modeling of olivine preferred orientations: Implications for the upper mantle seismic anisotropy, *J. Geophys., Res.,* 105, 7893-7908, 2000.

Toomey, D. R. W.S.D. Wilcock, J.A. Conder, D.W. Forsyth, J.D. Blundy, E.M. Parmentier, and W.C. Hammond, Asymmetric mantle dynamics in the MELT region of the East Pacific Rise, *Earth Planet. Sci. Lett.,* 200, 287-295, 2002.

Toramaru, A., and N. Fujii, Connectivity of melt phase in a partially molten peridotite, J. *Geophys. Res.,* 91 9239-9252, 1986.

Watson, E. B., Diffusion in fluid-bearing and slightly melted rocks; experimental and numerical approaches illustrated by iron transport in dunite, *Cont. Min. Petrol.,* 107, 417-434, 1991.

Weertman, J., The creep strength of the Earth's mantle, *Rev. Geophys. Space Phys.* 8, 145-168, 1970.

Xu, Y., M.E. Zimmerman, and D.L.Kohlstedt, Deformation behavior of partially molten mantle rocks, *MARGINS Theoretical and Experimental Earth Science Series. Volume I: Rheology and Deformation of the Lithosphere at Continental Margins,* eds. G.D. Karner, N.W. Driscoll, B. Taylor and D.L. Kohlstedt, Columbia University Press, in press, (2002).

Young, T.E., H.W. Green II, A.M. Hofmeister and D. Wallker, Infrared spectroscopic investigation of hydroxyl in β-$(Mg,Fe)_2SiO_4$ and coexisting olivine: Implications for mantle evolution and dynamics, *Phys. Chem. Miner.,* 19, 409-422, 1993.

Zener, C., Theory of lattice expansion introduced by cold-work, *Trans. AIME,* 147, 361-368, 1942.

Zhang, S., S.-I. Karato, J. FitzGerald, U.H. Faul, and Y. Zhou, Simple shear deformation of olivine aggregates, *Tectonophysics,* 316, 133-152, 2000.

Zhang, J., and C. Herzberg, Melting experiments on anhydrous peridotite KLB-1 from 5.0 to 22.5 GPa, *J. Geophys. Res.,* 99, 17,729-17,742, 1994.

Zimmerman, M.E., *The Structure and Rheology of Partially Molten Mantle Rocks,* Ph.D. Thesis University of Minnesota, 159 pp., 1999

Greg Hirth, Department of Geology and Geophysics, Woods Hole Oceanographic Institution, Woods Hole, MA 02543. (ghirth@whoi.edu)

D.L. Kohlstedt, Department of Geology and Geophysics, University of Minnesota-Twin Cities, Pillsbury Hall, Minneapolis, MN 55455. (dlkohl@umn.edu)

Experimental Constraints on Melt Generation in the Mantle Wedge

Glenn A. Gaetani

Department of Geology and Geophysics, Woods Hole Oceanographic Institution, Woods Hole, Massachusetts

Timothy L. Grove

Department of Earth, Atmospheric, and Planetary Sciences, Massachusetts Institute of Technology, Cambridge, Massachusetts

Experimental studies show that H_2O affects most aspects of melt generation in the sub-arc mantle wedge. For example, dissolved H_2O modifies the major element composition of peridotite partial melt by increasing the ratio of SiO_2 to MgO + FeO, mimicking the effect of decreasing pressure during anhydrous partial melting. Comparison of the normalized (anhydrous) compositions of experimentally produced hydrous and anhydrous melts shows that SiO_2 increases by ~1 wt% with addition of 3 to 6 wt% dissolved H_2O, while FeO + MgO decreases by ~2 wt%. Furthermore, mobility of partial melt in mantle peridotite may increase due to the influence of H_2O. Orthopyroxene-melt dihedral angles are ~70° under anhydrous conditions, trapping small amounts of melt at 4 grain junctions, but they decrease to ~52° under hydrous conditions, allowing connectivity down to very low melt fractions. Dissolved H_2O also decreases melt density and viscosity that, combined with enhanced connectivity, allows hydrous melt to segregate very efficiently from residual peridotite. Less melt is produced by hydrous peridotite, relative to anhydrous peridotite, for a given temperature increase or pressure decrease, because of the monotonic decrease of dissolved H_2O with increasing extent of melting. Primitive arc magmas with high pre-eruptive H_2O contents may form when a peridotite partial melt that is initially near fluid saturation percolates upward through the mantle wedge, maintaining equilibrium with hotter, overlying peridotite by dissolving the surrounding rock (reactive porous flow). Adiabatic decompression melting may occur in regions where hot mantle flows from the back arc into the wedge corner, generating nearly anhydrous partial melt.

1. INTRODUCTION

Thermal and chemical exchange between subducting oceanic lithosphere and the overlying mantle wedge at convergent margins produces what is arguably the most physically and compositionally complex melting regime in the upper mantle. For example, the transfer of heat from the hot, overlying mantle to the cold, subducting lithosphere produces an inverted thermal gradient, while viscous coupling between mantle and subducting lithosphere induces downward flow in the lower portion of the wedge [*McKenzie*, 1969; *Oxburgh and Turcotte*, 1970; *Toksöz et al.*, 1971; *Anderson et al.*, 1978; *Peacock*, 1991; *Davies and*

Inside the Subduction Factory
Geophysical Monograph 138

10.1029/138GM07

Stevenson, 1992; *Kincaid and Sacks,* 1997; *Peacock and Wang,* 1999]. Spatial and temporal variations in composition and flux of material transferred from subducting lithosphere to mantle wedge variably enrich the source regions of arc lavas in volatiles and incompatible elements [*Perfit et al.,* 1980; *McCulloch and Gamble,* 1991; *Plank and Langmuir,* 1993; *Stolper and Newman,* 1994; *Elliott et al.,* 1997; *Regelous et al.,* 1997; *Turner and Hawkesworth,* 1997; *Ayers,* 1998; *Eiler et al.,* 1998; *Johnson and Plank,* 1999; *Eiler et al.,* 2000]. It is, therefore, not surprising that subduction-related lavas are highly variable in composition, and that the processes by which they are generated within the mantle wedge and transported to the surface of the Earth are poorly understood.

It is generally accepted that partial melting of peridotite in subduction zones is initiated by an influx of volatiles from the subducted oceanic lithosphere [e.g., *Gill,* 1981; Tatsumi, 1986; *Davies and Stevenson,* 1992]. However, the mechanism by which partial melting proceeds is a matter of debate. Adiabatic decompression of peridotite along the mantle adiabat is thought to generate anhydrous partial melts beneath oceanic spreading ridges (path labeled anhydrous adiabat in Figure 1), and it has been proposed that a similar process operates in the sub-arc mantle [e.g., *Tatsumi et al.,* 1983; *Nye and Reid,* 1986; *DeBari et al.,* 1987; *Plank and Langmuir,* 1988]. Although there are examples of subduction-related magmas with very low pre-eruptive H_2O contents that are consistent with decompression partial melting at near-anhydrous conditions [*Tatsumi et al.,* 1983; *Bartels et al.,* 1991; *Sisson and Layne,* 1993; *Baker et al.,* 1994; *Bacon et al.,* 1997; *Sisson and Bronto,* 1998], this form of melt generation requires high temperatures and, therefore, may only be significant at convergent margins, such as Japan, where geophysical evidence suggests that the combination of rapid subduction, steeply dipping subducting lithosphere, and a young overriding plate produces enhanced mantle flow and advection of heat into the wedge, resulting in elevated temperatures [*Hasegawa et al.,* 1991; *Zhao and Hasegawa,* 1993; *Kincaid and Sacks,* 1997].

Partial melting of peridotite at low-temperature, fluid-saturated conditions near the base of the mantle wedge will generate a small amount of H_2O-rich partial melt, but the process of adiabatic decompression (path labeled hydrous adiabat in Figure 1) would cause this melt to cool, degas, and crystallize within the mantle wedge [e.g., Nicholls and Ringwood, 1972; *Tatsumi and Eggins,* 1995]. In order for melt generation to proceed, the mass of the ascending material must be small enough that it can be efficiently heated as it moves through the inverted thermal gradient that characterizes the mantle wedge. Given these constraints, potential modes for melt generation at convergent margins include the development of small, low-density diapirs that rise through the wedge and partially melt due to a combination of decompression and heating [*Tatsumi et al.,* 1983; *Tatsumi et al.,* 1986; *Kushiro,* 1990; *Pearce et al.,* 1995] or reaction between relatively cool, H_2O-rich magma and hotter peridotite as the melt percolates upward through the mantle wedge (path labeled hydrous flux melting in Figure 1) [e.g., *Kelemen,* 1986; 1990; *Grove et al.,* 2002].

Here we provide an overview of experimental constraints on peridotite partial melting under hydrous conditions in the sub-arc mantle. We begin with a discussion of the mechanism by which H_2O dissolves in molten silicate, and its effect on the composition, physical properties, and transport of partial melt, and on the amount of melt produced in response to heating or decompression. We then review the

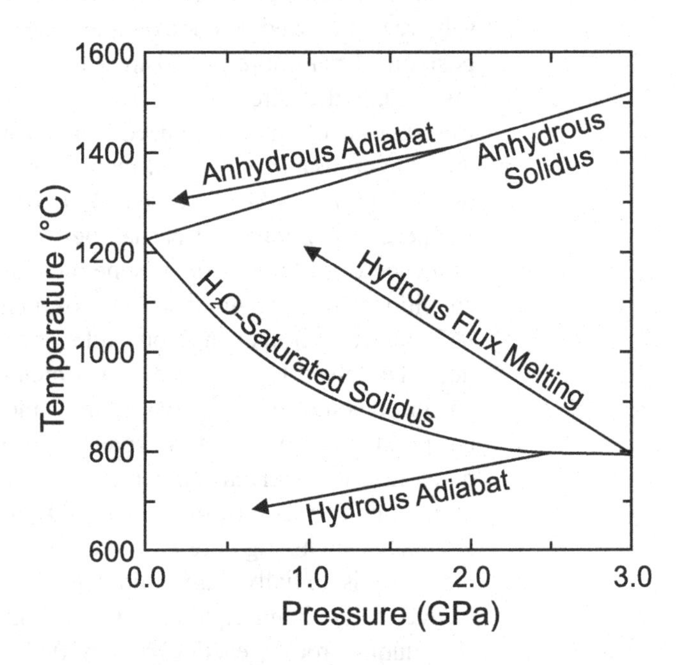

Figure 1. Schematic pressure-temperature diagram showing the anhydrous and H_2O-saturated solidii for mantle peridotite (curves) and illustrating several possible paths for melt generation in the mantle wedge (arrows). An adiabatic decompression path beginning at the H_2O-saturated solidus, near the base of the mantle wedge, will lead to solidification of melt to vapor + crystals. The hydrous flux melting path is a consequence of the inverted thermal gradient that is a unique property of the mantle wedge environment. This melting process requires the exchange of heat between the melt and its surroundings and, therefore, is not adiabatic. Anhydrous adiabatic decompression melting may occur when hot upwelling mantle ascends from depth and flows into the corner of the mantle wedge.

experimentally determined phase relations for primitive arc lavas, and discuss several alternative mechanisms for melt generation in the sub-arc mantle. We conclude that existing lines of evidence are consistent with a melt generation process in which cool, H_2O-rich partial melt generated near the base of the mantle wedge percolates upward through an inverted thermal gradient, assimilating peridotite and increasing magma mass as it ascends (hydrous flux melting), but that anhydrous adiabatic decompression melting may also occur in many sub-arc environments. The balance between buoyancy forces and thermal equilibration time limits the efficiency of hydrous decompression partial melting within ascending diapers.

2. DISSOLUTION AND SPECIATION OF H_2O IN SILICATE MELTS

The dissolution of H_2O in silicate melts influences peridotite partial melting and melt transport through its effects on phase equilibria and the physical properties of partial melts [e.g., *Kushiro,* 1972; *Kushiro et al.,* 1972; *Richet et al.,* 1996; *Schulze et al.,* 1996; *Ochs and Lange,* 1997; 1999]. Experimental and theoretical investigations of the speciation of H_2O in silicate melts provide a framework for interpreting these effects in terms of changes to the structure of the melt. Although there is not a clear consensus regarding the details of H_2O speciation in silicate liquids at magmatic temperatures [e.g., *Burnham,* 1975; *Stolper,* 1982a; *Nowak and Behrens,* 1995; Zhang, 1999; *Nowak and Behrens,* 2001], the thermodynamics that govern the dissolution of H_2O in silicate melts are well established [e.g., *Stolper,* 1982a; *Silver and Stolper,* 1985; 1989].

When an anhydrous silicate melt and an aqueous fluid are brought together at some pressure and temperature, they react until the chemical potential (μ_i^j) of each component is equivalent in the two phases. The concentration of molecular H_2O that dissolves in the silicate melt is governed by the equivalence of chemical potentials:

$$\mu_{H_2O}^{fluid} = \mu_{H_2O}^{melt} \quad (1)$$

which can be rearranged to give:

$$\mu_{H_2O}^{0,fluid} + RT\ \mathrm{In}\ a_{H_2O}^{fluid} = \mu_{H_2O}^{0,melt} + RT\ \mathrm{In}\ a_{H_2O}^{melt} \quad (2)$$

where $\mu_{H_2O}^{0,j}$ is the standard state chemical potential of H_2O in phase j (fluid or melt), R is the gas constant, T is absolute temperature, and $a_{H_2O}^{j}$ is the activity of H_2O in phase j. Rearranging Equation 2 gives:

$$\frac{-\left(\mu_{H_2O}^{0,melt} - \mu_{H_2O}^{0,fluid}\right)}{RT} = \mathrm{In}\frac{X_{H_2O}^{melt}\gamma_{H_2O}^{melt}}{f_{H_2O}^{fluid}/f_{H_2O}^{0,fluid}} \quad (3)$$

where $X_{H_2O}^{melt}$ is the mole fraction of molecular H_2O dissolved in the melt, $\gamma_{H_2O}^{melt}$ is the activity coefficient for H_2O dissolved in the melt, $f_{H_2O}^{fluid}$ is the fugacity of H_2O in the fluid, and $f_{H_2O}^{0,fluid}$ is the fugacity of H_2O in the standard state of the fluid. Activity-composition relationships are Henrian at low concentrations of dissolved molecular H_2O (i.e., $\gamma_{H_2O}^{melt}$ is independent of $X_{H_2O}^{melt}$), so it follows from Equation 3 that, at a given pressure and temperature, the solubility of molecular H_2O in silicate melt should increase as a linear function of $f_{H_2O}^{fluid}$.

Low-pressure solubility measurements demonstrate, however, that the dissolution mechanism for H_2O in silicate melts is more complicated than that described by Equations 1-3. The concentration of H_2O dissolved in a glass or melt varies with the square root of $f_{H_2O}^{fluid}$ rather than linearly [e.g., *Tomlinson,* 1956; *Russell,* 1957; *Kurkjian and Russell,* 1958; *Burnham and Davis,* 1974], indicating that when molecular H_2O (H_2O^{melt}) is introduced into an initially anhydrous silicate melt it reacts with oxygen (O^{melt}) to produce hydroxyl groups (OH^{melt}) [e.g., *Orlova,* 1962; *Ernsberger,* 1977; *Stolper,* 1982a; b]. The relative abundances of the three species in the melt are governed by the homogeneous equilibrium:

$$H_2O^{melt} + O^{melt} = 2OH^{melt} \quad (4)$$

and activities of the individual species are related by the equilibrium constant:

$$K_{eq} = \frac{(a_{OH}^{melt})^2}{a_O^{melt}\ a_{H_2O}^{melt}} \quad (5)$$

where a_{OH}^{melt} is the activity of hydroxyl groups in the melt, $a_{H_2O}^{melt}$ is the activity of molecular H_2O in the melt, and a_O^{melt} is the activity of oxygen in the melt that is not associated with hydrogen [*Silver and Stolper,* 1985]. The formation of OH^{melt} through a reaction between H_2O^{melt} and O^{melt} that had initially been bonded to two cations within the aluminosilicate network (a bridging oxygen) may involve several different equilibria, the simplest of which are:

$$H_2O + Si-O-Si = 2(Si-OH) \quad (6)$$

$$H_2O + Si-O-Al = Si-OH + Al-OH \quad (7)$$

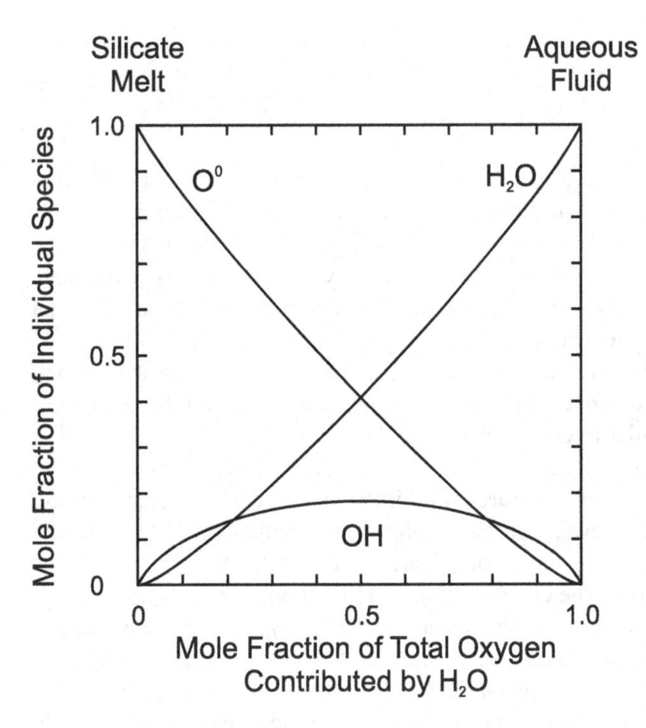

Figure 2. Plot of the mole fractions of molecular H_2O, hydroxyl groups, and oxygen versus the mole fraction of total oxygen contributed to the liquid by H_2O showing speciation calculated from Equation 5 by assuming ideal mixing and an equilibrium constant equal to 0.2. After Stolper [1982a].

$$H_2O + Al - O - Al = 2(Al - OH) \quad (8)$$

Reactions such as these decrease the length of individual polymers within the melt (degree of polymerization), affecting the activity of SiO_2 [*Kushiro,* 1975], as well as the viscosity of the melt [*Shaw,* 1963; *Kushiro,* 1978], and cation diffusivities [*Watson,* 1981].

Figure 2 shows how the concentrations of O^{melt}, OH^{melt}, and H_2O^{melt} will vary as functions of total dissolved H_2O if mixing of the three species is approximated as an ideal solution [*Silver and Stolper,* 1985]. At low concentrations of total dissolved H_2O, the concentration of OH^{melt} is higher than that of H_2O^{melt}. The abundance of H_2O^{melt} increases monotonically as a function of total dissolved H_2O, however, while the concentration of OH^{melt}. initially increases, then gradually levels off and passes through a maximum. As the bulk composition approaches pure H_2O, the concentrations of both OH^{melt} and O^{melt} decrease to zero. Although mixing between silicate melts and aqueous fluids is demonstrably non-ideal, spectroscopic measurements indicate that the curves in Figure 2 are a good approximation to H_2O speciation in quenched aluminosilicate glasses containing total dissolved H_2O concentrations of up to ~10 wt% [e.g., *Stolper,* 1982b; *Silver and Stolper,* 1989; *Dixon et al.,* 1995]. The crossover point, at which the concentrations of OH^{melt} and H_2O^{melt} in the glass are equal, occurs at ~4 wt% total dissolved H_2O (Figure 3). In detail, the shape of these curves will vary according to the temperature dependence of Equation 5 [*Stolper,* 1989; *Nowak and Behrens,* 1995; *Shen and Keppler,* 1995].

Spectroscopic measurements of H_2O speciation, like those shown in Figure 3, are typically carried out at room temperature on glasses quenched from magmatic conditions. This has led to questions regarding the appropriateness of these measurements for determining the speciation of H_2O dissolved in silicate melts at magmatic temperatures. It was noted by Dingwell and Webb [1990], for example, that the rate at which the structure of a silicate melt responds to temperature change (relaxation rate) is too rapid for speciation to be preserved at a temperature greater than that at which the melt transforms to a glass (the glass transition). It was subsequently shown by Zhang et al. [1997], on the basis of experimental data, that the kinetics of Reaction 4 are too rapid at temperatures ≥700 °C for H_2O speciation to be preserved during the quench. The presence of both OH^{melt} and H_2O^{melt} in molten silicate has been confirmed by near-infrared spectra collected in situ at magmat-

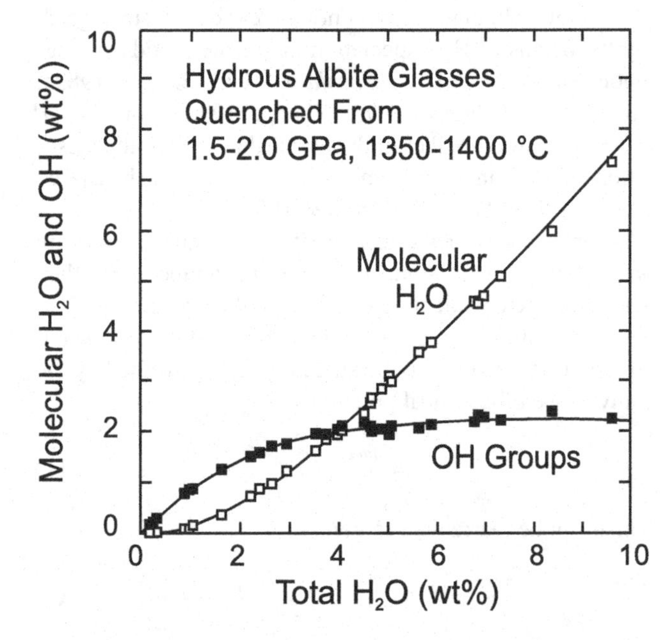

Figure 3. Plot of the weight concentrations of molecular H_2O (open symbols) and hydroxyl groups (filled symbols) dissolved in quenched albite glass versus total H_2O content of the glass. Curves represent best fit to the measured speciation calculated from Equation 5 using a regular solution model. After Silver and Stolper [1989].

ic temperatures [*Nowak and Behrens,* 1995; *Shen and Keppler,* 1995]. Although these measurements helped to resolve some of the questions regarding the speciation of H_2O in silicate melt, subsequent work has shown that the molar absorption coefficients used to determine species concentrations from IR absorption band intensities are temperature dependent [*Withers and Behrens,* 1999; *Withers et al.,* 1999; *Zhang,* 1999]. Developing a detailed understanding of the speciation of H_2O dissolved in silicate melt will, therefore, require analyses of silicate liquids at magmatic conditions using molar absorption coefficients calibrated over a range of temperatures and liquid compositions [*Nowak and Behrens,* 2001].

3. PHASE EQUILIBRIA OF HYDROUS PERIDOTITE PARTIAL MELTING

Our understanding of the phase equilibria that governs hydrous peridotite partial melting derives largely from experimental studies of systems such as $Mg_2SiO_4 - SiO_2 - H_2O$ and $Mg_2SiO_4 - CaMgSi_2O_6 - SiO_2 - H_2O$ [e.g., *Kushiro et al.,* 1968b; *Kushiro,* 1969]. These systems span the compositional range necessary to produce hydrous melts saturated with an upper mantle mineral assemblage of clinopyroxene, olivine, and orthopyroxene, while being comprised of a minimum number of components. The advantage of studying systems containing a small number of components is that they have low thermodynamic variance according to the Gibbs phase rule:

$$F = C + 2 - P \tag{9}$$

where *F* is the variance or degrees of freedom of the system, C is the number of components, and P is the number of phases [e.g., *Denbigh,* 1981]. For example, when forsterite (Mg_2SiO_4), enstatite ($Mg_2Si_2O_6$), silicate melt, and aqueous fluid coexist in the system $Mg_2SiO_4 - SiO_2 - H_2O$ the variance is equal to 1, so that only a single intensive variable is required to define the state of the system. At a given pressure, the 4 phases coexist at a single temperature so that fluid-saturated melting occurs at either a peritectic:

$$\text{enstatite} + \text{fluid} = \text{melt} + \text{forsterite} \tag{10}$$

or a eutectic:

$$\text{forsterite} + \text{enstatite} + \text{fluid} = \text{melt.} \tag{11}$$

For comparison, fluid-saturated partial melting of natural spinel lherzolite involves 6 phases (olivine, orthopyroxene, clinopyroxene, spinel, melt, and aqueous fluid) and approximately 11 components (SiO_2, TiO_2, Al_2O_3, Fe_2O_3, FeO, MnO, MgO, CaO, Na_2O, K_2O, H_2O), so that the variance is equal to 7. At a given pressure, 6 of the remaining intensive parameters can be varied without changing the assemblage of coexisting phases.

The minerals olivine and orthopyroxene comprise ~85% of a typical spinel lherzolite, and define the activity of SiO_2 in a coexisting silicate melt through the equilibrium:

$$Mg_2SiO_4^{\text{olivine}} + SiO_2^{\text{melt}} = Mg_2Si_2O_6^{\text{orthopyroxene}} \tag{12}$$

so that a comparison of phase relations involving forsterite, enstatite, and melt in the systems $Mg_2SiO_4 - SiO_2$ and $Mg_2SiO_4 - SiO_2 - H_2O$ provides a useful framework for understanding the influence of H_2O on the compositions of peridotite partial melts. At anhydrous conditions stoichiometric enstatite melts incongruently at pressures up to ~500 MPa to form forsterite and an SiO_2-rich partial melt. At pressures greater than ~500 MPa the melting of enstatite is congruent, as shown for a pressure of 2.0 GPa in Figure 4,

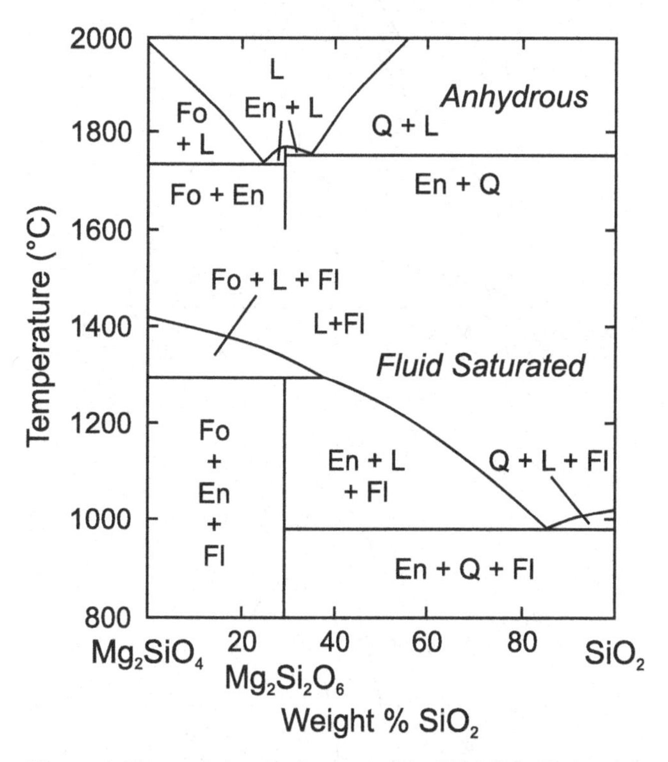

Figure 4. Phase relations in the system Mg_2SiO_4-SiO_2-H_2O at 2.0 GPa under anhydrous and H_2O-saturated conditions projected onto the Mg_2SiO_4-SiO_2 join. Abbreviations: Fo is forsterite; En is enstatite; Q is quartz; L is silicate liquid; Fl is aqueous fluid. After Morse [1980].

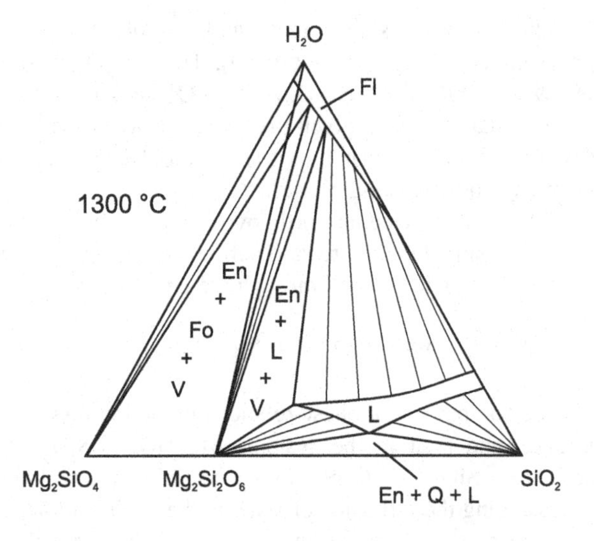

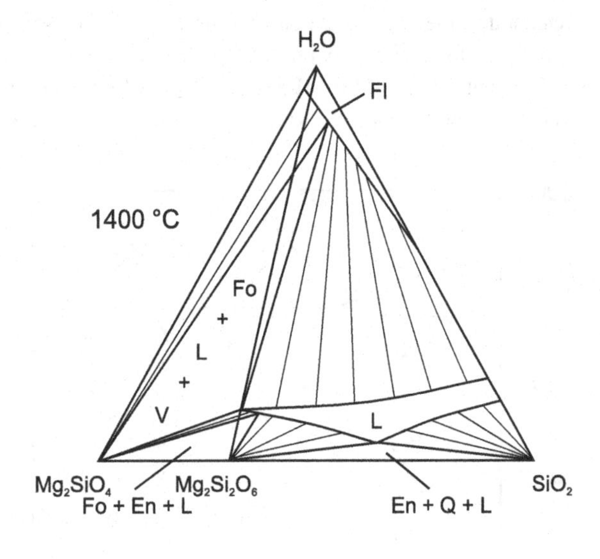

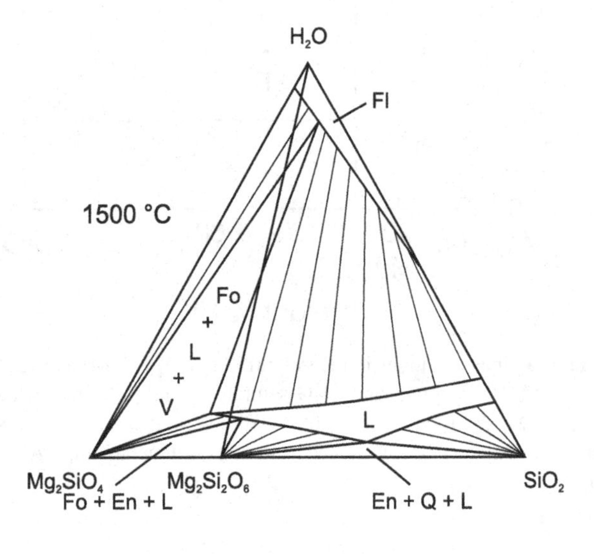

so that enstatite and forsterite coexist at a eutectic with melt of an intermediate composition [*Boyd et al.,* 1964]. The shift of anhydrous peridotite partial melt compositions toward higher normative olivine content with increasing pressure also occurs in experiments performed on natural systems [*Stolper,* 1980]. This compositional shift reflects a decrease in the activity of SiO_2 in peridotite partial melts with increasing pressure and temperature, although the change in SiO_2 concentration is mitigated by a decrease in the activity coefficient for SiO_2 in the melt [e.g., *Sack and Ghiorso,* 1989; *Hirschmann et al.,* 1998].

The addition of H_2O to the system Mg_2SiO_4 – SiO_2 changes melting relations significantly at upper mantle conditions (Figure 4, 5). The incongruent melting behavior of stoichiometric enstatite persists to at least 3.0 GPa, resulting in hydrous peridotite partial melts that have a lower normative olivine content than anhydrous partial melts produced at the same pressure [e.g., *Kushiro et al.,* 1968b; *Kushiro,* 1969; 1972]. At 2.0 GPa, for example, an anhydrous melt coexisting with forsterite and enstatite is olivine normative whereas hydrous melts contain normative quartz (Figure 4). Figure 5 contains schematic isothermal sections of the system Mg_2SiO_4 – SiO_2 – H_2O at 1.0 GPa, based on the experimental results of Kushiro et al. [1968b], summarizing phase relations for coexisting minerals, silicate melt, and aqueous fluid at temperatures of 1300 °C, 1400 °C, and 1500 °C. The 1300 °C isothermal section is analogous to conditions at which aqueous fluid, olivine, and orthopyroxene coexist below the peridotite solidus. Silicate melt coexists only with solid assemblages containing enstatite and/or quartz, due to the lower temperature of the $Mg_2Si_2O_6$ – SiO_2 – H_2O eutectic (Figure 4). At 1400 °C enstatite, forsterite, and hydrous melt coexist at fluid-undersaturated conditions, indicating that the fluid-saturated solidus for a mineral assemblage consisting of enstatite + forsterite occurs at a peritectic between 1300 °C and 1400 °C at 1.0 GPa. Although the liquid field extends to the $Mg_2Si_2O_6$ – H_2O join at this temperature, the melt that coexists with an assemblage of enstatite + forsterite is quartz normative. At 1500 °C, olivine-normative hydrous melts coexist with forsterite or with forsterite and fluid, but only quartz-normative melts coexist with forsterite and enstatite. As discussed below, a trend toward lower normative olivine in hydrous peridotite saturated melts is also observed in experiments performed on natural compositions.

Figure 5. Schematic isothermal, isobaric sections of the system Mg_2SiO_4-SiO_2-H_2O at 1.0 GPa and temperatures of 1300 °C, 1400 °C, and 1500 °C illustrating phase relations relevant to partial melting of hydrous mantle peridotite. Abbreviations as in Figure 3. After Kushiro et al. [1968b].

The temperature of the H_2O-saturated solidus for natural peridotite has been investigated experimentally a number of times [*Kushiro et al.,* 1968a; *Kushiro,* 1970; *Green,* 1973; *Millhollen et al.,* 1974; Mysen and Boettcher, 1975]. Results from experiments carried out in the pressure range of 2.0 to 3.0 GPa, which are summarized in Figure 6, fall into two categories: those that determined the H_2O-saturated solidus to be at 1000 °C [*Kushiro et al.,* 1968a; *Kushiro,* 1970; *Green,* 1973; *Millhollen et al.,* 1974] and those that found the first melt at a temperature of only ~800 °C [*Mysen and Boettcher,* 1975]. This discrepancy may result from the way in which the experiments were carried out. The studies that produced the higher H_2O-saturated solidus temperature either used capsules fabricated from Pt or Mo, which are prone to rapid loss of hydrogen, for short duration runs (0.5 to 3 hours) [*Kushiro et al.,* 1968a; *Millhollen et al.,* 1974] or

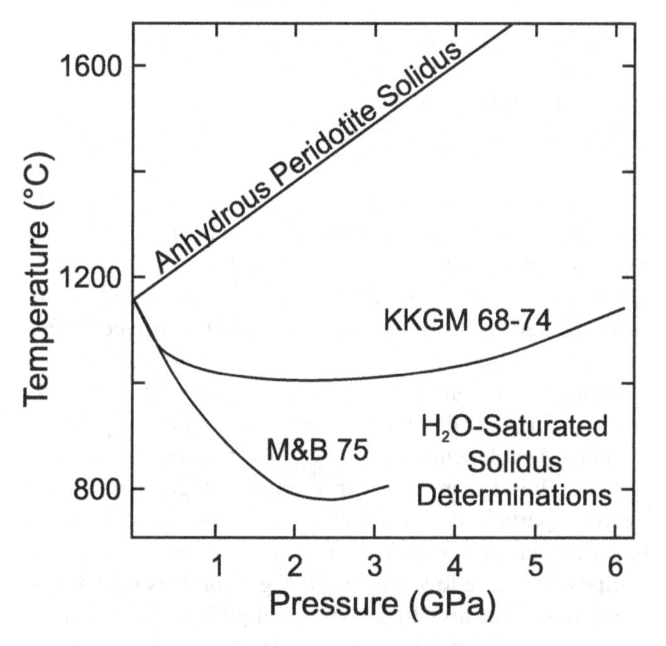

Figure 6. Pressure-temperature diagram summarizing experimental determinations of the fluid-saturated peridotite solidus. The anhydrous peridotite solidus is from Kushiro et al. [1968a]. Two curves representing the experimentally determined H_2O-saturated peridotite solidus are shown. The curve labeled KKGM 68-74 summarizes the experimental determinations of Kushiro et al. [1968a], Kushiro [1970], Green [1973], and Millhollen et al. [1974]. The Kushiro et al. [1968a] experiments extended to pressures of 6.0 GPa, but all other studies were carried out over the pressure range of 1.0 to 3.0 GPa. The curve labeled M&B 75 summarizes the results of melting experiments by Mysen and Boettcher [1975] on a natural peridotite (their composition C). These curves separate the stability fields for super-critical aqueous fluid + crystals from that of melt + crystals.

added relatively small amounts of H_2O (5.7 to 10 wt%) to the charge [*Green,* 1973; *Millhollen et al.,* 1974]. These techniques tend to hinder observation of the first experimentally produced melt at the H_2O-saturated solidus. Conversely, Mysen and Boettcher [1975] carried out long duration experiments (24 to 64 hours) in AgPd capsules and added large amounts of H_2O (20 to 30 wt%) to the starting peridotite. This experimental approach should have enhanced their ability to detect the first melt formed at the H_2O-saturated solidus. On the basis of recent experimental evidence for the durations required to produce near-equilibrium mantle melt compositions [*Baker and Stolper,* 1994], the Mysen and Boettcher [1975] experiments seem the most reliable of the early attempts at determining the H_2O-saturated solidus. It should also be noted that the solidus determined by Mysen and Boettcher [1975] is more consistent with the effects of H_2O on silicate melting as demonstrated in the systems Mg_2SiO_4 – H_2O and $NaAlSi_3O_8$ – H_2O [*Burnham and Davis,* 1974; *Hodges,* 1974]. Determination of the vapor-saturated peridotite solidus is worthy of further experimental investigation, as partial melting near the base of the mantle wedge is likely to begin there. Therefore, its location in pressure-temperature space is critical to a number of issues surrounding arc magma generation.

In recent years, many of the difficulties inherent in performing hydrous melting experiments on natural compositions have been overcome, providing high quality data relating to the compositions of partial melts of hydrous mantle peridotite [*Kushiro,* 1990; *Kawamoto and Hirose,* 1994; *Hirose and Kawamoto,* 1995; *Kawamoto and Holloway,* 1997; *Gaetani and Grove,* 1998]. The results from these studies demonstrate that, although the presence of H_2O expands the stability field of olivine, hydrous partial melts are compositionally similar to those produced by melting anhydrous peridotite. Figure 7 contains plots of wt% SiO_2 versus total FeO + MgO, in wt%, comparing the compositions of hydrous (3.3±0.3 to 6.26±0.10 wt% H_2O) and nominally anhydrous (≤1 wt% H_2O) melts in equilibrium with a spinel lherzolite mineral assemblage at 1.2 GPa from the experimental study of Gaetani and Grove [1998]. With H_2O included in the composition, the SiO_2 contents of the hydrous partial melts are lower than those of the anhydrous melts by ~2 to 3 wt%, producing a positive correlation between SiO_2 and total FeO + MgO (Figure 7a). This correlation does not, at first, appear to be consistent with an expansion of the olivine stability field. However, the hydrous melts are characterized by an elevated SiO_2/(MgO+FeO) ratio relative to the anhydrous melts and when compared on an anhydrous basis they fall along the

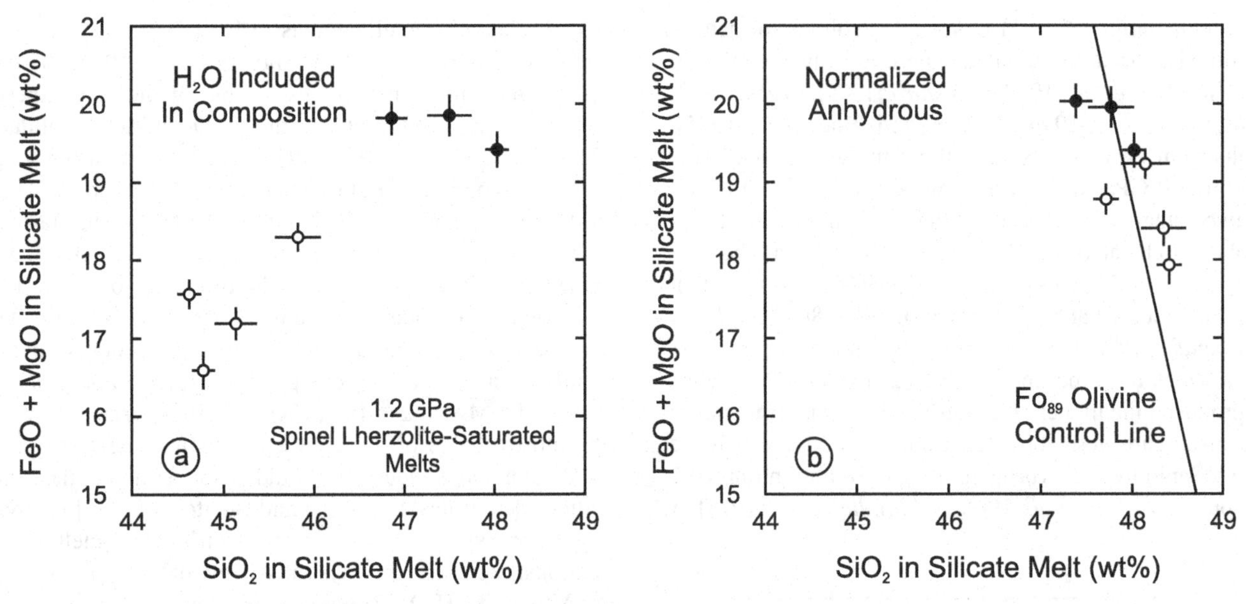

Figure 7. Plot of SiO_2 versus total FeO + MgO, in weight percent, comparing experimentally produced anhydrous (filled symbols) and hydrous (open symbols) silicate melts saturated with a spinel lherzolite assemblage at 1.2 GPa; (a) with H_2O included in the composition; (b) on an anhydrous basis. Solid line illustrates the effect of variations in normative olivine content of the melt. After Gaetani and Grove [1998].

same olivine control line as the nominally anhydrous melts, consistent with lower normative olivine contents (Figure 7b). These results also demonstrate that the presence of ~3 to 6 wt% H_2O dissolved in the melt increases the SiO_2 content by only ~1 wt% on an anhydrous basis.

Thermodynamic controls on the compositions of hydrous peridotite partial melts were examined by Gaetani and Grove [1998], who used the coexistence olivine and orthopyroxene to calculate activity coefficients for SiO_2 ($\gamma_{SiO_2}^{melt}$) and for MgO (γ_{MgO}^{melt}) in partial melts produced experimentally at pressures of 0.9 to 2.0 GPa. They used Equation 12 and the equilibrium:

$$Mg_2SiO_4^{olivine} = \frac{1}{2} Mg_2Si_2O_6^{orthopyroxene} + MgO^{melt} \quad (13)$$

to calculate the activities of SiO_2 and MgO, respectively, in each melt on the basis of the compositions of coexisting olivine and orthopyroxene. Activity coefficients were then calculated by dividing the activity of the component by its molar concentration. Results from the SiO_2 activity calculations are shown in Figure 8a. When the $\gamma_{SiO_2}^{melt}$ values are corrected for the effects of temperature, there is a positive correlation between ln $\gamma_{SiO_2}^{melt}$ and the mole fraction of H_2O dissolved in the liquid (Figure 8a). This trend is consistent with a smaller deviation from ideal mixing for SiO_2 with increasing H_2O, and is opposite to what would be expected from the formation of hydroxyl groups through the reaction of H_2O with bridging oxygens, as discussed above. One possible explanation for this apparent discrepancy is that the correlation is produced by variations in melt composition or the speciation of melt components and, therefore, is only indirectly related to the concentration of H_2O dissolved in the melt. Results from the MgO activity calculations are shown in Figure 8b. In contrast with the results for $\gamma_{SiO_2}^{melt}$, these calculations indicate that the dominant control on γ_{MgO}^{melt} is temperature. Any dependence of γ_{MgO}^{melt} on dissolved H_2O is negligible. The dominant control on the MgO content of hydrous partial melt is, therefore, temperature; hydrous melts are in equilibrium with residual mantle of a given composition have lower FeO and MgO contents than anhydrous melts because they exist as liquids at significantly lower temperatures. The difference in the mixing behaviors of SiO_2 and MgO with respect to H_2O is plausibly related to the role of the former as a network former and of the latter as a network modifier in silicate melts.

In summary, experimental studies demonstrate that the presence of H_2O profoundly affects the melt generation process, lowering the temperature of the peridotite solidus [e.g., *Kushiro et al.,* 1968a; *Green,* 1973; *Mysen and Boettcher,* 1975] and altering the compositions of peridotite partial melts [e.g., *Kushiro,* 1972; 1990; *Hirose and Kawamoto,* 1995; *Gaetani and Grove,* 1998]. The principal compositional difference between anhydrous and hydrous partial melts is that the latter are characterized by lower normative olivine. When compared on an anhydrous basis, hydrous partial melts contain higher SiO_2, lower FeO, and lower MgO than anhydrous melts produced at the same pressure. The compositional differences between hydrous and anhydrous peridotite partial melts are related both to the influence of dissolved H_2O on the activities of other melt components and to the difference in temperature between anhydrous and hydrous melting [e.g., *Kushiro,* 1975; *Gaetani and Grove,* 1998]. Expansion of the olivine stability field under hydrous conditions is a characteristic of melts generated in both analog (e.g., $CaO - MgO - Al_2O_3 - SiO_2 - H_2O$) and natural systems.

4. SEGREGATION OF HYDROUS PARTIAL MELT FROM MANTLE PERIDOTITE

Partial melting of peridotite is an intergranular process. At textural equilibrium, the grain-scale distribution of partial melt within the residual solid is determined by the magnitude of the interfacial energy between adjacent mineral grains (γ_{ss}) relative to that where a crystal is in contact with melt (γ_{sm}). For a partially molten polycrystalline aggregate in which crystal-melt interfaces have constant mean curvature, the distribution of partial melt under hydrostatic conditions is determined by the dihedral angle, θ, that forms where a pocket of melt is bounded by two mineral grains. For a monomineralic aggregate, the relationship between dihedral angle and interfacial energy is:

$$\cos\frac{\theta}{2} = \frac{\gamma_{ss}}{2\gamma_{sm}} \tag{14}$$

Given the simplifying assumptions discussed above, it can be shown geometrically that small amounts of melt will form isolated pockets at 4 grain junctions when $\theta > 60°$. For polycrystalline aggregates in which $\theta \leq 60°$, however, it becomes energetically favorable for the melt to wet three-grain junctions, so that an interconnected network is formed even for very small melt fractions [e.g., *Smith,* 1964; *Beere,* 1975; *Bulau et al.,* 1979; *von Bargen and Waff,* 1986; *Watson et al.,* 1990].

In nature, partially molten aggregates are characterized by a range of θ values due to the dependence of interfacial energy on crystallographic orientation. Use of a mean θ value to predict melt distribution in monomineralic rocks is supported by the experimental results of Daines and Richter [1988] that demonstrate the presence of an interconnected network in an olivine aggregate ($20° \leq \theta \leq 50°$) containing as little as 1% anhydrous partial melt. Melt distribution in

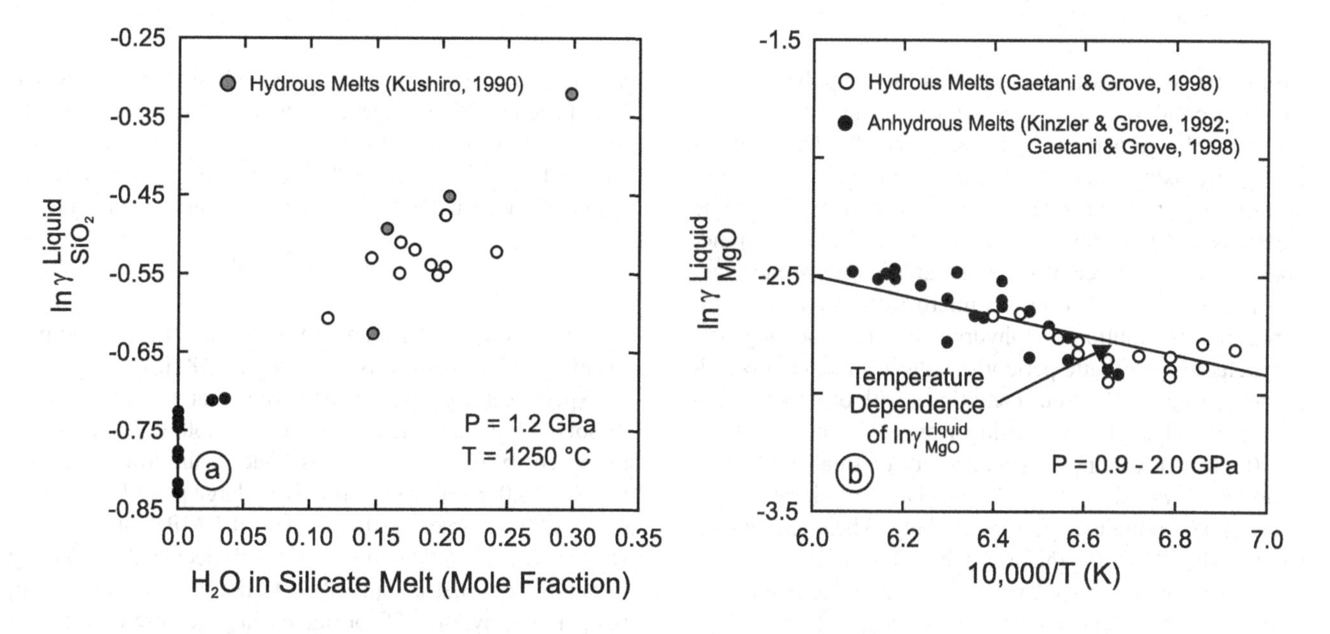

Figure 8. (a) Plot showing the relationship between the molar concentration of H_2O dissolved in a silicate melt and the activity coefficient for SiO_2 in silicate melt for nominally anhydrous (filled symbols) and H_2O-bearing (open and shaded symbols) melts saturated with a mantle peridotite mineral assemblage (Oliv + Opx ± Cpx ± Sp) at 1.2 GPa and 1250 °C. (b) Plot showing the relationship between temperature and the activity coefficient for MgO in silicate melt for nominally anhydrous (filled symbols) and H_2O-bearing (open and shaded symbols) saturated with a mantle peridotite mineral assemblage (Oliv + Opx ± Cpx ± Sp ± Gt) at 0.9 to 2.0 GPa and 1170 °C to 1370 °C. After Gaetani and Grove [1998].

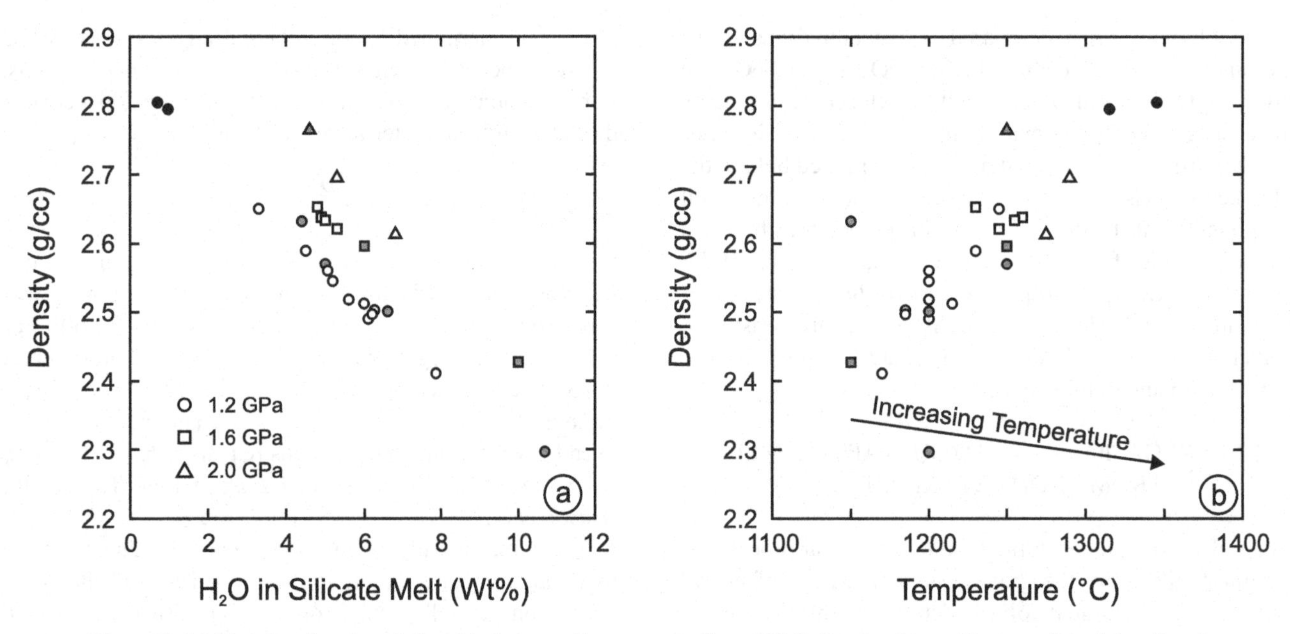

Figure 9. (a) Plot showing the relationship between the concentration of H_2O dissolved in a silicate melt and density of the melt, in g/cc, for nominally anhydrous (filled symbols) and H_2O-bearing (open and shaded symbols) melts saturated with a mantle peridotite mineral assemblage (Oliv + Opx ± Cpx ± Sp ± Gt) at 1.2 GPa (circles), 1.6 (squares), and 2.0 GPa (triangles) and temperatures of 1150 °C to 1350 °C calculated as discussed in the text. Open and filled symbols are data from Gaetani and Grove [1998]; shaded symbols are data from Kushiro [1990]. (b) Plot showing the relationship between temperature and melt density for experiments shown in (a). Arrow indicates the effect of increasing temperature on melt density.

the mantle wedge is also influenced by the polymineralic nature of peridotite (olivine + orthopyroxene ± clinopyroxene ± spinel ± garnet) and by the presence of H_2O. The presence of pyroxene tends to limit melt connectivity in anhydrous peridotite; melt is distributed along olivine-only triple junctions and four-grain junctions, and at four-grain junctions made up of three olivines and an orthopyroxene grain [*Toramaru and Fujii,* 1986]. A microstructural study comparing the distribution of anhydrous and H_2O-bearing partial melts in a synthetic peridotite indicates that connectivity increases under hydrous conditions, with orthopyroxene-melt dihedral angles decreasing from 70° (anhydrous) to 52° (hydrous), enhancing connectivity of small melt fractions [*Fujii et al.,* 1986]. On the basis of these results hydrous partial melts appear more likely to be interconnected in mantle lherzolite than are anhydrous melts.

Given an interconnected network of grain-edge channels, it is possible for partial melt to move independently of the residual solid via porous flow [e.g., *Watson,* 1982; *McKenzie,* 1984; *Scott and Stevenson,* 1986; *von Bargen and Waff,* 1986]. The extent to which melt remains in contact with residual solid during partial melting of hydrous peridotite is an especially important consideration, given that segregation of the melt removes most of the H_2O, limiting continued melting. The rate at which small melt fractions segregate from residual solid is dependent upon the viscosity (μ) of the melt and the density contrast ($\Delta\rho$) between the melt and residual solid [e.g., *McKenzie,* 1985]. The velocity of melt percolation driven by a matrix-melt density contrast is given by:

$$V = \frac{\Delta\rho \times g \times k}{\mu} \tag{15}$$

where g is the acceleration due to gravity and k is the permeability of the matrix [*Stolper et al.,* 1981].

Experimental studies demonstrate that the presence of dissolved H_2O has dramatic effects on both the density and the viscosity of silicate liquids. Due to the low molecular weight (18.016 g/mol) and relatively large partial molar volume (22.9 cc/mole at 1000 °C and 0.1 MPa) of dissolved H_2O, adding ~1 wt% to a basaltic melt decreases its density by an amount that is equivalent to the effect of increasing temperature by ~400 °C or decreasing pressure by 0.5 GPa [*Ochs and Lange,* 1999]. The effects of H_2O and of temperature on the densities of experimentally produced peridotite partial melts were calculated using the partial molar volumes and compressibilities determined by Lange and Carmichael [1987], Kress and Carmichael [1991], and Ochs and Lange [1999], and are presented in Figure 9. Although anhydrous partial melts exist as liquids at higher temperatures than hydrous melts, which tends to decrease density due to thermal expansion, the density change produced by the presence of dissolved H_2O is large enough to overwhelm the temperature effect (Figure 9b). Experimental determinations of the effect of H_2O on melt viscosity are limited to siliceous compositions, but they indicate that adding ~2 wt% H_2O can reduce melt viscosity by up to ~3 orders of magnitude, which is a decrease equivalent to that produced by increasing temperature by ~200 °C [*Lange,* 1994]. The viscosity decrease produced by dissolved H_2O in peridotite partial melts was estimated by using the model of Shaw [1972] to calculate the viscosities of experimentally produced peridotite partial melts. The results from these calculations suggest that the viscosity decrease produced by dissolved H_2O is not as large in mafic melts as that determined experimentally in felsic melts, but that partial melts containing ~10 wt% dissolved H_2O are less viscous that anhydrous melts by a factor of 25 (Figure 10). Overall, the combination of enhanced connectivity, decreased melt density, and lowered viscosity favor efficient melt segregation in the

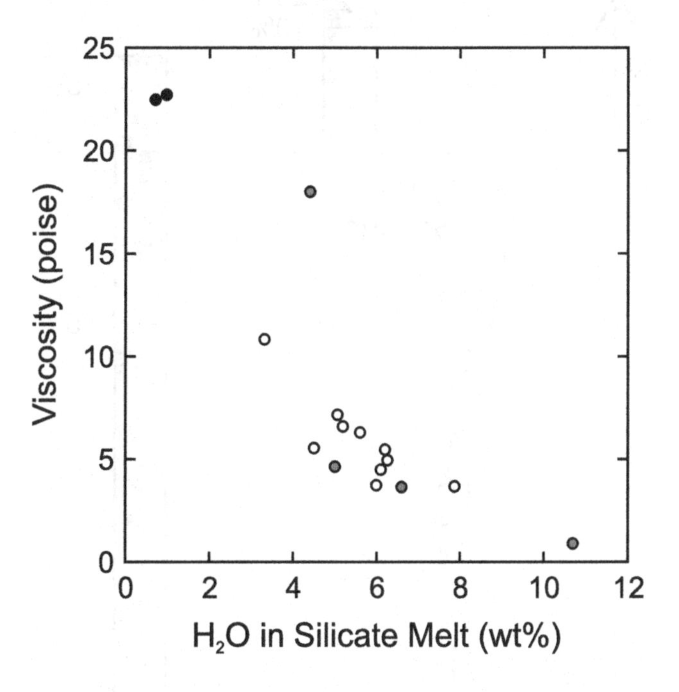

Figure 10. Plot showing the relationship between the concentration of H_2O dissolved in a silicate melt and viscosity of the melt, in poise, for nominally anhydrous (filled symbols) and H_2O-bearing (open and shaded symbols) melts saturated with a mantle peridotite mineral assemblage (Oliv + Opx ± Cpx ± Sp) at 1.2 GPa and temperatures of 1150 °C to 1350 °C calculated using the model of Shaw [1972]. Note that these calculations do not account explicitly for the effects of pressure on melt viscosity.

mantle wedge; the partial melting process is, therefore, likely to be near-fractional.

5. EFFECT OF H_2O ON MELT PRODUCTIVITY

Melt productivity refers to the amount of melt produced in response to a given change in temperature or pressure, and is an especially important consideration in the mantle wedge because the length of the melting column is limited to a region between the Benioff-Wadati zone and the base of the lithosphere [e.g., *Plank and Langmuir,* 1988]. Isobaric productivity is the increase in melt fraction (F) for a given temperature increase, $(\partial F/\partial T)_P$, whereas polybaric, isentropic productivity is the increase in melt fraction for a given pressure decrease, $(-\partial F/\partial P)_S$. Asimow et al. [1997] showed that the most significant factors controlling polybaric melt productivity during adiabatic ascent of mantle peridotite are the isobaric productivity, $(\partial F/\partial T)_P$, and the pressure-temperature slopes of constant melt fraction contours, $(\partial T/\partial P)_F$. The strong dependence of $(-\partial F/\partial P)_S$ on $(\partial F/\partial T)_P$ means that, although melting of hydrous mantle peridotite is likely to be a polybaric, near-fractional process, consideration of isobaric batch melting provides basic insights into melt generation processes in the mantle wedge.

Calculations carried out by Gaetani and Grove [1998], using a model for olivine-melt equilibrium, and by Hirschmann et al. [1999], using a thermodynamic model for mineral-melt equilibrium (MELTS), indicate that $(\partial F/\partial T)_P$ is lowest at small extents of melting, and increases continuously as hydrous partial melting proceeds; the relationship between temperature and melt fraction is strongly curved. Further, the results from both sets of calculations indicate that during isobaric heating $(\partial F/\partial T)_P$ is much lower for hydrous peridotite than for anhydrous peridotite. This is due to the monotonically decreasing concentration of dissolved H_2O in the melt with increasing extent of partial melting, which weakens the melting point depression effect. This can be seen in Figure 11, which is a comparison of peridotite partial melting experiments performed under anhydrous conditions [*Hirose and Kushiro,* 1993] and with 0.5 wt% H_2O added to the peridotite [*Hirose and Kawamoto,* 1995]. At 1.0 GPa a temperature increase of 50 °C increases F from 0.12 to 0.20 at anhydrous conditions, whereas a temperature increase of 100 °C is required to increase *F* from 0.13 to 0.20 at hydrous conditions. Although H_2O behaves as a flux and increases the extent to which peridotite partially melts at a given set of pressure-temperature conditions, closed-system melting at H_2O-undersaturated conditions is an inefficient process. If partial melting begins at the hydrous solidus, a significantly longer melting column is required to

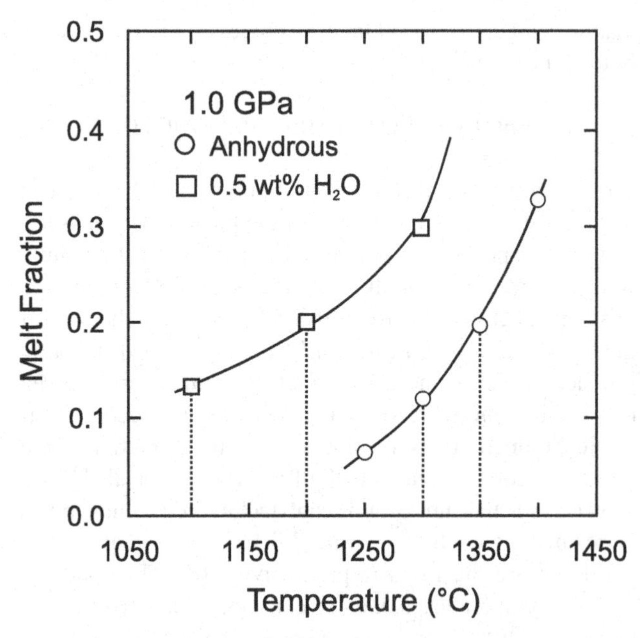

Figure 11. Plot of temperature versus melt fraction comparing results from anhydrous [*Hirose and Kushiro,* 1993] and hydrous [*Hirose and Kawamoto,* 1995] peridotite partial melting experiments carried out on KLB-1. After Hirose and Kawamoto [1995].

produce a given amount of partial melt under hydrous conditions relative to anhydrous partial melting (Figure 12).

The melting point lowering effect of H_2O can be quantified by examining the increase in melt fraction produced by increasing the concentration of H_2O in the peridotite at constant pressure and temperature, $(\partial F/\partial C_{H_2O})_{P,T}$. Figure 13 is a plot of melt fraction versus H_2O in the peridotite comparing the anhydrous peridotite melting experiments of Hirose and Kushiro [1993] with the hydrous experiments of Hirose and Kawamoto [1995]. From this plot it can be seen that the relationship between melt fraction and H_2O in the peridotite is approximately linear at a given pressure and temperature, and is similar to the relationship inferred from a suite of basaltic glasses from the Mariana trough by Stolper and Newman [1994]. It can also be seen that $(\partial F/\partial C_{H_2O})_{P,T}$ increases with increasing temperature, in agreement with the calculations of Gaetani and Grove [1998] and of Hirschmann et al. [1999].

6. CONSTRAINTS ON MANTLE MELTING DERIVED FROM PRIMITIVE ARC LAVAS

Petrologic studies of primitive arc lavas (i.e., those with high Mg/(Mg + ΣFe) that demonstrably represent liquids) are an important source of constraints on the process of peridotite partial melting and melt transport in the sub-arc mantle. When arc magmas segregate from the mantle and ascend through the overlying crust, their compositions are typically modified by processes, such as fractional crystallization, magma mixing, and assimilation of wall rock, that obscure the compositional signatures imparted by the mantle wedge and subducted lithosphere [*Gill,* 1981]. Further, the strong pressure-dependence of the solubility of H_2O in silicate melts causes degassing during the ascent and/or eruption of most magma, so that pre-eruptive H_2O contents can seldom be directly measured. Nevertheless, basalts, basaltic andesites, and andesites that have molar Mg/(Mg + ΣFe) > 0.7 and are saturated with Fo_{90} to Fo_{94} olivine erupt in arc settings, and these lavas are likely to preserve a record of processes that occur in the mantle wedge and subducted lithosphere. Here we discuss evidence from experimentally determined phase relations for these lavas, combined with

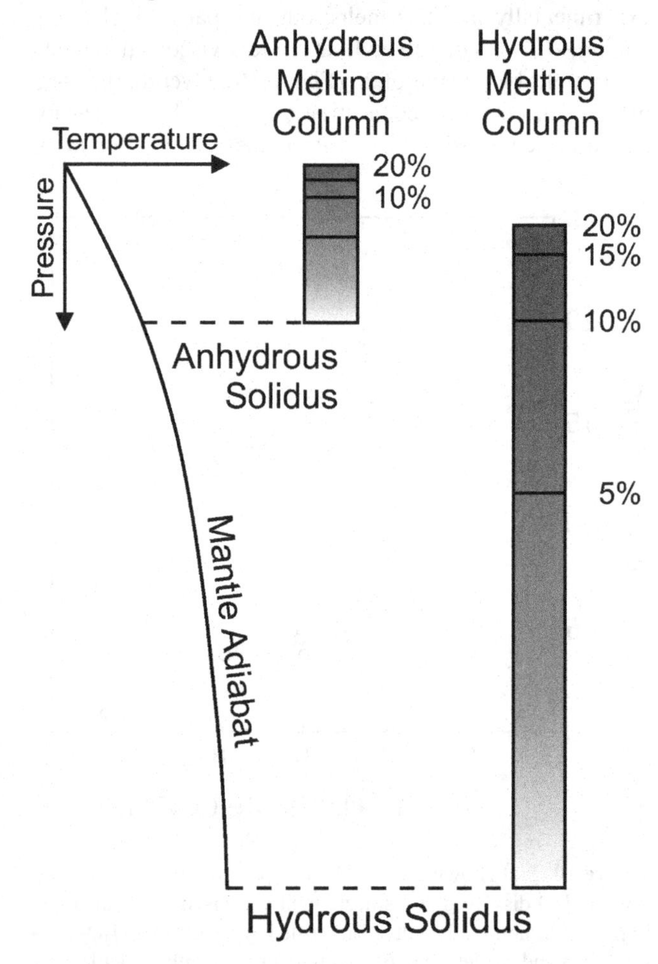

Figure 12. Schematic pressure-temperature diagram comparing the amount of melt produced by adiabatic decompression partial melting of mantle peridotite under anhydrous (left) and hydrous (right) conditions.

petrologically based estimates for pre-eruptive H_2O contents, for the conditions under which melt generation occurs in the mantle wedge.

There are a number of approaches that can be used to infer the pre-eruptive H_2O content of a magma on the basis of petrology and phase equilibria. Perhaps the most direct approach is to perform experiments to determine the liquidus phases for a primitive lava with different amounts of H_2O added. The correct pre-eruptive H_2O content will reproduce the phenocryst assemblage observed in the lava or, if little crystallization has occurred, may produce multiple saturation with peridotite minerals at the conditions of segregation from the mantle [e.g., *Baker et al.,* 1994; *Gaetani et al.,* 1994]. Thermodynamic models for mineral/melt equilibria under hydrous conditions can be used to estimate the concentration of H_2O in the magma from which a given mineral crystallized on the basis of the compositions of coexisting phases [e.g., *Housh and Luhr,* 1991; *Sisson and Grove,* 1993]. The presence of dissolved H_2O significantly affects the sequence in which minerals crystallize from magma, altering the compositional path followed by residual liquids (the "liquid line of descent"). This effect of H_2O on fractional crystallization can be identified in the compositional trends recorded in arc lava suites, and used to infer pre-eruptive H_2O contents [e.g., *Gaetani et al.,* 1993; *Sisson and Grove,* 1993]. Direct measurement of H_2O in

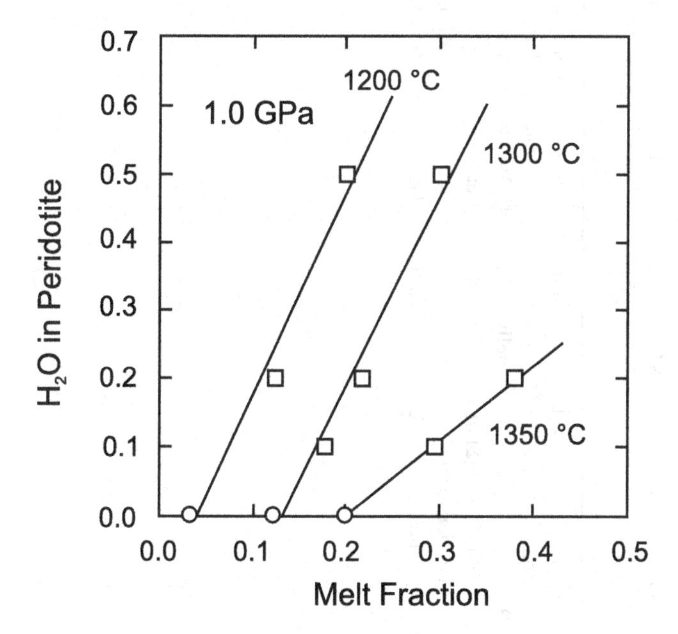

Figure 13. Plot of melt fraction versus bulk H_2O content comparing results from anhydrous [*Hirose and Kushiro,* 1993] and hydrous [*Hirose and Kawamoto,* 1995] peridotite partial melting experiments carried out on KLB-1.

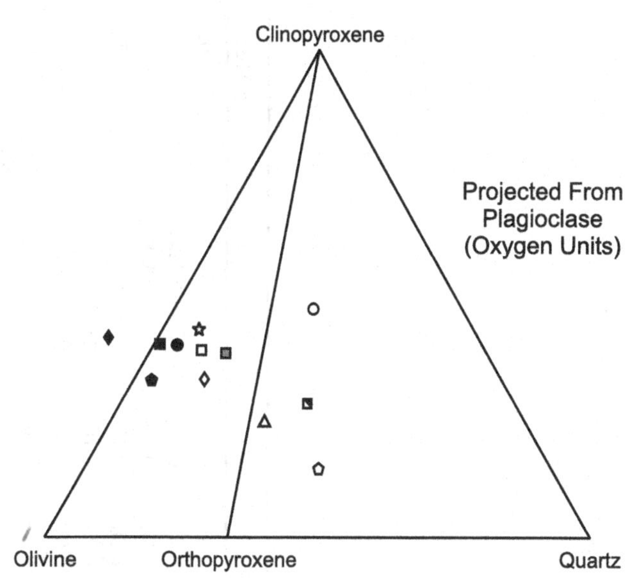

Figure 14. Compositions of primitive arc lavas (Table 1) projected into the Olivine – Clinopyroxene – Quartz pseudoternary of Grove [1993]. Projection uses oxygen units. The line from orthopyroxene to clinopyroxene separates quartz-normative from olivine-normative lavas. Squares represent lavas from the Cascades (filled: 82-72f; open: 95-15; shaded: 82-94a; diagonal fill: 85-41b); circles represent lavas from the Aleutians (filled: ID16; open: ADK-53); diamonds represent lavas from the Mexican arc (filled: M102; open: M108); pentagons represent lavas from Japan arc (filled: SD-438; open: TGI); star represents lava from the Indonesian arc (Avg-MGB); triangle represent lava from the Tongan arc (3-44). See Table 1 for compositions and references.

melt inclusions that form prior to magmatic degassing can also be used to infer pre-eruptive H_2O content [e.g., *Anderson,* 1973; 1974; *Sisson and Layne,* 1993; *Sisson and Bronto,* 1998]. Ideally, several of these methods are combined in order to provide the most reliable estimate of pre-eruptive magmatic H_2O content. In the following discussion, pre-eruptive H_2O contents have been estimated for primitive arc lavas on the basis of one or more of the above criteria.

Primitive lavas have been identified in the Aleutian, Cascade, Indonesian, Japan, Mexican, and Tongan arcs [*Anderson,* 1973; 1974; 1976; *Kay,* 1978; *Anderson,* 1979; *Tatsumi and Ishizaka,* 1982; *Ishizaka and Carlson,* 1983; *Tatsumi et al.,* 1983; *Nye and Reid,* 1986; *Falloon and Crawford,* 1991; *Baker et al.,* 1994; *Carmichael et al.,* 1996; *Bacon et al.,* 1997; *Shimoda et al.,* 1998; *Sisson and Bronto,* 1998]. The major element compositions of a subset of these lavas, including petrologic estimates for their pre-eruptive H_2O contents, are given in Table 1 and their anhydrous compositions are plotted on a pseudo-ternary projection in Figure 14. In the case of the Aleutian, Cascade,

Table 1. Major-element compositions of primitive lavas from select continental and oceanic arcs.

Sample(Arc)[Ref]	SiO_2	TiO_2	Al_2O_3	Fe_2O_3	FeO	MnO	MgO	CaO	Na_2O	K_2O	P_2O_5	LOI	Sum	H_2O	Mg#
ADK-53(A)[1]	55.5	0.86	15.5	4.70	1.98	0.07	5.58	9.51	3.22	1.47	0.32	—	98.71	—	0.62
ID16(A)[2]	48.94	0.70	16.01	1.06	7.95	0.17	11.42	10.89	2.21	0.52	0.12	0.33	100.32	—	0.70
95-15(C)[3]	50.93	0.66	15.54	8.18	—	0.14	10.65	9.94	2.73	0.76	0.28	0.21	99.83	4.5	0.72
82-94a(C)[4]	52.50	0.72	15.30	2.93	4.43	0.14	10.17	9.61	3.01	0.58	0.22	0.51	100.12	4.5	0.72
85-41b(C)[4]	57.87	0.60	14.67	—	5.69	0.11	8.88	8.13	3.18	0.72	0.16	1.28	100.01	6.5	0.74
82-72f(C)[5]	47.7	0.59	18.5	—	8.20	0.15	10.52	12.02	2.16	0.07	0.06	<0.01	100.72	<0.2	0.70
Avg-MGB(I)[6]	49.1	0.82	15.9	—	9.0	0.17	10.5	11.3	2.21	0.34	0.10	0.10	99.4	0.3	0.68
TGI(J)[7]	59.59	0.44	13.55	—	6.32	0.12	9.65	6.24	2.66	1.30	0.13	—	100.0	~4.0	0.73
SD-438(J)[7]	49.76	1.02	15.48	—	8.87	0.16	11.68	8.91	2.60	1.28	0.22	—	99.98	<0.2	0.70
M102(M)[8]	49.40	0.99	14.84	3.06	4.88	0.12	11.63	8.30	3.53	1.40	0.44	0.60	99.55	>3.5	0.73
M108(M)[8]	53.09	0.79	16.76	2.81	4.56	0.14	8.80	8.62	3.59	0.83	0.23	0.13	100.38	>3.5	0.69
3-44(T)[9]	54.35	0.20	10.67	—	9.41	0.19	14.99	8.66	1.14	0.35	0.03	0.05	100.01	—	0.74

Notes: The sample name is followed by a letter in parentheses indicating the arc where it was collected and a superscript indicating the reference from which the composition was taken. Arcs: A = Aleutians; C = Cascades; I = Indonesian; J = Japan; M = Mexican; T = Tonga. References: 1. Kay [1978]; 2. Nye and Reid [1986]; 3. Grove et al. [2001]; 4. Baker et al. [1994]; 5. Donnelly Nolan et al. [1991]; 6. Sisson and Bronto [1998] 7. Tatsumi [1982]; 8. Carmichael et al [1996]; 9. Falloon and Crawford [1991]. Mg# = molar Mg/[Mg + ΣFe] with all Fe as Fe^{2+}. A dash in the Fe_2O_3 column indicates that analysis reported all Fe as FeO. A dash in the FeO column indicates either that FeO was not determined or that analysis reported all Fe as FeO. All oxide concentrations are in wt%. The H_2O column reports the estimated pre-eruptive H_2O content of the magma cited in the reference. A dash in the H_2O column indicates that no information is available on pre-eruptive H_2O.

Japan, and Tongan arcs, the lavas have either been the subject of experimental investigations or evaluated on the basis of results from experimental studies in order to place constraints on the temperature and depth range over which melt generation occurs, and on the pre-eruptive H_2O content of arc magmas [*Tatsumi,* 1982; 1986; *Bartels et al.,* 1991; *Draper and Johnston,* 1992; *Baker et al.,* 1994; *Danyushevsky et al.,* 1997; *Elkins Tanton et al.,* 2001].

The primitive arc lavas reported in Table 1 span a compositional spectrum from silica-undersaturated shoshonite and absarokite, to tholeiite and high-alumina olivine tholeiite through basaltic andesite to magnesian andesite. When a measure of the degree of normative silica-saturation (i.e., the normative olivine/[olivine + quartz] ratio) is plotted against estimates for pre-eruptive H_2O content, it can be seen that the primitive tholeiites are low in H_2O and that the primitive basaltic andesites and magnesian andesites define a high-H_2O end-member at ~4 to 6 wt.% H_2O (Figure 15). Experiments performed at varying pressures and H_2O contents indicate that all of these magmas equilibrated with peridotite near the top of the mantle wedge, at depths of 30 to 60 km for the near-anhydrous tholeiites, and 30 to 45 km for the hydrous basaltic andesites, magnesian andesites and boninites (Figure 16). For example, Bartels et al. [1991] showed experimentally

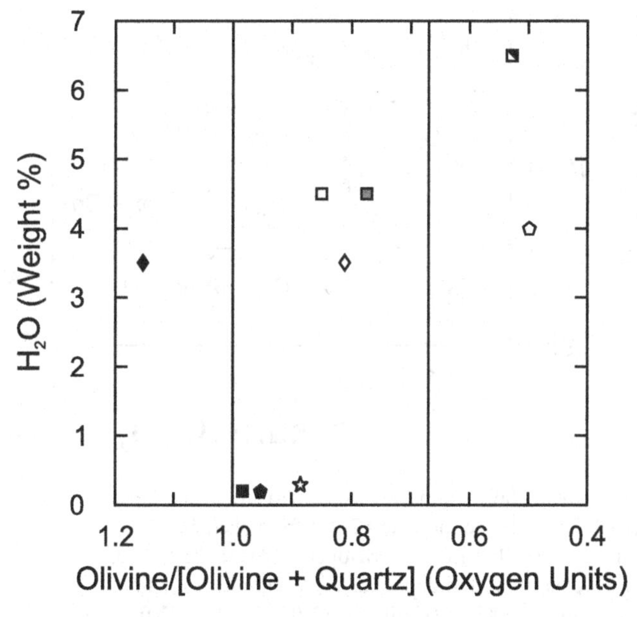

Figure 15. Plot of normative Olivine/[Olivine + Quartz] versus estimate for pre-eruptive H_2O content, in wt%, showing primitive arc lavas from Figure 14 and Table 1. Vertical line at Oliv/[Oliv + Qtz] = 1.0 separates silica-undersaturated from olivine-normative lava compositions. Vertical line at Oliv/[Oliv + Qtz] = 0.67 separates olivine-normative from quartz-normative lava compositions.

that primitive high-alumina olivine tholeiites (HAOT) from Medicine Lake volcano, directly east of the Mt. Shasta area in the S. Cascades, are multiply saturated under near-anhydrous conditions with olivine + orthopyroxene + high-Ca clinopyroxene + plagioclase + spinel at 1290 °C and 1.1 GPa (Figure 16a). Similarly, the primitive Aleutian tholeiite studied by Draper and Johnston [1992] last equilibrated with the same residual assemblage at ~1300 °C and 1.2 GPa under anhydrous conditions. Primitive basaltic andesites and magnesian andesites from the Cascade and Japan arcs are saturated with olivine + orthopyroxene at ~1.1 GPa and ~1200 °C, ~100 °C cooler than the mantle temperatures indicated by the tholeiites, and with H_2O contents of ~4 to 6 wt% [*Tatsumi,* 1982; 1986; *Baker et al.,* 1994]. At present, no experimental constraints exist on the depth of origin for primitive magmas from the Mexican arc, but petrologic estimates of pre-eruptive H_2O contents are ~3 to 6 wt.% and these lavas may record lower extents of partial melting in comparison to the associated andesitic magmas [*Carmichael et al.,* 1996].

Figure 17 shows the depth of melt segregation (filled circles) beneath the Cascade arc in the vicinity of Mt. Shasta and Medicine Lake Volcano inferred by Elkins Tanton et al. [2001] on the basis of pressures of multiple saturation for primitive HAOT lavas, calculated using the algorithm for spinel-lherzolite melting of Kinzler and Grove [1992a]. These near-anhydrous lavas indicate that there is a progressive increase in the depth at which melt segregates from its mantle residue along a 75-km-long, east-west transect across the arc. Beneath Mt. Shasta, magmas segregate near the base of the crust, at a depth of ~30 km and a temperature of ~1300 °C. That depth progressively increases to ~60 km and temperature increases to ~1450 °C at the eastern end of the transect. This pressure increase parallels the corner flow lines calculated by Furakawa [1993] and, therefore, may reflect hot, low-viscosity mantle flowing into the wedge corner. A similar temperature-depth structure has been suggested for the NE Japan arc on the basis of phase equilibrium experiments carried out on primitive lavas with varying amount of H_2O added [*Tatsumi et al.,* 1983].

7. MELTING PROCESSES IN THE SUB-ARC MANTLE WEDGE

To understand the nature of melt generation and transport within the mantle wedge, it is necessary to consider (1) the thermal structure of the wedge and subducted lithosphere, (2) the phase equilibria for mantle wedge and subducted lithosphere components, (3) the style of mantle flow in the wedge and its influence on melt transport, and (4) the chem-

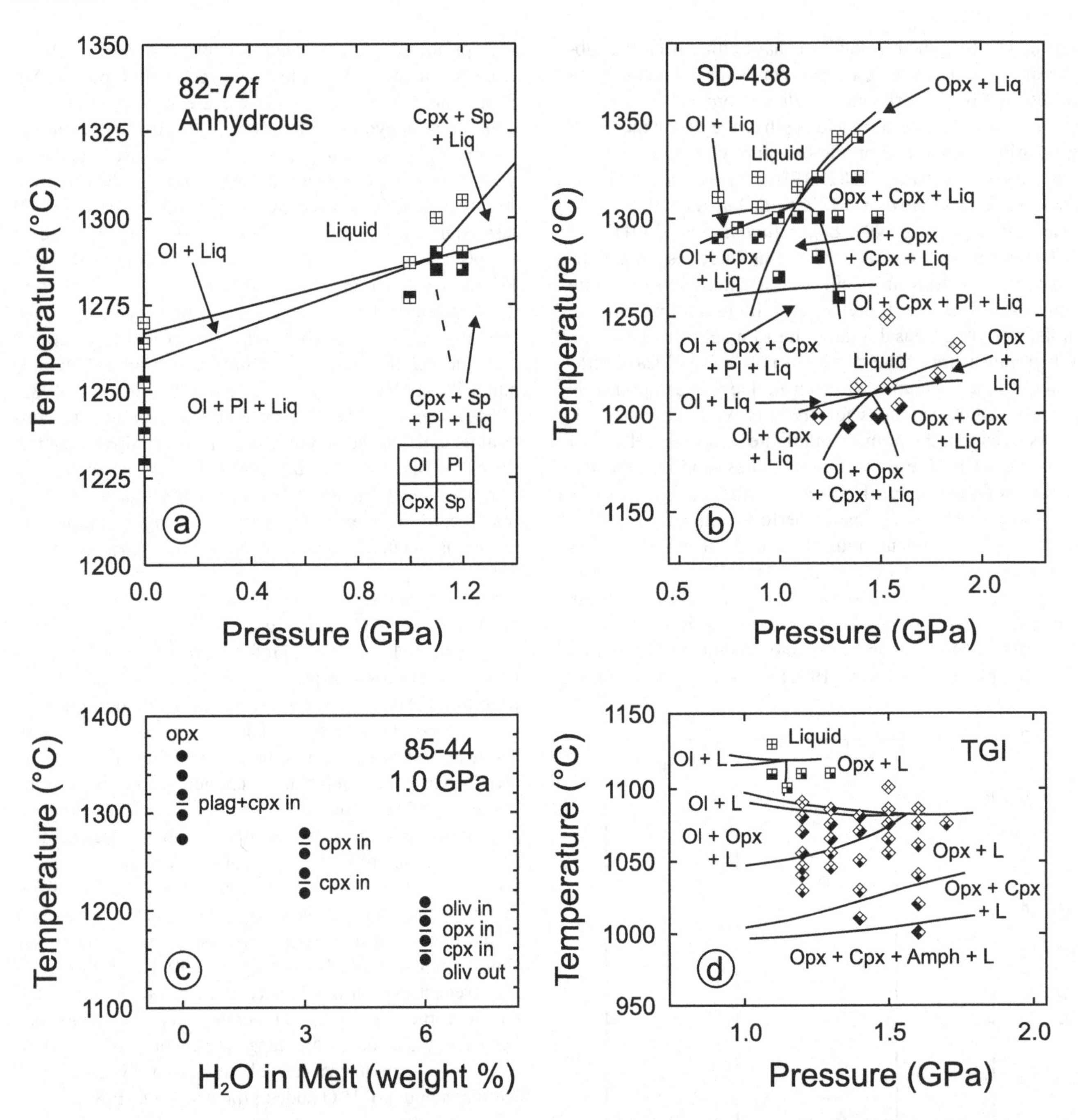

Figure 16. Elevated pressure, near-liquidus phase relations on primitive arc lavas from the Cascades (a and c) and Japan (b and d) arcs. (a) Results from anhydrous experiments on high-alumina olivine tholeiite 82-72f from Medicine Lake volcano studied by Bartels et al. [1991]. (b) Melting relations of basalt SD-438 from the Setouchi volcanic belt. After Tatsumi [1982]. Squares represent phase relations under anhydrous conditions and diamonds represent phase relations with 3.8 wt.% H_2O added. (c) Phase relations for primitive basaltic andesite 85-44 (similar to 82-94a in Table 1) from Mt. Shasta, N. California as a function of increasing H_2O content. After Baker et al. [1994]. (d) Phase relations of high magnesian andesite TGI with 7.5 wt% H_2O added (squares) and under H_2O-saturated conditions (diamonds). After Tatsumi [1981].

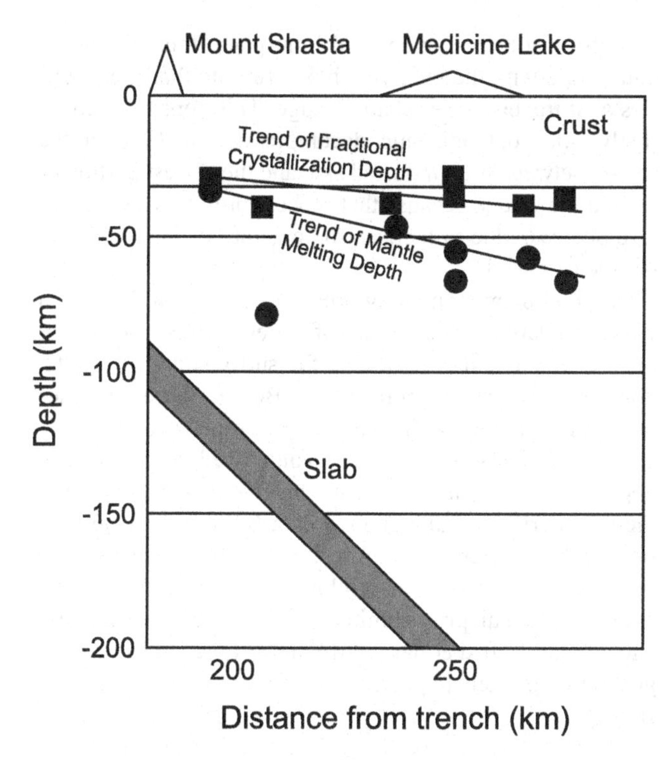

Figure 17. Cross section through the S. Cascade arc showing the position of the subducted lithosphere and the mantle segregation depths of primitive basalts. After Elkins Tanton et al. [2001]. See Tatsumi et al. [1983] for a similar view of the NE Japan arc.

ical characteristics of the mantle wedge and the volatile-rich inputs from the subducted lithosphere. Considering the potential complexity that may be introduced from the existence of so many independent variables, it seems surprising that a single-stage "slab melting" model continues to receive attention in the geochemical literature [e.g., *Drummond and Defant,* 1990; *Defant and Kepezhinskas,* 2001]. In the following discussion, constraints derived from the phase equilibria of primitive arc lavas, discussed above, are combined with experimentally determined physical properties of hydrous mantle peridotite and of H_2O-bearing melt, and numerical models for flow in the wedge and the thermal structure of the sub-arc mantle to constrain the melt generation processes that produce arc magmas. Three styles of melt generation are considered: (1) adiabatic decompression of anhydrous peridotite; (2) hydrous partial melting during ascent of buoyant diapers; (3) reactive porous flow involving percolation of hydrous partial melt through an inverted thermal gradient. The experimental evidence for the conditions at which arc magmas segregate from their mantle residue is consistent with more than one melting process operating simultaneously beneath some arcs.

7.1. Anhydrous Decompression Melting

Melt generation in the mid-ocean ridge tectonic regime occurs when hot, nearly anhydrous mantle peridotite ascends beneath spreading centers in response to extension of the lithosphere, decompresses along the mantle adiabat, intersects the solidus, and partially melts [e.g., *Klein and Langmuir,* 1987; *Kinzler and Grove,* 1992a; b; *Langmuir et al.,* 1992]. The pattern of flow in the sub-arc mantle, shown schematically in Figure 18, is driven by thermal and mechanical interactions between the subducting lithosphere and the mantle wedge, rather than extension of lithospheric plates, so that peridotite does not ascend vertically beneath the volcanic front. However, numerical models that include the temperature- and stress-dependence of mantle viscosity predict the ascent of a hot, low-viscosity tongue of mantle from the back arc into the wedge corner, where it overturns

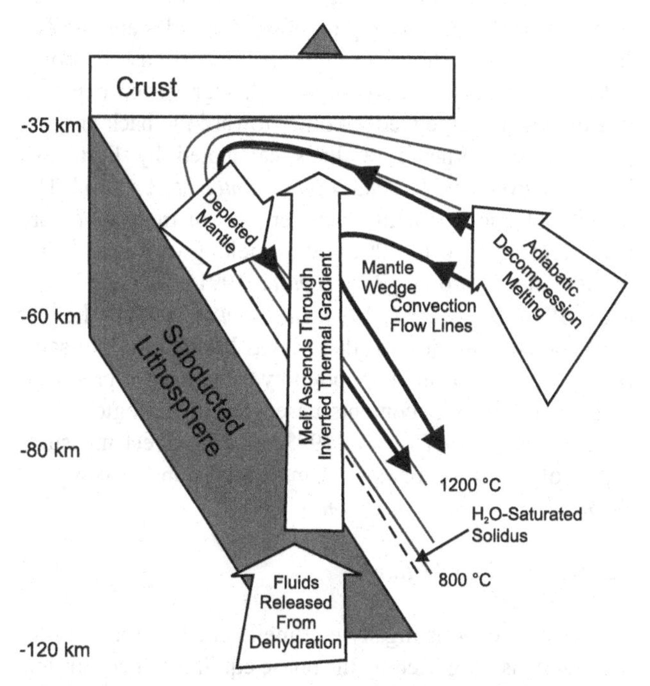

Figure 18. Schematic cross section through a convergent margin illustrating mantle flow and melt generation processes in the sub-arc mantle. First, near-anhydrous adiabatic decompression melting as it is advected into the wedge depletes the peridotite. Next, H_2O is released from the subducted lithosphere by dehydration of minerals in the subducted sediments, oceanic crust, and hydrated lithospheric mantle or by vapor-saturated melting of sediments and/or oceanic crust. Finally, an H_2O-rich phase rises into the overlying mantle wedge, initiating a second episode of partial melting in the inverted temperature gradient above the subducted slab. The H_2O content of the melt decreases as it encounters and dissolves mantle minerals in the hotter, overlying peridotite.

and descends through viscous coupling with the subducted lithosphere [*Andrews and Sleep,* 1974; *Bodri and Bodri,* 1978; *Furakawa,* 1993]. This pattern of mantle flow produces temperatures that are comparable to those required to partially melt anhydrous peridotite, suggesting that adiabatic decompression melting can occur on the upwelling limb of the mantle wedge.

The low pre-eruptive H_2O contents and high liquidus temperatures associated with some primitive lavas from the Japan, Cascades, and Indonesian arcs provide petrologic support for the occurrence of adiabatic decompression melting in the sub-arc mantle. For example, a plausible interpretation of the experimental results of Bartels et al. [1991], discussed above, is that primitive HAOT magmas are produced by adiabatic decompression melting and that melt continuously reequilibrates with peridotite as the mantle is advected into the wedge. Melt segregates from residual mantle at the base of the crust. Low-H_2O HAOT are found throughout the back-arc region of the Cascades and the NE Japan arc and, as shown in Figure 17 and discussed above, define a progressive increase of melt segregation depth as one moves from the wedge corner toward the back arc that parallels the corner flow lines calculated by Furakawa [1993] [*Tatsumi et al.,* 1983; *Elkins Tanton et al.,* 2001]. The major and trace element characteristics of these lavas are consistent with ~6 to 10% partial melting of a near-anhydrous spinel lherzolite [*Bartels et al.,* 1991; *Donnelly Nolan et al.,* 1991; *Baker et al.,* 1994; *Bacon et al.,* 1997]. The low H_2O contents implied for these lavas by the multiple-saturation experiments are supported by estimates for pre-eruptive magmatic H_2O contents based on mineralogical constraints [*Sisson and Grove,* 1993] and by direct measurements of the H_2O contents of melt inclusions [*Sisson and Layne,* 1993; *Sisson and Bronto,* 1998].

7.2. Hydrous Flux Melting

A second, contrasting style of melt generation in the sub-arc mantle is suggested by the phase equilibrium constraints and pre-eruptive H_2O contents of primitive basaltic andesites and magnesian andesites. In this process a cool, H_2O-rich component derived from the subducted slab ascends into the overlying mantle wedge, encounters higher temperature mantle peridotite, and initiates partial melting near the fluid-saturated solidus (hydrous flux melting). The important constraints that bear on the nature of this process are the existence of H_2O-rich primitive magmas that record temperatures of ~1200 °C near the crust-mantle boundary (~1 GPa; Figure 16). In order for a hydrous melting process to achieve these temperatures, material ascending from the vicinity of the cool, subducted lithosphere must thermally equilibrate with the relatively hot peridotite that it encounters as it traverses the mantle wedge. This constraint effectively rules out adiabatic decompression melting in the region between the subducted slab and the hottest portion of the mantle wedge because that process produces a temperature *decrease* due to the isentropic nature of the ascending mantle (Figure 1).

Two end-member melting processes can be envisioned in which a relatively small mass of material rises through the mantle wedge and is heated by the surrounding peridotite: diapirism and reactive porous flow. Both of these processes require that an H_2O-rich component be transferred from the subducted lithosphere to the mantle wedge in order to depress the peridotite solidus and initiate melting. In both of these scenarios partial melting proceeds through a combination of decompression *and* heating under hydrous conditions. In the discussion that follows, we show that melt production within diapirs is limited by a balance between buoyancy forces and thermal equilibration time, but that melt generation is likely to proceed efficiently by reactive porous flow.

7.2.1. Transfer of volatiles into the mantle wedge. The inception of hydrous partial melting in the sub-arc mantle is marked by the influx of a hydrous component from the subducted lithosphere. Recent experimental data on the stability of hydrous phases in the subducted lithosphere show that several H_2O-bearing minerals are stable to considerable pressures, and can transport H_2O to depths of > 200 km. For example, in the subducted sediment, phengite can transport ~5 wt% H_2O to depths of 300 km, while lawsonite and phengite in the oceanic crust contain ~10 and ~5 wt% H_2O, respectively, and are stable to a depth of 250 km [*Schmidt,* 1996; *Schmidt and Poli,* 1998]. Serpentine and chlorite present in the hydrated portion of the subducted lithospheric mantle and/or intermixed with the oceanic crust can transport up to ~10 wt% H_2O to depths > 200 km [Ulmer and Trommsdorff, 1995]. Thus, the peridotite above the subducted slab could be continuously supplied with H_2O to depths approaching 300 km [e.g., *Poli,* 1993; *Poli and Schmidt,* 1995; *Schmidt,* 1996; *Schmidt and Poli,* 1998].

The mechanism by which the hydrous component produced within the dehydrating subducted lithosphere is transported to the overlying mantle wedge is uncertain. The two most probable modes of transport are in fractures initiated by fluid overpressure within the subducted lithosphere or via porous flow along grain boundaries. Mibe et al. [1999] explored variations in the olivine-aqueous fluid dihedral angle over the pressure-temperature range relevant for the base of the mantle wedge, and determined that θ exceeds 60° at 3.0 GPa if the temperature is < 900 °C, indicating that small amounts of H_2O will become trapped at four-grain junctions, but is below 60° at higher temperatures, allowing aqueous fluid to form an interconnected network along olivine grain boundaries. This result led Mibe et al. [1999] to propose that the variation in θ may control the location of arc volcanism. A caveat to this model for transport of fluid into the mantle wedge is that Mibe et al. [1999] investigated connectivity for a pure H_2O fluid, and there are undoubtedly additional components dissolved in any fluid released from the subducted lithosphere, some of which may influence the pressure-temperature conditions over which θ is < 60° [*Watson and Brenan,* 1987].

The nature of the hydrous component (i.e., fluid versus melt) that initiates partial melting in the wedge remains controversial, largely due to uncertainty with respect to pressure-temperature conditions within the subducted lithosphere. Schmidt and Poli [1998] argued that dehydration reactions are responsible for most of the H_2O released from subducted lithosphere, leaving behind solid anhydrous reaction products. Given certain subduction parameters, numerical models for the thermal structure of subduction zones that use analytical solutions to describe corner flow in a constant viscosity mantle predict temperatures at the top of the subducted lithosphere (> 800 °C) that are high enough to allow fluid-undersaturated melting of the oceanic crust [*Peacock,* 1996; *Peacock and Wang,* 1999]. Models that use numerical solutions for flow and incorporate the temperature- and stress-dependence of mantle viscosity [*Furakawa,* 1993; *Furakawa and Tatsumi,* 1999] predict temperatures in the subducted lithosphere of "hot" subduction zones that are sufficient for melting of sediment and eclogite only at fluid-saturated conditions (~600 °C at 3.0 GPa) [*Lambert and Wyllie,* 1970]. Further, trace element models require a low extent of melting of the subducted slab (~2 to 5 wt%) [*Yogodzinski et al.,* 1995; *Grove et al.,* 2002]. Therefore, if melting does occur the most likely product is a near-fluid-saturated melt of sediment or oceanic crust. It is unlikely that slab temperatures are high enough to melt under anhydrous conditions.

7.2.2. Hydrous partial melting within buoyant diapirs. The ascent of buoyant diapirs through the mantle wedge is often invoked as a mechanism for melt generation in subduction zones [e.g., *Marsh,* 1979; *Tatsumi et al.,* 1983; *Kushiro,* 1990]. Although diapirism provides a means for decompressing mantle peridotite within a regime dominated by downward flow, the thermal requirements for melt generation must also be considered. For example, constraints derived from experimental petrology indicate that the temperature of a diapir must increase from ~600 °C at the beginning of ascent near the slab-wedge interface to >1200 °C once the diapir reaches the hottest portion of the wedge. Therefore, the size and ascent velocity of the diapir place limits on the conditions under which it can maintain thermal equilibrium with the surrounding mantle and, thereby, partially melt [*Davies and Stevenson,* 1992; *Kincaid and Sacks,* 1997; *Hall and Kincaid,* 2001].

Factors that control the velocity of a partially molten diapir's ascent through mantle peridotite are its size, mantle viscosity, and the density contrast between the diapir and surrounding peridotite. Stokes velocities calculated using the minimum viscosity for hydrous peridotite (10^{18} Pas) [*Hirth and Kohlstedt,* 1996] and the maximum density contrast between peridotite and a diapir comprised of peridotite + hydrous partial melt ($\Delta\rho$ =1100 kg/m$_3$) [*Ochs and Lange,* 1999], are plotted against diapir radius in Figure 19a. These should be considered maximum ascent velocities. The time required for thermal equilibration of a diapir is plotted against ascent time from the slab-wedge interface (120 km depth) to a depth of 60 km in Figure 19b. The combined results from these calculations demonstrate that as diapir size increases, both the ascent velocity and the time required for thermal equilibration increase. The vertical line at 50 cm/yr in Figure 19a ($v_A > 5v_S$) represents the ascent rate necessary for a diapir to rise vertically from just above the slab to the volcanic front. Diapirs with radii < 2 km will maintain thermal equilibrium with the surrounding peridotite, but their ascent rates are not rapid enough to overcome the downward flow of the mantle wedge. Therefore, any diapir small enough to thermally equilibrate with its surroundings will be subducted. Large diapirs ascend fast enough to escape the subduction flow, but their mass is too great to allow thermal equilibration on the timescales required for hydrous melting to proceed. Further, the heat required to raise the temperature of these large diapirs will significantly cool the surrounding mantle. Therefore, hydrous partial melting within buoyant diapirs does not appear to be a viable mechanism for producing arc magmas.

7.2.3. Melt generation by reactive porous flow. Melt generation via reactive porous flow begins when an H_2O-rich component derived from the subducted lithosphere infiltrates the mantle wedge along grain boundaries (Figure 20). The buoyant, H_2O-rich phase is heated as it ascends from the subducted lithosphere into the overlying peridotite, and partial melting begins when the ambient pressure-temperature conditions exceed the fluid-saturated solidus (Fig 20; P1, T1). The first melt will be extremely H_2O rich (~26 to 30 wt% at 3.0 GPa) [*Green,* 1973; *Hodges,* 1974; *Mysen and Boettcher,* 1975].

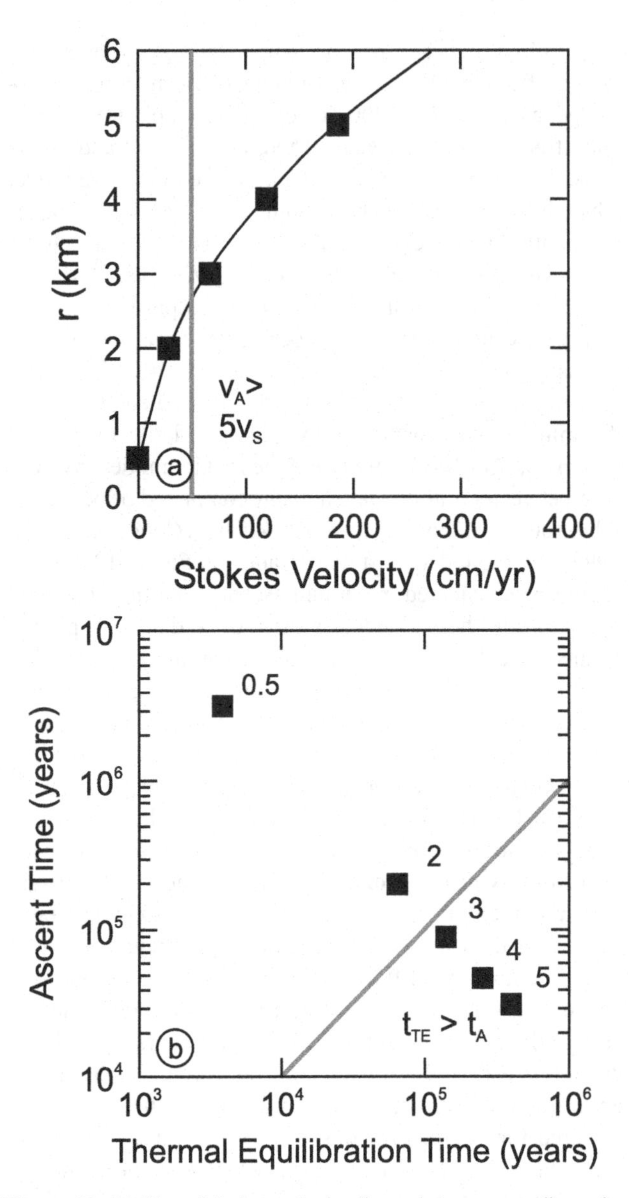

Figure 19. (a) Plot of Stokes velocity (in cm/yr) versus radius calculated for a diapir comprised of peridotite + hydrous partial melt. A hydrous mantle viscosity of 10^{18} Pas was assumed [*Hirth and Kohlstedt,* 1996]. Range of permeabilities is taken from von Bargen and Waff [1986]. The vertical line represents ascent velocities that exceed 5 times the maximum subduction rate. This is an arbitrary geometric factor that allows the diapir to ascend vertically in the convecting mantle wedge. After Grove et al. [2002]. (b) Ascent time (in years) for diapirs of variable radii (indicated next to symbol and shown in a) versus time required for thermal equilibration with the surrounding mantle [*Carslaw and Jaeger,* 1959]. A 60 km ascent path is assumed. Solid gray line separates the plot into a region where the diapir will come to thermal equilibrium with its surroundings (for the 0.5 and 2 km radius diapirs) and a region where the diapir will not in the time required for ascent ($t_{TE} > t_A$).

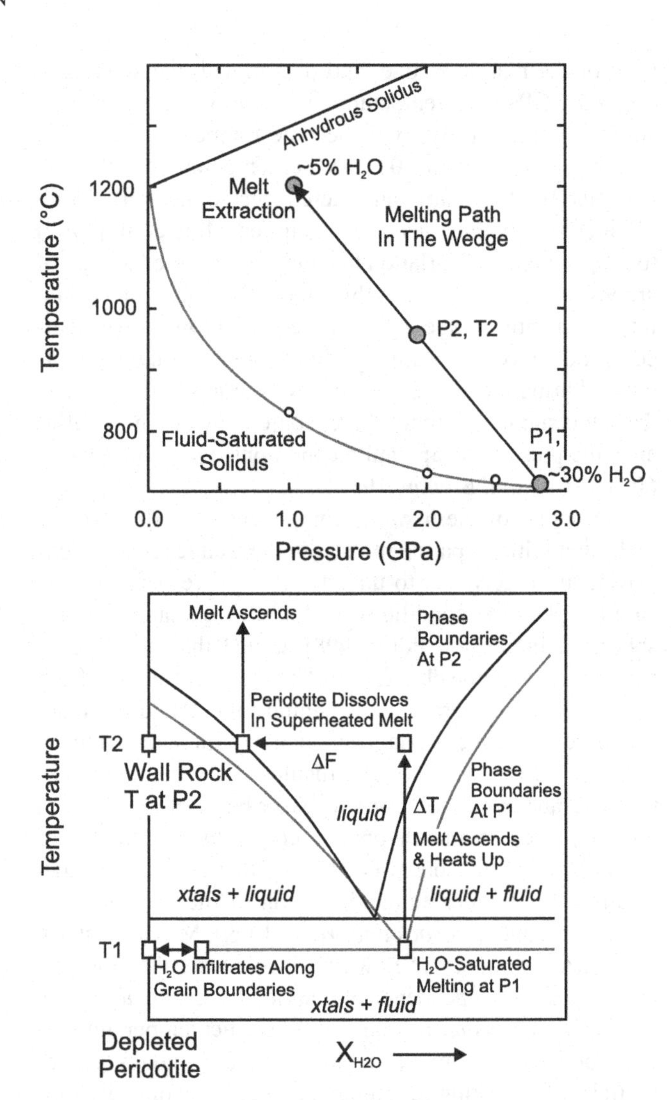

Figure 20. Pressure-temperature and temperature-composition diagrams illustrating hydrous flux melting of the mantle wedge by reactive porous flow through an inverted thermal gradient. Upper figure illustrates the path taken by a melt produced at the fluid-saturated solidus (P1, T1) as it ascends into shallow, hotter mantle (arrow). This path is determined by the thermal structure in the mantle wedge and the ascent trajectory taken by the melt. Lower figure illustrates the evolution of an initially fluid-saturated melt as it ascends into shallower, hotter mantle and equilibrates at a lower pressure (P2, T2). The melt is too H_2O-rich under these conditions and dissolves silicates to come into equilibrium with its surroundings. After segregation from the mantle, the melt records only the shallow conditions of final equilibration. After Grove et al. [2002].

Once it has formed, this hydrous partial melt ascends into overlying hotter mantle by porous flow and comes to thermal equilibrium with its new surroundings. The superheated melt is now out of chemical equilibrium, and must dissolve silicate minerals from the surrounding peridotite in order to equilibrate (Fig 20; P2, T2). This will cause the water content of the partial melt to decrease (ΔF) until it reaches the equilibrium H_2O content at the depth equivalent to P2. This process will continue as the melt ascends to shallower depths and encounters progressively hotter mantle peridotite.

In this scenario of hydrous flux melting, intergranular melt is heated by its surroundings as it ascends, and the latent heat of fusion is supplied by the surrounding rock. Mantle temperature ultimately determines the amount of melt produced, and its final H_2O content. Further, the melt will respond to the composition of the surrounding mantle at each stage of its ascent. If the same porous flow network experiences multiple melting episodes, the melt compositions produced by such a process should evolve over time [*Daines and Kohlstedt,* 1994; *Kelemen et al.,* 1997]. If the timescale of melt ascent is less than that required to achieve complete chemical equilibrium, elements with differing compatibilities may behave differently during mineral-melt exchange, and this could influence the melt composition as well [*Navon and Stolper,* 1987]. The final product, as evidenced by the experimental observation, is a melt that last equilibrated with a depleted mantle residue at shallow depths (Figure 16).

In the case of porous flow through an interconnected network, the dominant physical control on ascent velocity is the permeability of the mantle (Figure 21). To calculate the ascent rate of a hydrous partial melt, values of mantle permeability (K) were chosen that span the plausible range for a partially molten peridotite using the permeability model of von Bargen and Waff [1986]. Melt migration was assumed to occur by Darcy flow [*Ahern and Turcotte,* 1979]. Melt viscosity (10 Pas) and mantle/melt density contrast were the same as those used in the Stokes Law calculation, and the amount of trapped melt was assumed to be small (1 wt%). The ascent rates at this percentage of trapped melt span the range of values from 3 to 315 cm/yr. Only for values of K $> 10^{-12.7}$ m^2 will the melt migrate at a velocity rapid enough to escape the downward mantle subduction flow ($v_A > 5v_S$). Because porous flow along grain boundaries involves a small amount of melt relative to the mass of the surrounding solid, ascending hydrous melt rapidly achieves thermal equilibrium. Therefore, a flux melting process will proceed efficiently, with the surrounding peridotite providing enough energy to both raise the temperature of the migrating melt and supply the heat of fusion to create addition melt without significantly lowering the temperature of the mantle. As the melt ascends, increasing magma mass will, in turn, increase the ascent velocity. Thus, the calculated velocities shown in Figure 21 should be considered minimum values. If the melt segregates into channels ascent rates could be even faster [*Daines and Kohlstedt,* 1994]. Maximum melt ascent times to traverse the distance from the slab-mantle wedge interface to the base of the crust (~60 km) are on the order of 2 x 10^4 years. Estimates of the timescale for magma formation and ascent in subduction zones from U-Th disequilibrium are between 10^5 and 2 x 10^4 years and estimates of fluid addition from ^{226}Ra disequilibrium imply even shorter timescales (~10^3 years) [Turner et al., 2000]. Therefore, the calculations presented here indicate that melt transport involving a reactive porous flow mechanism could be quite rapid, and provide estimates for timescales that agree with independent geochemical estimates.

Grove et al. [2002] suggest that in the Cascade arc the melt generation process is comprised of several distinct steps. First, advection of hot, low-viscosity mantle into the wedge generates anhydrous HAOT, depleting the residual peridotite before it is overturned at the wedge corner. This depleted mantle travels along flow lines that parallel the subducted slab, losing heat as it descends. Next, an H_2O-rich component derived from dehydration of H_2O-bearing minerals in the subducted slab (sediments, altered oceanic crust and/or underlying hydrated mantle) ascends into the overlying mantle wedge. As this H_2O-rich component percolates into the wedge, it encounters mantle that is hot

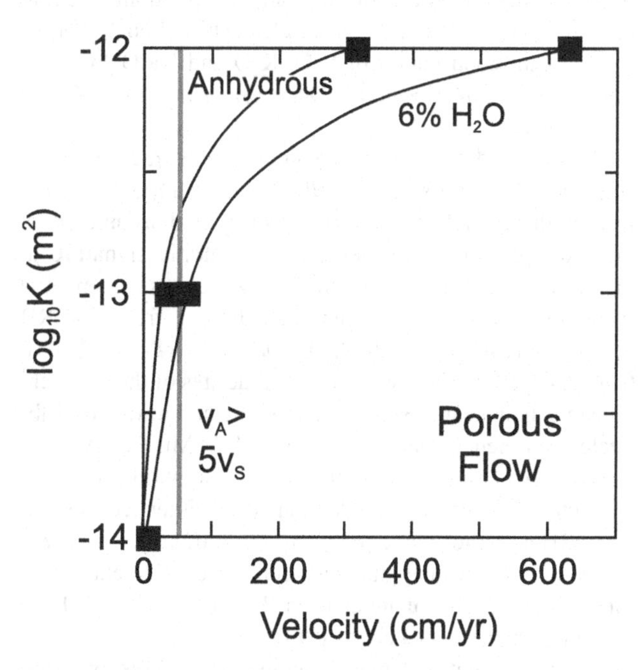

Figure 21. Ascent velocity versus permeability for melt percolation through an interconnected network within a crystal aggregate. Porous flow velocity was calculated using Darcy's law assuming the same density contrast as the Stoke's law calculation in Figure 19. Permeability range was taken from von Bargen and Waff [1986]. After Grove et al. [2002].

enough to melt at the fluid-saturated solidus. Buoyant, H_2O-rich partial melt of peridotite ascends to shallower depths where it encounters higher temperature mantle. The melt reacts and equilibrates with the hotter shallower mantle, thereby lowering its H_2O content. In other arc setting, such as the Mexican arc, mantle that has not been depleted by a prior melt extraction event may undergo flux melting by hydrous slab-derived fluid [*Carmichael et al.*, 1996].

7.2.4. Trace element characteristics of hydrous arc melts. Arc lavas are characterized by relative enrichments of the large ion lithophile elements (LILE; e.g., K, Sr and Ba) and light rare earth elements (LREE) and strong depletions in the high field strength elements (HFSE; e.g., Nb, Zr, Ti) and the heavy rare earths (HREE) relative to MORB [*Stern,* 1979; *Kay,* 1980; *Arculus and Johnson,* 1981; *Morris and Hart,* 1983; *Hawkesworth et al.*, 1993]. These trace element characteristics result from a combination of strong enrichments contributed by components delivered from the subducted lithosphere and the high extents of depletion experienced in the mantle source during hydrous flux melting. Mass balance calculations of the relative contributions from the slab-derived fluid and mantle peridotite to the composition of arc magmas show that the fluid-enriched component contributes >99 wt% of the incompatible trace elements and that peridotite melting contributes >90 wt% of the major elements, although there is a significant contribution of H_2O, K_2O and Na_2O from the slab [*Stolper and Newman,* 1994; *Eiler et al.,* 2000; *Grove et al.,* 2002].

There are indications that the incompatible trace elements contributed by the slab-derived hydrous component come from both subducted sediment and subducted oceanic crust. For example, the trace element and isotopic signature in primitive Adak-type magnesian andesites from the Aleutians indicates a strong contribution from MORB eclogite crust [*Kay,* 1978; *Yogodzinski et al.,* 1995]. Conversely, the primitive high-Mg andesites of the Setouchi volcanic belt bear a strong trace element imprint of subducted sediment [*Shimoda et al.,* 1998]. At Mt. Shasta in the Cascade arc, primitive basaltic andesite show trace element characteristics that may be dominated by either sediment or a MORB eclogite source [*Grove et al.,* 2002]. In all of these environments the magmas that retain trace element signatures characteristic of input from the subducted slab have high pre-eruptive H_2O contents (Table 1).

It is likely that any major element input from the subducted lithosphere has been strongly influenced by processes that occur as the fluid-rich component ascends through and reacts with the overlying mantle wedge. Some of these processes are illustrated in Figure 22. In the "hot slab" environments (~600 to 650 °C at 3.0 GPa) fluid-saturated, low-degree melts (2 to 5 wt%) of the slab may carry the trace element and isotopic signature of the subduction component [*Yogodzinski et al.,* 1995; *Grove et al.,* 2002]. It is likely that such melts compositionally resemble a rhyolite [*Lambert and Wyllie,* 1970]. As this SiO_2- and H_2O-rich melt ascends into the overlying mantle wedge, it will react with peridotite at conditions below the solidus, losing SiO_2 through a reaction of the form olivine + SiO_2 = orthopyroxene [*Carroll and Wyllie,* 1989; *Zhang and Frantz,* 2000]. At shallower mantle depths, this reaction operates in the reverse direc-

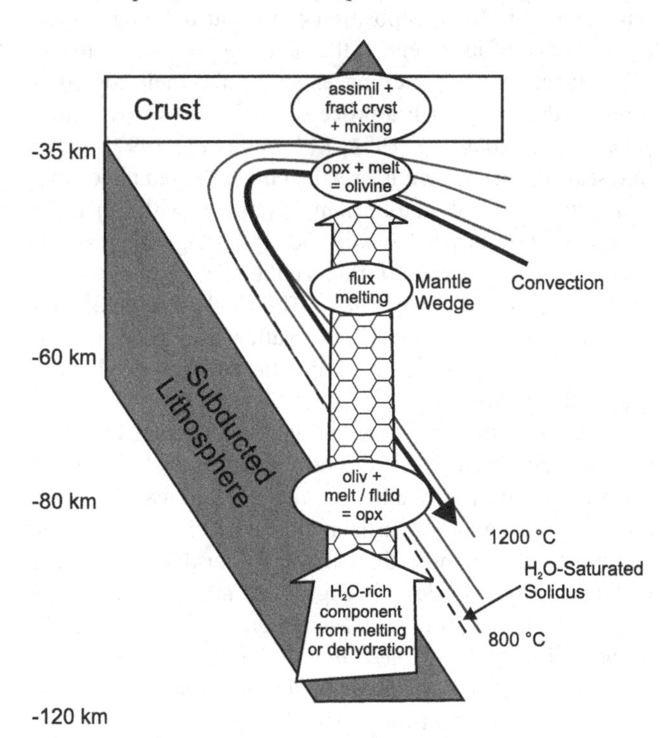

Figure 22. A summary of the processes of melt generation and subsequent modification of hydrous magmas produced in the mantle wedge. An H_2O-rich component is derived by dehydration of minerals in the subducted lithosphere and/or by low-degree vapor saturated melting of subducted sediment or basalt. This H_2O-rich component ascends into the overlying mantle wedge, where is modified by reaction with the descending mantle (SiO_2 is stripped from the H_2O-rich component). The modified H_2O-rich component ascends into shallower, hotter mantle that exceeds the vapor-saturated solidus of peridotite and melting begins. Melting continues into the hottest part of the mantle wedge by flux melting (Figure 20). As the melt continues its ascent into cooler shallowest part of the mantle wedge, it becomes SiO_2-rich by dissolving orthopyroxene and precipitating olivine. Finally the magma reaches the overlying crust where assimilation of deep and shallow crust, fractional crystallization, and magma mixing operate to modify most subduction zone magmas. After Grove et al. [2002].

tion, as the magma dissolves orthopyroxene and crystallizes olivine [*Kelemen,* 1995]. Such mantle–melt interaction may play a role in generating the high SiO_2 contents of primitive high-Mg andesites (Figure 22). Finally, arc magmas enter the crust where fractional crystallization, assimilation, and mixing processes occur prior to eruption. Therefore, the potential exists for significant processing of the fluid-rich component at the base of the wedge, in the shallower portion of the sub-arc mantle, and in the overlying crust.

8. CONCLUSIONS

In the preceding discussion we have shown that the influence of H_2O on peridotite partial melting in the sub-arc environment leads to hydrous flux melting, a process that occurs in the inverted geothermal gradient that is unique to the subduction zone tectonic regime. The melting process begins with the generation of a cool, hydrous, low degree melt of mantle peridotite. This melt ascends into shallower, hotter overlying wedge and maintains chemical equilibrium by dissolving the surrounding peridotite and diluting the H_2O content of the magma. Melting is controlled by the temperature-depth structure of the ascent path through the overlying mantle wedge. The presence of H_2O has a dramatic influence on the physical properties of the mantle and on the viscosity and density of the melt. The high H_2O content of the melt promotes rapid ascent by reactive porous flow. In some arc environments adiabatic decompression occurs along with hydrous flux melting. Decompression melting takes place under near-anhydrous conditions where hot mantle ascends into the wedge corner along an adiabat, intersects the peridotite solidus, and partially melts. These contrasting styles of melting occur in close spatial and temporal proximity in some arcs, and are examples of the complexity in melt generation processes that characterize the sub-arc mantle.

Acknowledgments. The authors are grateful to P. Asimow and M. Hirschmann for thoughtful reviews. We would be remiss if we did not thank J. Eiler for his editorial help and his (nearly) endless patience. This work was supported by the MARGINS program of NSF through grants no. EAR-0112013 and OCE-0001821 as well as by NSF grant no. EAR-9706214.

REFERENCES

Ahern, J.L., and D.L. Turcotte, Magma migration beneath an ocean ridge, *Earth Planet. Sci. Lett., 45,* 115-122, 1979.

Anderson, A.T., Jr., The before-eruption water content of some high-alumina magmas, *Bull. Volcanol., 37,* 530-552, 1973.

Anderson, A.T., Jr., Evidence for a picritic, volatile-rich magma beneath Mt. Shasta, California, *J. Petrol., 15,* 243-267, 1974.

Anderson, A.T., Jr., Magma mixing: petrological process and volcanological tool, *J Volcanol Geotherm Res, 1,* 3-33, 1976.

Anderson, A.T., Jr., Water in some hypersthenic magmas, *J. Geol., 87,* 509-531, 1979.

Anderson, R.N., S.E. DeLong, and W.M. Schwarz, Thermal model for subduction with dehydration in the downgoing slab, *J. Geol., 86,* 731-739, 1978.

Andrews, D.J., and N.H. Sleep, Numerical Modeling of tectonic flow behind island arcs, *Geophys. J. R. Astron. Soc., 38,* 237-251, 1974.

Arculus, R.J., and R.W. Johnson, Island arc magma sources: a geochemical assessment of the roles of slab-derived components and crustal contamination, *Geochem. J., 15,* 109-133, 1981.

Asimow, P.D., M.M. Hirschmann, and E.M. Stolper, An analysis of variations in isentropic melt productivity, *Philos. Trans. R. Soc. London Ser. A, 355,* 255-281, 1997.

Ayers, J.C., Trace element modeling of aqueous fluid - peridotite interaction in the mantle wedge of subduction zones, *Contrib. Mineral. Petrol., 132,* 390-404, 1998.

Bacon, C.R., P.E. Bruggman, R.L. Christiansen, M.A. Clynne, J.M. Donnelly Nolan, and W. Hildreth, Primitive magmas at five Cascade volcanic fields: melts from hot, heterogeneous sub-arc mantle, *Can. Mineral., 35,* 397-423, 1997.

Baker, M.B., T.L. Grove, and R. Price, Primitive basalts and andesites from the Mt. Shasta region, N. California: products of varying melt fraction and H_2O content, *Contrib. Mineral. Petrol., 118,* 111-129, 1994.

Baker, M.B., and E.M. Stolper, Determining the composition of high-pressure mantle melts using diamond aggregates, *Geochim. Cosmochim. Acta, 58,* 2811-2827, 1994.

Bartels, K.S., R.J. Kinzler, and T.L. Grove, High pressure phase relations of primitive high-alumina basalts from Medicine Lake volcano, northern California, *Contrib. Mineral. Petrol., 108,* 253-270, 1991.

Beere, W., A unifying theory of the stability of penetrating phases and sintering pores, *Acta Metall., 23,* 131-138, 1975.

Bodri, L., and B. Bodri, Numerical investigation of tectonic flow in island arc areas, *Tectonophys., 50,* 163-175, 1978.

Boyd, F.R., J.L. England, and B.T.C. Davis, Effects of pressure on the melting and polymorphism of enstatite, $MgSiO_3$, *J. Geophys. Res., 69,* 2101-2109, 1964.

Bulau, J.R., H.S. Waff, and J.A. Tyburczy, Mechanical and thermodynamic constraints on fluid distribution in partial melts, *J. Geophys. Res., 84,* 6102-6108, 1979.

Burnham, C.W., Water and magmas: a mixing model, *Geochim. Cosmochim. Acta, 39,* 1077-1084, 1975.

Burnham, C.W., and N.F. Davis, The role of H_2O in silicate melts II. Thermodynamic and phase relations in the system $NaAlSi_3O_8$-H_2O to 10 kilobars, 700 ° to 1100 °C, *Amer. J. Sci., 274,* 902-940, 1974.

Carmichael, I.S.E., R.A. Lange, and J.F. Luhr, Quaternary minettes and associated volcanic rocks of Mascota, western Mexico: a consequence of plate extension above a subduction modified mantle wedge, *Contrib. Mineral. Petrol., 124,* 302-333, 1996.

Carroll, M.R., and P.J. Wyllie, Experimental phase relations in the system tonalite-peridotite-H_2O at 15 kb: implications for assimilation and differentiation processes near the crust-mantle boundary, *J. Petrol., 30,* 1351-1382, 1989.

Carslaw, H.S., and J.C. Jaeger, *Conduction of Heat in Solids,* 510 pp., Clarendon Press, Oxford, 1959.

Daines, M.J., and D.L. Kohlstedt, The transition from porous to channelized flow due to melt/rock reaction during melt migration, *Geophys. Res. Lett., 21,* 145-148, 1994.

Daines, M.J., and F.M. Richter, An experimental method for directly determining the interconnectivity of melt in a partially molten system, *Geophys. Res. Lett., 15,* 1459-1462, 1988.

Danyushevsky, L.V., M. Carroll, and T.J. Falloon, Origin of high-An plagioclase in Tongan high-Ca boninites: implications for plagioclase-melt equilibria at low P(H_2O), *Can. Mineral., 35,* 313-326, 1997.

Davies, J.H., and D.J. Stevenson, Physical model of source region of subduction zone volcanics, *J. Geophys. Res., 97,* 2037-2070, 1992.

DeBari, S.S., S.M. Kay, and R.W. Kay, Ultramafic xenoliths from Adagdak volcano, Adak, Aleutian Islands, Alaska: deformed igneous cumulates from the moho of an island arc, *J. Geol., 95,* 329-341, 1987.

Defant, M.J., and P. Kepezhinskas, Evidence suggests slab melting in arc magmas, *EOS, Trans Amer Geophys Un, 82,* 65-69, 2001.

Denbigh, K., *The Principles of Chemical Equilibrium,* 494 pp., Cambridge University Press, Cambridge, 1981.

Dingwell, D.B., and S.L. Webb, Relaxation in silicate melts, *Eur. J. Mineral., 2,* 427-449, 1990.

Dixon, J.E., E.M. Stolper, and J.R. Holloway, An experimental study of water and carbon dioxide solubilities in mid-ocean ridge basaltic liquids. Part I: calibration and solubility models, *J. Petrol., 36,* 1607-1631, 1995.

Donnelly Nolan, J.M., D.E. Champion, T.L. Grove, M.B. Baker, J.E. Taggart, Jr., and P.E. Bruggman, The Giant Crater lava field: geology and geochemistry of a compositionally zoned, high alumina basalt to basaltic andesite eruption at Medicine Lake Highland, California, *J. Geophys. Res., 96,* 21843-21863, 1991.

Draper, D.S., and A.D. Johnston, Anhydrous PT phase relations of an Aleutian high-MgO basalt: an investigation of the role of olivine-liquid reaction in the generation of arc high-alumina basalts, *Contrib. Mineral. Petrol., 112,* 501-519, 1992.

Drummond, M.S., and M.J. Defant, A model for trondjhemite - tonalite - dacite genesis and crustal growth via slab melting, *J. Geophys. Res., 95,* 21503-21521, 1990.

Eiler, J.M., A.J. Crawford, T. Elliott, K.A. Farley, J.W. Valley, and E.M. Stolper, Oxygen isotope geochemistry of oceanic-arc lavas, *J. Petrol., 41,* 229-256, 2000.

Eiler, J.M., B. McInnes, J.W. Valley, C.M. Graham, and E.M. Stolper, Oxygen isotope evidence for slab-derived fluids in the sub-arc mantle, *Nature, 393,* 777-781, 1998.

Elkins Tanton, L.T., T.L. Grove, and J.M. Donnelly Nolan, Hot, shallow mantle melting under the Cascades volcanic arc, *Geology, 29,* 631-634, 2001.

Elliott, T., T. Plank, A. Zindler, W. White, and B. Bourdon, Element transport from slab to volcanic front at the Mariana arc, *J. Geophys. Res., 102,* 14991-15019, 1997.

Ernsberger, F.M., Molecular water in glass, *J. Amer. Ceram. Soc., 60,* 91-92, 1977.

Falloon, T.J., and A.J. Crawford, The petrogenesis of high-calcium boninites from the north Tonga ridge, *Earth Planet. Sci. Lett., 102,* 375-394, 1991.

Fujii, N., K. Osamura, and E. Takahashi, Effect of water saturation on the distribution of partial melt in the olivine-pyroxene-plagioclase system, *J. Geophys. Res., 91,* 9253-9259, 1986.

Furakawa, Y., Magmatic processes under arcs and formation of the volcanic front, *J. Geophys. Res., 98,* 8309-8319, 1993.

Furakawa, Y., and Y. Tatsumi, Melting of a subducting slab and production of high-Mg andesite magmas: unusual magmatism in SW Japan at 13-15 Ma, *Geophys. Res. Lett., 26,* 2271-2274, 1999.

Gaetani, G.A., and T.L. Grove, The influence of water on melting of mantle peridotite, *Contrib. Mineral. Petrol., 131,* 323-346, 1998.

Gaetani, G.A., T.L. Grove, and W.B. Bryan, The influence of water on the petrogenesis of subduction-related igneous rocks, *Nature, 365,* 332-334, 1993.

Gaetani, G.A., T.L. Grove, and W.B. Bryan, Experimental phase relations of basaltic andesite from Hole 839b under hydrous and anhydrous conditions, in *Proc ODP, Sci Results,* edited by J.W. Hawkins, Jr., L.M. Parson, J.F. Allan, et al., pp. 557-564, Ocean Drilling Program, College Station TX, 1994.

Gill, J.B., *Orogenic Andesites and Plate Tectonics,* 390 pp., Springer-Verlag, New York, NY, 1981.

Green, D.H., Experimental melting studies on a model upper mantle composition at high pressures under water-saturated and water-undersaturated conditions, *Earth Planet. Sci. Lett.,* 19, 37-53, 1973.

Grove, T.L., Corrections to expressions for calculating mineral components in "Origin of Calc - Alkaline Series Lavas at Medicine Lake Volcano by Fractionation, Assimilation and Mixing" and "Experimental Petrology of normal MORB near the Kane Fracture Zone:22° - 25°N, mid - Atlantic ridge, *Contrib. Mineral. Petrol., 114,* 422-424, 1993.

Grove, T.L., S.W. Parman, S.A. Bowring, R.C. Price, and M.B. Baker, The role of an H_2O-rich fluid component in the generation of primitive basaltic andesites and andesites from the Mt. Shasta region, N. California, *Contrib. Mineral. Petrol., 142,* 375-396, 2002.

Hall, P.S., and C. Kincaid, Diapiric flow at subduction zones: a recipe for rapid transport, *Science, 292,* 2472-2475, 2001.

Hasegawa, A., D. Zhao, S. Hori, A. Yamamoto, and S. Horiuchi, Deep structure of the northeastern Japan arc and its relationship to seismic and volcanic activity, *Nature, 352,* 683-689, 1991.

Hawkesworth, C.J., K. Gallagher, J.M. Hergt, and F. McDermott, Mantle and slab contributions in arc magmas, *Ann. Rev. Earth Planet. Sci., 21,* 175-204, 1993.

Hirose, K., and T. Kawamoto, Hydrous partial melting of lherzolite at 1 GPa: the effect of H_2O on the genesis of basaltic magmas, *Earth Planet. Sci. Lett., 133,* 463-473, 1995.

Hirose, K., and I. Kushiro, Partial melting of dry peridotites at high pressures: determination of compositions of melts segregated from peridotite using aggregates of diamond, *Earth Planet. Sci. Lett., 114,* 477-489, 1993.

Hirschmann, M.M., P.D. Asimow, M.S. Ghiorso, and E.M. Stolper, Calculation of peridotite partial melting from thermodynamic models of minerals and melts. III. Controls on isobaric melt production and the effect of water on melt production, *J. Petrol., 40,* 831-851, 1999.

Hirschmann, M.M., M.B. Baker, and E.M. Stolper, The effect of alkalis on the silica content of mantle-derived melts, *Geochim. Cosmochim. Acta, 62,* 883-902, 1998.

Hirth, J.G., and D.L. Kohlstedt, Water in the oceanic upper mantle: implications for rheology, melt extraction and the evolution of the lithosphere, *Earth Planet. Sci. Lett., 144, 93-108,* 1996.

Hodges, F.N., The solubility of H_2O in silicate melts, *Carnegie Inst. Wash. Yearbk., 73,* 251-255, 1974.

Housh, T.B., and J.F. Luhr, Plagioclase-melt equilibria in hydrous systems, *Amer. Mineral., 76,* 477-492, 1991.

Ishizaka, K., and R.W. Carlson, Nd-Sr systematics of the Setouchi volcanic rocks, southwest Japan: a clue to the origin of orogenic andesite, *Earth Planet. Sci. Lett., 64,* 327-340, 1983.

Johnson, M.C., and T. Plank, Dehydration and melting experiments constrain the fate of subducted sediments, *Geochem. Geophys. Geosyst., 1,* 1999GC000014, 1999.

Kawamoto, T., and K. Hirose, Au-Pd sample containers for melting experiments on iron and water bearing systems, *Eur. J. Mineral., 6,* 381-385, 1994.

Kawamoto, T., and J.R. Holloway, Melting temperature and partial melt chemistry of H_2O-saturated mantle peridotite to 11 Gigapascals, *Science, 276,* 240-243, 1997.

Kay, R.W., Aleutian magnesian andesites: melts from subducted Pacific ocean crust, *J Volcanol Geotherm Res, 4,* 497-522, 1978.

Kay, R.W., Volcanic arc magmas: implications of a melting-mixing model for element recycling in the crust - upper mantle system, *J. Geol., 88,* 497-522, 1980.

Kelemen, P.B., Assimilation of ultramafic rock in subduction-related magmatic arcs, *J. Geol., 94,* 829-843, 1986.

Kelemen, P.B., Reaction between ultramafic rock and fractionating basaltic magma I. Phase relations, the origin of calc-alkaline magma series, and the formation of discordant dunite, *J. Petrol., 31,* 51-98, 1990.

Kelemen, P.B., Genesis of high Mg# andesite and the continental crust, *Contrib. Mineral. Petrol., 120,* 1-19, 1995.

Kelemen, P.B., J.G. Hirth, N. Shimizu, M. Spiegelman, and H.J.B. Dick, A review of melt migration processes in the adiabatically upwelling mantle beneath oceanic spreading ridges, *Philos. Trans. R. Soc. London Ser. A, 355,* 283-318, 1997.

Kincaid, C., and I.S. Sacks, Thermal and dynamical evolution of the upper mantle in subduction zones, *J. Geophys. Res., 102,* 12295-12315, 1997.

Kinzler, R.J., and T.L. Grove, Primary magmas of mid-ocean ridge basalts 1. Experiments and methods, *J. Geophys. Res., 97,* 6885-6906, 1992a.

Kinzler, R.J., and T.L. Grove, Primary magmas of mid-ocean ridge basalts 2. Applications, *J. Geophys. Res., 97,* 6907-6926, 1992b.

Klein, E.M., and C.H. Langmuir, Global correlations of ocean ridge basalt chemistry with axial depth and crustal thickness, *J. Geophys. Res., 92,* 8089-8115, 1987.

Kress, V.C., and I.S.E. Carmichael, The compressibility of silicate liquids containing Fe_2O_3 and the effect of composition, temperature, oxygen fugacity and pressure on their redox states, *Contrib. Mineral. Petrol., 108,* 82-92, 1991.

Kurkjian, C.R., and L.E. Russell, Solubility of water in molten alkali silicates, *J. Soc. Glass Tech., 42,* 130-144T, 1958.

Kushiro, I., The system forsterite-diopside-silica with and without water at high pressures, *Amer. J. Sci., 267A,* 269-294, 1969.

Kushiro, I., Stability of amphibole and phlogopite in the upper mantle, *Carnegie Inst. Wash. Yearbk., 68,* 245-247, 1970.

Kushiro, I., Effect of water on the compositions of magmas formed at high pressures, *J. Petrol., 13,* 311-334, 1972.

Kushiro, I., On the nature of silicate melt and its significance in magma genesis: regularities in the shift of liquidus boundaries involving olivine, orthopyroxene, and silica minerals., *Amer. J. Sci., 275,* 411-431, 1975.

Kushiro, I., Density and viscosity of hydrous calc-alkalic andesite magma at high pressures, *Carnegie Inst. Wash. Yearbk., 77,* 675-677, 1978.

Kushiro, I., Partial melting of mantle wedge and evolution of island arc crust, *J. Geophys. Res., 95,* 15929-15939, 1990.

Kushiro, I., N. Shimizu, Y. Nakamura, and S. Akimoto, Compositions of coexisting liquid and solid phases formed upon melting of natural garnet and spinel lherzolite at high pressures: a preliminary report, *Earth Planet. Sci. Lett., 14,* 19-25, 1972.

Kushiro, I., Y. Syono, and S.-i. Akimoto, Melting of a peridotite nodule at high pressures and high water pressures, *J. Geophys. Res., 73,* 6023-6029, 1968a.

Kushiro, I., H.S. Yoder, Jr., and M. Nishikawa, Effect of water on the melting of enstatite, *Geol. Soc. Amer. Bull., 79,* 1685-1692, 1968b.

Lambert, I.B., and P.J. Wyllie, Melting in the deep crust and upper mantle and the nature of the low velocity layer, *Phys Earth Planet Int, 3,* 316-322, 1970.

Lange, R.A., The effect of H_2O, CO_2, and F on density and viscosity of silicate melts, in *Volatiles in Magmas*, edited by M.R. Carroll, and J.R. Holloway, Mineralogical Society of America, Washington D.C., 1994.

Lange, R.A., and I.S.E. Carmichael, Densities of Na_2O-K_2O-CaO-MgO-FeO-Fe_2O_3-Al_2O_3-TiO_2-SiO_2 liquids: new measurements and derived partial molar properties, *Geochim. Cosmochim. Acta, 51,* 2931-2946, 1987.

Langmuir, C.H., E.M. Klein, and T. Plank, Petrological systematics of mid-ocean ridge basalts: constraints on melt generation beneath ocean ridges, in *Mantle Flow and Melt Generation at Mid-Ocean Ridges,* edited by J. Phipps Morgan, D.K. Blackman, and J.M. Sinton, pp. 183-280, AGU, Washington D.C., 1992.

Marsh, B.D., Island arc development: some observations, experiments, and speculations, *Amer. J. Sci., 87,* 687-713, 1979.

McCulloch, M.T., and A.J. Gamble, Geochemical and geodynamical constraints on subduction zone magmatism, *Earth Planet. Sci. Lett., 102,* 358-374, 1991.

McKenzie, D.P., Speculations on the consequences and causes of plate motions., *Geophys. J. R. Astron. Soc., 18,* 1-32, 1969.

McKenzie, D.P., The generation and compaction of partially molten rock, *J. Petrol., 25,* 713-765, 1984.

McKenzie, D.P., The extraction of magma from the crust and mantle, *Earth Planet. Sci. Lett., 74,* 81-91, 1985.

Mibe, K., T. Fujii, and A. Yasuda, Control of the location of the volcanic front in island arcs by aqueous fluid connectivity in the mantle wedge, *Nature, 401,* 259-262, 1999.

Millhollen, G.L., A.J. Irving, and P.J. Wyllie, Melting interval of peridotite with 5.7 per cent water to 30 kilobars, *J. Geol., 82,* 575-587, 1974.

Morris, J.D., and S.R. Hart, Isotopic and incompatible element constraints on the genesis of island arc volcanics from Cold Bay and Okmok Island, Aleutians, and implications for mantle structure, *Geochim. Cosmochim. Acta, 47,* 2015-2030, 1983.

Morse, S.A., *Basalts and Phase Diagrams, 493* pp., Springer-Verlag, New York, New York, 1980.

Mysen, B.O., and A.L. Boettcher, Melting of a hydrous mantle: I. Phase relations of natural peridotite at high pressures and temperatures with controlled activities of water, carbon dioxide and hydrogen, *J. Petrol., 16,* 520-548, 1975.

Navon, O., and E.M. Stolper, Geochemical consequences of melt percolation: the upper mantle as a chromatographic column, *J. Geol., 95,* 285-307, 1987.

Nicholls, I.A., and A.E. Ringwood, Production of silica-saturated tholeiitic magmas in island arcs, *Earth Planet. Sci. Lett., 17,* 243-246, 1972.

Nowak, M., and H. Behrens, The speciation of water in haplogranitic glasses and melts determined by in situ near-infrared spectroscopy, *Geochim. Cosmochim. Acta, 59,* 3445-3450, 1995.

Nowak, M., and H. Behrens, Water in rhyolitic magmas: getting a grip on a slippery problem, *Earth Planet. Sci. Lett., 184,* 515-522, 2001.

Nye, C.J., and M.R. Reid, Geochemistry of primary and least fractionated lavas from Okmok Volcano, Central Aleutians: implications for arc magmagenesis, *J. Geophys. Res., 91,* 10271-10287, 1986.

Ochs, F.A., III, and R.A. Lange, The partial molar volume, thermal expansivity, and compressibility of H_2O in $NaAlSi_3O_8$ liquid: new measurements and an internally consistent model, *Contrib. Mineral. Petrol.*, 129, 155-165, 1997.

Ochs, F.A., III, and R.A. Lange, The density of hydrous magmatic liquids, *Science, 283,* 1314-1317, 1999.

Orlova, G.P., The solubility of water in albite melts, *Int. Geol. Rev., 6,* 254-258, 1962.

Oxburgh, E.R., and D.L. Turcotte, Thermal structure of island arcs, *Geol. Soc. Amer. Bull., 81,* 1665-1688, 1970.

Peacock, S.M., Numerical simulation of subduction zone pressure-temperature-time paths: constraints on fluid production and arc magmatism, *Philos. Trans. R. Soc. London Ser. A, 335,* 341-353, 1991.

Peacock, S.M., Thermal and petrologic structure of subduction zones, in *Subduction Top to Bottom,* edited by G.E. Bebout, D.W. Scholl, S.H. Kirby, et al., pp. 119-133, AGU, Washington D.C., 1996.

Peacock, S.M., and K. Wang, Seismic consequences of warm versus cool subduction metamorphism: examples from southwest and northeast Japan, *Science, 286,* 937-939, 1999.

Pearce, J.A., P.E. Baker, P.K. Harvey, and I.W. Luff, Geochemical evidence for subduction fluxes, mantle melting and fractional crystallization beneath the South Sandwich island arc, *J. Petrol., 36,* 1073-1109, 1995.

Perfit, M.R., D.A. Gust, A.E. Bence, R.J. Arculus, and S.R. Taylor, Chemical characteristics of island arc basalts: implications for mantle sources, *Chem. Geol., 30,* 227-256, 1980.

Plank, T., and C.H. Langmuir, An evaluation of the global variations in the major element chemistry of arc basalts, *Earth Planet. Sci. Lett., 90,* 349-370, 1988.

Plank, T., and C.H. Langmuir, Tracing trace elements from sediment input to volcanic output at subduction zones, *Nature*, 362, 739-743, 1993.

Poli, S., The amphibole-eclogite transformation: an experimental study on basalt, *Amer. J. Sci., 293,* 1061-1107, 1993.

Poli, S., and M.W. Schmidt, H_2O transport and release in subduction zones: experimental constraints on basaltic and andesitic systems, *J. Geophys. Res., 100,* 22299-22314, 1995.

Regelous, M., K.D. Collerson, A. Ewart, and J.I. Wendt, Trace element transport rates in subduction zones: evidence from Th, Sr and Pb isotope data for Tonga-Kermadec arc lavas, *Earth Planet. Sci. Lett., 150,* 291-302, 1997.

Richet, P., A.-M. Lejeune, F. Holtz, and J. Roux, Water and the viscosity of andesite melts, *Chem. Geol., 128,* 185-197, 1996.

Russell, L.E., Solubility of water in molten glass, *J. Soc. Glass Tech., 41,* 304-317T, 1957.

Sack, R.O., and M.S. Ghiorso, Importance of considerations of mixing properties in establishing an internally consistent thermodynamic database: thermochemistry of minerals in the system Mg_2SiO_4-Fe_2SiO_4-SiO_2, *Contrib. Mineral. Petrol., 102,* 41-68, 1989.

Schmidt, M.W., Experimental constraints on recycling of potassium from subducted oceanic crust, *Science, 272,* 1927-1930, 1996.

Schmidt, M.W., and S. Poli, Experimentally based water budgets for dehydrating slabs and consequences for arc magma generation, *Earth Planet. Sci. Lett., 163,* 361-379, 1998.

Schulze, F., H. Behrens, F. Holtz, J. Roux, and W. Johannes, The influence of H_2O on viscosity of a haplogranitic melt, *Amer. Mineral., 81,* 1155-1165, 1996.

Scott, D.R., and D.J. Stevenson, Magma ascent by porous flow, *J. Geophys. Res., 91,* 9283-9296, 1986.

Shaw, H.R., Obsidian-H_2O viscosities at 1000 and 2000 bars and in the temperature range 700-900 °C, *J. Geophys. Res., 68,* 6337-6343, 1963.

Shaw, H.R., Viscosities of magmatic silicate liquids: an empirical method of prediction, *Amer. J. Sci., 272,* 870-893, 1972.

Shen, A.H., and H. Keppler, Infrared spectroscopy of hydrous silicate melts to 1000 °C and 10 kbar: direct observations of H_2O

speciation in diamond anvil cells, *Amer. Mineral., 80,* 1335-1338, 1995.

Shimoda, G., Y. Tatsumi, S. Nohda, K. Ishizaka, and B.M. Jahn, Setouchi high-Mg andesites revisited: geochemical evidence for melting of subducted sediments, *Earth Planet. Sci. Lett., 160,* 479-492, 1998.

Silver, L., and E.M. Stolper, A thermodynamic model for hydrous silicate melts, *J. Geol., 93,* 161-178, 1985.

Silver, L., and E.M. Stolper, Water in albitic glasses, *J. Petrol., 30,* 667-709, 1989.

Sisson, T.W., and S. Bronto, Evidence for pressure-release melting beneath magmatic arcs from basalt at Galunggung, Indonesia, *Nature, 391,* 883-886, 1998.

Sisson, T.W., and T.L. Grove, Temperatures and H_2O contents of low-MgO high-alumina basalts, *Contrib. Mineral. Petrol., 113,* 167-184, 1993.

Sisson, T.W., and G.D. Layne, H_2O in basalt and basaltic andesite glass inclusions from four subduction-related volcanoes, *Earth Planet. Sci. Lett., 117,* 619-635, 1993.

Smith, C.S., Some elementary principles of polycrystalline microstructure, *Metall. Rev., 9,* 1-48, 1964.

Stern, R.J., On the origin of andesite in the northern Mariana island arc: implications from Agrigan, *Contrib. Mineral. Petrol., 68,* 207-219, 1979.

Stolper, E.M., A phase diagram for mid-ocean ridge basalts: preliminary results and implications for petrogenesis, *Contrib. Mineral. Petrol., 74,* 13-27, 1980.

Stolper, E.M., The speciation of water in silicate melts, *Geochim. Cosmochim. Acta, 46,* 2609-2620, 1982a.

Stolper, E.M., Water in silicate melts: an infrared spectroscopic study, *Contrib. Mineral. Petrol., 81,* 1-17, 1982b.

Stolper, E.M., Temperature dependence of the speciation of water in rhyolitic melts and glasses, *Amer. Mineral., 74,* 1247-1257, 1989.

Stolper, E.M., and S. Newman, The role of water in the petrogenesis of Mariana trough magmas, *Earth Planet. Sci. Lett., 121,* 293-325, 1994.

Stolper, E.M., D. Walker, B.H. Hager, and J.F. Hays, Melt segregation from partially molten source regions: the importance of melt density and source region size, *J. Geophys. Res., 86,* 6261-6271, 1981.

Tatsumi, Y., Melting experiments on a high-magnesian andesite, *Earth Planet. Sci. Lett., 54,* 357-365, 1981.

Tatsumi, Y., Origin of high-magnesian andesites in the Setouchi volcanic belt, southwest Japan, II. Melting phase relations at high pressures, *Earth Planet. Sci. Lett., 60,* 305-317, 1982.

Tatsumi, Y., Formation of the volcanic front in subduction zones, *Geophys. Res. Lett., 13,* 717-720, 1986.

Tatsumi, Y., and S.M. Eggins, *Subduction Zone Magmatism,* 211 pp., Blackwell Science, Cambridge, Massachusetts, 1995.

Tatsumi, Y., D.L. Hamilton, and R.W. Nesbitt, Chemical characteristics of fluid phase released from a subducted lithosphere and origin of arc magmas: evidence from high-pressure experiments and natural rocks, *J Volcanol Geotherm Res, 29,* 293-309, 1986.

Tatsumi, Y., and K. Ishizaka, Origin of high-magnesian andesites in the Setouchi volcanic belt, southwest Japan, I. Petrographical and chemical characteristics, *Earth Planet. Sci. Lett., 60,* 305-317, 1982.

Tatsumi, Y., M. Sakuyama, H. Fukuyama, and I. Kushiro, Generation of arc basalt magmas and thermal structure of the mantle wedge in subduction zones, *J. Geophys. Res., 88,* 5815-5825, 1983.

Toksöz, M.N., J.W. Minear, and B.R. Julian, Temperature field and geophysical effects of a downgoing slab, *J. Geophys. Res., 76,* 1113-1138, 1971.

Tomlinson, J.W., A note on the solubility of water in a molten sodium silicate, *J. Soc. Glass Tech., 40,* 25-31T, 1956.

Toramaru, A., and N. Fujii, Connectivity of melt phase in a partially molten peridotite, *J. Geophys. Res., 91,* 9239-9252, 1986.

Turner, S.P., R.M.M. George, P.J. Evans, C.J. Hawkesworth, and G.F. Zellmer, Timescales of magma formation, ascent and storage beneath subduction-zone volcanoes, *Philos. Trans. R. Soc. London Ser. A, 358,* 1443-1464, 2000.

Turner, S.P., and C.J. Hawkesworth, Constraints on flux rates and mantle dynamics beneath island arcs from Tonga-Kermadec lava geochemistry, *Nature, 389,* 568-573, 1997.

Ulmer, P., and V. Trommsdorff, Serpentine stability to mantle depths and subduction-related magmatism, *Science, 268,* 858-860, 1995.

von Bargen, N., and H.S. Waff, Permeabilities, interfacial areas and curvatures in partially molten systems: results of numerical computations of equilibrium microstructures, *J. Geophys. Res., 91,* 9261-9276, 1986.

Watson, E.B., Diffusion in magmas at depth in the Earth: the effects of pressure and dissolved H_2O, *Earth Planet. Sci. Lett., 52,* 291-301, 1981.

Watson, E.B., Melt infiltration and magma evolution, *Geology, 10,* 236-240, 1982.

Watson, E.B., and J.M. Brenan, Fluids in the lithosphere, 1. Experimentally determined wetting characteristics of CO_2-H_2O fluids and their implications for fluid transport, host-rock physical properties, and fluid inclusion formation, *Earth Planet. Sci. Lett., 85,* 497-515, 1987.

Watson, E.B., J.M. Brenan, and D.R. Baker, Distribution of fluids in the continental mantle, in *Continental Mantle,* edited by M.A. Menzies, pp. 111-125, Claredon press, Oxford, 1990.

Withers, A.C., and H. Behrens, Temperature-induced changes in the NIR spectra of hydrous albitic and rhyolitic glasses between 300 and 100 K, *Phys. Chem. Min., 27,* 119-132, 1999.

Withers, A.C., Y. Zhang, and H. Behrens, Reconciliation of experimental results on H_2O speciation in rhyolitic glass using in-situ and quenching techniques, *Earth Planet. Sci. Lett., 173,* 343-349, 1999.

Yogodzinski, G.M., R.W. Kay, O.N. Volynets, A.V. Koloskov, and S.M. Kay, Magnesian andesite in the western Aleutian Komandorsky region: implications for slab melting and processes in the mantle wedge, *Geol. Soc. Amer. Bull., 107,* 505-519, 1995.

Zhang, Y., H_2O in rhyolitic glasses and melts: measurement, speciation, solubility, and diffusion, *Rev. Geophys., 37,* 493-516, 1999.

Zhang, Y., J. Jenkins, and Z. Xu, Kinetics of the reaction $H_2O + O = 2OH$ in rhyolitic glasses upon cooling: geospeedometry and comparison with glass transition, *Geochim. Cosmochim. Acta, 61,* 2167-2173, 1997.

Zhang, Y.-G., and J.D. Frantz, Enstatite-forsterite-water equilibria at elevated temperatures and pressures, *Amer. Mineral., 85,* 918-925, 2000.

Zhao, D., and A. Hasegawa, P wave tomographic imaging of the crust and upper mantle beneath the Japan islands, *J. Geophys. Res., 98,* 4333-4353, 1993.

G.A. Gaetani, MS#8, Woods Hole Oceanographic Institution, Woods Hole, MA 02543 (e-mail: ggaetani@whoi.edu).

T.L. Grove, Department of Earth, Atmospheric, and Planetary Sciences, Massachusetts Institute of Technology, Cambridge MA 02139.

Mapping Water Content in the Upper Mantle

Shun-ichiro Karato

Yale University, Department of Geology and Geophysics, New Haven, Connecticut

Variations in water (hydrogen) content in Earth's (upper) mantle can be inferred from geophysical observations if the relationship between water content and relevant physical properties is known and if high-resolution geophysical measurements are available. This paper reviews the current status of mineral physics understanding of the effects of water on elastic and non-elastic deformation of minerals such as olivine and its influence on seismologically measurable properties. Important effects of water on seismic wave propagation are through indirect effects due to hydrogen-related defects in nominally anhydrous minerals as opposed to the direct effects caused by the formation of hydrous minerals. Two cases of indirect effects are reviewed: (i) effects through the enhancement of anelasticity and (ii) effects through the modifications of lattice preferred orientation. The former causes enhanced attenuation (low Q) and low velocities by the increase of water content and the latter modifies the nature of seismic anisotropy. Experimental data are reviewed to formulate ways to infer water content from seismological data and analytical equations are derived that relate velocity and attenuation anomalies to anomalies in temperature and/or water content. The results are applied to infer the distribution of water in Earth's upper mantle using seismological observations. In subduction zones, the regions of high water content in the shallow upper mantle (200 km) are inferred to be localized to the mantle beneath current or recent volcanoes although wider distribution is hinted in the deeper portions (200 km). In the upper mantle beneath hot spot volcanoes such as Hawaii and Iceland, both seismic wave attenuation and anisotropy measurements suggest the presence of a column of material with a high water content, indicating that Hawaii and Iceland are not only "hot" spots but also "wet" spots.

INTRODUCTION

Water in the solid Earth is known to affect a number of geochemical and geophysical processes including melting and resultant chemical differentiation (e.g., [*Kushiro,* 1972]) and plastic flow associated with mantle convection [*Mackwell et al.,* 1985; *Karato et al.,* 1986; *Mei and Kohlstedt,* 2000a,b; *Karato and Jung,* 2003]. These processes also re-distribute water because strong partitioning of water occurs during phase transformations, particularly melting [*Karato,* 1986; *Hirth and Kohlstedt,* 1996; *Karato and Jung,* 1998] and the wadsleyite to olivine transformation [*Young et al.,* 1993; *Inoue,* 1994; *Kawamoto et al.,* 1996]. Consequently, it is likely that the water content in Earth's mantle varies both laterally and radially.

Several studies of mantle and volcanic rocks show that water contents in Earth's mantle vary significantly with

Inside the Subduction Factory
Geophysical Monograph 138

10.1029/138GM08

geological settings [*Sakuyama,* 1979; *Bell and Rossman,* 1992; *Thompson,* 1992; *Stolper and Newman,* 1994; *Green and Falloon,* 1998; *Ingrin and Skogby,* 2000; *Jamtveit et al.* 2001]. These direct measurements provide estimates of water contents in Earth's mantle, but the data are limited to volcanically regions. In addition, when water content is inferred from mantle minerals, one can question whether the original water content is preserved because of the known high diffusion coefficients of water (hydrogen) in minerals (e.g., [*Mackwell and Kohlstedt,* 1990]). Therefore it would be valuable if one could estimate water contents from geophysical measurements because geophysical measurements have much wider spatial coverage and are intrinsically *in-situ.* Many geophysically observable properties are known to be affected by water (electrical conductivity: *Karato* [1990]; plastic deformation (viscosity): *Blacic* [1972], *Chopra and Paterson* [1984], *Karato et al.* [1986]; seismic wave propagation: *Karato* [1995]), but I will focus on seismic wave propagation in this paper.

The purpose of this paper is to review the current status of our understanding of the effects of water on seismologically observable properties, to formulate a strategy toward mapping water content in Earth's mantle, and to apply it to certain regions. I will emphasize elastic and non-elastic properties including seismic wave attenuation and deformation fabrics because these properties can be determined by seismological measurements with higher resolution than other physical properties such as electrical conductivity. Earlier attempts along this line include studies by *Nolet and Zielhuis* [1994] *and Sato* [1994]. *Nolet and Zielhuis* [1994] found unusually low velocity regions in the deep upper mantle (300–500 km) of Europe and suggested the presence of higher than usual water content. Given the small amount of mineral physics data available at that time, they speculated that the cause of low velocities might be water-assisted partial melting. However, a significant degree of partial melting at this depth is highly unlikely in view of the low heat flow (~40 mW/m^2 [*Cermak and Hurtig,* 1979]) in this regions. *Sato* [1994] investigated the variation in water content in the upper mantle beneath island arcs using a combination of seismological observations and mineral physics data. *Sato* [1994] used the homologous temperature (T/T_m, T: temperature, T_m: melting temperature) formulation of anelastic properties to infer the water content assuming that attenuation changes with water content through the change in the solidus of peridotite. However, *Karato* [1998a] showed that the experimental observations on olivine do not support homologous temperature scaling using the solidus as T_m. Therefore neither of these previous studies has strong support from mineral physics. In this paper, I will develop a new method to infer water content from the combination of seismological observations with mineral physics.

2. HYDROGEN-RELATED DEFECTS IN NOMINALLY ANHYDROUS MINERALS

Experimental studies under high-pressures and analyses of naturally occurring minerals have established that a significant amount of water can be dissolved in nominally anhydrous minerals such as olivine (for review see [*Bell and Rossman,* 1992; *Ingrin and Skogby,* 2000; *Williams and Hemley,* 2001]). The maximum amount of water that can be dissolved in olivine increases with pressure (water fugacity) and is ~10^4 ppm H/Si (~0.1 wt%)* at ~14 GPa (e.g., [*Kohlstedt et al.,* 1996]). Although some hydrous minerals can contain a much higher amount of water than olivine or pyroxenes (e.g., 4.3wt% in phlogopite), these minerals are not stable under high temperature conditions (T>1100 K) in most of Earth's mantle. Therefore I consider only the effects of dissolved water in nominally anhydrous minerals on seismic wave propagation.

Water is incorporated in anhydrous minerals as hydrogen-related point defects (e.g., [*Karato,* 1989a; *Paterson,* 1989; *Bai and Kohlstedt,* 1992; *Mei and Kohlstedt,* 2000a,b; *Bolfan-Casanova et al.,* 2000]). Infrared (IR) spectroscopy shows that hydrogen occurs as protons, H^+, and closely associated with a nearby oxygen ions, O^{2-}, to make an OH^- dipole. Protons are likely to occupy lattice sites that are usually occupied by positively charged species. For instance, Si vacancies (V_{Si}'''' using the Kröger-Vink notation of point defects; [*Kröger and Vink,* 1956]) and Mg vacancies (V_{Mg}'') are among the possible sites where protons might reside [*Karato,* 1989a; *Kohlstedt et al.,* 1996; *Mei and Kohlstedt,* 2000a,b]. For example, when one proton occupies Mg vacancy site, then one creates a defect $(H_{Mg})'$ which has a −1 effective charge relative to the perfect lattice. It must also be noted that the concentration of hydrogen-related defects is much higher than the concentration of point defects in olivine under "dry" environments (~10^{-6}–10^{-5}) (Fig. 1). Because of the high concentration of hydrogen-related defects, the incorporation of hydrogen in olivine will modify the charge balance of a crystal. Consequently the concentration of other charged point defects such as V_{Mg}'', $Si_I^{\bullet\bullet\bullet\bullet}$ and V_{Si}'''' will also be modified [*Karato,* 1989a; *Yan,* 1992]. It is for this reason the incorporation of hydrogen not only directly affects properties that are controlled by the most abundant and fastest diffusing species (e.g., electrical conductivity) but also indirectly modifies properties

* I use the unit H/Si, i.e., the molar ratio of hydrogen to silicon. 1 ppm H/Si~10^{-5} wt% of water (H_2O) in olivine.

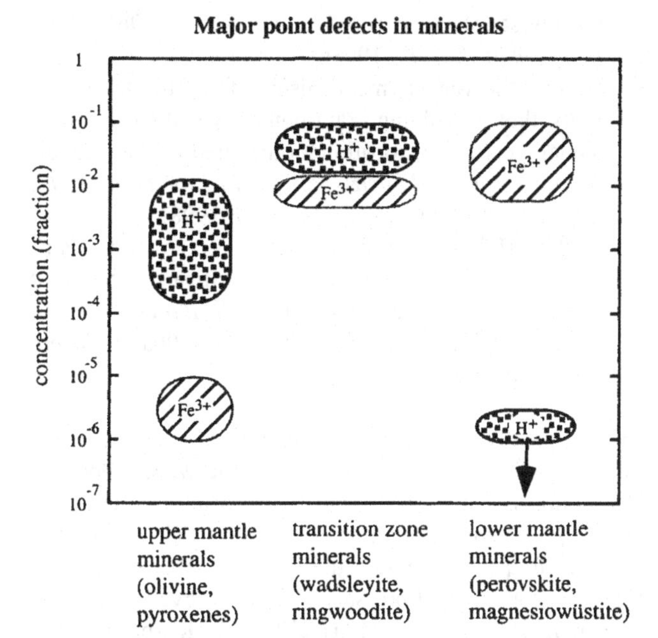

Figure 1. Major point defects in mantle minerals. The solubility of ferric iron (Fe^{3+}) and hydrogen H (H^+) are shown. The actual concentrations of these defects depend on the chemical environment including the fugacity of oxygen (for Fe^{3+}) and fugacity of water (for H^+). The excess electrostatic charge due to these defects must be compensated for by some negatively charged defects, such as vacancies at M-site or Al^{3+} at Si-site. The data sources: ferric iron [*O'Neill et al.*, 1993], hydrogen [*Kohlstedt et al.*, 1996; *Hirth and Kohlstedt*, 1996; *Karato and Jung*, 1998; *Bolfan-Bosanova et al.*, 2000]. Note that the concentration of ferric iron was determined under "dry" conditions. Under "wet" conditions, a high solubility of hydrogen (in olivine) will increase the concentration of ferric iron.

which are controlled by the motion of less abundant, slowest diffusing species (e.g., creep and anelasticity).

Finally, water or hydrogen may also be dissolved at grain-boundaries. Evidence for this includes a much higher concentrations of hydroxyls in polycrystalline olivine than single crystals [*Karato et al.,* 1986] which likely causes enhanced mobility of grain-boundaries by water [*Karato,* 1989b; *Jung and Karato,* 2001a].

3. ENHANCEMENT OF INTRA-GRANULAR AND GRAIN-BOUNDARY PROCESSES IN OLIVINE BY WATER

3.1. *Intra-granular Processes*

Enhancement of dislocation creep (plastic deformation due to dislocation motion) in olivine by the addition of water is well known [*Blacic,* 1972; *Chopra and Paterson,* 1984; *Mackwell et al.,* 1985; *Karato et al.,* 1986; *Mei and Kohlstedt,* 2000b]. *Karato et al.* [1986] showed that water also enhances diffusion creep (plastic deformation due to atomic diffusion) in olivine. However, the exact functional relationship between water content (water fugacity) and creep strength has only been investigated recently [*Mei and Kohsltedt,* 2000a,b; *Karato and Jung,* 2003]. Based on experiments at low pressures (<0.45 GPa), *Mei and Kohlstedt* [2000a,b] found that creep rate in olivine under "wet" (water-saturated) conditions *increases* with pressure. However, *Karato and Jung* [2003] showed, by the experimental study at higher pressures (to 2 GPa), that creep rate under "wet" conditions is a non-monotonous function of pressure [*Karato,* 1989a]: when a system is saturated with water, olivine first becomes weaker with pressure and then gets stronger with pressure at higher pressures. This behavior can be described by the following relationship [*Karato and Jung,* 2003],

$$\dot{\varepsilon} = [A_d exp\left(-\frac{H_d^*}{RT}\right) + A_w C_{OH}^r(T,P) exp\left(-\frac{H_w^*}{RT}\right)]\sigma^n \quad (1)$$

where $\dot{\varepsilon}$ is strain-rate, A_w and A_d are constants that are independent of P (pressure) and T (temperature), H_d^* and H_w^* are activation enthalpies at dry (d) or wet (w) conditions ($H_{d,w}^* = E_{d,w}^* + PV_{d,w}^*$, where $E_{d,w}^*$ is activation energy and $V_{d,w}^*$ is activation volume), σ is stress and C_{OH} is the concentration of OH. When the system is open then C_{OH} is a function of activity of water (a_{H2O}) and pressure (P) and temperature (T). At low pressures, the effect of activation volume ($exp(-PV_{d,w}^*/RT)$ term) is not important. Consequently, the effect of water fugacity dominates and strain-rate increases with increasing pressure because water fugacity increases with pressure. However, at higher pressures, the activation volume term becomes important and the strain-rate decreases with pressure.

3.2. *Grain-boundary Processes*

Karato [1989b] showed that rates of grain-growth in olivine are enhanced by water. *Chopra and Paterson* [1984] and *Jung and Karato* [2001a] found that dynamic recrystallization in olivine is significantly enhanced by water. *Jung and Karato* [2001a] also found that the size of dynamically recrystallized grains increases with water content. Both of these observations strongly suggest water enhances grain-boundary mobility.

Karato et al. [1986] determined that the solubility of water in fine-grained (~10–40 μm) olivine is about ~1000–1500 ppm H/Si at P=300 MPa and T=1573K (OH content measured from IR spectroscopy corresponding to absorption peaks at around 3500-3650 cm^{-1}) which is about 10 times larger than the solubility of water in olivine single crystals. *Jung and Karato* [2001a] found that the grain-boundary mobility is more enhanced by water than is intragranular dislocation motion. They also noted that the grain-boundaries in olivine hot-pressed under high water fugacity conditions are very weak. I conclude that water weakens chemical bonding at grain-boundaries and enhances grain-boundary processes such as grain-boundary migration and sliding.

4. EFFECTS OF WATER ON SEISMIC WAVE PROPAGATION IN OLIVINE

Effects of water on seismic wave propagation in upper mantle minerals have been reviewed by Karato [1995]. There I concluded that the most important effects are (i) enhancement of anelasticity, leading to higher attenuation of seismic waves and lower seismic wave velocities and (ii) modification of lattice preferred orientation, leading to changes in seismic anisotropy. This earlier analysis is further supported by subsequent experimental studies. In the following section I briefly summarize these developments and will provide a quantitative assessment of these effects.

4.1. Effects of Water on Anelasticity

The important points that have been well established include:

1. The propagation of seismic waves (velocities and attenuation) is significantly affected by anelasticity [e.g., *Kanamori and Anderson,* 1977; *Karato and Spetzler,* 1990; *Karato,* 1993; *Jackson,* 2000].
2. Anelasticity in minerals (and ceramics) at high homologous temperatures and low frequencies follow a behavior called *high-temperature background* viz.,

$$Q^{-1} = B \cdot \omega^{-\alpha} exp(-\frac{\alpha H^*}{RT}) \tag{2}$$

   where Q is the Q-factor defined by $Q^{-1} \equiv \Delta E / 2\pi E$ (ΔE: energy loss per unit cycle, E: elastic energy stored in a system), B is a constant, ω is the frequency, α is a constant, and H^* is the activation enthalpy. For these temperatures and frequencies, attenuation is closely related to plastic deformation [e.g., *Karato and Spetzler,* 1990] especially transient micro-creep [*Karato,* 1998c; *Gribb and Cooper,* 1998; *Jackson,* 2000] which is itself closely connected to macro-(steady-state) creep [*Lifshitz and Shikin,* 1965; *Amin et al.,* 1970; *Karato,* 1998b].
3. Among the various mechanisms of solid-state anelasticity, those involving grain-boundary sliding (or diffusional creep) or dislocation motion are the likely candidates [*Karato and Spetzler,* 1990; *Jackson et al.,* 1992; *Tan et al.,* 1997; *Gribb and Cooper,* 1998].
4. Both diffusional and dislocation creep and grain-boundary mobility in olivine are significantly enhanced by hydrogen-related defects [e.g., *Karato et al.,* 1986; *Karato,* 1989a; *Mei and Kohlstedt,* 2000a,b; *Karato and Jung,* 2003].

Consequently, one expects that the addition of water enhances anelasticity and reduces seismic wave velocities. This notion is borne out by the results of a reconnaissance study by *Jackson et al.* [1992] on anelasticity of Åheim dunite at frequencies within the range of seismic waves at high temperatures and pressures. They observed a factor of ~2 increase in Q at P=300 MPa and T= 1273 K when samples are dried prior to measurements at high-temperature (1473 K). The frequency dependence of attenuation (Q^{-1}) given by equation (1) can be translated into, $Q^{-1} \propto (\omega\tau)^{-\alpha}$ where τ is the relaxation time [*Jackson et al.,* 1992]. Microscopic theory of seismic wave attenuation indicates that the relaxation time for anelasticity is related to strain-rate ($\dot{\varepsilon}$) as $\tau \propto \dot{\varepsilon}^{-1}$ [*Karato and Spetzler,* 1990] hence $Q(wet)/Q(dry) = [\dot{\varepsilon}(dry)/\dot{\varepsilon}(wet)]^{\alpha}$. This means that $Q(wet)/Q(dry) \sim 2$ translates to $\dot{\varepsilon}(wet)/\dot{\varepsilon}(dry) \sim 20$ for α=0.23 [*Jackson et al.,* 1992], which is in excellent agreement with the experimental data on creep in the same material [*Chopra and Paterson,* 1984], suggesting that micro-creep responsible for anelasticity is enhanced by water to a similar degree as is macro-creep.

Therefore I conclude that water enhances anelasticity and that the effects of water on anelasticity can be quantified through the effects of water on creep that modifies the relaxation time. It follows that Q^{-1} depends on temperature, pressure and water content as,

$$Q^{-1}(T,P,C_{OH}) \propto [A_d \cdot exp(-\frac{H_d^*}{RT}) + A_w \cdot C_{OH}^r \cdot exp(-\frac{H_w^*}{RT})]^{\alpha} \tag{3}$$

hence the lateral variation of Q^{-1} is related to the lateral variation in water content and temperature as,

$$log\frac{Q^{-1}}{Q_0^{-1}} \cong \alpha r \cdot log(\frac{C_{OH}}{C_{OH_0}}) - \frac{\alpha H^*}{R}(\frac{1}{T} - \frac{1}{T_0}) \tag{4}$$

where the quantities with suffix 0 are those of reference values and the second term in equation (3) is assumed to dominate (as in most cases in Earth). Here I use H^* for activation enthalpy, ignoring a small difference between H^*_w and H^*_d. The variation of Q as a function of temperature (T) and water content (C_{OH}) calculated using equation (3) is shown in Fig. 2. This relation allows us to infer the water content C_{OH} if temperature T is known.

A high water content has an effect similar to a high temperature. Therefore a region of a high water content would look like seismologically a region of high temperature. To quantify this point, it is convenient to introduce a quantify called "rheologically effective temperature" T_{eff} defined as,

$$Q^{-1}(\omega, T, P) \propto exp(-\alpha \frac{H^*}{RT_{eff}}) \quad (5)$$

where

$$T_{eff} = T \cdot \xi \quad (6)$$

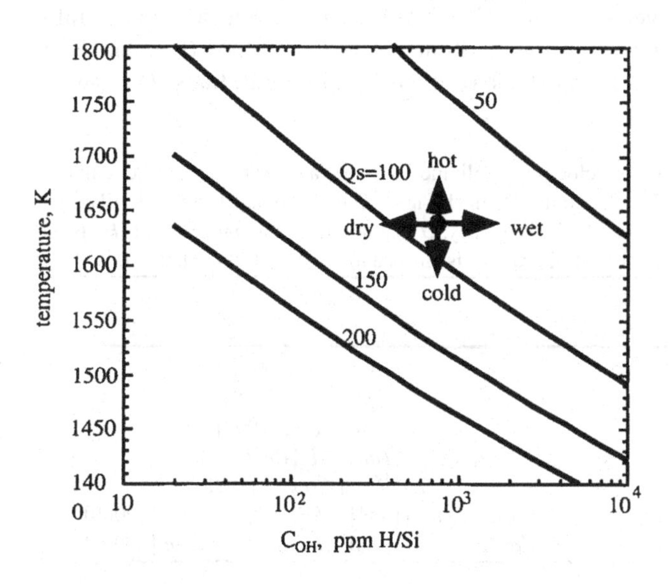

Figure 2. Calculated dependence of Q_s (Q for S-waves: Q^{-1} is seismic wave attenuation) on temperature and water content. A solid circle represents values corresponding to the asthenosphere (~200 km depth) of an average Earth: Q_s=80; [*Dziewonski and Anderson,* 1981]), C_{OH} (olivine water content) = 800 ppm H/Si; [*Hirth and Kohlstedt,* 1996], T~1650 K. The frequency dependence, α, and the activation enthalpy, H^*, are from *Jackson et al.* [1992] (Table 1). The pre-exponential factor is adjusted to yield Q_s=80 at 200 km depth. Arrows indicate directions of change in Q_s due to the change in temperature and/or water content.

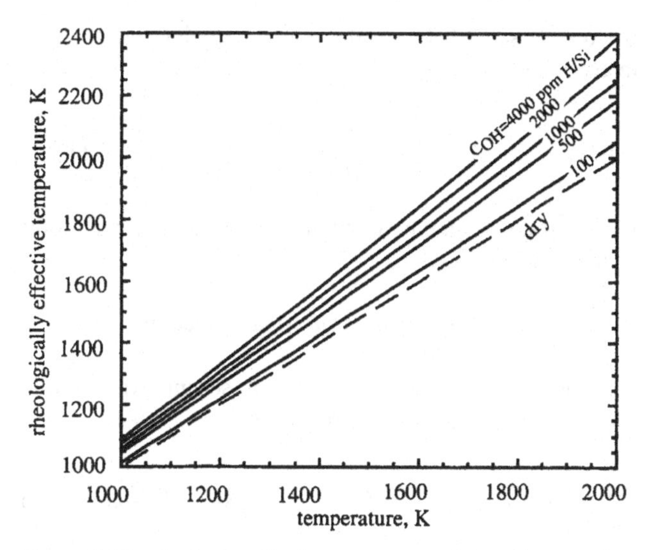

Figure 3. Rheologically effective temperature defined by equation (6) in the text, T_{eff}, vs. temperature relation for various water content, C_{OH} (ppm H/Si).

with

$$\xi \equiv \frac{1}{[1-(RT/H^*) \cdot log(A_w \cdot C^r_{OH}/A_d)]} \quad (7)$$

T_{eff} is slightly modified from actual temperature incorporating the effects of water (Fig. 3). ξ is only weakly dependent on temperature and therefore the difference between real temperature, T, and rheologically effective temperature, T_{eff}, is primarily dependent on water content. Therefore the observed Q values may be first interpreted as a variation in T_{eff}, and if T can be inferred from some other sources, then C_{OH} can be estimated.

Rheologically effective temperature is directly related to viscosity and hence the variation in Q is directly related to the variation in viscosity. The strain-rate is now written as,

$$\dot{\varepsilon} \propto \sigma^n \cdot exp(-\frac{H^*}{RT_{eff}}) \quad (8)$$

and hence the viscosity η is given by,

$$\eta \propto \sigma^{1-n} exp(\frac{H^*}{RT_{eff}}) \propto \dot{\varepsilon}^{(1-n)/n} exp(\frac{H^*}{nRT_{eff}}) \quad (9)$$

or

$$\frac{\eta}{\eta_0} = (\frac{Q}{Q_0})^{\frac{1}{\alpha}} \quad (10\text{-a})$$

if stress is same between the reference point and point of interest, and

$$\frac{\eta}{\eta_0} = \left(\frac{Q}{Q_0}\right)^{\frac{1}{\alpha n}} \tag{10-b}$$

if strain-rate is same between the reference point and point of interest.

4.2. Effects of Water on Seismic Wave Velocities

A change in Q results in a change in seismic wave velocities V through velocity dispersion viz.,

$$V(\omega,T,P,C_{OH}) = V_\infty(T,P)[1 - F \cdot Q^{-1}(\omega,T,P,C_{OH})] \tag{11}$$

with

$$V_\infty(T,P) \cong \bar{V}_\infty(P) \cdot [1 - A(P) \cdot T] \tag{12}$$

where $V(\omega, T, P, C_{OH})$ is seismic wave velocity corrected for the anelastic dispersion, $\bar{V}_\infty(T,P)$ is the velocity at infinite frequency where only anharmonic effect is important ($A(P)$ is a parameter related to thermal expansion [*Anderson,* 1995]; see Table 1) and $F = (1/2)cot(\pi\alpha/2)$ [*Minster and Anderson,* 1981]. Water has little effect on $\bar{V}_\infty(T,P)$ and therefore the most important effect of water on seismic wave velocity is through anelastic dispersion [*Karato,* 1995]. Using (4), equation (12) can be rewritten as,

$$\frac{V - V_0}{V_0} \cong -A(P) \cdot \delta T - F \cdot Q_o^{-1} \cdot \left[\left(\frac{C_{OH}}{C_{OH_o}}\right)^{r\alpha} exp\left(\frac{\alpha H^* \delta T}{RT_0^2}\right) - 1\right] \tag{13a}$$

or equivalently,

$$\frac{V - V_0}{V_0} \cong -A(P) \cdot \delta T_{eff} - F \cdot Q_o^{-1} \cdot \left[exp\left(\frac{\alpha H^* \delta T_{eff}}{RT_0^2}\right) - 1\right] \tag{13-b}$$

where $\delta T_{eff} \equiv T_{eff} - T_o$ where V is the actual seismic wave velocity and V_o is the seismic wave velocity in a laterally homogeneous Earth model such as PREM (Fig. 4). Using equation (9), the relations (13a,b) can be translated into,

Table 1. Parameters controlling the temperature dependence of seismic wave velocities in olivine. A is a parameter related to the temperature derivative of seismic wave velocities at high-frequencies defined by equation (12) in the text. The functional form $Q^{-1} = B \cdot \omega^{-\alpha} \cdot exp(-\alpha H^*/RT)$ is assumed for seismic wave attenuation where B is a constant, ω is frequency, α is a non-dimensional constant, H^* is the activation enthalpy, and the functional form $\dot{\varepsilon} \propto \sigma^n$ is used for creep where is $\dot{\varepsilon}$ strain rate, σ is stress and n is the stress exponent.

A (K^{-1})	B (s^{α})	α	H^*(kJ/mol)	n	source
$6x10^{-5}$ [a], $7x10^{-5}$ [b]					*Anderson* [1995]
	18	0.20	440		*Gueguen et al.* [1989][c]
	160[d], 80[e]	0.23	400		*Jackson et al.* [1992][f]
	$3.3x10^3$	0.31	420		*Tan et al.* [1997][g]
	$3.7x10^7$	0.35	700		*Gribb and Cooper* [1998][h]
				3.0-3.5[i]	*Karato et al.* [1986], *Mei and Kohlstedt* [2000b]
				1.0[j]	*Karato et al.* [1986], *Mei and Kohlstedt* [2000a]

[a]: for P wave
[b]: for S-wave
[c]: single crystal forsterite at room pressure
[d]: for "dry" conditions
[e]: for "wet" conditions
[f]: coarse-grained dunite (~1 μm) at P=300 MPa confining pressure
[g]: fine-grained dunite (~50 μm) at P=300 MPa confining pressure
[h]: fine-grained forsterite (~3 μm) at room pressure
[i]: dislocation creep
[j]: diffusion creep

$$\frac{V-V_o}{V_o} \cong (ART_o^2 / H^*) \cdot log\frac{\eta}{\eta_o} -$$

$$(FQ_o^{-1}) \cdot [(\frac{\eta_o}{\eta})^{\alpha} - 1] \quad (14\text{-a})$$

if stress is same between the reference point and point of interest, and

$$\frac{V-V_o}{V_o} \cong (nART_o^2 / H^*) \cdot log\frac{\eta}{\eta_o} -$$

$$(FQ_o^{-1}) \cdot [(\frac{\eta_o}{\eta})^{\alpha n} - 1] \quad (14\text{-b})$$

Using equation (10) or (14), the lateral variation in viscosity can be inferred from the lateral variation in seismic wave velocities or attenuation from a model for T_o and Q_o and material parameters A, H^*, n and α. The results are shown in Fig. 5.

In this formulation, the important parameters are a, n, A, α and E^*+PV^* $(=H^*)$ all of which are constrained by laboratory studies (Table 1). A small variation in major element chemistry (such as Fe/(Fe+Mg)) has small effects on creep (e.g., *Karato* [1989c]) and is likely to have little effect on anelasticity. Therefore we ignore the effects of variation in major element chemistry on anelasticity. However, a variation in major element chemistry has important effects on $\bar{V} \propto (T,P)$(Table 2).

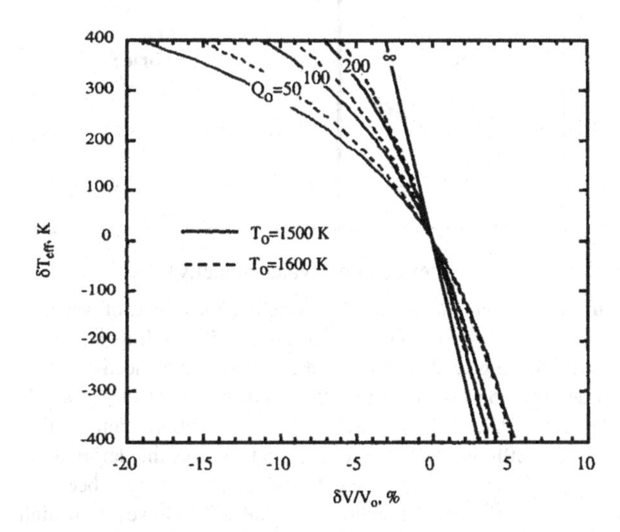

Figure 4. Relation between velocity anomalies and anomalies in effective temperature. Q_o and T_o are the background Q and temperature respectively. Difference between S and P waves is mainly from the difference in background Q_o. Q_o for S waves is ~0.4-0.5 of Q_o for P waves.

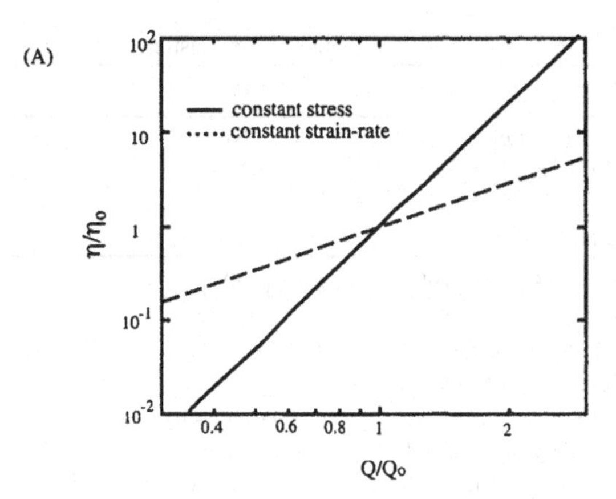

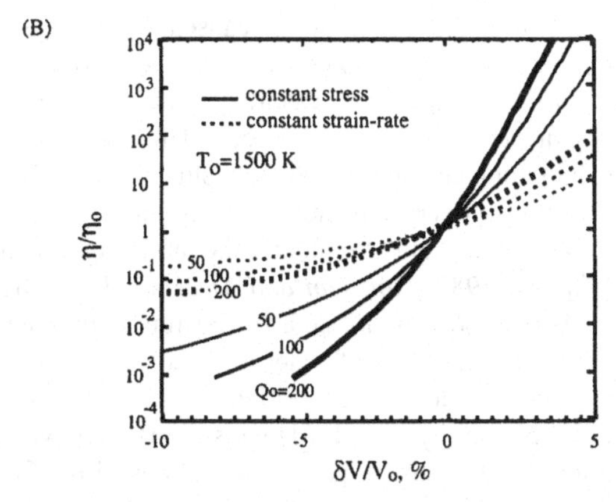

Figure 5. Anomalies in viscosity can be calculated from anomalies in seismic wave attenuation or velocities using the relationships shown in these figures. (A) Relation between attenuation anomalies and viscosity anomalies. (B) Relation between velocity anomalies and viscosity anomalies. Difference between S and P waves is mainly from the difference in background Q_o.

4.3. Effects of Water on Seismic Anisotropy

Seismic anisotropy in the upper mantle has traditionally been interpreted using the knowledge of deformation fabrics of peridotites from the shallow upper mantle or deformed in the laboratory at low water fugacity [*Nicolas and Christensen,* 1987; *Karato,* 1989d]. Although upper mantle peridotites show a remarkably consistent deformation fabric [*Ismail and Mainprice,* 1998], this does not mean that upper mantle rocks result in a similar fabric under all conditions. Rocks considered in these previous studies are mostly from the lithospheric mantle, which is likely to be depleted of water (as well as other "incompatible" elements) due to previous partial melting. Similarly, most deformation experiments have been conducted under low water fugacity conditions.

Table 2. Importance of various factors on elastic and anelastic properties of mantle minerals

	Q^{-1}	$V_\infty - V_0$	V_∞	$\delta logV_s/\delta logV_p$
major element chemistry[a]	No	No	Yes	1.2-1.4
partial melting[b]	No	Yes	Small	1.0-1.2
water	Yes	Yes	No	1.5-2.0
temperature[c]	Yes	Yes	Yes	1.5-2.0

a: based on *Karato* [1989c] *and Bass* [1995]
b: based on *Takei* [2000]
c: including anelastic effects [*Karato,* 1993]
Q^{-1}: seismic wave attenuation
V_0 : velocity at zero frequency (relaxed velocity)
V_∞: velocity at infinite frequency (unrelaxed velocity)
$\delta logV_s/\delta logV_p$: ratio of lateral variation of S-wave velocity to that of P wave velocity

Changes in deformation fabric caused by changes in deformation conditions are known in many materials (e.g., quartz [*Tullis et al.,* 1973], calcite [*Wenk et al.,* 1973]) which are due to changes in microscopic deformation mechanisms, including changes in dominant slip systems of crystal dislocations. Recrystallization that occurs in most of mantle conditions could also modify the fabric significantly (e.g., [*Karato,* 1987; *Gottstein and Mecking,* 1985; *Wenk and Tomé,* 1999; *Lee et al.,* 2002]). Existing experimental data show that the effects of water to enhance deformation are anisotropic. Briefly, the addition of water selectively enhances the mobility of **b**=[001] (**b** is the Burgers vector) dislocations more than **b**=[100] dislocations (Fig. 6). In addition, water enhances grain-boundary mobility relative to intra-granular dislocation motion [*Jung and Karato,* 2001a]. These observations led to a hypothesis that a high water fugacity might modify the deformation fabrics of olivine-rich rocks [*Karato,* 1995].

We have recently tested this hypothesis by conducting shear deformation experiments under relatively high water fugacity conditions (water fugacity to ~13 GPa; *Jung and Karato* [2001b]). The results indicate that the deformation fabrics of olivine under high water-fugacity ("wet") conditions are different from those under water-poor ("dry") conditions: rather than a maximum of [100] orientation subparallel to the shear direction, there is a maximum of [001] orientation subparallel to the shear direction [*Jung and Karato,* 2001b] (plate 1).

Note that the transition conditions between different fabrics are not directly dependent on strain-rates, although they do depend on temperature, stress and water content. Therefore the results shown in plate 1 can be directly applied to hot regions in Earth (i.e., the asthenosphere) where the temperatures are similar to the experimental conditions of *Jung and Karato* (2001b). The transition from "dry" fabric to "wet" fabric under these high temperature conditions occurs when the olivine water content exceeds ~1000 ppm H/Si at low stresses. The olivine water content in a typical oceanic asthenosphere is estimated to be ~800 ppm H/Si [*Hirth and Kohlstedt,* 1996], but olivine in the

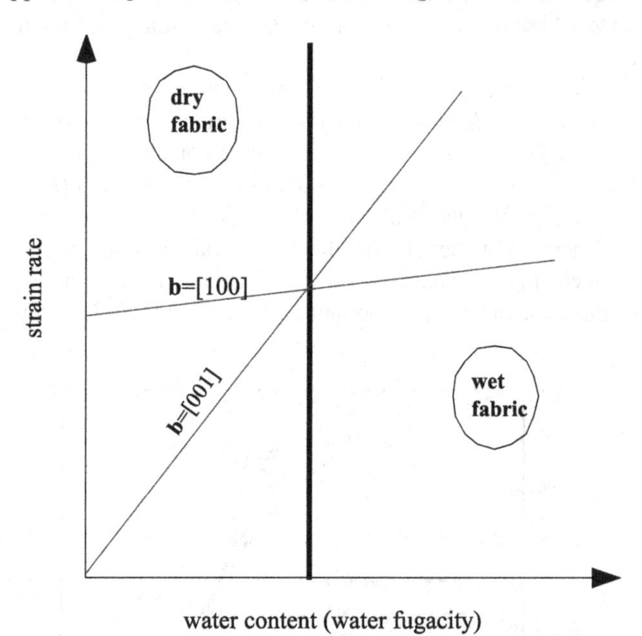

Figure 6. A schematic diagram showing that the effects of water on plastic deformation of olivine are anisotropic. Plastic flow involving **b**=[001] (**b**: Burgers vector) dislocations is more enhanced by water than deformation involving **b**=[100] dislocations [*Mackwell et al.,* 1985; *Yan,* 1992]. Under low water fugacity conditions, deformation due to **b**=[100] dislocations is easiest and controls the deformation fabric, causing a fabric in which [100] direction of olivine becomes subparallel to the flow direction ("dry fabric"). However, at high water fugacity conditions, deformation by **b**=[001] dislocations could become easier than deformation using **b**=[100] dislocations and a new type of fabric may dominate in which [001] direction of olivine becomes subparallel to the flow direction ("wet fabric") [*Karato,* 1995; *Jun and Karato,* 2001b].

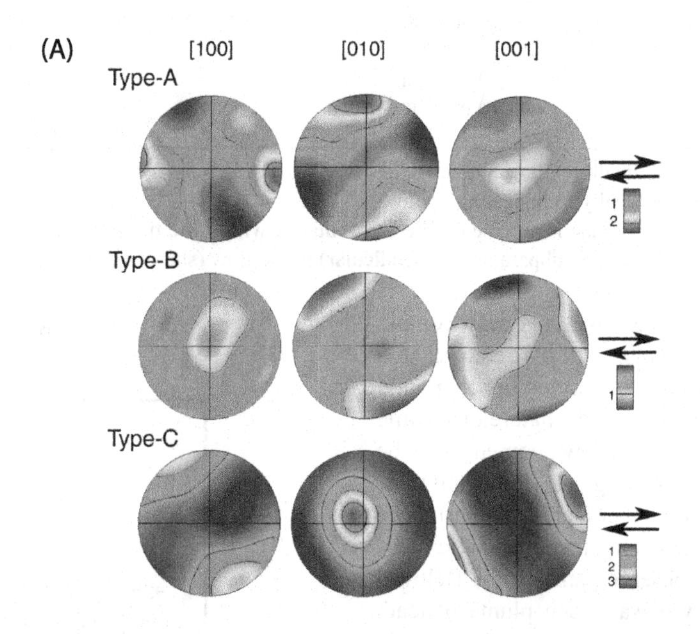

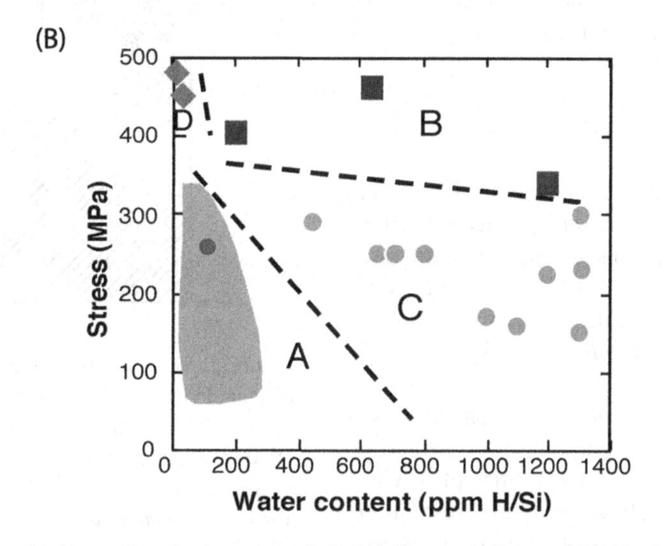

Plate 1. Effects of water on lattice preferred orientation of olivine (*Jung and Karato,* 2001b). (A) Lattice preferred orientation of olivine deformed in simple shear. Pole figures corresponding to three types of lattice preferred orientation are shown. Shear direction is shown by arrows. (B) A fabric diagram showing the dominant types of lattice preferred orientation as a function of stress and water content. All the data are at relatively high temperatures (T=1400-1570K).

Table 3. Seismic anisotropy due to lattice preferred orientation of olivine (after *Jung and Karato* [2001b]). Transverse isotropy (the vertical axis is the axis of symmetry) is assumed when calculating the SH/SV, PH/PV anisotropy (see *Jung and Karato* [2001b] for details).

	dry	wet
horizontal shear	$V_{SH}>V_{SV}$	$V_{SV}>V_{SH}$
	$V_{PH}>V_{PV}$	$V_{PV}>V_{PH}$
vertical shear	$V_{SV}>V_{SH}$	$V_{SH}>V_{SV}$
	$V_{PV}>V_{PH}$	$V_{PH}>V_{PV}$
fast S-wave[a]	// flow	// flow (low stress), ⊥ flow (high stress)

$V_{SH\ (SV)}$: velocity of horizontally (vertically) polarized S-wave
$V_{PH\ (PV)}$: velocity of horizontally (vertically) propagating P wave
[a]: polarization of faster S-waves. Flow plane is assumed to be horizontal or with some tilt, but not vertical. "// (⊥) flow" means that the direction of the faster S-wave polarization is subparallel (perpendicular) to the flow (shear) direction projected to the surface.

source region of subduction zone volcanoes may contain ~3000 ppm H/Si [*Karato and Jung,* 1998]. Therefore, in a relatively water-rich environment in Earth's upper mantle such as subduction zone upper mantle, the nature of seismic anisotropy corresponding to a given flow geometry will be different from that in water-poor regions (for details see Table 3). Therefore if one knows other factors that influence anisotropy, such as the flow geometry, then one could identify water-rich regions from observed anisotropy. Below I use this approach to identify a water-rich plume beneath hotspots, where one can assume a vertical flow geometry in the deep upper mantle. I will also briefly discuss some implications of the present results on the anisotropy of subduction zones.

5. SOME APPLICATIONS

In this section I apply the set of relations developed in previous sections to infer the distribution of water in Earth's upper mantle. I focus on two regions: subduction zones and hotspots, both of which are settings where a high water content is expected based on geochemical evidence [*Kushiro,* 1972; *Iwamori,* 1998; *Green and Falloon,* 1998].

5.1. Inferences from Velocity and Attenuation Tomography

Both seismic wave attenuation and velocities of olivine are sensitive to water content. However, spatial resolution of attenuation measurements is more limited than that of velocity measurements. Therefore, I first use velocity anomalies to map water content. High resolution tomographic images have been obtained for the upper mantle beneath Japan where the densest seismic network on Earth is in place [*Hasegawa et al.,* 1991, 1994; *Zhao et al.,* 1992] (a similar result was obtained for Lau basin, *Zhao et al.* [1997]). I use the results by *Zhao et al.* [1992] to map anomalies in rheologically effective temperature, $\delta T_{eff,}$ using

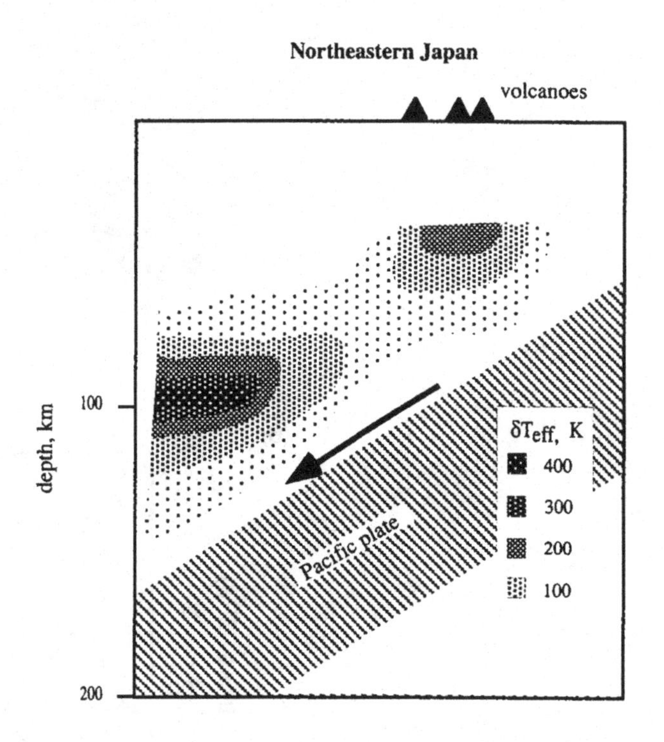

Figure 7. Variation of anomalies in rheologically effective temperature in the northeastern Japan subduction zone estimated from anomalies of seismic wave velocities [*Hasegawa et al.,* 1991; *Zhao et al.,* 1992] (no vertical exaggeration). The background Q_o is assumed to be Q_o=200 (Q for P waves) [*Umino and Hasegawa,* 1984; *Takanami et al.,* 2000] and the background temperature model by *Furukawa* [1993] was used. Anomalously high effective temperatures correspond either to anomalously high temperatures and/or high water content. The maximum anomaly of δT_{eff} ~400 K is too high to be attributed only to temperature anomaly. A combination of modestly high temperatures (δT~200K) and high water content (δC_{OH}~3000 ppm H/Si) can explain anomalies in these regions. The maximum anomaly in this region corresponds to the viscosity anomalies of ~1/10 (for constant strain rate) to ~1/300 (for constant stress) of average viscosity.

equation (13-b). The results are shown in Fig. 7. It is seen that the regions with significantly high T_{eff} are highly localized. Inferred maximum anomaly in rheologically effective temperature is δT_{eff} ~400 K, which is too high to be attributed solely to temperature anomaly that is constrained by heat flow anomalies [*Furukawa,* 1993]. Therefore some enrichment of water may also be considered. A possible combination is an excess temperature of δT~200 K and an excess water content of δC_{OH}~3000 ppm H/Si, which corresponds to δT_{eff}~400 K (see Fig. 3). Regions just above the subducting Pacific plate show small δT_{eff} despite significantly lower than typical temperatures [*Furukawa,* 1993], suggesting a higher than normal water content in olivine. These regions with high water content agree well with the theoretical model of water transport by *Iwamori* [1998].

Long wavelength high attenuation anomalies are identified in many subduction zones [*Romanowicz,* 1994; *Romanowicz and Durek,* 2000; *Billien et al.,* 2000]. Similarly, *Umino and Hasegawa* [1984] and *Takanami et al.* [2000], and *Flanagan and Wiens* [1990, 1994] and *Roth et al.* [1999] made attenuation measurements in the relatively shallow portions (≤ 200 km) of Japan and Tonga-Kermadec (Lau basin) subduction zones respectively, both of these studies showed that enhanced attenuation occurs in the upper mantle beneath active volcanoes. However, the degree to which subduction zone upper mantle is enriched with water is not well constrained by attenuation measurements. *Flanagan and Wiens* [1994] and *Roth et al.* [1999] for example, obtained significantly lower over-all Q_s values in several back arc basins than *Umino and Hasegawa* [1984] and *Takanami et al.* [2000]. A somewhat different picture is found for the deeper portions (≥200 km) of subduction zones. Both long wavelength surface wave/normal mode studies [*Romanowicz,* 1994] and global scale body wave studies [*Billien et al.,* 2000] suggest broad regions of low Q in the deep upper mantle, hinting at a wide distribution of water in that depth range. This conclusion is consistent with broad low velocity anomalies in the deep upper mantle found by *Nolet and Zielhuis* [1994] and *Revenaugh and Sipkin* [1994].

Another notable region is the upper mantle beneath Hawaii (and other hotspots). Low velocity and high attenuation anomalies beneath Hawaii and other hotspots have been identified by *Zhang and Tanimoto* [1993], *Romanowicz* [1994] and *Li et al.* [2000]. These anomalies can be attributed to high temperature anomalies associated with a hot plume. However, the petrological studies by *Takahashi et al.* [1998], *Green and Falloon* [1998] and *Takahashi and Nakajima* [2002] suggest that the temperature anomalies associated with hotspots are ~150 K or less, less than a simple core-mantle-boundary originated plume hypothesis would imply (~300 K, e.g., [*Watson and McKenzie,* 1991]). A higher water content could contribute to the observed higher attenuation and lower seismic wave velocities in these regions. A higher water content in olivine from hotspot magmas has been documented by Jamtveit et al. [2001].

5.2. Inference from Seismic Anisotropy

Additional insights into the regional variation in water content comes from the observations on seismic anisotropy. The pattern of shear wave splitting in subduction zones is rather complicated [e.g., *Buttles and Olson,* 1998; *Fischer et al.,* 2000; *Smith et al.,* 2001]. However, there is a general trend that the direction of fast S-wave polarization tends to be subparallel to trench near trenches and becomes high angle away from them (Plate 2b). Similarly, the direction of polarization of fast S-waves near the collision zones in the continents is generally parallel to the collision zones (e.g., *Silver,* 1996). In the past, such a trend was interpreted as due to the complication in flow pattern [e.g., *Nicolas,* 1993; *Silver,* 1996; *Buttles and Olson,* 1998; *Smith et al.,* 2001], but the present study provides an alternative and simpler explanation: the change in the direction of the faster S-wave polarization is due to the spatial variation in water content (and stress). Both water content and stress magnitude are expected to be higher near trenches leading to a change in the pattern of shear wave splitting [see also *Jung and Karato,* 2001b].

Using high density seismic records, *Gaherty* [2001] revealed the detailed anisotropic structure beneath Iceland. If transverse isotropy is assumed, (V_{SH} > V_{SV} ($V_{SH(SV)}$: velocity of horizontally (vertically) polarized S-wave) anisotropy is inferred in the deeper portion (>100 km) and V_{SV} > V_{SH} in the shallow portion (<100 km). Assuming a vertical flow beneath Iceland, such an observation can be interpreted by lattice orientation of olivine under "wet" conditions (type-C or B) in the deeper portion and under "dry" conditions (type-A) in the shallower portion, suggesting dewatering due to partial melting at ~100 km. This implies a potential temperature of ~1750 K for this region which is consistent with an independent estimate by *Schilling* [1991]. The results in the central Pacific around Hawaii are consistent with this picture although the interpretation is more ambiguous because most of the seismic observations reveal only long wavelength structures. Using a global data set, *Ekström and Dziewonski* [1998] and *Montagner and Guillot* [2000] found anomalous anisotropy in the Pacific upper mantle. Upper mantle beneath Hawaii is characterized by

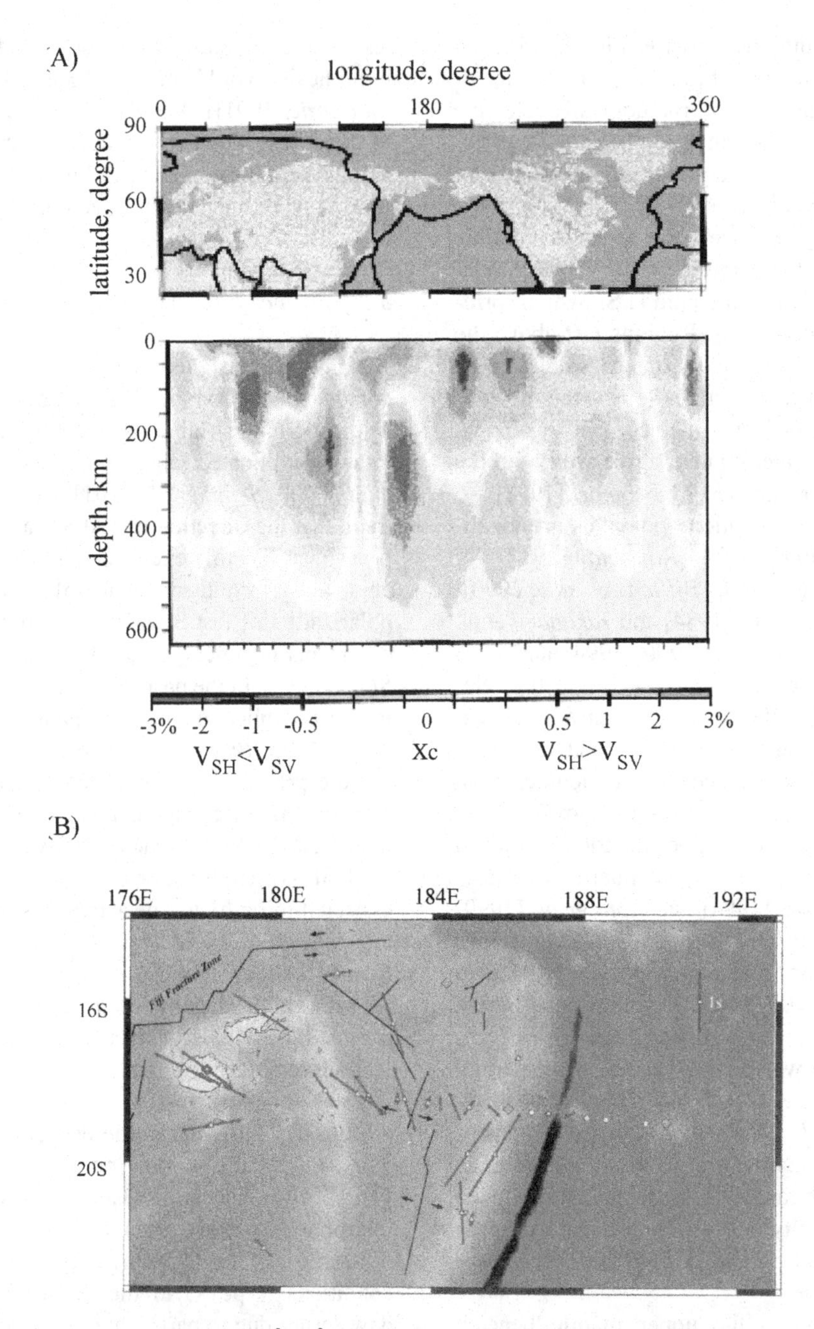

Plate 2. (A) Polarization anisotropy ($X_i \equiv V_{SH}^2/V_{SV}^2 - 1$) in the upper mantle beneath the Pacific (from *Montagner and Guillot* [2000]). The lower figure is a cross section along the 20N. Hawaii is located near 205E, 20N. Strong $V_{SH}>V_{SV}$ anisotropy is identified beneath Hawaii (similar results were obtained by *Ekström and Dziewonski* [1998]). Note also a marked "anisotropy dipole": the upper mantle of the western Pacific shows $V_{SH}>V_{SV}$ anisotropy toward the west of the subducting slab, whereas $V_{SV}>V_{SH}$ anisotropy in the eastern side of the subducting slab. Flow direction in these regions is expected to be nearly vertical. Therefore the observed $V_{SH}>V_{SV}$ anisotropy beneath Hawaii and in the western portions of the subduction zone suggests that these regions are enriched with water. (B) Shear wave splitting observed in the Lau back-arc basin [*Smith et al.,* 2001]. Red arrows indicate the direction of fast S-wave polarization and the magnitude of shear wave splitting. The fast S-wave polarization is nearly parallel to the trench near the trench, but becomes high angle to the trench away from it. Such a trend is observed in other backarc regions [e.g., *Margheriti et al.,* 1996].

strong $V_{SH} > V_{SV}$ anisotropy [*Ekström and Dziewonski,* 1998; *Montagner and Guillot,* 2000]. $V_{SH} > V_{SV}$ in the deeper upper mantle in the central Pacific may be due to "wet" olivine fabric corresponding to a vertical flow, whereas the anisotropy in the shallow portion may reflect strong horizontal flow of "dry" olivine after dewatering. This model implies a change in anisotropy in the Hawaiian plume conduit at a certain depth (~100 km), the detection of which requires high-resolution local studies (Plate 3).

6. DISCUSSION

6.1. How to Distinguish Effects of Water from Those of Partial Melting

Geophysical anomalies such as anomalously low velocities and high attenuation have often been attributed to partial melting [e.g., *Shankland et al.,* 1981; The MELT Seismic Team, 1998]. Qualitatively, much of the effects of high water (hydrogen) content are similar to those due to the presence of partial melting. Therefore one must ask how to distinguish the effects of partial melting from those of high water content. Two observations are critical to this point.

The anelastic relaxation caused by partial melting can be evaluated through the estimate of the relaxation time. The characteristic time of anelastic relaxation depends on the geometry of melt pocket and viscosity of melt [*O'Connell and Budianski,* 1977]. Using a realistic melt geometry, *Hammond and Humphreys* [2000a,b] investigated the effects of partial melting on seismic wave propagation. They concluded that the characteristic frequencies of anelastic relaxation are higher than seismic frequencies and therefore Earth materials would behave like a relaxed body and there is little effect of partial melting on attenuation (Q^{-1}) (Table 2) (*Takei* [2000] reached the same conclusion). This conclusion is consistent with *Sato et al.* [1989]'s observation that there is little change in attenuation in a peridotite across the solidus.

In the relaxed regime, attenuation is not affected by partial melting, but seismic wave velocities are affected. The degree to which velocities are affected by partial melting depends on the melt fraction and geometry. *Takei* [2000] (see also *Hammond and Humphreys* [2000a,b]) showed that for a reasonable range of geometries (corresponding to a dihedral angle of ~30-50°), partial melting has an important effect on seismic wave velocity only when the melt fraction exceeds ~1%. The melt fraction for a given material is controlled by the degree of partial melting and the efficiency of melt transport. The degree of partial melting is determined by the temperature and water content. At the global scale of the source region of MORB, temperature is likely to be lower than the dry solidus of peridotite and in this case the degree of partial melting is determined by the amount of water and is less than ~0.2% in the source region of MORB (e.g., *Plank and Langmuir* [1992]). A higher degree of partial melting (say >3%) occurs only when temperature exceeds the dry solidus, which would occur only in the shallow portions at the vicinity of upwelling current. A higher degree of partial melting is possible in regions where a larger amount of water is available (e.g., wedge mantle). In these cases, however, high permeability of melt through peridotites makes it difficult to keep a high degree of melt fraction (>1%) for a geologically long time scale: although the total degree of melting can be higher, the melt fraction contained in a given piece of peridotite at a given time will be limited to a small value by high permeability (e.g., *Shankland et al.,* 1981; *Kohlstedt,* 1992). Therefore I conclude that partial melting is unlikely to have an important effect on seismic wave propagation in most cases.

Another useful parameter to distinguish effects of partial melting from those of high temperature and/or of water is the ratio of lateral variation in shear and compressional wave velocities, $R_{s/p} \equiv \delta \log V_S / \delta \log V_P$. This ratio can be calculated using the formula given by Takei [2000]. For equilibrium geometry of melt pockets appropriate for peridotite this ratio is estimated to be $R_{s/p}$ ~1.0–1.2 (*Hammond and Humphreys* [2000a] used unreasonably large values of compressibility of melt to estimate large values of $R_{s/p}$ ~1.5–2.5). In contrast, when effects of anelasticity are large, this ratio becomes larger, $R_{s/p}$ ~1.5–2.0 (see Table 2). Consequently, the observed $R_{s/p}$ can be used to evaluate the importance of partial melting. In most cases, the ratio $R_{s/p}$ in the upper mantle is ~1.5–2.0 (e.g., *Shankland et al.* [1981], *Zhao et al.* [1992], *Masters et al.* [2000]) indicating that partial melting is not important in most of Earth's upper mantle.

6.2. Some Remarks on the Use of Seismological Observations to Infer Water Content

In using velocity anomalies to infer rheologically effective temperature or water content, the influence of changes in major element chemistry must be corrected. A combination with geochemical studies on major element distribution (e.g., [*Gaul et al.,* 2000]) will be one way to make correction for this effect. In contrast to velocity anomalies, anomalies in attenuation (Q^{-1}) are insensitive to major element chemistry such as the Fe/(Fe+Mg) and is free from this complication. However, currently available data on

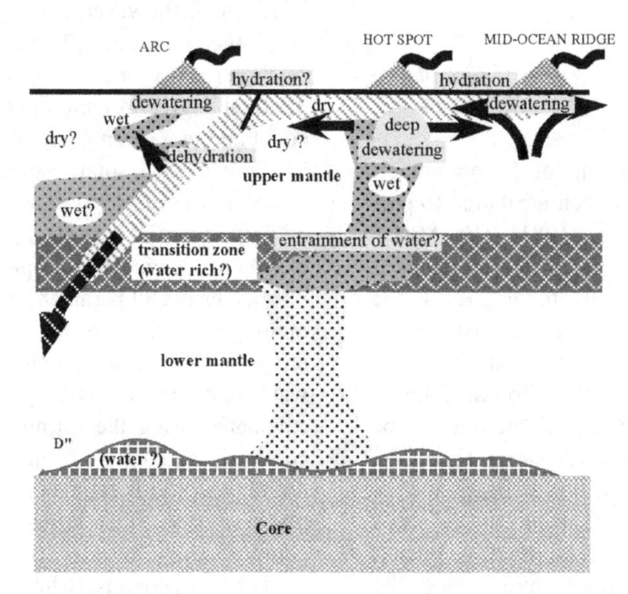

Plate 3. A cartoon showing possible water distribution in the upper mantle suggested from seismological observations. Dewatering at mid-ocean ridges can cause rheological stratification of the upper mantle resulting in a dry lithosphere. The top ~ 5km of the lithosphere is hydrated by hydrothermal circulation. A subduction zone is likely to be enriched with water, but water-rich regions in the shallow upper mantle could be localized (Figure 7). A wider distribution of water is suggested in the deeper upper mantle both from attenuation tomography and anisotropy. The deep upper mantle beneath some hot spots such as Hawaii or Iceland appears to be water-rich due probably to the entrainment of water through the interaction of upwelling plume with the water-rich transition zone. These plumes inject materials into the asthenosphere but the water content in these materials depends critically on the depth of dewatering due to partial melting. The currently available data suggest deep dewatering in plumes and materials injected by plumes are largely "dry."

attenuation have less spatial resolution and at best semi-quantitative.

Interpretation of seismic anisotropy is also not always unique. There is a trade-off between flow geometry and physical mechanisms of anisotropic structure formation. For example, the observed $V_{SH} > V_{SV}$ anisotropy beneath Hawaii could well be caused by horizontal flow under "dry" conditions. Similarly, anisotropy in the subduction zone upper mantle might reflect a complicated three- dimensional flow pattern (e.g., *Smith et al.* [2001]). Seismic observations with higher spatial resolution are needed to resolve these issues.

7. CONCLUDING REMARKS

Due to the recent progress in both mineral physics and seismology, mapping water content in Earth's mantle has become a realistic target. I have shown that currently available experimental and seismological data can be used to infer regions of high water content such as the upper mantle above subducting lithospheric plates and the upper mantle beneath hotspots (Plate 3).

Subduction zone upper mantle is expected to have a high water content, but currently available data on velocity and attenuation tomography suggest that a high water content is limited to the vicinity of active or recently active volcanoes in the shallow portions (≤200 km). The degree to which the deeper portions of upper mantle (≥200 km) in the subduction zone upper mantle is enriched with water is, however, not well resolved. There is a hint that deeper portions of subduction zones are enriched with water to a broader extent. This issue is closely related to the process of water transport [*Iwamori,* 1998; *Mibe et al.,* 1999] and has an important influence on the dynamics of subduction zones [*Billen and Gurnis,* 2001].

Inferred high water content in the mantle beneath Hawaii and Iceland is consistent with petrological and geochemical inference (e.g., [*Schilling et al.,* 1983; *Green and Falloon,* 1998; *Jamtveit et al.,* 2001]). The source regions of hotspots (ocean island basalts) likely contain higher than normal water content. The influence of "wet" plumes on global water circulation depends critically on the nature of dewatering due to partial melting. Current seismological observations favor a view that the asthenospheric materials around Hawaii are "dry" suggesting a deep dewatering in plume conduit. High-resolution seismic mapping is needed to delineate the details of dewatering processes in plumes.

In the past, the distribution of water has been inferred only through geochemical approach in which chemical analysis of materials from Earth's interior provided a basic data set. The present study demonstrates that an alternative approach is possible to infer distribution of water in which geophysical (seismological) data are interpreted in terms of variation in water content. One of the big advantages of this new approach is its wide spatial coverage. Although unique interpretation of such a data set is often difficult, growing understanding of mineral physics on the effects of water and the improvements in seismological studies gives a premise that a great progress toward the better understanding of distribution of water in this planet will be possible in near future as a result of this truly interdisciplinary effort.

Acknowledgments. This research was supported by NSF. My interests in the effects of water on seismic properties were initiated by the stimulating conversation with Guust Nolet in 1994. This paper was written in late 2000, and accepted for publication early 2001. Some changes have been made to update the references. Haemyeong Jung conducted much of the experiments on olivine. Jean-Paul Montagner, Simon Peacock, Barbara Romanowicz, Paul Silver, Yasuko Takei, Peter van Keken and Dapeng Zhao provided useful data and discussions. John Eiler, Mike Gurnis, Marc Hirschmann and Greg Hirth provided detailed comments that helped improve the presentation of this paper. Marc Hirschmann invited me to a MARGINS workshop in which this paper was presented. Thanks all.

REFERENCES

Amin, K.E., Mukherjee, A.K. and Dorn, J.E., A universal law for high-temperature diffusion controlled transient creep, *J. Mech. Phys. Solids,* 18, 413-426, 1970.

Anderson, O.L., *Equations of States of Solids for Geophysics and Ceramic Sciences*, Oxford University Press, Oxford, pp. 405, 1995.

Bai, Q. and Kohlstedt, D.L., Substantial hydrogen solubility in olivine and implications for water storage in the mantle, *Nature,* 357, 672-674, 1992.

Bass, J.D., Elasticity of minerals, glasses, and melts, In *Mineral Physics & Crystallography, A Handbook of Physical Constants* (ed. T.J. Ahrens), Amer. Geophys. Union, pp. 45-63, 1995.

Bell, D.R., and G.R. Rossman, Water in Earth's mantle: The role of nominally anhydrous minerals, *Science,* 255, 1391-1397, 1992.

Billen, M.I. and Gurnis, M., A low viscosity wedge in subduction zones, *Earth Planet. Sci. Lett.,* 193, 227-236, 2001.

Billien, M., Lévêque, J-J. and Trampert, J., 2000. Global maps of Rayleigh wave attenuation for periods between 40 and 150 seconds s, *Geophys. Res. Lett.,* 27, 3619-3622, 2000.

Blacic, J.D., Effects of water in the experimental deformation of olivine, In *Flow and Fracture of Rocks* (eds., H.C. Heard, I.Y. Borg, N.L. Carter and C.B. Raleigh), Amer. Geophys. Union, pp. 109-115, 1972.

Bolfan-Casanova, N., Keppler, H. and Rubie, D.C., Water partitioning between nominally anhydrous minerals in the $MgO–SiO_2–H_2O$ system up to 24 GPa: implications for the distribution of water in the Earth's mantle, *Earth Planet Sci. Lett.*, 182, 209-221, 2000.

Buttles, J. and Olson, P., A laboratory model of subduction zone anisotropy, *Earth Planet. Sci. Lett.*, 164, 245-262, 1998.

Cermak, V. and Hurtig, H., Heat flow map of Europe, In *Terrestrial Heat Flow I Europe* (eds. V. Cermak and L. Rybach), Springer-Verlag, enclosure, New York, 1979.

Chopra, P.N. and Paterson, M.S., The role of water in the deformation of dunite, *J. Geophys. Res.*,89, 7861-7876, 1984.

Ekström, G. and Dziewonski, A.M., The unique anisotropy of the Pacific upper mantle, *Nature*, 394, 168-172, 1998.

Fischer, K.M., Parmentier, E.M., Stine, A.R. and Wolf, E.R., Modeling anisotropy and plate-driven flow in the Tonga subduction back arc, *J. Geophys. Res.*, 105, 16,181-16,191, 2000.

Flanagan, M.P. and Wiens, D.A., Attenuation structure beneath the Lau back arc spreading from teleseismic phases, *Geophys. Res. Lett.*, 17, 2117-2120, 1990.

Flanagan, M.P. and Wiens, D.A., Radial upper mantle attenuation structure of inactive back arc basins from differential shear wave attenuation measurements, *J. Geophys. Res.*, 99, 15,469-15,485, 1994.

Furukawa, Y., Magmatic processes under arcs and formation of the volcanic front, *J. Geophys. Res.*, 98, 8309-8319, 1993.

Gaherty, J.B., Seismic evidence for hot spot-induced buoyant flow beneath the Reykjanes ridge, *Science*, 293, 1645-1647, 2001.

Gaul, O.F., Griffin, W.L., O'Reilly, S.Y. and Pearson, N.J., Mapping olivine composition in the lithospheric mantle, *Earth Planet. Sci. Lett.*, 182, 223-235, 2000.

Gottstein, G. and Mecking, H., Recrystallization, In: *Preferred Orientation in Deformed Metals and Rocks: An Introduction to Modern Texture Analysis* (ed. H-R. Wenk), Academic Press, New York, pp. 183-232, 1985.

Green, D.H. and Falloon, T.J., Pyrolite: a Ringwood concept and its current expression, In: *The Earth's Mantle* (I. Jackson, Ed.), Cambridge University Press, Cambridge, pp. 311-378, 1998.

Gribb, T.T. and Cooper, R.F., Low-frequency shear attenuation in polycrystalline olivine: grain boundary diffusion and the physical significance of the Andrade model for viscoelastic rheology, *J. Geophys. Res.*, 103, 27,267-27,279, 1998.

Gueguen, Y., Darot, M., Mazot, P. and Woirgard, J., Q^{-1} of forsterite single crystals, *Phys. Earth Planet. Inter.*, 55, 254-258, 1989.

Hammond, W.C. and Humphreys, E.D., Upper mantle seismic wave velocity: Effects of realistic partial melt geometries, *J. Geophys. Res.*, 105, 10,975-10,986, 2000a.

Hammond, W.C. and Humphreys, E.D., Upper mantle seismic wave velocity: Effects of realistic partial melt distribution, *J. Geophys. Res.*, 105, 10,987-10,999, 2000b.

Hasegawa, A., Zhao, D., Hori, S., Yamamoto, A. and Horiuchi, S., Deep structure of the northeastern Japan arc and its relationship to seismic and volcanic activity, *Nature*, 352, 683-689, 1991.

Hasegawa, A., Horiuchi, S. and Umino, N., Seismic structure of the northeastern Japan convergent margin: A synthesis, *J. Geophys. Res.*, 99, 22295-22311, 1994.

Hirth, G. and Kohlstedt, D.L., Water in the oceanic upper mantle – implications for rheology, melt extraction and the evolution of the lithosphere, *Earth Planet. Sci. Lett.*, 144, 93-108, 1996.

Ingrin, J. and H. Skogby, Hydrogen in nominally anhydrous upper-mantle minerals: concentration levels and implications, *Eur. J. Mineral.*, 12, 543-570, 2000.

Inoue, T., Effect of water on melting phase relations and melt composition in the system Mg_2SiO_4–$MgSiO_3$–H_2O up to 15 GPa, *Phys. Earth Planet. Inter.*, 85, 237-263, 1994.

Ismail, B.W. and Mainprice, D., An olivine fabric database: an overview of upper mantle fabrics and seismic anisotropy, *Tectonophysics*, 296, 145-157, 1998.

Iwamori, H., Transportation of H_2O and melting in subduction zones, *Earth Planet. Sci. Lett.*, 160, 65-80, 1998.

Jackson, I., Laboratory measurement of seismic wave dispersion and attenuation: recent progress, In *Earth's Deep Interior: Mineral Physics and Tomography from the Atomic to the Global Scale*, (eds. S. Karato, A.M. Forte, R.C. Liebermann, G. Masters and L. Stixrude), Amer. Geophys. Union, pp. 265-289, 2000.

Jackson, I., Paterson, M.S. and Fitz Gerald, J.D., Seismic wave dispersion and attenuation in Åheim dunite, *Geophys. J. Int.*, 108: 517-534, 1992.

Jamtveit, B., Brooker, R., Brooks, K., Larsen, L.M. and Pedersen, T., The water content of olivines from the North Atlantic volcanic province, *Earth Planet. Sci. Lett.*, 186, 401-415, 2001.

Jung, H. and Karato, S., Effects of water on the size of dynamically recrystallized grains in olivine, *J. Struct. Geol.*, 23, 1337-1344, 2001a.

Jung, H. and Karato, S., Water-induced fabric transitions in olivine, *Science*, 293, 1460-1463, 2001b.

Kanamori, H. and Anderson, D.L., Importance of physical dispersion in surface wave and free oscillation problems, *Rev. Geophys. Space Phys.*, 15, 105-112, 1977.

Karato, S., Does partial melting reduce the creep strength of the upper mantle? *Nature*, 319, 309-310, 1986.

Karato, S., Seismic anisotropy due to lattice preferred orientation of minerals: kinematic or dynamic? In *High Pressure Research in Mineral Physics* (eds. M.H. Manghnani and Y. Syono), Terra Pub/AGU, pp. 455-471, 1987.

Karato, S., Defects and plastic deformation of olivine, In: *Rheology of Solids and of the Earth* (S. Karato and M. Toriumi, eds.), Oxford University Press, Oxford, pp. 176-208, 1989a.

Karato, S., Grain-growth kinetics in olivine aggregates, *Tectonophysics*, 168, 255-273, 1989b.

Karato, S., Plasticity-crystal structure systematics in dense oxides and its implications for the creep strength of Earth's deep mantle: A preliminary result, *Phys. Earth Planet. Inter.*, 55, 234-240, 1989c.

Karato, S., Seismic anisotropy: mechanisms and tectonic implications, In: *Rheology of Solids and of the Earth* (S. Karato and M. Toriumi, eds.), Oxford University Press, Oxford, pp. 393-422, 1989d.

Karato, S., The role of hydrogen in the electrical conductivity of the upper mantle, *Nature*, 347, 272-273, 1990.

Karato, S., 1993. Importance of anelasticity in the interpretation of seismic tomography, *Geophys. Res. Lett.*, 20, 1623-1626, 1993.

Karato, S., 1995. Effects of water on the seismic wave velocity in the Earth's upper mantle, *Proc. Japan Academy,* Ser. B., 71, 61-66, 1995.

Karato, S., Effects of pressure on plastic deformation of polycrystalline solids: Some geological applications, *Mat. Res. Soc. Symp. Proc.,* vol. 499, 3-14, 1998a.

Karato, S., 1998b. Micro-physics of postglacial rebound, In *Dynamics of the Ice Age Earth,* (ed. P. Wu), Trans Tech Pub., Zürich, pp.351-364, 1998b.

Karato, S., A dislocation model of seismic wave attenuation and micro-creep in the Earth: Harold Jeffreys and rheology of the solid Earth, *PAGEOPH,* 153, 239-256, 1998c.

Karato, S. and Jung, H., Water, partial melting and the origin of the seismic low velocity and high attenuation zone in the upper mantle, *Earth Planet. Sci. Lett.,* 157, 193-207, 1998.

Karato, S. and Jung, H., Effects of pressure on high-temperature dislocation creep in olivine, *Phil. Mag.,* 83, 401-414, 2003.

Karato, S. and Spetzler, H.A., Defect microdynamics and physical mechanisms of seismic wave attenuation and velocity dispersion in the Earth's mantle, *Rev. Geophys.,* 28, 399-421, 1990.

Karato, S., Paterson, M.S. and Fitz Gerald, J.D., Rheology of synthetic olivine aggregates: influence of grain size and water, *J. Geophys. Res.,* 91, 8151-8176, 1986.

Kawamoto, T, Hervig, R.L. and Holloway, J.R., Experimental evidence for a hydrous transition zone in the early Earth's mantle, *Earth Planet. Sci. Lett.,* 142, 587-592, 1996.

Kohlstedt, D.L., Keppler, H. and Rubie, D.C., Solubility of water in the α, β and γ phases of $(Mg,Fe)_2SiO_4$, *Contrib. Mineral. Petrol.,* 123, 345-357, 1996.

Kröger F.A. and Vink, H.J., Relation between the concentration of imperfections in crystalline solids, In: *Solid State Physics vol. 3* (eds., F. Seitz and D. Turnbull), pp. 367-435, 1956.

Kushiro, I., Effect of water on the composition of magmas formed at high pressures, *J. Petrol.,* 13, 311-334, 1972.

Lee, K-H., Jiang, Z. and Karato, S., A scanning electron microscope study on the effects of dynamic recrystallization on lattice preferred orientation in olivine, *Tectonophysics*, 351, 331-341, 2002.

Li, X., Kind, R., Priestly, K., Sobolev, S.V., Tilman, F., Yuan, X. and Weber, M., Mapping the Hawaiian plume conduit with converted seismic waves, *Nature,* 405, 938-941, 2000.

Lifshitz, I.M. and Shikin, V.B., The theory of diffusional viscous flow of polycrystalline solids, *Soviet Phys. Solid State,* 6, 2211-2218, 1965.

Mackwell, S.J. and Kohlstedt, D.L., Diffusion of hydrogen in olivine: implications for water in the mantle, *J. Geophys. Res.,* 95, 5079-5088, 1990.

Mackwell, S.J., Kohlstedt, D.L. and Paterson, M.S., The role of water in the deformation of olivine single crystals, *J. Geophys. Res.,* 90, 11,319-11,333, 1985.

Margheriti, L., Nostro, C., Cocco, M. and Amato, A., Seismic anisotropy beneath the Northern Apennines (Italy) and its tectonic implications, *Geophys. Res. Lett.,* 23, 2721-2724, 1996.

Masters, G., Laske, G., Bolton, H. and Dziewonski, A.M., The relative behavior of shear velocity, bulk sound speed, and compressional velocity in the mantle: implications for chemical and thermal structure, In *Earth's Deep Interior: Mineral Physics and Tomography from the Atomic to the Global Scale* (eds. S. Karato, A.M. Forte, R.C. Liebermann, G. Masters and L. Stixrude), Amer. Geophys. Union, pp. 63-87, 2000.

Mei, S. and Kohlstedt, D.L.,. Influence of water on plastic deformation of olivine aggregates 1. Diffusion creep regime, *J. Geophys. Res.,* 105, 21,457-21,469, 2000a.

Mei, S. and Kohlstedt, D.L., Influence of water on plastic deformation of olivine aggregates 2. Dislocation creep regime, *J. Geophys. Res.,* 105, 21,471-21,481, 2000b.

Mibe, K., Fujii, T. and Yasuda, A., Control of the location of the volcanic front in island arcs by aqueous fluid connectivity in the mantle wedge, *Nature,* 401, 259-262, 1999.

Minster, B.J. and Anderson, D.L., A model for dislocation-controlled rheology for the mantle, *Philos. Trans. Roy. Soc. London,* 299, 319-356, 1981.

Montagner, J-P. and Guillot, L., Seismic anisotropy in the earth's mantle, In *Problems in Geophysics for the New Millenium* (by E. Boschi, G. Ekström and A. Morelli), Editrice Compostiori, Rome, pp. 217-253, 2000.

Nicolas, A., Why fast polarization directions of SKS seismic waves are parallel to mountain belts, *Phys. Earth Planet. Inter.,* 78: 337-344, 1993.

Nicolas, A. and Christensen, N.I., Formation of anisotropy in upper mantle peridotites: A review, In *Composition, Structure and Dynamics of the Lithosphere-Asthenosphere System* (eds., K. Fuchs and C. Froidevaux), Amer. Geophys. Union, pp. 111-123, 1987.

Nolet, G. and Zielhuis, A., Low S velocities under the Tornquist-Teisseyre zone: evidence for water injection into the transition zone by subduction, *J. Geophys. Res.,* 99, 15,813-15,820, 1994.

O'Connell, R.J. and Budianski, B., Viscoelastic properties of fluid-saturated cracked solids, *J. Geophys. Res.,* 82, 5719-5735, 1977.

O'Neill, H.C.St., McCammon, C.A., Canil, D., Rubie, D.C., Ross, C.R., II. and Seifert, F., Mössbauer spectroscopy of mantle transition zone phases and determination of minimum Fe^{3+} content, *Amer. Mineral.,* 78, 456-460, 1993.

Paterson, M.S., The interaction of water with quartz and its influence in dislocation flow—an overview, In *Rheology of Solids and of the Earth* (eds. S. Karato and M. Toriumi), Oxford University Press, Oxford, pp. 107-1542, 1989.

Plank, T. and Langmuir, C.H., Effects of the melting regime on the composition of the oceanic crust, *J. Geophys. Res.,* 97, 19749-19770, 1992.

Revenaugh, J. and Sipkin, S.A., Seismic evidence for silicate melt atop the 410-km mantle discontinuity, *Nature*, 369, 474-476, 1994.

Romanowicz, B., Anelastic tomography: a new perspective on upper-mantle thermal structure, *Earth Planet. Sci. Lett.,* 128, 113-121, 1994.

Romanowicz, B. and Durek, J.J., Seismological constraints on attenuation in the Earth: a review, In *Earth's Deep Interior: Mineral Physics and Tomography from the Atomic to the Global*

Scale, (eds. S. Karato, A.M. Forte, R.C. Liebermann, G. Masters and L. Stixrude), Amer. Geophys. Union, pp. 161-179, 2000.

Roth, E.G., Wiens, D.G., Dorman, L.M., Hildebrand, J. and Webb, S.C., Seismic attenuation tomography of the Tonga-Fiji region using phase pair methods, *J. Geophys. Res.,* 104, 4795-4809, 1999.

Sakuyama, M., Lateral variation of H_2O contents in quarternary magmas of Northeastern Japan, *Earth Planet. Sci. Lett.,* 43, 103-111, 1979.

Sato, H., Sacks, I.S., Murase, T., Muncill, G. and Fukuyama, H., Q_P-melting temperature relation in peridotite at high pressure and temperature: attenuation mechanism and implications for the mechanical properties of the upper mantle, *J. Geophys. Res.,* 94, 10,647-10,661, 1989.

Sato, H., H_2O and magmatism in island arc mantle inferred from seismic anelasticity and heat flow data, *J. Phys. Earth,* 42, 439-453, 1994.

Schilling, J-G., Fluxes and excess temperatures of mantle plumes inferred from their interaction with migrating mid-oceanic ridges, *Nature,* 352, 397-403.

Schilling, J-G., Zajac, M., Evans, R., Johnston, T. and White, W., Petrologic and geochemical variations along the mid-Atlantic Ridge from 29°N to 73°N, *Amer. J. Sci.,* 283, 510-586, 1983.

Shankland, T.J., O'Connell, R.J. and Waff, H.S., Geophysical constraints on partial melt in the upper mantle, *Rev. Geophys. Space Phys.,* 19, 394-406, 1981.

Silver, P.G., Seismic anisotropy beneath the continents: probing the depths of geology, *Ann. Rev. Earth Planet. Sci.,* 24: 385-432, 1996.

Smith, G.P., Wiens, D.A., Fischer, K.M., Dorman, L.M., Webb, S.C. and Hildebrand, J.A., A complex pattern of mantle flow in the Lau backarc, *Science,* 292, 713-716, 2001.

Stolper, E.M. and Newman, S., The role of water in the petrogenesis of Mariana trough magmas, *Earth Planet. Sci. Lett.,* 121, 293-325, 1994.

Takahashi, E. and Nakajima, K., Melting process in the Hawaiian plume: An experimental study, In *Hawaiian Volcanoes: Recent Progress in Deep Underwater Researches* (eds. E. Takahashi, P. Lipman, M. Garcia, J. Naka), Amer. Geophys. Union, 403-418, 2002.

Takahashi, E., Nakajima, K. and Wright, T.L., Origin of the Columbia River basalts: melting model of a heterogeneous plume head, *Earth Planet. Sci. Lett.,* 162, 63-80, 1998.

Takanami, T., Sacks, S.I. and Hasegawa, A., Attenuation structure beneath the volcanic front in northeastern Japan from broadband seismograms, *Phys. Earth Planet. Inter.,* 121, 339-357, 2000.

Takei, Y., Acoustic properties of partially molten media studied on a simple binary system with a controllable dihedral angle, *J. Geophys. Res.,* 105, 16,665-16,682, 2000.

Tan, B., Jackson, I. and Fitz Gerald, J.D., Shear wave dispersion and attenuation in fine-grained synthetic olivine aggregates: preliminary results, *Geophys. Res. Lett.,* 24, 1055-1058, 1997.

The MELT Seismic Team, Imaging the deep seismic structure beneath a mid-oceanic ridge: The MELT experiment, *Science,* 280, 1215-1218, 1998.

Thompson, A.B., Water in the Earth's upper mantle, *Nature,* 358, 295-302, 1992.

Tullis, J., Christie, J.M. and Griggs, D.T., Microstructures and preferred orientations of experimentally deformed quartzites, *Bull. Geol. Soc. Amer.,* 84, 297-314, 1973.

Umino, N. and Hasegawa, A., Three-dimensional Qs structure in the northeastern Japan arc, *J. Seism. Soc. Japan,* 37, 217-228, 1984.

Wenk, R-H. and Tomé, C.N., Modeling dynamic recrystallization of olivine aggregates deformed in simple shear, *J. Geophys. Res.,* 104, 25,513-25,527, 1999.

Wenk, R-H., Venkitasubramanayan, C.S., Baker, D.W. and Turner, F.J., Preferred orientation in experimentally deformed limestone, *Contrib. Mineral. Petrol.,* 38, 81-114, 1973.

Watson, S. and McKenzie, D., Melt generation by plumes: a study of Hawaiian volcanism, *J. Petrol.,* 32, 501-537, 1991.

Williams, Q. and Hemley, R.J., Hydrogen in the deep Earth, *Ann. Rev. Earth Planet. Sci.,* 29: 365-418, 2001.

Yan, H., *Dislocation Recovery in Olivine,* MSc. Thesis, University of Minnesota, pp. 98, 1992.

Young, T.E., Green, H.W., Hofmeister, A.M. and Walker, D., Infrared spectroscopic investigation of hydroxyl in β-$(Mg,Fe)_2SiO_4$ and coexisting olivine: implications for mantle evolution and dynamics, *Phys. Chem. Mineral.,* 19, 409-422, 1993.

Zhang, Y. and Tanimoto, T., High-resolution global upper mantle structure and plate tectonics, *J. Geophys. Res.,* 98, 9793-9823, 1993.

Zhao, D., Hasegawa, A. and Horiuchi, S., Tomographic imaging of P and S wave velocity structure beneath northeastern Japan, *J. Geophys. Res.,* 97, 19909-19928, 1992.

Zhao, D., Xu, Y., Wiens, D.G., Dorman, L., Hildebrand, J. and Webb, S., Depth extent of the Lau back-arc spreading center and its relation to the subduction processes, *Science,* 278, 254-257, 1997.

S. Karato, Yale University, Department of Geology and Geophysics, New Haven, CT 06520 shun-ichiro.karato@yale.edu

Volcanism and Geochemistry in Central America: Progress and Problems

M. J. Carr and M. D. Feigenson

Department of Geological Sciences, Rutgers University

L. C. Patino

Department of Geological Sciences, Michigan State University

J. A. Walker

Department of Geology and Environmental Geosciences, Northern Illinois University

Most Central American volcanoes occur in an impressive volcanic front that trends parallel to the strike of the subducting Cocos Plate. The volcanic front is a chain, made of right-stepping, linear segments, 100 to 300 Km in length. Volcanoes cluster into centers, whose spacing is random but averages about 27 Km. These closely spaced, easily accessible volcanic centers allow mapping of geochemical variations along the volcanic front. Abundant back-arc volcanoes in southeast Guatemala and central Honduras allow two cross-arc transects. Several element and isotope ratios (e.g. Ba/La, U/Th, B/La, $^{10}Be/^{9}Be$, $^{87}Sr/^{86}Sr$) that are thought to signal subducted marine sediments or altered MORB consistently define a chevron pattern along the arc, with its maximum in Nicaragua. Ba/La, a particularly sensitive signal, is 130 at the maximum in Nicaragua but decreases out on the limbs to 40 in Guatemala and 20 in Costa Rica, which is just above the nominal mantle value of 15. This high amplitude regional variation, roughly symmetrical about Nicaragua, contrasts with the near constancy, or small gradient, in several plate tectonic parameters such as convergence rate, age of the subducting Cocos Plate, and thickness and type of subducted sediment. The large geochemical changes over relatively short distances make Central America an important margin for seeking the tectonic causes of geochemical variations; the regional variation has both a high amplitude and structure, including flat areas and gradients. The geochemical database continues to improve and is already adequate to compare to tectonic models with length scales of 100 Km or longer.

1. INTRODUCTION

The selection of Central America as a focus area by the National Science Foundation's Margins Program should increase the amount of research carried out in Central America and lead to improved understanding of all aspects

Inside the Subduction Factory
Geophysical Monograph 138

10.1029/138GM09

of the subduction process. The primary goal of this paper is to facilitate future research in Central America; first, by updating regional data bases and second, by surveying current ideas with emphasis on assumptions, caveats, inconsistencies, flaws, unexplained observations and important problems. There are several reviews of Central American volcanology [e.g. *Stoiber and Carr*, 1973; *Carr et al.*, 1982; Carr and Stoiber, 1990] and progressively refined models attempting to explain the geochemistry of the arc [e.g. *Carr*, 1984; *Feigenson and Carr*, 1986; *Plank and Langmuir*, 1988; *Carr et al.*, 1990; *Plank and Langmuir*, 1993; *Leeman et al.*, 1994; *Reagan et al.*, 1994; *Walker et al*, 1995; *Noll et al.*, 1996; *Chan et al.*, 1999; and *Patino et al.*, 2000].

This paper is structured as follows; first, an updated database of physical and geochemical parameters of Central American volcanoes; second, review of the Cocos Plate section that is critical for understanding the volcanic geochemistry; third, a volcanological framework for the diverse magma types in the region; fourth, a summary of the regional, cross-arc, and local geochemical zoning; and finally, a summary of segmentation, including a proposed first-order geochemical segmentation of the arc.

2. DATA

Data are in three files located at M. J. Carr's web site: http://www-rci.rutgers.edu/~carr/index.html.

CAVolcFront.zip is Table 1. It lists physical parameters of 39 Quaternary volcanic centers that comprise the volcanic front of Central America. Figure 1 shows the tectonic setting of the volcanic front and Figure 2 names the volcanic centers. With only a few exceptions, the volcanic front centers are groups of vents, including explosive vents, ranging in size from maars to calderas, and including constructive morphologies, such as composite cones, domes, shields and cinder cones.

Deciding which volcanoes to group into a center is not always clear-cut. In Guatemala, the Atitlán center is based

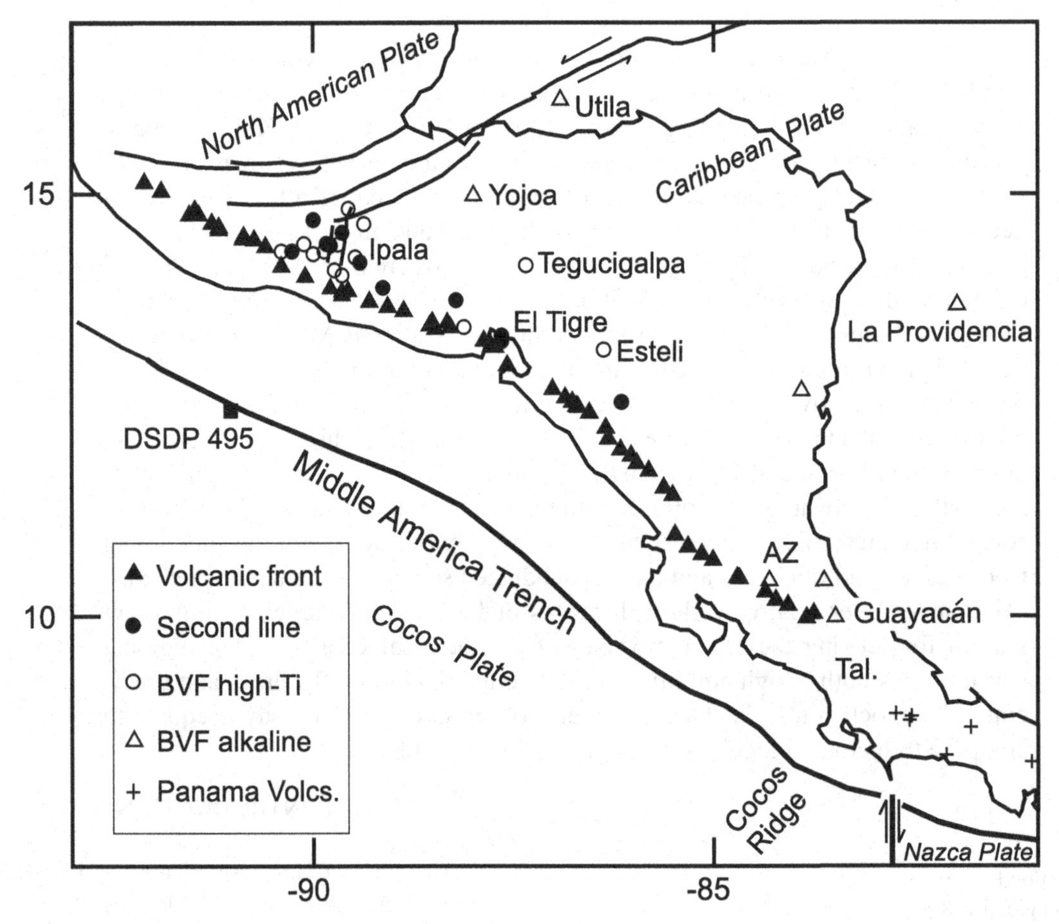

Figure 1. Volcanological and tectonic framework of Central America. AZ is Aguas Zarcas. Tal. is the Talamanca range in southern Costa Rica.

Table 1. Physical parameters of Central American volcanoes

Center No.	Name	Distance Km	Back Km	Latitude	Longitude	Volcano Volume Km^3	Center Volume Km^3
1	Tacaná	22.3	154.1	15.13	–92.11	20	20
2	Tajumulco	46.7	155.9	15.04	–91.90	45	45
3	Santa María	93.9	147.0	14.76	–91.55	18	77
	Santiaguito	93.9	145.0	14.75	–91.57	2	
	Cerro Quemado	94.8	152.5	14.80	–91.52	5	
	Siete Orejas	84.7	148.7	14.81	–91.62	40	
	Chicabál	82.7	144.1	14.79	–91.66	12	
4	Atitlán	137.1	148.9	14.58	–91.18	33	330
	Tolimán	135.3	152.1	14.62	–91.18	18	
	San Pedro	125.5	151.5	14.66	–91.27	27	
	Los Chocoyos	137.6	148.7	14.58	–91.18	250	
5	Fuego	170.1	154.8	14.48	–90.88	73	135
	Acatenango	169.2	156.4	14.50	–90.88	62	
6	Agua	183.2	160.1	14.47	–90.75	68	68
7	Pacaya	201.6	159.8	14.38	–90.60	17	17
8	Tecuamburro	231.0	147.0	14.15	–90.42	39	39
9	Moyuta	266.5	152.3	14.03	–90.10	15	15
10	Santa Ana	319.4	159.2	13.85	–89.63	220	351
	Apaneca	304.5	149.6	13.84	–89.80	125	
	Izalco	321.2	156.1	13.82	–89.63	2	
	Cerro Verde	321.1	157.5	13.83	–89.63	2	
	Conejo	320.0	157.5	13.82	–89.60	1	
	San Marcelino	318.0	158.0	13.81	–89.58	1	
11	Boquerón	358.0	166.4	13.73	–89.28	63	63
12	Ilopango	383.0	172.4	13.67	–89.05	29	30
	Islas Quemadas	383.0	172.4	13.67	–89.05	1	
13	San Vicente	404.1	178.1	13.62	–88.85	59	60
	Apastapeque	405.9	192.1	13.72	–88.77	1	
14	Tecapa	442.7	185.5	13.50	–88.50	65	168
	Usulután	449.8	179.4	13.42	–88.47	15	
	Berlin	440.5	183.6	13.49	–88.53	60	
	Tigre	441.2	180.8	13.47	–88.53	20	
	Taburete	442.7	178.1	13.44	–88.53	8	
15	San Miguel	467.8	191.4	13.43	–88.27	58	68
	Chinameca	460.4	193.1	13.48	–88.32	10	
	*Aramuaca	480.7	198.6	13.43	–88.13	<1	
16	Conchagua	514.4	199.3	13.28	–87.85	27	27
17	Conchaguita	524.9	198.7	13.23	–87.77	1	1
18	Meanguera	531.8	198.9	13.20	–87.71	3	3
19	Cosigüina	556.9	185.9	12.98	–87.57	33	33
20	San Cristóbal	624.7	189.4	12.70	–87.00	65	110
	Casita	627.7	191.2	12.70	–86.97	45	
21	Telica	644.0	188.0	12.60	–86.85	28	30
	Santa Clara	648.8	186.6	12.57	–86.82	2	
22	Rota	655.0	187.0	12.55	–86.75	12	12
23	Las Pilas	664.9	187.5	12.50	–86.68	14	28
	Cerro Negro	663.3	186.6	12.50	–86.70	<1	
	El Hoyo	667.0	187.1	12.49	–86.67	14	
24	Momotombo	683.3	187.7	12.42	–86.53	18	18
	*Momotombito	692.1	184.6	12.35	–86.48	<1	

Table 1. (continued)

Center No.	Name	Distance Km	Back Km	Latitude	Longitude	Volcano Volume Km^3	Center Volume Km^3
25	Apoyeque	711.2	182.8	12.25	−86.33	6	9
	Nejapa	720.0	170.0	12.11	−86.32	3	
26	Masaya	742.7	167.3	11.98	−86.15	168	178
	Apoyo	754.8	167.7	11.93	−86.05	10	
27	Mombacho	766.4	162.2	11.83	−85.98	19	20
	Granada	762.2	165.7	11.88	−86.00	1	
28	Zapatera	784.8	161.2	11.74	−85.84	5	5
29	Concepción	816.9	153.7	11.53	−85.62	19	19
30	Maderas	834.0	149.4	11.42	−85.50	22	22
31	Orosí	861.5	109.9	10.98	−85.48	50	100
	Cacao	863.0	110.0	10.96	−85.45	50	
32	Rincón de la Vieja	882.3	103.1	10.83	−85.33	201	201
33	Miravalles	903.9	105.3	10.75	−85.15	132	132
34	Tenorio	920.4	104.9	10.67	−85.02	95	95
35	Arenal	958.3	101.5	10.47	−84.73	13	15
	Chato	960.0	100.0	10.45	−84.69	2	
36	Platanar	1001.7	105.9	10.30	−84.37	32	48
	Porvenir	1003.8	103.3	10.27	−84.36	16	
37	Poás	1021.0	104.9	10.20	−84.22	168	168
38	Barba	1037.2	106.0	10.13	−84.08	326	326
39	Irazú	1067.3	104.8	9.98	−83.85	227	378
	Turrialba	1072.1	114.1	10.03	−83.77	151	

* volcano too small to be considered a separate center and not on the flank of the nearest center.

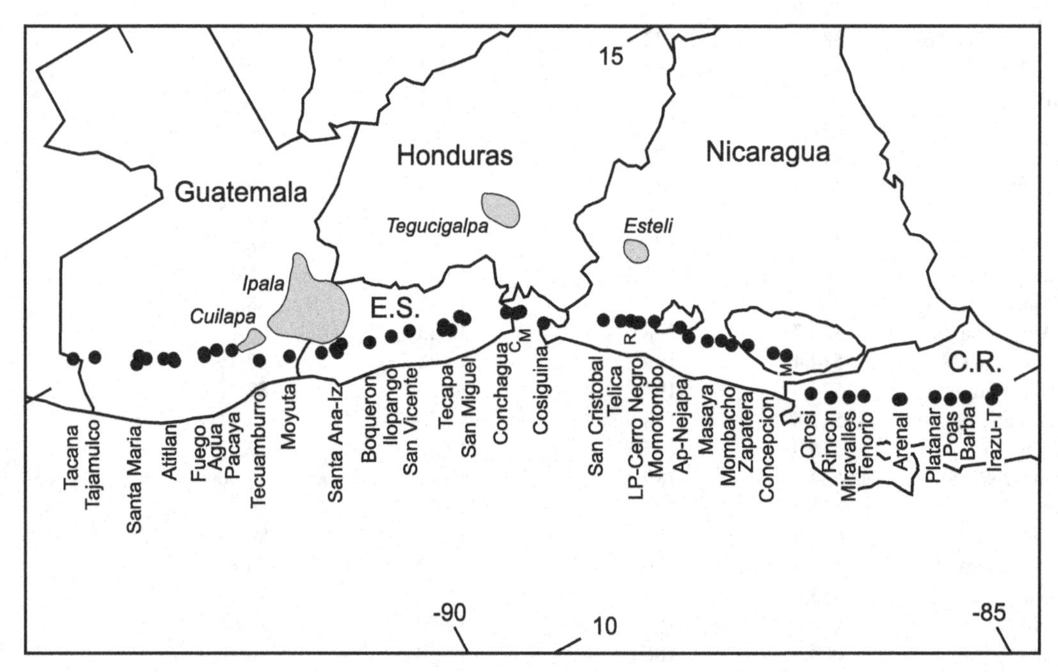

Figure 2. Locations of volcanic centers and BVF fields. E.S. is El Salvador, C.R. is Costa Rica. The abbreviations are: in El Salvador, Iz for Izalco, C for Conchaguita, M for Meanguera; in Nicaragua, R for Rota, M for Maderas; in Costa Rica, T for Turrialba. The stippled areas show approximate extents of the BVF volcanic fields with high-Ti character.

on the huge Atitlán caldera (25 Km in diameter). The caldera now includes two pairs of composite cones, Atitlán-Tolimán and San Pedro-Santo Tomás that, absent the caldera, would be grouped into two separate centers. The ring of composite cones near Apaneca, El Salvador, merges with the huge Santa Ana volcanic complex; separating them or not is arbitrary. The same is true for the Tecapa and San Miguel centers in El Salvador. Geochemistry has not proven helpful in defining dividing lines between nearby centers because the geochemical variation within a center is commonly as great as that between adjacent centers. For example, in northern Central America there is strong cross-arc chemical zonation among vents in the same center [Halsor and Rose, 1988]. The volcanoes nearest to the trench have more mafic character and generally lower LIL element contents and higher Ba/La.

Another flaw in the assumption that it is wise to group volcanoes into centers is the presence of two small volcanoes along the volcanic front at some distance from adjacent centers. Momotombito is a small composite cone in Nicaragua that is arbitrarily grouped with Momotombo even though it is 9 Km distant. Aramuaca is a maar located 13 Km southeast of San Miguel in El Salvador. It's chemistry is not known because no fresh samples were found. Most of Central America is volcanic and it is problematic to decide how eroded and old a volcanic structure should be to be excluded from the list. Until there is more extensive dating, it seems best to be conservative in deciding which volcanoes to include among the Quaternary centers.

CALine2.zip is Table 2. It lists the locations and relative ages of eleven volcanoes that comprise a secondary volcanic belt, landward of the volcanic front. These volcanoes are analogous to the double chains seen in parts of the Japanese arcs and other areas. Central America has few of these volcanoes and none have historic activity. This list should be modified and improved, as the ages of these volcanoes are determined.

CAGeochem.zip provides geochemical data for samples from Central American volcanoes. Most of the data, including nearly all the Guatemalan data, were previously published as CENTAM [*Carr and Rose*, 1987]. The new file contains many new columns of data for the Salvadoran, Honduran, Nicaraguan and Costa Rican samples. To first order, all the samples from volcanoes northwest of Tecuamburro volcano in southeast Guatemala are from the work of W. I. Rose Jr. at Michigan Technological University and his colleagues. M. J. Carr and colleagues at Rutgers University collected most of the rest of the samples. Many others contributed samples (e.g. G. E. Alvarado and several students from Dartmouth College) and powders (e.g. A. R. McBirney, W. G. Melson, T. N. Donnelly, M. K. Reagan). The column headed, 'collector-location', notes the collector and the type of sample available at Rutgers University (ru) or Michigan Technological University (mtu). If a reference is given, the data are from the literature. If the entry is ru or mtu then the entire sample resides at that location. If the entry says 'ru powder' then only rock powder is available. The column titled 'Quality' is included to allow a quick sort of the samples with isotopic data and high quality REE and other trace element data. Quality is zero if the only data available are major elements and a few trace elements or if the trace element data are less reliable or less consistent with the rest of the database.

Table 2. Volcanoes of the secondary front.

Name	Country	Latitude	Longitude	Age
Jumatepeque	Guatemala	14.33	–90.27	H
Jumay	Guatemala	14.70	–90.00	Q
Suchitán	Guatemala	14.40	–89.78	P
Ipala	Guatemala	14.55	–89.63	H
Retana caldera	Guatemala	14.42	–89.83	P
Masahuat	El Salvador	14.20	–89.40	P
Guazapa	El Salvador	13.90	–89.11	P
Cacaguatique	El Salvador	13.75	–88.20	P
El Tigre	Honduras	13.27	–87.63	H
Zacate Grande	Honduras	13.33	–87.63	P
Ciguatepe	Nicaragua	12.55	–86.15	Q

Age estimates; H is Holocene, Q is Quaternary. P is Plio-Quaternary

The locations of the samples in CAGeochem are most accurately given by the easting and northing columns that refer to the map grid in Km units found on 1:50,000 scale topographic maps in Central America. Guatemala, Honduras and Nicaragua conveniently use the UTM grid system, but El Salvador and Costa Rica have local grids based on a Lambert projection. Other columns provide the latitude and longitude of the vent, not the sample locations. Two odd parameters are called 'Distance' and 'Back'. These units are in Km and refer to a Lambert conical conformal projection of the volcano locations and a subsequent 30° counterclockwise rotation. The origin of this reference frame is a spot near the Middle America Trench south of the northwesternmost volcano, Tacaná in Guatemala. Distance, measured parallel to the volcanic front, is a good estimate of distance along the arc. Back is the cross-arc direction, but it is not a good estimate of distance from the trench because the trench is not a great circle. 'Center volumes' were estimated from 1:50,000 scale maps and 100-meter contour intervals, whose areas were determined with a digitizer. The volumes of volcanic centers in Table 1 and CAGeochem are more precise than previous ones, which were made using simple geometric models.

However, the accuracy is not much improved because most of the error is caused by pre-volcanic topography hidden by the volcanoes. The column entry, 'volcano volume', is incomplete because the partition of the volume of volcanic centers into the constituent volcanoes that comprise them has not been done systematically.

Silicic rocks are inadequately represented in CAGeochem for El Salvador, Nicaragua and Costa Rica because the Rutgers group has been biased toward the collection of mafic rocks. This data file should not be used to estimate unbiased average compositions. Current work in El Salvador initiated by W. I. Rose Jr. and others at Michigan Technological University, but now including several other universities, is refining the tephra stratigraphy and doing justice to the silicic rocks. Similarly, active research by T. A. Vogel and L. C. Patino of Michigan State University is rapidly establishing the tephra stratigraphy and geochemistry of silicic rocks in Costa Rica. In Nicaragua, the geology of all the volcanoes has been mapped and tephra stratigraphy established through a program of the Geological Survey of the Czech Republic. Maps are available in Managua, Nicaragua at the Instituto Nicaragüense de Estudios Territoriales (INETER). Geologic reports should be available soon [P. Hradecky pers. comm.]. Geochemical work remains to be done on many large silicic deposits in Nicaragua, including Cosigüina, the Monte Galan caldera at Momotombo, Apoyeque and the Las Sierras section of Masaya. Williams [1983b] and Sussman [1985] describe the Mafic and silicic tephra from the youngest parts of the Masaya complex.

Panama has several Quaternary volcanic centers related to active subduction of the Nazca Plate [*de Boer et al.*, 1991; *Defant et al.*, 1992]. Adakites occur in young Panamanian volcanoes [e.g. *Defant et al.*, 1991]. Small bodies of adakite lavas and dikes occur in the Talamanca range (see Figure 1) in southern Costa Rica [*Drummond et al.*, 1995]. These dikes and domes are dated between 1.9 and 3.5 Ma [*Abratis and Wörner*, 2001] and occur above the subducting Cocos Ridge. In contrast, adakites appear to be absent along the Central American volcanic front. Panamanian volcanism is separated from the Central American volcanic front by plate boundaries cutting both the upper plate and the subducting plate. There is also a volcanic gap of about 175 Km, located above the subducting Cocos Ridge. Young and hot lithosphere of the Nazca Plate is subducting beneath Panama. The profound tectonic and magmatic differences between Panama and Central America make it appropriate to separate these volcanic belts if one seeks to explain variations within a single convergent plate margin. However, *Harry and Green* [1999] group the volcanic belts of Central America and Panama and relate the large geochemical contrasts between Central America and Panama to age variations among the subducting lithosphere segments.

3. COCOS PLATE SLAB SIGNALS

One of the fascinating characteristics of Central America is the pronounced regional variation in the geochemical ratios that define slab signals. A slab signal is a trace element or isotopic ratios that is enriched (e.g. Ba/La) or depleted (e.g. Nb/La) in arc magma, relative to the mantle, because of additions of hydrous fluids or silicic melts derived from a subducted slab. The wide range in ages and geologic histories of subducted slabs prevent a uniform global slab signal. Similarities in the subduction process and the relatively uniform composition of the basalt section of oceanic lithosphere allow many common features among arcs but the sediment input to subduction is variable and the resulting slab signal varies along with the sediment input (*Plank and Langmuir*, 1993). In Central America the slab provides several signals or ratios that trace different parts of the Cocos Plate stratigraphy.

Patino et al., [2000] showed that Central American magmas have special geochemical characteristics derived mainly from the two distinctly different sediments that form a 400 M thick veneer at the top of the Cocos Plate. The discussion here stresses how different element and isotope ratios can provide a variety of slab signals that have different resolution and focus on different parts of the Cocos stratigraphy.

The oceanic crust and sediments input into the subduction system appear to have low variation along strike of the trench. From Guatemala to northwestern Costa Rica the age, source and stratigraphy of the Cocos Plate crustal section are similar, suggesting a near uniform input [*Aubouin et al.*, 1982; *Kimura et al.*, 1997]. The tectonic processes along the Cocos-Caribbean convergent margin result in the subduction of most of the Cocos sediment section, which is clearly imaged tens of kilometers landward of the trench [*von Huene et al.*, 2000]. Deeper processes that cannot be resolved with seismic images may remove some sediment but, to first order, the sediment section from DSDP 495, analyzed by *Patino et al.*, [2000], characterizes what is subducted into the arc to melt generation depths.

The crustal section of the Cocos Plate consists of three stratigraphic units, a basal MORB/altered MORB, a middle carbonate unit and an upper hemipelagic unit. The MORB section has not been directly characterized and so, in Figure 3 the NMORB and EMORB of *Sun and McDonough* [1989] (diamonds) represent the MORB section. The carbonate sediments are filled triangles and the hemipelagic sediments are open triangles. Figure 3 shows several trace element ratios that track different parts of the stratigraphy. These ratios are plotted on a log scale because of their wide variation in the Cocos Plate section.

Ba/La has little variation in the sediment section but the sediments have much higher Ba/La compared to MORB. This distribution makes Ba/La the best tracer of slab signal for the sediment section as a whole.

U/Th variation is large, especially in the carbonate sediments, which have low contents of both U and Th. The base of the hemipelagic section has values slightly higher than MORB and there is a progressive increase up section to a value of nearly 2.0. The mean values of U/Th for the two sediments are statistically indistinguishable and so, U/Th is a tracer for the entire sediment section. However, the much higher dispersion of this ratio indicates that U/Th might provide a less clearly resolved view of regional variation along the arc than Ba/La.

Ba/Th is exceptionally enriched in the carbonate section and provides a first order tracer of the carbonate section.

U/La is similar in MORB and carbonates but much higher in the hemipelagic section. The U content is especially high in the upper part of the Cocos section, apparently trapped by organic matter that also increases up section [*Patino et al.*, 2000]. The unusual distribution of U in the Cocos section make the U/La ratio a useful slab signal in Central America but this ratio may not be useful in other areas.

Because ^{10}Be decays with a half-life of about 1.5 Ma, it is concentrated in the upper part of the hemipelagic sediments and is essentially zero below 150 M in the Cocos Plate section [*Reagan et al.*, 1994]. The $^{10}Be/^{9}Be$ ratio therefore provides a unique fingerprint of the top of the section.

The slab signals found in Central American magmas provide different information. The $^{10}Be/^{9}Be$ data provide the most precise depth control because ^{10}Be is present only at the top of the sediment section. Ba/La and U/Th represent the entire sediment section but Ba/La has the least dispersion, suggesting it is the better of the two whole sediment signals. Ba/Th and U/La trace the carbonate and hemipelagic sediments, respectively.

4. FRAMEWORK

Central American volcanoes can be subdivided on the basis of location or tectonics, geochemistry and activity. The active system, the volcanic front, is further subdivided into segments defined both by location and by size of volcanic centers. The segmentation of the volcanic front is discussed below in section 6. This section focuses on three volcanic systems that can be geographically or tectonically separated: the volcanic front, the second line and behind the front volcanism (BVF) (see Figure 1). Each volcanic system has a typical geochemistry or magma type but there is some intermingling of magma types as magmas take opportunistic paths to the surface. Adding to the geochemical complexity is the existence of regional variation in both the volcanic front and BVF systems. This zoning is discussed in section 5.

4.1. Volcanic Front

The volcanic front is the source of all the historic volcanic activity in Central America and most of the Quaternary

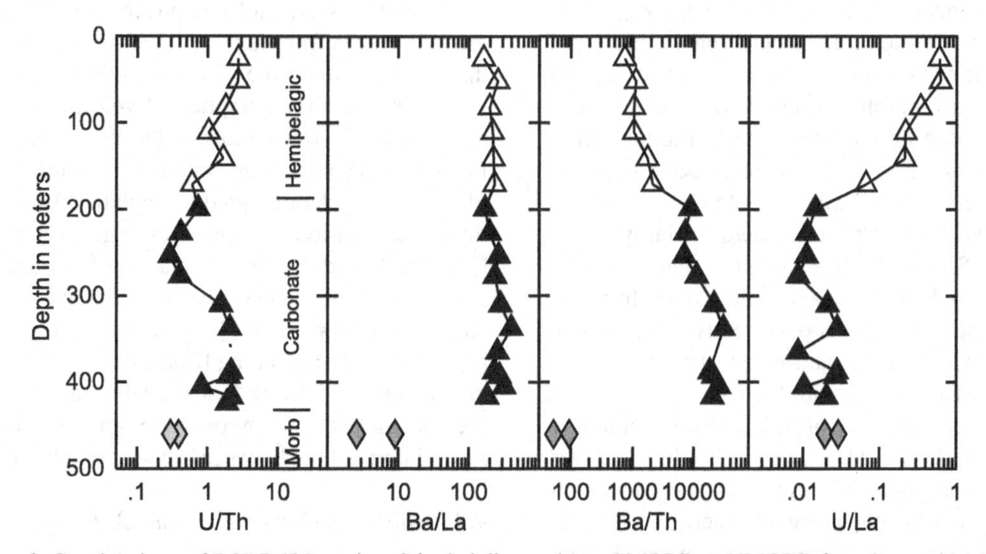

Figure 3. Geochemistry of DSDP 495 section. Stippled diamonds are EMORB an NMORB from Sun and McDonough [1989]. Filled triangles are carbonate sediments. Open triangles are hemipelagic sediments.

volcanic rocks. Several narrow lines of active volcanic centers (Figures 1 and 2) define the volcanic front. These volcanic lines or segments are 165 to 190 Km landward of the axis of the Middle America Trench but the depth to the seismic zone beneath them is much more variable. The dip of the seismic zone appears to steepen toward the center of the volcanic front and so depths to the seismic zone range from less than 100 Km in Guatemala and central Costa Rica to about 200 Km in eastern Nicaragua (*Carr et al.*, 1990).

With few exceptions (e.g. Agua in Guatemala) each volcanic center is a group of several distinct vents of various types. The structure of the centers varies along the arc [*Stoiber and Carr,* 1973]. In Guatemala, there are prominent transverse lineaments, five to ten Km in length, with two or more overlapping composite volcanoes or domes. In El Salvador, the two largest volcanic centers are made from several composite volcanoes in arcuate or circular arrays (the Santa Ana center and the Tecapa center, respectively). In western Nicaragua, most centers are clusters of small composite cones, shields, domes and cinder cones aligned in a grid-like pattern [*van Wyk de Vries*, 1993].

Throughout Central America, with the possible exception of central Costa Rica, the volcanic front coincides with a shallow seismic zone created by right-lateral, strike-slip faulting parallel to the volcanic lines and associated N-S normal faults and grabens [*Carr and Stoiber*, 1977; *White and Harlow*, 1993]. In this transtensional setting many volcanoes are associated with the N-S extensional structures.

The volcanic front has many calderas that erupt silicic tephra. The geology and geochemistry of the calderas in Guatemala and El Salvador are well investigated [*e.g. Rose et al.*, 1999]. Most of these calderas occur on the landward side of the volcanic centers. *Rose et al.* [1999] established a regional tephra stratigraphic framework for northern Central America based on the voluminous eruptions from the calderas. In Nicaragua, small silicic calderas occur in the following centers: Momotombo (Monte Galan caldera), Apoyeque and Masaya (Apoyo caldera). Masaya also includes the Las Sierras caldera complex that produced large volumes of predominantly andesitic tephra. In Costa Rica, large silicic tephra deposits occur in association with Rincón del la Vieja, Miravalles and Barva and a small silicic tephra deposit is associated with Platanar.

The individual vents within volcanic centers commonly have more than one lava field or magma batch. The lavas range from basalt to rhyolite, but few lavas have MgO contents greater than 6.0 weight percent. Where the crust is thinner, as in Nicaragua, mafic basalts are abundant. In central and western Guatemala, where thick, old continental crust occurs, basaltic lavas are usually present but their abundance and their MgO contents are low. Throughout Central America, large and extensively zoned plagioclase and pyroxene phenocrysts are typical. Magnetite is ubiquitous and olivine is common in basalts. An important weakness in the Central American database is the lack of regional scale mineralogical studies. The only systematic regional mineralogical variation, known to the authors, is the occurrence of hornblende in basalts and basaltic andesites that have Na_2O contents greater than about 3.5 weight percent. These high Na_2O contents are restricted to central and western Guatemala and central Costa Rica (*Carr et al.*, 1982).

The geochemistry of most volcanic front magmas is marked by a strong enrichment in trace element and isotopic ratios, associated with hydrous fluids or silicic melts derived from the subducted Cocos Plate (e.g. Ba/La >50; $^{10}Be/^{9}Be > 2$). Most magmas have strong depletions in high field strength (HFS) elements and low TiO_2 contents. High water contents were measured in melt inclusions in tephra from Fuego volcano in Guatemala [*Sisson and Layne*, 1993] and from Cerro Negro in Nicaragua [*Roggensack et al.*, 1997] but not from Masaya in Nicaragua [*Bosenberg and Lindsay, pers. comm.*].

4.2. Second Line

A weakly developed second line of composite volcanoes sporadically occurs parallel to and 20 to 75 Km behind the volcanic front (filled circles in Figure 1). The literature of Central American volcanism largely ignores this group because it lacks historic activity and most of the volcanoes are moderately to deeply eroded. However, El Tigre volcano in Honduras is a small composite cone that is minimally eroded and has fresh lavas and tephra. This cone is no older than many cones at the volcanic front.

In contrast to the complicated volcanic centers common at the volcanic front, most of the volcanoes in the second line grew as single composite cones. Younger cinder cones of the BVF system erupted through the flanks of several of these older composite volcanoes (e.g. Ipala and Suchitán in Guatemala and Zacate Grande in Honduras). However, these late cinder cones have a distinctively different geochemistry [*Walker*, 1981]. Systematic study of the secondary front is just beginning [*Patino et al.*, 1997]. The limited data available indicate that the lavas are calc-alkaline and are plagioclase and pyroxene phyric. Lavas have Ba/La ratios intermediate between the volcanic front and the BVF.

4.3. Behind the Front Volcanism (BVF)

Widespread back-arc volcanism occurs in Central America (open symbols in Figure 1). This volcanic system overlaps the second line and extends to the volcanic front, but has distinctive structural, morphological, and geochemical characteristics. *Walker* [1981] called this behind the front volcanism because it extends from the volcanic front to more than 200 Km behind it. Volcanism occurs with back-arc spreading in many circum-Pacific arcs and the BVF volcanism in Central America is analogous to it, but there is an important structural difference: back-arc rifts are commonly sub parallel to the volcanic arc whereas the N-S striking extensional structures in Central America trend at a high angle to the volcanic front, which strikes about N60W. The transverse nature of the rifting allows the back-arc magmas in Central America in some cases to extend all the way into the flanks of the volcanic front centers. In northern Central America, BVF volcanism extends to the left-lateral, strike-slip faults that mark the Caribbean-North America plate boundary (Figures 1 and 2). The extensional tectonics that gives rise to the BVF in southeastern Guatemala is likely dominated by the transform faulting of the nearby plate boundary [*Burkart and Self*, 1985].

Behind the volcanic front volcanism occurs in clusters of cinder cones, small shields and lava fields in strongly extensional settings, the largest of which is the Ipala graben in southeastern Guatemala. There are also a small number of rhyolite obsidian domes and small calderas that produce silicic tephra. Only a few of these numerous vents are shown in Figure 1. The approximate extents of the major subalkaline BVF fields are shown in Figure 2. There has been no historic activity in the back-arc but Holocene activity is certain given the youthful morphology of many southeast Guatemalan and Salvadoran cones and lavas. In several cases, morphologically young cinder cones occur on the flanks of the second line of composite volcanoes. In El Salvador, cinder cones with typical BVF geochemistry occur on the flanks of the historically active volcanic front centers Santa Ana [*Pullinger*, 1998] and Boquerón [*Fairbrothers et al.*, 1978]. Contrarily, a recent maar, erupted 12 Km behind Boquerón, has geochemistry more like Boquerón than the BVF cones. Clearly the plumbing is opportunistic. There is a clear geographic overlap between the volcanic front and the BVF in western El Salvador. In contrast, there is a clear separation in southeastern Guatemala [*Walker et al.*, 2000].

Geochemically, most BVF samples have Ba/La ratios in the range 15 (typical mantle) to 45 (moderate slab signal). Many samples are depleted in Nb but Zr and Ti depletions are rare. Compared with volcanic front basalts, the BVF basalts have high TiO_2 contents and are good examples of high-Ti magma (see discussion in section 4.5). These geochemical characteristics indicate a low to moderate input of fluids from the subducting plate. Typical lavas are nearly aphyric with rare olivine phenocrysts, although more evolved, phyric lavas are present, especially at the larger shield volcanoes. In a few exceptional cases (Volcán Culma northeast of Jutiapa, Guatemala and some young lavas north of Estelí, Nicaragua) there are BVF lavas with very large (>1 cm) phenocrysts. BVF lavas from southeast Guatemala to Tegucigalpa, Honduras to central Nicaragua (Estelí) are geochemically similar, at least as far as has been sampled. Very different alkaline lavas occur in central and northern Honduras (Yojoa and Utila in Figure 1). Lavas from these volcanoes have no apparent contribution from the subducted slab, no HFS depletion and minimal crustal contamination. They appear to be the clearest window into the geochemistry of the local asthenosphere [*Patino et al.*, 1997; *Walker et al.*, 2000]. They are similar to the EMORB of *Sun and McDonough* [1989]. In Costa Rica, the BVF lavas are substantially different from the BVF lavas in the rest of Central America. Instead, they are similar to ocean island basalts (OIB). Pliocene lavas at Guayacán are alkaline and are derived by low and variable degrees of melting from an OIB-like source [*Feigenson et al.*, 1993]. At Aguas Zarcas (AZ in Figure 1), BVF cones and lavas extend from the back-arc to the flanks of the Platanar center on the front. The lavas are very rich in incompatible elements, plot near the alkaline/subalkaline discrimination lines and have shoshonitic affinities [*Malavassi*, 1991; *Alvarado and Carr*, 1993].

4.4. Magma Genesis

Two different melt generation processes occur in Central America, flux melting and decompression melting. Most magmas from the volcanic front and the second line are depleted in HFS elements and enriched in the slab signals described above. Recent studies of Central America [*e.g. Patino et al.*, 2000] infer that these magmas form because a hydrous or silicic flux, derived from the subducted slab, enters the mantle wedge, lowers the melting point of mantle peridotite and causes melting. In the back-arc, magmas have negligible to moderate input of elements from the subducting slab but they occur in an extensional setting. Mantle upwelling, similar to what occurs at mid-ocean ridges but on a much smaller scale, appears sufficient to cause decompression melting that generates BVF magma [*Walker et al.*, 1995].

Cameron et al. [in press] identify magmas from the volcanic front that have geochemical characteristics compatible with decompression melting. It is probable that the dichotomy of flux melting at the volcanic front and decompression melting in the back-arc is oversimplified.

4.5. Low-Ti and High-Ti Lavas on the Volcanic Front

Along most of the volcanic front, there is a bimodal distribution of TiO_2 in basalts and basaltic andesites. Exceptions include eastern Nicaragua, where the TiO_2 distribution is unimodal, and northwestern Costa Rica, where all samples have low TiO_2 contents. There is no regionally consistent TiO_2 value that separates high-Ti and low-Ti lavas because the modal values of the high-Ti lavas range from 1.15 weight percent TiO_2 in El Salvador to 1.5 weight percent in western Nicaragua. The low-Ti lavas are more consistent with modes between 0.7 to 0.9 weight percent. The high-Ti and low-Ti groups usually define overlapping distributions on histograms, so precise separation is impossible. Part of the overlap is the result of magma mixing because many vents erupt both high-Ti and low-Ti magmas. The low-Ti and high-Ti basalts are separate magma types in the sense that they are sufficiently geochemically distinct that it is not possible to derive one from the other by low-pressure assimilation-fractional crystallization processes (AFC).

High-Ti is a misleading term because the TiO_2 contents of the lavas appear high only relative to the abundant lavas with strong HFS depletions and low TiO_2. Thus, basaltic lavas called high-Ti in Central America have normal TiO_2 contents compared to the global basalt population. The high-Ti lavas differ substantially along the length of the arc and therefore, they are discussed in separate, local contexts below.

In northern Central America, *Halsor and Rose* [1988] pointed out several examples of paired volcanoes or short, cross-arc volcanic lineaments (e.g. Santa María-Cerro Quemado, Atitlán-Tolimán, Fuego-Acatenango, and Izalco-Santa Ana). In each case, the seaward volcano is more active, more explosive and more mafic and has steeper slopes and generally lower incompatible element contents. At the Fuego and Santa Ana centers, the more seaward volcano also has lower TiO_2 and HFS element contents. The cross-arc volcanological and geochemical gradients in the paired volcanoes can qualitatively be explained by mixing a magma derived from flux melting (high water content and strong HFS depletion) with a BVF magma derived from decompression melting (low water content and weak HFS depletion). Quantitative tests of this mixing hypothesis have been inconclusive so far. Establishing the cause of the gradients seen in the paired volcanoes remains an important problem.

Pacaya volcano is at the southeast end of the central Guatemala segment, adjacent to Agua and Fuego volcanoes (Figure 2). The next volcanic segment to the SE, which consists of Tecuamburro and Moyuta volcanoes, is 13 Km closer to the trench. Behind Tecuamburro is a Holocene BVF cinder cone field called Cuilapa. The westernmost cones of this field sit on the flanks of Pacaya. The basaltic lava, currently erupting at Pacaya, has a TiO_2 content of about 1.15 weight percent, a level similar to that of BVF lavas in the Cuilapa field and substantially higher than is found at the adjacent volcanic front centers. Pacaya is on the volcanic front, but its current eruption has characteristics between the low-Ti volcanic front and the high-Ti back-arc. No cross-arc geochemical gradient has been described at Pacaya. However, the volcano does appear to be receiving a back-arc input from its SE flank.

The most unusual examples of high-Ti lavas occur in the generally N-S oriented lines of rifts, explosion pits and cinder cones that cross the volcanic front in several places in Nicaragua but most notably at Nejapa and Granada. Granada is on the northwest flank of the Mombacho center (Figure 2). *Ui* [1972] first described the mafic and LIL-poor lavas from these alignments. *Walker* [1984] and *Walker et al.* [1990] discovered a mixture of high-Ti, low-Ti and hybrid magmas at both the Nejapa and Granada volcanic alignments and called these NG basalts. Walker suggested that these two small rift zones fortuitously expose the real complexity of magma genesis in Central America. In contrast, large volcanic centers may partially or completely obscure the complexity through mixing processes in large magma chambers, as appears to be the case at Masaya volcano [*Walker et al.*, 1993]. *Carr et al.* [1990] discovered up-bowed REE patterns in NG basalts, consistent with sequential batch melting, which depletes the more incompatible light rare earth elements and allows an up-bowed REE pattern in the later melts. These same lavas had lower Nd isotope ratios than the adjacent low-Ti lavas, suggesting they came from a more enriched source. *Carr et al.* [1990] explained the low light REE and LIL element contents, coupled with the relatively enriched Nd isotopes, by proposing an initially enriched source that had recently been depleted of its incompatible elements by an episode of low degree melting. *Reagan et al.* [1994] prefer to derive the high-Ti lavas from the same source as the low-Ti lavas. *Walker et al.* [2001] explain the high variability in the NG basalts through variable contributions from the subducting slab.

In central Costa Rica, high-Ti and low-Ti lavas are present in each volcanic center. The high-Ti lavas here differ only slightly from the low-Ti lavas; they have higher REE and HFS contents and lower Ba, Sr and Pb, indicating an origin involving lower degrees of melting and a weaker slab input [*Reagan and Gill*, 1989].

4.6 Non-uniform Distribution of High-Ti Magmas

The presence of two or more magma types at the same volcanic center should be considered normal in Central America. Many volcanic front centers erupt all the magma types found in the region, while others do not. Field observation suggest some controls on where the magmas come from and their ascent paths. For example, the volcanoes Moyuta and Tecuamburro (Figure 2) have only low-Ti lavas and are separated from the Cuilapa high-Ti cinder cone field by a gap of about 15 Km. The immediately adjacent volcanoes, Santa Ana in El Salvador and Pacaya in Guatemala, are overlapped by BVF cinder cone fields. Both of these centers have some lavas with high-Ti, BVF characteristics. In this region high-Ti lavas appear to migrate to the front from the back-arc in some cases but not in others. These field observations suggest separate locations for the generation of high-Ti and low-Ti magmas, followed by migrations to the same vent. The low-Ti magmas are likely generated in the mantle wedge by a slab-derived flux. The back-arc magmas occur in extensional structures forming behind the volcanic front. Extensional structures in the crust can facilitate the along strike movement of magma. Where these structures extend into the volcanic front, as they do at Pacaya and Santa Ana, back-arc magmas have easy access to the volcanic front centers.

Nicaraguan volcanoes have the clearest examples of distinct high-Ti and low-Ti magmas. The relatively thin crust in Nicaragua and the N-S striking extensional structures that cross the volcanic front promote rapid transit of magmas through the crust, allowing less fractionated basalts to erupt and minimizing hybridization of the different magma types. The high-Ti magmas erupt in apparently random locations. It seems each vent has the potential to erupt either type of magma, regardless of distance from the trench. At the extreme, these field observations suggest the high-Ti and low-Ti magmas follow the same flow paths and share a common locale for the melting processes. The presence of both high-Ti and low-Ti lavas with minimal soil development at the Las Pilas and Telica centers argue that there is minimal time lag and random sequencing in the eruptions of these magma types. The apparently low level of hybridization in these young lavas is surprising if they indeed share an extensive subcrustal plumbing system.

The intimate association of high-Ti and low-Ti magmas at Nicaraguan centers could be due to tectonic factors. First, there are no large extensional structures extending north of the volcanic front, limiting decompression melting due to extension in the back-arc. Second, the Nicaraguan seismic zone progressively steepens and becomes nearly vertical below about 150 Km (*Protti et al.*, 1995). Steep descent of the slab may lead to steep counterflow in the mantle that rises into the arc to replace mantle drawn down by the slab. In this model, decompression melting occurs without extension in the back-arc and the sites of magma generation for the low-Ti and high-Ti magmas would be closer than they are in northern Central America but still separate.

The only relationship between the physical characteristics of the volcanoes and the variety of erupted magmas is that many large volcanic centers have a more restricted range of chemistry. The clearest example of this is Masaya, a shield volcano, with a volume of about 180 Km^3, which makes it the largest center in the eastern Nicaragua segment. *Walker et al.* [1993] explained the compositional homogeneity of Masaya's lavas by magma mixing and AFC processes in a large shallow magma chamber. San Cristobal, the largest volcanic center in western Nicaragua, is not as homogeneous as Masaya, but does have the smallest range in Ba/La and U/Th ratios among the centers in this segment. The tendency for large centers to be more extensively mixed argues that the best places to find the widest range of magma types are small to medium sized volcanic centers with multiple distinct vents. In Nicaragua, the widest range of magmas occurs at Telica, a moderate sized center made from as many as six small, overlapping volcanoes [*Patino et al.*, 2000].

5. GEOCHEMICAL ZONING

Geochemical zoning occurs both along and across Central America and within volcanic centers. These chemical variations arise from changes in the mantle and crust, changes in the strength of the slab signal and changes in the type of slab signal, primarily the extent of the hemipelagic sediment component. Zoning will be briefly reviewed here because it is well defined in a series of recent papers. Regional zoning is extensively described [*e.g. Carr*, 1984; *Plank and Langmuir*, 1988; *Carr et al.*, 1990; *Leeman et al.*, 1994; and *Patino et al.*, 2000]. Similarly, cross arc zoning is thoroughly described by *Patino et al.*, [1997] and *Walker et al.*, [2000].

5.1. Mantle Zoning

The first order mantle zoning is the presence of unusual isotopic ratios and trace element contents in basalts from central Costa Rica and northern Panama. Geochemically, these basalts are similar to the basalts produced by the Galapagos hot spot. Although the unusual nature of central Costa Rican volcanics was apparent in their steep REE patterns and high LIL element contents, *Carr et al.* [1990] tried to integrate this group into the rest of the arc. Tournon [1984], *Malavassi* [1991], *Kussmaul et al.*, [1994] and *Leeman et al.*, [1994] showed that the magmas of central Costa Rica were distinct from those of the rest of Central America and had an ocean-island basalt (OIB) character. More recently, *Reagan et al.* [1994] showed that the distinction between the central Costa Rican lavas and the lavas in western Costa Rica and Nicaragua was clear in U-series isotopes and that the boundary was gradational across western Costa Rica and possibly into eastern Nicaragua. In Pb isotope space the Quaternary and Tertiary lavas of central Costa Rica plot in the OIB field and, like the Galapagos hot spot, extend from the MORB field toward the high mantle uranium (HIMU) variety of OIB [*Feigenson et al.*, 1996].

Figure 4 shows $^{206}Pb/^{204}Pb$ values along the volcanic front. $^{206}Pb/^{204}Pb$ values are sharply higher in central Costa Rica, with values of 18.8 to 19.3. Northwest of central Costa Rica, the $^{206}Pb/^{204}Pb$ values are less than 18.7, indicating a source similar to EMORB-source mantle. Refining the location and nature of the boundary between the two mantle domains in Costa Rica is an important problem.

There is little agreement on the origin of the unusual magmas in central Costa Rica. Most studies agree that the mantle source has Galapagos hot spot characteristics, but there are many ideas on how that source gets into the present volcanic system. *Abratis and Wörner* [2001] cite a window in the subducting Cocos Plate, inferred from plate reconstructions [*Johnston and Thorkelson*, 1997], that allows Galapagos mantle to rise into central Costa Rica. *Herrstrom et al.* [1995] cite S-wave splitting evidence for mantle flow parallel to the Andes that brings unusual mantle from the southeast. *Feigenson et al.* [1996] show that the Galapagos signature is present in Eocene to Quaternary lavas in easternmost Nicaragua and on islands in the Caribbean, including La Providencia (Figure 1), well behind the arc. Wherever the Galapagos signature occurs in active volcanoes, it coincides with the track of the hotspot during the last 90 my. They conclude that the hotspot added its geochemical signature to the mantle that passed over it and Galapagos-like magma erupts where this mantle is currently melting.

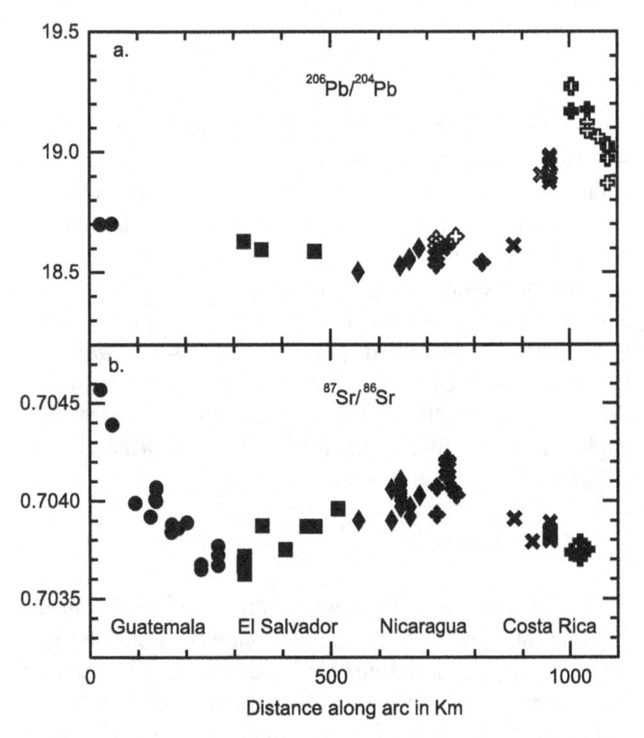

Figure 4. Crustal and mantle variations along the strike of the Central American volcanic front.

A second aspect of mantle zoning consists of possible variations in the EMORB-like mantle that seems to be the primary source for Central American magmas [*Patino et al.*, 2000]. Very little is known about possible variations because inputs from the slab prevent clear views of this source. One exception is the alkali basalt field near Lake Yojoa, Honduras. These back-arc basalts are close to and likely related to the transform fault system that separates the Caribbean and North American plates [*Walker et al.*, 2000]. The isotopic and trace element characteristics of Yojoa lavas are consistent with derivation by low degree melting of a source like the EMORB-source of *Sun and McDonough* [1989]. *Chan et al.* [1999] found δ^6 Li values in two high-Ti Nicaraguan lavas and one back-arc lava at Aguas Zarcas in Costa Rica that are higher than MORB, suggesting the mantle beneath much of Central America may have isotopically light composition, consistent with a source, less depleted than MORB.

5.2. Crustal Zoning

The crust along the volcanic front of Central America is thicker at both ends of the arc. In central and western Guatemala the volcanoes sit on the edge of a plateau comprising Paleozoic schists through Tertiary volcanics, whereas in Costa Rica, the basement appears to be Cretaceous and younger oceanic crust, sediments and volcanics [*Weyl*, 1980]. The Costa Rican crust contrasts with neighboring Nicaragua. In Costa Rica, most Tertiary and Quaternary volcanics appear to be superposed, suggesting voluminous intrusive and extrusive arc magmas created the thick crust in central Costa Rica. In Nicaragua, the Tertiary and

Cretaceous volcanic deposits are, for the most part, progressively further inland [*McBirney*, 1985; *Ehrenborg*, 1996], resulting in a relatively thin crust beneath the Nicaraguan volcanic lines. These crustal variations have isotopic and major element consequences.

From El Salvador to western Costa Rica, the Sr and Nd isotopes of lavas define an array with an unusual positive correlation [*Feigenson and Carr*, 1986]. The high Nd and high Sr end of the array has isotopic values consistent with derivation from EMORB-like mantle after addition of Sr from the slab. Sr derived from the subducted Cocos crust has a higher isotopic ratio because of interaction with sea water. A crustal overprint disrupts the unusual positive correlation of Sr and Nd isotopes in central and western Guatemala. In this region, Paleozoic rocks crop out along strike with the volcanic lines and Nd ratios become progressive lower as Sr ratios become progressively higher (Figure 4). These relationships are common in arcs with continental crustal contamination. Although the thick crust in Guatemala most likely increases the amount of assimilation, the main reason crustal contamination is obvious here is the highly radiogenic nature of the older crust found only in this area. Crustal contamination occurs all along the arc, but outside of central and western Guatemala, the assimilant is young enough and similar enough to present magmas that contamination is a minor consideration.

The most interesting zoning, related to the crust, occurs in major elements and physical parameters. *Carr* [1984] and Plank and Langmuir [1988] related volcano heights, maximum magma density, minimum SiO_2 contents and Na_2O and FeO contents to estimated crustal thickness (Figure 5). *Carr* [1984] explained the correlations between crustal thickness, volcano heights, maximum magma density and minimum SiO_2 content through a model that used ponding at the base of the crust and magma compressibility to set maximum magma densities along the arc. Magma density then controlled the other parameters, except for Na_2O. Although thicker crust should increase fractionation and moderate pressures, coupled with high water contents, should suppress plagioclase crystallization, these two effects increase Na_2O only a modest amount. The positive correlation between Na_2O and crustal thickness was not explained.

Plank and Langmuir [1988] argued that Na_2O and FeO contents in arc lavas, both in Central America and globally, are controlled by the extent of melting and the mantle potential temperature, a model well established for mid-ocean ridges. The thickness of the crust was assumed to control the extent of melting because the sharp density contrast at the base of the crust will stop rising diapirs that are undergoing decompression melting. This model easily explains the regional variations in major elements and is consistent the magma ponding model of *Carr* [1984]. However, this major element derived model is very different from the magma genesis model derived from trace elements and described below in section 5.4. Reconciling these different models is an important issue.

5.3. Zoning of Slab Signals

Across the arc, the intensity of slab signal, as estimated from Ba/La, U/Th, $^{87}Sr/^{86}Sr$, B/La or $^{10}Be/^{9}Be$, decreases behind the front but not in a consistent manner [*Walker et al.*, 2000]. Cross-arc transects more than 100 Km in length occur across southeast Guatemala and central Honduras, both the result of extensional tectonics related to the strike-slip Caribbean-North America plate boundary. In southeast Guatemala, Ba/La abruptly drops to mantle level just 10 to 30 Km behind the volcanic front [*Walker et al.*, 1995]. In Honduras, Ba/La, U/Th and $^{10}Be/^{9}Be$ decrease with distance across the arc in a more or less progressive manner [*Patino et al.*, 1997].

Along the volcanic front, the primary zoning arises from changes in the strength or intensity of the slab signal, which varies in a symmetric pattern centered on western to central Nicaragua. Ba/La and U/Th, which vary only slightly down the Cocos Plate sediment stratigraphy, most clearly show this regional variation but other ratios are also useful (see Figures 4b, 5c, 5d and 7c). Maxima occur in Nicaragua between Telica volcano, which has the maximum Ba/La, and Masaya volcano, which has the maximum $^{10}Be/^{9}Be$ and maximum $^{87}Sr/^{86}Sr$ outside of Guatemala. The intensity of the slab signal varies by at least a factor of four in Ba/La, so the signal is robust. Ba/La and U/Th correlate well enough ($r>0.80$) with $^{10}Be/^{9}Be$ to be proxies for it and therefore are unambiguous, easily measured indicators of subducted sediment. Ba/La is a superior slab signal because of its lower variation in the Cocos Plate stratigraphy (see section 3).

Since 1990, the working model to explain the regional variation in intensity of slab signal has been based on the positive correlation between slab signal and apparent degree of melting, estimated from the overall slope of REE patterns. For magmas derived from the same source, higher La/Yb equals steeper REE slope and implies lower degree of melting. The mirror image in the along strike variations of La/Yb and U/Th (Figure 5c and 5d) shows the regionally consistent, positive correlation between slab signal and degree of melting. The La/Yb plot has a log scale to allow for the anomalously high La/Yb values in central Costa Rica derived from the Galapagos-like mantle in this area. Even excluding the central Costa Rican data (crosses at distances of 1000 Km and greater in Figure 5), there is a convincing

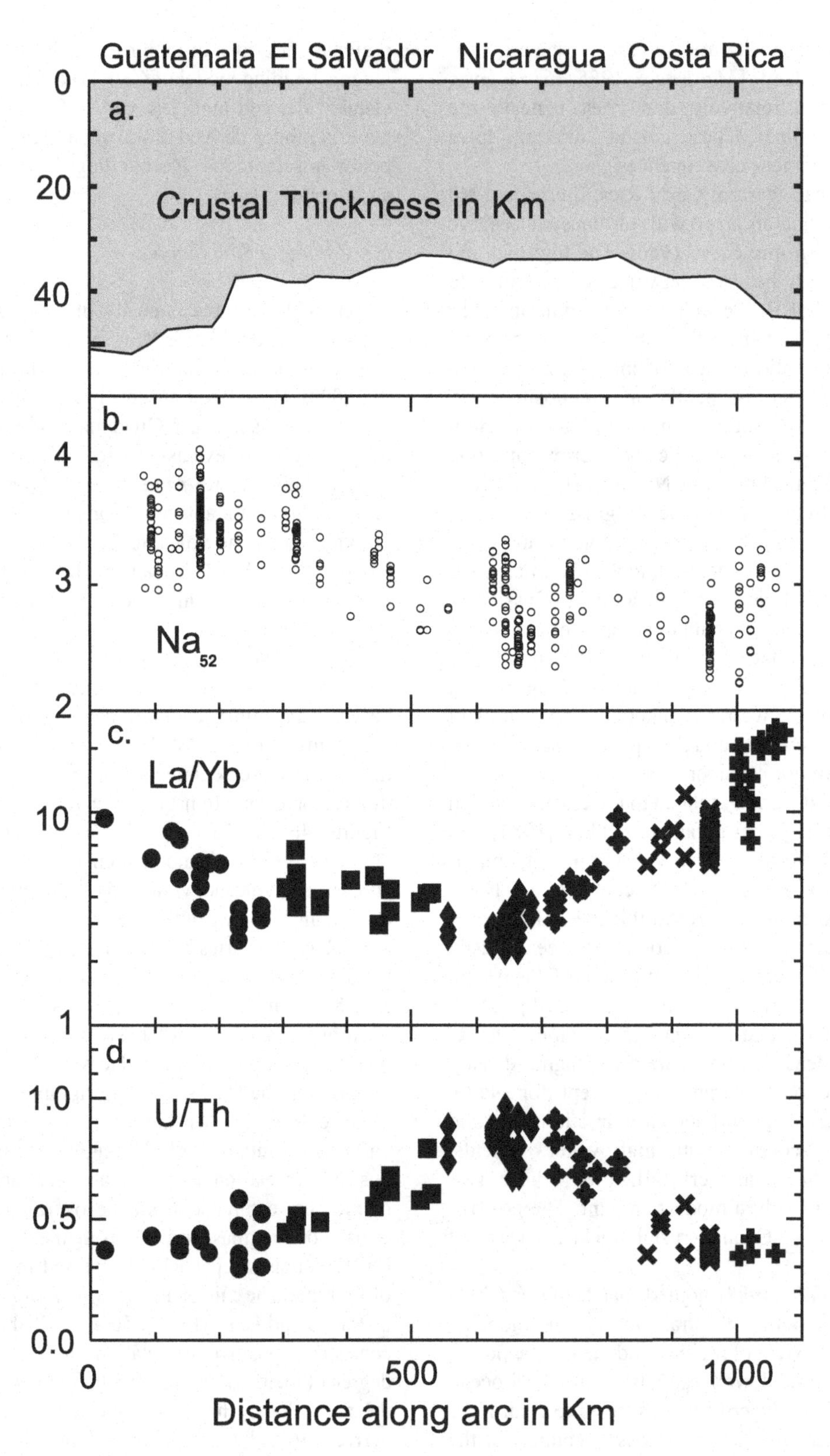

Figure 5. Regional variation in crustal thickness, Na_{52}, U/Th and La/Yb. Na_{52} is Na_2O content of lavas with SiO_2 contents between 48 and 55, corrected to 52% SiO_2 via $Na_{52} = Na_2O - (SiO_2-52)*0.14$.

mirror image between La/Yb and U/Th. A further constraint on models attempting to explain the regional variation in slab signal arises from a crude negative correlation between degree of melting and volumes of erupted volcanics. Nicaragua, which has the lowest La/Yb (or highest degree of melting), also has smaller volcanoes, just the opposite of what would be expected. Although there are huge variations in volcano size over short distances, Figure 7a shows that there has been less magma production in Nicaragua than in other segments of the arc.

[*Carr et al.*, 1990] reconciled the positive correlation between slab signal and apparent degree of melting and the negative correlation between degree of melting and volumes of erupted volcanics by proposing a constant slab flux that is more and less focused depending on external tectonic factors, such as the dip of the slab. Regions with concentrated flux produce small volumes of high degree melt magmas with high slab signal. Regions with diffuse flux produce large volumes of low degree melt magmas with low slab signal. A key assumption is that higher flux concentration leads to higher degrees of melting but not necessarily in a linear manner. Equal increments of flux concentration result in progressively smaller increments of melting, allowing elements, carried in the flux, to be enriched in the lavas. Measurements of H_2O and incompatible elements in Mariana Trough lavas indicate this assumption is valid [*Stolper and Newman*, 1994].

The regional variations in slab signal reflect an unresolved combination of changes in slab flux and changes in magma production rate. A high slab signal is not the same as a high slab flux. The ratios used as slab signals are dimensionless, whereas a flux should be in units of mass per unit time per arc length. A very high slab signal, such as a Ba/La ratio of 100, could be either a high flux of Ba into an average sized magma batch or an average flux of Ba into a small magma batch. The relative amounts of slab flux and magma are the same, but the fluxes and batch sizes are different. It is well established that there are strong regional variations in slab signal along the volcanic front. These may reflect changes in flux or a linkage between flux focusing and degree of melting, as suggested by *Carr et al.*, [1990]. This problem will not be unraveled until there are well-determined estimates of magmatic flux along the arc, or reliable geochemical methods of relating a signal to a flux.

A local (intravolcano) variation adds more complexity to the slab signal [*Patino et al.*, 2000]. Separate magma batches at a volcanic center differ in the major and trace element ratios that most emphasize the contrast between the carbonate and hemipelagic sections of the subducted Cocos Plate, which are Ba/Th and U/La (Figure 3). Lavas with apparently low hemipelagic content (high Ba/Th and low U/La) also have slightly lower K_2O contents, as would be expected if the K_2O-rich hemipelagic muds do not contribute to the flux. At well-sampled volcanoes, there are binary mixing arrays in Ba/Th versus U/La, two of which, Telica and Arenal, are shown in Figure 6, along with all the data for the eastern Nicaragua segment. The upper left end of the Telica array (the high Ba/Th end) can be modeled as a mixture of mantle + altered oceanic crust + carbonate. The other end of the array (the high U/La end) is reached just by adding hemipelagic component to the previous mix. The Arenal array differs from the Telica array primarily by having smaller sediment amounts. In general, the distance of a mixing array from the mantle point, EM, corresponds with the strength of the slab signal, with western Nicaragua (Telica) being the maximum. These apparent binary mixing hyperbolae occur in all segments of the arc, except for eastern Nicaragua. In eastern Nicaragua available data (Figure 6) indicate that all the magmas are near the right hand side of the data array, suggesting that the hemipelagic component is strongly represented in all the lavas.

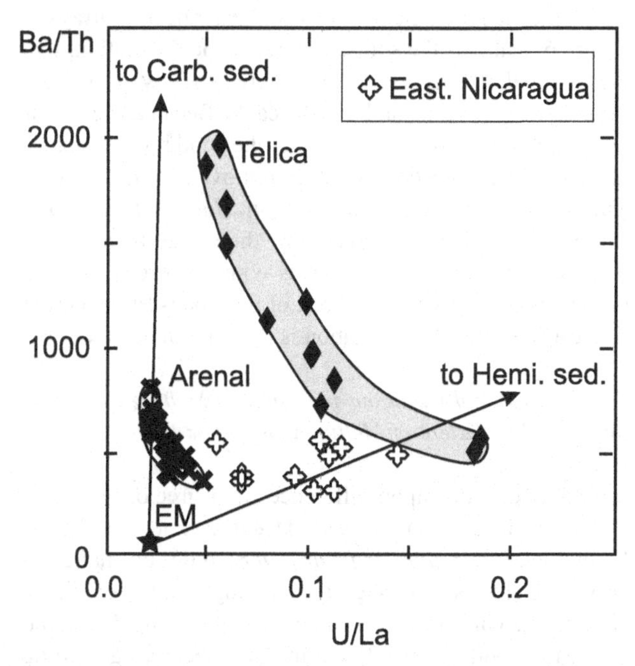

Figure 6. Intravolcano variation at Arenal and Telica volcanoes. Balloons enclose low-Ti lavas at Arenal and Telica volcanoes, which define arrays, called the local variation by Patino et al. (2000). The open symbols are low-Ti lavas from eastern Nicaragua, where no local variation is present. The eastern Nicaragua array extends from near the mantle (EM) to progressively closer to the sediment array that is parallel to the Telica array but well outside the diagram. The arrows point toward the locations of the ends of the sediment array.

The regularity in the local mixing arrays is perplexing because it implies repeated similar magma generation and mixing events. The order of mixing, derived from the array geometry, is first; create two separate melts; one from asthenosphere and the entire slab section (high U/La); the other from asthenosphere and the entire slab section, except for the hemipelagic sediments (high Ba/Th). These two melts are then mixed, creating the array. For most of the arc, this presumably complicated process generates parallel arrays, implying some unknown process that generates constant proportionality. Similarly, Reagan et al. [1994) discovered that slab tracers, presumed the result of fluid transport, correlated well with Th addition, even though Th should be immobile in a hydrous fluid. Delivering the correct amount of Th in a separate melt is possible only if the fluid-melt proportions are just right. These aspects of Central American geochemistry have a suspicious just right quality, which strongly suggests an important process that is not understood.

The hemipelagic sediments carry the bulk of the incompatible elements in the sediment section. Removing some or all of the hemipelagic section can generate the arrays in Figure 6. If this is the sole cause of the local variation, then the hemipelagic section is sequestered, removed or redistributed on short time and length scales. Because the oceanic crust offshore Nicaragua has a graben and horst structure, the hemipelagic section could be removed from the horsts and doubled into the grabens [*von Huene et al.*, 2000; *Patino et al.*, 2000]. Alternatively, there may be a melt of hemipelagic sediments that mixes with a hydrous flux that mobilizes elements from the rest of the Cocos Plate section. The cause of the local variation is not well understood.

5.4. Are Regional Variations in Degree of Melting Controlled by Slab Flux, Extent of Melting Column or Both?

Two models call upon differences in degree of melting to explain regional variations across Central America. The earlier model, *Plank and Langmuir* [1988], relies on changes in crustal thickness to change the melting heights of diapirs feeding the volcanoes. This model has global application to arcs and is widely accepted in the mid-ocean ridge setting. The later model [*Carr et al.*, 1990] attempts to explain the correlation of slab signal (e.g. U/Th in Figure 5d) with degree of melting (inverse to La/Yb in Figure 5c). In this model, an external physical control, slab dip, controls the focusing of slab flux, which, in turn, controls degree of melting. The crustal thickness model has a realistic physical basis (variation in crustal thickness) but does not explain the regional variation in slab signals like Ba/La or $^{10}Be/^{9}Be$. The slab flux model explains the slab signals but lacks a reliable physical control. It is based on an assumption that fluid movements in the mantle wedge vary with dip of the slab. The two models could complement each other if slab dip and crustal thickness are linked by a physical process that causes a negative correlation between them. The seismic data used to determine slab dip are only fair in coverage and quality but most reviewers of the data [*see Protti et al.*, 1995] find the steepest dips in Nicaragua, as is required to generate the negative correlation. Therefore, the two models seem to be complementary, but whether by cause or by accident is not known.

6. SEGMENTATION OF THE VOLCANIC FRONT: GEOGRAPHY, VOLUME AND GEOCHEMISTRY

The volcanic front can be segmented in two ways. The locations of the active volcanic centers define eight lineaments or volcanic segments that are separated by changes in strike and or dextral steps of as much as 40 Km. The boundaries between the segments are the stippled bars in Figure 7b. The distribution of volumes of erupted volcanics in Figure 7a defines a less obvious segmentation. The volumes of volcanic centers are lognormally distributed, but this distribution does not appear to be spatially random because there are several progressions starting from a large volcanic center and proceeding to successively smaller ones. There are seven very large volcanic centers (named in Figure 7a) that define volume-segments. The boundaries between the volume segments are roughly located at the small volcanoes out on the tails of the volume progressions. These two methods of segmenting the arc do not give exactly the same results but there is considerable overlap. One difference is that the volume distribution is continuous across the two northwesternmost volcanic segments in Guatemala, apparently unaffected by the step in the volcanic front. The other two differences are minima in volumes in central El Salvador and eastern Nicaragua where there is no volcanic segmentation.

Most of the eight geographic lineaments were recognized by *Dollfus and Montserrat* [1868] but, despite their long residence in the literature, their origin is not explained. *Stoiber and Carr* [1973] and *Carr and Stoiber* [1977] pointed out numerous geological features that were discontinuous at the same places as the volcanic lineaments and suggested that segmentation of the upper plate was initiated by breaks in the lower plate. However, this was not proved and the question of whether lower plate irregularities broke the upper plate or upper plate structures imposed a structural pattern on the descending plate was not resolved. Extensive marine geologic investigations offshore Costa Rica discovered structures in the subducting Cocos Plate, such as the

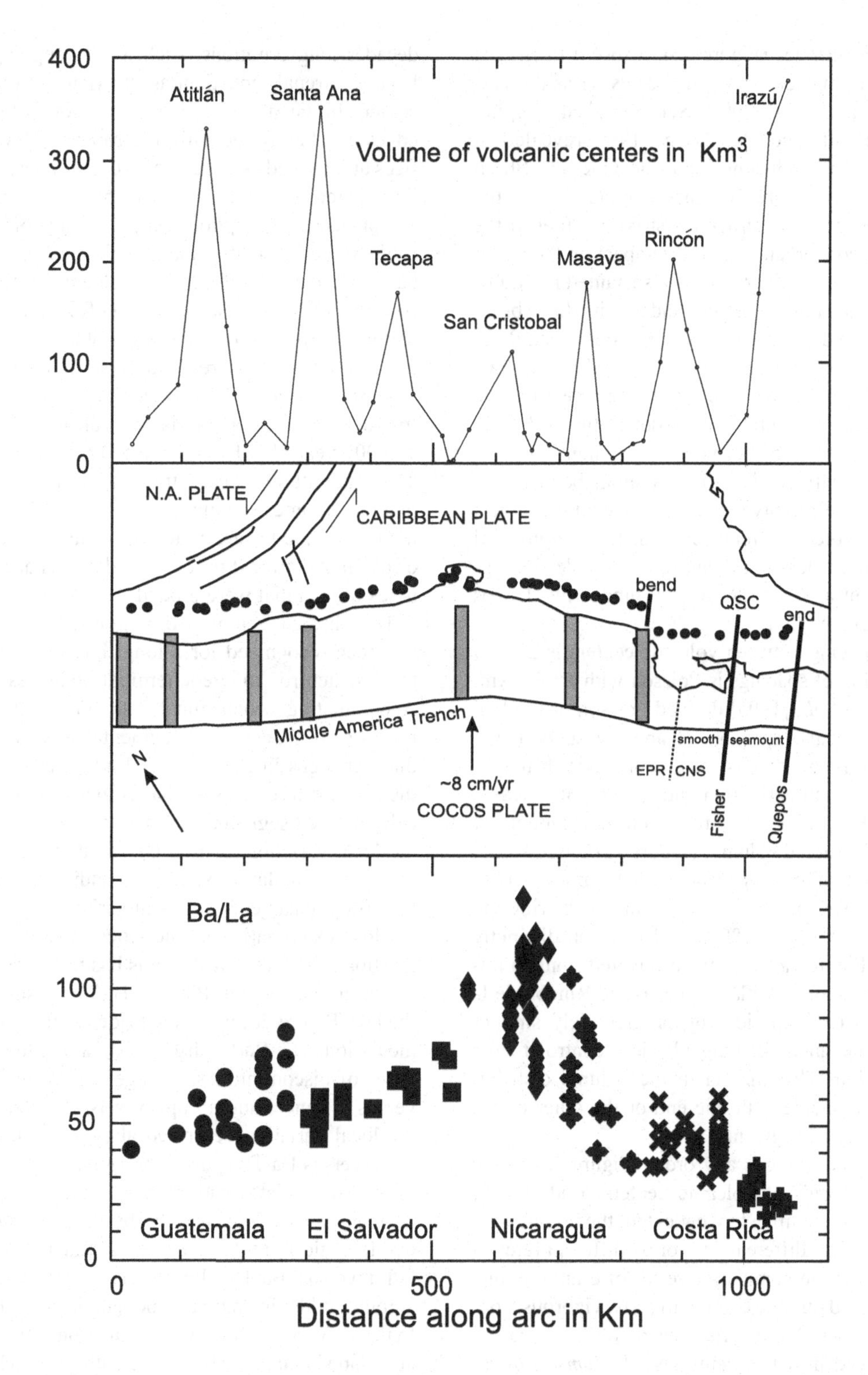

Figure 7. Segmentation of the volcanic front. Panel 7a shows the volumes of the 39 Central America volcanic centers. Panel 7b shows the plate boundaries, the volcanic front (circles), boundaries of volcanic segments (stippled bars) and structural boundaries in the Cocos Plate [von Huene et al., 2000] and inclined seismic zone; bend, QSC, and end [Protti et al., 1995].

Quepos and Fisher Ridges (Figure 7b), whose influence is clearly traceable to the coastline and even as far as the volcanic front [*von Huene et al.*, 2000]. Seismological mapping of the Wadati-Benioff zone [*Protti et al.*, 1995] revealed an apparent tear in the subducting slab, the Quesada Sharp Contortion (QSC in Figure 7b) that coincides with the Fisher Ridge. The offshore structures strike parallel to the direction of plate convergence, so their impact on the structure of the upper plate is the result of a sustained period of subduction. The Quepos Ridge coincides with the abrupt end on intermediate depth seismicity in Central Costa Rica. The boundary between oceanic crust generated by the East Pacific Rise (EPR) and oceanic crust generated by the Cocos-Nazca spreading center (CNS) (dotted line in Figure 7b) does not generate an obvious volcanic segment. *Protti et al.*, [1995] also tentatively identified a small bend in the intermediate depth seismicity that coincides with the largest right step in the volcanic front. Overall, the geophysical data are beginning to identify the causes of the volcanic segmentation. In central Costa Rica, the subducting Cocos Plate initiates segmentation.

The average spacing between volcanic centers is 27 Km and the distribution of spacings is Poisson with $\lambda = 24$ Km. *D'Bremond d'Ars et al.*, [1995] derived Poisson or random distributions of volcano spacings at arcs by superposing several generations of diapirs originating via Raleigh-Taylor instability. The distribution and spacing of volcanic centers in Figures 1, 2 and 7b agrees with their model but the distribution of erupted volumes (Figure 7a) suggests that earlier generations of diapirs influence later ones. Along Central America, seven peaks in the volume distribution occur at intervals of 120 to 180 Km. The seven distinctly larger centers in Figure 7a may have originated from an initial generation of diapirs. Atitlán, Tecapa, and Rincon de la Vieja are flanked on both sides by progressively smaller centers, suggesting an underlying physical control to the volume distribution. The minima in the volume distributions commonly coincide with one end or the other of the geographically defined segments.

It is surprising that there is any order (Figure 7a) in the distribution of volumes of volcanic centers. Radiometric dating is needed to determine to what extent the volume differences are caused by different ages or by different rates of eruption. Volume is the cumulative result of eruptions that are large enough and durable enough to resist vigorous tropical erosion processes. Tephra eruptions of 10 Km^3 size can be largely removed in a few centuries [*Williams*, 1983a]. Furthermore, substantial volume can be instantaneously removed by caldera-forming events. Persisting volume is added primarily by lava fields and domes, created in decades-long eruptions, such as the ongoing eruptions of Pacaya, Arenal, and Santiaguito (part of the Santa María center. In Guatemala, there are morphologically young cones and barely recognizable roots but few volcanic edifices at intermediate stages of erosion. There is either a continuing recent pulse of activity or very rapid reduction of volcanoes when eruptions cease. The age of the current volcanic front in Guatemala is estimated as <84,000 years, based on the age of the last caldera-forming eruption of Atitlán caldera. The large Atitlán-Tolimán center postdates and partly fills the caldera. Most other Guatemalan centers are smaller so it is reasonable to estimate their ages as <84,000 years as well. In Nicaragua, parts of the Las Sierras formation that underlies Masaya volcano correlates with the 135,000 year old J1 marine ash layer [*Walker et al.*, 1993]. These admittedly weak stratigraphic relationships suggest that the volumes of volcanic centers are the sum of roughly the last 100,000 years of activity minus substantial volumes of distal and eroded pyroclastic deposits and moderate volumes of lava that were eroded.

The physical segmentation shown by volcano locations has been recognized for a long time but attempts to relate this structure to geochemical features have failed. Accumulating geochemical data (Figure 7c) now allow a provisional geochemical segmentation of the arc based on three changes in gradients of Ba/La versus distance along the arc and three offsets that coincide with the geographically defined segment boundaries. The recent discovery of the local variation, caused by variation in the hemipelagic component of slab flux, adds an additional tool for measuring abrupt changes in regional variation.

Close examination of the Cocos Plate sediment section (section 3) shows that Ba/La is has the most uniform distribution in the section. Plotting Ba/La versus distance along the arc (Figure 7c) provides evidence for geochemical segmentation. The Ba/La distribution along the arc shows two types of discontinuities; changes in the gradient of Ba/La versus distance; and abrupt offsets. An additional factor is the local variation described above, which is best seen in U/La versus Ba/Th space. The signal of the uppermost unit of the Cocos Plate stratigraphy (U/La) is reduced or missing in many lavas, presumably by loss or redistribution of hemipelagic muds as they are subducted. Examination of U/La versus Ba/Th allows qualitative description of the amount of hemipelagic component present in each segment. In the summary below, which runs from SE to NW along the arc, slab signal is synonymous with Ba/La (Figure 7c).

Central Costa Rica (crosses in Figure 7c) has a constant, low slab signal with hemipelagic component present and variable.

Western Costa Rica (Xs) has a constant, moderate slab signal with hemipelagic component present and variable.

Eastern Nicaragua (pointed crosses) has a very strong gradient in slab signal from low in the SE to high in the NW. Masaya volcano near the NW end of this segment is the global maximum in Be isotope ratio. The hemipelagic component is not only present but, in contrast to other areas, apparently always present (Figure 6). Further sampling is in progress to test this observation.

Western Nicaragua (diamonds) has the highest Ba/La in Central America. This ratio has very high dispersion, especially at the smaller volcanic centers, but it does not vary significantly along the length of the segment. The hemipelagic component is present and variable.

El Salvador (squares) has a sharp decrease in Ba/La at the border with Nicaragua. There is a strong gradient in slab signal from high (SE) to low (NW). The hemipelagic component is present and variable.

Guatemala (circles) has an increase in Ba/La at the border with El Salvador. This offset coincides with the onset of extensive back arc volcanism in southeast Guatemala. There is a strong gradient in slab signal from high (SE) to low (NW) across Guatemala. The hemipelagic component is present and variable. Central and western Guatemala have Sr and Nd isotope systematics that are perturbed by assimilation of Paleozoic crust [*Carr et al.*, 1990].

One possible explanation for the offsets in Ba/La could be that the steps in the volcanic front change the depth to a smoothly varying Wadati-Benioff zone and that depth to the seismic zone controls Ba/La. The cross-arc zoning in Honduras [*Patino et al.*, 1997] does show decreasing Ba/La with increasing depth to the seismic zone. If depth to the seismic zone controls Ba/La, then large geographic steps by the volcanic front should produce large jumps in Ba/La and the segment to the NW (always further from the trench because the steps are all dextral) should have lower Ba/La. At the boundary between central and western Costa Rica there is no step in the volcanic front but there is a tear in the Cocos Plate allowing the western Costa Rica segment to have a greater depth to the seismic zone. However, that segment has higher Ba/La, not lower. The largest step along the volcanic front occurs at the boundary between western Costa Rica and eastern Nicaragua, but instead of a sharp drop in Ba/La into Nicaragua, there is no obvious change. The step between eastern and western Nicaragua is at least 10 Km, but Ba/La does not drop, instead it increases. The step between western Nicaragua and El Salvador is on the order of 10 Km and here the Ba/La ratio sharply decreases. The boundary between El Salvador and Guatemala is a change in strike rather than a rightward step and a large increase occurs not a decrease. The two assumptions, that the Wadati-Benioff zone has a simple geometry and that depth to the seismic zone controls Ba/La, fail to explain the Ba/La offsets and lack of offsets seen in Central America.

7. CONCLUSIONS

Central American volcanoes provide a rich physical and geochemical data set. Several layers of geochemical insight result from selecting isotopic and trace element ratios that maximize the differences between different sources. Although a rich framework of data and interpretation made Central America a selection for focused study in the NSF Margins Subduction Factory Initiative, much remains to be done and many important problems are essentially untouched. A partial list of problems, unexplained observations and remaining work follows.

1. The first-order geochemistry of the mafic volcanic front is known, but few volcanic centers have been comprehensively studied.
2. Silicic volcanic centers have been or are being studied in Guatemala, El Salvador and Costa Rica, but, in Nicaragua, there are several small silicic calderas that have not been investigated.
3. The secondary front of isolated composite volcanoes, located 20 to 75 Km behind the main front, should be integrated into regional volcanological and geochemical investigations.
4. Important aspects of the magma genesis of the high-Ti lavas in Nejapa and Granada, Nicaragua remain unresolved.
5. Are the short, cross arc gradients in the paired volcanoes of northern Central America the result of mixing between low-Ti basalts and back-arc, high-Ti basalts?
6. The existence of cross-arc geochemical variation within most volcanic front centers in Guatemala contrasts with the lack of such variation in Nicaragua. This discrepancy emphasizes how little is known about magma flow paths and the loci of melting.
7. The transition between EMORB mantle beneath Nicaragua and OIB mantle beneath central Costa Rica occurs in easternmost Nicaragua and western Costa Rica. Is the transition, sharp, gradual or intermingled?
8. Do the regional variations in slab signal mean there are regional variations in slab flux or constant flux with regional variations in degree of melting caused by variable focusing of the flux? If the volcanic output rate were constant along the arc, the former would be the case, but currently available volume data and age estimates favor

the latter because there is low magma output in Nicaragua, where the slab signal is at the maximum.

9. Some of the detailed systematics in isotopic and trace element ratios are difficult to explain unless the sources involved are repeatedly tapped in just the right proportions. This just-right phenomenon is not understood.
10. Are regional variations in degree of melting controlled by flux focusing, extent of melting column or both? In other words is the *Plank and Langmuir* [1988] model correct, the *Carr et al.* [1990] model, or both?
11. The variation in the sizes of volcanic centers along the arc is an observation that needs to be integrated into models of magma genesis.
12. The breaks between the volcanic lines that comprise the Central American volcanic front corresponds with discontinuities in the regional slab signal, including three changes in the gradient and three abrupt offsets in the slab signal.

Acknowledgments. R.E. Stoiber and his numerous students and colleagues did much of the work on Central America described here. The chain of inspiration started by Dick is long and rich and we will miss his challenging insights. This work was partially supported by National Science Foundation Grants EAR 9628251, EAR 9905167, OCE-9521716 and EAR 9406624. We thank J. Whitlock for her careful measurements of the volumes of volcanic centers. We thank Louise Bolge for use of new data from Arenal volcano. R. Stern, J. Eiler and G. Alvarado provided useful criticisms that greatly improved this work. R. Harwood and B. Cameron made exceptional contributions to the field work.

REFERENCES

Abratis, M. and G. Wörner, Ridge collision, slab-window formation, and the flux of Pacific asthenosphere into the Caribbean realm. *Geology, 29*, 127-130, 2001.

Alvarado, G. E. and M. J. Carr, The Platanar-Aguas Zarcas volcanic centers, Costa Rica: Spatial-temporal association of Quaternary calc-alkaline and alkaline volcanism. *Bull. Volcanol., 55*, 443-453, 1993.

Aubouin, J., J. Azema, J-Ch. Carfantan, A. Demant, C. Rangin, M. Tardy and J. Tournon, The Middle America trench in the geological framework of Central America, *in Initial Rep. Deep Sea Drill. Proj. 67*, 747-755, 1982.

Burkart, B. and S. Self, Extension and rotation of crustal blocks in northern Central America and effect on the volcanic arc, *Geology, 13*, 22-26, 1985.

Cameron, B. I., J. A. Walker, M. J. Carr, L.C. Patino and O. Matías, Contrasting melt generation processes at front stratovolcanoes in southeastern Guatemala, *J. Volcanol. Geotherm. Res.*, in press.

Carr, M. J., Symmetrical and segmented variation of physical and geochemical characteristics of the Central American volcanic front, *J. Volcanol. Geotherm. Res., 20*, 231-252, 1984.

Carr, M. J., and W. I. Rose Jr., CENTAM-a data base of analyses of Central American volcanic rocks, *J. Volcanol. Geotherm. Res., 33*, 239-240, 1987.

Carr, M. J., and R. E. Stoiber, Geologic setting of some destructive earthquakes in Central America. *Geol. Soc. Amer. Bull., 89*, 151-156, 1977.

Carr, M. J., and R. E. Stoiber, Volcanism, in *The Geology of North America, Volume H, The Caribbean* Region, edited by G. Dengo and J.E. Case, Boulder, Colorado, Geol. Soc. America, pp. 375-391, 1990.

Carr, M. J., M. D. Feigenson and E. A. Bennett, Incompatible element and isotopic evidence for tectonic control of source mixing and melt extraction along the Central American arc. *Contribs. Mineral. Petrol., 105*, 369-380, 1990.

Carr, M. J., W. I. Rose Jr., and R. E. Stoiber, Central America, in *Orogenic Andesites*, edited by R.S. Thorpe, pp. 149-166, John Wiley, 1982.

Chan, L. H., W. P. Leeman and C.-F. You, Lithium isotope composition of Central American volcanic arc lavas: implications for modifications of subarc mantle by slab-derived fluids, *Chemical Geology, 160*, 255-280, 1999.

de Boer, J. Z., M. J. Defant, R.H. Stewart and H. Bellon, Evidence for active subduction below western Panama, *Geology, 19*, 649-652, 1991.

de Bremond D'ars, J., C. Jaupart and R.S.J. Sparks, Distribution of volcanoes in active margins, *J. Geophys. Res., 100*, 20421-20432, 1995.

Defant, M. J., T. E. Jackson, M. S. Drummond, J. Z. de Boer, M. D. Feigenson, R. C. Maury, and R. H. Stewart, The geochemistry of young volcanism throughout western Panama and southeastern Costa Rica, an overview, *J. Geol. Soc. Lond., 149*, 569-579, 1992.

Dollfus, A., and E. Montserrat, *Voyage geologique dans les republiques de Guatemala et de Salvador: France, Mission Scientifique au Mexique at dans l'Amerique Centrale, Geologie, 9*, 539 pp, Imprimerie imperiale, Paris, 1868.

Drummond, M. S., M. Bordelon, J. Z.de Boer, M. J. Defant, H. Bellon and M.D. Feigenson, Igneous petrogenesis and tectonic setting of plutonic and volcanic rocks of the Cordillera de Talamanca, Costa Rica-Panama, Central American arc: *American Journal of Science, 295*, 875-919, 1995.

Ehrenborg, J., A new stratigraphy for the Tertiary volcanic rocks of the Nicaraguan highland, *Geol. Soc. America. Bull., 108*, 830-842, 1996.

Fairbrothers, G. E., M. J. Carr, and D. G Mayfield, Temporal magmatic variation at Boquerón Volcano, El Salvador, *Contrib. Mineral. Petrol., 67*, 1-9, 1978.

Feigenson, M. D. and M. J. Carr, The source of Central American lavas: inferences from geochemical inverse modeling, *Contrib. Mineral. Petrol., 113*, 226-235, 1993.

Feigenson, M. D. and M. J. Carr, Positively correlated Nd and Sr isotope ratios of lavas from the Central American volcanic front, *Geology, 14*, 79-82, 1986.

Feigenson, M. D., M. J. Carr, L. C. Patino, S. Maharaj and S. Juliano, Isotopic identification of distinct mantle domains beneath Central America, *Geol. Society of America, Abstracts with Programs, 28*, 380, 1996.

Halsor, S. P. and W. I. Rose Jr., Common characteristics of active paired volcanoes in northern Central America, *J. Geophys. Res., 93*, 4467-4476, 1988.

Harry, D. L. and N. L. Green, Slab dehydration and slab petrogenesis in subduction systems involving very young oceanic lithosphere, *Chemical Geol., 160*, 309-333, 1999.

Herrstrom, E. A., M. K. Reagan, and J. D. Morris, Variations in lava composition associated-with flow of asthenosphere beneath southern Central America, *Geology 23*, 617-620, 1995.

Johnston, S.T., and D. J. Thorkelson, Cocos-Nazca slab window beneath Central America, *Earth Planet. Sci. Lett., 146*, 465-474, 1997.

Kussmaul, S., Tournon, J. , Alvarado, G. E., Evolution of the Neogene to Quaternary igneous rocks of Costa Rica. *Profil 7*, 97-123, Stuttgart, 1994.

Kimura, G., E. A. Silver, P Blum, and Shipboard Scientific Party, *Proceedings of Ocean Drilling Project, Initial Reports , 170.* Ocean Drilling Program, College Station Texas, 1997.

Leeman, W. P., M. J. Carr, and J. D Morris, Boron geochemistry of the Central American arc: Constraints on the genesis of subduction-related magmas, *Geochim. Cosmochim. Acta 58,* 149-168, 1994.

Malavassi, E., Magma sources and crustal processes at the southern terminous of the Central American volcanic front. PhD. Thesis, 435pp., Univ. Calif. Santa Cruz, 1991.

McBirney, A. R., Volcanic evolution of Central America. *Boletín Vulcanología, Heredia, Costa Rica. 14*, 21-23, 1985.

Noll, P. D., H. E. Newsom, W. P. Leeman and J. G. Ryan, The role of hydrothermal fluids in the production of subduction zone magmas: Evidence from siderophile and chalcophile trace elements and boron. *Geochim Cosmochim Acta, 60*, 587-612, 1996.

Patino, L. C., M. J. Carr and M. D. Feigenson, Cross-arc geochemical variations in volcanic fields in Honduras, C. A.: Progressive changes in source with distance from the volcanic front, *Contrib. Mineral. Petrol., 129*, 341-351, 1997.

Patino, L. C., M. J. Carr and M.D. Feigenson, Local and regional variations in Central American arc lavas controlled by variations in subducted sediment input, *Contrib. Mineral. Petrol.,138*, 265-283, 2000.

Plank, T. and C.H. Langmuir, An evaluation of the global variations in the major element chemistry of arc basalts, *Earth Planet. Sci. Lett., 90*, 349-370, 1988.

Plank, T. and C.H. Langmuir, Tracing trace elements from sediment input to volcanic output at subduction zones, *Nature, 362*,.739-743, 1993.

Pullinger, C., Evolution of the Santa Ana volcanic complex, El Salvador. M.S. Thesis, 251 pp, Michigan Technological University, Houghton, MI, 1998.

Protti, M., F. Gundel, and K. McNally, Correlation between the age of the subducting Cocos plate and the geometry of the Wadati-Benioff zone under Nicaragua and Costa Rica, *Geol. Soc. Am. Spec. Pap., 295*, p. 309-326, 1995.

Reagan, M. K. and J. B. Gill, Coexisting calc-alkaline and high-niobium basalts from Turrialba volcano, Costa Rica: Implications for residual titanates in arc magma sources, *J. Geophys. Res., 94*, 4619-4633, 1989.

Reagan, M. K., J. D. Morris, E. A. Herrstrom and M. T. Murrell, Uranium series and beryllium isotope evidence for an extended history of subduction modification of the mantle below Nicaragua, *Geochim. Cosmochim. Acta 58*, 4199-4212, 1994.

Roggensack, K., R. L. Hervig, S. B. McKnight, S. N. Williams, Explosive basaltic volcanism from Cerro Negro Volcano: Influence of volatiles on eruptive style, *Science 277*, 1639-1642, 1997.

Rose Jr., W. I., F. M. Conway, C. R. Pullinger, A. Deino and W C Mclntosh, A more precise age framework for late Quaternary silicic eruptions in northern Central America, *Bull. Volcanol., 61,* 106-120, 1999.

Sisson, T and G. D. Layne, H_2O in basalt and basaltic andesite glass inclusions from four subduction-related volcanoes, *Earth Planet. Sci. Lett., 117*, 619-635. 1993.

Stoiber, R. E., and M. J. Carr, Quaternary volcanic and tectonic segmentation of Central America, *Bull. Volcanol., 37*, 304-325, 1973.

Stolper, E. and S. Newman, The role of water in the petrogenesis of Mariana Trough magmas, *Earth Planet. Sci. Lett. 121,* 293-325, 1994.

Sun S., and W. F. McDonough, Chemical and isotopic systematics of oceanic basalts: implications for mantle composition and processes, in *Magmatism in the Ocean Basins* edited by A. D. Saunders and M. J. Norry, *Geological Soc. Special Publ., 42*, 313-345, 1989.

Sussman, D., Apoyo Caldera, Nicaragua: A major Quaternary silicic eruptive center, *J. Volcanol Geotherm. Res. 24*, 249-282, 1985.

Tournon, J., Magmatismes du mesozoique à l' actuel en Amerique Centrale: L'exemple de Costa Rica, des ophiolites aux andesites, Ph.D. thesis, 335 pp., Mém. Sc. Terre, Univ. Curie, Paris, 84-49, 1984.

Ui, T., Recent volcanism in the Masaya-Granada area, Nicaragua, *Bull. Volcanol. 36*, 174-190, 1972.

van Wyk de Vries, B., Tectonics and magma evolution of Nicaraguan volcanic systems, Ph.D. Thesis, 328pp., Open University, Milton Keynes, UK, 1993.

von Huene, R., C. R. Ranero, W. Weinrebe and K. Hinz, Quaternary convergent margin tectonics of Costa Rica: Segmentation of the Cocos Plate, and Central American volcanism, *Tectonics, 19*, 314-334, 2000.

Walker, J. A., Petrogenesis of lavas from cinder cone fields behind the volcanic front of Central America, *J. Geol., 87,* 721-739, 1981.

Walker, J. A., Volcanic rocks from the Nejapa and Granada cinder cone alignments, Nicaragua, *J. Petrol., 25*, 299-342, 1984.

Walker, J. A., M. J. Carr, M. D. Feigenson and R. I. Kalamarides, The petrogenetic significance of High-and Low-Ti basalts in Central Nicaragua, *J. Petrol., 31,* 1141-1164, 1990.

Walker, J. A., M. J. Carr, L. C. Patino, C. M. Johnson, M. D. Feigenson and R. L. Ward, Abrupt change in magma generation processes across the Central American arc in southeastern Guatemala: flux-dominated melting near the base of the wedge to decompression melting near the top of the wedge, *Contrib. Mineral. Petrol., 120,* 378-390, 1995.

Walker, J. A, L. C. Patino, B. I. Cameron and M. J. Carr, Petrogenetic insights provided by compositional transects across the Central American arc: Southeastern Guatemala and Honduras, *J. Geophys. Res., 105*, 18,949-18,963, 2000.

Walker, J. A, L. C. Patino, M. J. Carr and M. D. Feigenson, Slab control over HFSE depletions in central Nicaragua, *Earth Planet. Sci .Lett., 192*, 533-543, 2001.

Walker, J. A., S. N. Williams, R. I. Kalamarides, and M. D. Feigenson, Shallow open-system evolution of basaltic magma beneath a subduction zone volcano: the Masaya caldera complex, Nicaragua, *J. Volcanol. Geotherm. Res., 56*, 379-400, 1993.

Weyl, R., *Geology of Central America*, 372 p., Gebrüder Borntraeger, Berlin, 1980.

White, R. A., and D. Harlow, Destructive upper-crustal earthquakes of Central America since 1900, *Bull. Seismol. Soc. America, 83*, 1115-1142, 1993.

Williams, S. N., The October 1902 plinian eruption of Santa María volcano, Guatemala, *J. Volcanol. Geotherm. Res., 16*, 33-56, 1983a.

Williams, S. N., Plinian airfall deposits of basaltic composition, *Geology, 11*, 211-214, 1983b.

An Overview of the Izu-Bonin-Mariana Subduction Factory

Robert J. Stern

Geosciences Department, University of Texas at Dallas, Richardson, Texas

Matthew J. Fouch

Carnegie Institution of Washington, Department of Geological Sciences, Arizona State University, Tempe, Arizona

Simon L. Klemperer

Department of Geophysics, Stanford University, Stanford, California

The Izu-Bonin-Mariana (IBM) arc system extends 2800km from near Tokyo, Japan to Guam and is an outstanding example of an intraoceanic convergent margin (IOCM). Inputs from sub-arc crust are minimized at IOCMs and output fluxes from the Subduction Factory can be more confidently assessed than for arcs built on continental crust. The history of the IBM IOCM since subduction began about 43 Ma may be better understood than for any other convergent margin. IBM subducts the oldest seafloor on the planet and is under strong extension. The stratigraphy of the western Pacific plate being subducted beneath IBM varies simply parallel to the arc, with abundant off-ridge volcanics and volcaniclastics in the south which diminish northward, and this seafloor is completely subducted. The Wadati-Benioff Zone varies simply along strike, from dipping gently and failing to penetrate the 660 km discontinuity in the north to plunging vertically into the deep mantle in the south. The northern IBM arc is about 22km thick, with a felsic middle crust; this middle crust is exposed in the collision zone at the northern end of the IBM IOCM. There are four Subduction Factory outputs across the IBM IOCM: (1) serpentinite mud volcanoes in the forearc, and as lavas erupted from along (2) the volcanic front of the arc and (3) back-arc basin and (4) from arc cross-chains. This contribution summarizes our present understanding of matter fed into and produced by the IBM Subduction Factory, with the intention of motivating scientific efforts to understand this outstanding example of one of earth's most dynamic, mysterious, and important geosystems.

1. INTRODUCTION

The Izu-Bonin-Mariana arc system (IBM) stretches over 2800 km, from near Tokyo, Japan, to beyond Guam, U.S.A. (Figure 1), and is an excellent example of an intraoceanic convergent margin (IOCM). IOCMs are built on oceanic crust and contrast fundamentally with arcs built on continental crust, such as Japan or the Andes. Because IOCM crust is thinner, denser, and more refractory than that beneath Andean-type margins, study of IOCM melts and fluids allows more confident assessment of mantle-to-crust fluxes and processes than is possible for Andean-type convergent margins. Because IOCMs are far removed from continents they are less affected by the large volume of alluvial and glacial sediments. The consequent thin sedimentary cover makes it much easier to study arc infrastructure and determine the mass and composition of subducted sediments. Active hydrothermal systems found on the submarine parts of IOCMs give us a chance to study how many of earth's important ore deposits formed.

Inside the Subduction Factory
Geophysical Monograph 138

10.1029/138GM10

IBM presents an outstanding opportunity to study the operation of the Subduction Factory at an IOCM, for several reasons: 1) the history of IBM evolution is one of the best known of any convergent margin; 2) there are four opportunities across the arc to sample products of the Subduction Factory—the forearc, the active magmatic arc, arc cross-chains, and back-arc basins—more than any other convergent margin; 3) Subducted sediments are simple, diagnostic, and completely subducted; 4) IBM is an excellent example of convergent margins undergoing extension; and 5) IBM is large and diverse. Furthermore, a collision zone in the north provides an unparalleled opportunity to study the composition of middle IBM crust and so better infer products and processes leading to the formation of this crust. Additional information about the advantages of IBM for the Subduction Factory experiment can be found in the Subduction Factory Science Plan (http://www.ldeo.columbia.edu/margins/SF_Sci_Plan_revised.pdf).

The following overview of the IBM arc system is presented in the context of the Subduction Factory experiment. Emphasis is on the active components and on the nature of inputs and outputs. Although we try to summarize the most important points, it is important to appreciate that the Subduction Factory is one of earth's largest and most complex geosystems, planetary in scale and extending thousands of kilometers into the earth. This is certainly true for IBM. While we have a good understanding of what goes in and what comes out of the IBM Subduction Factory, we are just beginning to understand how it operates.

2. NOMENCLATURE AND GEOGRAPHY

The IBM arc system lies along the eastern margin of the Philippine Sea Plate in the Western Pacific. In this paper we will use different terms to describe all or part of the subducting Pacific Plate, including 'subducting components', 'subducted slab', and 'incoming plate', and the overriding crust and mantle, including 'mantle wedge', 'arc crust', and 'overriding plate'. Subducted and overriding materials are the two key elements of the IBM Subduction Factory. We define the Subduction Factory as that part of a convergent margin/subduction zone system where material is transferred from subducted lithosphere and sediments to the overlying mantle wedge for processing, from which some products are transferred to the surface. At one time, all of the eastern Philippine Sea Plate—from New Guinea to Japan—was an active convergent plate boundary responsible for subducting the Pacific Plate, but active-arc volcanism has ceased in the region south of the Challenger Cusp (Figure 1). Because we are concerned here only with that part of the IBM system associated with an active Subduction Factory, we focus on the IBM arc system east and north of the Challenger Cusp. We consider the width of the Subduction Factory to correspond to the active part of the IBM arc system, which lies between the trench—entry slot for all non-mantle contributions to the Subduction Factory—and a western limit that is defined by the most distant point from the trench where Subduction Factory products issue, such as lavas from back-arc spreading ridges or most distal parts of cross-chain volcanism. So defined, the IBM Subduction Factory is no more than 350 km wide. However, products of IBM generated during its Tertiary history extend as far west as the Palau-Kyushu Ridge, up to 1,000 km from the trench. We cannot define the depth limit of the IBM Subduction Factory, because we don't know at what depth transfer of material from subducted plate to overlying mantle ceases.

The northern boundary of the active IBM arc system follows the Nankai Trough northeastward and onto southern Honshu, joining up with a complex system of thrusts that continue offshore eastward to the Japan Trench. The intersection of the IBM, Japan, and Sagami trenches is the only trench-trench-trench triple junction on earth. Thus defined, the pertinent portion of the IBM arc system spans over 25° of latitude, from 11°N to 35°20'N. It is bounded on the east by a great trench, which ranges from almost 11km deep in the Challenger Deep (deepest spot on Earth's surface) to less than 3 km where the Ogasawara Plateau enters the trench (Figures 2, 3). The Ogasawara Plateau [*Smoot,* 1983] separates a deeper portion of the IBM Trench to the north—where depths of 8-9 km are common—from a shallower trench in the south, characterized by depths of 6-8 km. The difference in trench depths can be reasonably ascribed to the different histories of the Pacific Plate north and south of the Ogasawara Plateau, which was built near the boundary between Jurassic seafloor to the south and Cretaceous seafloor to the north (Figure 3). The presence of a generally deeper trench to the north and shallower trench to the south contrasts with the expected greater trench depth for the older, southern seafloor [*Stein and Stein,* 1992]. This may reflect the fact that seafloor to the south was rejuvenated by mid- and late- Cretaceous off-ridge volcanism whereas seafloor to the north was not.

Variations in age and bathymetry on the incoming plate north and south of the Ogasawara Plateau do not correspond in any obvious way with morphological variations along the arc system, which can be reasonably divided into thirds along strike. Boundaries are defined by the Sofugan Tectonic Line (~29°30'N) separating the Izu and Bonin segments, and by the northern end of the Mariana Trough back-

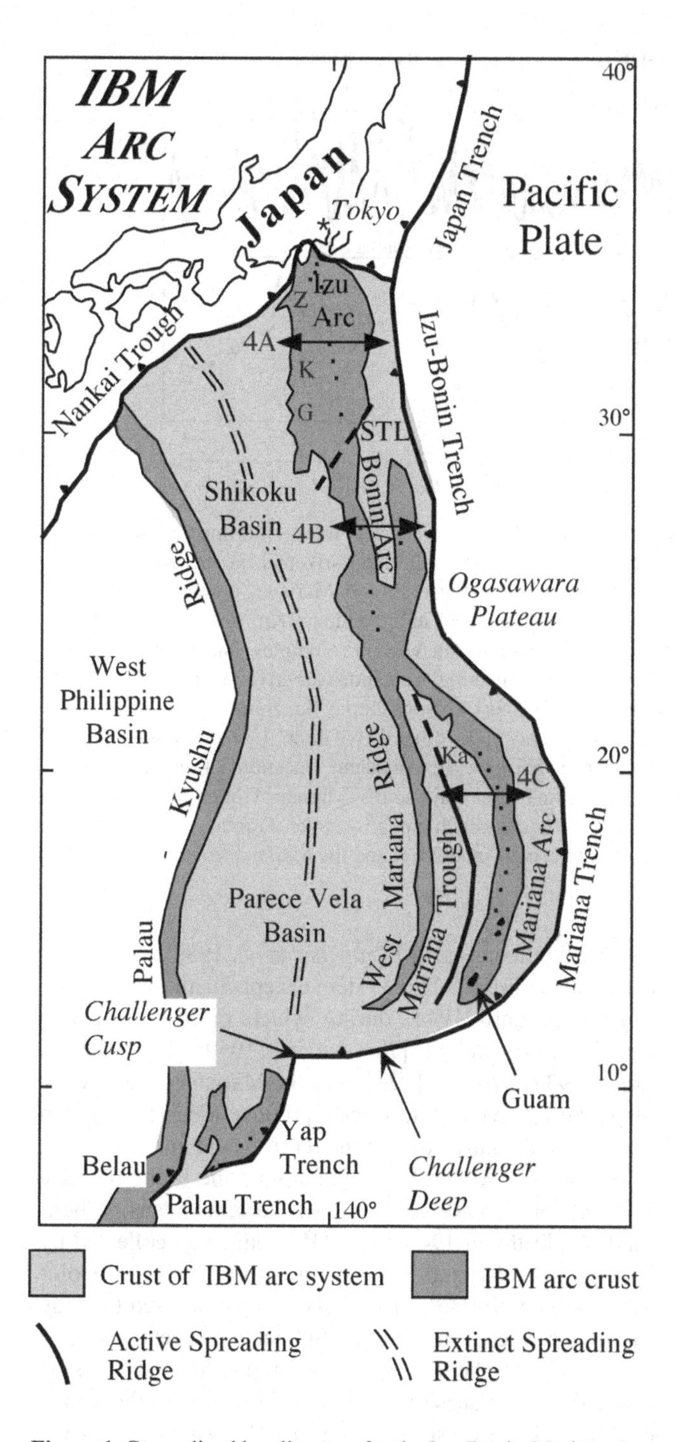

Figure 1. Generalized locality map for the Izu-Bonin-Mariana Arc system. Dashed line labeled STL = Sofugan Tectonic Line. Traverse shown in Figure 2 follows the magmatic arc from Japan through the Izu, Bonin, and Mariana magmatic arcs. Locations of profiles shown in Figure 4 are shown and labeled 4A, 4B, and 4C. Location of cross-chains is shown, with Zenisu (Z) at the far north, Kan'ei (K) and Genroku (G) farther south in the Izu arc, and Kasuga (Ka) in the Marianas.

arc basin (~23°N), that we take to define the boundary between the Bonin and Mariana segments. Forearc, active arc, and back arc are expressed differently on either side of these boundaries (Figure 4).

The forearc is that part of the arc system between the trench and the line of arc volcanoes that define the magmatic front. The IBM forearc is very narrow in the south, between the Challenger Cusp and Guam. Elsewhere, from Guam to Honshu, it is about 200 km wide. Uplifted sectors of the forearc situated near the magmatic front are called the 'frontal arc' [*Karig* 1971b]. The frontal arc is composed of Eocene igneous basement surmounted in the south by reef terraces of Eocene and younger age. Where exposed above sealevel, the frontal arc is expressed as the island chain from Guam north to Ferdinand de Medinilla in the Marianas and as the Bonin or Ogasawara Islands farther north. There is no accretionary prism associated with the IBM forearc or trench.

The magmatic axis of the arc is well defined from Honshu to Guam (Figure 2). This 'magmatic arc' is often submarine, reflecting fluctuations between 1 and 4 km water depth for the platform on which the arc volcanoes are built. Volcanic islands dominate the magmatic front from Oshima south to Sofugan (the Izu Islands). The Izu segment farther south also contains several submarine felsic calderas [*Yuasa and Nohara,* 1992]. The Izu segment is punctuated by inter-arc rifts [*Klaus et al.,* 1992; *Taylor et al.,* 1991]. The Bonin segment to the south of the Sofugan Tectonic Line contains mostly submarine volcanoes and a few island volcanoes, such as Nishino-shima. The highest elevations in the IBM arc (not including Fujiyama on the Izu Peninsula, where IBM comes onshore in Japan) are found in the southern part of the Bonin segment, where the extinct volcanic islands of Minami and Kita Iwo Jima rise to almost 1000 meters above sealevel. The bathymetric high associated with magmatic arc of the Izu and Bonin segments is often referred to as the Shichito Ridge in Japanese publications, and the Bonins are often referred to as the Ogasawara Islands. Volcanoes erupting lavas of unusual composition—the shoshonitic province—are found in the transition between the Bonin and Mariana arc segments. The magmatic arc in the Marianas is submarine to the north of Uracas (Northern Seamount Province), is mostly defined by volcanic islands in the Central Island Province, and again becomes submarine south of Anatahan (Southern Seamount Province) [*Bloomer et al.,* 1989b; *Meijer,* 1982].

The back-arc regions of the three segments vary markedly (Figure 4). The Izu segment is marked by several volcanic cross-chains which extend SW away from the magmatic front [*Ishizuka et al.,* 1998]. The Bonin arc segment has no back-arc basin, inter-arc rift, or rear-arc cross chains.

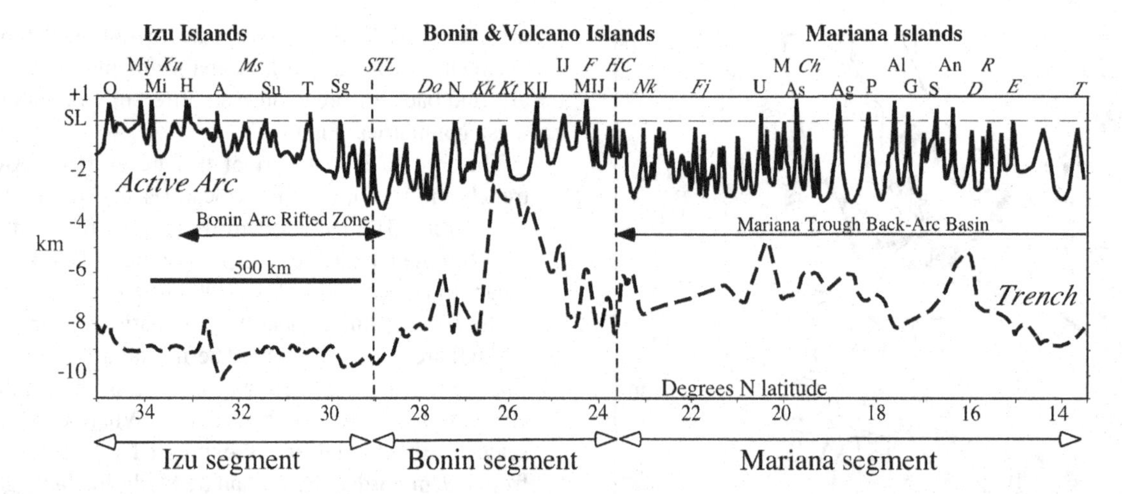

Figure 2. Along-strike profiles of the IBM Arc System, from Japan (left) to Guam (right). The thick solid line shows the bathymetry and topography along the volcanic axis of the active arc, with the thin dashed horizontal line marking sealevel. The approximate locations of the principal island groups (Izu, Bonin-Volcano, and Mariana) are shown. Submarine volcanoes (and the Sofugan Tectonic Line, STL) are given as italicized abbreviations: Ku, Kurose;Ms, Myojin-sho; Do, Doyo; Kk, Kaikata; Kt, Kaitoku;F, Fukutoku-oka-no-ba; HC, Hiyoshi Volcanic Complex, Nk, Nikko; Fj, Fukujin, Ch, Chamorro, D, Diamante; R, Ruby, E, Esmeralda; T; Tracy. Subaerial volcanoes are given as normal abbreviations: O, Oshima; My, Miyakejima; Mi, Mikurajima; H, Hachijojima; A, Aogashima; Su, Sumisujima, T, Torishima; Sg, Sofugan; N, Nishinoshima; KIJ, Kita Iwo Jima; IJ, Iwo Jima; MIJ, Minami Iwo Jima; U, Uracas; M, Maug; As, Asuncion; Ag, Agrigan; P, Pagan; Al, Alamagan; G; Guguan; S, Sarigan; An, Anatahan. Locations of important zones of intraarc and back-arc extension in the north (Bonin Arc Rifted Zone) and south (Mariana Trough Back-Arc Basin) are marked. The thick dashed line shows the maximum depth in the trench along its strike. Frontal arc elements are not shown, but consist of the Bonin or Ogasawara Islands between 26° and 28°N and the Mariana frontal arc islands between 13° and 16°N.

The Mariana segment is characterized by an active back arc basin [*Fryer,* 1995] known as the Mariana Trough. The Mariana Trough shows marked variations along strike, with seafloor spreading south of 19°15' and rifting farther north [*Gribble et al.,* 1998; *Martinez et al.,* 1995].

The IBM arc system south and west of Guam is markedly different than the region to the north. The forearc region is very narrow and the intersection of backarc basin spreading axis with the arc magmatic systems is complex. Earlier ideas that this was a region dominated by dextral strike-slip faults which trend E-W are not supported by more recent geophysical data sets [*Martinez et al.,* 2000].

3. EARLY STUDIES OF THE IBM ARC SYSTEM

Study of the IBM arc system began with geologic studies of the islands and has gone down from this. An excellent history of work in the Marianas dating back to 1792 is given by *Cloud et al.* [1956]. Modern geologic studies began with the identification of boninite lavas from the Ogasawara Islands by *Petersen* [1890]. The term 'boninite' was neglected for 90 years, but the rocks that Petersen described are now recognized as among the most interesting and important of all igneous rocks [*Crawford et al.,* 1989]. Japan controlled the entire IBM arc system except Guam for 30 years, from 1914 until 1944, during which time the volcanic islands were studied [*Tanakadate,* 1940; *Tsuya,* 1936; *Tsuya,* 1937]. *Hobbs* [1923] used the Marianas as an example of how 'asymmetric forces' produce arcuate mountain ranges, incidentally coining the term 'island arc'.

The Second World War raged around the Philippine Sea Plate and particularly along the IBM arc, with tens of thousands of deaths in 1944 and 1945. Soundings collected by U.S. warships led to the first overview of the marine geology of the region [Hess, 1948]. This paper showed the way for modern research on the Subduction Factory (*Hess, 1948,* p.422): "The trenches are the topographic expression of the dominant structural feature of the region, the down-buckling of the crust or tectogene. This structure...is the core and essence of mountain building, and all other major structures as well as the volcanic activity and the seismic activity of the region, are subordinate to and related to it." Hess mistakenly identified a peridotite belt along the IBM islands but would have been gratified to learn that ultramafic rocks are common, both as part of the inner trench wall and as part of serpentine mud volcanoes in the forearc.

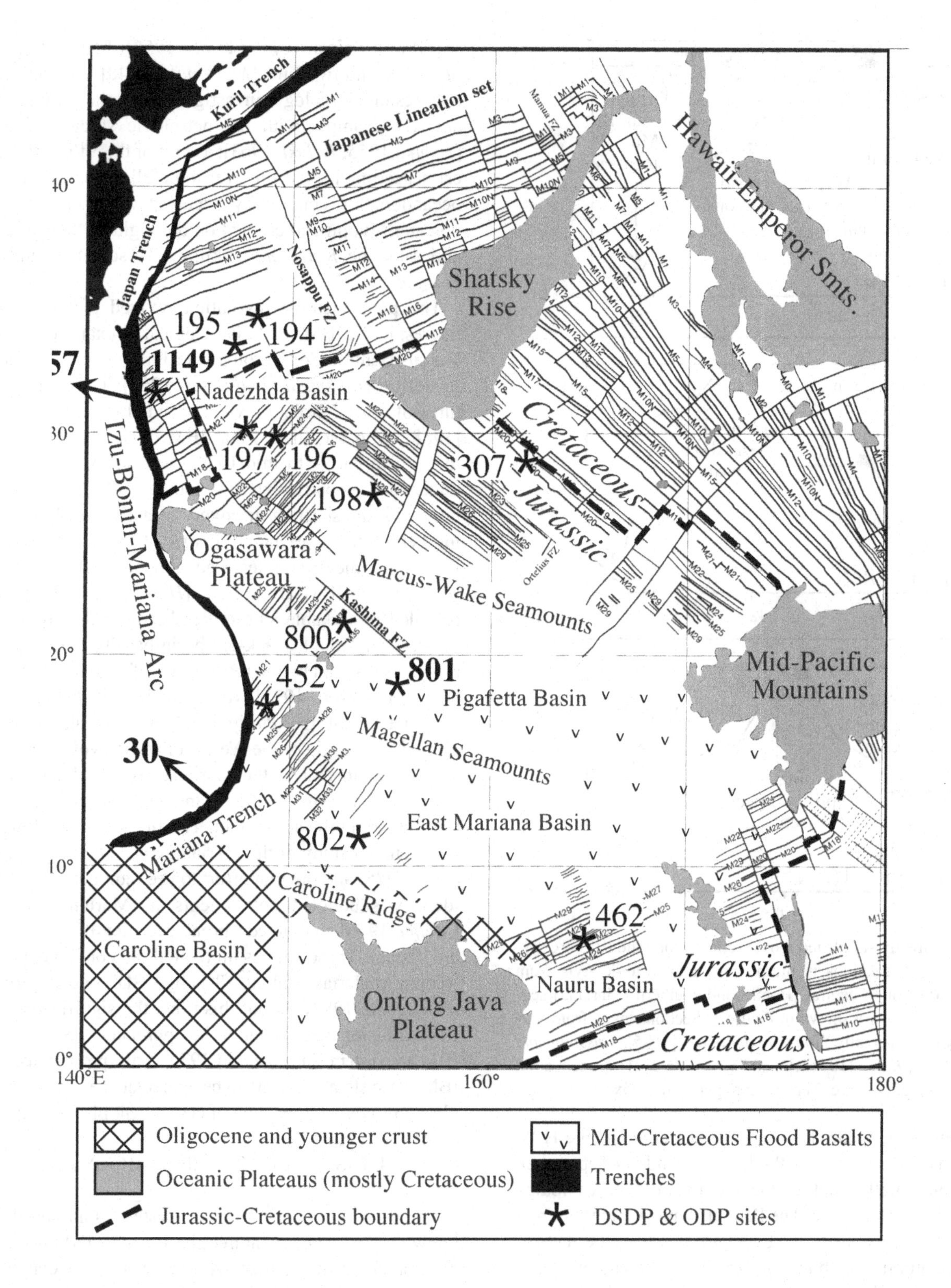

Figure 3. Simplified geologic and magnetic map of the western Pacific, modified after *Nakanishi et al.* [1992]. Relative motion of the Pacific Plate with respect to the Philippine Sea Plate is shown with arrows, numbers correspond to velocities (mm/year), after *Seno et al.* [1993].

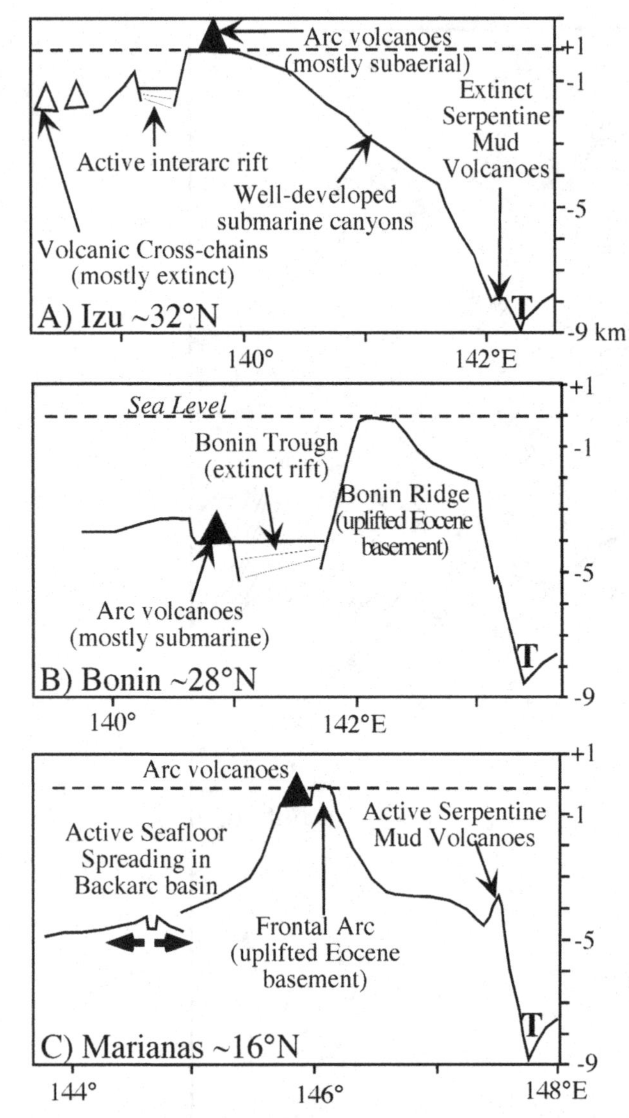

Figure 4. Greatly simplified cross sections along the IBM arc system, showing variations in morphology and tectonic style from north (Izu) through central (Bonin) and southern (Mariana) segments. Locations are shown in Figure 1. Sections taken from various reports, especially *Honza and Tamaki* [1985], *Hussong and Uyeda* [1982] *Karig* [1971b;], and *Taylor* [1992]. T= trench, solid triangle = magmatic arc. Vertical exaggeration ~ 25x.

As an historical aside, former U.S. President George H.W. Bush was shot down over the Bonin Islands in September, 1944, one small event in the bloody military campaign fought on this island arc. The IBM arc's most important role in world history may have occurred when the U.S.A. launched atomic bomb attacks on Hiroshima and Nagasaki from Tinian in the Marianas to end the Second World War.

Geologic studies of some of the major islands occurred during the post-war period, particularly on Guam [*Tracey et al.,* 1964], Saipan [*Cloud et al.,* 1956], Pagan [*Corwin et al.,* 1957] and Iwo Jima [*McDonald,* 1948]. Japanese scientists resumed geologic investigations in the northern IBM arc, for example with the petrologic studies of Isshiki [*Isshiki,* 1955; *Isshiki,* 1963]; as part of this effort, the greatest peacetime loss of life ever to befall the earth science or oceanographic community occurred when the submarine volcano Myojin-sho erupted on the night of Sept. 23, 1952 and sank the *Kaiyo-maru,* killing 31 scientists and crew [*Jaggar,* 1952].

The plate tectonic revolution revived global scientific interest in the IBM arc. Seismological re-examination of the Wadati-Benioff Zone around the Philippine Sea plate led this effort [*Katsumata and Sykes,* 1969]. The principal features of crustal structure were outlined by *Murauchi et al.* [1968], who established that IBM crust was about 20 km thick. The early- to mid-1970s was an exciting time for understanding particularly the southern IBM system, beginning with the startling discovery of active seafloor spreading behind the Mariana arc [*Karig,* 1971b]. This was confirmed by the first integrated geochemical and isotopic study of the IBM arc system [*Hart et al.,* 1972], which revealed remarkable similarities between the composition of Mariana Trough 'back-arc basin basalts' and mid-ocean ridge basalts, and by the recognition of high heatflow in the basin [*Anderson,* 1975]. Karig realized that many of the deep basins of the Philippine Sea Plate and elsewhere around the margin of the Western Pacific evolved in a manner similar to that of the Mariana Trough. *Karig* [1971a] established a firm historical and tectonic perspective for understanding the arc system. The first modern whole-rock geochemical study of the Mariana volcanic islands [*Larson et al.,* 1975] and the first effort to use radiogenic isotopes to identify subducted components in Mariana Arc lavas [*Meijer,* 1976] completed a broad geoscientific sketch of the arc system. These studies motivated the convergent margin 'biopsy' undertaken along 18°N by DSDP Leg 60 [*Hussong and Uyeda,* 1982]. About this time, the significance of the IBM collision zone began to be appreciated. Studies of the IBM arc system since the mid-1970's have proliferated, as the IBM arc system has come to be appreciated as an outstanding place to study the operation of convergent plate margins.

4. HISTORY OF THE IBM ARC SYSTEM

The evolution of the IBM arc system is among the best known of any convergent margin. Because IBM has always been an arc system under strong extension, its components encompass a broad area, from the Palau-Kyushu Ridge to the IBM trench (Figure 1). In general, the oldest components are preserved farther west, and a complete record of evolution is preserved in the forearc. The early history has been outlined by *Bloomer et al.* [1995] and *Stern and Bloomer* [1992], who concluded that the IBM subduction zone began as part of a hemispheric-scale foundering of old, dense lithosphere in the Western Pacific (Figure 5A). The beginning of large-scale lithospheric subsidence (not true subduction, but its precursor) is constrained by the age of igneous basement of the IBM forearc, including boninites, to have begun by about 50 Ma [*Bloomer et al.,* 1995; *Cosca et al.,* 1998]. During this stage, the forearc was the site of igneous activity, including the eruption of depleted tholeiites, boninites and associated low-K rhyodacites [*Hickey and Frey,* 1982; *Stern et al.,* 1991; *Taylor et al.,* 1994]. Magmatic activity localized along the present magmatic arc and allowed forearc lithosphere to cool in Late Eocene or Early Oligocene time. This marked the transition from lithospheric subsidence to subduction. True subduction—that is, the beginning of down-dip motion of the lithosphere—probably began about 43 Ma, when the Pacific Plate suddenly—and otherwise inexplicably—changed from a northerly to more westerly motion [*Richards and Lithgow-Bertelloni,* 1996].

This simple model of how subduction began depends in part on the orientation of spreading centers and fracture zones, needed to localize large-scale lithospheric foundering. *Stern and Bloomer* [1992] preferred an E-W spreading 'Tethyan' spreading regime and a N-S fracture zone in which the IBM subduction zone nucleated, but paleomagnetic data suggest that elements of the IBM arc system have rotated ~90° clockwise since the early Oligocene [*Hall et al.,* 1995]. It is difficult to reconcile the simple model of subduction initiation via fracture zone collapse with paleogeographic models inferred from paleomagnetic data. Large-scale rotations seemingly required by paleomagnetic data have recently been questioned because the IBM collision zone has been at about the same place for the last 13 Ma, an interval during which paleomagnetic data suggest the Philippine Sea Plate (and the collision zone) should have migrated several hundred km to the NE zone (see discussion and reply by *Ali and Moss* [1999] and *Takahashi and Saito* [1999]).

The beginning of true subduction localized the magmatic arc close to its present position, about 200 km away from the trench, and allowed the sub-forearc mantle to stabilize and cool (Figure 5B). Note that because of a lack of accretionary prism, the distance from the magmatic arc to the trench is thought not to have changed much since the magmatic arc first became established. The arc stabilized until about 30 Ma, when it began to rift to form the Parece Vela Basin. Spreading propagated north and south from this point, resulting in the 'bowed-out' appearance of the Parece Vela basin [*Taylor*, 1992]. Spreading began in the northernmost part of the IBM arc about 25Ma and propagated south to form the Shikoku Basin [*Kobayashi et al.,* 1995]. These two spreading systems met about 20 Ma and spreading continued until about 15 Ma (Figure 5C). The arc was disrupted during rifting but began to build again as a distinct magmatic system once seafloor spreading began. Arc volcanism, especially explosive volcanism, waned during much of this episode, with a resurgence beginning about 20 Ma in the south and about 17 Ma in the north [*Lee et al.,* 1995; *Taylor,* 1992]. Tephra from northern and southern IBM show that strong compositional differences observed for the modern arc have existed over most of the arc's history, with northern IBM being more depleted and southern IBM being relatively enriched [*Bryant et al.,* 1999]. A similar conclusion resulted from studies of coarser volcaniclastic materials [*Gill et al.,* 1994]. About 15 Ma, the northernmost IBM

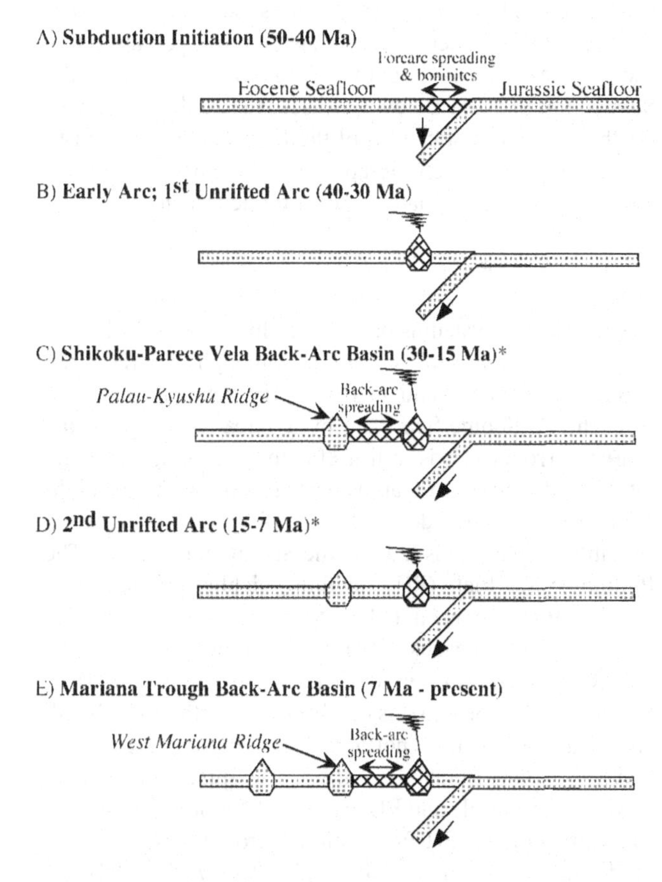

Figure 5. Simplified history of the IBM arc system. Shaded areas are magmatically inactive, cross-hatched areas are magmatically active.

began to collide with Honshu, probably as a result of new subduction along the Nankai Trough.

A new episode of rifting to form the Mariana Trough back-arc basin began sometime after 10 Ma, with seafloor spreading beginning about 3-4 Ma [*Bibee et al.,* 1980; *Yamazaki and Stern,* 1997] (Figure 5D). Because the old arc is physically removed from above the site of melting in the mantle when a back-arc basin forms, the present Mariana arc volcanoes cannot be older than 3-4 Ma but the Izu-Bonin volcanoes could be as old as ~25 Ma. The Izu inter-arc rifts began to form about 2 Ma [*Taylor,* 1992].

5. INPUTS INTO THE SUBDUCTION FACTORY

Everything delivered to the IBM trench, including the entire sedimentary section, is subducted. In addition, *Cloos* [1992] argued that the subduction channel shear zone beneath the Marianas widens with depth in the shallow parts of the subduction zone, so that subcretion is unlikely and the entire subducted assemblage—minus perhaps a few accreted seamounts—is delivered to and processed in the Subduction Factory. Below we discuss the delivery vectors for this input, modifications of the lithosphere just prior to its descent, and the stratigraphy and composition of sediments on the Pacific plate adjacent to the trench.

5.1. Plate Motions

The IBM arc system is part of the Philippine Sea Plate, at least to the first approximation. Although the IBM arc deforms internally—and in fact in the south is separated from the Philippine Sea Plate by a spreading ridge in the Mariana Trough—it is still useful to discuss approximate rates and directions of that plate with its lithospheric neighbors, because these define how rapidly and along what streamlines material is fed into the Subduction Factory. The Philippine Sea Plate (PH) has four neighboring plates: the Pacific (PA), Eurasian (EU), North America (NA), and Caroline (CR). There is minor relative motion between PH and CR; furthermore, CR does not feed the IBM Subduction Factory, so it is not discussed further. The North America plate includes northern Japan, but relative motion between it and Eurasia is sufficiently small that relative motion between PH and EU explains the motion of interest. The Euler pole for PH-PA as inferred from the NUVEL-1A model for current plate motions [*DeMets et al.,* 1994] lies about 8°N, 137.3°E, near the southern end of the Philippine Sea Plate. PA rotates around this pole CCW ~1°/Ma with respect to PH. The PH-PA Euler pole inferred from earthquake slip vectors lies very close to the NUVEL-1A pole [*Seno et al.,* 1993]. This means that relative to the southernmost IBM, PA is moving NW at about 20-30 mm/y, whereas relative to the northernmost IBM, PA is moving WNW and twice as fast (Figure 3).

The NUVEL-1A Euler Pole for EU-PH lies about 51°N, 160.5°E, off the coast of Kamchatka. Earthquake slip vectors [*Seno et al.,* 1993] indicate a more southerly pole location, NE of Hokkaido. Recent GPS (global positioning satellite) campaigns [*Kotake et al.,* 1998] indicate that the EU-PH Euler Pole lies ~42°N, 152.5°E, a bit south of Seno's slip vector pole, with a CW rotation of 1.5°/Ma for PH relative to EU. PH is moving NW at about 40 mm/year relative to Hokkaido, so velocities increase to the south along the EU-PH plate boundary. To a first approximation, the GPS, NUVEL-1A, and earthquake slip vector data agree, except for Guam, which GPS data indicate is moving much more slowly to the NW (~10mm/year) than predicted from NUVEL-1A. This is due to the opening of the Mariana Trough, for which full spreading rates of 3-5 cm/year are inferred [*Bibee et al.,* 1980; *Hussong and Uyeda,* 1982; *Yamazaki and Stern,* 1997].

It should be noted that the IBM arc is not experiencing trench 'roll-back', that is, the migration of the trench towards the ocean. The trench is moving towards Eurasia, although a strongly extensional regime is maintained in the IBM arc system because of rapid PH-EU convergence. The nearly vertical orientation of the subducted plate beneath southern IBM exerts a strong "sea-anchor" force that strongly resists its lateral motion. Back-arc basin spreading is thought to be due to the combined effects of the sea-anchor force and rapid PH-EU convergence [*Scholz and Campos,* 1995].

The obliquity of convergence between PA and the IBM arc system change markedly along the IBM arc system. Figure 6 shows the plate convergence inferred from earthquake slip vectors [*McCaffrey,* 1996]. This is nearly strike-slip in the northernmost Marianas, adjacent to and south of the northern terminus of the Mariana Trough, where the arc has been 'bowed-out' by back-arc basin opening, resulting in a trench which strikes approximately parallel to the convergence vectors. Convergence is strongly oblique for most of the Mariana Arc system but is more nearly orthogonal for the southernmost Marianas and most of the Izu-Bonin segments. Figure 6 also shows the arc-parallel slip rate in the forearc, which reaches a maximum of 30mm/yr in the northern Marianas. According to McCaffrey, this is fast enough to have produced geologically significant effects, such as unroofing of high-grade metamorphic rocks, and provides one explanation for why the forearc in southern IBM is tectonically more active than that in northern IBM.

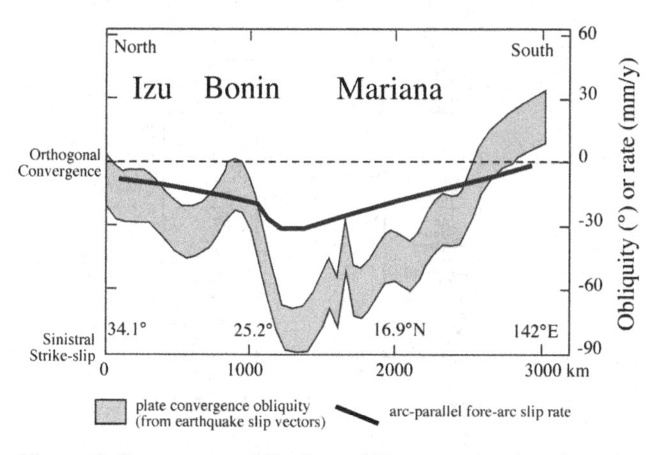

Figure 6. Convergence obliquity and fore-arc extension along the IBM arc, modified after *McCaffrey* [1996]. Arc segments Izu, Bonin, and Mariana are shown. Note that the northernmost Mariana arc is characterized by nearly strike-slip motion, whereas much of the Izu-Bonin and southern Mariana Arc is characterized by plate convergence that is almost orthogonal.

5.2. Trench and Forearc Bulge

The trench and the associated forearc bulge mark where raw materials are fed into the Subduction Factory and where the lithosphere is first bent or broken to begin its downward trajectory. The phenomenon of the trench is well known and will not be further discussed, other than to note that the IBM trench lacks sediment fill [*Bellaiche,* 1980]. Flexure of lithosphere about to enter a subduction zone is commonly observed just outboard of the trench, where it is referred to as the "outer rise". The flexural bulge rises to about 300m above the surrounding seafloor just before the trench, and this flexure can be used to constrain the state of stress across the convergent margin. Based on application of an elastic-plastic model, *Bodine and Watts* [1979] concluded that horizontal stresses are low across the northern Mariana Trench, intermediate across the southern Mariana and Bonin trench segments, and high across the Izu trench segment. Vertical stress was modeled by *Harry and Ferguson* [1991], who noted larger vertical stresses for the southern Marianas than the southern Izu segment. *Harry and Ferguson* [1991] also concluded that the load inducing flexure results from a narrow zone of interplate coupling located 15-20 km arcward of the trench axis, and that the weight of the subducted slab does not contribute to the support of the trench-outer rise bathymetry. Faulting associated with development of the flexural bulge provides one last opportunity for fluids to be introduced deep into the about-to-be-subducted plate. These occur as 'stair-stepping' normal faults which trend parallel to the trench [*Bellaiche,* 1980]. Open cracks or tension fractures showing no displacement are associated with the normal faults [*Ogawa et al.,* 1997]. These fractures are thought to allow seawater to penetrate deeply into mantle lithosphere, forming serpentinites that carry substantial water into the Subduction Factory [*Peacock,* 2001].

5.3. Geology and Composition of the Westernmost Pacific Plate

The IBM arc system now subducts mid-Jurassic to Early Cretaceous lithosphere, with younger lithosphere in the north and older lithosphere in the south. It is not possible to directly know the composition of subducted materials presently being processed by the IBM Subduction Factory—what is now 130 km deep in the subduction zone entered the trench 4-10 million years ago. However, the composition of the Western Pacific seafloor—sediments, crust, and mantle lithosphere—varies sufficiently systematically that, to a first approximation, we can understand what is now being processed by studying what now lies on the seafloor east of the IBM trench.

The seafloor east of the IBM arc system can be subdivided into a northern 'smooth' portion and a southern 'rough' portion, separated by the Ogasawara Plateau (Figure 3). These large-scale variations mark distinct geologic histories to the north and south. The featureless north is dominated by the Nadezhda Basin. In the south, crude alignments of seamounts, atolls, and islands allows define three great, WNW-ESE trending chains to be identified [*Winterer et al.,* 1993]. The Marcus-Wake-Ogasawara Plateau extends westward from the Mid-Pacific Mountains for 3,000 km, culminating at the trench as the great Ogasawara Plateau. A second ridge intersects the trench about 20°N, and is known as the Magellan Chain. The Magellan chain is complex, with the Dutton Ridge in the north and the Magellan seamounts in the south. The southernmost ridge is the Caroline Islands Ridge, which intersects the Yap trench to the south of the Marianas, at about 9°N, and includes the only high volcanic islands on the NW Pacific plate—the islands of Truk, Ponape, and Kosrae. The first two chains formed by off-ridge volcanism during Cretaceous time, whereas the Caroline Islands chain formed over the past 20 million years. last 20 million years by the westward passage of the Pacific Plate over a mantle hot-spot. The Marcus-Wake and Magellan chains are typified by flat-topped guyots, with summit depths on the order of 2000 m, whereas the Caroline Islands chains are much shallower. Between these chains lie two important abyssal plains or basins: the Pigafetta Basin lies between the Marcus-Wake and Magellan chains, and the East Mariana Basin lies between the Magellan and Caroline chains. The East Mariana Basin is a flat-floored basin, with a maximum water depth of 6.1 km [*Whitman,* 1986], and the deepest part of the Pigafetta Basin is only slightly shallower.

The age of Western Pacific seafloor has been interpreted from magnetic anomalies [*Nakanishi et al.,* 1992] and confirmed by drilling. Three major sets of magnetic anomalies have been identified in the area of interest [*Larson and Chase,* 1972]. Each of these lineation sets comprises M-series anomalies (mid-Jurassic to mid-Cretaceous) that are essentially Pacific Plate growth rings. These anomaly sets indicate that the small, roughly triangular Pacific plate grew by spreading along three ridges [*Bartolini and Larson,* 2001]. The oldest identifiable lineations are M33 to M35 [*Nakanishi,* 1993] or perhaps even M38 [*Handschumacher et al.,* 1988]. It is difficult to say how old these lineations and the older crust might be; the oldest magnetic lineations for which ages have been assigned are M29 (157 Ma; [*Channell et al.,* 1995]). Magnetic lineations as old as M29 are not known from other oceans, and the area in the Western Pacific that lies inside the M29 lineation—that is, crust older than M29—is on the order of $3x10^6$ km^2, about a third of the size of the United States. ODP site 801 lies on crust considerably older than M29 and the MORB basement there yields Ar-Ar ages of 167±5 Ma [*Pringle,* 1992]. The oldest sediments at site 801C are Callovian or latest Bathonian (~162 Ma [*Harland et al.,* 1990]).

Seafloor spreading in the Pacific during the Cretaceous evolved from a more E-W 'Tethyan' orientation to the modern N-S trend. This occurred during mid-Cretaceous time, a ~35-40 Ma interval characterized by a lack of magnetic reversals known as the Cretaceous Superchron. Subsequently, the location of N-S trending spreading ridges relative to the Pacific Basin migrated progressively to the east throughout Cretaceous and Tertiary time, resulting in the present marked asymmetry of the Pacific, with very young seafloor in the Eastern Pacific and very old seafloor in the Western Pacific.

Sediments being delivered to the IBM trench are about 500m thick (Figure 7). Away from seamounts, the pelagic sequence is dominated by chert and pelagic clay, with little carbonate. Carbonates are important near guyots, common in the southern part of the region. Interestingly, Cenozoic sediments are unimportant except for deposits of volcanic ash adjacent to Japan and carbonate associated with the relatively shallow Caroline Ridge and plate. Strong seafloor currents are probably responsible for this erosion or non-deposition. The compositions of sediments being subducted beneath the northern and southern parts of the IBM arc are significantly different, because of the Cretaceous off-ridge volcanic succession in the south that is missing in the north.

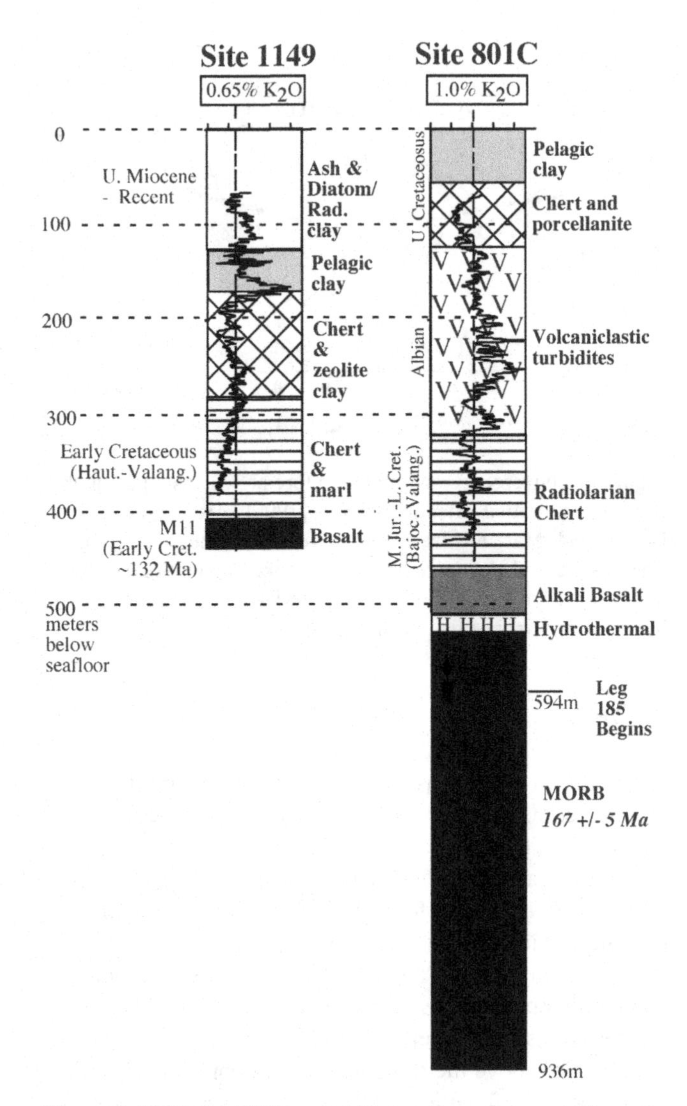

Figure 7. ODP sites 801C and 1149, locations shown on Figure 3. 801C is thought to be representative of sediments being subducted beneath the Marianas and 1149 of those going down beneath Izu-Bonin. Superimposed on simplified lithological columns are K contents determined using the natural gamma down hole logging tool. Note that enrichments in K and other incompatible elements in sediments being subducted beneath the Mariana arc is largely due to the thick sequence of Albian (Cretaceous) volcaniclastics http://www-odp.tamu.edu/publications/185_IR/VOLUME/CHAPTERS/IR185_04.PDF

Lavas and volcaniclastics associated with an intense episode of intraplate volcanism correspond in time closely to the Cretaceous Superchron. Off-ridge volcanism became increasingly important approaching the Ontong-Java Plateau. There are 100-400m thick tholeiitic sills in the East Mariana Basin and Pigafetta Basin [*Abrams et al.,* 1993], and at least 650m of tholeiitic flows and sills in the Nauru

Basin, near ODP Site 462 [*Shipley et al.,* 1993]. *Castillo et al.* [1994] suggest that this province may reflect the formation of a mid-Cretaceous spreading system in the Nauru and East Mariana basins. Farther north, deposits related to this episode consist of thick sequences of Aptian-Albian volcaniclastic turbidites shed from emerging volcanic islands, such as preserved at DSDP site 585 and ODP sites 800 and 801. A few hundred meters of volcaniclastic deposits probably characterizes the sedimentary succession in and around the East Mariana and Pigafetta basins. Farther north still, at DSDP sites 196 and 307 and ODP site 1149, there is little evidence of mid-Cretaceous volcanic activity. It appears that the Aptian-Albian volcanic episode was largely restricted to the region south of present 20°N latitude. Paleomagnetic and plate kinematic considerations place this extensive region of off-ridge volcanism in the present vicinity of Polynesia, a broad region of off-ridge volcanism, shallow bathymetry, and thin lithosphere known as the 'Superswell' [*McNutt et al.,* 1990; *Menard,* 1984].

The compositions of sediments being subducted beneath the IBM arc have been examined by various workers, most recently by *Plank and Langmuir* [1998]. The mean composition for Izu-Bonin sediments may soon be superseded by data from ODP site 1149 (see Initial Reports for the Leg: http://www-odp.tamu.edu:80/publications/185_IR/185ir.htm). Relative to MORB, both the Izu-Bonin and Mariana sediment composites show elevated concentrations of large-ion lithophile elements Rb, Ba, Th, K, La Pb (Figure 8). Relative to the Global Subducting Sediment mean (GLOSS; [*Plank and Langmuir,* 1998]), IBM sediments are depleted in most LIL and moderately incompatible elements, although the sedimentary column being subducted beneath the Marianas is slightly enriched in incompatible High Field Strength Elements (HFSE) Nb and Ta. Sediments subducted beneath the IBM arc have significant Ce anomalies (Ce/Ce*=0.33 to 0.9; [*Lin,* 1992] and high Pb/Ce (~0.2) and Th/U (2-5) but low Ba/La (10-20) and Sr/Nd (<10) [*Plank and Langmuir,* 1998]). These sediments contain much higher $^{18}O/^{16}O$ and more radiogenic Sr than mantle-derived lavas or those from the IBM arc system (Figure 9A). Subducted sediments are therefore expected to have ^{18}O-rich compositions ($\delta^{18}O$ = 15-30‰; [*Stern and Ito,* 1983]), although the sedimentary section subducted beneath the Marianas should have somewhat lower $\delta^{18}O$ because of the contribution of Cretaceous off-ridge lavas and volcaniclastics. Western Pacific sediments have quite radiogenic Sr and nonradiogenic Nd compared to mantle-derived lavas, including both those from Cretaceous off-ridge volcanoes now being subducted beneath the southern IBM, and those of IBM arc system lavas (Figure 9B). In contrast, the Pb isotopic compositions of the sediments define a relatively tight cluster whereas W. Pacific volcanics show a very wide range (Figure 9C,D). IBM arc lavas most resemble subducted volcanics on a Sr-Nd isotopic diagram but are more similar to sediment compositions on Pb isotopic diagrams.

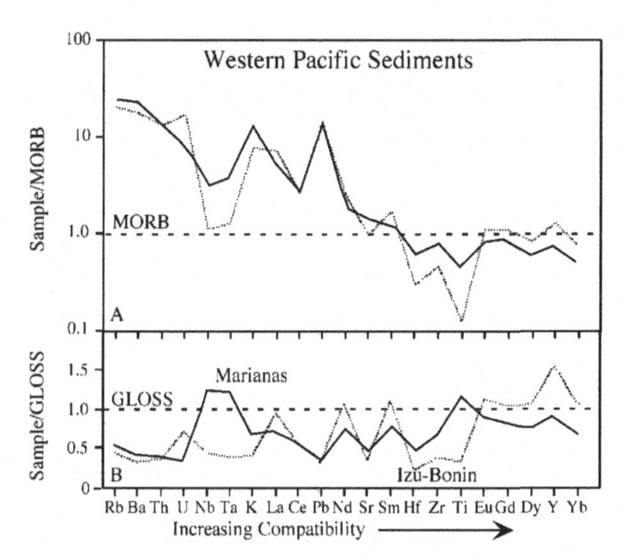

Figure 8. Chemical composition of sediments being fed into the IBM Subduction Factory. Composites for sediments outboard of Izu-Bonin (dotted line) and Mariana (solid line) are after *Plank and Langmuir* [1998], (A) normalized to abundances in MORB [*Hofmann,* 1988] and (B) normalized to Global Subducting Sediments (GLOSS; [*Plank and Langmuir,* 1998]).

About 470m of oceanic crust was penetrated at 801C during Legs 129 and 185 of the Ocean Drilling Program (Figure 6). These are typical MORB that were affected by low-temperature hydrothermal alteration. The isotopic compositions of Sr, Nd and Pb are also typical of Pacific MORB (Figure 9; [*Castillo et al.,* 1992]). This crust is overlain by a 3m thick, bright yellow hydrothermal deposit [*Alt et al.,* 1992] and about 60 m of alkali olivine basalt, 157.4±0.5 Ma old [*Pringle,* 1992], enriched in light REE and other incompatible elements.

6. GEOPHYSICS OF THE SUBDUCTED SLAB AND MANTLE

The deep structure of the IBM system has been imaged using a variety of geophysical techniques. This section provides an overview of these data, including a discussion of mantle structure at depths (>200 km) below the typical definition of the Subduction Factory. Although deep structure may not seem to directly influence shallow subduction processes, deep mantle features exert important controls on features coupled to the shallow subduction factory, such as slab descent, slab rheology, mantle circulation, and trench migration.

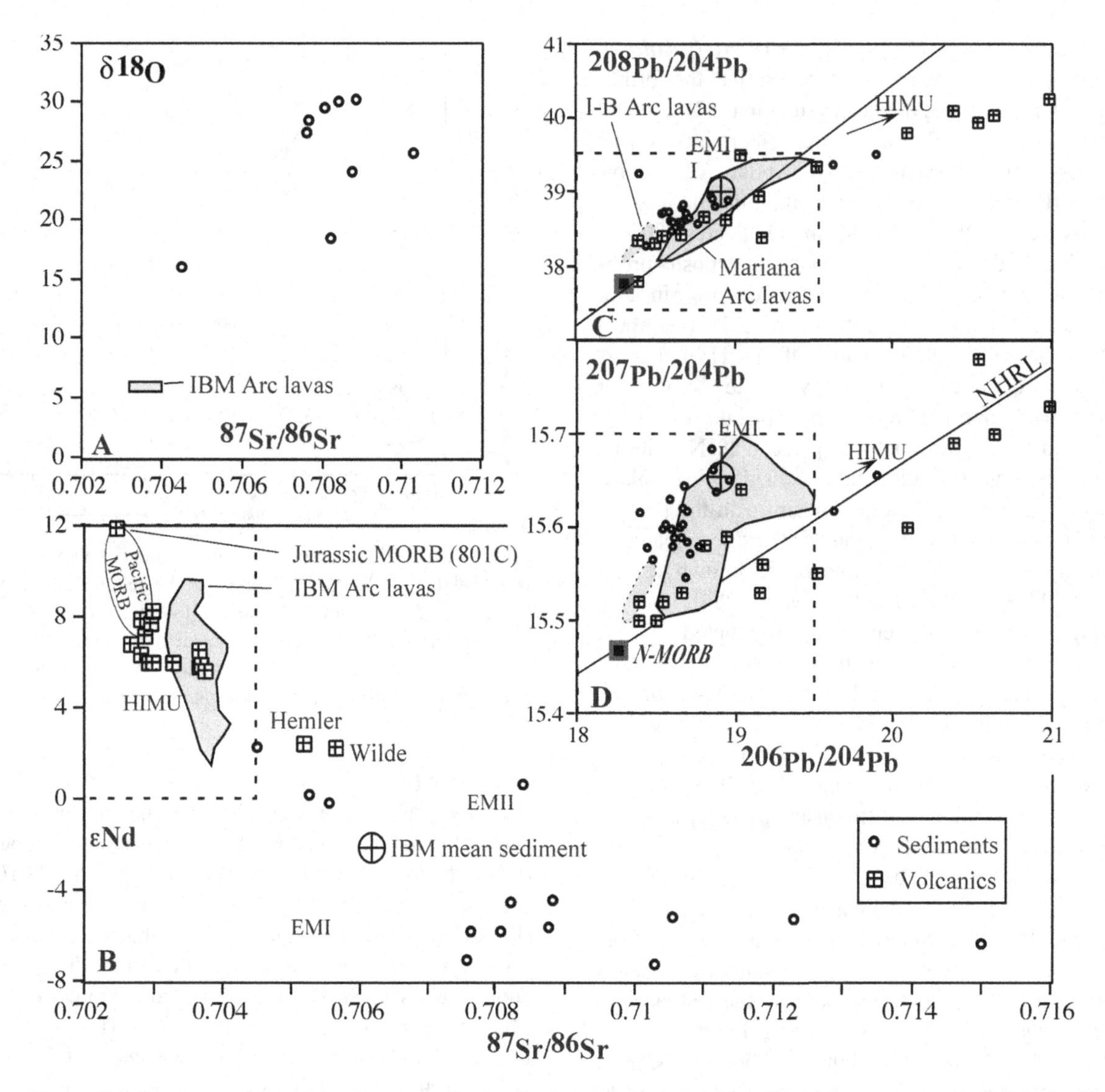

Figure 9. Isotopic composition of sediments and volcanics being fed into the IBM Subduction Factory. A) O-Sr isotopic systematics for sediments (data from *Stern and Ito* [1983], *Woodhead and Fraser* [1985], and *Woodhead* [1989]). B) Sr-Nd isotopic systematics for W. Pacific sediments (data from *Lin* [1992], *Stern and Ito* [1983], and *Woodhead* [1989] and means calculated for Jurassic seafloor and Jurassic alkali basalt at ODP 801C [Castillo et al., 1992]; alkalic sill at ODP 800 [*Castillo et al.,* 1992]; Cretaceous off-ridge tholeiites at ODP 802 [*Castillo et al.,* 1994], Nauru Basin (ODP 462; [*Castillo et al.,* 1991], and the Ontong Java Plateau [*Mahoney and Spencer,* 1991]; Magellan seamounts (Himu, Hemler, Golden Dragon); Wake seamounts (Wilde, Miami, Lamont, and Scripps) [*Staudigel et al.,* 1991]; Marshall Islands [*Davis et al.,* 1989]; and Caroline Islands [*Hart,* 1988]. Notice that data for lavas clusters about εNd ~+6 to +8, $^{87}Sr/^{86}Sr$ ~ 0.7027- 0.7040 except for the Jurassic MORB at 801C and Hemler and Wilde seamounts. Approximate locations of mantle reservoirs EMI, EMII, and HIMU are also shown. Sediment samples with $^{87}Sr/^{86}Sr < 0.706$ contain a high proportion of volcaniclastic material. Location of weighted mean for sediment being subducted beneath IBM [*Plank and Langmuir,* 1998] is shown, along with fields for fresh MORB and IBM arc lavas. C) $^{208}Pb/^{204}Pb$ vs. $^{206}Pb/^{204}Pb$ for W. Pacific sediments and volcanics, data sources as above. D) $^{207}Pb/^{204}Pb$ vs. $^{206}Pb/^{204}Pb$. Also shown is the estimated composition of IBM mean sediment from *Plank and Langmuir,* [1998] and the Northern Hemisphere Reference Line (NHRL) after *Hart* [1984]. Dashed rectangles in B-D show the areas plotted in Figure 21.

6.1. Seismicity

Spatial patterns of seismicity are essential for locating and understanding the morphology and rheology of subducting slabs, and this is particularly true for the IBM Wadati-Benioff Zone. Pioneering efforts [*Katsumata and Sykes,* 1969] outlined the most important features of the IBM Wadati-Benioff Zone. Their study detected a zone of deep earthquakes beneath the southern Marianas and provided some of the first constraints on the deep, vertical nature of subducting Pacific lithosphere beneath southern IBM. They also found a region of reduced shallow seismicity (≤70 km) and a lack of deep (≥300 km) events beneath the Volcano Islands adjacent to the junction of the Izu-Bonin and Mariana trenches, where the trench is nearly parallel to the convergence vector.

More recently, Engdahl et al. [1998] provided an earthquake catalog containing improved locations (Figure 10). This data set shows that, beneath northern IBM, the dip of the Wadati-Benioff Zone steepens smoothly from ~40° to ~80° southwards, and seismicity diminishes between depths of ~150 km and ~300 km (Figures 11a-c). The subducted slab beneath central IBM (near 25°N; Figure 11c) is delineated by reduced seismic activity that nevertheless defines a more vertical orientation that persists southward (Figures 11d-f).

Deep earthquakes, here defined as seismic events ≥300 km deep, are common beneath parts of the IBM arc system (Figures 10, 11). Deep events in the IBM system are less frequent than for most other subduction zones with deep seismicity, such as Tonga/Fiji/Kermadec and South America. Beneath northern IBM, deep seismicity extends southward to ~27.5°N, and a small pocket of events between 275 km and 325 km depth exists at ~22°N. There is narrow band of deep earthquakes beneath southern IBM between ~21°N and ~17°N, but south of this there are extremely few deep events. Although early studies assumed that seismicity demarcated the upper boundary of the slab, more recent evidence has shown that many of these earthquakes occur within the slab. For instance, a study by *Nakamura et al.* [1998] showed that a region of events beneath northernmost IBM region occur ~20 km beneath the top of the subducting plate. They propose that transformational faulting, which occurs when metastable olivine changes to a more compact spinel structure, produces this zone of seismicity. Indeed, the faulting mechanism for deep earthquakes is a hotly debated topic (e.g., [*Green and Houston,* 1995]), and has yet to be resolved.

Double seismic zones have been detected in the IBM subduction zone, but their significance is not understood. Beneath southern IBM, *Samowitz and Forsyth* [1981] found a double seismic zone at 80 to 120 km depth, with the two zones separated by 30-35 km. Earthquake focal mechanisms indicate that the more seismic upper zone is in downdip compression, while the lower zone is in downdip extension. This double seismic zone is located at a depth where the curvature of slab is greatest; at greater depths it unbends into a more planar donfiguration. *Samowitz and*

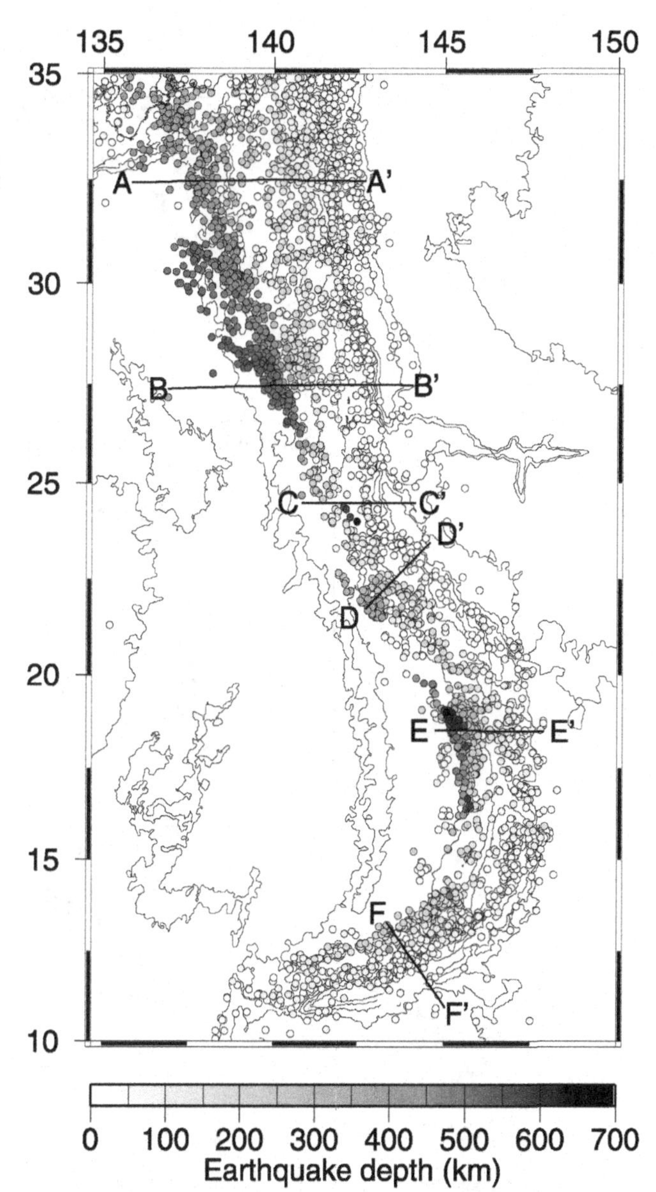

Figure 10. Map view of bathymetry and seismicity in the Izu-Bonin-Mariana Subduction Factory using the earthquake catalog of *Engdahl et al.* [1998]. Circles denote epicentral locations; lighter circles represent shallower events, darker circles represent deeper events. Black lines denote cross-sectional areas depicted in Figure 11. Large variations in the seismic structure of the region are evident.

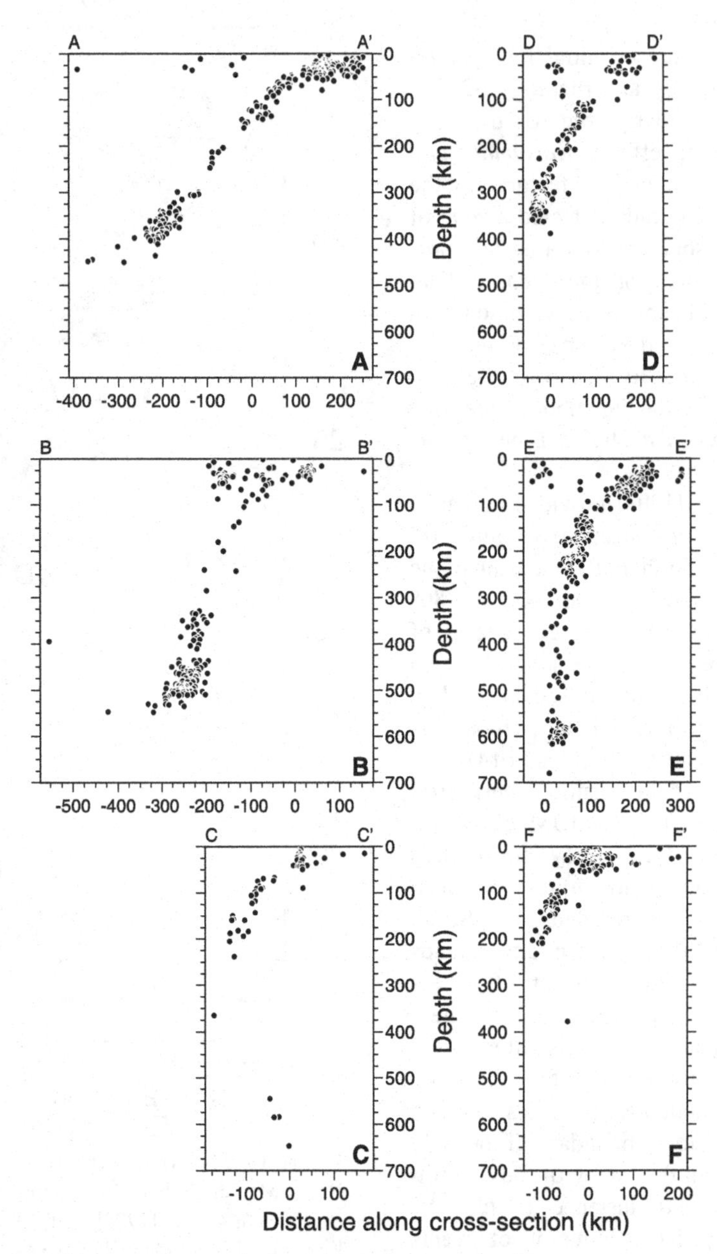

Figure 11. Cross-sectional views of the IBM system using the earthquake catalog of *Engdahl et al.* [1998]. Black circles represent hypocentral locations in volume ~60 km to each side of the lines shown in Figure 10. Large variations in slab dip and maximum depth of seismicity are apparent. Distance along each section is measured from the magmatic arc. A) Northern Izu-Bonin region. Slab dip is ~45°; seismicity tapers off from ~175 km to ~300 km but increases around 400 km, and terminates at ~475 km. B) Central Izu-Bonin region. Slab dip is nearly vertical; seismicity tapers off from ~100 km to ~325 km but increases in rate and extends horizontally around 500 km, and terminates at ~550 km. C) Southern Izu-Bonin region. Slab dip is ~50°; seismicity is continuous to ~200 km, but a very few anomalous events are evident down to ~600 km. D) Northern Mariana region. Slab dip is ~60°; seismicity is continuous to ~375 km and terminates at ~400 km, but a very few anomalous events are evident down to ~600 km. E) Central Mariana region. Slab dip is vertical; seismicity tapers off slightly between ~275 km and ~575 km, but is essentially continuous. A pocket of deep events around 600 km exists, as well as 1 deep event at 680 km. F) Southern Mariana region. Slab dip is ~55°; seismicity is continuous to ~225 km, with an anomalous event at 375 km.

Forsyth [1981] suggested that unbending or thermal stresses in the upper 150 km of the slab may cause the seismicity. For northern IBM, *Iidaka and Furukawa* [1994] detected a double seismic zone between depths of 300 km and 400 km, which also is separated by 30-35 km between the upper and lower zones. They concluded that the double seismic zone results from transformational faulting of a metastable olivine wedge in the slab. Recent work suggests that compositional variations in the subducting slab may also contribute to double seismic zones [*Abers,* 1996], or that double seismic zones represent the locus of serpentine dehydration in the slab [*Peacock,* 2001].

6.2. Seismic Anisotropy and Mantle Flow

For nearly two decades, seismic anisotropy in the form of shear wave splitting has been used to examine deformation in subduction zones [*Ando et al.,* 1983; *Fouch and Fischer,* 1996; *Peyton et al.,* 2001]. Two principal goals of mapping seismic anisotropy are a) to determine the history of strain and deformation in the mantle, and b) to evaluate coupling between lithosphere and asthenosphere. The distribution of seismicity in most subduction zones provides an ideal opportunity to examine the lateral and depth distribution of subducting and overriding plate fabric, mantle flow, and possible slab-mantle coupling.

Evaluation of shear wave splitting in phases such as S and SKS are routinely performed for most regions using standard data analysis techniques [*Silver and Chan,* 1991]. These measurements produce two independent pieces of information: the azimuth of propagation of the faster wave (the fast direction) and the delay between the fast and slow phases (the splitting time). The primary assumption is that olivine, the principal component of the upper mantle, is responsible for the splitting, and that the fast direction (*a*-axis of olivine) is aligned with either the flow direction or the direction of maximum extension [*Ribe,* 1989; *Zhang and Karato,* 1995] in the upper 200-400 km of the mantle, where dislocation creep is the dominant deformation mechanism. The magnitude of the splitting time reflects how completely olivines are aligned and how thick is the layer of aligned olivines.

Studies of seismic anisotropy near the IBM system have allowed workers to examine the distribution of strain in the mantle wedge and the subducting slab. *Fouch and Fischer* [1996] evaluated shear wave splitting in the northern IBM subduction zone and found fast directions of NW-SE to WNW-ESE (i.e., roughly convergence-parallel) and splitting times of ~0.5 s. These splitting values for local events are comparable to observations for most convergent margins. The geometry of ray paths did not allow them to determine the relative anisotropy of the slab, mantle wedge, and overriding plate; however, most of the ray paths sample mantle wedge structure. Modeling of the splitting data indicated that anisotropy may exist to depths no greater than ~410 km. These results indicate a relatively small strength of anisotropy, and are consistent with a model in which induced flow in the back-arc wedge is the primary cause of anisotropy beneath Izu-Bonin (Figure 12).

Xie [1992] and *Fouch and Fischer* [1998] examined mantle anisotropy in the southern IBM subduction zone. Using short-period waveforms and a qualitative analysis technique, *Xie* [1992] determined E-W fast directions and splitting times of ≤0.35 s beneath Guam. *Fouch and Fischer* [1998] utilized broadband waveforms and a more accurate method of waveform analysis, and found ~NW-SE fast directions and splitting times ranging from 0.1 s to 0.4 s. The steep dip of the Marianas slab makes it difficult to distinguish anisotropy in the slab from that in the mantle wedge, but the use of broadband data facilitated the analysis of frequency dependence in fast directions. Modeling of these data suggested that a combination of fossil anisotropy in the subducting slab, slab-induced flow in the mantle wedge, and anisotropy possibly due to back-arc extension in the overriding plate contribute to the shear wave splitting (Figure 12).

6.3. Slab Morphology and Lower Mantle Penetration

Although first-order slab characteristics can be constrained using earthquake locations (section 6.1), more detailed information regarding lateral and vertical variations in slab morphology is required for a complete analysis of the Subduction Factory. Additionally, the depth to which subducting slabs penetrate is fundamental for determining the nature of mantle convection in the earth, as it helps constrain details of mass and heat exchange between the upper and lower mantle [*Silver et al.,* 1988; *Anderson,* 1989]. Using a variety of analysis techniques, several studies have focused on the IBM system as an end-member example of the large variations in deep seismic structure found worldwide. Most of these studies conclude that the slab beneath northern IBM stagnates near the upper-lower mantle boundary (~660 km depth; Figure 12). In contrast, the slab beneath southern and central IBM descends steeply through the upper mantle, and appears to penetrate into the lower mantle.

Residual sphere images of the southern IBM subduction zone provided some of the first conclusive evidence of slab penetration into the lower mantle [*Creager and Jordan,* 1986]. These images showed that the slab beneath southern IBM penetrates to depths of 900-1000 km, which can be

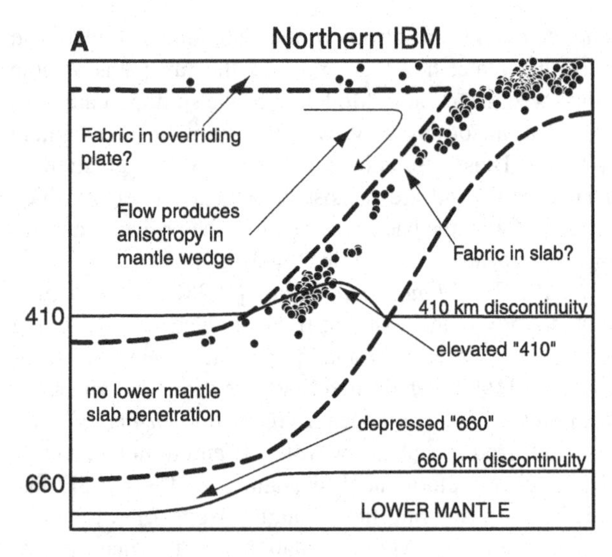

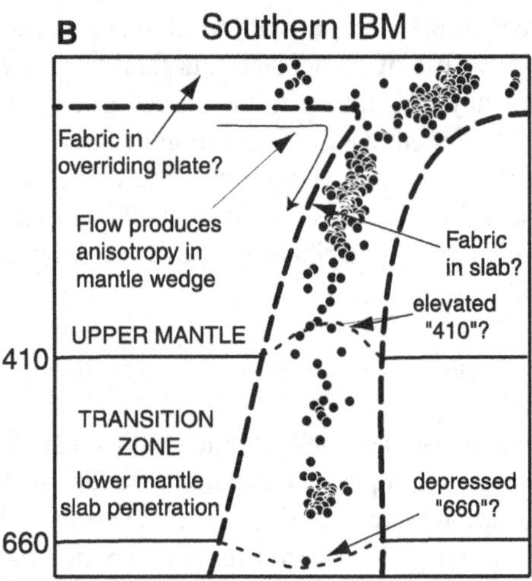

Figure 12. Cartoon representation of seismic observations in the IBM system. Black circles show representative seismicity for each region. Thick dashed lines show rough interpretation of tomographic images for each area. A) Northern IBM (near section A in Figure 11). Slab dip is variable, but averages around 50°. Tomographic images suggest that the slab does not penetrate the lower mantle, but extends horizontally to the west. The 410 km seismic discontinuity (thin line at 410) appears to be locally elevated; the 660 km seismic discontinuity (thin line at 660) appears to be depressed for an extended distance to the west. Seismic anisotropy indicates the presence of fabric in the mantle wedge, as well as possibly in the overriding plate and subducting slab. B) Southern IBM (near section E in Figure 11). Slab dip is slightly variable, but generally becomes vertical. Tomographic images suggest that the slab penetrates the lower mantle. Regional variations of upper mantle discontinuities have not yet been documented, but a locally elevated "410" and a locally depressed "660" (thin dashed lines) might be expected. Seismic anisotropy indicates the presence of fabric in the mantle wedge and subducting slab, as well as possibly in the overriding plate.

explained solely by thermal effects [*Fischer et al.,* 1988]. Waveform modeling [*Brudzinski et al.,* 1997] indicates that a horizontal high-velocity feature near 660 km depth exists beneath the northern IBM arc, consistent with the hypothesis that cold, seismically fast slab material rests on top of the upper-lower mantle boundary here. *Castle and Creager* [1999] used a migration method to detect scatterers at ~1000 km depth beneath the central IBM arc, which they interpreted to be produced by subducted crust.

In the past decade, high-resolution images resulting from travel-time tomography have shown that the deep subducted slab beneath northern and southern IBM behaves differently. The first of these images was produced by *Van der Hilst et al.* [1991], who showed that the slab beneath northern IBM deflects at the upper-lower mantle boundary near 660 km, while the slab beneath southern IBM descends through the boundary into the lower mantle (Figure 12). These results have been confirmed by *Fukao et al.* [1992] using slightly different data analysis and inversion techniques. *Van der Hilst and Seno* [1993] suggested that whether or not the slab penetrates the 660 km discontinuity may be due to the rate of trench migration: the Izu-Bonin trench appears to have migrated northwestward, while the Mariana trench appears to have remained stationary. In addition, *Widiyantoro et al.* [1999] used P- and S-wave data to show that the rigidity of the subducting slab beneath southern IBM is highly variable. Bulk-sound (a combination of P- and S-wave velocity) images beneath northern IBM show a clear spreading of fast-velocity material along the upper-lower mantle discontinuity, while the shear-wave images do not reveal such a layer. In contrast, both bulk-sound and shear-wave images of the southern IBM region clearly show that fast-velocity material penetrates into the lower mantle. The tomography results of *Widiyantoro et al.* [1999] for northern IBM are somewhat surprising given the implications of recent work by *Takenaka et al.* [1999], which suggest that shear-wave anomalies due to temperature variations should be more easily detected than bulk sound anomalies. However, *Widiyantoro et al.* [1999] suggest that rheological variations within slabs may help determine whether or not these are able to penetrate the 660-km discontinuity, and that temperature and compositional variations may not be the only explanation for the seismic results.

6.4. Upper Mantle Structure

Analysis of seismic discontinuities in the mantle transition zone between depths 410 km and 660 km (hereafter referred to as the "410" and the "660") provide important constraints on the thermal, chemical, and rheological properties of subducting slabs. Additionally, constraints on the depth of the "660" help resolve the extent of slab penetration into the lower mantle. In general, these discontinuities are believed to be due to mineralogical phase changes in olivine due to variations in pressure and temperature. The 410 discontinuity marks the exothermic olivine—β-spinel (wadsleyite) transformation and the 660 km discontinuity marks the endothermic γ-spinel (ringwoodite)—perovskite + magnesiowüstite structural transformation [*Lay,* 1994]. Several studies have shown that near subduction zones, upper mantle discontinuities are generally anticorrelated, with relatively shallower "410" and deeper "660". These observations suggest that thermal anomalies due to subducting slab material are responsible for topography on the upper mantle discontinuities.

Studies of the transition zone beneath the IBM system corroborate the tomographic and other seismic results discussed in section 6.3. In general, global studies exhibit a thickened transition zone and/or a depressed "660" near the IBM system [*Flanagan and Shearer,* 1998; *Gu et al.,* 1998; *Revenaugh and Jordan,* 1991; *Shearer and Masters,* 1992; *Vidale and Benz,* 1992], consistent with a slab-induced thermal anomaly. Detailed studies targeting the northern IBM region have yielded results similar to the global studies. *Collier and Helffrich* [1997] utilized data from several seismic networks to show that the transition zone in this region may be thickened by as much as 100 km, corroborating the results of *Revenaugh and Jordan* [1991] and *Shearer* [1993]. Work by *Castle and Creager* [1998] showed that the "660" beneath northern IBM mimics the high-resolution tomographic results. The discontinuity is depressed westward from the trench, again suggesting that the slab lies flat on the upper-lower mantle boundary. Beneath central IBM the depressed portion of the "660" is narrow, again suggesting penetration of slab material into the lower mantle. Other studies focusing on the "660"—beneath northern IBM have shown similar depressions of the discontinuity, between 30 km and 80 km [*Tajima and Grand,* 1998; *Wicks and Richards,* 1993], but these studies do not possess the lateral resolution to illuminate details of slab penetration into the lower mantle.

7. IBM SUBDUCTION FACTORY OUTPUTS

Outputs from the Subduction Factory are found as low-temperature fluids released in the forearc and as melts erupted and emplaced along the magmatic front of the arc, in arc cross-chain volcanoes, and in back-arc basin rifts and spreading centers. These outputs define the width of the IBM Subduction Factory, which reaches a maximum of ~300km at 18°N. In the following sections, we focus on the outputs of the IBM Subduction Factory. We first present what we know about the crustal structure of the IBM arc system, then focus on fluid outputs from the forearc before describing melts from along the magmatic front, cross chain volcanoes, and back-arc basin rifts and spreading systems.

7.1. IBM Crustal Structure

In 1992, Japanese scientists performed an ambitious wide-angle OBS experiment at 32°15'N [*Suyehiro et al.,* 1996; *Takahashi et al.,* 1998](Figure 13A). This defined a relatively thin (~20 km) arc crust, that may be simplistically viewed as composed of four layers of approximately equal thickness. The upper layer with Vp=1.5-5.8 km/s represents sediments and volcanics and overlies a middle-crustal layer with Vp=6.2±0.2 km/s. The lower crust has an upper half with Vp=6.6-7.2 km/s (presumably mafic plutonics) overlying a layer with Vp=7.3±0.1 km/s. The 6.2 km/s and the 7.3 km/slayers have excited most comment because layers of these velocities are not prominent in seismic data from other intraoceanic arcs.

The presence of the 6.2 km/s layer is exciting because this velocity corresponds to a wide range of intermediate-composition plutonic rocks and is close to the mean velocity of the continental crust (6.4 km/s) [*Christensen and Mooney,* 1995], thereby suggesting that continental crust can be directly created in an IOCM. Insight into the lithology of the IBM middle crust is provided by exposures in the Izu Collision Zone (Section 7.6). Dredges from fault scarps bounding the rifted eastern edge of the Palau-Kyushu Ridge (Figure 1) (arc crust rifted from the IBM by formation of the Shikoku and Parece-Vela basins) at 30°N also reveal extensive submarine exposures of tonalite [*Taira et al.,* 1998]. Similar silicic rocks were dredged by P. Fryer (pers. comm., 2000) from major fault scarps in the Mariana forearc southeast of Guam. Given these exposures, *Taira et al.* [1998] interpret the 6.0-6.4 km/sec layer to be tonalitic. A similar 6 km/sec layer was also found in the IBM arc near 31°30'N & 33°N by *Hino* [1991] and in the northern Palau-Kyushu Ridge, suggesting its regional presence in at least the northern part of IBM system [*Taira et al.,* 1998].

Velocities > 7.2 km/s, at < 20 km depth in an active arc where velocities are reduced by the high lower-crustal temperatures, are best interpreted as due to mafic granulites, pyroxenites or dunites [*Christensen and Mooney,* 1995], or some mixture of mafic and ultramafic cumulates as exposed in the roots of some arc complexes [*Burns,* 1995]. In the forearc, these same velocities may correspond to variably-serpentinized upper mantle.

Although *Suyehiro et al.* [1996] provide the best cross-section yet of IBM arc crust, important geologic differences

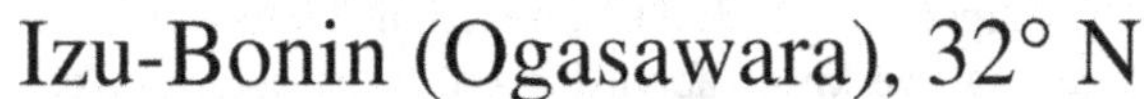

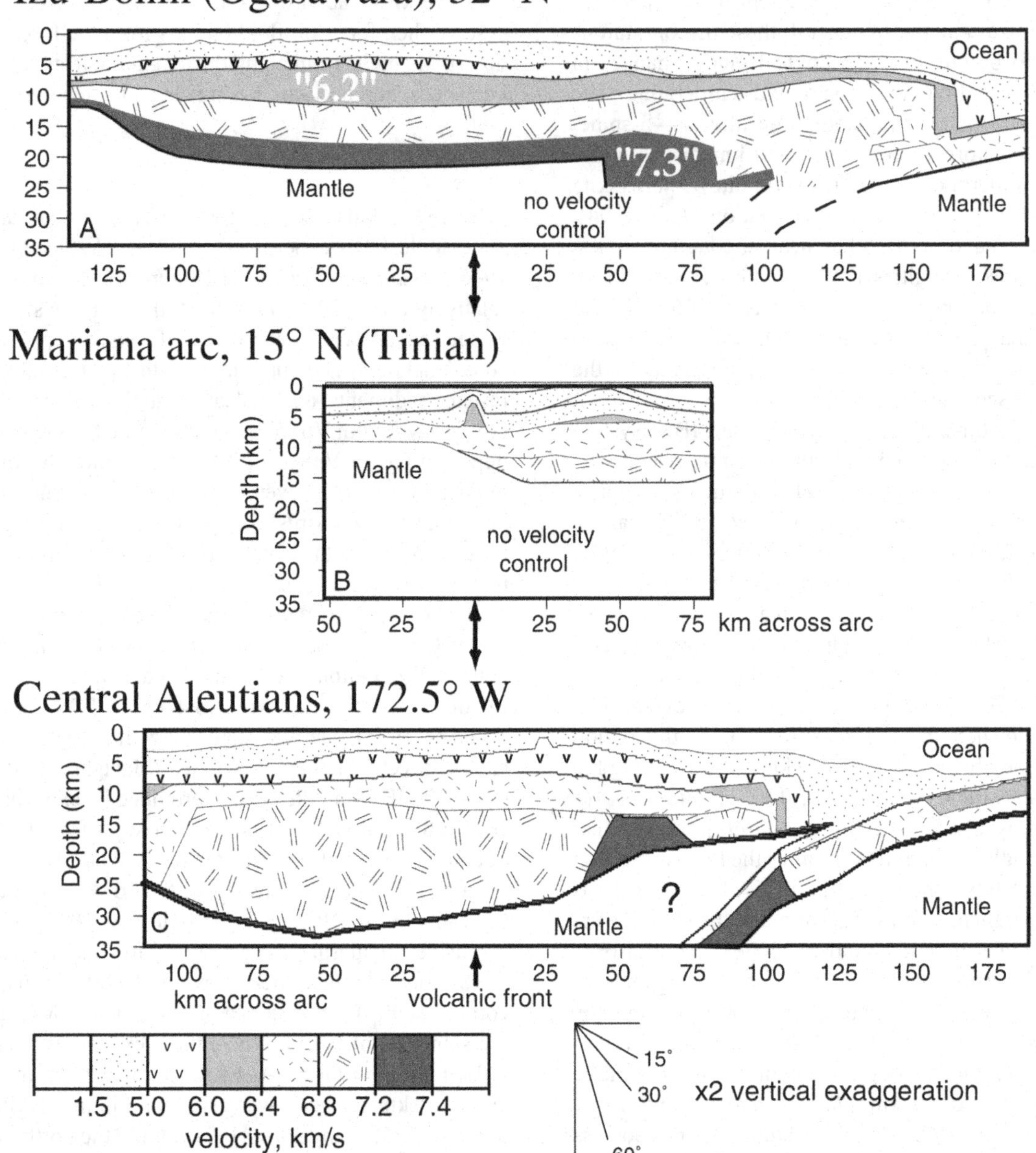

Figure 13. Arc-perpendicular cross-sections aligned along the volcanic front, redrawn to a uniform vertical exaggeration (x2) and scale, and with uniform velocity shading. "6.2" and "7.3" identify the 6.2±0.2 km/s middle crustal layer and 7.3±0.1 km/s deepest crustal layer of the IBM arc. Neither layer is prominent in the other two models. Redrawn after A: [*Suyehiro et al.,* 1996]; B: [*Lange,* 1992]; C: [*Holbrook et al.,* 1999].

along strike make it questionable whether this structure can be extrapolated to the crust of southern IBM. A German OBS study across the Mariana arc-backarc at 15°N *Lange* [1992] provides a crude velocity model and a crustal thickness of only 16 km (Figure 13b). It is intriguing that neither the 6.2 km/sec nor the 7.3 km/sec layers seen at 32°15'N were detected beneath the Marianas. This section broadly agrees with earlier sonobuoy measurements at 18°N [*LaTraille and Hussong,* 1980] but gravity data have been used to infer a ≥20-km thick crust [*Sager,* 1980]. This leaves open the question of whether a deeper mafic layer might be present but not recognized because of limited penetration of these seismic experiments. Recognizing high-velocity lower-crustal layers is a particular problem for older surveys in which only first-arrival times were modeled [*Hyndman and Klemperer,* 1989]. Even in modern surveys, the lower-crustal velocity and thickness of such layers can be hard to constrain tightly [*Suyehiro et al.,* 1996; *Takahashi et al.,* 1998]. However, comparison with the Aleutians, the only other arc for which such high-quality data exist [*Fliedner and Klemperer,* 1999, 2000; *Holbrook et al.,* 1999](Figure 13C) suggests that very fast (7.3 km/sec) layers are not ubiquitous beneath arcs. Layers with similar velocities are known from the lower crust of rift provinces [*Christensen and Mooney,* 1995] and it is possible that the high velocities observed for the IBM lower crust represent mafic/ultramafic underplating related to the active rifting in the Izu-Bonin segment of the arc [*Klaus et al.,* 1992], rather than to arc magmatism. Layers in the upper and middle crust are easier to recognize than those in the lower crust, so it seems very likely that, if a similar 5-km-thick 6.2 km/s layer was present, it would have been detected by the seismic experiments in all three regions (Figure 13), suggesting that this layer in particular may be restricted to the northern part of the IBM chain.

The crustal structure of the IBM arc at 32°15'N is well enough known that *Taira et al.* [1998] were able to calculate a mean crustal composition of 56% SiO2, and a mean crustal growth rate of 80 km^3/km/Ma. This calculation required interpreting which parts of the seismically-defined arc crust are pre-existing oceanic crust, or back-arc basin crust, or (in the forearc) hydrated mantle. As discussed previously, important uncertainties remain in extrapolating these results along the full length of the IBM system. Until a new generation of multiscale seismic experiments is undertaken, we can only approximate the composition and volume of the arc.

The Mariana Trough back-arc basin has a crustal structure that is similar to normal oceanic crust. Three independent efforts in the late 1970s used seismic refraction to examine the structure of Mariana Trough crust at 18°N. *Bibee et al.* [1980] examined crustal structure within 50 km of the ridge, in the region underlain by new oceanic crust. Another study [*LaTraille and Hussong,* 1980] examined regional crustal structure, from the W. Mariana Ridge across the Mariana Trench and eastward onto the Pacific Plate. The first study concluded that the crust was 5 to 7 km thick, with a velocity structure typical of other slow-spreading ridges. Seismically slow mantle was not found beneath the spreading axis. In contrast, the study of *LaTraille and Hussong* [1980] only imaged the Moho in the western part of the Mariana Trough, more than 80 km away from the spreading axis and beneath the region underlain by rifted arc crust. They concluded that the crust was about 6 km thick in the westernmost Trough and thinned towards the spreading axis. They also noted that the crust was extremely heterogeneous and that it was associated with low-velocity upper layers similar to those found on very young crust of other slow-spreading ridges. These two results are thus somewhat contradictory.

The third refraction study focused on the structure of the Pagan Fracture Zone, near 18°N. This study used 6 ocean-bottom seismometers to generate two orthogonal refraction crustal profiles, with one line parallel and one normal to the ridge axis [*Sinton and Hussong,* 1983]. They found that the crust along the spreading ridge away from the fracture zone is between 4 and 6 km thick, while the crust beneath the fracture zone is 1 to 2 km thinner. They also found a thick section of mid- to lower crust with P wave velocities of 6.6 to 6 km/sec, similar to that of normal oceanic crust. The results of *Sinton and Hussong* [1983] also suggest that slower mantle underlies the region east of the spreading ridge relative to that in the west (Vp = 7.5 vs. 8.0 km/sec).

Mariana Trough crust thickens north of 21°N. Gravity data were interpreted by Ishihara and Yamazaki [1991] to indicate that the crust thickens from 6 to 9 km near 21°20'N (Region III, Figure 14A&C) to about 15 km near 23°30' to 24°N (Region I, Figure 14A&C). This is also the part of the Mariana Trough where the extension axis shoals from about 3.5 km depth to about 2.0 km depth. Recovery of mantle peridotite from deep graben near 20°N [Stern et al., 1996] shows that the crust is thin at least this far north, and the fact that the extension axis shoals north from this point also suggest crustal thickening to the north.

7.2. The Forearc

The IBM forearc is about 200 km wide and differs from many other forearcs by not having an accretionary prism. This is because IBM is isolated from continent-derived sediments. In addition, because the supply of sediments to the IBM forearc is limited to pelagic 'rain' and volcaniclastics shed from the volcanic arc, the infrastructure of especially the outer forearc is not deeply buried. The relatively naked IBM forearc provides unique opportunities to sample the rocks that formed when subduction began, to track the evolution of Subduction Factory products, and to examine fluids squeezed and sweated out of the downgoing slab.

There are important differences along the IBM forearc that reflect its evolution and relative tectonic stability. The threefold subdivision proposed for the IBM arc system in Section 2 provides a useful subdivision for the forearc region as well. The Izu section has no well-developed frontal arc ridge comparable to those of the Bonin and

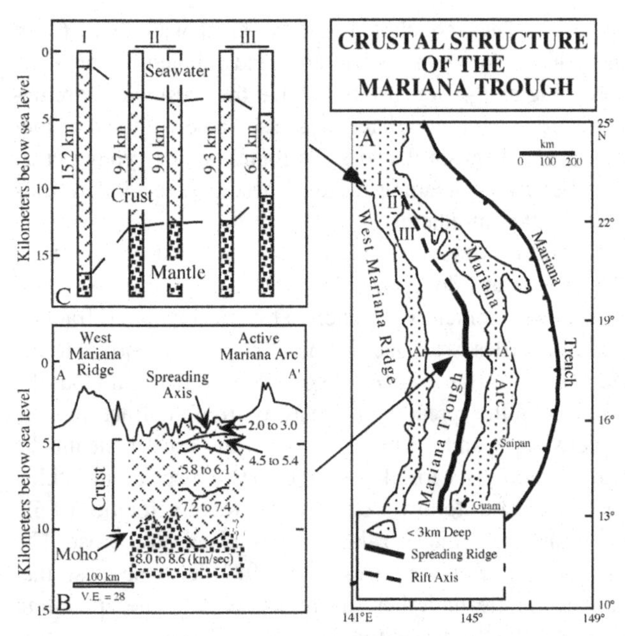

Figure 14. Crustal structure of the Mariana Trough. (A) Shows the location of B and C. (B) Crustal structure beneath the central Mariana Trough is combined from the seismic refraction results of *Bibee et al.* [1980] in the east, centered beneath the spreading ridge, and the results of *LaTraille and Hussong* [1980] in the west. Numbers correspond to P wave seismic velocities, in km/sec. The Moho is defined by velocities greater than 8 km/sec, although the 7.4 km/sec velocity in the east may be anomalously slow mantle, perhaps due to the presence of partial melt. (C) Crustal structure beneath the northern Mariana Trough is based on interpretation of gravity data [*Ishihara and Yamazaki,* 1991]. Note the rapid thinning southward to crustal thicknesses corresponding to normal oceanic crust just south of 22°N.

Mariana segments (Figure 4). The long stability of the Izu forearc is demonstrated by several well-developed submarine canyons up to 150 km long across the forearc which began to form in early Miocene time [*Klaus and Taylor,* 1991]. The few conical features near the trench that are likely to be serpentine mud volcanoes are covered with pelagic sediments and seem to be inactive, for example Torishima seamount drilled during ODP Leg 125. The Bonin forearc is characterized by a well-developed forearc basin (Bonin Trough), which formed as a result of Oligocene rifting [*Taylor,* 1992]. The forearc basin becomes progressively more filled or poorly developed away from the Bonin segment, and the free-air gravity expression of the Bonin Trough becomes muted [*Honza and Tamaki,* 1985]. The Bonin Ridge is capped by the Bonin or Ogasawara islands and is associated with one of the highest free-air gravity anomalies on Earth (up to 360 mGal) [*Honza and Tamaki,* 1985].

Dredging of the IBM inner trench wall has recovered mostly igneous rocks, including peridotites, boninites, and arc tholeiites [*Bloomer,* 1983; *Bloomer and Hawkins,* 1987; Ishii, 1985; *Tararin et al.,* 1987]. Similar sequences of boninite and arc tholeiite were penetrated during DSDP and ODP drilling [*Hickey and Frey,* 1982; *Pearce et al.,* 1992]. These rocks are interpreted as forming during the initial stages in the development of the IBM subduction zone during Eocene time [*Stern and Bloomer,* 1992]. There is also evidence that pre-Eocene rocks make up a small portion of the forearc basement, including Pacific MORB crust [*DeBari et al.,* 1999] and accreted Cretaceous seamounts and pelagic sediments [*Bloomer and Hawkins,* 1983; *Johnson et al.,* 1991].

The frontal arc islands of the southern Marianas (Guam, Saipan, etc.) and the Bonin Islands (Chichijima, Hahajima, etc.) provide an opportunity to study forearc evolution in detail. These islands show the same sort of tholeiitic/boninitic igneous basement as that found in the submerged forearc [*Hickey-Vargas and Reagan,* 1987; *Reagan and Meijer,* 1984; *Taylor et al.,* 1994; *Umino,* 1985], although the island sequences also have abundant low-K rhyodacites [*Meijer,* 1983; *Taylor et al.,* 1994]. The northern Bonin Islands are dominated by boninite series volcanics whereas the southern Bonin Islands consist of tholeiite and calc-alkaline lavas [*Taylor et al.,* 1994], demonstrating an intermingling of magma types similar to that observed downhole in DSDP Site 458. The Eocene volcanic sequences in the Bonin Islands are overlain by shallow water limestones of Oligocene-Early Miocene age [*Taylor et al.,* 1994]. Mariana frontal arc islands have a more complete record of the interplay between sealevel and tectonics in a well-developed sequence of reefal limestones of Oligocene to Holocene age [*Cloud et al.,* 1956; *Tracey et al.,* 1964]. The Balanos volcanic member of mid-Miocene age (13.5 Ma) on Guam [*Meijer et al.,* 1983] is part of a thicker sequence of volcanic and sedimentary rocks known as the Umatac Formation. This is the on-land equivalent of the thick Miocene ash blanket drilled at DSDP Site 60 and corresponds to the mid-Miocene magmatic maximum identified by *Lee et al.* [1995]

The Mariana forearc is tectonically unstable, reflecting the combined effects of stretching parallel to it due to being 'bowed-out' by back-arc extension and because of oblique convergence (Figure 6; [*McCaffrey,* 1996; *Stern and Smoot,* 1998]). A detailed survey of the inner forearc near 22°N [*Wessel et al.,* 1994] confirmed that the northern Mariana forearc is dominated by sinistral shear [*McCaffrey,* 1996]. This tectonic instability is an important reason why the Mariana forearc has no identifiable submarine canyons and is also the main reason why fluids and serpentinized mantle rise to the surface at several places here. Because an important motivation for selecting IBM as a MARGINS focus area was that this is the only place known where we can sample fluids released at shallow depth from a subduction zone, we concentrate in the following discussion on the Mariana forearc (Figure 15).

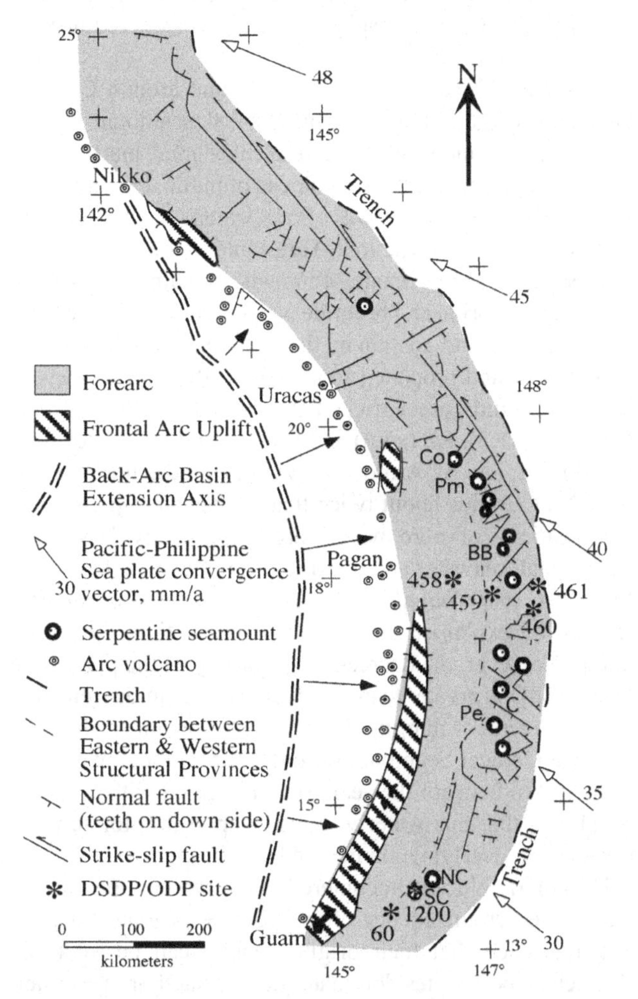

Figure 15. The IBM forearc, between 13° and 25°N, shown in gray. Generalized structural features and location of serpentine mud volcanoes are taken from *Stern and Smoot* [1998]. Dashed line marks the boundary between rough and smooth terrains of *Stern and Smoot* [1998], or between Eastern and Western Structural Provinces of *Mrozowski and Hayes* [1980]. Arrows radiating out from the back-arc basin extension axis show how differential extension causes decreasing radius of curvature of the arc, leading to stretching of the forearc along strike. Location of DSDP sites 60, 458, 459, 460, 461, and 1200 are shown. ODP sites 778-781 are centered on Chamorro Seamount. Names of serpentine mudvolcanoes, after *Fryer et al.* [1999] from south to north: South Chamorro, SC; North Chamorro, NC; Blue Moon, BM; Peacock, Pe; Celestial, C; Turquoise, T; Big Blue, BB; Pacman, Pm; Conical, Co.

Mariana forearc seafloor is relatively smooth in its western 2/3 and rugged seafloor in its eastern 1/3. The inner, smooth forearc slopes gently eastward and lies mostly 3 to 4 km below sealevel, reflecting deposition of distal volcaniclastic turbidites, ash, and pelagic sediments. In contrast, the outer, rugged forearc has a mountainous local relief of over 2 km. Rough and smooth forearc sectors correspond to the eastern and western structural provinces of *Mrozowski and Hayes* [1980]. The western structural province is typified by a thick sequence of well-stratified, relatively undeformed sediments above an eastward-shoaling basement reflector, comprising a forearc basin. This basin is over 3km thick in the flexural moat east of the active volcanic arc [*Mrozowski and Hayes,* 1980] and is filled with relatively undeformed sediments as old as latest Eocene or early Oligocene [*Hussong and Uyeda,* 1982]. A basaltic sill of Late Miocene or younger age was penetrated at ODP site 781A, near Conical Seamount (Figure 15,16) [*Marlow et al.,* 1992], and forearc volcanic flows and intrusions likely comprise a minor part of the Mariana forearc basin sequence. Although the Mariana forearc basin is structurally much simpler than the rugged region to the east, it is cut by numerous high-angle normal faults that parallel the arc and trench, with surface offsets that range from 10 to 250m. Intersection of acoustic basement with the seafloor defines the eastern limit of the forearc basin and the boundary between the western and eastern structural provinces.

The IBM forearc has been drilled during DSDP Legs 6 and 60 and ODP legs 125, 126, and 195 (Figure 15 & 16). DSDP Site 60 penetrated 350 m without reaching basement, with the interval from 50-350m consisting of Miocene ash [*Party,* 1971]. DSDP Leg 60 drilled four holes in the forearc. Site 458 penetrated almost 500m, with the upper half consisting pelagic sediments and ashes reaching back to the early Oligocene and the lower half consisting of pillowed tholeiites and boninites of probable Eocene age [*Party,* 1982a]. Site 459 penetrated almost 700m, with the upper 560 m consisting of pelagic sediments and ash beds as old as middle Eocene, underlain by pillowed tholeiites of middle Eocene and older age [*Party,* 1982b]. Site 460/460A penetrated about 100m of Pleistocene diatomaceous ooze, Oligo-Miocene conglomerates, and vitric muds of Eocene to Oligocene age. Site 461, near the trench axis, penetrated 21m of Quaternary conglomerate. ODP sites 779-781 were all drilled within 15 km of the summit of Conical Seamount (Figure 16). ODP site 1200 penetrated 200m into the South Chamorro mud volcano and installed a borehole geophysical observatory. In addition to these sites in the Mariana forearc, ODP sites 782-793 were drilled near 31-32°N in the northern IBM forearc.

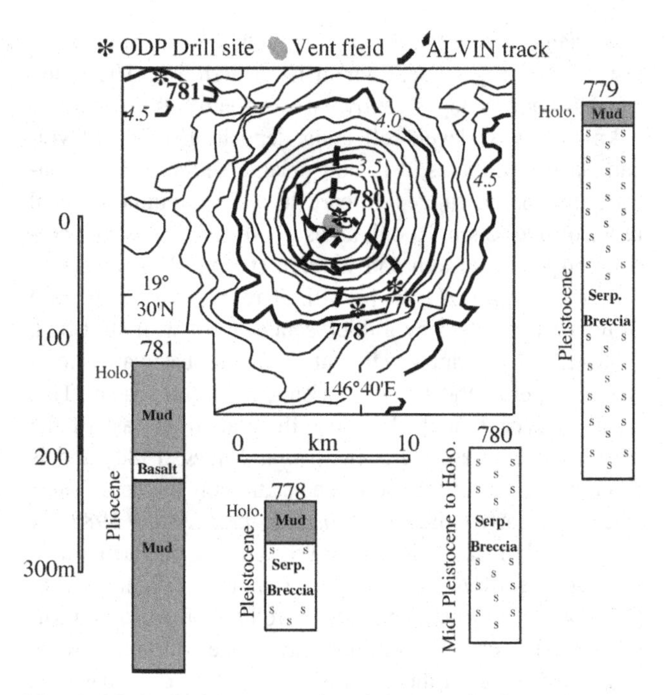

Figure 16. Conical Seamount, location shown in Figure 15. Map shows the location of active vents and chimney structures [*Fryer et al.,* 1995], and location of ODP drill sites and ALVIN dive tracks [*Fryer et al.,* 1990a]. Tracks for five dives (1851-1855) on the flanks are shown; four dives (1859-1862) on the summit are not shown. Rock sequences penetrated by ODP sites 778-781 are portrayed very schematically, along with age assignments (from ODP Leg 125 Initial Reports). Note the presence of a basalt sill at ODP site 781. Bathymetry in km.

The rugged, outer Mariana forearc contains several serpentinite mudvolcanoes, some of which are active (Figure 15). Conical Seamount is a good representative of these. This edifice is 10-15 km in diameter and 1500m tall (Figure 16). Young flows mantle its flanks and extend as far as 18 km from the summit area [*Fryer et al.,* 1990a]. The youngest flows are light green, have contorted surfaces, and are not covered with pelagic sediments. They contain numerous cobbles and boulders, mostly of serpentinized peridotites and dunites, with small amounts of metabasalt and metagabbro. Vent chimneys composed of carbonate (aragonite and calcite) and hydrated Mg-silicate are found near the summit region (Figure 16). The carbonate chimneys and concretions are unusual in being composed of aragonite well below the aragonite compensation depth in this region (~400m). *Haggerty* [1991] concluded that these carbonates were not precipitated in equilibrium with seawater; furthermore, isotopically light C indicated equilibrium with methane. Compared to ambient seawater, fluids emanating from vents in the summit area are slightly cooler (0.03°C) and have higher pH (9.3 vs. 7.7) and alkalinity (5.5 vs. 2.4 meq/L). The vent waters are enriched in methane (1000 vs. 2.1 nM), silica (0.75 vs. 0.12 mM), and H_2S (2.1 mM vs. none detected) [*Fryer et al.,* 1990a]. A biogenic origin for the methane enrichments was discounted by *Fryer et al.* [1990a] because of the lack of a sedimentary substrate where bacteria could thrive, but the possible role of a deep biosphere was not considered at that time.

ODP Leg 125 drilled four holes in and around Conical Seamount (Figure 16). Sites on Conical Seamount recovered cores of unconsolidated serpentine mud matrix containing aragonite needles and blocks of mafic and ultramafic lithologies [*Fryer,* 1992]. Pore waters from Site 780 (summit of Conical Seamount) are some of the most unusual ever sampled in oceanic sediments, containing less than half of the chloride and bromide concentrations of seawater, pH up to 12.6, and contain methane up to 37mmol/kg along with ethane and propane [*Mottl,* 1992]. Relative to seawater, these fluids are enriched in alkalinity (x26), sulfate (x1.7), K (x1.5), Rb (x5.6) and B (x10) and are depleted in Li, Mg, Ca, and Sr. Although low in Na, these pore waters have Na/Cl ratios about twice that in seawater. Sr isotopic compositions of pore waters are significantly less radiogenic than seawater and become less radiogenic downcore [*Haggerty and Chaudhuri,* 1992]. Some aspects of the fluids can be explained as a result of interaction with peridotite. However, the decrease in chlorinity of the pore fluid at site 780D is greater than observed in pore fluids from any other convergent margin. *Mottl* [1992] concluded that the fluids at Conical Seamount originated from the downgoing slab, 30 km beneath the seafloor, and agreed with *Sakai et al.* [1990] that this water was also responsible for serpentinizing the overlying mantle wedge.

Ultramafic blocks encountered in the serpentine mud are largely serpentinized but often preserve original igneous minerals such that their original nature can be inferred. Refractory peridotites dominate (mostly harzburgite, minor dunite) [*Ishii et al.,* 1992]. Orthopyroxene and olivine have Mg# ~ 92. Spinels in the harzburgite are generally more Cr-rich than those found in MORB-type abyssal peridotites [*Dick and Bullen,* 1984], indicating that these are residues left after extensive melting. These peridotites are similar to those recovered from the Mariana inner trench wall. *Parkinson and Pearce* [1998] interprets the trace element signature of Conical Seamount harzburgites as residual MORB mantle (15-20% fractional melting) subsequently modified by interaction with boninitic melt. Intriguingly, recent Os isotopic data give a mixture of modern and ~1Ga model ages for both Conical and Torishima seamount harzburgites [*Parkinson et al.,* 1998].

Mafic crustal blocks are about 10% as common as peridotite blocks in ODP cores drilled at Conical Seamount [*Johnson,* 1992]. Basalt compositions have both arc- and MORB- affinities. Metamorphism in most samples ranges from low-T zeolite to lower greenschist facies. At site 778A, *Maekawa et al.* [1993] documented clasts containing blueschist-facies mineralogy (lawsonite, aragonite, sodic pyroxene, and blue amphibole) and used these to estimate metamorphic conditions of 150-250°C and 5-6 kbar. This is the first direct link documented between an active subduction zone and blueschist-facies metamorphism.

It is not clear what routes solids and fluids follow upward to feed the mud volcanoes. Some authors infer diapiric upwelling [*Phipps and Ballotti,* 1990], whereas others infer that ascent is localized along fault intersections [*Fryer,* 1996a]. Regardless, the serpentine muds are weak plastic materials with low density and strengths and are easily mobilized [*Phipps and Ballotti,* 1990]. These authors also noted that, although these muds can transport blocks of serpentinized peridotite as large as 20m upward, such blocks would sink unless the mud continues to well up. Fluids appear to be upwelling faster than a few mm/yr [*Mottl,* 1992]. The identification of slab-derived fluids and solids requires that ascent paths must reach down to the slab itself.

Recent attention has been focused on South Chamorro seamount, the southernmost of IBM serpentine mud volcanoes (Figure 15). This is the site of active venting, which supports a vigorous biological community of mussels, snails, tube worms, and crabs, that is far richer than the Conical Seamount community [*Fryer and Mottl,* 1997]. The muds are dominantly serpentine but also contain blue, sodic amphiboles (crossite), jadeite, and hydrated garnet. South Chamorro is also the 'southern forearc seamount' where *Haggerty* [1991] reported very light C from chimney materials and interpreted this to indicate equilibrium with organic methane. South Chamorro seamount was drilled during ODP Leg 195 in March, 2001 (http://www.odp.tamu.edu:80/publications/prelim/195_prel/195toc.html).

Recent samplings of fluids from serpentine mud volcanoes built at different heights above the subduction zone suggest that systematic variations in fluid chemistry can be identified. These are interpreted to manifest progressive decarbonation reactions in the downgoing slab [*Fryer et al.,* 1999]. Pacman Seamount, situated ~15km above the subduction interface, vents fluids with low alkalinity and is associated with brucite chimneys, whereas vent fields on S. Chamorro (~20km above the subduction interface) and Conical seamounts (~25 km above the subduction interface) issue fluids with much higher alkalinity and have chimneys of carbonate and silicates. *Fryer et al.* [1999] infer from this relationship that important decarbonation reactions happen in the slab when the subduction interface above it lies at 15-20 km depth.

The IBM forearc may host an ultra-deep biosphere. If the ultimate limit on life is high temperature, then life may exist to greater depth in cold subduction zone environments than anywhere else in the planet. Intense fluid mediation of the relatively cool forearc may support the deepest biosphere in the planet. No data exists to support this speculation.

7.3. The Magmatic Front

The magmatic front lies at a height of about 150 km above the Wadati-Benioff zone, with no systematic differences along the IBM arc system. Estimates of magmatic production rates vary widely, largely because these are based on different models. Looking just at volcanics, *Sample and Karig* [1982] estimated a production rate of 12.4 km^3/Ma/km for the Mariana segment. Plutonic equivalents and cumulates need to be included to obtain a magmatic production rate, but the ratio of intrusive to extrusive igneous products is poorly constrained. This uncertainty is shown by differing estimates of crustal growth rates of 70-80 km^3/Ma per km of arc estimated for the northern IBM arc [*Taira et al.,* 1998] and 30 km^3/Ma per km of arc estimated for Phanerozoic arcs [*Reymer and Schubert,* 1984]. Resolving the magmatic addition rate to the IBM arc is a critical constraint needed to understand IBM Subduction Factory processes.

Both submarine and subaerial volcanoes define the IBM magmatic front (Figure 2). On this basis, the Mariana segment has been subdivided into Northern Seamount Province, Central Island Province, and Southern Seamount Province [*Bloomer et al.,* 1989b]. It is still not clear where the southern IBM magmatic arc ends [*Martinez et al.,* 2000]. Photographs of individual subaerial volcanoes of the IBM arc can be seen on-line at Volcano World (*Izu islands:* http://volcano.und.nodak.edu/vwdocs/volc_images/north_asia/Izu-Volcanic-Islands/IZU1.html ; Iwo Jima:: http://volcano.und.nodak.edu/vwdocs/volc_images/north_asia/iwo-jima.html ; Mariana islands: http://volcano.und.nodak.edu/vwdocs/volc_images/southeast_asia/mariana/basic_geology.html).

One problem in studying IBM (and other) arc lavas is that these are predominantly porphyritic so that bulk compositions generally do not correspond to magmatic liquids. Such samples should not be studied using the techniques routinely used for studying aphyric or glassy samples, which are appropriate for studying backarc basin basalts (BAB). Accumulation of plagioclase phenocrysts in particular has led to a misperception that mafic members are dominantly high-Al basalts when in fact aphyric samples or glass inclusions in phenocrysts are tholeiites [*Jackson,* 1993; *Lee and Stern,* 1998]. Primitive compositions (Mg#>65) are uncommon among IBM arc lavas, so fractionation conditions and history need to be resolved.

A related issue that awaits resolution is the abundance of felsic material in the IBM arc. Like all IOCM, IBM is commonly thought to have a basaltic bulk composition. Recent evidence from geophysics [*Suyehiro et al.,* 1996], exposures in the IBM collision zone [*Kawate and Arima,* 1998], glass inclusions in phenocrysts [*Lee and Stern,* 1998], the abundance of felsic tephra in DSDP cores [*Lee et al.,* 1995], and the presence of submarine silicic calderas in the Izu segment [*Iizasa et al.,* 1999] indicates that felsic rocks are a more important part of the IBM arc system than previously thought.

7.3.1. Elemental abundances. A key element for understanding operation of the IBM Subduction Factory is knowing the typical compositions and regional variability in magma compositions along the magmatic front of the IBM arc. An EXCEL spreadsheet, available upon request to the first author, was constructed for 62 volcanoes from along the magmatic front of the IBM arc system, from Oshima (34.73°N) to an unnamed seamount west of Rota (14.33°N). This compilation is not exhaustive; instead data were entered for samples which contained information considered to be especially important, such as high-quality data for REE, HFSE abundances (Y, Zr, Nb, Ta), or isotope data (B, Li, Be, O, Sr, Nd, Pb, Hf). Data were assembled from the literature [*Bloomer et al.,* 1989a; *Dixon and Stern,* 1983; *Eiler et al.,* 2000; *Elliott et al.,* 1997; *Hamuro et al.,*1983; *Hochstaedter et al.,* 2001; *Ishikawa and Nakamura,* 1994; *Ishikawa and Tera,* 1999; *Lin et al.,* 1989; *Lin et al.,* 1990; *Morimoto and Ossaka,* 1970; *Notsu et al.,* 1983; *Pearce et al.,* 1999; *Stern and Bibee,* 1984; *Stern et al.,* 1989; *Stern et al.,* 1993; *Stern et al.,* 1984; *Sun and Stern,* 2001; *Sun et al.,* 1998a; *Sun et al.,* 1998b; *Taylor and Nesbitt,* 1998; *White and Patchett,* 1984; *Woodhead,* 1989; *Yuasa and Nohara,* 1992; *Yuasa and Tamaki,* 1982]. Means were calculated for each of these volcanoes (volcano means) despite big differences in the number of samples from which means are calculated, from a minimum of 1 to a maximum of 20. Means for the entire IBM system assume equal weighting for each volcano mean. These means are the basis for the following overview.

IBM lavas from along the magmatic front of the arc show a wide range of compositions. For example, volcano means plot in the low-K, medium-K, high-K, and shoshonitic fields on potassium-silica diagrams (Figure 17). Low-K suites characterize the Izu and most of the Bonin segment (as far south as 25°N), a medium-K suite typifies most of the Mariana segment (as far north as 23°N), and a largely

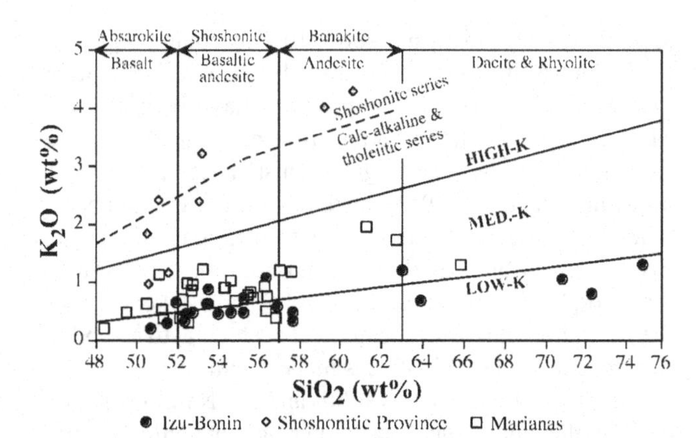

Figure 17. Potassium-silica diagram for the mean composition of 62 volcanoes along the magmatic front of the IBM arc system. Data sources discussed in text. Names for rocks indicated for calc-alkaline and tholeiitic series below the double arrows, above the double arrows for the shoshonitic series.

shoshonitic province is found between the Bonin and Mariana segments. This subdivision will be used in subsequent discussions of along-arc compositional variations, and symbols on figures correspond to this first order variation. Volcano means scatter unsystematically around a mean for the volcanic front of 55% SiO_2 (Figure 18B), indicating that these lavas are generally quite fractionated. Although the major element variations along the arc are systematic, the significant variations in composition do not correspond exactly to the along-arc subdivisions shown on Figures 1 and 2. One variation that does seem consistent with this segmentation is the abundance of large submarine silicic calderas in the Izu segment; where nine are found [*Iizasa et al.,* 1999], in contrast to the absence of this volcanic style in the rest of the IBM arc. Along-arc variations in potash concentrations (Figure 18C) show that this variation is not related to fractionation, with high K lavas in a different place than the high SiO_2 lavas. The low potassium contents of Izu segment lavas is reflected in the fact that some of the felsic volcanoes have mean potash contents below the IBM arc mean of 0.98% K_2O.

Determining water and CO_2 contents of arc lavas has proved challenging because arc lavas are extensively degassed. Water contents determined for ten glass inclusions in phenocrysts from three volcanoes average 2.2% H_2O accompanied by no detectable CO_2 (except for 400-600 ppm in one shoshonite). Water contents in glass inclusions in a gabbroic xenolith from Agrigan gives a mean of 5.1% H_2O and negligible CO_2 [Newman et al., 2000]. It is not clear whether the much higher value for the xenolith better reflects magmatic water contents or post-entrapment crystallization, but an estimate of 3% H_2O in IBM magmas

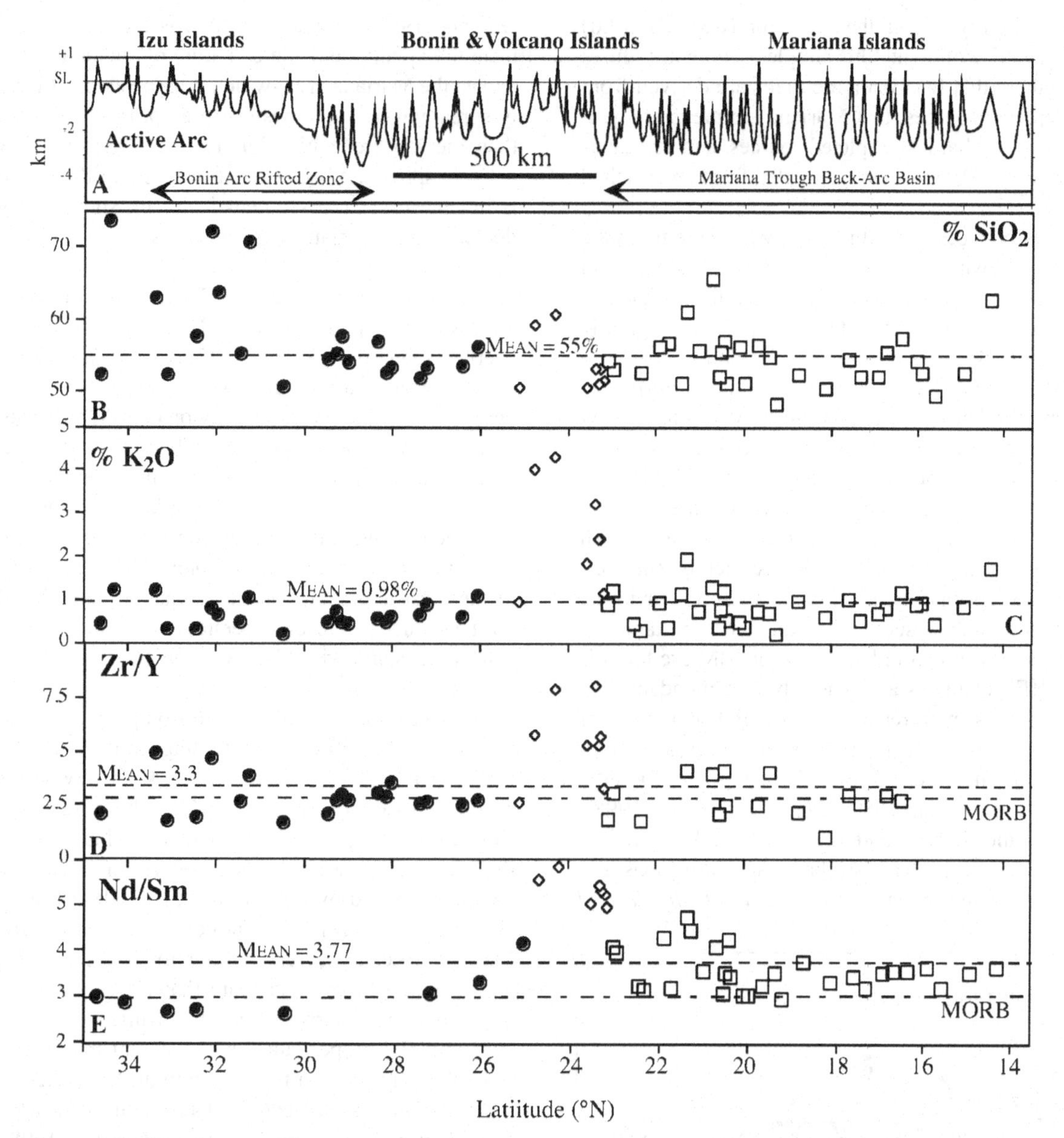

Figure 18. Profiles of mean silica and potash contents and Zr/Y and Nd/Sm for volcanoes along the magmatic front of the IBM arc. Data sources as described in text, same symbols as in Figure 17. A) Variations in bathymetry along the same profile, similar to Figure 2. Names of individual volcanoes can be found in Figure 2. Along-strike variations in silica B) and potash C) are shown, using symbols as in Figure 17. Mean silica and potash contents are shown for the IBM magmatic front as a dashed line. Along-strike variations in Zr/Y D) and Nd/Sm E) are shown, using symbols as in Figure 17. Mean values for Zr/Y and Nd/Sm are shown for the IBM magmatic front as a dashed line labeled Mean and also for typical MORB [*Hofmann,* 1988].

prior to degassing seems conservative at this time. Water contents of northern IBM arc magmas are unknown.

There are important differences in source depletion observed along the arc. Because Zr behaves more incompatibly than Y for most mantle minerals, the ratio Zr/Y is relatively high in melts generated from undepleted mantle sources and decreases for melts from depleted sources. A similar relationship holds for Nd/Sm. The mean Zr/Y for the arc system is 3.3, slightly higher than the MORB mean of 2.9 [*Hofmann,* 1988]. Similarly, the Nd/Sm ratio monitors source depletion, with higher Nd/Sm indicating a less-depleted source. The mean Nd/Sm for the arc system of

3.77 is also slightly higher than that for N-MORB (3.0) [*Hofmann*, 1988]. Following this simple argument, Figures 18D and E show that Izu-Bonin arc melts are derived from the most depleted sources, that Mariana arc melts are generated from somewhat less-depleted sources, and that lavas of the Shoshonitic Province are generated from an enriched source.

Differences between typical mafic lavas of the three arc segments are shown another way in Figure 19, along with the field of subducted sediment generalized from Figure 8. The three arc patterns are broadly similar in being greatly enriched in LIL elements (Rb, Ba, K, Pb, and Sr) and significantly depleted in HFSE Nb and Ta relative to MORB. The patterns for Izu and Mariana are very similar except that the Izu lavas are generally more depleted for elements to the left of (more incompatible than) Sm. The shoshonitic pattern is similar to the features shown by the Izu and Mariana patterns, albeit at generally elevated levels. All three patterns broadly parallel the field defined by subducted sediment except that the positive anomaly for Sr in the lavas is not seen for the sediments. Another important feature of the pattern shown in Figure 19 for IBM arc lavas is that heavy REE elements are generally less abundant than in MORB. This is interpreted to reflect higher degrees of melting for IBM arc melts relative to MORB; it is estimated that IBM arc melts represent 25-30% melting of mantle peridotite [*Bloomer et al.*, 1989a; *Peate and Pearce,* 1998]. Alternatively, the source peridotite may have been melted previously, beneath the back-arc basin spreading axis and then been remelted beneath the arc [*McCulloch and Gamble,* 1991]. However, if this is the dominant mechanism to explain the depleted nature of arc melts, one would expect the Mariana arc—which is associated with an actively spreading back-arc basin—to erupt more depleted lavas than the Izu arc, which has not been associated with an actively spreading back-arc basin for 25 million years. This is not the case: Izu segment lavas are significantly more depleted than Mariana segment lavas.

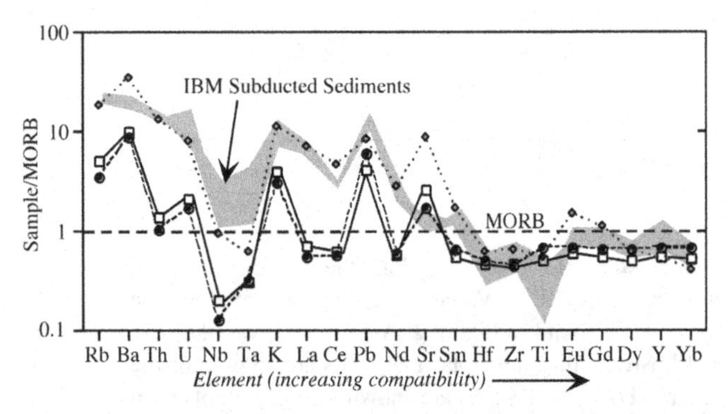

Figure 19. Incompatibility plot for lavas from along the front of the IBM arc. Typical compositions for the Mariana arc (GUG-9 of *Elliott et al.* [1997]; 51% SiO_2) and Shoshonitic province (54H of *Peate and Pearce* [1998]; 47.8% SiO_2) are plotted, along with a mean calculated for Izu arc mafic lavas (51.6% SiO_2 [*Taylor and Nesbitt,* 1998]). Symbols are the same as in Figure 17 (filled circles: Izu-Bonin; diamonds: Shoshonitic Province; open squares: Mariana). Composition of MORB and element order is after N-MORB of *Hofmann* [1988]. Grey field outlines the composition of IBM subducted sediment, from Figure 8.

7.3.2. Isotopic characteristics A wide range of isotopic tracers have been applied to the IBM arc system since the pioneering studies of *Hart et al.* [1972] and *Meijer* [1976]. Some isotopic systems allow subducted components to be unequivocally identified in the source region of IBM arc melts, while others do not allow such contributions to be identified. Uncertainties in our understanding of the nature of the mantle source as it existed before the subduction component was added looms over all interpretations of the isotopic datasets. Regardless of these issues, the isotopic database provides one of the most important constraints that we have for how the IBM Subduction Factory operates. Accordingly, the isotopic database is summarized below, in order of increasing atomic weight.

There has not been a lot of He isotopic work, principally because of the difficulty of finding samples that have not been extensively degassed, but work on olivine separates promises to overcome this problem because glass inclusions in these retain volatile elements. This also applies to the other rare gasses (Ar, Ne, Xe). A small amount of He data is presented by *Poreda and Craig* [1989], who report $^3He/^4He$ of 7.22-7.65 x atmospheric values for three Mariana volcanoes, very similar to values for MORB of 8±1 x atmospheric. It appears that He in the arc lavas is overwhelmingly derived from the mantle.

Li and B isotope data for convergent margin lavas are powerful ways to identify contributions from altered oceanic crust. The first Li isotope data for the IBM arc system were reported by *Moriguti and Nakamura* [1998]. They noted that lavas from Oshima in the northernmost IBM arc have significantly heavier Li than do lavas generated by unmodified mantle, indicating significant participation of altered oceanic crust in the source region of melts beneath the magmatic front. They also noted that Li isotopic compositions decrease away from the trench, indicating decreasing involvement of altered oceanic crust. There is a larger database for B isotopic compositions of lavas from along the IBM arc magmatic front, which is significantly heavier (mean $\delta^{11}B = +4.8‰$) than expected for unmodified mantle ($\delta^{11}B<-1‰$; Figure 20A). This isotopic system is especially powerful because compositions expected for sediments, altered oceanic crust, and unmodified mantle are

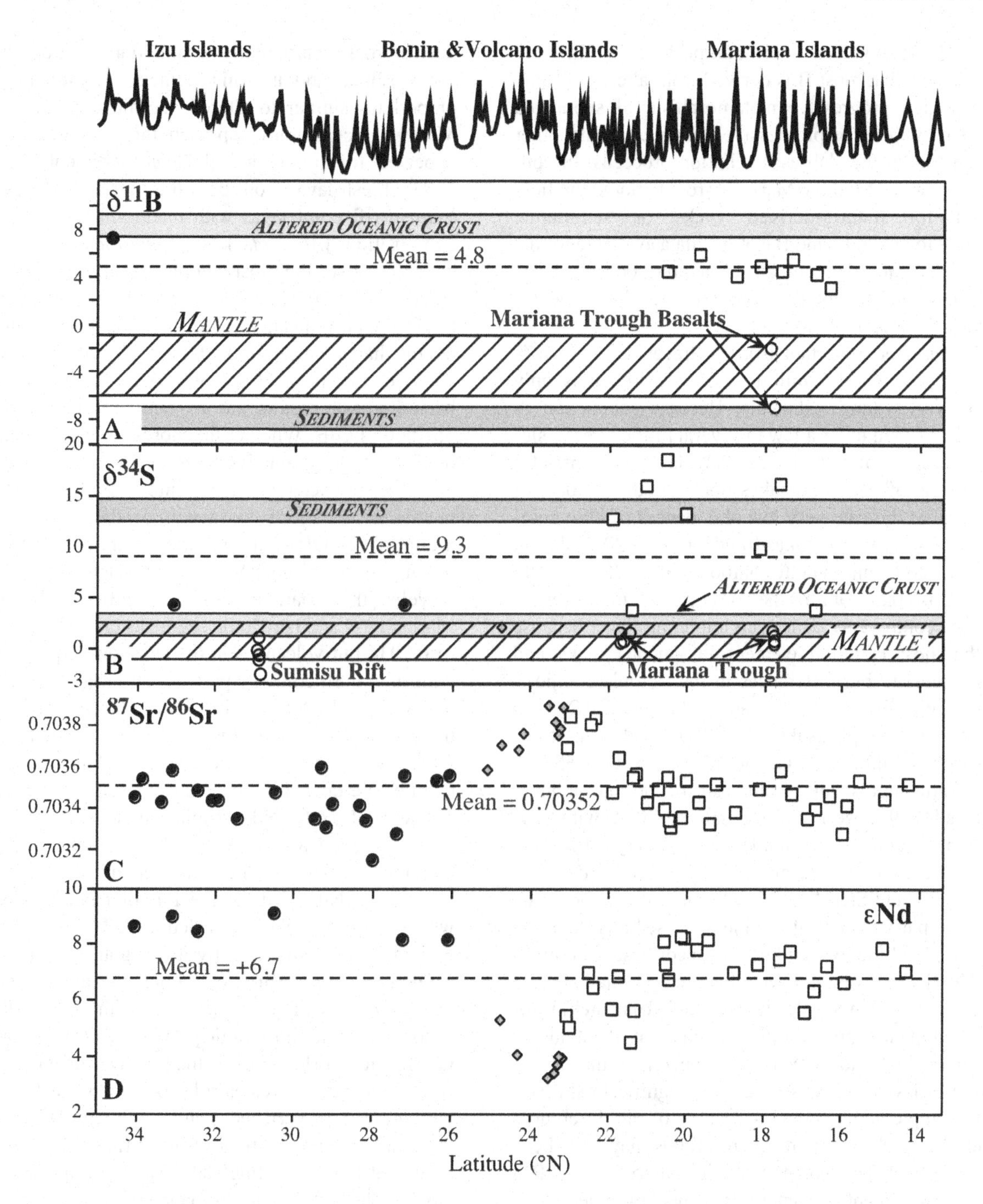

Figure 20. Isotopic variations along the IBM arc. Symbols are the same as in Figure 17 (filled circles: Izu-Bonin; diamonds: Shoshonitic Province; open squares: Mariana). (A) B isotopic variations; data for IBM arc after *Ishikawa and Nakamura* [1994] and *Ishikawa and Tera* [1999]; data for Mariana Trough basalts after *Chaussidon and Jambon* [1994]. Isotopic composition of altered oceanic crust, sediment, and mantle are from *Ishikawa and Tera* [1999]. (B) S isotopic composition of IBM arc lavas, Mariana Trough, and Sumisu rift lavas. Data for IBM arc lavas from *Alt et al.* [1993], *Ueda and Sakai* [1984], and *Woodhead et al.* [1987]; Data for ten Mariana Trough basalts are from *Alt et al.* [1993]; for 7 Sumisu Rift lavas from *Hochstaedter et al.* [1990a]. Fields for sediments, altered oceanic crust, and mantle are from *Alt et al.* [1993]. (C) Sr isotopic data for IBM arc lavas, normallized to E&A $SrCO_3$; data sources described in text. (D) Nd isotopic data for IBM arc lavas, adjusted to a common value for the La Jolla Nd standard; data sources described in text.

distinct. The heavy B isotopic composition of IBM arc lavas indicates that most B is derived from altered oceanic crust, although progressive metamorphism of subducted components will leave progressively heavier B and other light elements in the slab as distillation proceeds. In contrast, two samples from the Mariana Trough plot in the field expected for mantle-derived rocks or sediments. Considered together, Li and B isotopic data available for the IBM arc indicate that subducted altered oceanic crust makes important contributions to the source of melts generated along the magmatic front.

^{10}Be is another important isotopic tracer. Because it is cosmogenic, with a 1.5 million year half-life, it is only found in very young sediments. Because subduction to depths of ~130 km takes a few to several million years, significant amounts of ^{10}Be in fresh arc lavas demonstrates that recycling of sediments takes place. Studies to date of IBM arc lavas indicate very low abundances of ^{10}Be compared to other arcs [Woodhead and Fraser, 1985], due in part to low sedimentation rates outboard of the IBM trench.

Oxygen isotopic data for the IBM arc system has resulted in some controversy. This is an important tracer because O is the most common element in igneous rocks and because there are big differences in the isotopic composition of mantle ($\delta^{18}O$ ~+5.7 ‰) and sediments ($\delta^{18}O$ ~+20 ‰). The first study [Matsuhisa, 1979] concluded that fresh basalts from Oshima and Hachijojima had $\delta^{18}O$ = +5.3 to +6.1‰, indistinguishable from that of normal mantle. Matsuhisa [1979] observed a covariation of $\delta^{18}O$ with SiO_2 that could be modeled by simple fractional crystallization. Similar observations were made in the southern IBM arc system by Ito and Stern [1986], who interpreted their data to limit the participation of sediments or sediment-derived fluids to <1% of the oxygen budget. In contrast, Woodhead et al. [1987] concluded that Mariana arc lavas had slightly heavier oxygen than would be expected from melting of unmodified mantle, and that the participation of subducted sediment could be identified. One concern is that minor alteration of lavas can cause slight but significant shifts in oxygen isotopic composition, and some of the differences between the results of the different groups may have been due to this effect. The oxygen isotopic composition of IBM arc lavas has recently been re-examined by Eiler et al. [2000], who analyzed olivines and concluded that the oxygen isotopic composition of pristine Mariana arc lavas—including shoshonites—could not be distinguished from that of MORB or OIB melts. They concluded that no more than 1% or so of the oxygen in these rocks could be recycled from subducted sediments.

In contrast to the mantle-like oxygen in arc lavas, sulphur isotopic compositions are often elevated relative to mantle-derived rocks, particularly for the Mariana arc. Sparse data for Iwo Jima and Izu-Bonin segments are more mantle-like, or perhaps similar to altered oceanic crust (Figure 20b). Mariana arc samples approach the isotopic composition expected for sulfate in subducted sediments [*Alt et al.*, 1993]. Fresh lavas from both the Mariana Trough and the Sumisu rift erupt lavas with mantle-like sulphur isotopic compositions [*Alt et al.*, 1993; *Hochstaedter et al.*, 1990a].

Sr comprises the most complete isotopic data set for the IBM arc (Figure 20C). $^{87}Sr/^{86}Sr$ varies between 0.7030 and 0.7040, with a mean value of 0.70352. This is more radiogenic than the mean of 0.7028 found for MORB [*Ito et al.*, 1987] or the mean of 0.70304 found for Mariana Trough back-arc basin basalts, but it is similar to the isotopic composition of OIB. Whether the more radiogenic composition of IBM arc lavas relative to MORB reflects the mixing of Sr from subducted materials into a depleted MORB-like mantle or manifests extraction of these melts from an unmodified OIB-like mantle remains controversial. Because of the completeness of the Sr-isotope dataset, long-wavelength variations can be identified. Most of the Mariana Arc and Izu-Bonin segments cluster around a mean of 0.7034 but volcanoes in the shoshonitic province along with the northernmost part of the Mariana arc cluster around 0.7038. Whether the elevated $^{87}Sr/^{86}Sr$ of lavas from the shoshonitic province is due to the involvement of proportionately more or different subducted material or reflects mantle heterogeneities unrelated to subduction is unknown.

The database for Nd isotopic compositions of lavas from along the IBM arc is reasonably complete (Figure 20D). Isotopic data from the literature were adjusted to a common value of $^{143}Nd/^{144}Nd$ for the La Jolla Nd standard, from which values of εNd were calculated. The mean for the arc system is +6.7, significantly lower than typical values for MORB of ~+10. Variations along the arc mostly mirror variations seen for Sr isotopes, with the northernmost Mariana arc and shoshonitic province volcanoes having less radiogenic Nd (lower εNd) than the rest of the arc. This is consistent with a greater role for subducted sediments or derivation from enriched mantle, although Nd is generally thought not to be mobile in hydrous fluids like those sweated out of the subducting slab. Izu-Bonin arc lavas mostly have higher εNd than Mariana arc lavas. Compared to Mariana arc lavas, the higher εNd of Izu-Bonin lavas coupled with similar $^{87}Sr/^{86}Sr$ results in the latter lavas lying significantly farther to the right than other IBM arc lavas, well to the right of where the mantle array would lie on Figure 21A if it were shown. This is strong evidence that subducted components (altered seafloor or sediments) help control the Sr systematics of the arc source region. For the most part, there is a clear break between the space occupied

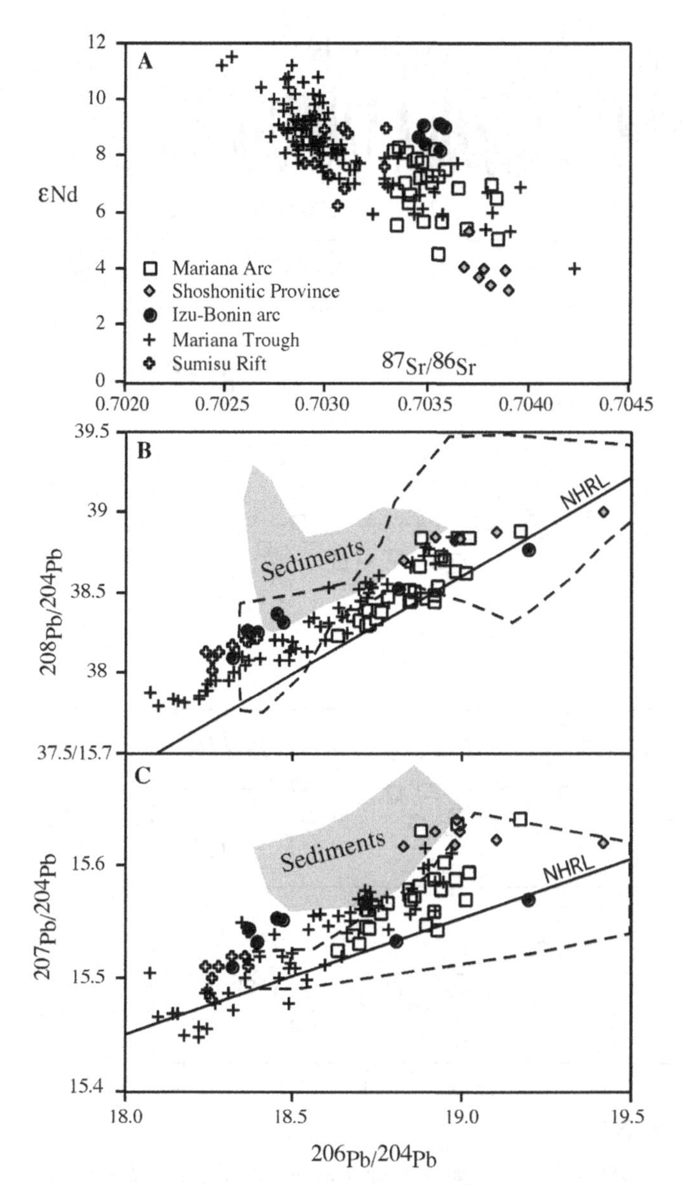

Figure 21. Isotopic composition of Sr, Nd, and Pb for IBM arc lavas, including Sumisu Rift and Mariana Trough basalts. Data for IBM arc are averages for individual volcanoes, data for Sumisu Rift and Mariana Trough are individual analyses. Fields for sediments (shaded) and W. Pacific off-ridge volcanics (dashed), taken from Figure 9, are shown in B and C. Location of these diagrams is shown with dashed boxes on Figure 9. The Northern Hemisphere Reference Line (NHRL [*Hart,* 1984]) is shown for comparison.

by back-arc basin (Mariana Trough) and interarc basin (Sumisu Rift) lavas and those of the arc (Mariana arc, Shoshonitic province, Izu-Bonin arc). Most of the former have $^{87}Sr/^{86}Sr$ <0.7032 and most of the latter have $^{87}Sr/^{86}Sr$ >0.7032. A few back-arc basin (BAB) samples are more radiogenic and are isotopically indistinguishable from arc samples. These BAB samples are invariably from the northern rifted part of the Mariana Trough where the BAB extension axis is rifting, not spreading, and has captured the arc magmatic supply [*Gribble et al.,* 1998].

A growing body of Hf isotopic data has been reported for the IBM arc system. The arc system has a mean εHf of 14 (Figure 22A). Variations in Hf isotopic composition along the arc are similar to those seen for Nd, with εHf decreasing markedly in the Shoshonitic Province. The data from the entire arc fall in the Nd-Hf mantle array of Vervoort and coworkers (Figure 23A). *Pearce et al.* [1999] used coupled Hf-Nd isotopic data to infer that the mantle source for IBM arc melts has always been of Indian Ocean provenance and that Pacific volcanogenic sediments make the most important contribution to the subduction component.

There are also considerable variations in Pb isotopic composition within the arc, most of which varies systematically along strike of the arc. Mariana arc lavas cluster around the mean $^{206}Pb/^{204}Pb$ for the IBM arc of 18.85 (Figure 22B). Most of the shoshonitic edifices also cluster around this mean, with the exception of very radiogenic Iwo Jima data. This volcano defines a trend toward 'HIMU', possibly due to the participation of subducted HIMU off-ridge volcanoes (Figure 9C). Pb isotopic variations in the southern Izu-Bonin segment are radiogenic in the south but fall off rapidly to non-radiogenic values ($^{206}Pb/^{204}Pb$ ~18.4) at the north end of the IBM arc. Figure 21B and C shows the Pb isotopic data on a conventional Pb isotope diagram. The field of Western Pacific sediments (from Figure 9) is also shown (shaded), as well as the field defined by the products of W. Pacific off-ridge volcanism (dashed line). The trajectory of oceanic volcanic rocks erupted in the northern hemisphere which are unrelated to subduction is shown as the 'Northern Hemisphere Reference Line (NHRL [Hart, 1984]). Significant deviations from the NHRL, especially on the $^{207}Pb/^{204}Pb$ vs. $^{206}Pb/^{204}Pb$ plot (Figure 21B) are commonly interpreted as due to the mixing of Pb from subducted sediments into the mantle source region. It is also noteworthy that especially the Mariana Arc lavas plot in the field of off-ridge volcanics and that subducted off-ridge volcanics and volcaniclastics should also be considered as a source of Pb for IBM arc melts.

Hart [1984] used Δ7/4 and Δ8/4 to quantify deviations of $^{207}Pb/^{204}Pb$ and $^{208}Pb/^{204}Pb$ from NHRL, Δ>0 lying above NHRL and Δ<0 lying below. Because Western Pacific sediments lie above the NHRL, involvement of Pb from subducted sediments may be monitored with Δ7/4. Figure 22C plots this parameter along strike of the IBM arc. The IBM arc system is characterized by a mean value of 4.4 for Δ7/4, with still higher values characterizing the northernmost

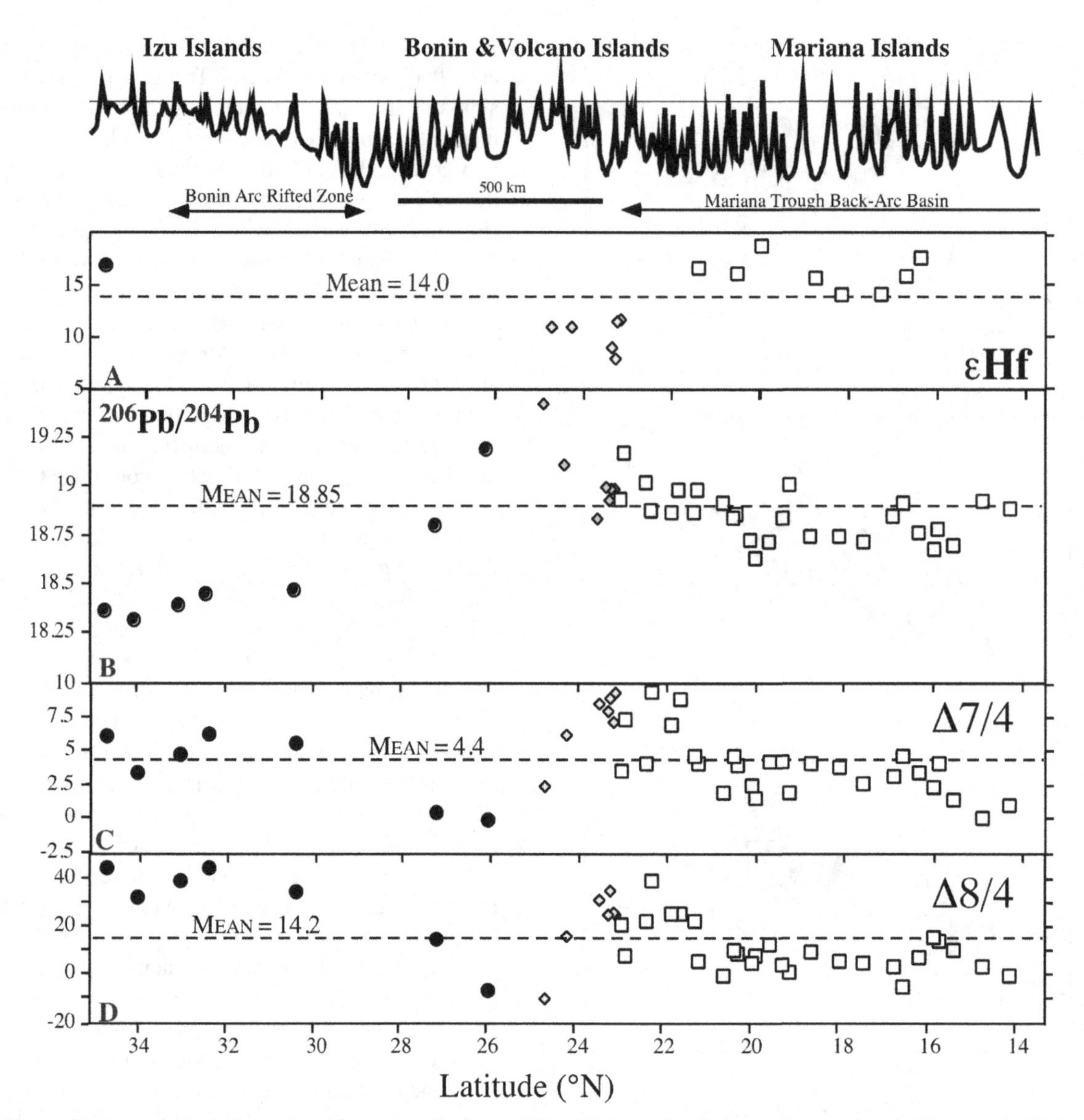

Figure 22. Isotopic variations along the magmatic front of the IBM arc. Symbols are the same as in Figure 17 (filled circles: Izu-Bonin; diamonds: Shoshonitic Province; open squares: Mariana). A) εHf ; B) $^{206}Pb/^{204}Pb$ isotopic variations; C) Δ7/4; and D) Δ8/4. Hf isotopic data are from *Pearce et al.* [1999], *White and Patchett* [1984], *Woodhead* [1989], and *Vervoort* (personal communication, 2000). Delta notation for Pb follows *Hart* [1984]; Epsilon notation for Hf follows *Pearce et al.* [1999].

Marianas and Shohsonitic Province. Pelagic sediments also lie above the NHRL on a plot of $^{208}Pb/^{204}Pb$ vs. $^{206}Pb/^{204}Pb$ plot (Figure 21C) and so have strongly positive Δ8/4. The complementary parameter Δ8/4 is not usually used to track sediment involvement, however, the IBM arc system shows systematic variations in Δ8/4 that differ in detail from those shown by Δ7/4 (Figure 22D). The arc system has a mean Δ8/4 ~14 and is even higher at the northern end of the Marianas and in the southern shoshonitic edifices. However, Δ8/4 drops precipitously for Iwo Jima and the southernmost Izu-Bonin segment before rising to a regional high over the northern Izu-Bonin segment. The strongly positive Δ8/4 over the northernmost IBM arc may be due to migration into the mantle beneath the northernmost IBM arc of Eurasian or Indian Ocean-type mantle that is distinguished by elevated Δ8/4 [*Hickey-Vargas et al.,* 1995].

Uranium-thorium studies of IBM arc magmas provide important constraints on timescales and processes of fluid fractionation, melt generation, and migration. These elements have similar distribution coefficients between mantle minerals and melt, but U is much more mobile than Th in oxidizing fluids. Such fluids can strip U preferentially from the subducted slab or the mantle and can correspondingly enrich the arc source region. Fractionation of U and Th is detectable in young lavas through the use of short-lived radiogenic daughter products. Striking confirmation that fluid-mediated fractionation has been important was demonstrated by the first U-Th disequilibrium study of IBM arc lavas [*Newman et al.,* 1984]. Subsequent studies confirm that Mariana lavas generally contain large U-excesses [*Elliott et al.,* 1997; *Gill and Williams,* 1990; *McDermott and Hawkesworth,* 1991]. These excesses vary systematically with composition. Lavas that have been interpreted to be derived from a depleted source to which a hydrous fluid was added (e.g., depleted arc tholeiites with high Ba/La) show large excesses in U, whereas shoshonites and lavas from relatively enriched sources show small to significant U-excesses (Figure 23B). Some samples show ($^{238}U/^{230}Th$) <1, which suggests small degrees of melting or long-term dynamic melting of sources with residual garnet or clinopyroxene [*Turner et al.,* 2000]. Those samples showing strong fractionation of radionuclides lie near a 30 kyr isochron. This constrains the amount of time that has elapsed between fluid-mediated fractionation and eruption. Whether the timing of fluid mediated fractionation corresponds to dehydration of the slab, breakdown of amphiboles in the convecting mantle wedge, or some other event is not yet known.

Elliott et al. [1997] recently suggested that material was added from the subducted slab in two stages. Lavas with ($^{238}U/^{230}Th$) ~1 have incompatible trace element abundances suggesting that their source has been metasomatized with a component from subducted sediments, possibly sediment melts. These workers also suggest that a late stage, water-rich component, with ($^{238}U/^{230}Th$) >1, was derived from altered crust and transferred to the mantle source region <30,000 years ago.

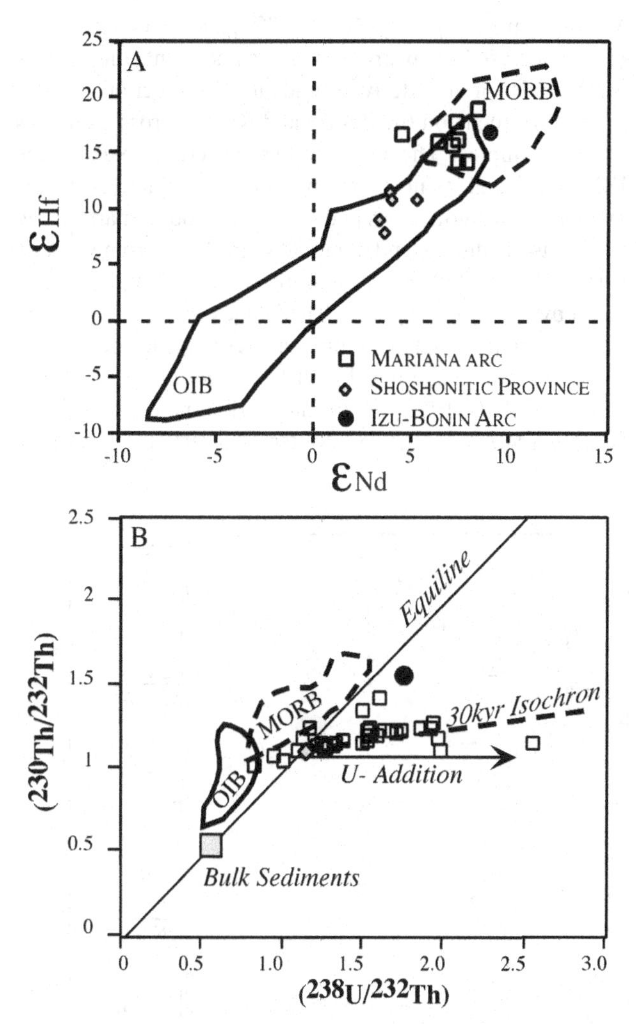

Figure 23. (A) Hf-Nd isotopic variations for the IBM arc system, data from [*Pearce et al.,* 1999; *White and Patchett,* 1984; *Woodhead,* 1989] (Vervoort, personal communication, 2000). Fields for OIB and MORB are from *Vervoort et al.* [1999] and define the mantle array. (B) U-Th disequilibrium data for IBM arc lavas. Data are not averaged for each edifice. Data from *Elliott et al.* [1997], *Gill and Williams* [1990], *McDermott and Hawkesworth* [1991, *Newman et al.* [1984], and *Reagan* (personal communication, 2000).

7.4. Cross-chains

Another perspective on IBM Subduction Factory outputs comes from considering chains of volcanoes that trend at high angles to the arc front, known as cross-chains. Because the Wadati-Benioff zone lies at different depths beneath cross-chain volcanoes while mantle and subducted material lie approximately on flowlines, the products of these volcanoes allow us to track output of the Subduction Factory across its breadth. Cross-chains are best developed in the Izu segment, where several extinct and one active cross chain are known. The Kan'ei, Manji, Enpo, and Genroku cross-chains extend WSW between 30°30' and 32°N (Figure 1) and give K-Ar ages between 3 and 12 Ma [*Ishizuka et al.,* 1998]. Detailed geochemical and isotopic studies of these cross-chains are given by *Hochstaedter et al.* [2000; 2001], who concluded that magmas generated beneath the magmatic front had the strongest relative enrichments of fluid-mobile elements in spite of having the lowest concentrations of incompatible elements. Interestingly, a similar study by [*Tatsumi et al.*, 1992] concluded that Rb/Zr ratios increased away from the volcanic front, implying a greater role for slab-derived fluids in that direction.

Here we focus on the Zenisu and Kasuga cross-chains as active examples. The most impressive cross-chain in the IBM arc lies at its northern end, where five active or dormant subaerial volcanoes and submarine counterparts along the Zenisu Ridge extend about 300 km WSW from Oshima (Figure 24A). Volcanism along the Zenisu Ridge is dominated by rhyolitic eruptions. The islands of Kozushima and Niijima are dominated by medium-K rhyolitic rocks, with subordinate andesite and basalt [*Aramaki and Itoh,* 1993; *Aramaki et al.,* 1993]. Mafic lavas dominate on Toshima [*Taylor and Nesbitt,* 1998]. Systematic variations in Li, B, Sr, and Pb isotopic compositions are consistent with decreasing participation of a slab-derived component away from the trench (Figure 24B-E; [*Ishikawa and Nakamura,* 1994; *Moriguti and Nakamura,* 1998; *Notsu et al.,* 1983]). Moreover, the subducted component that dominates along the magmatic front can be identified on the basis of Li and B isotopes as subducted altered oceanic crust.

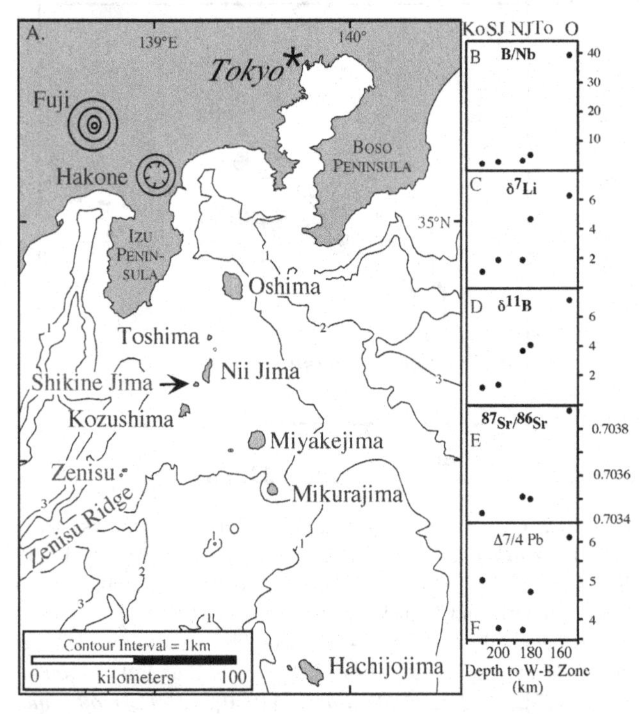

Figure 24. A) Locality map for the northern IBM arc system and the Zenisu cross-chain. B-F plot volcano-means for parameters sensitive to fluid input (B) or contributions from the subducted slab (C-F) vs. depth to the Wadati-Benioff zone. B) Variations of B/Nb vs. depth to the Wadati-Benioff Zone. Islands Oshima (Os), Toshima (To), Nii Jima (NJ), Shikine Jima (SJ); Kozushima (Ko); C) Variations of δ^7Li vs. depth to the Wadati-Benioff Zone; D) Variations of δ^{11}B vs. depth to the Wadati-Benioff Zone; E) Variations of ^{87}Sr/^{86}Sr vs. depth to the Wadati-Benioff Zone; F) Variations of Δ7/4 Pb vs. depth to the Wadati-Benioff Zone. Notice the decrease in all parameters sensitive to subducted inputs with greater depth to the seismic zone. Data are from *Ishikawa and Nakamura,* [1994], *Moriguti and Nakamura* [1998], and *Notsu et al.* [1983].

The Kasuga cross-chain, near 22°N, (Figure 1, 26B) is the best studied of any submarine arc volcano in the IBM arc system. It has been mapped with SeaMARC II imagery and sampled several times by dredging and 8 Alvin dives [*Fryer et al.,* 1997]. The cross-chain consists of three volcanoes, Kasuga 1, 2, and 3, numbered with increasing distance from the magmatic front and increasing depth to the Wadati-Benioff zone. Kasuga 1 is extinct but Kasuga 2 and Kasuga 3 are active, with fresh lava exposures and hydrothermal activity at their summits. Kasuga 2 hydrothermal fluids are compositionally unique among submarine hydrothermal systems [*McMurtry et al.,* 1993]. In contrast to lavas from the IBM magmatic front, Kasuga 2 and 3 lavas are often glassy and primitive; Mg# >65 is common. These lavas range from medium-K to shoshonitic suites and the basalts range from nepheline-normative to hypersthene-normative [*Stern et al.,* 1993]; this appears to be the only IBM cross-chain that has shoshonitic lavas. Much of the compositional range may be due to magma mixing [*Meen et al.,* 1998]. Glassy margins of three Kasuga basalts contain 1.45-1.69% H_2O [*Newman et al.,* 2000]. Felsic volcanism is also important in the Kasuga cross-chain, although lavas with >66% SiO_2 are not reported [*Fryer et al.,* 1997]. U-Th disequilibrium studies show no U excess or even Th excess [*Gill and Williams,* 1990], indicating that fractionation of U-Th by hydrous fluids was not important. The range of δ^{18}O for basaltic glasses (5.7 to 6.0‰) and clinopyroxene separates (5.23-5.57‰) indicates a mantle source, and the range for andesites and dacites (6.0-6.2‰; [*Stern et al.,* 1993]) is consistent with their evolution by fractionation of basalt. These lavas do not show the elevated ^{87}Sr/^{86}Sr for a given εNd that is characteristic of lavas erupted along the magmatic front, and Pb isotopic variations and mantle-like Ce/Pb (10-23) limit the role of subducted sediment in modifying the mantle source region. Instead, the inverse variation of magma production rate and isotopic heterogeneity was interpreted by *Stern et al.* [1993] to indicate diminished melting with distance from the magmatic front due to diminished water supply from the subducted slab.

7.5. Magmatism Related to Extension

Extension-related igneous activity provides a third perspective on the high-temperature products of the IBM

Subduction Factory. This perspective is important not only because these volcanoes are located farther from the trench than the magmatic axis of the arc, but also because the melt generation process itself is different. Magma generation beneath the magmatic front is probably caused by fluid fluxing of downwelling mantle, whereas beneath back-arc basin spreading centers melting is probably caused by adiabatic decompression of upwelling mantle. Back-arc Basin melt generation is thus similar to that beneath mid-ocean ridges but aided by addition of water from the subduction zone (Figure 25). Evolution of an unrifted arc to one with a back-arc basin thus must be accompanied by reorganization of the underlying mantle flow, a problem that has not yet been explored. Also, mantle flow beneath an arc system associated with active back-arc basin spreading may carry mantle previously depleted by melting beneath the back-arc basin spreading center (1 in Figure 25B) to be remelted beneath cross-chains and the magmatic front (2,3 in Figure 25B) [*McCulloch and Gamble,* 1991].

Extension is manifested as inter-arc rift basins in the Izu segment [*Klaus et al.,* 1992] and as an active back-arc basin (Mariana Trough) in the Mariana segment [*Fryer,* 1995]. Izu rifts may be earlier stages of tectonic evolution than that of a mature back-arc basin like the Mariana Trough. Here we briefly consider the magmatic products of the Sumisu Rift inter-arc basin and the Mariana Trough back-arc basin.

The Izu rift zone spans most of the Izu segment, from 33° to 27°30' N, just south of the Sofugan Tectonic Line, a distance of about 700km. The rift zone parallels the arc magmatic front and lies just to the west of it, except near 29°N where magmatic arc and rift coincide (Figure 26A). Six basins have been named in the rift zone: Hachijo, Aogashima, Sumisu, Torishima, Sofugan, and Nishinoshima. The rift zone deepens southward, with flat, sediment covered basins lying at depths of a little deeper than 1 km in the north to little over 3km in the south [*Klaus et al.,* 1992; *Taylor et al.,* 1991]. The Sumisu Rift is by far the best known of the Izu rifts and is representative. Studies of the Sumisu Rift include SeaMARC II and Seabeam swathmapping and Alvin diving [*Taylor,* 1992], multichannel seismic reflection profiling [*Taylor et al.,* 1991] and ODP drilling (sites 790 and 791 in the basin, sites 788 and 789 on the flanking uplift to the east [*Taylor,* 1992]. This geophysical background provides a firm foundation for igneous rock geochemical studies [*Fryer et al.,* 1990b; *Hochstaedter et al.,* 1990a; *Hochstaedter et al.,* 1990b; *Ikeda and Yuasa,* 1989; *Gill et al.,* 1992]. Three oblique transfer zones divide the rift along strike into four segments with different fault trends and uplift/subsidence patterns. Differential strain is accommodated by interdigitating, rift-parallel faults and cross-rift volcanism, which is often concentrated at cross-rift

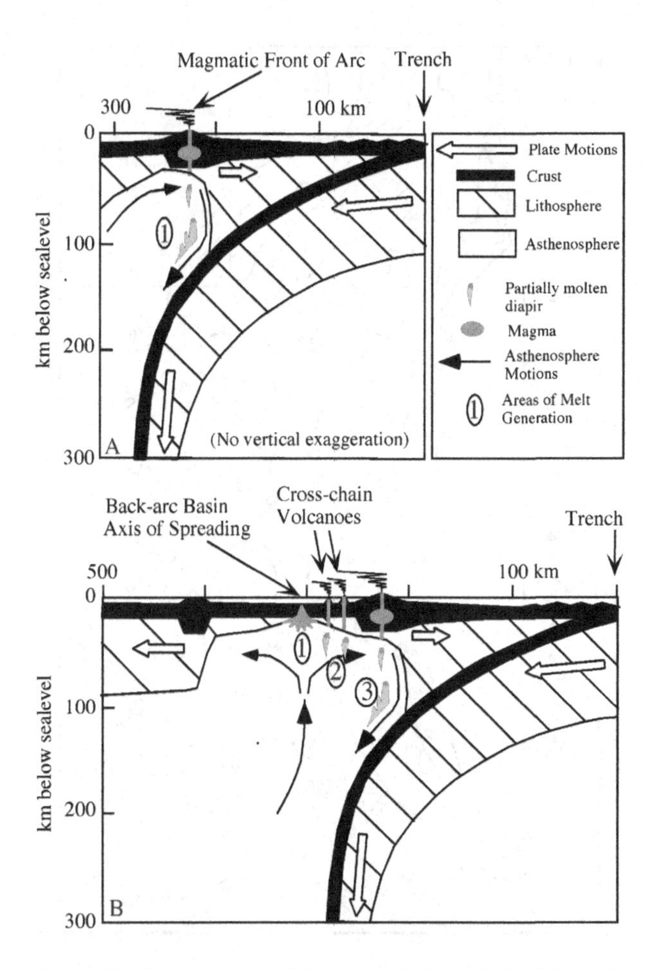

Figure 25. Conceptual models for evolution of the Subduction Factory from one lacking an actively spreading back-arc basin (A) to one associated with back-arc spreading (B). Note that magma generation and migration beneath the magmatic front occurs in a region of mantle downwelling, whereas that beneath the back-arc basin spreading axis occurs in a region of mantle upwelling. Note in (B) the multiple melting opportunities for mantle rising beneath the back-arc basin, with melt generation beneath the spreading axis (1), beneath cross-chains (2) and beneath the magmatic front (3). The IBM arc system presents all of these stages, with the Bonin segment now acting as an unrifted arc (A) whereas the Mariana segment is an outstanding example of an arc associated with back-arc spreading (B). The Izu segment represents an intermediate situation, where rifting is underway but has not proceeded to seafloor spreading.

transfer zones. The locus of maximum sediment thickness (>1km) and basin subsidence occurs along an inner rift adjacent to the arc margin. Volcanism is also concentrated at highs between the basins, most of which mimic cross-chains extending W to SW from the major arc volcanoes, indicating the dominance of magmatic point sources. From estimates of extension (2-5km), the age of the rift (~2 Ma), and accel-

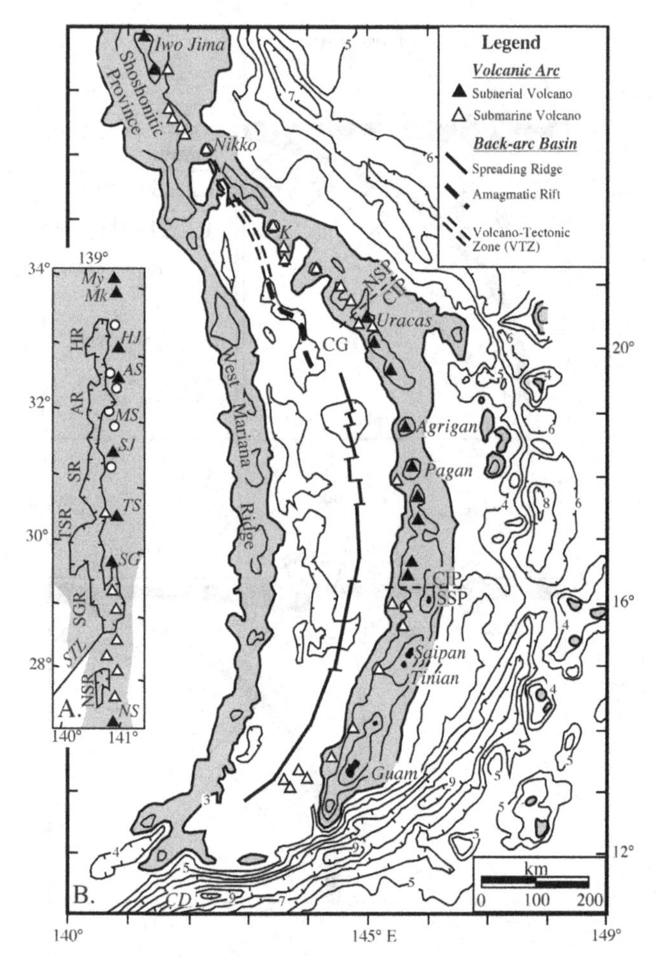

Figure 26. Actively extending systems of the IBM arc. A) Tectonic sketch map of Izu rifts, modified after *Taylor* [1992]; shaded where shallower than 3km. Note this is rotated clockwise about 10° from true north. Calderas shown with open circles. Abbreviations for frontal arc volcanoes (from the north): *My,* Miyake Jima; *Mk,* Mikura Jima; *HJ,* Hachijo Jima; *AS,* Aoga Shima; *MS,* Myojin Sho; *TS,* Tori Shima; *SJ,* Sofu Gan; *NS,* Nishino Shima. *STL* = Sofugan Tectonic Line. Abbreviations for rifts (from the north): HR, Hachijo Rift; AR, Aogashima Rift; SR, Sumisu Rift; TSR, Torishima Rift, SGR, Sofugan Rift; NSR, Nishinoshima Rift. B) Tectonic sketch map of the Mariana Arc and back-arc basin. Bathymetric contours are in km, region shallower than 3km is shaded. CD= Challenger Deep, SSP = Southern Seamount Province, CIP=Central Island Province, NSP=Northern Seamount Province, *K*=location of Kasuga cross-chain, CG=Central Graben. Note that seafloor spreading characterizes the Mariana Trough south of 19°45'N and rifting is characteristic from this latitude north to Nikko, where the extension axis intersects the arc.

erating subsidence, *Taylor et al.* [1991] infer that the Sumisu rift is in the early stages of developing as a back-arc basin.

The Mariana Trough stretches 1300 km from north to south (Figure 26B), about the distance from Los Angeles to Portland, Tokyo to Seoul, or London to Rome, and has roughly the dimensions and areal extent of Japan or California. The Trough is crudely crescent-shaped, opening on the south; it is bounded to the east by the active Mariana arc, to the west by the remnant arc of the West Mariana Ridge [*Karig,* 1972], and to the south by the Challenger Deep part of the Mariana Trench. It narrows northward until the Mariana arc and West Mariana Ridge meet at about 24°N. It is widest in the middle, at 18°N, where it is about 240 km wide, and narrows to about half this at its southern, open end. Depths in the basin are distributed asymmetrically, being greater adjacent to the West Mariana Ridge than next to the active arc, due to a westward-thinning wedge of volcaniclastic sediments derived from the active arc, and also less thermal buoyancy of the mantle. Where not covered by sediments, the seafloor is deeper and bathymetry more rugged than normal. Zero-age seafloor of the Philippine Sea, including the Mariana Trough, lies at a mean depth of 3200 m compared to normal zero-age seafloor depths of 2500 m [*Park et al.,* 1990].

The extension axis can be subdivided along strike into a southern two-thirds characterized by slow seafloor spreading and a northern third characterized by rifting (Figure 26B). From 19°45'N south to 13°10'N, the axis has the typical morphology of a slow-spreading ridge, with an axial graben that is sometimes occupied by a neovolcanic axial ridge [*Hawkins et al.,* 1990; *Martinez et al.,* 2000; *Stüben et al.,* 1998]. South of this the ridge resembles a fast spreading ridge, probably because magma supply is enhanced by proximity to the arc [*Martinez et al.,* 2000]. Spreading half-rates in the region between 16° and 18°N are estimated at 1.5 to 2.2cm/year [*Bibee et al.,* 1980; *Hussong and Uyeda,* 1982; *Yamazaki and Stern,* 1997]. The ridge becomes punctiform north of 18°30', and true seafloor spreading does not occur north of 19°45'N (but see the different conclusion of *Yamazaki et al.* [1993]). Rifting forms a series of amagmatic deeps between 19°45'N and 21°10'N called the 'Central Graben' by *Martinez et al.* [1995]. These basins have low heatflow, lack igneous activity, and contain the greatest depths in the Mariana Trough (>5400m). The deepest part of the Central Graben is also unique among active back-arc basins in exposing mantle peridotites along the extension axis [*Stern et al.,* 1996]. Extension north of the Central Graben occurs by combined tectonic and magmatic processes that are distinct from seafloor spreading, in a region known as the Volcano-Tectonic Zone (VTZ) [Martinez et al., 1995]. The VTZ corresponds with a part of the Mariana Trough where the crust thickens from 6 to 15 km (Figure 14). The southern VTZ is dominated by fissure eruptions associated with a ridge-like feature, ~30 km long, which rises to less than 2800 m water depth and which is similar to the inflated segment at the southern terminus of the spreading ridge. The northern VTZ is dominated by point-source volcanism, with edifices spaced 50-60 km apart alternating with rift basins. There is no volcanic activity along the adjacent arc segment and it appears that the extension axis has captured the arc magma supply between the Kasuga cross-chain at 22°N and Nikko near 23°N, where the extension axis intersects the arc. North of this point, incipient rifting is magmatically manifested by the shoshonitic lavas of the Hiyoshi complex, Fukutoku-okano-ba (or Sin Iwo Jima), and Iwo Jima (Figure 26B). Rifting and spreading propagate northward at a rate of 10 to 40 cm/year [*Martinez et al.*, 1995; *Stern et al.*, 1984], so the variations in tectonic and magmatic style seen along-strike north of 18°N reveal the sequence of events that occur as the back-arc basin rift evolves from updoming through rifting to seafloor spreading.

The following geochemical discussion and data for associated figures is extracted from the following literature sources: [*Hochstaedter et al.*, 1990a; *Ikeda and Yuasa,* 1989; *Alt et al.*, 1993; *Chaussidon and Jambon,* 1994; *Fryer et al.*, 1990b; *Garcia et al.*, 1979; *Gribble et al.*, 1996; *Gribble et al.*, 1998; *Hart et al.*, 1972; *Hawkins et al.*, 1990; *Hawkins and Melchior,* 1985; *Ikeda et al.*, 1998; *Ito and Stern,* 1986; *Jackson,* 1989; *Poreda,* 1985; *Sano et al.*, 1998; *Sinton and Fryer,* 1987; *Stern et al.*, 1990; *Volpe et al.*, 1987; *Eiler et al.*, 2000; *Hochstaedter et al.*, 1990b; *Macpherson et al.*, 2000; *Newman et al.*, 2000]. Data for the Mariana Trough samples are entered in an annotated EXCEL database that is available from the first author upon request. Before summarizing the geochemical and isotopic signatures of extension-related IBM lavas, three points need to be emphasized. First, even though the Sumisu Rift is in the early stages of extension and seafloor spreading is not occurring, these basalts have geochemical and isotopic affinities with back-arc basin basalts such as erupt from spreading segments of the Mariana Trough [*Hochstaedter et al.*, 1990b]. Second, and in contrast to the absence of felsic volcanism in the Mariana Trough spreading ridge, volcanism in the Sumisu Rift is compositionally bimodal, with abundant felsic (>70% SiO_2) material as well as basalt. The bimodal nature of Sumisu Rift igneous activity makes it more similar to the northern VTZ than to the spreading portion of the Mariana Trough (Figure 27A). Extension-related igneous activity in many tectonic environments is compositionally bimodal during rifting and becomes entirely mafic when seafloor spreading begins. Finally, extension-related lavas are wet, generally containing much more water than found in MORB or OIB. Glassy Mariana Trough basalts contain 0.2-2.78% H_2O, with a mean of 1.59%. This range does not seem to due to degassing, because glass inclusions

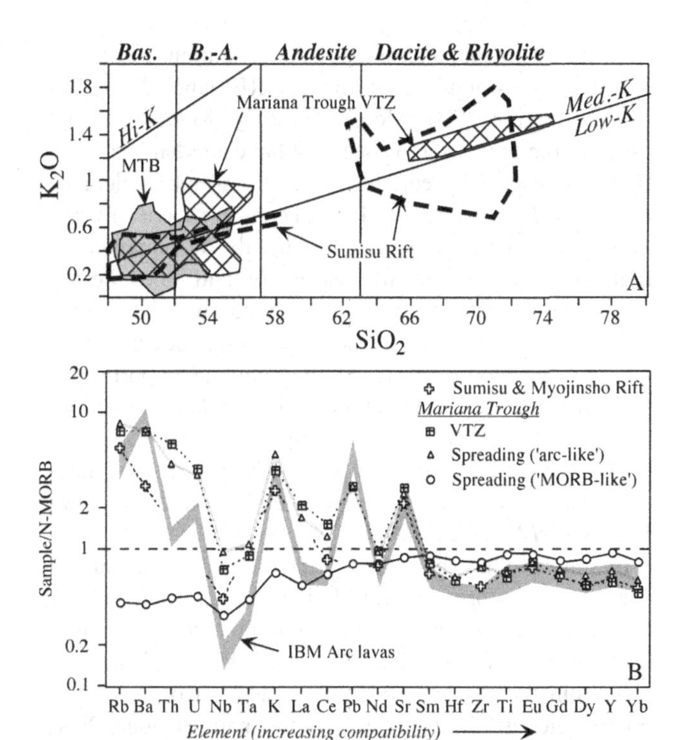

Figure 27. Chemical features of IBM rift-related lavas. A) Potash-silica diagram for IBM rift-related igneous rocks. Fields are shown for Sumisu and Myojinsho rifts (66 samples), Mariana Trough VTZ (41 samples), and Mariana Trough spreading segment (labeled MTB, 167 samples). Note the bimodal nature of lavas erupted in rifts and strictly mafic nature of lavas erupted in the region undergoing spreading. B) Incompatibility plot for extension-related IBM lavas. Sumisu rift data is mean from data set of *Hochstaedter et al.*, 1990b]; note that this pattern is incomplete. Mariana Trough data are for glasses analyzed by Pearce (unpublished ICP-MS data). VTZ sample is T7-54:1-1 (1.69% H_2O). Two representative samples are given for the Mariana Trough spreading segment: 'arc-like' sample GTVA75:1-1 (2.21% H_2O) and 'MORB-like' sample DS84:2-1 (0.21% H_2O) [*Gribble et al.*, 1996]. Composition of N-MORB and element order is after *Hofmann* [1988]. Grey field outlines the composition of Mariana and Izu segment lavas, from Figure 19.

contain similar water abundances as found in glassy lava rims.

Extension-related basalts generally show chemical signatures of a 'subduction component' that is dominantly carried by a hydrous fluid. Trace element abundances vary strongly with how wet the melt was. Figure 27B presents trace element abundances for the two rifting segments, Mariana Trough VTZ and Sumisu Rift, along with end-member examples of basalts erupted along the Mariana Trough spreading ridge. As can be seen from this figure, some basalts in the Mariana Trough spreading segment have trace element patterns—and concentrations of water—which are indistinguishable from MORB (labeled 'MORB-like' on Figure 27B). More commonly, Mariana Trough basalts from the spreading segment have an identifiable arc signature, especially enrichments in fluid-mobile elements Rb, Ba, K, Pb, and Sr, and elevated concentrations of water. This is particularly perplexing because the Mariana Wadati-Benioff zone dives nearly vertically and does not lie beneath most of the Mariana Trough, so how slab-derived fluids are transported to the melt generation zone is unclear. Arc-like geochemical features are ubiquitous in portions of the IBM arc system undergoing rifting. Melting models indicate that the degree of melting in spreading segment basalts ('spread' basalts) can be directly linked to the abundance of water in the mantle source [*Gribble et al.,* 1998; *Stolper and Newman,* 1994].

In spite of the geochemical indications for a subduction component, isotopic data for Mariana Trough spread basalts mostly indicate a mantle source. Helium isotopes data indicate a mean R/R_A = 7.87±1.34 (1 std. dev), indistinguishable from the value for MORB of 8±1. Two analyses of boron isotopes fall in the fields for the mantle (Figure 20A). Mean $\delta^{13}C$ of –4.15±1.34‰ is close to mantle values as well. Oxygen isotopic data mostly on glasses gives a mean $\delta^{18}O$ of +5.85±0.15‰, overlapping the range of 5.35-6.05‰ for MORB glasses [*Ito et al.,* 1987], although glasses with strong arc-like signatures sometimes have $\delta^{18}O$ of 6.0-6.1‰. Indications that Mariana Trough spread basalts have oxygen isotopic compositions that are indistinguishable from MORB is confirmed by analysis of olivines [*Eiler et al.,* 2000]. Sulphur isotopic data for samples from both the spreading and VTZ segments of the Mariana Trough spread and the Sumisu Rift fall in the field for mantle-derived basalts (Figure 20B).

Isotopic data for Sr, Nd, and Pb indicate that the mantle source of rift volcanism is very similar to that of associated arc segments whereas spreading segments have distinct mantle sources. Figure 21A shows that the εNd of Sumisu Rift lavas is similar to Izu-Bonin arc lavas but that they have lower $^{87}Sr/^{86}Sr$. Mariana Trough VTZ and spread basalts are not distinguished in Figure 21, but spread basalts have significantly lower $^{87}Sr/^{86}Sr$ (mean = 0.70291±0.00015) and higher εNd (8.9±1.1) than lavas erupted along the associated magmatic front (see Figure 20C,D: 0.70352, +6.7). Lavas erupted from the Mariana Trough VTZ have Sr- and Nd-isotopic compositions that overlap those of the magmatic front (0.70343±0.00032; +6.8±1.0). Pb isotopic compositions tell a similar story. On Pb isotopic diagrams, Sumisu Rift and Izu-Bonin arc lavas plot in restricted areas that overlap, and both show elevated $^{208}Pb/^{204}Pb$ relative to the NHRL. Mariana Trough VTZ lavas have mean $^{206}Pb/^{204}Pb$ (18.76±0.13) that is indistinguishable from the mean for the IBM arc of 18.85. In contrast, Mariana Trough spread basalts have less radiogenic and more variable $^{206}Pb/^{204}Pb$ (18.24±0.27) and show elevated $^{208}Pb/^{204}Pb$ relative to the NHRL. The radiogenic isotope data strongly indicate that IBM rift basalts are produced by parts and processes of the Subduction Factory that generally produce basalts along the magmatic front, but that spreading basalts are produced differently.

In summary, although arc lavas have affinities with continental crust and back-arc basin lavas are in most ways indistinguishable from MORB and oceanic crust, the source regions of these two magma stems are similar in many respects. In particular, both mantle source regions are strongly affected by a 'subduction component' that travels with water extracted from the subducted slab or sediments. Although arc and back-arc basin melts have very different modes of formation and tectonic environments of eruption, arc and back-arc basin lavas show similar enrichments in LIL elements, depletions of HFSE, and water contents. In spite of these affinities, significantly different isotopic compositions indicate that back-arc basin melts have less of an identifiable component from subducted materials than do arc lavas. Fluids from the slab that interact with the back-arc basin source are more equilibrated with asthenospheric mantle than have those delivered to the arc source. There is a great need for quantitative chemical modeling in tandem with development of realistic models for mantle, fluid, and melt flow to explain these similarities and differences.

7.6. IBM Collision Zone

The IBM arc system in the north terminates against Honshu in the Izu Collision Zone (ICZ; Figure 28). This is one of the few examples on Earth of active collision and terrane accretion, and it provides outstanding opportunities to study strain partitioning during terrane accretion and the nature of middle and deep crust of an IOCM. The ongoing collision also presents a first-order geohazard; the Great Kanto Earthquake of 1923 had an estimated surface wave magnitude of 8.2. This earthquake had an epicenter SW of Tokyo and it devastated southeastern Japan, including the cities of Tokyo and Yokohama. The earthquake killed about 140,000 people and was probably caused by the collision. Uplift associated with the collision zone has resulted in the highest topographic relief in Japan; if submarine relief is considered, the total relief in the region is greater than that of the Himalayas [*Niitsuma,* 1989]. Collision-related uplift is active and rapid, as shown by the steep tilts of very young sedimentary beds.

The ICZ is caused by the attempted subduction of thickened arc crust at the northeastern corner of the Philippine Sea plate. The north end of the IBM arc enters the Nankai Trough subduction zone at a rate of ~4cm/y but the ~22km-thick IBM crust is too thick and buoyant to be readily subducted. Instead, the northern end of IBM plows end-on into southern Honshu. This collision has been occurring in about the same place since Middle Miocene time, ~15 Ma ago [*Itoh,* 1986]. The deformation is partly accommodated by oroclinal bending of margin-parallel lithotectonic belts in southern Honshu, such as the Cretaceous-Paleogene Shimanto Belt accretionary prism [*Niitsuma,* 1989]. The collision has rotated the Shimanto Belt from its normal ENE trend elsewhere in Honshu to a N-S and NNW-SSE trend to the west and east of the ICZ respectively. This cuspate rotation of trends that were originally more E-W is shown by the deflection of the Nankai Trough into the NNE-trending Suruga Trough on the west side of the Izu Peninsula and the NW-trending Sagami Trough on the east side (Figure 28A).

Even though the Izu Peninsula has acted for the most part as a rigid indenter during the collision, it is also being deformed and thickened. Cumulative thickening has about doubled the crustal thickness, from ~22km for the arc at 32°15'N to >40km [*Soh et al.,* 1998] in the collision zone. Collision has caused the incremental migration of the subduction zone to the south with time, manifested as major thrust faults which young from north to south. The three thrusts exposed on land (labeled 1, 2, and 3 on Figure 28) are each associated with deep, foreland-type basins which also young southwards, ranging in age from Middle Miocene in the north to Pleistocene in the south [*Soh et al.,* 1998]. The Ashigara Basin (Figure 28A) is a good example; it filled with about 5km of shallowing- and coarsening-upwards sediments deposited between 1.6 Ma and 0.5 Ma [*Soh et al.,* 1998]. The southernmost and youngest thrust (5 in Figure 28) lies parallel to the Zenisu Ridge on its southern flank and is associated with a broad zone of crustal seismicity [*Le Pichon et al.,* 1987].

Uplift has exhumed the deep infrastructure of the arc in the Tanzawa Mountains (Figure 28A). Exposures are made up of Tanzawa Group volcanics and sediments and the Tanzawa plutonic complex, both of Miocene age. The Tanzawa Group comprises altered basaltic and andesitic tuff, volcanic breccia and lavas, and shallow marine limestone, which is considered to be a shallow submarine volcanic sequence erupted on the IBM arc and later accreted to Honshu. The Tanzawa Group has been metamorphosed from zeolite to amphibolite facies

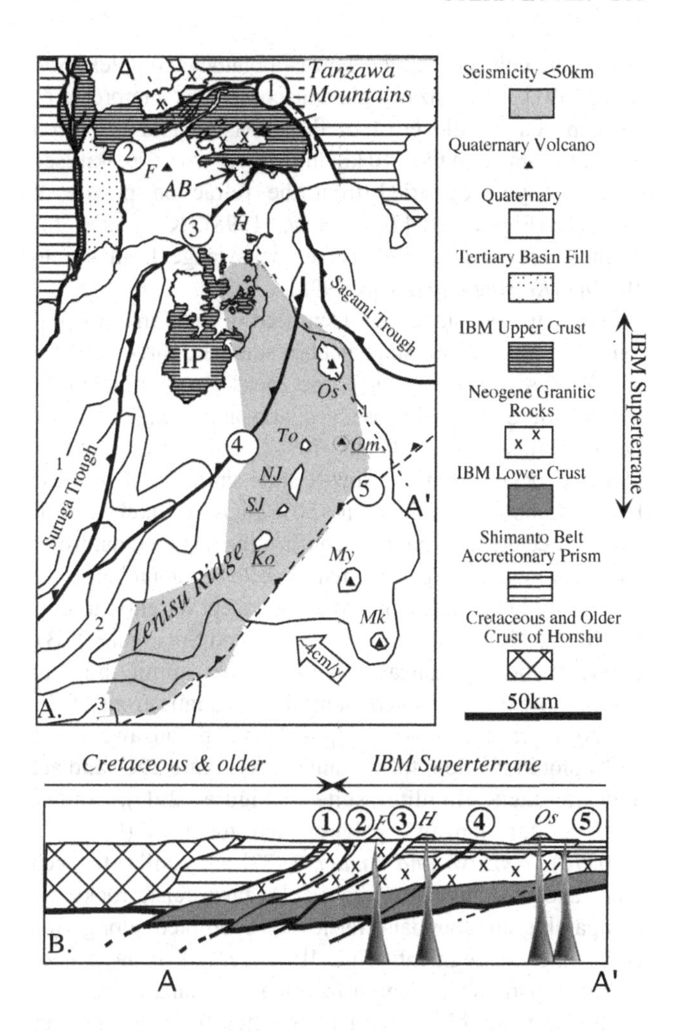

Figure 28. The IBM collision zone. A) Geologic map showing the location of major thrust faults and tectonic lines (1-5 circled) and the Tanzawa Mountains. 1= Mineoka/Tonoki-Aikawa/Itoigawa-Shizuoka Thrust; 2=TonokiAikawa/Nishikatsura/ Minobu Thrust; 3=Kouzu-Matsuda/Kannawa Thrust; 4=Izu-Toho Tectonic Line; 5=Zenisu-Oki Tectonic Line. Dashed line indicates incipient thrust fault. IP = Izu Peninsula; AB = Ashigara Basin. Water depth contours are labeled in km. Volcanoes are abbreviated F=Fuji, H=Hakone, Os = Oshima, Om = Omurodashi (submarine), To = Toshima, NJ = Nii Jima, SJ = Shikine Jima, Ko = Kozushima, My = Miyakejima, Mk = Mikurajima. Underlining indicates dominantly felsic volcano. Arrow labeled '4 cm/y' indicates convergence direction and rate between IBM arc and Honshu. Note the generalized location of seismicity shallower than 50km. Dashed line A-A' shows approximate location of section shown in B. Figure modified after *Kawate and Arima* [1998] and *Taira et al.* [1998]. B) Along-arc cross-section A-A', modified after *Taira et al.* [1998], showing sequence of imbrications of the IBM Superterrane. No vertical exaggeration. Scales for A and B are identical.

toward the contact with the Tanzawa plutonic complex [*Soh et al.,* 1991]. The Tanzawa plutonic complex is interpreted as the exposed middle crust of the IBM arc and has been explicitly linked to the ~10 km thick layer with 6-6.3km/sec velocity layer identified from the refraction profile at 32°15'N (Figure 13) [*Taira et al.,* 1998]. K-Ar ages for biotites, hornblende, and whole rocks range from 4.6 to 10.7Ma [*Kawate and Arima,* 1998].

Four intrusive stages are identified in the Tanzawa plutonic complex, a first-stage gabbro suite and three younger stages of a tonalite suite. These rocks are dominated by hornblende and plagioclase ± quartz and contain a wide range in silica, from ~44% to >70% (Figure 29A). Tanzawa plutonic rocks comprise a low-K suite similar to that of the Izu Arc; gabbroic rocks with <50% SiO_2 are likely cumulates. Trace element patterns confirm the depleted nature of Tanzawa magmas, which often have lower abundances of incompatible elements than MORB, except for enrichments in Rb, Ba, Th, and K and spikes in Pb and Sr (Figure 29B). Fields defined by Tanzawa plutonic rocks mimic those of basalts now being erupted along the magmatic front of the Izu Arc. Trace element compositions for Tanzawa felsic rocks plot in the field for volcanic arcs (Figure 29C) and arc andesite-dacite-rhyolite suites (Figure 29D). Initial $^{87}Sr/^{86}Sr$ for Tanzawa gabbros and tonalites (0.70331-0.70370) [*Ishizaka and Yanagi,* 1977] are similar to each other and to other Izu arc lavas. Tanzawa felsic rocks are comparable to abundant felsic lavas erupted along and across the Izu segment of the IBM arc. All of these geochemical parameters demonstrate that the Tanzawa igneous suite is comparable to what is expected from the igneous infrastructure of an IOCM such as the northern IBM arc.

The Tanzawa plutonic complex is of special interest because of the insights it provides on the evolution of middle crust in the northern IBM arc. This has been modeled as taking place in two stages [*Kawate and Arima,* 1998]. The first stage was melting of amphibolite-facies low-K tholeiite in the lower crust of the IBM arc to form an intermediate melt (~60% silica). About 50% melting of amphibolite is required, at temperatures of ~1100°C and P<1.2 GPa [*Nakajima and Arima,* 1998]. This corresponds to conditions expected for arc crust that is less than 40 km thick; garnet is stabilized at higher pressures. They further proposed delamination of pyroxenitic residue after generation of the anatectic melts. More siliceous melts were derived from the parental intermediate melt by fractionation of hornblende, plagioclase, and magnetite.

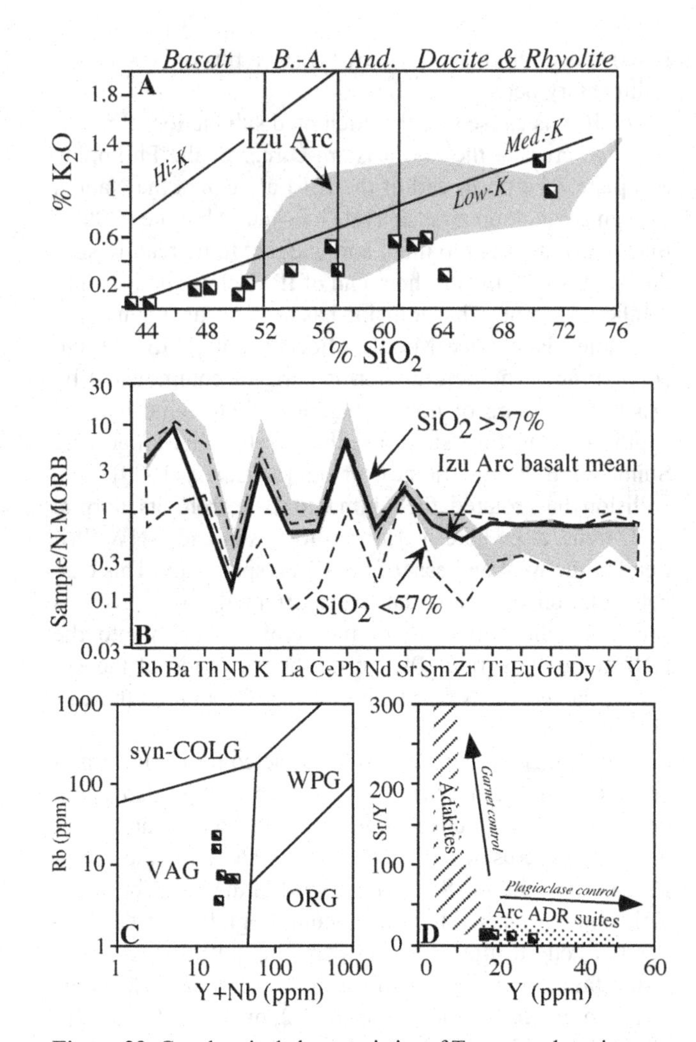

Figure 29. Geochemical characteristics of Tanzawa plutonic complex, all data from *Kawate and Arima* [1998]. A) K-Si diagram. Field occupied by volcano means from the Izu Arc (Figure 17) is shown in gray. Italicized volcanic rock names across the top (B.A. = basaltic andesite; And. = andesite) are intended as compositional guide. B) Incompatibility plot, composition of N-MORB and element order is after *Hofmann* [1988]. Field defined by mafic rocks (<57% SiO_2) is outlined with dashed line and that for felsic rocks (>57% SiO_2) is shaded. Izu Arc Basalt Mean is from Figure 19, note absence of data for U, Ta, and Hf. C) Tectonic discriminant diagram after *Pearce et al.* [1984]. Syn-COLG = syn-collisional granitoid; WPG = within-plate granitoid; ORG = ocean-ridge granitoid; VAG = volcanic arc granitoid. D) Petrogenetic diagram for adakites vs. arc andesite-dacite-rhyolite (ADR) suites, showing the composition of felsic rocks generated in equilibrium with garnet vs. those generated in equilibrium with plagioclase. Figure modified after *Drummond and Defant* [1990].

8. CONCLUSIONS

The IBM Subduction Factory opened for business about 50 million years ago and is now in an equilibrium state of operation. We understand the most important parts of the IBM arc system, including raw materials fed into and finished products issuing from this Subduction Factory. Inputs into the IBM Subduction Factory define an end-member assemblage of the oldest seafloor on earth which youngs northward. Sedimentary sequences on this crust are dominated by a late Mesozoic pelagic succession with a southward-thickening wedge of mid-Cretacous volcanic and volcaniclastic rocks. Seafloor and associated supracrustal sequences are entirely subducted. Plate convergence increases from about 3 cm/year in the south to about 6 cm/year in the north. Relative convergence between the Pacific and the IBM arc system is oblique along much of the IBM trench and varies from orthogonal to margin-parallel along strike. In contrast to the situation for Pacific-Philippine Sea plate convergence, which has a pole of rotation near the southern end of the IBM arc, Philippine Sea-Eurasian plate convergence has a pole of rotation north of the IBM arc system. The clockwise rotation of the Philippine Sea plate coupled with the relatively slow subduction of the oldest and most dense oceanic lithosphere on the planet results in vertical sinking of the Pacific plate in the southern IBM, producing a sea-anchor force that is responsible for back-arc extension in the southern IBM. Combined effects of back-arc extension and oblique convergence result in stretching of the forearc in the south.

Along-strike variations in seafloor age and convergence rate control the shape of the subduction zone. The Wadati-Benioff zone smoothly steepens dip from 40° to 80° southwards, although earthquakes deeper than 300km are common only beneath the Izu segment and central Marianas. Double seismic zones exist beneath the northern and southern IBM. The trajectory of subducted material can be traced with earthquakes down to a depth of about 700 km and tomographically can be outlined well into the lower mantle. In contrast, the more rapid subduction of younger, less dense oceanic lithosphere beneath northern IBM has resulted in a gentler angle of subduction and the stagnation of the subducted slab above the 660 km discontinuity. Shear-wave splitting data indicate that olivines in the mantle beneath IBM are oriented NW-SE to E-W and this is about what is expected for flow induced by NW plate convergence.

The IBM arc system is naturally divided into Izu, Bonin, and Mariana segments characterized by different morphologies, tectonic histories, and locus of outputs. The crust is about 22 km thick beneath the Izu segment and has a distinctive middle crust composed of felsic igneous rocks. The thickness and velocity structure for this arc segment is quite different from that of the central Aleutians, the only other arc for which we have a similar level of geophysical understanding. The middle crust of the IBM arc system can be studied on-land in the Izu Collision Zone, where IBM has been colliding end-on with Honshu for the past 15 Ma. The bulk crust of northern IBM contains about 56% SiO_2, much more siliceous than generally thought for IOCM, which have been thought to have basaltic bulk compositions. Further work is needed to evaluate the crustal structure beneath other parts of the arc. Back-arc basin crust of the Mariana Trough is similar in thickness to crust produced by spreading at mid-ocean ridges.

Outputs from the IBM Subduction Factory can be examined at forearc mud volcanoes and cold-water vents, along the magmatic arc, at cross-chains and in extension-related back-arc magmatic systems. Low sedimentation rates over most of the IBM arc system allow most of the forearc to be studied directly, simplifying assessments of the source of fluids issuing from vents and, incidentally providing one of the best places on the planet to study how subduction begins. The forearc mud volcanoes of southern IBM result from strong extension of this forearc and provide unique opportunities to sample slab-derived fluids, blueschist-facies subducted crust, and entrained upper mantle. We speculate that the cool mantle beneath the IBM forearc may support the deepest biosphere in the planet.

The 2500-km long line of submarine and subaerial arc volcanoes provide the most spectacular output from the IBM Subduction Factory. These volcanoes erupt porphyritic and fractionated lavas, the study of which demands new techniques and approaches. There are important compositional variations along strike of the IBM arc system. The Izu segment erupts some of the most depleted arc lavas on the planet but is flanked southward by the only active shoshonitic province found along the magmatic front of an IOCM. Mariana arc lavas are less depleted than Izu arc lavas, and this may reflect an important role for subducted volcaniclastics in the south. The Izu segment is also distinctive in having abundant submarine felsic volcanoes. Mariana arc lavas contain about 2-3% magmatic water. These lavas are characterized by distinctive depletions in Ta and Nb and enrichments in Ba, Rb, K, Pb and Sr. Isotopic tracers allow some subducted components to be unequivocally identified (e.g., water, ^{11}B, ^{34}S) but not others (e.g., ^{4}He, ^{18}O). Uncertainties regarding the affinities of the unmodified mantle source (MORB-like or OIB-like) complicate interpretations of other isotopic tracers. Lavas from arc cross-chains such as the Zenisu Ridge and the Kasuga seamounts indicate that the role of subducted components decreases across the arc, leading to diminished melt production with distance from the trench. Extension-related igneous activity, localized in the Mariana Trough and the Izu rifts, provides a final perspective on IBM Subduction Factory outputs. Depending on the speed or duration of extension, IBM extensional zones may be rifting or spreading. Lavas associated with extension are primitive, glassy, and relatively straightforward to study. Rifting is associated with bimodal volcanism whereas only basalts are erupted from regions undergoing spreading. Lavas associated with backarc basin seafloor spreading contain variable amounts of water, with an average of about 1.6%; those with little water are indistinguishable from MORB whereas those with abundant water also carry a subduction component that is readily identified in trace element or isotopic signatures.

We hope that this summary and synthesis serves to motivate and guide students and scientists to help elucidate the workings of the large, complex, and deeply hidden Subduction Factory. Development of a quantitative model that explains all aspects of Subduction Factory products is our ultimate goal. A truly interdisciplinary and international effort will be required to achieve this for IBM.

Acknowledgments. We appreciate the opportunity to synthesize our understanding of the IBM arc system provided by the conveners of the Subduction Factory workshop at Eugene. We thank those who let us use their unpublished data in this synthesis: Julie Morris, Julian Pearce, Mark Reagan, and Jeff Vervoort, and to Jim Gill for pointing out some of our mistakes. We are also grateful for thoughtful reviews by John Eiler and Mark Reagan. This is UTD Geosciences contribution # 951.

REFERENCES

Abers, G.A., Plate structure and the origin of double seismic zones, in *Subduction: Top to Bottom,* edited by G.E. Bebout, D.W. Scholl, S.H. Kirby, and J.P. Platt, pp. 223-228, AGU Geophysical Monograph 96, American Geophysical Union, Washington DC, 1996.

Abrams, L.J., R.L. Larson, T.H. Shipley, and Y. Lancelot, Cretaceous Volcanic Sequences and Jurassic Oceanic Crust in the East Mariana and Pigafetta Basins of the Western Pacific, in *The Mesozoic Pacific: Geology, Tectonics, and Volcanism,* edited by M.S. Pringle, W.W. Sager, W.V. Sliter, and S. Stein, pp. 77-101, AGU Geophysical Monograph 97, American Geophysical Union, Washington DC, 1993.

Ali, J.R., and S.J. Moss, Miocene intraarc bending at an arc-arc collision zone, central Japan: Comment, *The Island Arc,* 8, 114-117, 1999.

Alt, J.C., C. France-Lanord, P.A. Floyd, P. Castillo, and A. Galy., Low-temperature hydrothermal alteration of Jurassic ocean crust, Site 801, in *Proceedings of the Ocean Drilling Project, Scientific Results,* edited by R.L. Larson, Y. Lancelot, et al., pp. 415-427, Ocean Drilling Project, College Station, TX, 1992.

Alt, J.C., W.C. Shanks III, and M.C. Jackson, Cycling of sulfur in subduction zones; the geochemistry of sulfur in the Mariana island arc and back-arc trough, *Earth Planet. Sci. Lett.,* 119, 477-494, 1993.

Anderson, D.L., *Theory of the Earth,* 366 pp., Blackwell Scientific, Cambridge, Mass., 1989.

Anderson, R.N., Heat flow in the Mariana marginal basin, *J. Geophys. Res.,* 80, 4043-4048, 1975.

Ando, M., Y. Ishikawa, and F. Yamazaki, Shear wave polarization anisotropy in the upper mantle beneath Honshu, Japan, *J. Geophys. Res.,* 88, 5850-5864, 1983.

Aramaki, S., and J. Itoh, Rocks of Niijima volcano, in *Characterization of Volcanic rocks in Izu Islands: Data set for Prevention of Volcanic Hazard,* pp. 31-45, Bureau of General Affair, Disaster Prevention Division, Tokyo Metropolitan Government, 1993.

Aramaki, S., S. Oshima, K. Ono, S. Sudoh, and J. Itoh, Rocks of Kozushima volcano, in *Characterization of Volcanic rocks in Izu Islands: Data set for Prevention of Volcanic Hazard,* pp. 7-29, Tokyo Metropolitan Government, 1993.

Bartolini, A., and R.L. Larson, Pacific microplate and the Pangea supercontinent in the Early to Middle Jurassic, *Geology,* 29, 735-738, 2001.

Bellaiche, G., Sedimentation and structure of the Izu-Ogasawara (Bonin) Trench off Tokyo: New lights on the results of a diving campaign with the Bathyscape "Archimede", *Earth Planet. Sci. Lett.,* 47, 124-130, 1980.

Bibee, L.D., G.G.S. Jr., and R.S. Lu, Inter-arc spreading in the Mariana Trough, *Marine Geol.,* 35, 183-197, 1980.

Bloomer, S.H., Distribution and origin of igneous rocks from the landward slopes of the Mariana Trench: Implications for its structure and evolution, *J. Geophys. Res.,* 88, 7411-7428, 1983.

Bloomer, S.H., and J.W. Hawkins, Gabbroic and ultramafic rocks from the Mariana Trench: An Island Arc ophiolite, in *The Tectonic and Geologic Evolution of Southeast Asian Seas and Islands: Part 2. Geophysical Monograph 27,* edited by D.E. Hayes, pp. 294-317, American Geophysical Union, Washington D.C., 1983.

Bloomer, S.H., and J.W. Hawkins, Petrology and geochemistry of boninite series volcanic rocks from the Mariana Trench, *Contrib. Mineral. Petrol., 97,* 361-377, 1987.

Bloomer, S.H., R.J. Stern, E. Fisk, and C.H. Geschwind, Shoshonitic Volcanism in the Northern Mariana Arc. 1. Mineralogic and Major and Trace Element Characteristics, *J. Geophys. Res.,* 94, 4469-4496, 1989a.

Bloomer, S.H., R.J. Stern, and N.C. Smoot, Physical volcanology of the submarine Mariana and Volcano Arcs, *Bull. Volcanol., 51,* 210-224, 1989b.

Bloomer, S.H., B. Taylor, C.J. MacLeod, and et. al., Early Arc volcanism and the Ophiolite problem: A perspective from drilling in the Western Pacific, in *Active Margins and Marginal Basins of the Western Pacific,* edited by B. Taylor, and J. Natland, pp. 67-96, AGU Geophysical Monograph 88, American Geophysical Union, Washington D.C., 1995.

Bodine, J.H., and A.B. Watts, On Lithospheric Flexure seaward of the Bonin and Mariana Trenches, *Earth Planet. Sci. Lett.*, 43, 132-148, 1979.

Brudzinski, M.R., W.-P. Chen, R.L. Nowack, and B.S. Huang, Variations of P wave speeds in the mantle transition zone beneath the northern Philippine Sea, *J. Geophys. Res.*, 102 (6), 11,815-11,827, 1997.

Bryant, C.J., R.J. Arculus, and S.M. Eggins, Laser ablation—inductively coupled plasma mass spectrometry and tephras: A new approach to understanding arc-magma genesis, *Geology, 27,* 1119-1122, 1999.

Burns, L.E., The Border Ranges ultramafic and mafic complex, south-central Alaska: cumulate fractionates of island arc volcanics, *Can. J. Earth Sci., 22,* 1020-1038, 1995.

Castillo, P.R., R.W. Carlson, and R. Batiza, Origin of Nauru Basin igneous complex: Sr, Nd, and Pb isotope and REE constraints, *Earth Planet. Sci. Lett., 103,* 200-213, 1991.

Castillo, P.R., P.A. Floyd, and C. France-Lanord, Isotope geochemistry of Leg 129: Implications for the origin of the widespread Cretaceous volcanic event in the Pacific, in *Proceedings of the Ocean Drilling Project, Scientific Results,* edited by R.L. Larson, Y. Lancelot, et al., pp. 405-413, Ocean Drilling Project, College Station, TX, 1992.

Castillo, P.R., M.S. Pringle, and R.W. Carlson, East Mariana Basin tholeiites: Cretaceous intraplate basalts or rift basalts related to the Ontong Java plume?, *Earth Planet. Sci. Lett., 123,* 139-154, 1994.

Castle, J.C., and K.C. Creager, Topography of the 660-km seismic discontinuity beneath Izu-Bonin: Implications for tectonic history and slab deformation, *J. Geophys. Res., 103,* 12,511-12,527, 1998.

Castle, J.C., and K.C. Creager, A steeply dipping discontinuity in the lower mantle beneath Izu-Bonin, *J. Geophys. Res., 104,* 7279-7292, 1999.

Channell, J.T., E. Erba, M. Nakanishi, and K. Tamaki, Late Jurassic-Early Cretaceous timescales and oceanic magnetic anomaly block models, in *Geochronology, timescales and global stratigraphic correlation,* edited by W.A. Berggren, D.V. Kent, M.-P. Aubry, and J. Hardenbol, pp. 51-63, SEPM Special Publication, Tulsa, 1995.

Chaussidon, M., and A. Jambon, Boron content and isotopic composition of oceanic basalts: Geochemical and cosmochemical implications, *Earth Planet. Sci. Lett., 121,* 277-291, 1994.

Christensen, N.J., and W.D. Mooney, Seismic velocity structure and composition of the continental crust: A global view, *J. Geophys. Res., 100,* 9761-9788, 1995.

Cloos, M., Thrust-type subduction earthquakes and seamount asperities: A physical model for seismic rupture, *Geology, 20,* 601-604, 1992.

Cloud, P.E., Jr., R.G. Schmidt, and H.W. Burke, Geology of Saipan, Mariana Islands. Part 1. General Geology, pp. 126, U.S. Geological Survey Professional Paper 280, Washington DC, 1956.

Collier, J.D., and G.R. Helffrich, Topography of the "410" and "660" km seismic discontinuities in the Izu-Bonin subduction zone, *Geophys. Res. Lett., 24,* 1535-1538, 1997.

Corwin, G., L.D. Bonham, M.J. Terman, and G.W. Viele, Military Geology of Pagan, Mariana Islands, pp. 259, United States Army, Washington, D.C., 1957.

Cosca, M.A., R.J. Arculus, J.A. Pearce, and J.G. Mitchell, $^{40}Ar/^{39}Ar$ and K-Ar geochronological age constraints for the inception and early evolution of the Izu-Bonin-Mariana arc system, *The Island Arc, 7,* 1998.

Crawford, A.J., T.J. Falloon, and D.H. Green, Classification, petrogenesis and tectonic setting of boninites, in *Boninites and Related Rocks,* edited by A.J. Crawford, pp. 1-49, Unwin Hyman, Boston, 1989.

Creager, K.C., and T.H. Jordan, Slab penetration into the lower mantle beneath the Mariana and other island arcs of the Northwest Pacific, *J. Geophys. Res.*, 91, 3573-3589, 1986.

Davis, A.S., M.S. Pringle, L.G. Pickthorn, and D.A. Clague, Petrology and age of Alkalic Lava from the Ratak Chain of the Marshall Islands, *J. Geophys. Res.*, 94, 5757-5774, 1989.

DeBari, S.M., B. Taylor, K. Spencer, and K. Fujioka, A trapped Philippine Sea plate origin for MORB from the inner slope of the Izu-Bonin trench, *Earth Planet. Sci. Lett., 174,* 183-197, 1999.

DeMets, C., R.G. Gordon, D.D. Argus, and S. Stein, Effect of recent revisions to the geomagnetic reversal timescale on estimates of current plate motions, *Geophys. Res. Lett., 21,* 2191-2194, 1994.

Dick, H.J.B., and T. Bullen, Chromian spinels as a petrogenetic indicator in abyssal and alpine-type peridotites and spatially associated lavas, *Contrib. Mineral. Petrol., 86,* 54-76, 1984.

Dixon, T.H., and R.J. Stern, Petrology, chemistry, and isotopic composition of submarine volcanoes in the southern Mariana arc, *Geol. Soc. Am., Bull., 94,* 1159-1172, 1983.

Drummond, M.S., and M.J. Defant, A model for trondhjemite-tonalite-dacite genesis and crustal growth via slab melting: Archean to modern comparisons, *J. Geophys. Res., 95,* 503-521, 1990.

Eiler, J.M., A. Crawford, T. Elliott, K.A. Farley, J.W. Valley, and E.M. Stolper, Oxygen Isotope Geochemistry of Oceanic-Arc lavas, *J. Petrol., 41,* 229-256, 2000.

Elliott, T., T. Plank, A. Zindler, W. White, and B. Bourdon, Element transport from slab to volcanic front at the Mariana arc, *J. Geophys. Res., 102,* 14,991-15,019, 1997.

Engdahl, E.R., R.D. van der Hilst, and R. Buland, Global teleseismic earthquake relocation with improved travel times and procedures for depth determination, *Bull. Seismol. Soc. Am., 88,* 722-743, 1998.

Fischer, K.M., T.H. Jordan, and K.C. Creager, Seismic constraints on the morphology of deep slabs, *J. Geophys. Res., 93,* 4773-4783, 1988.

Flanagan, M.P., and P.M. Shearer, Global mapping of topography on transition zone velocity discontinuities by stacking SS precursors, *J. Geophys. Res., 103,* 2673-2692, 1998.

Fliedner, M.M., and Klemperer, S.L., Structure of an island arc; wide-angle seismic studies in the eastern Aleutian Islands, Alaska, *J. Geophys. Res., 104,* 10,667-10,694.

Fliedner, M.M., and S.L. Klemperer, The transition from oceanic arc to continental arc in the crustal structure of the Aleutian Arc, *Earth and Planetary Science Letters, 179,* 567-579, 2000.

Fouch, M.J., and K.M. Fischer, Mantle anisotropy beneath northwest Pacific subduction zones, *J. Geophys. Res.,* 101, 15,987-16,002, 1996.

Fouch, M.J., and K.M. Fischer, Shear wave anisotropy in the Mariana subduction zone, *Geophys. Res. Lett., 25,* 1221-1224, 1998.

Fryer, P., A synthesis of Leg 125 Drilling of Serpentine Seamounts on the Mariana and Izu-Bonin forearcs, in *Proceedings of the Ocean Drilling Project, Scientific Results,* edited by P. Fryer, J. Pearce, L.B. Stokking, et al., pp. 3-11, Ocean Drilling Program, College Station, 1992.

Fryer, P., Evolution of the Mariana Convergent Margin, *Rev. Geophys., 34,* 89-125, 1996a.

Fryer, P., Geology of the Mariana Trough, in *Back-arc Basins: Tectonics and Magmatism,* edited by B. Taylor, Plenum Press 1995.

Fryer, P., J.B. Gill, and M.C. Jackson, Volcanologic and tectonic evolution of the Kasuga seamounts, northern Mariana Trough: Alvin submersible investigations, *J. Volcanol. Geotherm. Res., 79,* 277-311, 1997.

Fryer, P., M. Mottl, J. L., S. Phipps, and H. Maekawa, Serpentine bodies in the Forearcs of Western Pacific Convergent Margins: Origin and Associated Fluids, in *Active Margins and Marginal Basins of the Western Pacific Convergent Margins,* edited by B. Taylor, and J. Natland, pp. 259-270, AGU Geophysical Monograph 88, American Geophysical Union, Washington DC, 1995.

Fryer, P., and M.J. Mottl, "Shinkai 6500" investigations of a resurgent mud volcano on the Southeastern Mariana forearc, *JAMSTEC Journal of Deep Sea Research, 13,* 103-114, 1997.

Fryer, P., K.L. Saboda, L.E. Johnson, M.E. Mackay, G.F. Moore, and P. Stoffers, Conical Seamount: SeaMARC II, ALVIN Submersible, and seismic-reflection studies, in *Proceedings of the Ocean Drilling Project, Scientific Results,* edited by P. Fryer, and J. Pearce, pp. 3-11, Ocean Drilling Project, College Station, 1990a.

Fryer, P., B. Taylor, C.H. Langmuir, and A.G. Hochstaedter, Petrology and geochemistry of lavas from the Sumisu and Torishima backarc rifts, *Earth Planet. Sci. Lett., 100,* 161-178, 1990b.

Fryer, P., C.G. Wheat, and M. Mottl, Mariana blueschist mud volcanism; implications for conditions within the subduction zone, *Geology, 27,* 103-106, 1999.

Fukao, Y., M. Obayashi, H. Inoue, and M. Nenbai, Subducting slabs stagnant in the mantle transition zone, *J. Geophys. Res., 97,* 4809-4822, 1992.

Garcia, M.O., N.W.K. Liu, and D.W. Muenow, Volatiles in submarine volcanic rocks from the Mariana Island arc and trough, *Geochim. Cosmochim. Acta, 43,* 305-312, 1979.

Gill, J.B., R.N. Hiscott, and Ph. Vidal, Turbidite geochemistry and evolution of the Izu-Bonin arc and continents, *Lithos, 33,* 135-168, 1994.

Gill, J.B., C. Seales, P. Thompson, A.G. Hochstaedter, and C. Dunlap, Petrology and Geochemistry of Pliocene-Pleistocene volcanic rocks from the Izu Arc, Leg 126, in *Proceedings of the Ocean Drilling Program, Scientific Results,* edited by B. Taylor, K. Fujioka, et al., pp. 383-404, Ocean Drilling Program, College Station, 1992.

Gill, J.B., and R.W. Williams, Th isotope and U-series studies of subduction-related volcanic rocks, *Geochim. Cosmochim. Acta, 54,* 1427-1442, 1990.

Green, H.W., and H. Houston, The mechanics of deep earthquakes, *Ann. Rev. Earth Planet Sci., 23,* 169-213, 1995.

Gribble, R.F., R.J. Stern, S.H. Bloomer, D. Stüben, T. O'Hearn, and S. Newman, MORB Mantle amd subduction components interact to generate basalts in the southern Mariana Trough back-arc basin, *Geochim. Cosmochim. Acta, 60,* 2153-2166, 1996.

Gribble, R.F., R.J. Stern, S. Newman, S.H. Bloomer, and T. O'Hearn, Chemical and Isotopic Composition of Lavas from the Northern Mariana Trough: Implications for Magmagenesis in Back-arc basins, *J. Petrol., 39,* 125-154, 1998.

Gu, Y., A.M. Dziewonski, and C.B. Agee, Global de-correlation of the topography of transition zone discontinuities, *Earth Planet. Sci. Lett., 157,* 57-67, 1998.

Haggerty, J.A., Evidence from fluid seeps atop serpentine seamounts in the Mariana Forearc: Clues for emplacement of the seamounts and their relationship to forearc tectonics, *Marine Geol., 102,* 293-309, 1991.

Haggerty, J.A., and S. Chaudhuri, Strontium isotopic composition of the interstitial waters from Leg 125: Mariana and Bonin forearcs, in *Proceedings of the Ocean Drilling Project, Scientific Results,* edited by P. Fryer, J. Pearce, L.B. Stokking, and et. al., pp. 397-400, Ocean Drilling Program, College Station, 1992.

Hall, R., M. Fuller, J.R. Ali, and C.D. Anderson, The Philippine Sea Plate: Magnetism and Reconstructions, in *Active Margins and Marginal Basins of the Western Pacific,* edited by B. Taylor, and J. Natland, pp. 371-404, AGU Geophysical Monograph 88, American Geophysical Union Monograph, Washington DC, 1995.

Hamuro, K., S. Aramaki, K. Fujioka, T. Ishii, and T. Tanaka, The Higashi-Izu-Oki submarine volcanoes, Part 2, and the submarine volcanoes near the Izu Shoto Islands (in Japanese), *Bull. Earthquake Res. Inst., 58,* 527-557, 1983.

Handschumacher, D.W., W.W. Sager, T.W.C. Hilde, and D.R. Bracey, Pre-Cretaceous tectonic evolution of the Pacific plate and extension of the geomagnetic polarity reversal timescale with implications for the origin of the Jurassic "Quiet Zone", *Tectonophy., 155,* 365-380, 1988.

Harland, W.B., R.L. Armstrong, A.V. Cox, L.E. Craig, A.G. Smith, and D.G. Smith, *A Geologic Timescale,* 263 pp., Cambridge University Press, Cambridge, UK, 1990.

Harry, D.L., and J.F. Ferguson, Bounding the state of stress at oceanic convergent zones, *Tectonophys., 187,* 305-314, 1991.

Hart, S.R., A large scale isotope anomaly in the Southern Hemisphere mantle, *Nature,* 309, 753-757, 1984.

Hart, S.R., Heterogeneous mantle domains: Signatures, genesis and mixing chronologies, *Earth Planet. Sci. Lett., 90,* 273-296, 1988.

Hart, S.R., W.E. Glassley, and D.E. Karig, Basalts and seafloor spreading behind the Mariana Island Arc, *Earth Planet. Sci. Lett., 15,* 12-18, 1972.

Hawkins, J.W., P.F. Lonsdale, J.D. Macdougall, and A.M. Volpe, Petrology of the axial ridge of the Mariana Trough backarc spreading center, *Earth Planet. Sci. Lett., 100,* 226-250, 1990.

Hawkins, J.W., and Melchior, Petrology of Mariana Trough and Lau Basin basalts, *J. Geophys. Res., 90,* 11,431-11,468, 1985.

Hess, H.H., Major structural features of the western North Pacific, An interpretation of H.O. 5485, Bathymetric chart, Korea to New Guinea, *Geol. Soc. Am. Bull., 59,* 417-445, 1948.

Hickey, R.L., and F.A. Frey, Geochemical characteristics of boninite series volcanics: Implications for their source, *Geochim. Cosmochim. Acta, 45,* 2099-2115, 1982.

Hickey-Vargas, R., J.M. Hergt, and P. Spadea, The Indian Ocean-type isotopic signature in Western Pacific marginal basins: Origin and significance, in *Active Margins and Marginal Basins of the Western Pacific,* edited by B. Taylor, and J. Natland, pp. 175-197, AGU Geophysical Monograph 88, American Geophysical Union, Washinton DC, 1995.

Hickey-Vargas, R.L., and M. Reagan, Temporal variations of isotope and rare earth element abundances from Guam: Implications for the evolution of the Mariana island arc, *Contrib. Mineral. Petrol., 97,* 497-508, 1987.

Hino, R., A study on the crustal structure of the Izu-Bonin arc by airgun-OBS profiling, Ph.D. thesis, Tohoku University, Sendai, Japan, 1991.

Hobbs, W.H., The Asiatic Arcs, *Bull. Geol. Soc. Am., 34,* 243-252, 1923.

Hochstaedter, A.G., J.B. Gill, M. Kusakabe, S. Newman, M. Pringle, B. Taylor, and P. Fryer, Volcanism in the Sumisu Rift, I. Major element, volatile, and stable isotope geochemistry, *Earth and Planetary Science Letters, 100,* 179-194, 1990a.

Hochstaedter, A.G., J.B. Gill, and J.D. Morris, Volcanism in the Sumisu Rift, II. Subduction and non-subduction related components, *Earth and Planetary Science Letters, 100,* 195-209, 1990b.

Hochstaedter, A.G., J. B. Gill, R. Peters, P. Broughton, P. Holden, and B. Taylor, Across-Arc Geochemical trends in the Izu-Bonin Arc: Contributions from the Subducting Slab, *Geochem. Geophys. Geosyst., 2* (Article), 2000GC000105, 2001.

Hochstaedter, A.G., J.B. Gill, B. Taylor, O. Ishizuka, M. Yuasa, and S. Morita, Across-arc geochemical trends in the Izu-Bonin Arc; Constraints on source composition and mantle melting. *J. Geophys. Res., 105,* 495-512, 2000.

Hofmann, A.W., Chemical differentiation of the Earth: The relationship between mantle, continental crust, and oceanic crust, *Earth Planet. Sci. Lett., 90,* 297-314, 1988.

Holbrook, W.S., D. Lizarralde, S. McGeary, N. Bangs, and J. Deibold, Structure and composition of the Aleutian island arc and implications for continental crustal growth, *Geology, 27,* 31-34, 1999.

Honza, E., and K. Tamaki, The Bonin Arc, in *The Ocean Basins and Margins,* edited by A.E. Nairn, F.G. Stehli, and S. Uyeda, pp. 459-502, Plenum Publishing Co,, 1985.

Hussong, D.M., and S. Uyeda, Tectonic processes and the history of the Mariana Arc, A synthesis of the results of Deep Sea Drilling leg 60, in *Initial Reports of the Deep Sea Drilling Project,* edited by D.M. Hussong, et al., pp. 909-929, U.S. Government Printing Office, Washington D.C., 1982.

Hyndman, R.D., and S.L. Klemperer, Lower-crustal porosity from electrical measurements and inferences about composition from seismic velocities, *Geophys. Res. Lett., 16,* 255-258, 1989.

Iidaka, T., and Y. Furukawa, Double seismic zone for deep earthquakes in the Izu-Bonin subduction zone, *Science, 263* (5150), 1116-1118, 1994.

Iizasa, K., R.S. Fiske, O. Ishizuka, M. Yuasa, J. Hashimoto, J. Ishibashi, J. Naka, Y. Horii, Y. Fujiwara, A. Imai, and S. Koyama, A Kuroko-type polymetallic sulfide deposit in a submarine silicic caldera, *Science, 283,* 975-977, 1999.

Ikeda, Y., K. Nagao, R.J. Stern, M. Yuasa, and S. Newman, Noble gases in pillow basalt glasses from the northern Mariana Trough back-arc basin, *The Island Arc, 7,* 471-478, 1998.

Ikeda, Y., and M. Yuasa, Volcanism in nascent back-arc basins behind the Shichito Ridge and adjacent areas in the Izu-Ogasawara arc, northwest Pacific: Evidence for mixing between E-type MORB and island arc magmas at the initiation of back-arc rifting, *Contrib. Mineral. Petrol., 101,* 377-393, 1989.

Ishihara, T., and T. Yamazaki, Gravity anomalies over the Izu-Ogasawara (Bonin) and northern Mariana Arcs, *Bull. Geol. Surv. Japan, 42,* 687-701, 1991.

Ishii, T., Dredged samples from the Ogasawara fore-arc seamount of "Ogasawara Paleoland"—"Fore-arc Ophiolite", in *Formation of Active Ocean Margins,* edited by N. Nasu, and e. al., pp. 307-342, Terra Scientific Publishing, Tokyo, 1985.

Ishii, T., P.T. Robinson, H. Maekawa, and R. Fiske, Petrological studies of peridotites from diapiric Serpentinite Seamounts in the Izu-Ogasawara-Mariana forearc, Leg 125, in *Proceedings of the Ocean Drilling Project, Scientific Results,* edited by P. Fryer, J. Pearce, L.B. Stokking, et al., pp. 445-485, Ocean Drilling Program, College Station, 1992.

Ishikawa, T., and E. Nakamura, Origin of the slab component in arc lavas from across-arc variation of B and Pb isotopes, *Nature, 370,* 205-208, 1994.

Ishikawa, T., and F. Tera, Two isotopically distinct fluid components involved in the Mariana Arc; evidence from Nb/B ratios and B, Sr, Nd, and Pb isotope systematics, *Geology,* 27, 83-86, 1999.

Ishizaka, K., and T. Yanagi, K, Rb, and Sr abundances and Sr isotopic composition of the Tanzawa granitic and associated gabbroic rocks, Japan: Low-potash island arc plutonic complex, *Earth Planet. Sci. Lett., 33,* 345-352, 1977.

Ishizuka, O., K. Uto, M. Yuasa, and A.G. Hochstaedter, K-Ar ages from seamount chains in the back-arc region of the Izu-Ogasawara arc, *The Island Arc, 7,* 408-421, 1998.

Isshiki, N., Ao-ga-sima Volcano, *Jap. J. Geol. Geog., 27,* 209-218, 1955.

Isshiki, N., Petrology of Hachijo-Jima volcano group, Seven Izu Islands, Japan, *J. Fac. Sci., U. Tokyo, 15,* 91-134, 1963.

Ito, E., and R.J. Stern, Oxygen- and strontium-isotopic investigations of subduction zone volcanism: The case of the Volcano Arc and the Marianas Island Arc, *Earth Planet. Sci. Lett., 76,* 312-320, 1986.

Ito, E., W.M. White, and C. Göpel, The O, Sr, Nd, and Pb isotope geochemistry of MORB, *Earth Planet. Sci. Letters, 62,* 157-176, 1987.

Itoh, Y., Differential rotation of the northern part of southwest Japan: Paleomagnetism of Early to Late Miocene rocks from Yatsuo Area in Chichibu District, *J. Geomag. Geo-elec., 38,* 325-334, 1986.

Jackson, M.C., Petrology and Petrogenesis of Recent Submarine Volcanics from the Northern Mariana Arc and Back-Arc Basin, Ph.D. thesis, University of Hawaii, Honolulu, 1989.

Jackson, M.C., Crystal accumulation and Magma Mixing in the Petrogenesis of Tholeiitic Andesites from Fukujin Seamount, Northern Mariana Island Arc, *J. Petrol., 34,* 259-289, 1993.

Jaggar, T., Myojin Reef, *Volcano Notes* (518), 13-14, 1952.

Johnson, L.E., Mafic clasts in serpentine seamounts: Petrology and geochemistry of a diverse crustal suite from the outer Mariana forearc, in *Proceedings of the Ocean Drilling Project, Scientific Results,* edited by P. Fryer, J. Pearce, L.B. Stokking, et al., pp. 401-413, Ocean Drilling Program, College Station, 1992.

Johnson, L.E., P. Fryer, B. Taylor, M. Silk, D.L. Jones, W.V. Sliter, T. Itaya, and T. Ishii, New evidence for crustal accretion in the outer Mariana forearc: Cretaceous Radiolarian cherts and MORB-like lavas, *Geology,* 19, 811-814, 1991.

Karig, D.E., Origin and development of marginal basins in the western Pacific, *J. Geophys. Res.,* 76, 2542-2561, 1971a.

Karig, D.E., Structural history of the Mariana island arc system, *Geol. Soc. Am. Bull.,* 82, 323-344, 1971b.

Karig, D.E., Remnant arcs, *Geol. Soc. Am. Bull., 83,* 1057-1068, 1972.

Katsumata, M., and L.R. Sykes, Seismicity and tectonics of the western Pacific: Izu-Mariana-Caroline and Ryukyu-Taiwan regions, *J. Geophys. Res., 74,* 5923-5948, 1969.

Kawate, S., and M. Arima, Petrogenesis of the Tanzawa plutonic complex, central Japan: Exposed felsic middle crust of the Izu-Bonin-Mariana arc, *The Island Arc, 7,* 342-358, 1998.

Klaus, A., and B. Taylor, Submarine canyon development in the Izu-Bonin forearc: A SeaMARC II and Seismic Survey of Aoga Shima, *Marine Geophys. Res., 13,* 131-152, 1991.

Klaus, A., B. Taylor, G.F. Moore, F. Murakami, and Y. Okamura, Back-arc rifting in the Izu-Bonin Island Arc: Structural evolution of Hachijo and Aoga Shima rifts, *The Island Arc, 1,* 16-31, 1992.

Kobayashi, K., S. Kasuga, and K. Okino, Shikoku Basin and Its Margins, in *Backarc Basins: Tectonics and Magmatism,* edited by B. Taylor, pp. 381-405, Plenum Press, New York, 1995.

Kotake, Y., T. Kato, S. Miyazaki, and A. Sengoku, Relative motion of the Philippine Sea Plate derived from GPS observations and tectonics of the south-western Japan (in Japanese), *Zisin, J. Seismol. Soc., Jap., 51,* 171-180, 1998.

Lange, K., Auswertung eines weitwinkelreflexions- und refraktionsseismischen Profils im suedlichen Marianen Backarc Becken, Diplomarbeit thesis, Hamburg U., Hamburg, 1992.

Larson, E.E., R.L. Reynolds, R. Merrill, S. Levi, M. Ozima, Y. Aoki, H. Kinoshita, S. Zasshu, N. Kawai, T. Nakajima, and K. Hirooka, Major-element Petrochemistry of some Extrusive Rocks from the Volcanically Active Mariana Islands, *Bull. Volcanol., 38,* 361-377, 1975.

Larson, R.L., and C.G. Chase, Late Mesozoic evolution of the western Pacific Ocean, *Geol. Soc. Am. Bull., 83,* 3627-3644, 1972.

LaTraille, S.L., and D.M. Hussong, Crustal structure across the Mariana Island Arc, in *The Tectonic and Geologic Evolution of Southeast Asian Seas and Islands. Geophysical Monograph 23,* edited by D.E. Hayes, pp. 209-221, American Geophysical Union, Washington DC, 1980.

Lay, T., *Structure and Fate of Subducting Slabs,* 185 p. pp., Academic Press, San Diego, 1994.

Le Pichon, X., T. Iiyama, H. Chamley, M. Faure, H. Fujimoto, T. Furuta, Y. Ida, H. Kagami, S. Lalemant, J. Leggett, A. Murata, H. Okada, C. Rangin, V. Renard, A. Taira, and H. Tokuyama, The eastern and western ends of Nankai Trough, results of Box 5 and Box 7 Kaiko survey, *Earth Planet. Sci. Lett., 83,* 199-213, 1987.

Lee, J., and R.J. Stern, Glass inclusions in Mariana Arc phenocrysts; A new perspective on magmatic evolution in a typical intraoceanic arc, *J. Geol., 106,* 19-33, 1998.

Lee, J., R.J. Stern, and S.H. Bloomer, Forty million years of magmatic evolution in the Mariana Arc: The tephra glass record, *J. Geophys. Res., 100,* 17,671-17,687, 1995.

Lin, P.-N., Trace element and isotopic characteristics of western Pacific pelagic sediments: Implications for the petrogenesis of Mariana Arc magmas, *Geochim. Cosmochim. Acta, 56,* 1641-1654, 1992.

Lin, P.N., R.J. Stern, and S.H. Bloomer, Shoshonitic volcanism in the Northern Mariana Arc. 2. Large-ion lithophile and rare earth element abundances: Evidence for the source of incompatible element enrichments in Intraoceanic Arcs, *J. Geophys. Res., 94,* 4497-4514, 1989.

Lin, P.N., R.J. Stern, J. Morris, and S.H. Bloomer, Nd- and Sr-isotopic compositions of lavas from the northern Mariana and southern Volcano arcs: Implications for the origin of island arc melts, *Contrib. Mineral. Petrol., 105,* 381-392, 1990.

Macpherson, C.G., D.R. Hilton, D.P. Mattey, and J.M. Sinton, Evidence for an ^{18}O-depleted mantle plume from contrasting $^{18}O/^{16}O$ ratios of back-arc lavas from the Manus Basin and Mariana Trough, *Earth Planet. Sci. Lett., 176,* 171-183, 2000.

Maekawa, H., M. Shozuni, T. Ishii, P. Fryer, and J.A. Pearce, Blueschist metamorphism in an active subduction zone, *Nature, 364,* 520-523, 1993.

Mahoney, J.J., and K.J. Spencer, Isotopic evidence for the origin of the Manihiki and Ontong Java oceanic plateaus, *Earth Planet. Sci. Lett., 104,* 196-210, 1991.

Marlow, M.S., L.E. Johnson, J.A. Pearce, P. Fryer, L.B.G. Pickhorn, and B.J. Murton, Upper Cenozoic volcanic rocks in the Mariana forearc recovered from drilling at Ocean Drilling Program Site 781: Implications for forearc magmatism, *J. Geophys. Res., 97,* 15,085-15,098, 1992.

Martinez, F., P. Fryer, N.A. Baker, and T. Yamazaki, Evolution of Backarc Rifting: Mariana Trough, 20°-24°N, *J. Geophy. Res., 100,* 3807-3827, 1995.

Martinez, F., P. Fryer, and N. Becker, Geophysical characteristics of the southern Mariana Trough, 11°50'N-13°40'N, *J. Geophys. Res., 105,* 16,591-16,607, 2000.

Matsuhisa, Y., Oxygen Isotopic compositions of Volcanic Rocks from the East Japan Island Arcs and their Bearing on Petrogenesis, *J. Volcanol. Geotherm. Res., 5,* 271-296, 1979.

McCaffrey, R., Estimates of modern arc-parallel strain rates in forearcs, *Geology, 24,* 27-30, 1996.

McCulloch, M.T., and A.J. Gamble, Geochemical and geodynamical constraints on subduction zone magmatism, *Earth Planet. Sci. Lett., 102,* 358-374, 1991.

McDermott, F., and C. Hawkesworth, Th, Pb, and Sr isotope variations in young island arc volcanics and oceanic sediments, *Earth Planet. Sci. Lett., 104,* 1-15, 1991.

McDonald, G.A., Petrography of Iwo Jima, *Geol. Soc. Am. Bull., 59,* 1009-1018, 1948.

McMurtry, G.M., P.N. Sedwick, P. Fryer, D.L. Vonderhaar, and H.-W. Yeh, Unusual geochemistry of Hydrothermal Vents on Submarine Arc Volcanoes: Kasuga Seamounts, Northern Mariana Arc, *Earth Planet. Sci. Lett., 114,* 517-528, 1993.

McNutt, M.K., E.L. Winterer, W.W. Sager, J.H. Natland, and G. Ito, The Darwin Rise: Cretaceous Superswell?, *Geophys. Res. Lett., 17,* 1101-1108, 1990.

Meen, J.K., R.J. Stern, and S.H. Bloomer, Evidence for magma mixing in the Mariana arc system, *The Island Arc, 7,* 443-459, 1998.

Meijer, A., Pb and Sr isotopic data bearing on the origin of volcanic rocks from the Mariana island arc system, *Geol. Soc. Am. Bull., 87,* 1358-1369, 1976.

Meijer, A., Mariana-Volcano Islands, in *Andesites,* edited by R.S. Thorpe, pp. 293-306, John Wiley and Sons, 1982.

Meijer, A., The origin of low-K rhyolites from the Mariana frontal arc, *Contrib. Mineral. Petrol., 83,* 45-51, 1983.

Meijer, A., M. Reagan, H. Ellis, M. Shafiqullah, J. Sutter, P. Damon, and K. S., Chronology of Volcanic Events in the Eastern Philippine Sea, edited by D.E. Hayes, pp. 349-359, The Tectonic and Geologic Evolution of Southeast Asian Seas and Islands Part 2, AGU Geophysical Monograph 27, American Geophysical Union, Washington DC, 1983.

Menard, H.W., Darwin Reprise, *J. Geophys. Res., 89,* 9960-9968, 1984.

Moriguti, T., and E. Nakamura, Across-arc variation of Li isotopes in lavas and implications for crust/mantle recycling at subduction zones, *Earth Planet. Sci. Lett., 164,* 167-174, 1998.

Morimoto, R., and J. Ossaka, The 1970-activity of Myojin reef, *J. Earthquake Stud. (in Japanese), 79,* 301-320, 1970.

Mottl, M.J., Pore waters from Serpentinite Seamounts in the Mariana and Izu-Bonin forearcs, Leg 125: Evidence for volatiles from the Subducting Slab, in *Proceedings of the Ocean Drilling Project, Scientific Results,* edited by P. Fryer, J. Pearce, L.B. Stokking, et al., pp. 373-385, Ocean Drilling Program, College Station, 1992.

Mrozowski, C.L., and D.E. Hayes, A seismic reflection study of faulting in the Mariana Forearc, in *The Tectonic and Geologic Evolution of Southeast Asian Seas and Islands,* edited by D.E. Hayes, pp. 223-234, AGU Geophysical Monograph 23, American Geophysical Union, Washington DC, 1980.

Murauchi, S., N. Den, S. Asano, H. Hotta, T. Yoshii, T. Asanuma, K. Hagiwara, K. Ichikawa, T. Sato, W.J. Ludwig, J.I. Ewing, N.T. Edgar, and R.E. Houtz, Crustal structure of the Philippine Sea, *J. Geophys. Res., 73,* 3143-3171, 1968.

Nakajima, K., and M. Arima, Melting experiments on hydrous low-K tholeiite: Implications for the genesis of tonalitic crust in the Izu-Bonin-Mariana arc, *The Island Arc, 7,* 359-373, 1998.

Nakamura, M., M. Ando, and T. Ohkura, Fine structure of deep Wadati-Benioff zone in the Izu-Bonin region estimated from S-to-P converted phase, *Phys. Earth Planet. Int., 106,* 63-74, 1998.

Nakanishi, M., Topographic Expression of Five Fracture Zones in the Northwestern Pacific Ocean, in *The Mesozoic Pacific: Geology, Tectonics, and Volcanism,* edited by M.S. Pringle, Sager, W.W., Sliter, W.V., Stein S., pp. 121-136, AGU Geophysical Monograph 97, American Geophysical Union, Washington D.C, 1993.

Nakanishi, M., K. Tamaki, and K. Kobayashi, Magnetic anomaly lineations from Late Jurassic to Early Cretaceous in the west-central Pacific Ocean, *Geophys. J. Internat., 109,* 701-719, 1992.

Newman, S., J.D. Macdougall, and R.C. Finkel, ^{230}Th-^{238}U disequilibrium in island arcs; evidence from the Aleutians and Marianas, *Nature, 308,* 268-270, 1984.

Newman, S., E. Stolper, and R.J. Stern, H2O and CO2 in magmas from the Mariana arc and back arc system, *Geochem. Geophys. Geosyst., 1* (Article), 1999GC000027, 2000.

Niitsuma, N., Collision tectonics in the southern Fossa Magna, *Modern Geol., 14,* 3-18, 1989.

Notsu, K., N. Isshiki, and M. Hirano, Comprehensive strontium isotope study of Quaternary volcanic rocks from the Izu-Ogasawara arc, *Geochem. J.* (17), 289-302, 1983.

Ogawa, Y., K. Kobayashi, H. Hotta, and K. Fujioka, Tension gashes on the oceanward slopes of the northern Japan and Mariana Trenches, *Marine Geol., 141,* 111-123, 1997.

Park, C.-H., K. Tamaki, and K. Kobayashi, Age-depth correlation of the Philippine Sea back-arc basins and other marginal basins in the world, *Tectonophys., 181,* 351-371, 1990.

Parkinson, I.J., C.J. Hawkesworth, and A.S. Cohen, Ancient mantle in a modern arc: Osmium isotopes in Izu-Bonin-Mariana forearc peridotites, *Science, 281,* 2011-2013, 1998.

Parkinson, I.J., and J.A. Pearce, Peridotites from the Izu-Bonin-Mariana forearc (ODP Leg 125); evidence for mantle melting and melt-mantle interaction in a supra-subduction zone setting, *J. Petrol., 39,* 1577-1618, 1998.

Peacock, S.M., Are the lower planes of double seismic zones caused by serpentine dehydration in subducted oceanic mantle?, *Geology, 29,* 299-302, 2001.

Pearce, J.A., N.B.W. Harris, and A.G. Tindle, Trace element discrimination diagrams for the tectonic interpretation of granitic rocks, *J. Petrol., 25,* 956-983, 1984.

Pearce, J.A., P.D. Kempton, G.M. Nowell, and S.R. Noble, Hf-Nd Element and Isotope Perspective on the Nature and Provenance of Mantle and Subduction Components in Western Pacific Arc-Basin systems, *J. Petrol., 40,* 1579-1611, 1999.

Pearce, J.A., S.R. Van Der Laan, R.J. Arculus, B.J. Murton, T. Ishii, J.A. Peate, and I.J. Parkinson, Boninite and Harzburgite from Leg 125 (Bonin-Mariana Forearc): A Case Study of Magma Genesis during the Initial Stages of Subduction, in *Proceedings of the Ocean Drilling Program,* edited by P. Fryer, J.A. Pearce, I.J. Stokking, et al., pp. 623-659, College Station, 1992.

Peate, D.W., and J.A. Pearce, Causes of spatial compositional variations in Mariana arc lavas: Trace element evidence, *The Island Arc, 7,* 479-495, 1998.

Petersen, J., Der Boninit von Peel island, *Jahrbuch der Hamburischen Wissenschafteftlichen Anstalten, Hamburg, 8,* 341-349, 1890.

Peyton, V., V. Levin, J. Park, M. Brandon, J. Lees, E. Gordeev, and A. Ozerov, Mantle flow at a slab edge: Seismic anisotropy in the Kamchatka region, *Geophys. Res. Lett., 28,* 379-382, 2001.

Phipps, S.P., and D. Ballotti, Rheology of Serpentine Muds in the Mariana-Izu-Bonin forearc, in *Proceedings of the Ocean Drilling Project, Scientific Results,* edited by P. Fryer, J.A. Pearce, L.B. Stokking, et al., pp. 363-372, Ocean Drilling Project, College Station, 1990.

Plank, T., and C. Langmuir, The chemical composition of subducting sediment and its consequence for the crust and mantle, *Chem. Geol., 145,* 325-394, 1998.

Poreda, R., Helium-3 and deuterium in backarc basalts: Lau and Mariana Trough, *Earth Planet. Sci. Lett., 73,* 244-254, 1985.

Poreda, R., and H. Craig, Helium isotope ratios in circum-Pacific volcanic arcs, *Nature, 338,* 473-478, 1989.

Pringle, M.S., Radiometric ages of basaltic basement recovered at sites 800, 801, and 802, Leg 129, Western Pacific Ocean, in *Proceedings of the Ocean Drilling Project, Scientific Results,* edited by R.L. Larson, Y. Lancelot, et al., pp. 389-404, Ocean Drilling Project, College Station, TX, 1992.

Reagan, M.K., and A. Meijer, Geology and geochemistry of early arc-volcanic rocks from Guam, *Geol. Soc. Am. Bull., 95,* 701-713, 1984.

Revenaugh, J., and T.H. Jordan, Mantle layering from ScS reverberations: 2. The transition zone, *J. Geophys. Res., 96,* 19,763-19,780, 1991.

Reymer, A., and G. Schubert, Phanerozoic addition rates to the continental crust and crustal growth, *Tectonics, 3,* 63-77, 1984.

Ribe, N.M., Seismic anisotropy and mantle flow, *J. Geophys. Res., 94,* 4213-4223, 1989.

Richards, M.A., and C. Lithgow-Bertelloni, Plate motion changes, the Hawaiian-Emperor Bend, and the apparent success and failure of geodynamic models, *Earth Planet. Sci. Lett., 137,* 19-27, 1996.

Sager, W.W., Mariana Arc structure inferred from gravity and seismic data, *J. Geophys. Res., 85,* 5382-5388, 1980.

Sakai, R., M. Kusakabe, M. Noto, and T. Ishii, Origin of waters responsible for serpentinization of the Izu-Ogaswara-Mariana forearc seamounts in view of hydrogen and oxygen isotope ratios, *Earth Planet. Sci. Lett., 100,* 291-303, 1990.

Samowitz, I.R., and D.W. Forsyth, Double seismic zone beneath the Mariana island arc, *J. Geophys. Res., 86,* 7013-7021, 1981.

Sample, J.C., and D.E. Karig, A volcanic production rate for the Mariana Island Arc, *J. Volcanol. Geotherm. Res., 13,* 73-82, 1982.

Sano, Y., Y. Nishio, T. Gamo, A. Jambon, and B. Marty, Noble gas and carbon isotopes in Mariana Trough basalt glasses, *Appl. Geochem., 13,* 441-449, 1998.

Scholz, C.H., and J. Campos, On the mechanism of seismic decoupling and back arc spreading at subduction zones, *J. Geophys. Res., 100,* 22,103-22,115, 1995.

Seno, T., S. Stein, and A.E. Gripp, A model for the motion of the Philippine Sea Plate consistent with NUVEL-1 and geological data, *J. Geophys. Res., 98,* 17,941-17,948, 1993.

Shipboard Scientific Party, Site 60, in *Initial Reports of the Deep Sea Drilling Project,* edited by A.G. Fischer, E.L. Winterer, W.R. Riedel, et al., pp. 587-629, U.S. Government Printing Office, Washington, D.C., 1971.

Shipboard Scientific Party, Site 458: Mariana Fore-Arc, in *Initial reports of the Deep Sea Drilling Project, Leg 60,* edited by D.M. Hussong, S. Uyeda, et al., pp. 263-307, U.S. Government Printing Ofice, Washington DC, 1982a.

Shipboard Scientific Party, Site 459: Mariana Fore-Arc, in *Initial Reports of the Deep Sea Drilling Project, Leg 60,* edited by D.M. Hussong, S. Uyeda, et al., pp. 309-369, U.S. Government Printing Office, Washington D.C., 1982b.

Shearer, P.M., Global mapping of upper mantle reflectors from long-period SS precursors, *Geophys. J. Int., 115,* 878-904, 1993.

Shearer, P.M., and T.G. Masters, Global mapping of topography on the 660-km discontinuity, *Nature, 355,* 791-796, 1992.

Shipley, T.H., L.J. Abrams, Y. Lancelot, and R.L. Larson, Late Jurassic-Early Cretaceous oceanic crust and Early Cretaceous volcanic sequences of the Nauru Basin, Western Pacific, in *The Mesozoic Pacific: Geology, Tectonics, and Volcanism,* edited by M.S. Pringle, W.W. Sager, W.V. Sliter, and S. Stein, pp. 103-119, AGU Monograph 97, American Geophysical Union, Washington, D.C., 1993.

Silver, P.G., R.W. Carlson, and P. Olson, Deep slabs, geochemical heterogeneity and the large-scale structure of mantle convection, *Ann. Rev. Earth Planet Sci., 16,* 477-541, 1988.

Silver, P.G., and W.W. Chan, Shear wave splitting and subcontinental mantle deformation, *J. Geophys. Res., 96,* 16,429-16,454, 1991.

Sinton, J.B., and D.M. Hussong, Crustal structure of a Short Length Transform Fault in the Central Mariana Trough, in *The Tectonic and Geologic Evolution of Southeast Asian Seas and Islands: Part 2. Geophysical Monograph 27,* edited by D.E. Hayes, pp. 236-254, American Geophysical Union, Washington DC, 1983.

Sinton, J.H., and P. Fryer, Mariana Trough lavas from 18°N: Implications for the origin of back arc basin basalts, *J. Geophys. Res., 92,* 12,782-12,802, 1987.

Smoot, N.C., Ogasawara Plateau: Multibeam sonar bathymetry and possible tectonic implications, *J. Geol., 91,* 591-598, 1983.

Soh, W., K. Nakayam, and T. Kimura, Arc-arc collision in the Izu collision zone, central Japan, deduced from the Ashigara Basin and adjacent Tanzawa Mountains, *The Island Arc, 7,* 330-341, 1998.

Soh, W., K.T. Pickering, A. Taira, and H. Tokuyama, Basin evolution in the arc-arc Izu Collision Zone, Mio-Pliocene Miura Group, central Japan, *J. Geol. Soc., Lond., 148,* 317-330, 1991.

Staudigel, H., K.-H. Park, M. Pringle, J.L. Rubenstone, W.H.F. Smith, and A. Zindler, The longevity of the South Pacific isotopic and thermal anomaly, *Earth Planet. Sci. Lett., 102,* 24-44, 1991.

Stein, C., and S. Stein, A model for the global variation in oceanic depth and heat flow with lithospheric age, *Nature, 359,* 123-129, 1992.

Stern, R.J., and L.D. Bibee, Esmeralda Bank: Geochemistry of an active submarine volcano in the Mariana Island Arc, *Contrib. Mineral. Petrol., 86,* 159-169, 1984.

Stern, R.J., and S.H. Bloomer, Subduction zone infancy : Examples from the Eocene Izu-Bonin-Mariana and Jurassic California, *Geol. Soc. Am. Bull., 104,* 1621-1636, 1992.

Stern, R.J., S.H. Bloomer, P.-N. Lin, and N.C. Smoot, Submarine arc volcanism in the southern Mariana Arc as an ophiolite analogue, *Tectonophys., 168,* 151-170, 1989.

Stern, R.J., S.H. Bloomer, F. Martinez, T. Yamazaki, and T.M. Harrison, The composition of back-arc basin lower crust and upper mantle in the Mariana Trough: A first report, *The Island Arc, 5,* 354-372, 1996.

Stern, R.J., and E. Ito, Trace element and isotopic constraints on the source of magmas in the active Volcano and Mariana island arcs, Western Pacific, *J. Volcanol. Geotherm. Res., 18,* 461-482, 1983.

Stern, R.J., M.C. Jackson, P. Fryer, and E. Ito, O, Sr, Nd, and Pb isotopic composition of the Kasuga Cross-Chain in the Mariana Arc: A new perspective on the K-H relationship, *Earth Planet. Sci. Lett., 119,* 459-475, 1993.

Stern, R.J., P.-N. Lin, J.D. Morris, M.C. Jackson, P. Fryer, S.H. Bloomer, and E. Ito, Enriched back-arc basin basalts from the northern Mariana Trough: Implications for the magmatic evolution of back-arc basins, *Earth Planet. Sci. Lett., 100,* 210-225, 1990.

Stern, R.J., J. Morris, S.H. Bloomer, and J.W. Hawkins, The source of the subduction component in convergent margin magmas: Trace element and radiogenic isotope evidence from Eocene boninites, Mariana forearc, *Geochim. Cosmochim. Acta, 55,* 1467-1481, 1991.

Stern, R.J., and N.C. Smoot, A bathymetric overview of the Mariana forearc, *The Island Arc, 7,* 525-540, 1998.

Stern, R.J., N.C. Smoot, and M. Rubin, Unzipping of the Volcano Arc, Japan, *Tectonophys., 102,* 153-174, 1984.

Stolper, E., and S. Newman, The role of water in the petrogenesis of Mariana Trough magmas, *Earth Planet. Sci. Lett., 121,* 293-325, 1994.

Stüben, D., T. Neumann, N.E. Taibi, and G.P. Glasby, Segmentation of the southern Mariana back-arc spreading center, *The Island Arc, 7,* 513-524, 1998.

Sun, C.H., and R.J. Stern, Genesis of Mariana Shoshonites: Contribution of the Subduction Component, *J. Geophys. Res., 106,* 589-608, 2001.

Sun, C.H., R.J. Stern, J. Naka, I. Sakamoto, and M. Arima, Geological and geochemical studies with Dolphin 3K on North Hiyoshi seamount, Izu-Bonin-Mariana arc, *JAMSTEC Journal of Deep-Sea Research, 14,* 139-156, 1998a.

Sun, C.H., R.J. Stern, T. Yoshida, and J.-I. Kimura, Fukutoku-oka-no-ba Volcano: A new perspective on the Alkalic Volcano Province in the Izu-Bonin-Mariana arc, *The Island Arc, 7,* 432-442, 1998b.

Suyehiro, K., N. Takahashi, Y. Ariie, and e. al., Continental crust, crustal underplating, and low-Q upper mantle beneath an oceanic island arc, *Science, 272,* 390-392, 1996.

Taira, A., S. Saito, and others, Nature and growth rate of the Northern Izu-Bonin (Ogasawara) arc crust and their implications for continental crust formation, *The Island Arc, 7,* 395-407, 1998.

Tajima, F., and S.P. Grand, Variation of transition zone high-velocity anomalies and depression of 660 km discontinuity associated with subduction zones from the southern Kuriles to Izu-Bonin and Ryukyu, *J. Geophys. Res., 103,* 15,015-15,036, 1998.

Takahashi, M., and K. Saito, Miocene intraarc bending at an arc-arc collision zone, central Japan: reply, *The Island Arc, 8,* 117-123, 1999.

Takahashi, N., K. Suyehiro, and M. Shinohara, Implications from the seismic crustal structure of the northern Izu-Bonin arc, *The Island Arc, 7,* 383-394, 1998.

Takenaka, S., H. Sanshadokoro, and S. Yoshioka, Velocity anomalies and spatial distributions of physical properties in horizontally lying slabs beneath the Northwestern Pacific region, *Phys. Earth Planet. Int., 112,* 137-157, 1999.

Tanakadate, H., Volcanoes in the Japanese Mandated South Seas, *Bulletin Volcanologique, Ser. 2, 6,* 199-223, 1940.

Tararin, I.A., I.N. Govorov, and B.I. Vasil'yev, Boninites of the Izu Trench, *Doklady of USSR Acadamey of Sciences, 295,* 117-120, 1987.

Tatsumi, Y., M. Murasaki, and S. Nohda, Across-arc variation of lava chemistry in the Izu-Bonin Arc: Identification of subduction components, *J. Volcanol. Geotherm. Res., 49,* 179-190, 1992.

Taylor, B., Rifting and the volcanic-tectonic evolution of the Izu-Bonin-Mariana Arc, in *Proceedings of the Ocean Drilling Program, Scientific Results,* edited by B. Taylor, K. Fujioka, et al., pp. 627-651, Ocean Drilling Program, College Station, 1992.

Taylor, B., A. Klaus, G.R. Brown, G.F. Moore, Y. Okamura, and F. Murakami, Structural development of Sumisu Rift, Izu-Bonin Arc, *J. Geophys. Res., 96,* 113-129, 1991.

Taylor, R.N., and R.W. Nesbitt, Isotopic characteristics of subduction fluids in an intraoceanic setting, Izu-Bonin Arc, Japan, *Earth Planet. Sci. Lett., 164,* 79-98, 1998.

Taylor, R.N., R.W. Nesbitt, P. Vidal, R.S. Harmon, B. Auvray, and I.W. Croudace, Mineralogy, chemistry, and genesis of the Boninite Series Volcanics, Chichijima, Bonin Islands, Japan, J. *Petrol., 35,* 577-617, 1994.

Tracey, J.I., S.O. Schlanger, J.T. Stark, D.B. Doan, and H.O. May, General Geology of Guam, pp. 104, U.S. Geological Survey Professional Paper P-0403A, Washington DC, 1964.

Tsuya, H., Geology and petrography of Io-sima (Sulphur Island), Volcano Islands Group, *Bull. Earthquake Res. Inst., Jap., 14,* 453-480, 1936.

Tsuya, H., On the Volcanism of the Huzi Volcanic Zone, with special reference to the Geology and Petrology of the Idu and the Southern islands, *Bull. Earthquake Res. Inst., 15,* 215-357, 1937.

Turner, S.P., J. Blundy, B. Wood, and M. Hole, Large ^{230}Th-excesses in basalts produced by partial melting of spinel lherzolite, *Chem. Geol., 162,* 127-136, 2000.

Ueda, A., and H. Sakai, Sulfur isotope study of Quaternary volcanic rocks from the Japanese Islands Arc, *Geochim. Cosmochim. Acta, 48,* 1837-1848, 1984.

Umino, S., Volcanic geology of Chichijima, the Bonin Islands (Ogasawara Islands), *J. Geol. Soc., Jap., 91,* 505-523, 1985.

Van der Hilst, R., and T. Seno, Effects of relative plate motion on the deep structure and penetration depth of slabs below the Izu-Bonin and Mariana island arcs, *Earth Planet. Sci. Lett., 120,* 395-407, 1993.

Van der Hilst, R.D., E.R. Engdahl, W. Spakman, and G. Nolet, Tomographic imaging of subducted lithosphere below northwest Pacific island arcs, *Nature, 353,* 37-43, 1991.

Vervoort, J.D., P.J. Patchett, J. Blichert-Toft, and F. Albarede, Relationships between Lu–Hf and Sm–Nd isotopic systems in the global sedimentary system, *Earth Planet. Sci. Lett., 168,* 79-99, 1999.

Vidale, J.E., and H.M. Benz, Upper-mantle seismic discontinuities and the thermal structure of subduction zones, *Nature, 356* (6371), 678-683, 1992.

Volpe, A.M., J.D. Macdougall, and J.W. Hawkins, Mariana Trough basalts (MTB): Trace element ad Sr-Nd isotopic evidence for mixing between MORB-like and arc-like melts, *Earth Planet. Sci. Lett., 82,* 241-254, 1987.

Wessel, J.K., P. Fryer, P. Wessel, and B. Taylor, Extension in the northern Mariana inner forearc, *J. Geophys. Res., 99,* 15,181-15,203, 1994.

White, W.M., and J. Patchett, Hf-Nd-Sr isotopes and incompatible element abundances in island arcs; implications for magma origins and crust-mantle evolution, *Earth Planet. Sci. Lett., 67,* 167-185, 1984.

Wicks, C.W., and M.A. Richards, A detailed map of the 660-kilometer discontinuity beneath the Izu-Bonin subduction zone, *Science, 261* (5127), 1424-1427, 1993.

Widiyantoro, S., B.L.N. Kennett, and R.D. van der Hilst, Seismic tomography with P and S data reveals lateral variations in the rigidity of deep slabs, *Earth Planet. Sci. Lett., 173,* 91-100, 1999.

Winterer, E.L., J.H. Natland, R.J. Van Waasbergen, R.A. Duncan, M.K. McNutt, C.J. Wolfe, I.P. Silva, W.W. Sager, and W.V. Slither, Cretaceous Guyots in the Northwest Pacific: An overview of their Geology and Geophysics, in *The Mesozoic Pacific: Geology, Tectonics, and Volcanism,* edited by M.S. Pringle, W.W. Sager, W.V. Sliter, and S. Stein, pp. 307-334, AGU Geophysical Monograph 97, American Geophysical Union, Washington D.C., 1993.

Woodhead, J.D., Geochemistry of the Mariana arc (western Pacific): Source compositions and processes, *Chem. Geol., 76,* 1-24, 1989.

Woodhead, J.D., and D.G. Fraser, Pb, Sr, and ^{10}Be isotopic studies of volcanic rocks from the Northern Mariana Islands. Implications for magma genesis and crustal recycling in the Western Pacific, *Geochim. Cosmochim. Acta, 49,* 1925-1930, 1985.

Woodhead, J.D., R.S. Harmon, and D.G. Fraser, O, S, Sr, and Pb isotope variations in volcanic rocks from the Northern Mariana Islands: Implications for crustal recycling in intraoceanic arcs, *Earth Planet. Sci. Lett., 83,* 39-52, 1987.

Xie, J., Shear-wave splitting near Guam, *Phys. Earth Planet. Int., 72,* 211-219, 1992.

Yamazaki, T., F. Murakami, and E. Saito, Mode of seafloor spreading in the northern Mariana Trough, *Tectonophys., 221,* 207-222, 1993.

Yamazaki, T., and R.J. Stern, Topography and magnetic vector anomalies in the Mariana Trough, *JAMSTEC Journal of Deep Sea Research, 13,* 1997.

Yuasa, M., and M. Nohara, Petrographic and geochemical along-arc variations of volcanic rocks on the volcanic front of the Izu-Ogasawara (Bonin) Arc, *Bull. Geol. Surv. Japan, 43,* 421-426, 1992.

Yuasa, M., and K. Tamaki, Basalt from Minami-Iwojima Island, Volcano Islands (in Japanese), *Bull. Geol. Surv. Japan, 33,* 531-540, 1982.

Zhang, S., and S. Karato, Lattice preferred orientation of olivine aggregates deformed in simple shear, *Nature, 375,* 774-777, 1995.

M. J. Fouch, Carnegie Institution of Washington, Department of Terrestrial Magnetism, 5241 Broad Branch Road NW, Washington, DC 20015, now at: Department of Geological Sciences, Arizona State University, P.O. Box 871404, Tempe, Arizona 85287-1404

S. L. Klemperer, Department of Geophysics, Stanford University, Stanford California 94305-2215.

R. J. Stern, Geosciences Department, University of Texas at Dallas, Box 830688, Richardson Texas 75083-0688

and Genroku (G) farther south in the Izu arc, and Kasuga (Ka) in the Marianas.

Along-Strike Variation in the Aleutian Island Arc: Genesis of High Mg# Andesite and Implications for Continental Crust

Peter B. Kelemen

Dept. of Geology & Geophysics, Woods Hole Oceanographic Institution, Woods Hole, Massachusetts

Gene M. Yogodzinski

Dept. of Geological Sciences, University of South Carolina, Columbia, South Carolina

David W. Scholl

Dept. of Geophysics, Stanford University, Stanford, California

Based on a compilation of whole rock geochemistry for approximately 1100 lava samples and 200 plutonic rock samples from the Aleutian island arc, we characterize along-strike variation, including data for the western part of the arc which has recently become available. We concentrate on the observation that western Aleutian, high Mg# andesite compositions bracket the composition of the continental crust. Isotope data show that this is not due to recycling of terrigenous sediments. Thus, the western Aleutians can provide insight into genesis of juvenile continental crust. The composition of primitive magmas (molar Mg# > 0.6) varies systematically along the strike of the arc. Concentrations of SiO_2, Na_2O and perhaps K_2O increase from east to west, while MgO, FeO, CaO decrease. Thus, primitive magmas in the central and eastern Aleutians (east of 174°W) are mainly basalts, while those in the western Aleutians are mainly andesites. Along-strike variation in Aleutian magma compositions may be related to a westward decrease in sediment input, and/or to the westward decrease in down-dip subduction velocity. $^{206}Pb/^{204}Pb$, $^{207}Pb/^{204}Pb$, $^{208}Pb/^{204}Pb$ and $^{87}Sr/^{86}Sr$ all decrease from east to west, whereas $^{143}Nd/^{144}Nd$ increases from east to west. These data, together with analyses of sediment from DSDP Site 183, indicate that the proportion of recycled sediment in Aleutian magmas decreases from east to west. Some proposed trace element signatures of sediment recycling in arc magmas do not vary systematically along the strike of the Aleutians, and do not correlate with radiogenic isotope variations. Thus, for example, Th/Nb and fractionation-corrected K concentration in Aleutian lavas are not related to the flux of subducting sediment. Th/La is strongly correlated with Ba/La, rendering it doubtful that Ba/La is a

Inside the Subduction Factory
Geophysical Monograph 138

10.1029/138GM11

proxy for an aqueous fluid component derived from subducted basalt. Ce/Pb > 4 is common in Aleutian lavas west of 174°W, in lavas with MORB-like Pb, Sr and Nd isotope ratios, and is also found behind the main arc trend in the central Aleutians. Thus, Ce/Pb in Aleutian lavas with MORB-like isotope ratios is not always low, and may be affected by a component derived from partial melting of subducted basalt in eclogite facies. Enriched, primitive andesites, with high Sr/Y, steep REE patterns, and low Yb and Y, are an important lava type in the Aleutians west of 174°W. High Sr/Y and Dy/Yb, indicative of abundant garnet in the source of melting, are correlated with major element systematics. Lavas with a "garnet signature" have high SiO_2, Na_2O and K_2O. Enriched, primitive Aleutian andesites did not form via crystal fractionation from primitive basalt, melting of primitive basalt, mixing of primitive basalt and evolved dacite, or partial melting of metasomatized peridotite. Instead, as proposed by *Kay* [1978], they formed by partial melting of subducted eclogite, followed by reaction with the mantle during ascent into the arc crust. In the eastern Aleutians, an eclogite-melt signature is less evident, but trace element systematics have led earlier workers to the hypothesis that partial melts of subducted sediment are an important component. Thus, partial melting of subducted sediment and/or basalt is occurring beneath most of the present-day Aleutian arc. This is consistent with the most recent thermal models for arcs. Enriched primitive andesites are observed mainly in the west because the mantle is relatively cold, whereas in the east, a hotter wedge gives rise to abundant, mantle-derived basalts which obscure the subduction zone melt component. Enriched primitive andesites, partial melts of eclogite, and products of small amounts of reaction between eclogite melts and mantle peridotite under conditions of decreasing magma mass, all have middle to heavy REE slopes that are steeper than those in typical Aleutian andesite and continental crust. Thus, direct partial melts of eclogite—without magma/mantle interaction—do not form an important component in the continental crust. Extensive reaction, with gradually increasing melt mass and melt/rock ratios ~ 0.1 to ~ 0.01, is required to increase heavy REE concentrations to the levels observed in most Aleutian andesites and in continental crust.

1. INTRODUCTION

In this paper, we present results of a compilation of whole rock geochemistry for approximately 1100 lava samples and 200 plutonic rock samples from the Aleutian island arc. Aleutian lava compositions have been compiled and analyzed in several previous studies [e.g., *Kelemen,* 1995; *Kay and Kay,* 1994; *Myers,* 1988]. We combine these previous compilations, and add recent data. These data characterize along-strike variation, including data for the western part of the arc that has recently become available. We evaluate current ideas about arc magma genesis in light of the spatial-geochemical patterns in the Aleutians.

We focus on evaluating hypotheses for the origin of andesites with high Mg/(Mg+Fe), or Mg#, which are abundant at and west of Adak (~ 174°W; Figure 1). These lavas are important because they overlap and bracket the major and trace element composition of the continental crust. Such lavas are rare or absent in intraoceanic island arcs other than the Aleutians. Also, the western Aleutians show the smallest influence of a subducted sediment component of any part of the Aleutians, so that it is unlikely that enrichments in elements such as U, Th, K and light rare earths are due to recycling of subducted, continental sediments. Thus, juvenile continental crust is being produced in the western Aleutians. Understanding this process forms the main focus of our paper.

In addition, along-strike variation in isotope ratios is systematic, and clearly related to variation in sediment input. This allows us to test trace element "proxies" for sediment recycling in arc lavas. These proxies are in current use in studies of global geochemical cycling and investigations of the Central American and Marianas arcs, so our results are timely and of broad relevance.

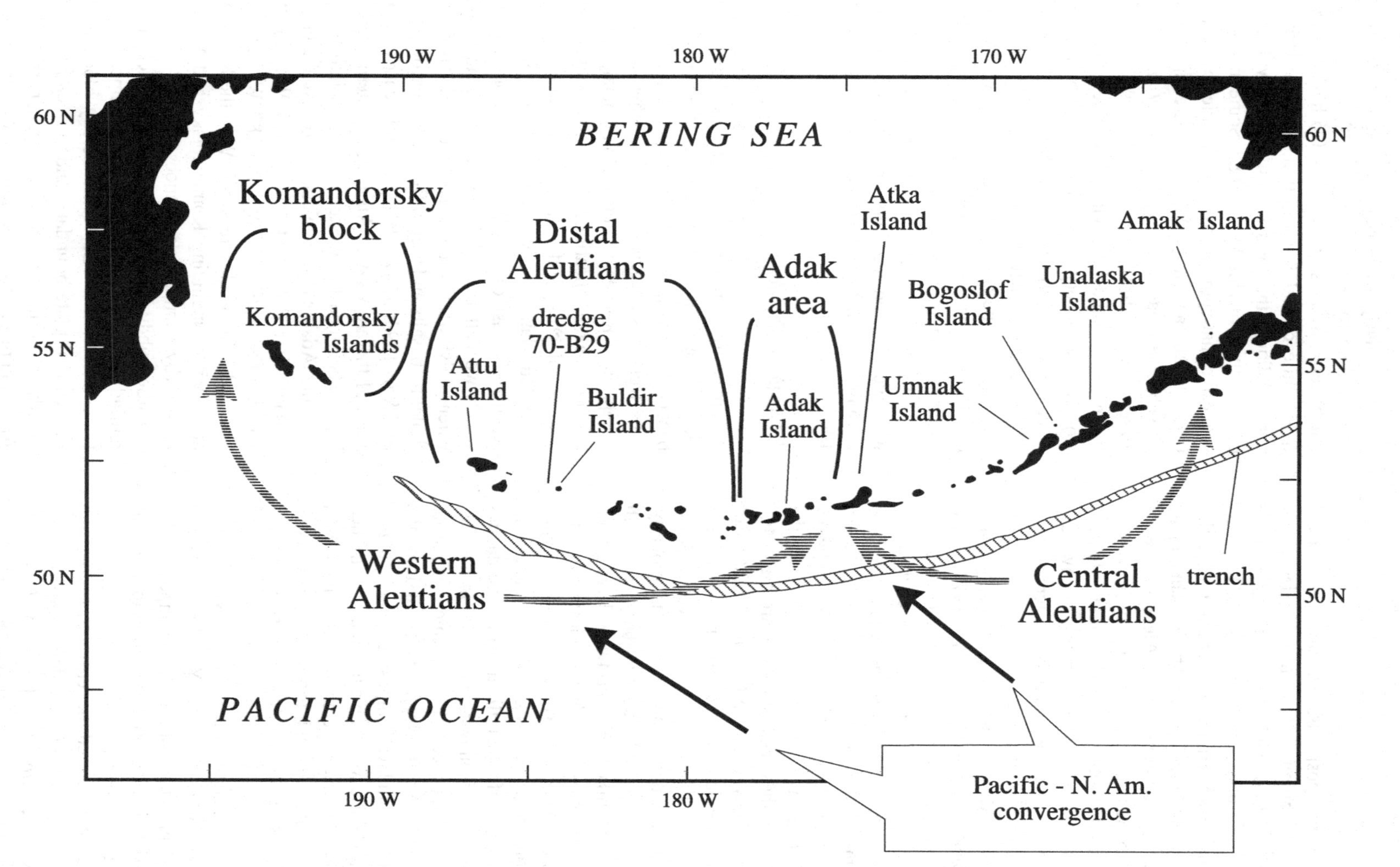

Figure 1. Overview of the Aleutian island arc, highlighting the geographic features mentioned in the text, and showing the approximate convergence vector near Adak (N42W) and Attu (N49W) from *Engebretson et al.* [1985].

1.1 Terminology

In describing lava compositions, we have used some terms that require definition. "High Mg# andesite" is defined as lava with 54 to 65 wt% SiO_2 and Mg# > 0.45. This could, in principle, include some boninites. However, classical boninites in the western Pacific generally have flat to light rare earth element (REE) depleted trace element patterns, whereas high Mg# andesites in the Aleutians are all light REE enriched. Note that MgO content plays no role in our definition of high Mg# andesite. Some Aleutian andesites with Mg# > 0.6 have MgO contents as low as 4 wt%. As a result, we have not used the familiar terms "high Mg andesite" or "magnesian andesite".

Some Aleutian high Mg# andesites—particularly some lavas on Adak Island first reported by *Kay* [1978]—also have been called "adakites" [e.g., *Defant and Drummond,* 1990]. The term adakite is used in a variety of contexts by different investigators, but generally refers to andesites and dacites with extreme light REE enrichment (e.g., La/Yb > 9), very high Sr/Y (e.g., Sr/Y > 50), and low Y and heavy REE concentrations (e.g., Y < 20 ppm, Yb < 2 ppm). In the Aleutians, all lavas with these characteristics are high Mg# andesites and dacites. However, note that the definition of "adakite" outlined above does not specify a range of Mg#. In arcs other than the Aleutians, many highly evolved lavas with low Mg# have been termed adakites. Thus, globally, not all adakites are high Mg# andesites. Similarly, most high Mg# andesites, in the Aleutians and worldwide, have La/Yb < 9 and Sr/Y < 50, so most high Mg# andesites are not adakites. Finally, we note that for some workers "adakite" has a genetic connotation as well as a compositional definition. Some investigators infer that all andesites and dacites with extreme light REE enrichment, very high Sr/Y ratios, and low Y and heavy REE concentrations formed via partial melting of subducted basalt in eclogite facies, and use the term adakite to refer to both lava composition and lava genesis interchangeably. While we believe that many "enriched andesites" may indeed include a component derived from partial melting of eclogite, we feel it is important to separate rock names, based on composition, from genetic interpretations. For this reason, we have not used the term "adakite" in this paper. Aleutian andesites and dacites with Mg# > 0.45, La/Yb > 9, and Sr/Y > 50, plus Y < 20 ppm and/or Yb<1 ppm, form an important end-member composition on most chemical variation diagrams. We refer to these as "enriched, high Mg# andesites".

More informally, we have used the terms "tholeiitic" and "calc-alkaline". For our purposes, the definitions of these terms proposed by either *Miyashiro* [1974] or *Irvine and Baragar* [1971] are approximately equivalent and equally useful. Similarly, we have informally used the term "primitive" to refer to lavas with Mg# > 0.6. This usage reflects our belief that lavas with Mg# > 0.6 have undergone relatively little crystal fractionation, and are derived from a parental liquid that was in equilibrium with mantle peridotite with an olivine Mg# > 0.88. Some investigators have suggested that some lavas and plutonic rocks with Mg# < 0.6—globally and in the Aleutians—are direct partial melts of subducted basalts [e.g., *Schiano et al.,* 1997; *Hauri,* 1996; *Myers,* 1988; *Myers et al.,* 1985; *Marsh,* 1976]. Indeed, this could be a viable hypothesis in some cases. However, it is very difficult to evaluate such hypotheses since many (most?) lavas with Mg# < 0.6 have been affected by crustal differentiation processes. In this paper, we have adopted the convention that lavas with Mg# < 0.6 are considered to be the products of differentiation from primitive magmas until proven otherwise.

Finally, for simplicity in data presentation, in this paper we have reported longitude in terms of degrees west of Greenwich so that, for example, 170°E is referred to here as 190°W.

1.2 Regional Divisions

In presenting our results, we have used regional divisions that are somewhat different from those used in previous papers (Figure 1). For our current purposes, Aleutian volcanoes from Atka eastward are all rather similar, with the exception of Rechesnoi volcano on Umnak Island. In order to avoid the complications of potential contamination from continental crust and overlying sediments, we have not compiled data for volcanic centers in the "eastern Aleutians", east of 164°W. The arc from 164 to 174°W, which we call the "central Aleutians", has been studied extensively. Most of it is a classic, oceanic arc dominated by tholeiitic basalts and their differentiation products [e.g., *Kay and Kay,* 1994; *Myers,* 1988, and references cited therein]. We call the arc west of Atka the "western Aleutians". We have subdivided the western Aleutians into three regions: (1) "the Adak area" (volcanoes on Adak, Great Sitkin, Kanaga, and Bobrof at 174 to 177°W), (2) "the distal Aleutians" from 177°W to 187°W, and (3) "the Komandorsky block" from 188°W to the Kamchatka Strait at ~ 195°W, including Komandorsky and Medny Islands and nearby, submarine volcanoes.

These subdivisions of the western Aleutians serve to remind readers of some important compositional and tectonic distinctions. The Komandorsky block lies within a transcurrent plate boundary, and may currently be moving mainly with the Pacific plate rather than with North American plate [*Avé Lallemant and Oldow,* 2000; *Geist and*

Scholl, 1994; *Geist and Scholl,* 1992]. For this reason, it is not entirely clear how the Komandorsky block is related to the modern Aleutian arc. We believe that there is a compositional continuum between the Komandorsky lavas and the main part of the active island arc further east, and this is important to our interpretations of Aleutian petrogenesis. However, it is crucial to note that this apparent continuum is dependent on very limited sampling of the distal Aleutians (Figure 2).

Adak and neighboring islands are heterogeneous, including both tholeiitic and calc-alkaline lavas [e.g., *Myers et al.,* 1985; *Kay et al.,* 1982]. Lavas in the Komandorsky block are predominantly calc-alkaline andesites, though basalts are present [*Yogodzinski et al.,* 1995; *Yogodzinski et al.,* 1994]. Distal Aleutian lavas are relatively homogeneous, compared to the Adak area, and are transitional between the dominantly tholeiitic, basaltic lavas of the eastern and central arc and primitive, calc-alkaline andesites and dacites with extreme trace element signatures that predominate in the Komandorsky block (Figure 3). As a result of this continuum in compositional variation, we presume that the Komandorsky lavas are genetically related to the rest of the western Aleutians. In fact, recent work supports this idea, and proposes that the Komandorsky block originated in the forearc region of the distal Aleutians, and later was transported ~ 700 km WNW along the transcurrent plate boundary [*Scholl et al.,* 2001; *Rostovtseva and Shapiro,* 1998].

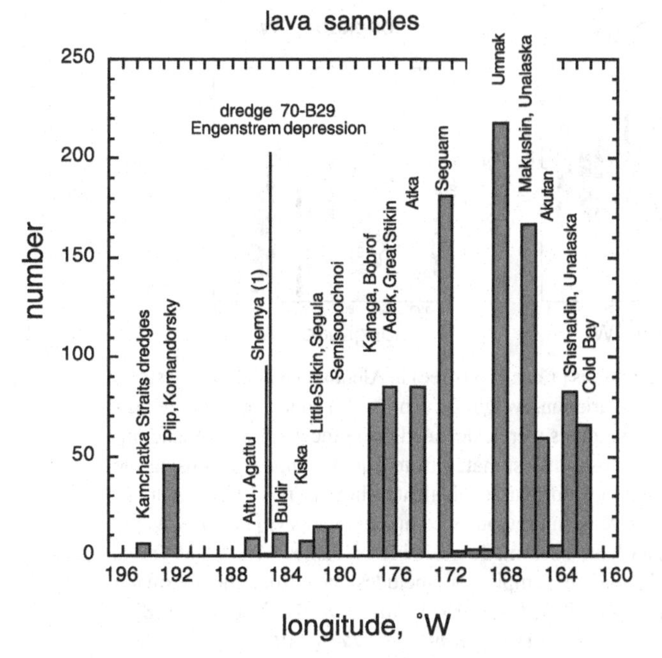

Figure 2. Histogram illustrating the number of Aleutian lava samples in our compilation as a function of longitude along the arc. Sources of data are given in the caption for Figure 3.

1.3 Andesite Paradox: Why Does Continental Crust Resemble arc Lavas?

A long-standing problem in Earth Science is to resolve the following, apparent paradox. It is well documented that the continental crust has a bulk composition very similar to high Mg# andesite lavas and plutonic rocks in subduction-related magmatic arcs, with molar Mg/(Mg+Fe), or Mg#, from 0.45 to 0.54 at 57 to 65 wt% SiO_2, 8000 to 24,000 ppm K, and La/Yb of ~ 5 to 20 (see Figures 4, 5 and 6, and [*Christensen and Mooney,* 1995; *Kelemen,* 1995; *Rudnick,* 1995; *Rudnick and Fountain,* 1995] for reviews). Lavas having these properties occur almost exclusively in arcs [*Gill,* 1981]. Thus, it is commonly inferred that continental crust formed mainly as a result of arc processes [e.g., *Kelemen,* 1995; *Rudnick,* 1995; *Kay and Kay,* 1991; *Taylor,* 1977]. However, most currently active, oceanic arcs are dominated by basalts and low Mg# andesites rather than high Mg# andesite lavas (Figure 5B).

Various hypotheses have been proposed to resolve this paradox. (1) At some times and places, the net magmatic flux through the Moho beneath arcs has been andesitic rather than basaltic [e.g., *Defant and Kepezhinskas,* 2001; *Martin,* 1999; *Kelemen et al.,* 1998; *Kelemen,* 1995; *Drummond and Defant,* 1990; *Martin,* 1986; *Ringwood,* 1974]. (2) High Mg# andesites in arcs form as a result of intracrustal crystal fractionation and magma mixing, involving mantle-derived basalts and their differentiates. Later, a dense mafic or ultramafic plutonic layer delaminates and returns to the mantle, leaving an andesitic crust [e.g., *Jull and Kelemen,* 2001; *Tatsumi,* 2000; *Kay and Kay,* 1993; *Kay and Kay,* 1991]. (3) High Mg# andesites in arcs form via fractionation of ultramafic cumulates from primary, mantle-derived basalts. The ultramafic cumulates remain, undetected, below the seismic Moho in arcs and continents [e.g., *Fliedner and Klemperer,* 1999; *Kay and Kay,* 1985b]. (4) Arcs may not have been involved at all. Processes of intracrustal differentiation and delamination have yielded a crustal composition whose resemblance to high Mg# andesites in arcs is coincidental [e.g., *Stein and Hofmann,* 1994; *Arndt and Goldstein,* 1989]. In this paper, we use data from the Aleutians to evaluate and extend hypotheses (1) through (3).

1.4 Calc-Alkaline Lavas in the Western Aleutians: Juvenile Continental Crust?

In the Aleutian island arc, particularly in the western Aleutians, high Mg# andesite lavas with compositions similar to continental crust are abundant (Figures 4, 5 and 6). Though similar lavas are also found in the Cascades, Baja

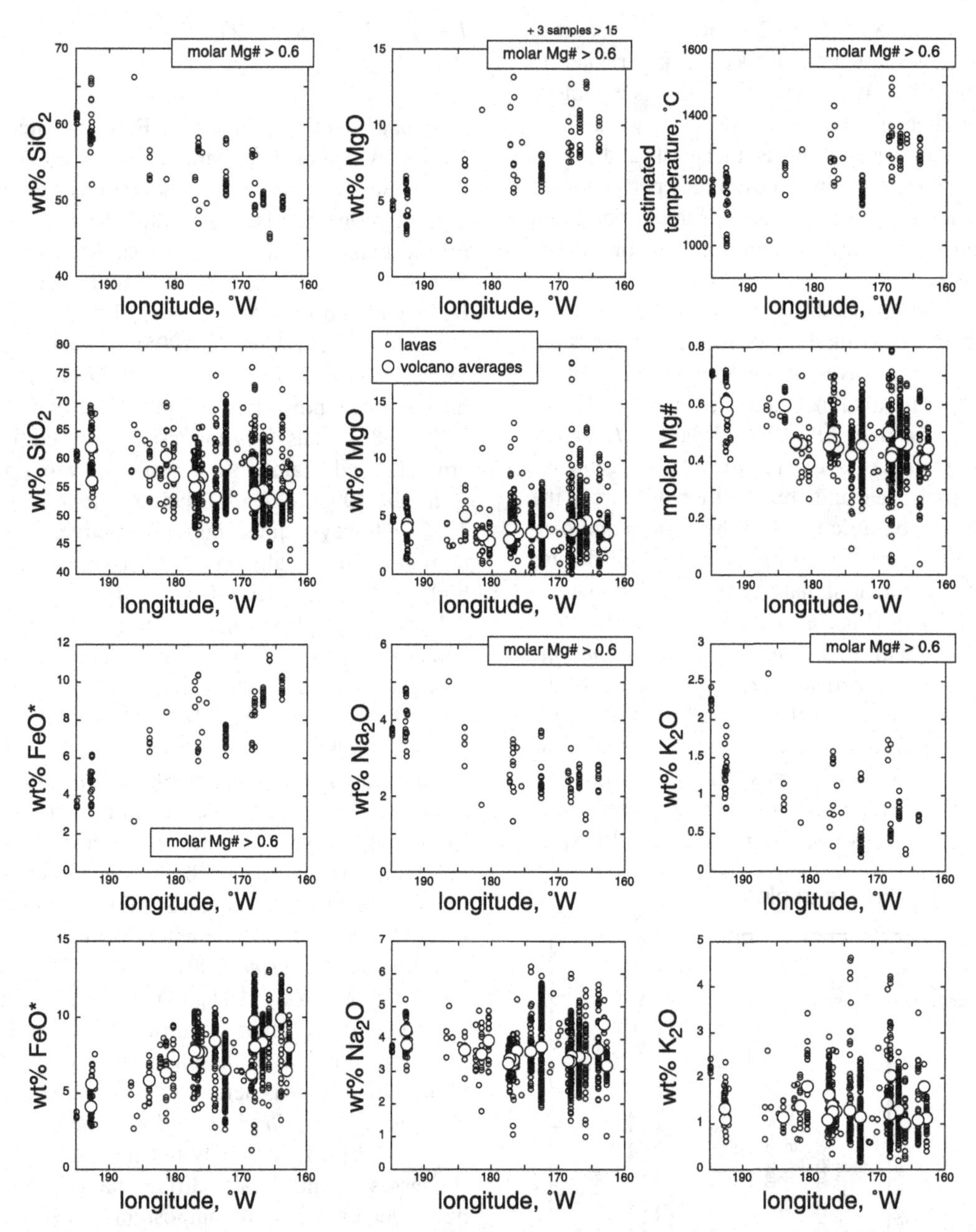

Figure 3. Major element variation and estimated magmatic temperatures (1 bar, H_2O-free) in Aleutian lavas along the strike of the arc. In this and all subsequent plots of major element oxide variation, weight percent (wt%) concentrations are normalized to 100% volatile free, with all Fe as FeO. Magmatic temperatures were calculated using the empirical olivine/liquid thermometer of [*Gaetani and Grove,* 1998], with olivine compositions estimated from liquid compositions using the olivine/liquid Fe/Mg Kd of [*Baker et al.,* 1996], assuming lava compositions are equivalent to liquids, no H_2O, pressure of 1 bar, and 80% of Fe is ferrous. Compositions of Aleutian lavas have been previously compiled by *Myers* [1988], *Kay and Kay* [1994], and *Kelemen* [1995]. More recently, compiled data have been made available by James Myers and Travis McElfrish at http://www.gg.uwyo.edu/aleutians/index.htm. The online compilation includes data from the following sources: [*Brophy,* 1986; *Myers,* 1986; *Myers,* 1985; *Morris and Hart,* 1983; *Perfit,* 1983; *Kay et al.,* 1982; *Marsh,* 1982b; *Romick,* 1982; *McCulloch and Perfit,* 1981; *Perfit et al.,* 1980a; *Kay et al.,* 1978a; *Kay,* 1977; *Perfit,* 1977; *Drewes et al.,* 1961;]. In our current compilation, these have been supplemented with additional data from: [*Class et al.,* 2000; *Yogodzinski et al.,* 1995; *Miller et al.,* 1994; *Yogodzinski et al.,* 1994; *Yogodzinski et al.,* 1993; *Tsvetkov,* 1991; *Goldstein,* 1986; *Rubenstone,* 1984; *Sun,* 1980; *Gates et al.,* 1971; *Fraser and Barrett,* 1959; *Fraser and Snyder,* 1956].

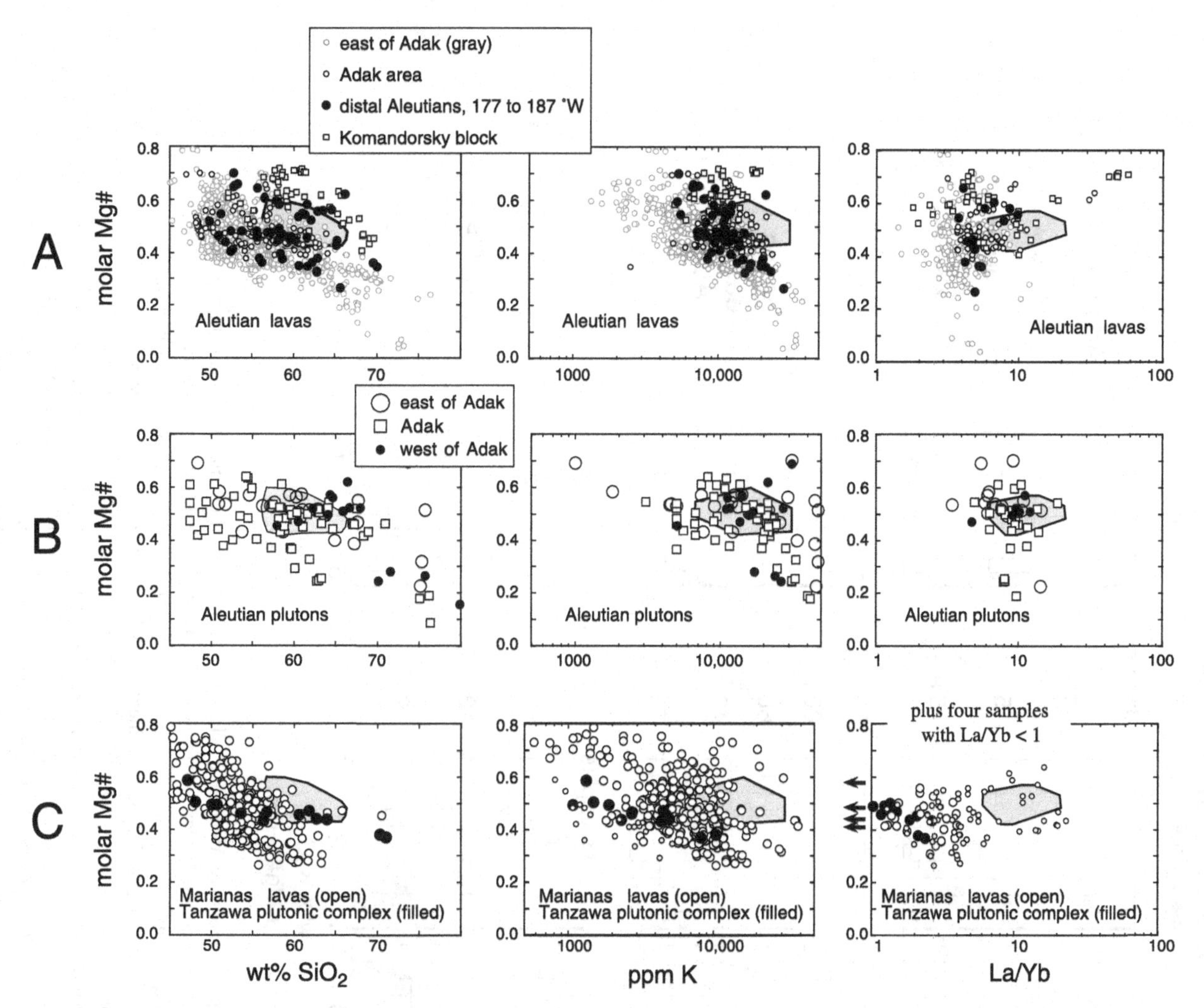

Figure 4. Compositions of Aleutian lavas (A) and plutonic rocks (B) compared to estimates of the composition of continental crust (shaded regions in each plot). Panel (C) shows compositions of lavas and plutonic rocks of the Izu-Bonin-Mariana arc system. Aleutian lavas from Adak and west of Adak, and plutonic rocks throughout the Aleutians, commonly have the SiO_2, Mg#, K, and La/Yb ratios comparable to estimates for bulk continental crust. Lavas and plutonic rocks of the Izu-Bonin-Mariana system generally do not have these characteristics. Sources of data for Aleutian lavas in caption for Figure 3. Aleutian plutonic rock data from [*Tsvetkov,* 1991; *Kay et al.,* 1990; *Romick et al.,* 1990; *Kay et al.,* 1986; *Kay et al.,* 1983; *Citron,* 1980; *Perfit et al.,* 1980b; *Gates et al.,* 1971; *Drewes et al.,* 1961; *Fraser and Barrett,* 1959; *Fraser and Snyder,* 1956], and Sue Kay, pers. comm. 2001. Estimates of continental crust composition have been recently compiled and reviewed in [*Christensen and Mooney,* 1995; *Kelemen,* 1995; *Rudnick,* 1995; *Rudnick and Fountain,* 1995]. Marianas data are from [*Elliott et al.,* 1997; *Bloomer et al.,* 1989; *Woodhead,* 1989; *Stern and Bibee,* 1984; *Wood et al.,* 1981; *Dixon and Batiza,* 1979; *Stern,* 1979]. Data on the Tanzawa plutonic complex (northern end of the Izu-Bonin Marianas arc system) are from [*Kawate and Arima,* 1998].

California, Central America, the southern Andes, the Philippines, and SW Japan [e.g., *Defant et al.,* 1991; *Defant et al.,* 1989; *Luhr et al.,* 1989; *Hughes and Taylor,* 1986; *Rogers et al.,* 1985; *Puig et al.,* 1984; *Tatsumi and Ishizaka,* 1982], these other localities are underlain by older, continental basement and/or sediment derived from continental crust. Light REE enriched, high Mg# andesite compositions are rare or absent in intraoceanic, island arcs other than the Aleutians (e.g., the Marianas, Figure 4). Also, radiogenic Pb isotope ratios in most intraoceanic arcs suggest the presence of a recycled terrigenous sediment component, whereas the western Aleutians have depleted Pb isotopes (Figure 5B). As a consequence, the western Aleutians offer the best opportunity to study the genesis of high Mg# andesites—and, by

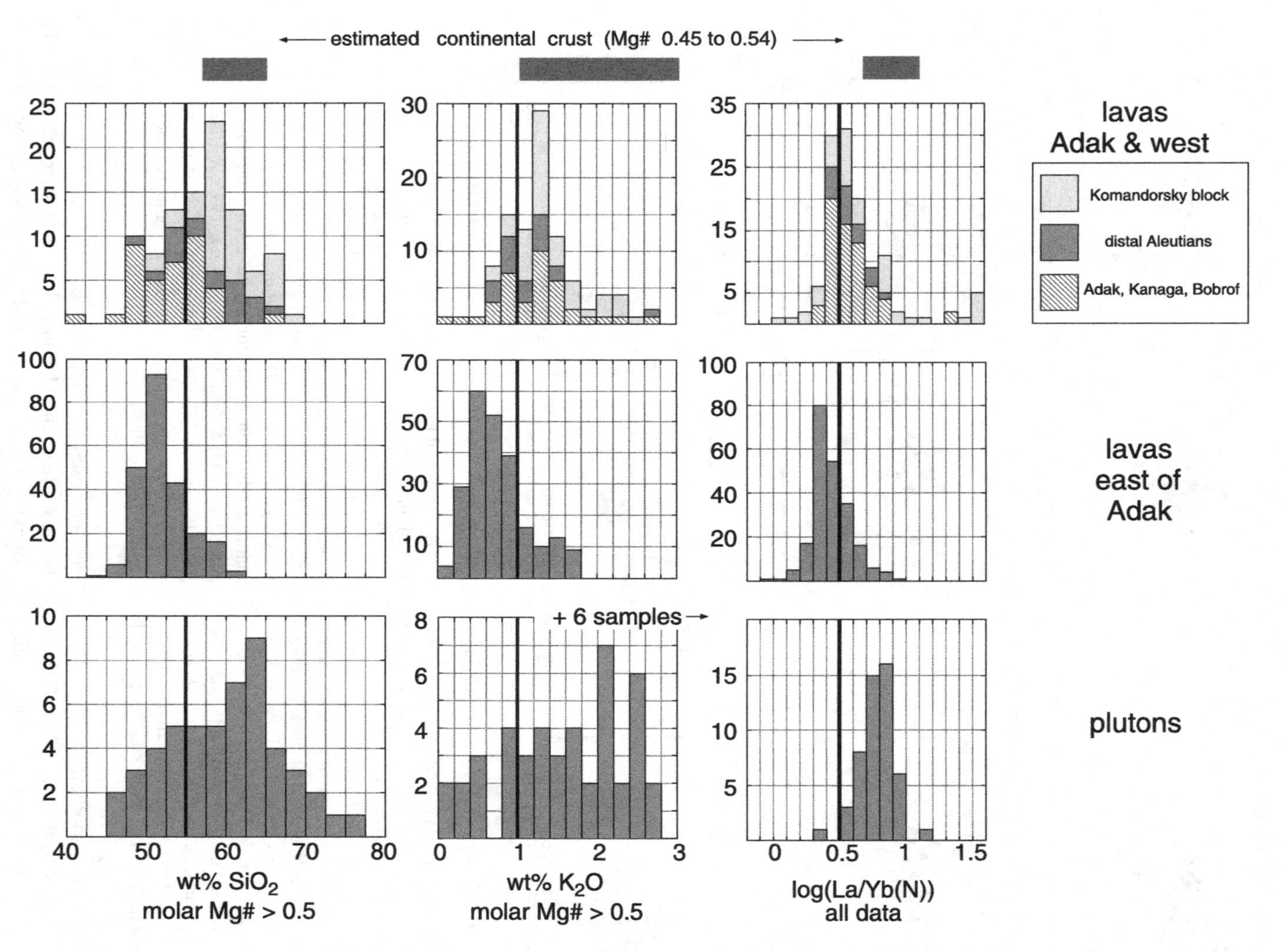

Figure 5. A. Histograms of rock compositions in the Aleutian arc. Top bars show range of estimates for the composition of continental crust. Please note that the histograms for lavas from Adak and west are cumulative, so that the maximum value for each bin refers to the sum of all samples from the Adak area + distal Aleutians + Komandorsky block. Aleutian data sources in caption for Figure 3. Range of compositions for continental crust reviewed in [*Kelemen,* 1995]. Section 1.2 and Figure 1 give definitions of regional terms. B. Histograms of wt% SiO_2, wt% K_2O, and $^{207}Pb/^{204}Pb$ in intraoceanic arc lavas. To our knowledge, the western Aleutians are unique among island arcs in having higher SiO_2 and K_2O, and lower $^{207}Pb/^{204}Pb$, in primitive magmas. Sources of Aleutian data are in caption for Figure 3. All other arc data were downloaded from the GEOROC database at Max Plank. Continental crust estimate has molar Mg# of 0.54 [*Rudnick and Fountain,* 1995]. The composition of sample 70B29, the only available dredge sample from the western Aleutians, is indicated.

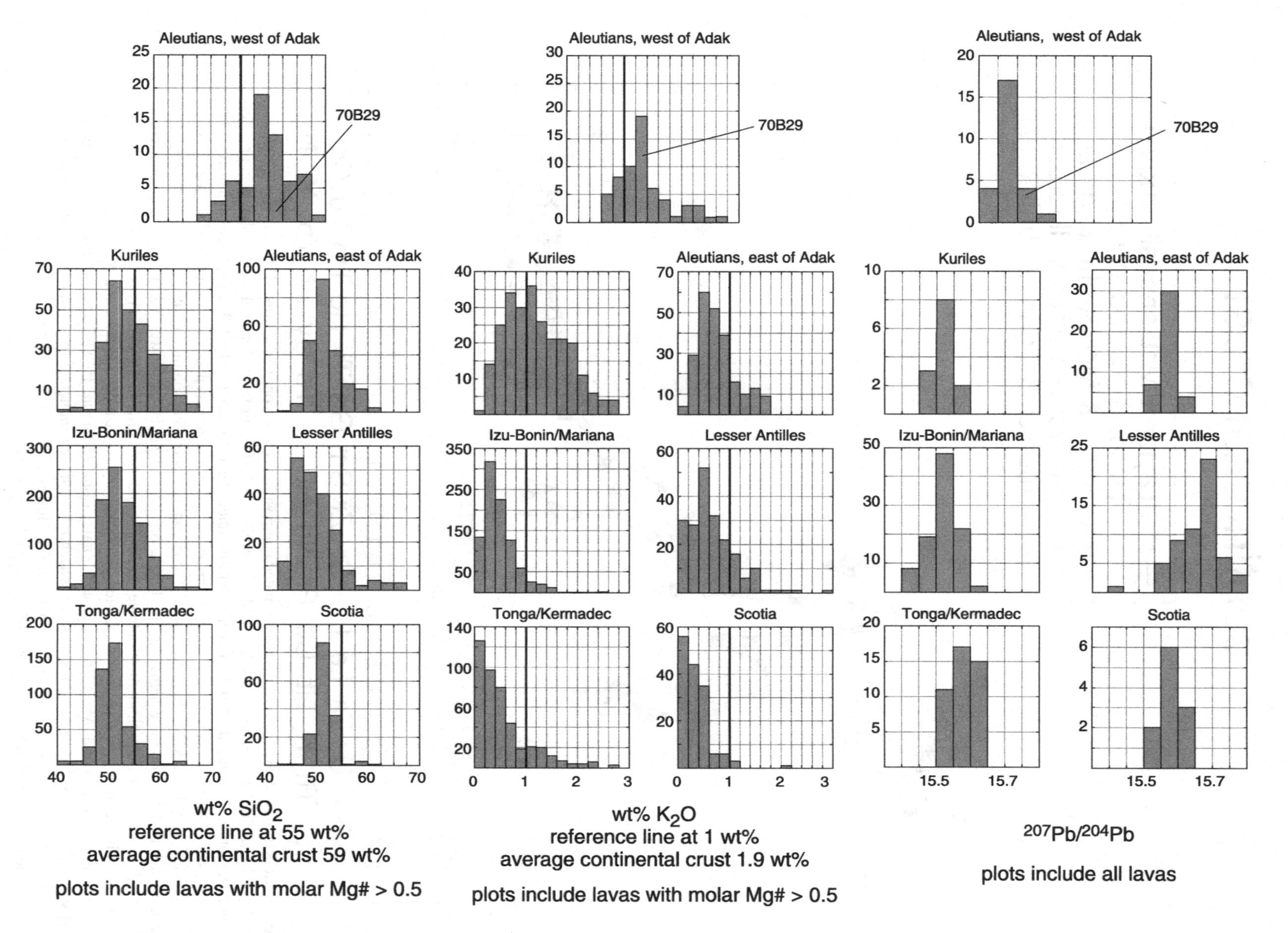

Figure 5. Continued.

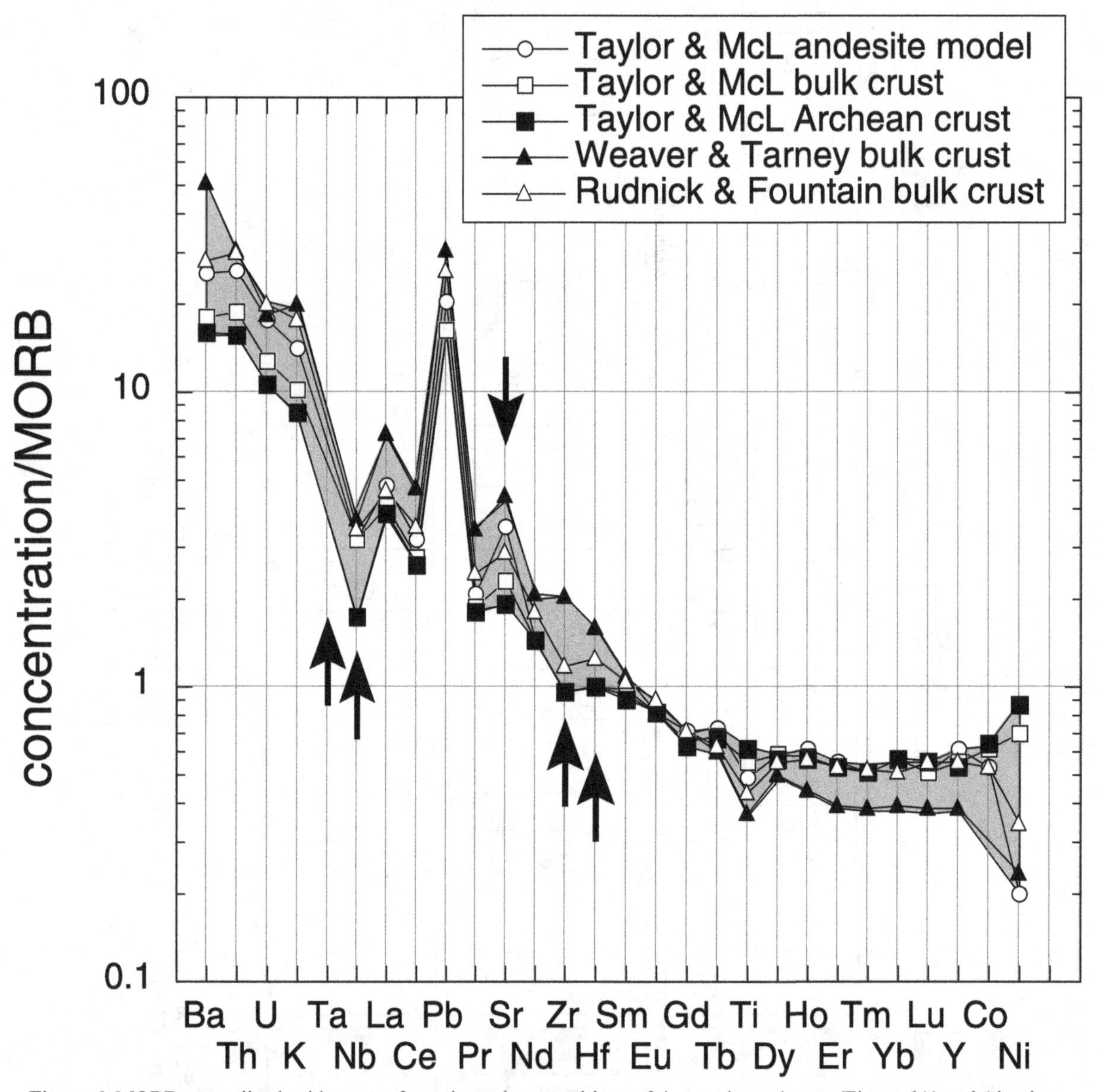

Figure 6. MORB-normalized spidergrams for estimated compositions of the continental crust (Figure 6A) and Aleutian lavas (Fig. 6b). In the Aleutian lava plots, the field of compiled estimates for the continental crust is shown in gray for reference. In plot for the Adak area, two enriched, primitive andesites with strong light REE enrichment and heavy REE depletion cross the other lava patterns. In Figure 6A, arrows emphasize elements in continental crust which may be concentrated in detrital sediments (Nb, Ta in rutile, Zr, Hf in zircon) or removed in solution during surficial weathering (Sr). In general, Aleutian magmas have higher La/Nb, La/Ta, Sm/Zr, and Sm/Hf and lower Nd/Sr than continental crust. Other than that, we feel that the similarities in trace element abundance between Aleutian lavas and continental crust is striking, and indicates that similar processes operated during formation of continental crust and Aleutian magmas. Data sources in caption for Figure 3. MORB normalization values from [*Hofmann,* 1988]. Note that *Plank and Langmuir* [1998] suggested that some estimates of Nb and Ta concentration in the continental crust [*McLennan and Taylor,* 1985; *Rudnick and Fountain,* 1995] are too high by a factor of ~ 2. However, because these estimates provide neither an upper nor a lower bound in Figure 6A, we did not modify the published values.

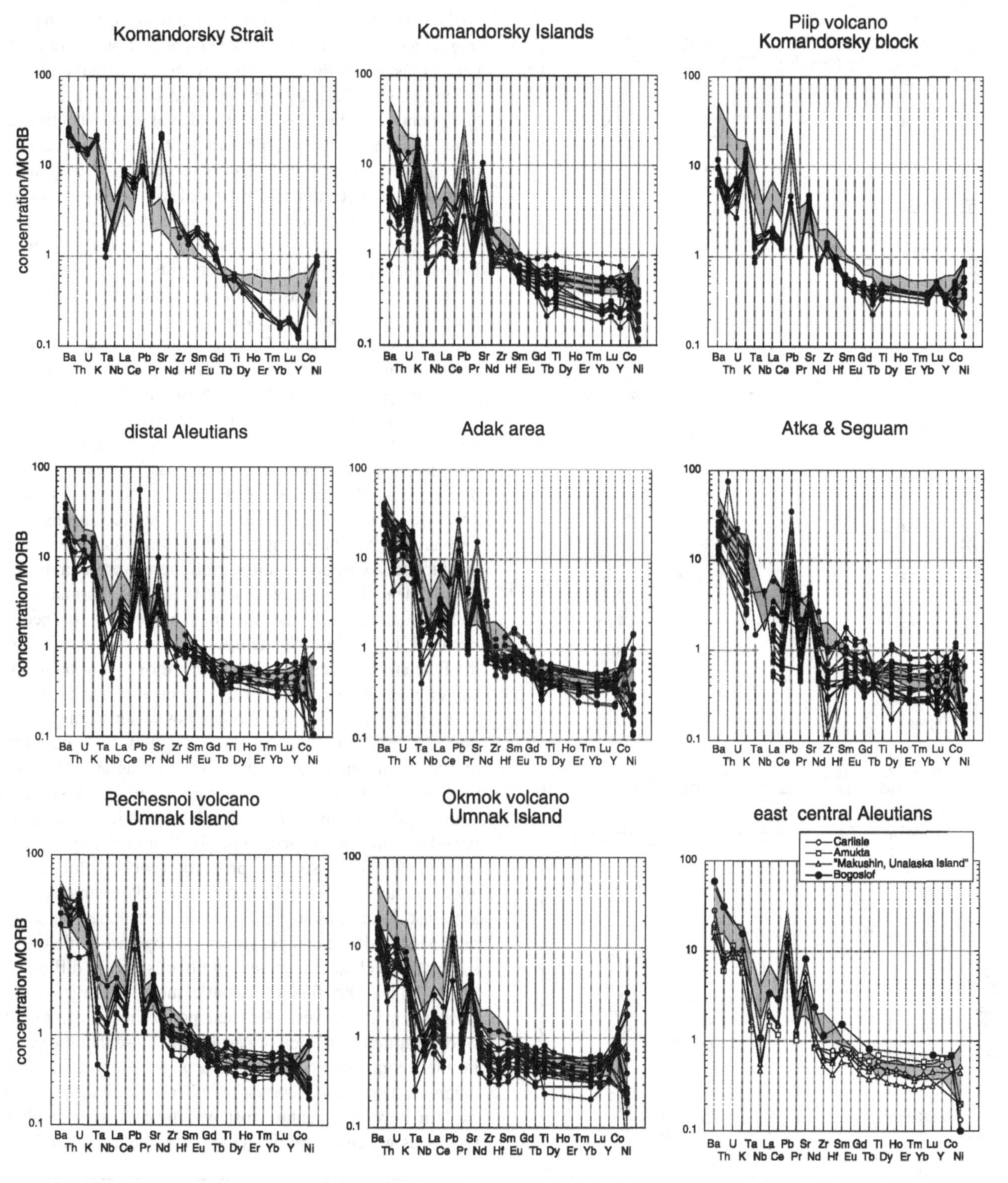

Figure 6. Continued.

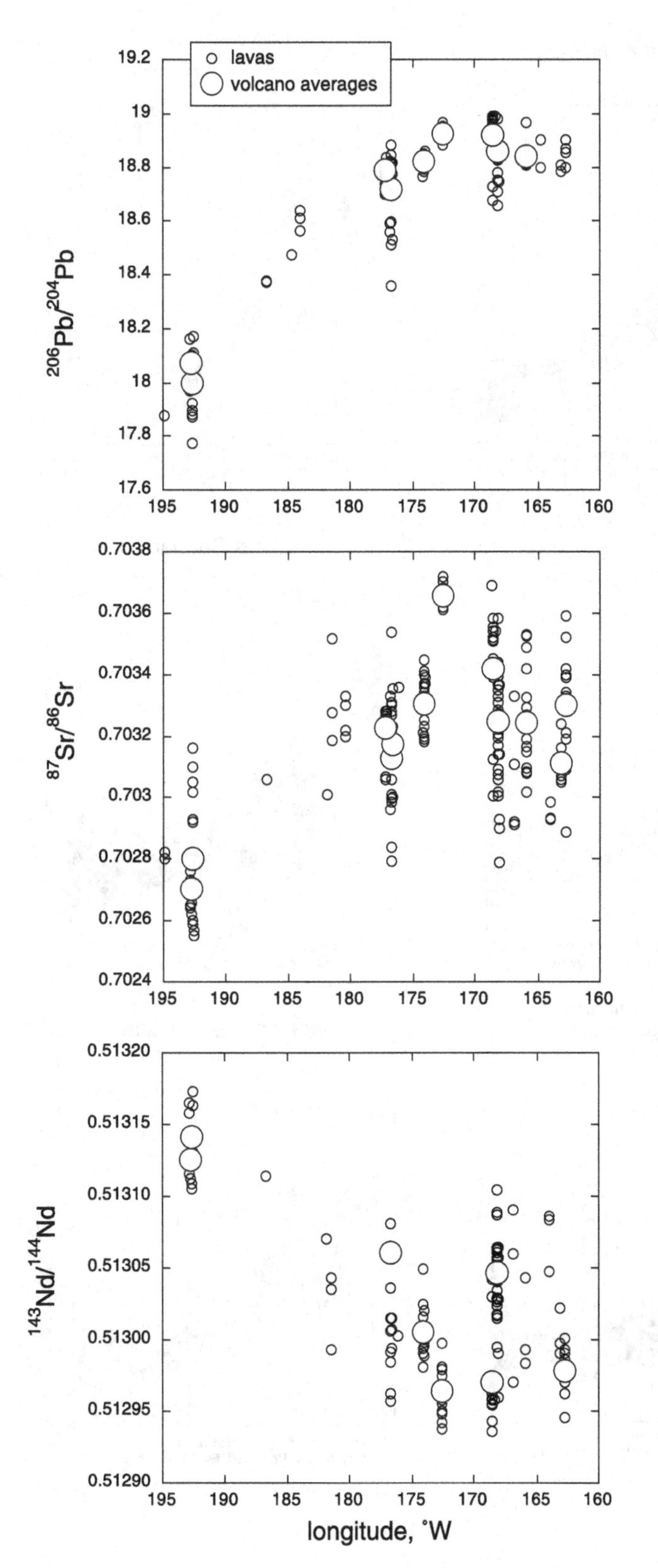

Figure 7. $^{206}Pb/^{204}Pb$, $^{87}Sr/^{86}Sr$ and $^{143}Nd/^{144}Nd$ in Aleutian lavas versus longitude, illustrating systematic along-strike variation in these ratios. Data sources in caption for Figure 3, plus new Pb isotope data for western Aleutian lavas in Table 1.

analogy, the formation of juvenile continental crust—in a setting which is clearly free of contamination from older continental crust and sediment derived from continental crust.

Enrichment of arc lavas in some elements that are important in understanding the genesis and evolution of continental crust, such as K, U, Th, Pb and light rare earth elements (REE), is sometimes attributed to recycling of components from subducted, continentally derived sediments [e.g., *Elliott et al.,* 1997; *Hawkesworth et al.,* 1997; *Hochstaedter et al.,* 1996; *Miller et al.,* 1994; *Plank and Langmuir,* 1993]. Isotopic and geological evidence suggests that sediment input to the source of arc lavas is common in the central Aleutians, and in most arc lavas worldwide, but is absent or minimal in the west (Figures 5B and 7). Thus, if the processes that form *juvenile* continental crust in an oceanic arc can be documented anywhere on Earth, it is in the western Aleutians.

High Mg# andesites are rare in oceanic arcs worldwide. How could such lavas accumulate to form large volumes of continental crust? There are several possible answers. First, the genesis of high Mg# andesites in arcs may involve processes that were more common in the Precambrian than they are today. For example, perhaps high temperatures in subduction zones are required, and these were common in the Archean [e.g., *Martin,* 1986]. Second, formation of abundant high Mg# andesite may be related to specific events. If high subduction temperatures are required, during the Phanerozoic these may have been most common in special circumstances such as "ridge subduction" [e.g., *Rogers et al.,* 1985], in which very young, hot oceanic crust is subducted. Third, hydrous andesite magmas become much more viscous than basalts when they degas at mid-crustal pressures, and may commonly form plutonic rocks rather than erupting as lavas [*Kelemen,* 1995], as is suggested by data from the Aleutians (Figures 4B and 5A, plus [*Kay et al.,* 1990]). If so, the proportion of basaltic versus andesitic lavas is not indicative of the bulk composition of arc crust. Fourth, as suggested later in this paper, perhaps a high Mg# andesite component formed by partial melting of subducting eclogite is present in all arcs, but is difficult to detect where the flux of basaltic melts is large; in the Archean, perhaps very depleted peridotite in the mantle wedge limited formation of basaltic melts, so that the eclogite melt components comprised a larger proportion of arc magmas. These four factors, together or individually, may explain how continental crust was formed as a result of processes which are rarely evident in present-day arc lavas.

Table 1: New Pb concentrations and isotope data for Aleutian lavas.

lat (°N)	lon (°W)	location	sample	Ce ppm *	Pb ppm	$^{206}Pb/^{204}Pb$	$^{207}Pb/^{204}Pb$	$^{208}Pb/^{204}Pb$	data from
52.370	184.020	Buldir	BUL6B	17.2	3.02	18.565	15.486	38.028	this paper
52.370	184.020	Buldir	BUL6A	20.5	3.48	18.641	15.514	38.132	this paper
52.370	184.020	Buldir	BUL4D	22.7	2.95	18.611	15.480	38.054	this paper
52.533	184.745	dredge near Buldir	70-B29	26.6	4.31	18.477	15.484	37.941	this paper
55.455	192.832	dredge, Piip volc.	V35G5B	15.0	1.92	18.089	15.431	37.535	1
55.425	192.732	dredge, Piip volc.	V35G1A	17.2	2.32	18.098	15.430	37.520	1
55.343	192.550	Komandorsky Isl.	V35G8B	10.7	1.33	18.030	15.411	37.537	2
55.469	192.892	Komandorsky Isl.	V35G7C	14.7	2.98	18.161	15.438	37.594	2

Pb concentrations determined by isotope dilution on hand picked rock chips leached in 2.5N HCl at 80°C for 1/2 hour, dissolved and analyzed on a Finnigan MAT Element I ICPMS. Estimated precision is ± 1% relative.
Pb isotope ratios determined on powdered, hand picked rock chips leached for 1h in 6.2N HCl at 100°C analyzed on a VG 354 mass spectrometer. Estimated precision, based on standard reproducibility, is ± 0.15% relative.
All results are corrected against NBS981 (Todt et al, 1996).
*: Ce concentrations from Kay & Kay, 1994; Yogodzinski et al., 1994, 1995
1.Yogodzinski et al., 1994
2 Yogodzinski et al., 1995

1.5 Along-Strike Variation in Convergence Rate

Systematic, along-strike variation in Aleutian lava compositions may be related to along-strike variation in the rate of "down dip" convergence between the Pacific and North American plates. Because of the arcuate shape of the Aleutians, convergence is nearly orthogonal to the trench in the eastern and central arc, and strongly oblique in the west. In addition, oblique convergence leads to strain partitioning in which absolute plate motions within the western arc are intermediate between those of the North American and Pacific plates [*Avé Lallemant and Oldow,* 2000; *Geist and Scholl,* 1994; *Geist and Scholl,* 1992]. As a result, the trench orthogonal convergence velocity, which is ~ 60 to 75 mm/year beneath the arc from Adak eastward, decreases to < 40 mm/year beneath the distal Aleutians and the Komandorsky block (Figure 8). (Orthogonal convergence rates beneath the arc have been projected from values of orthogonal convergence rate versus longitude along the trench [*Fournelle et al.,* 1994] along the plate convergence vector [*Engebretson et al.,* 1985]).

In contrast to convergence rate, the age of subducting oceanic crust entering the Aleutian trench is ~ 50 to 60 Ma and does not vary systematically along-strike [*Atwater,* 1989; *Geist et al.,* 1988; *Lonsdale,* 1988; *Scholl et al.,* 1987]. According to the plate reconstructions of *Lonsdale* [1988], the dead Kula-Pacific spreading ridge (which ceased spreading at ~ 43 Ma) was subducted between 15 Myr ago at the longitude of Umnak Island (~ 168°W) and 3 Myr ago at the longitude of Attu Island (~ 187°W). During this event, the subducting crust was 28 to 40 Myr old, and at any given time it was progressively *older* to the west. For example, using Lonsdale's "simplest reconstruction" of the present down-slab position of the dead Kula-Pacific spreading ridge and a down dip convergence rate of 65-70 km/Myr, and assuming that the subducted spreading ridge had a NE-SW (not an E-W) trend when it went into the trench, we find that the minimum age of the oceanic crust consumed during subduction of the extinct ridge varied from ~ 28 Myr at 168°W to ~ 40 Myr at 187°W. Alternatively, using Lonsdale's "more plausible" reconstruction, which considers age offsets across subducted fracture zones, we find that the minimum age of the subducting crust varied from ~32 Myr at 168°W to ~ 39 Myr at 187°W.

At any particular time, the lateral gradient in average slab temperature along the ridge was not large, and in general the hottest subducting crust was in the central Aleutians. Taking the middle Miocene (~10-12 Ma), for example, when most exposed plutons were emplaced and many enriched, high Mg# andesites were erupted in the Komandorsky block, the oceanic crust entering the trench at and east of Adak (~ 177°W) was roughly 32 Myr. At the same time, west of 180°W, the thermal age of the crust entering the subduction zone was on the order of 49 Myr.

Convergence rate could affect lava composition in a variety of ways. With decreasing convergence rate, the down dip flux of subducted sediment is smaller (Figure 9). Slow convergence leads to both increased conductive cooling of the convecting mantle wedge beneath an arc and increased heating of the subducting plate [e.g., *Kincaid and Sacks,* 1997, their Figure 10]. There are many possible consequences of these thermal effects: (1) dehydration reactions in the subducting plate might occur at shallower depths

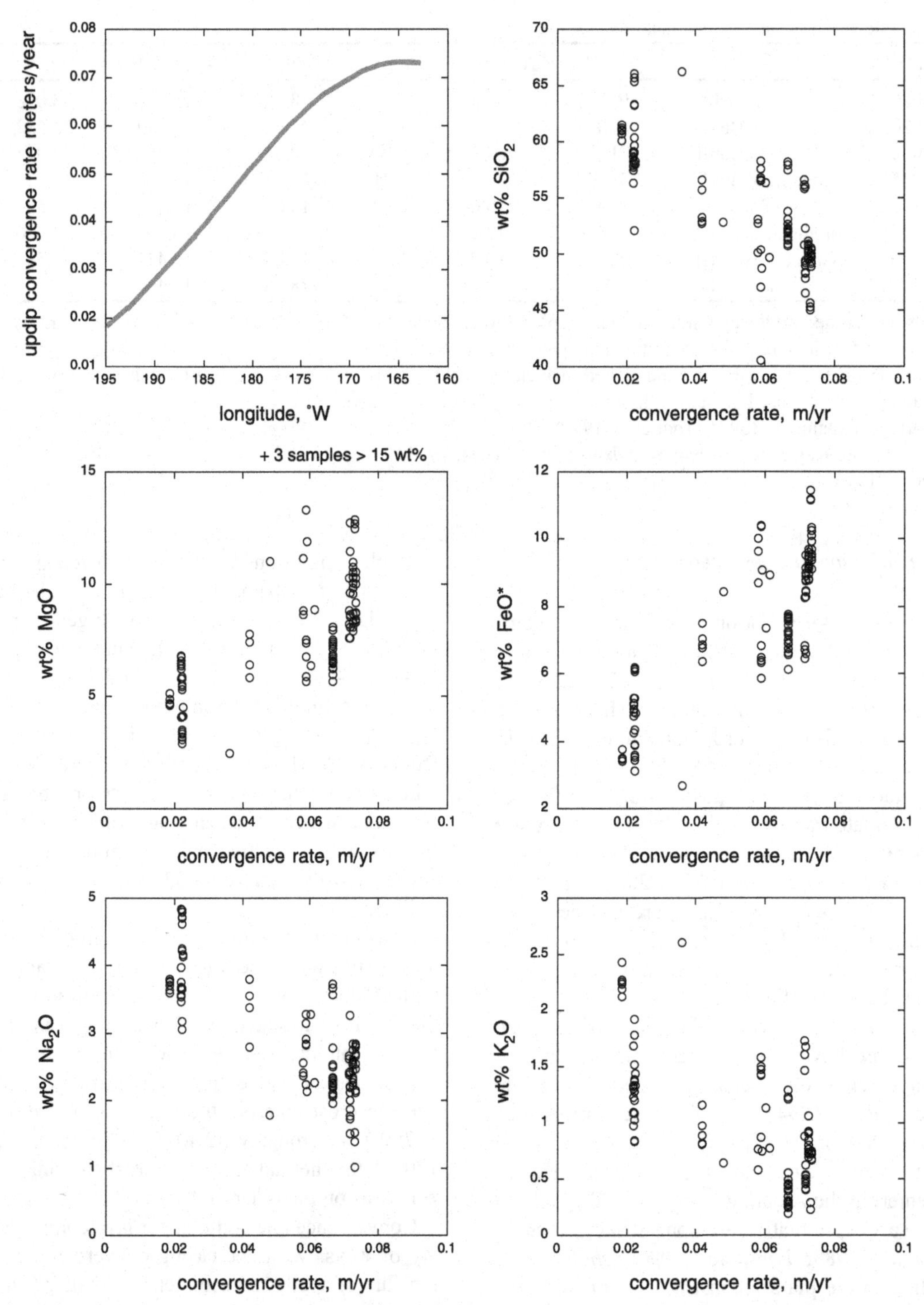

Figure 8. Some major element abundances in Aleutian lavas versus trench-orthogonal convergence rate beneath Aleutian volcanoes. Convergence rate from [*Fournelle et al.,* 1994] is plotted versus the longitude of each volcano. However, the convergence rate for each volcano is determined at a position in the trench that is updip of the volcano, projected along the plate convergence vectors shown in Figure 1 [*Engebretson et al.,* 1985]. Longitudes of positions in trench updip of each volcano were calculated using the following empirical function:

$$\text{longitude } (°\text{W}) = -393.71+5.6627*\text{volcano longitude } (°\text{W}) - 0.013863*\text{ volcano longitude } (°\text{W})^2$$

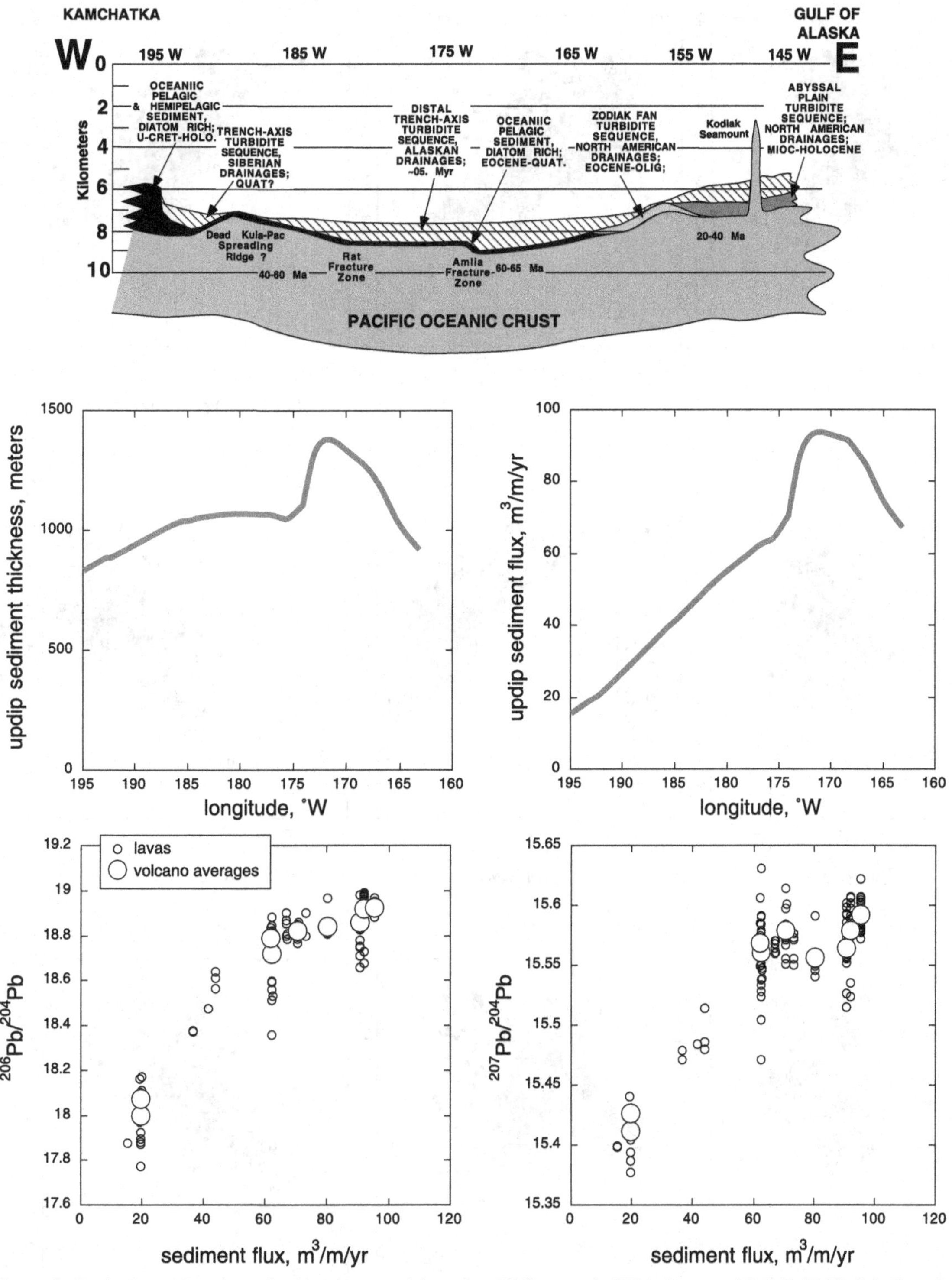

Figure 9. Geologic section along the Aleutian trench based on [*Vallier et al.,* 1994; *Ryan and Scholl,* 1989; *Scholl et al.,* 1987; *von Huene,* 1986; *McCarthy and Scholl,* 1985] and our unpublished data, sediment thickness subducted beneath Aleutian volcanoes, based on this section, sediment flux beneath Aleutian volcanoes (product of convergence rate from Figure 8 and sediment thickness), and Pb isotope ratios versus sediment flux. Sources of Pb isotope data in caption for Figure 3, and new data in Table 1.

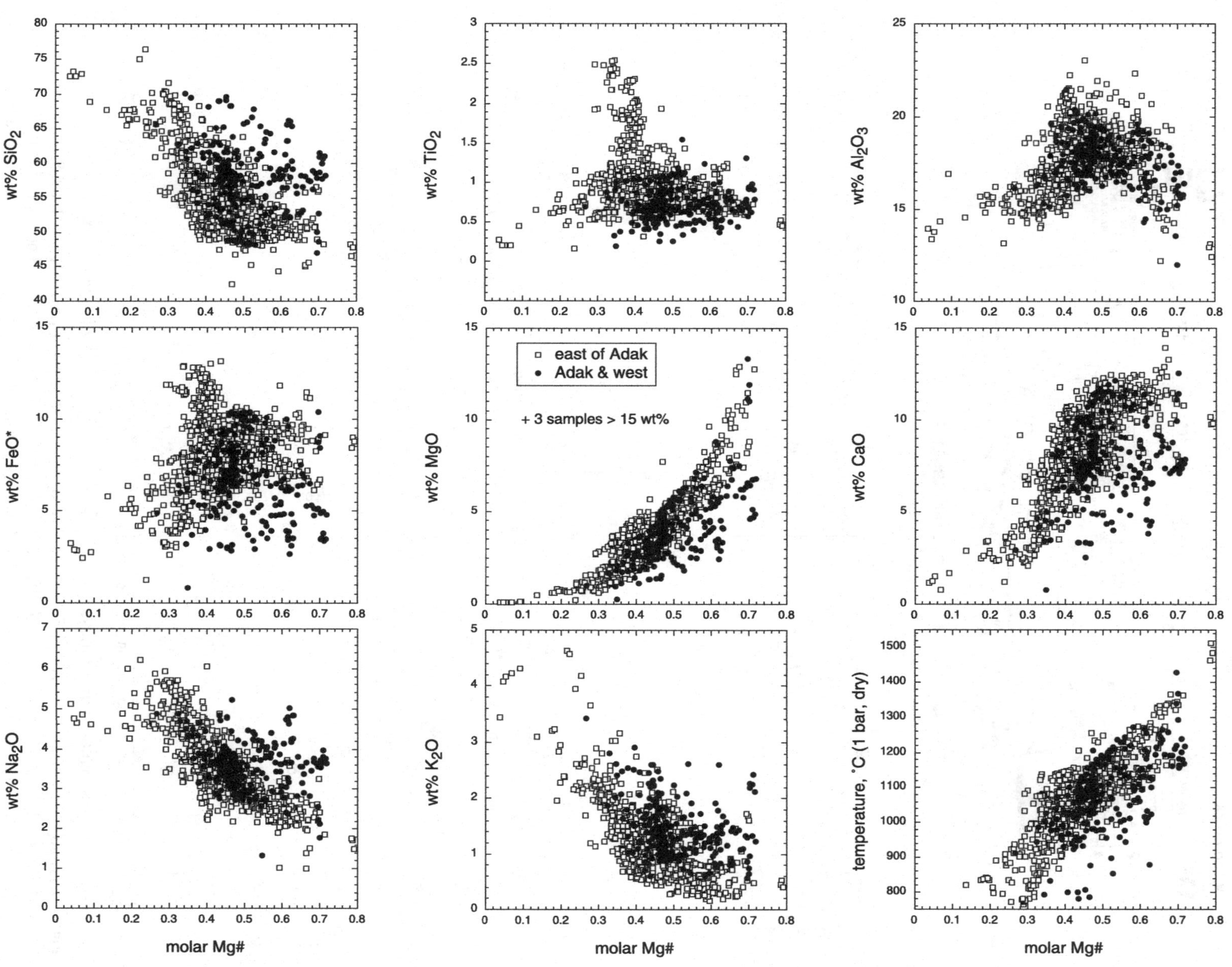

Figure 10. Variation of major element oxides versus molar Mg# in Aleutian lavas. Sources of data in caption for Figure 3.

beneath the forearc, limiting the amount of H_2O available to flux melting in the mantle wedge, (2) anatexis of subducted material might begin at shallower depths, extend to higher degrees of melting, and/or become more common or extensive, (3) there might be a lower extent of melting in the colder mantle, and (4) cooling of the shallow mantle and lower crust could lead to crystal fractionation at higher pressure, perhaps within the shallow mantle below the seismic Moho. Assuming that volcanic output is proportional to the magmatic flux through the Moho, it seems evident that some combination of (1), (3) and (4) limits the overall magma flux from the mantle to the crust in the western Aleutians. With the exception of Tanaga volcano, none of the emergent volcanoes in the western Aleutians are as large as the twenty largest volcanoes in the central Aleutians [Fournelle et al., 1994].

1.6 Along-Strike Variation in Sediment Thickness and Composition

Variation in Aleutian lava compositions might be related to the composition and volume of subducted sediment. Based on geophysical data, moderate size (20-40 km wide and as much as 6 km thick) prisms of tectonically accreted trench floor sediment form the lower part of the landward side of the Aleutian trench from 164 to 185°W. West of about 185°W, the wedge seems to rapidly dwindle in size and opposite Attu (188°W), the prism is quite small, perhaps just a few km wide and thick. Geophysical imaging, DSDP drilling, and rock dredging, imply that the mass of the modern accretionary prism began to form in latest Miocene time, i.e., 5-6 Myr ago [*Vallier et al.,* 1994; *Ryan and Scholl,* 1989; *Scholl et al.,* 1987; *von Huene,* 1986; *McCarthy and Scholl,* 1985].

Mass balance calculations imply that, during the past 5 Myr or so, no more than 20 percent of the volume of trench-floor sediment swept into the Aleutian subduction zone has contributed to the mass of the accretionary prism [*von Huene and Scholl,* 1991]. Based on more recent geophysical and drilling data, compiled during the past decade, it is now supposed that no more than 15 percent of the mass of ocean floor sediment that enters the Aleutian subduction zone is retained in an accretionary prism [*Scholl and Huene,* 1998].

Thus, it is possible to approximate the flux of sediment beneath a given part of the arc as the product of the trench-orthogonal convergence rate (Section 1.5), and the thickness of sediment in the trench. Figure 9 illustrates the thickness and inferred composition of sediments in the Aleutian trench, based on [*Vallier et al.,* 1994; *Ryan and Scholl,* 1989; *Scholl et al.,* 1987; *von Huene,* 1986; *McCarthy and Scholl,* 1985] and our unpublished data.

About half of the sediment derived from erosion of the Aleutian arc itself is shed southward toward the Aleutian Trench, with the other half accumulating in the Bering Sea. Very little of this material appears to make its way to the trench floor because the bulk of the sediment pools in the fore arc, which is underlain by a trough of arc-derived detritus, ash debris, and pelagic (mainly diatoms) sediment as thick as 3- 4 km. This fore arc basin runs along virtually the entire length of the Aleutian island arc.

In general, sediments in the trench are comprised largely of turbidites composed of sediments shed off the Alaska Range and older sediments derived from the Chugach, Wrangell, and St Elias Mountains. The trench axis slopes westward continuously to about 180°W. Sediments thicken gradually from east to west between 160 and ~172°W. West of 172°W, they thin gradually from east to west, and then more abruptly from about 1 km thick at ~182°W where the Rat Fracture Zone intersects the trench to just a few hundred meters at ~ 190°W. Even further west they thicken again as a result of turbidite sedimentation from Kamchatka. The thinning of the sedimentary section in the Aleutian trench from 182°W to 190°W is accompanied by an increasing proportion of pelagic to continentally-derived sediments.

The major variations in the thickness and relative proportions of different sediment types occur west of 182°W. Because the convergence vector is strongly oblique in the western Aleutians, any sediments beneath, e.g., Buldir volcano in the distal Aleutians (184°W) would have entered the trench at a longitude of ~ 179°W. Sediments overlying the Pacific - North America plate boundary west of ~184°W may not be subducted at all. Thus, there is little systematic, along-strike variation in the thickness or composition of sediment being subducted beneath the central and western Aleutian arc. If these estimates are correct, the main control on sediment flux beneath the Aleutian arc is the trench-orthogonal convergence rate.

2. ALONG-STRIKE VARIATION IN LAVA COMPOSITION

In this part of the paper, we emphasize systematic east-to-west trends in lava composition. Previous studies of along-strike variation in the Aleutians have not emphasized systematic trends along the entire arc. Instead, spatial variation in lava composition has been attributed to (1) varying stress regimes within a series of tectonic blocks along the arc [e.g., *Kay and Kay,* 1994; *Singer and Myers,* 1990; *Kay et al.,* 1982], or (2) the availability of volatiles and fluid associated with subduction of thick sedimentary sections deposited over ancient oceanic fracture zones [e.g., *Miller et al.,* 1992; *Singer et al.,* 1992a; *Singer et al.,* 1992b; *Singer et al.,*

1992c]. In keeping with (2), it has long been noted that some of the largest Aleutian volcanic centers overlie subducted oceanic fracture zones [*Kay et al.,* 1982; *Marsh,* 1982a].

An exception is the observation of *Fournelle et al.* [1994] that, in general, the size of Aleutian volcanoes decreases from east to west, which they ascribed to decreasing subduction rate and magma flux from east to west. Myers and co-workers noted that calc-alkaline lavas tend to be erupted from smaller volcanic centers, whereas tholeiitic lavas predominate at larger centers [*Myers,* 1988; *Myers and Marsh,* 1987; *Myers et al.,* 1986a; *Myers et al.,* 1986b; *Myers et al.,* 1985].

2.1 Along-Strike Variation in Mg# and MgO

Figure 3 illustrates along-strike variation in major element composition of lavas in the Aleutians. Prior to discussing these results, we would like to reiterate an important caveat. Lavas from the Komandorsky block form an important end-member in most of the trends apparent in Figure 3. If the Komandorsky data were omitted, the apparent correlations of major element composition with longitude would be much weaker. The Komandorsky block is tectonically anomalous, as described in Section 1.2, and it is reasonable to question whether processes there are related to magma genesis in the tectonically "normal" parts of the western and central Aleutians. We believe that a compositional continuum is evident in Figure 3, from the Komandorsky block through the distal Aleutians to the Adak area and the central Aleutians. A continuum is also evident in similar plots of isotope ratios as a function of distance along the arc, discussed later in this paper. As a result of this interpretation, we believe that the Komandorsky lavas are genetically related to the rest of the arc, and form an important compositional end-member that is present in gradually diminishing proportions to the east. This is important to our interpretation of magma genesis in the Aleutian arc. Thus, we wish to remind readers that the apparent compositional continuum is dependent on a limited number of samples from the distal Aleutians (Figure 2), which have compositions intermediate between the Komandorsky samples and lavas further to the east. Because it is relatively unknown but very important, we hope that the distal Aleutians will be an area of intensive sampling in the future.

With this caveat, we turn to Figure 3. Although the data show considerable scatter, due mainly to crystal fractionation, the average Mg# for volcanic centers apparently increases from east to west. In the western Aleutians, where samples are few, most of them are primitive. This is fortunate, because we rely heavily on the composition of primitive lavas to infer the nature of mantle-derived melts in the Aleutians. (Note that primitive lavas can be affected by magma mixing and assimilation processes, which—if ignored—could lead to erroneous conclusions concerning primary magma compositions; such effects are discussed in Section 4.2 of this paper).

MgO contents in primitive Aleutian lavas decrease from east to west. Primitive magmas in the Aleutians have a variety of MgO contents, ranging from 4 to 18 wt%, and most of this variation is not due to crystal fractionation. As a result, it is inappropriate to constrain the nature of Aleutian primary magmas by comparing concentrations of incompatible elements, such as Na, K or various trace elements, in lavas with a given MgO content. One could, instead, correct incompatible element concentrations for the effects of fractionation by extrapolating to a given Mg#. However, because the slopes of variation trends for, e.g., Na_2O and K_2O versus Mg# vary from one volcanic center to another in the Aleutians, comparisons of fractionation-corrected compositions of lavas with Mg# < 0.6 are fraught with peril. For this reason, we have chosen to concentrate on comparing uncorrected, incompatible element concentrations in primitive lavas with Mg# > 0.6.

The MgO content of mantle-derived magmas depends strongly on temperature. For example, many workers have developed geothermometers which depend on equilibrium Mg partitioning between olivine and liquid [e.g., *Gaetani and Grove,* 1998; *Roeder and Emslie,* 1970, and references cited therein]. For the most recent of these [*Gaetani and Grove,* 1998], we can use an olivine/liquid Fe/Mg Kd = 0.35 - 0.013 (mol% Na_2O + K_2O), based on the data of [*Baker et al.,* 1996], and the assumption that 80% of all Fe is ferrous to estimate equilibrium olivine compositions for primitive Aleutian lavas, and then derive a temperature estimate.

Based on the assumptions outlined in the previous paragraph, estimated temperatures of equilibration between olivine and primitive lavas at 1 bar vary from ~ 1500°C for three picritic lavas with 16 to 18 wt% MgO[1], or ~ 1350°C for more common basalts with ~ 10 wt% MgO, to 1000°C for lavas with 4 wt% MgO. Because Fe^3/Fe^2 and water contents for Aleutian melts prior to possible crustal degassing, crystallization and eruption are not yet known, it is not possible to quantify magmatic temperatures more accurately. Probably, the low MgO, high $Na_2O + K_2O$, andesite lavas in the western Aleutians also include abundant H_2O, whereas the primitive basalts in the eastern Aleutians have much less

[1] *Nye and Reid* [1986] argued that these high MgO lavas do, in fact, represent liquid compositions even though they include olivine phenocrysts.

H_2O. If so, then the total variation in Aleutian primitive magma temperatures is on the order of 300°C or more. Well-studied volcanic centers in the Aleutians exhibit a ~ 200°C range of estimated temperatures for primitive lavas which is not correlated with Mg# and therefore appears to be a primary feature rather than the result of crystal fractionation. This variability makes it difficult to be sure, but in general there seems to be a systematic, along-strike variation in magmatic temperature. For example, the maximum estimated temperature varies from ~1350°C (or even 1500°C) in the central Aleutians to ~1200°C in the western Aleutians.

2.2 Along-Strike Variation in Silica and Alkali Contents

Figure 3 illustrates along-strike variation in SiO_2, FeO, CaO, Na_2O, and K_2O. SiO_2 and Na_2O show a systematic increase from east to west. There may be a similar trend for K_2O, but it is less clear. CaO, FeO and molar Ca/Al (not shown) show a systematic decrease from east to west. These trends are clearer if lava compositions are plotted versus trench-orthogonal convergence rate (Figure 8). This is because the central Aleutians show relatively little variation in both lava composition and convergence rate, and plot as a group when convergence rate rather than longitude is considered. TiO_2 and Al_2O_3 show no clear correlation with longitude or convergence rate, and we believe this is important though we have omitted both oxides from Figure 3 to save space.

Plots of major element oxides versus Mg# for Aleutian lavas (Figure 10) can be used to evaluate the amount of variation which can be produced by crystal fractionation from a parent with an Mg# of, e.g., 0.7 to produce a derivative liquid with an Mg# of 0.6. It is clear from these plots that the east-to-west variation in Aleutian lava compositions is not the result of systematically increasing degrees of crystal fractionation from east to west.

The high alkali contents of primitive lavas in the western Aleutians probably arise as a result of either (a) small degrees of partial melting of the mantle, (b) variation in mantle source composition, possibly including a component derived from subducted metabasalt and/or metasediment, (c) combined crystal fractionation and melt/rock reaction in the shallow mantle, or (d) mixing of relatively alkali-poor, primitive basalt and alkali-rich, evolved dacites or rhyolites. These possibilities will be discussed further in Section 4. However, we note here that the lack of systematic variation in TiO_2 and Al_2O_3 in primitive lavas along strike suggests that the variation of alkali element concentrations cannot be explained as the result of different degrees of melting of a common source composition, or different degrees of crystal fractionation from a common parental magma (i.e., a variety of primitive magma types are required).

High alkali and H_2O contents could explain the high SiO_2 contents in primitive lavas in the western Aleutians. Increases in Na, K and H contents increase the size of the olivine primary phase volume, thereby increasing the SiO_2 contents of magmas that can be in equilibrium with mantle olivine [*Hirschmann et al.,* 1998; *Ryerson,* 1985; *Kushiro,* 1975]. Like continental crust, Aleutian high Mg# andesites are quartz normative. After decades of debate, it remains controversial whether quartz-normative andesites can be formed by partial melting of mantle peridotites [*Hirose,* 1997; *Baker et al.,* 1996; *Falloon et al.,* 1996; *Kushiro,* 1990; *Mysen et al.,* 1974]. However, in our opinion, experimental data [*Baker et al.,* 1994; *Tatsumi and Ishizaka,* 1982] illustrate that quartz-normative, high Mg# andesites similar to those in the western Aleutians can *equilibrate* with mantle peridotite, provided they have sufficiently high H_2O contents. In Section 4, we discuss whether such liquids could have formed by small degrees of melting of mantle peridotite, or whether their high alkali contents arise via reaction of partial melts of subducted metabasalt and/or metasediment with mantle peridotite.

2.3 Along-Strike Variation in Heavy, Radiogenic Isotope Ratios

There are systematic along-strike variations in Pb, Nd and Sr isotope ratios in Aleutian lavas. $^{206}Pb/^{204}Pb$, $^{207}Pb/^{204}Pb$, $^{208}Pb/^{204}Pb$ and $^{87}Sr/^{86}Sr$ all decrease from east to west, whereas $^{143}Nd/^{144}Nd$ increases from east to west (Figure 7), as previously documented [e.g., *Yogodzinski et al.,* 1995]. Again, the apparent strength of the correlations between isotope ratios and longitude depends heavily on data from the Komandorsky block and the western Aleutians.

In the central Aleutians, high Pb isotope ratios have been ascribed to recycling of Pb from subducted sediments into Aleutian lavas [e.g., *Class et al.,* 2000; *Kay and Kay,* 1994; *Miller et al.,* 1994; *Singer et al.,* 1992a; *Myers and Marsh,* 1987; *Kay et al.,* 1978b, and references therein]. Indeed, very high Pb isotope ratios ($^{207}Pb/^{204}Pb$ ~ 19.2) were reported for voluminous, terrigenous sediments recovered in DSDP Site 183 at 52.57°N, 161.20°W, SE of the eastern Aleutians [*Miller et al.,* 1994]. However, pelagic and metalliferous sediments from just above the basaltic basement in Site 183 have Pb isotope ratios more similar to mid-ocean ridge basalts (MORB) [*Peucker-Ehrenbrink et al.,* 1994]. In fact, some metalliferous sediments from elsewhere in the Pacific have Pb isotope characteristics indistinguishable from MORB. Metalliferous sediments are not abundant compared to continentally derived turbidites in DSDP Site 183, but they have high Pb concentrations, and the relative abundance of pelagic sediment to terrigenous turbidites on the subducting Pacific oceanic crust may increase westward. Thus, without further evaluation, Pb isotope ratios alone cannot be used as an unambiguous tracer for a recycled sediment component in Aleutian lavas.

Pb isotope ratios in Aleutian lavas are positively correlated with $^{87}Sr/^{86}Sr$ and negatively correlated with $^{143}Nd/^{144}Nd$ (Figure 11). In contrast, pelagic and metalliferous sediments have high $^{87}Sr/^{86}Sr$ and low $^{143}Nd/^{144}Nd$. Like Pb, Sr is likely to be mobilized in both aqueous fluids and melts derived from subducted sediment. Thus, the positive correlation of Pb and Sr isotope ratios in Aleutian lavas suggests that only the Aleutian lavas with high Pb isotope ratios include a substantial, recycled sediment component. Otherwise, we should observe high $^{87}Sr/^{86}Sr$ in lavas that include unradiogenic Pb from pelagic or metalliferous sediments. In support of this, Figure 9 shows that $^{207}Pb/^{204}Pb$ in Aleutian lavas is correlated with trench-orthogonal sediment flux (the product of sediment thickness and trench-orthogonal convergence rate).

The unusually low Pb and Sr isotope ratios in lavas from the Komandorsky block deserve special mention. A few samples of MORB from the North Pacific have similarly low values, but they are rare. It would be of great interest to obtain samples of lavas formed at the extinct Kula-Pacific spreading ridge to see if such depleted MORBs were common along that ridge, particularly near its western end.

2.4 Along-Strike Variation in Proposed Indicators of a Sediment Melt Component

As a result of the along-strike variation in trench-orthogonal sediment flux (Section 1.6 and Figure 9), and the related variation in radiogenic isotope ratios in lavas (Section 2.3 and Figure 7), the Aleutian arc represents an ideal testing ground for proposed relationships between the flux of trace elements in subducted sediment and their concentration in arc lavas. For example, *Plank and Langmuir* [1993] demonstrated that there is a worldwide correlation between the flux of subducted K and Th, and the concentration of these elements in primitive arc lavas. Similarly, *Elliott et al.* [1997] hypothesized that a vector toward high La/Sm, Th/Nb, low Ba/La in lavas from the Marianas arc represented addition of a component derived from partial melting of subducted sediment. They contrasted this with a vector toward high Ba/La in Marianas lavas, which they inferred was due to addition of an aqueous fluid derived by dehydration of subducted basalt. And, *Turner et al.* [1996] and *Hawkesworth et al.* [1997] found a correlation between Th/Ce (nearly equivalent to Th/La) and isotopic indicators of a recycled sediment component in lavas from the Lesser Antilles. Noting that Th is relatively immobile in aqueous fluid/solid equilibria, compared to silicate melt/solid equilibria, all of these groups suggested that the Th-enriched, sediment component must be a partial melt of subducted sediment, in the Marianas [*Elliott et al.,* 1997], the Lesser Antilles [*Hawkesworth et al.,* 1997; *Turner et al.,* 1996], and worldwide [*Plank and Langmuir,* 1993].

Figure 12 shows the variation in La and Th in primitive lavas, and La/Sm, Th/La, Th/Nb, and Ba/La in all lavas along the strike of the Aleutian arc. Note that variation in K along strike, in primitive lavas and all lavas, is shown in Figure 3. With the exception of Ba/La and Th/La, none of these variables show systematic variation along the arc, and in fact K may be highest in the west where the sediment component is minimal or absent. Th/La is correlated with Pb isotope ratios in Aleutian lavas (Figure 13), so it probably is a good proxy for the presence of recycled sediment, as proposed by *Hawkesworth et al.* [1997].

2.5 Evaluation of Proposed Indicators of a Fluid Component From Subducted Material

2.5.1 Barium/Lanthanum. As noted in the previous section, *Elliott et al.* [1997] and many other workers have proposed that Ba is mobile in aqueous fluids, whereas La, Nb and Ta are relatively immobile, so that Ba concentration in primitive lavas, and Ba/La, Ba/Nb and Ba/Ta in more evolved lavas, can be used to detect the presence of an aqueous fluid component derived from dehydration of subducted oceanic crust and/or sediments. Except for Ba/La (Figure 12), we see no systematic variation of these factors along the Aleutian arc (Figure 14).

Interestingly, as previously noted [*Kay and Kay,* 1994; *Yogodzinski et al.,* 1994; *Kay,* 1980], Ba/La is positively correlated with Th/La and Pb isotope ratios in Aleutian lavas (Figure 13). Aleutian lavas with high Th/La have high $^{207}Pb/^{204}Pb$ (derived from sediment) and high Ba/La. Recycling of Pb and other components in the Aleutians cannot be clearly separated into a Ba-rich fluid component and a Th-rich sediment melt, perhaps because Ba/La is very high in Aleutian sediments [Plank, pers. comm., 2001; Turner et al., pers. comm. 2001] In this way, the pattern of Ba/La variation is strikingly different from many other intraoceanic arcs worldwide [e.g., *Elliott et al., this volume*]. The relatively high solubility of Ba and Pb in aqueous fluids, compared to relatively insoluble Th, is well-documented [e.g., *Johnson and Plank,* 1999; *Ayers,* 1998; *Ayers et al.,* 1997;

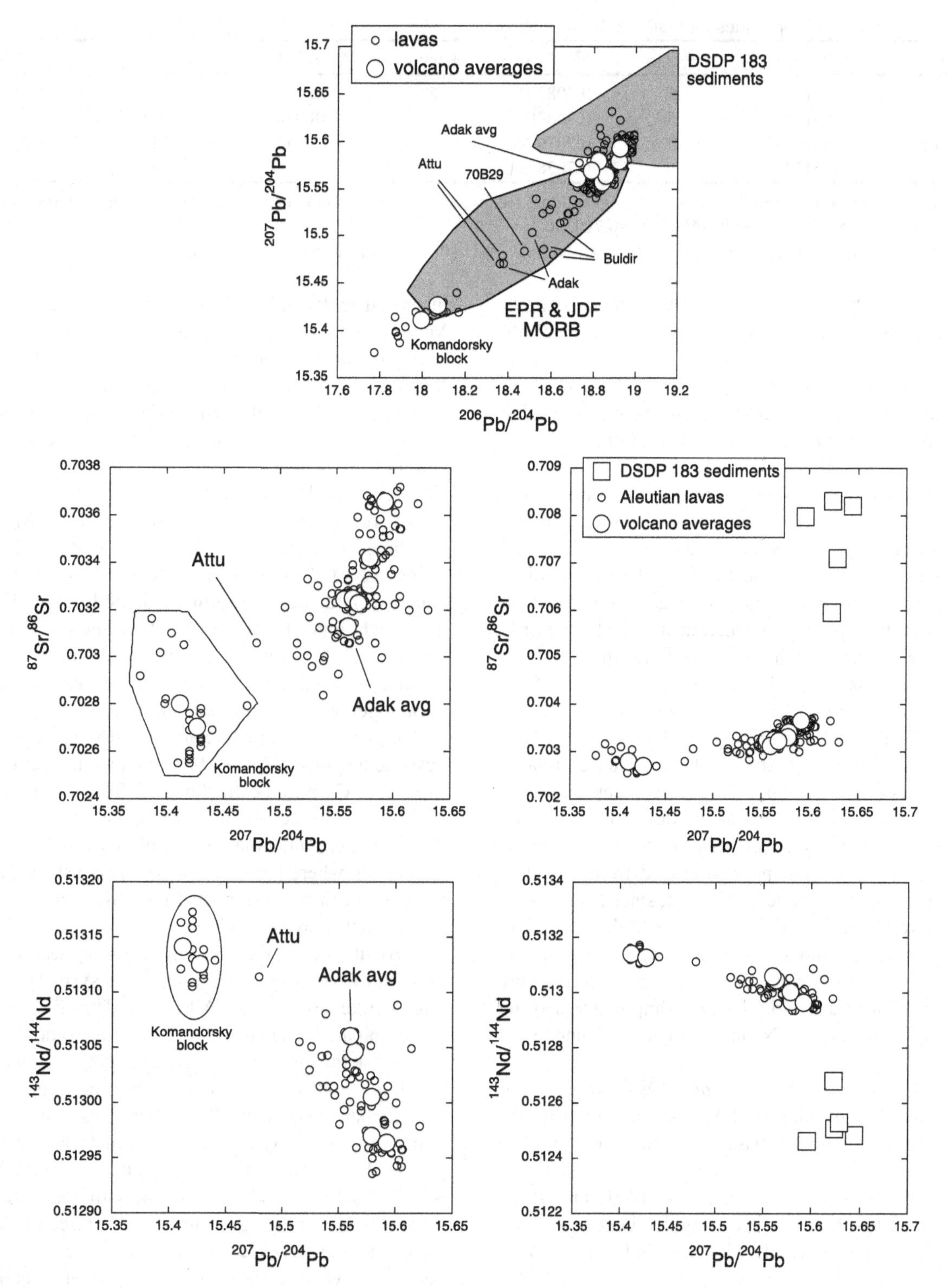

Figure 11. Isotope variation in Aleutian lavas and in sediments from DSDP Site 183 (52.57°N, 161.20°W) southeast of the Aleutian trench. Sources of data in caption for Figure 3, [*Plank and Langmuir,* 1998; *Peucker-Ehrenbrink et al.,* 1994; *von Drach et al.,* 1986], new Pb isotope data for western Aleutian lavas in Table 1, and new Sr and Nd isotope data for DSDP Site 183 sediments in Table 2.

Table 2: New Sr and Nd isotope ratios for DSDP Hole 183 sediments

sample	$^{143}Nd/^{144}Nd$	± ppm	$^{87}Sr/^{86}Sr$	± ppm	$^{206}Pb/^{204}Pb$*	$^{207}Pb/^{204}Pb$*	$^{208}Pb/^{204}Pb$*
38-3, 20-21	0.512485	5	0.708201	7	19.093	15.645	38.814
39-1, 83-84	0.512510	5	0.708309	8	18.934	15.624	38.734
39-1, 108-109	0.512462	5	0.707979	7	18.546	15.596	38.546
39-1, 108-109	0.512416	6	0.707850	7			

Sr and Nd isotopic analysis were carried out with conventional techniques (Hauri and Hart, 1993). All results are corrected against LaJolla std $^{143}Nd/^{144}Nd=0.511847$ and NBS987=$^{87}Sr/^{86}Sr=0.710240$.

*: Pb isotope data from Peucker-Ehrenbrink et al., 1994. Sample 39-1, 108-109 is a carbonate sample without Pb data

Brenan et al., 1996; *Brenan et al.,* 1995a; *Brenan et al.,* 1995b]. In contrast, Ba, Pb and Th all behave incompatibly during melting. Thus, the correlation of Ba, Pb and Th concentrations in Aleutian lavas suggests that transport of all these elements from subducted crust into the mantle wedge is mainly via a silicate melt, not an aqueous fluid.

2.5.2 Strontium/Neodymium. Virtually all primitive lavas in the Aleutians have superchondritic Sr/Nd ratios, and this ratio shows no systematic variation along-strike. Elevated Sr/Nd ratios (> 17; [*Anders and Grevesse,* 1989]), could be indicative of Sr addition to the arc magma source via an aqueous fluid, because experimental data indicate that Sr is more soluble in high P, high T aqueous fluids than the REE [*Johnson and Plank,* 1999; *Ayers,* 1998; *Stalder et al.,* 1998; *Ayers et al.,* 1997; *Kogiso et al.,* 1997; *Brenan et al.,* 1996; *Brenan et al.,* 1995a; *Brenan et al.,* 1995b; *Tatsumi et al.,* 1986]. Sr/Nd in primitive Aleutian lavas is not correlated with Ba/La, as might be expected if enrichments of Sr, Ba and Pb were all the result of aqueous fluid addition to the arc source[2]. Alternatively, superchondritic Sr/Nd could indicate the presence of a component derived from cumulate gabbro in subducted, oceanic lower crust, since plagioclase-rich cumulates are enriched in Sr relative to the REE. And finally, trace element modeling (Section 4) shows that partial melts of mid-ocean ridge basalt in eclogite facies may have super-chondritic Sr/Nd. Thus, we simply note that the explanation for elevated Sr/Nd in the Aleutians is uncertain.

2.5.3 Cerium/Lead. Miller et al. [1994] argued that unradiogenic Pb in the source of Aleutian arc magmas was transported from subducted basalts into the mantle wedge in an aqueous fluid. They suggested that Pb is relatively mobile in aqueous fluids, while Ce is relatively immobile, so that Ce/Pb can be used to distinguish between (a) Pb transported in aqueous fluids with low Ce/Pb, and (b) Pb in partial melts of the Aleutian mantle and/or subducted MORB with Ce/Pb greater than 10. This hypothesis is supported by more recent experimental data on fluid/rock partitioning [*Brenan et al.,* 1995a]. Sediments have low Ce/Pb, so Miller et al. could not use Ce/Pb to distinguish between melt and fluid transport of a sediment Pb component in Umnak lava with radiogenic Pb isotopes. However, since Ce/Pb is low (less than ~ 4) in all lavas from Umnak Island, regardless of their Pb isotope ratios, Miller et al. concluded that the unradiogenic, MORB-like Pb isotope ratios in some Umnak lavas were due to transport of Pb in aqueous fluids derived from subducted basalt. This led to the much repeated aphorism, "sediments melt, basalts dehydrate". On this basis, one could infer that Ce/Pb should be low in all Aleutian lavas, especially those with unradiogenic Pb isotope ratios.

Complicating this picture are the trace element models of possible igneous processes beneath arcs, in Section 4. They show that Ce may be fractionated from Pb during partial melting of eclogite, and during reaction of melts with mantle peridotite, producing a range of possible Ce/Pb ratios in melts even where the source Ce/Pb is constant. The resulting Ce/Pb can be greater than or substantially less than in the original source of melting.

In Aleutian data, Ce/Pb shows a weak, negative correlation with Pb isotope ratios, and shows systematic variation along-strike (Figure 15). All of the data discussed in this paragraph are for lavas whose Pb isotope ratios are known, and whose Pb concentration has been determined by isotope dilution. Ce/Pb is greater than 4 and ranges up to ~ 18 for all Aleutian lavas with $^{207}Pb/^{204}Pb$ less than 15.5, all of which are west of Adak. On Adak, 6 of 15 lavas have Ce/Pb greater than 4, ranging up to ~ 15. In the central Aleutians, ~ 50 lavas have Ce/Pb less than 4, while three have Ce/Pb greater than 4, ranging up to ~ 7. All three of the central Aleutian lavas with Ce/Pb from 4 to 7 are from volcanoes behind the volcanic front; two from Amak Island and one from Bogoslof.

To summarize, unradiogenic Pb isotope ratios, reflecting Pb derived from subducted oceanic crust and/or the sub-arc

[2]Note that we use only primitive lavas in this discussion, in an attempt to avoid the effects of plagioclase crystallization on Sr concentration.

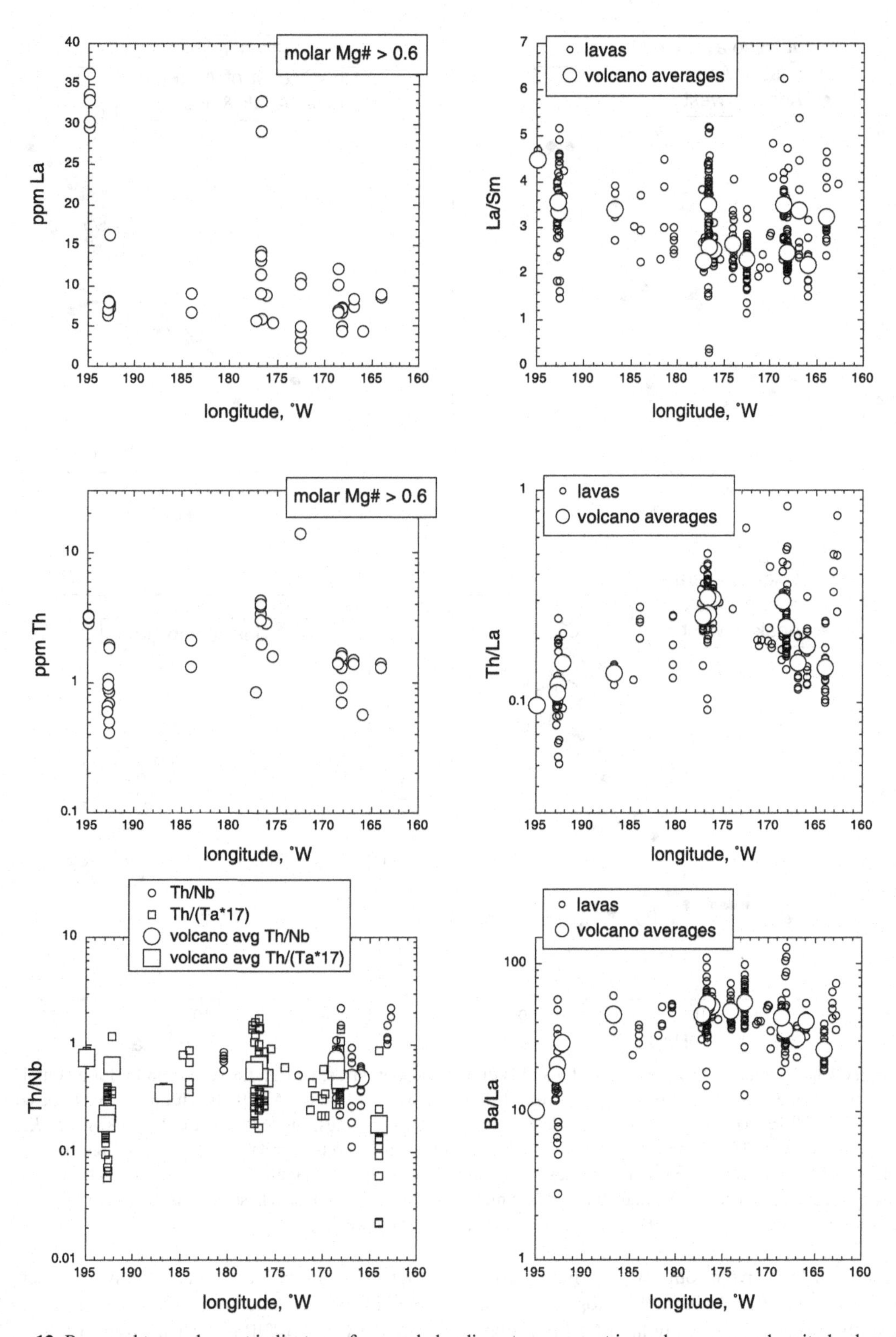

Figure 12. Proposed trace element indicators of a recycled sediment component in arc lavas, versus longitude along the Aleutian arc. Of these, only Th/La and perhaps Ba/La seem to show systematic variation along-strike. Sources of data in caption for Figure 3.

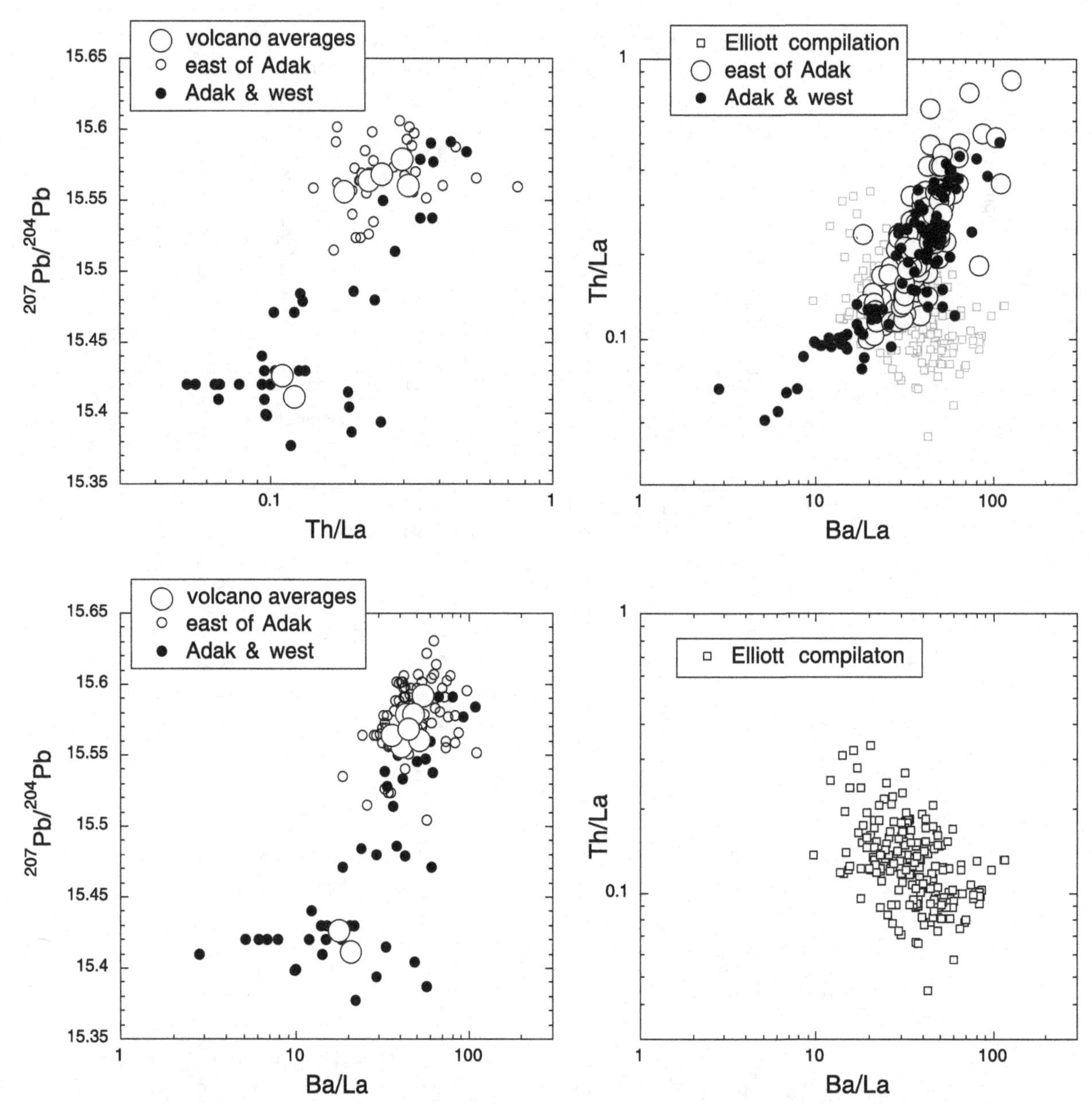

Figure 13. Variation of $^{207}Pb/^{204}Pb$, Th/La and Ba/La in Aleutian lavas, and Th/La versus Ba/La for a compilation of arcs worldwide—excepting the Aleutians—from Elliott (this volume). In the Aleutians, Th/La appears to be correlated with $^{207}Pb/^{204}Pb$. Th/La is definitely correlated with Ba/La, as previously noted by [*Kay,* 1980; *Kay and Kay,* 1994; *Yogodzinski et al.,* 1994]. In the "worldwide" data set, Ba/La is negatively correlated with Th/La. High Ba/La has been interpreted to reflect addition of an aqueous fluid component to the arc magma source, whereas high Th/La has been interpreted to reflect addition of a distinct, partial melt of subducted sediment. These components cannot be easily separated in the Aleutians. Sources of Aleutian data in caption for Figure 3.

mantle, are found in lavas with both low and high Ce/Pb ratios. In the western Aleutians, where Pb isotope ratios are lowest, almost all lavas have relatively high Ce/Pb, probably indicative of Pb transport in a melt rather than an aqueous fluid. In the central Aleutians, high Ce/Pb together with low $^{207}Pb/^{204}Pb$ are observed only in volcanoes situated behind the volcanic front. Other central Aleutian samples with low $^{207}Pb/^{204}Pb$, such as some of the Umnak lavas studied by Miller et al., have low Ce/Pb. These low Ce/Pb ratios *could* reflect transport of Pb from subducted basalt in

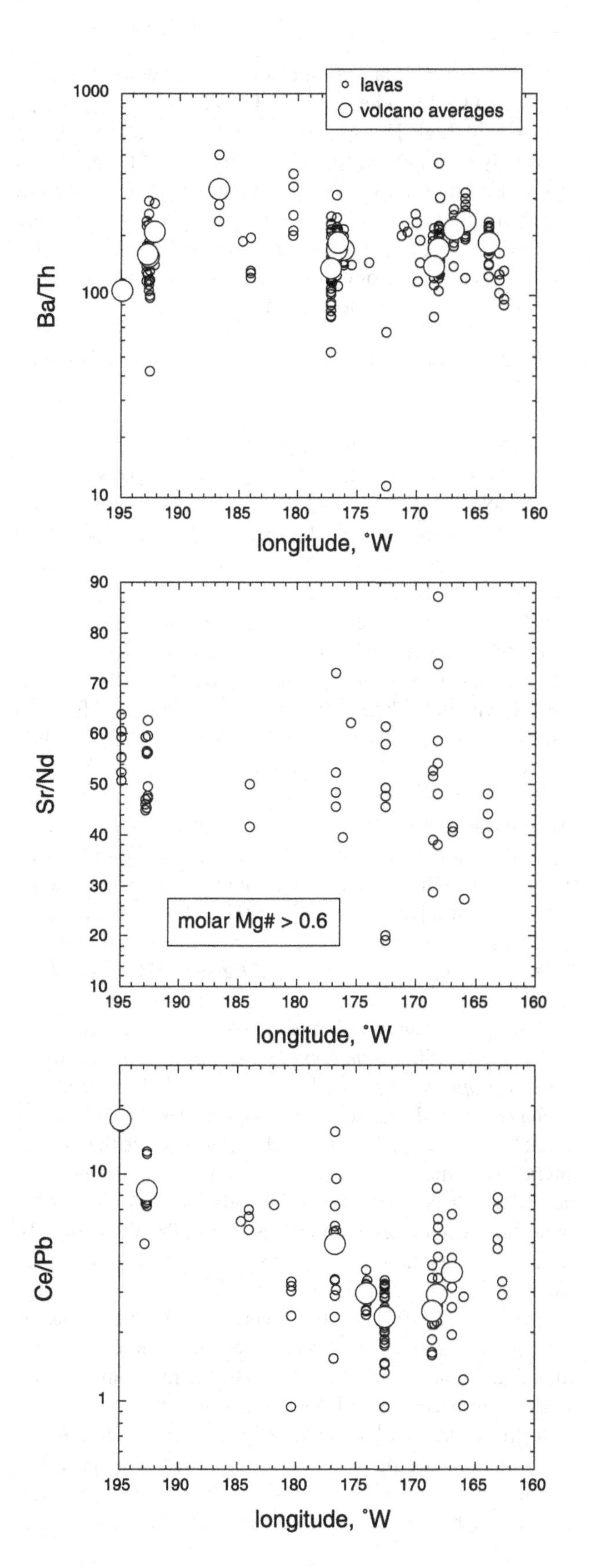

an aqueous fluid, but are *also consistent* with Pb transport in a partial melt of subducted basalt (see Section 4).

2.6 Along-Strike Variation in Proposed Indicators of an "Eclogite Melt" Component

Kay [1978] proposed that a few highly light REE enriched, heavy REE depleted, high Mg# andesites—from the base of Moffett volcano on Adak Island and two samples dredged from the Western Aleutian seafloor—were produced by partial melting of subducted basalt in eclogite facies, followed by reaction of this melt with the mantle during transport from the subduction zone to the surface. Because we feel use of the term "adakite" for these lavas is problematic (see Section 1.1), we refer to these lavas as "enriched, high Mg# andesites".

Kay noted that enriched, high Mg# andesites had such high Sr/Y, steep REE patterns, and low Yb and Y concentrations, that they seemed to require partial melting of a source with abundant garnet, in which Yb and Y were compatible and La and Sr were highly incompatible. The source must have contained little or no plagioclase, in which Sr is compatible and Yb and Y are highly incompatible. Many igneous processes can fractionate light REE from heavy REE whereas, among minerals that are abundant in metabasalt and mantle peridotite, only garnet can produce a strong fractionation of middle REE, such as Dy from heavy REE such as Yb. Thus, Dy/Yb is better than La/Yb as an indicator of abundant garnet in the source of arc magmas. Figure 16 shows that Sr/Y correlates with Dy/Yb in Aleutian lavas, reinforcing the idea that both high Sr/Y and high Dy/Yb reflect fractionation between melt and abundant, residual garnet. For further discussion of the origin of the high Sr/Y, Dy/Yb and La/Sm component in western Aleutian lavas, and particularly whether it really requires a component produced by partial melting of subducted basalt in eclogite facies, please see Sections 4.5 and 4.7.

We can look for along-strike variation in the abundance of a component with abundant, residual garnet in our data compilation using Sr/Y in primitive lavas and Dy/Yb ratios in all lavas. These should be high in magmas including a substantial component produced by small amounts of partial melting of basalt in eclogite facies followed by reaction with mantle peridotite. These ratios show little systematic variation as a function of longitude, because moderate Sr/Y and Dy/Yb ratios are found in lavas throughout the arc. However, the highest ratios are found in lavas at and west of

Figure 14. Proposed trace element indicators of an aqueous fluid component in arc lavas, versus longitude along the Aleutian arc. Sources of data in caption for Figure 3.

Adak Island. This is more easily seen in plots of Sr/Y versus La/Sm and Dy/Yb, in which it is clear that highly enriched lavas are found only in the western Aleutians (Figure 16).

Figure 16 shows that the high Sr/Y, La/Sm, Dy/Yb component in primitive Aleutian lavas is enriched in SiO_2, Na_2O, K_2O, Ce/Pb, and Th/Nb compared to the lower Sr/Y lavas in the eastern and central Aleutians. In many of these plots, lavas from Piip volcano in the Komandorsky block [*Yogodzinski et al.,* 1994] form an important, third end-member. The Piip lavas share the major element characteristics of the enriched, high Mg# andesites, and their low Ba/La ratios, but they lack high Sr/Y, La/Yb, and Dy/Yb. As pointed out by *Yogodzinski et al.* [1994], many of the characteristics of Piip lavas can be explained as the result of a small degree of partial melting of spinel peridotite.

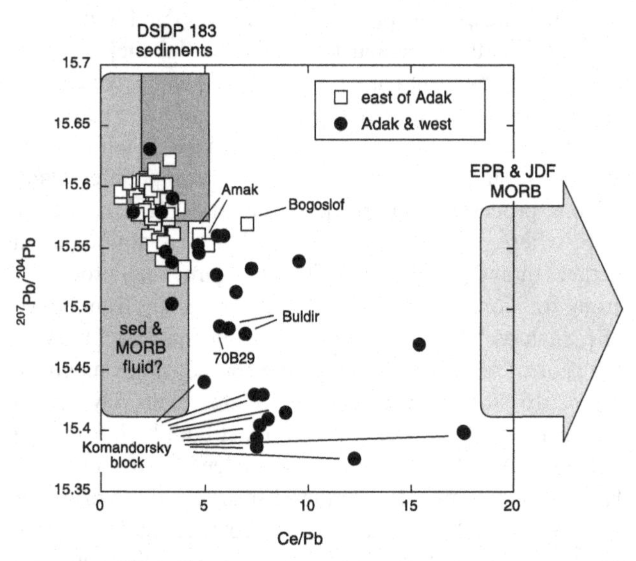

Figure 15. $^{207}Pb/^{204}Pb$ versus Ce/Pb in Aleutian lavas. All analyses by isotope dilution. Closed circles indicate data from volcanoes at and west of Adak. All closed circles that are not labeled with a location are from Adak Island. This plot is similar to Figure 4 of [*Miller et al.,* 1994], but includes data for volcanoes other than Okmok and Rechesnoi, plus updated fields for DSDP Site 183 sediments and MORB. The plot demonstrates that relatively high Ce/Pb is common in Aleutian lavas with unradiogenic Pb isotopes. Sources of Aleutian data in caption for Figure 3, plus new Pb concentrations and Pb isotope ratios for western Aleutian lavas in Table 1. Field for MORB glasses from the East Pacific Rise and Juan de Fuca Ridge based on data in the RIDGE petrological database at http://petdb.ldeo.columbia.edu/. Field for DSDP Site 183 sediments from data in [*Plank and Langmuir,* 1998; *Miller et al.,* 1994; *Peucker-Ehrenbrink et al.,* 1994; *von Drach et al.,* 1986]. Field for "aqueous fluid" is based on the reasoning of [*Brenan et al.,* 1995a; *Miller et al.,* 1994], but expanded to include fluid derived from subducted sediment as well as altered MORB.

It is noteworthy that Th/La is low in lavas with high Sr/Y, Dy/Yb and Ce/Pb, whereas Sr/Y, Dy/Yb and Ce/Pb are low in lavas with high Th/La (Figure 16). Also, recall that Ba/La is strongly correlated with Th/La (Figure 13). If high Ba/La and Th/La ratios are proxies for a component derived from partial melting of subducted sediment (Section 2.4), and high Sr/Y, Dy/Yb and Ce/Pb are caused by addition of a melt of subducted eclogite, these two components can clearly be distinguished in the Aleutians.

2.7 Proposed Indicators of Other Melt Components Derived From Subducted Basalt

Marsh, Myers, Brophy, Fournelle and their co-workers [e.g., *Brophy and Marsh,* 1986; *Myers et al.,* 1986a; *Myers et al.,* 1986b; *Myers et al.,* 1985] have proposed that more typical, low Mg#, high alumina basalt magmas in the Aleutians were also produced by partial melting of subducted basalt. Because these lavas do not have high Dy/Yb and Sr/Y, this group called upon diapiric ascent of partially molten, subducted basalt into the sub-arc mantle. Decompression within these diapirs would lead to garnet breakdown, and subsequent separation of the melt from its residue would produce high alumina basalt with the isotope signature of subducted basalt but without the trace element signature indicative of an eclogite melt. However, the hypothesis that high alumina basalts are partial melts of subducted basalts is controversial and difficult to verify, because high alumina basalts may also be produced by mantle melting [*Bartels et al.,* 1991] and by crustal crystal fractionation processes [e.g., *Sisson and Grove,* 1993; *Kelemen et al.,* 1990b; *Baker and Eggler,* 1983; *Kay et al.,* 1982].

Similarly, Drummond and Defant [e.g., *Defant and Drummond,* 1990; *Drummond and Defant,* 1990], Rapp and Watson [*Rapp and Watson,* 1995; *Rapp et al.,* 1991] and others have inferred that some low Mg# andesites and dacites might be direct partial melts of subducted eclogite that did not interact with mantle peridotite. Again, this hypothesis is difficult to verify for arcs worldwide because low Mg# andesites—even those with high Sr/Y and Dy/Yb—could be produced by crystal fractionation involving garnet, or by partial melting with residual garnet, within thick crust [e.g., *Müntener et al.,* 2001]. In the Aleutians such ambiguities do not arise, because the crust is relatively thin—at least at present—and because high Sr/Y and Dy/Yb ratios in Aleutian lavas are restricted to high Mg# andesites.

In this paper, we have adopted the assumption that low Mg# magmas are produced by differentiation of primitive magmas which were in Fe/Mg equilibrium with residual mantle olivine, unless proven otherwise.

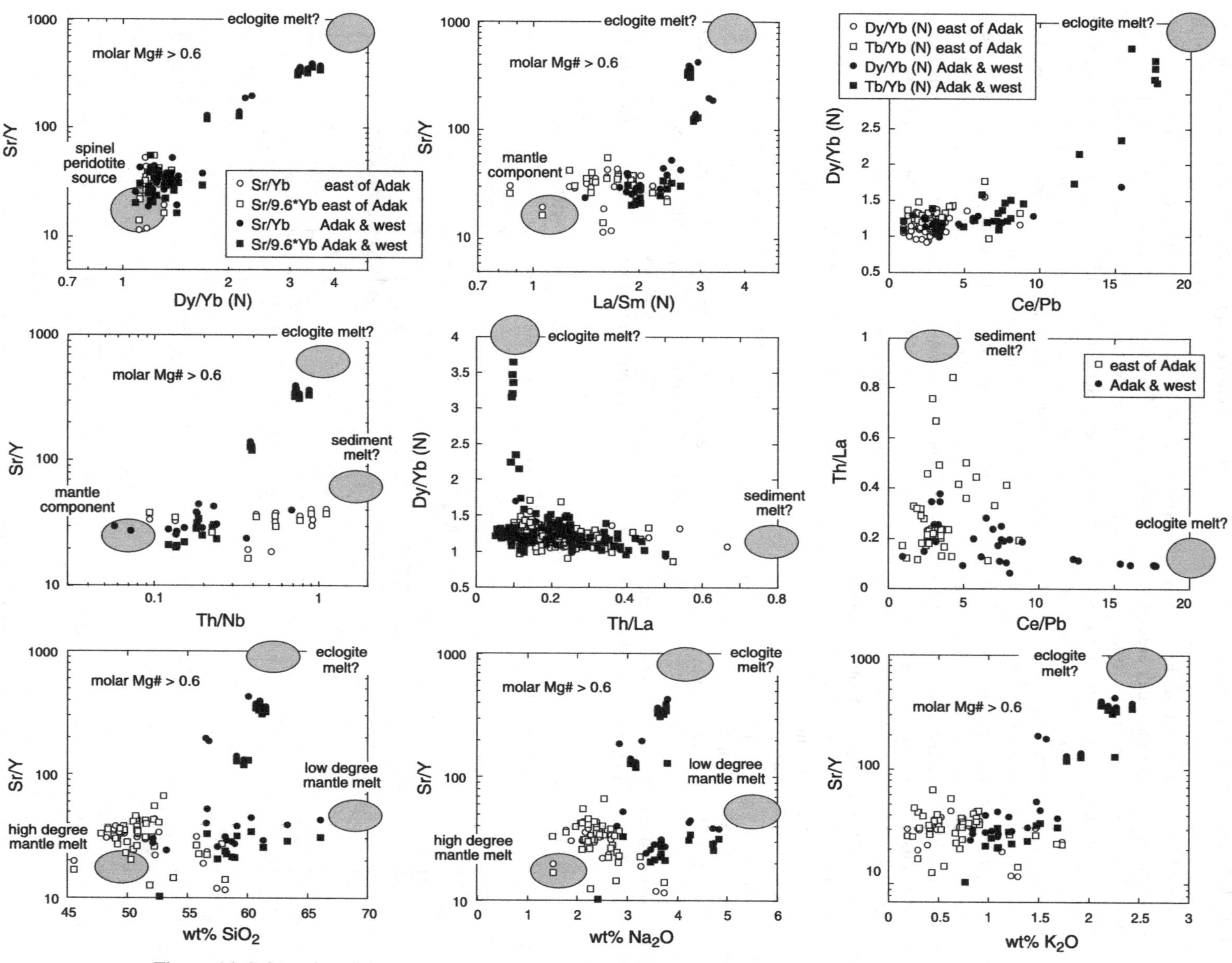

Figure 16. Sr/Y and Dy/Yb, proposed indicators of an eclogite melt component, versus other compositional data for Aleutian lavas. Sources of data in caption for Figure 3. Where Y data were not available, Sr/(9.6*Yb) was used instead, based on the chondritic Y/Yb ratio. Similarly, normalized Tb/Yb is used interchangeably with Dy/Yb, depending on what data are available. For REE, (N) indicates that concentrations are normalized to C1 chondrite values [*Anders and Grevesse,* 1989].

3. PLUTONIC ROCKS IN THE ALEUTIANS

Plutonic rocks in the Aleutians have been studied mainly on Unalaska Island in the east and in the Adak area [e.g., *Kay et al.*, 1990; *Kay et al.*, 1983; *Perfit et al.*, 1980b, and references therein]. There are also some small, hypabyssal intrusions in the western Aleutians and on the Komandorsky Islands [e.g., *Yogodzinski et al.*, 1995; *Yogodzinski et al.*, 1993; *Tsvetkov*, 1991]. Though some of the most mafic Aleutian gabbros may be "cumulate" (products of partial crystallization, from which remaining liquid was later removed), the high K contents of intermediate to felsic plutonic rocks in the Aleutians suggest that most are close to magma compositions.

Light REE enriched, K-rich, high Mg# andesite compositions similar to continental crust are more common among Aleutian plutonic rocks than they are among lavas (Figures 4 and 5). Although *Kay et al.* [1990] noted that plutonic rocks in the Adak area were somewhat more K-rich at a given Mg# than those on Unalaska Island, these variations are small compared to the along-strike variation in lava compositions.

Plutonic rocks analyzed to date—all Tertiary—may reflect an earlier phase of Aleutian magmatism. However, there is no identified change in convergence rate, age of subducting plate, or sediment source and abundance, which would explain a change in magma composition [*Atwater*, 1989; *Lonsdale*, 1988; *Engebretson et al.*, 1985]. Instead, the plutonic rocks may represent hydrous andesite magmas that were emplaced in the mid-crust after degassing of H_2O left them too viscous to erupt, as previously suggested by *Kay et al.* [1990] and *Kay and Kay* [1994]. In contrast, low viscosity, H_2O-poor, basalts may erupt readily. Thus, the proportion of high Mg# andesites among lavas may be less than their proportion within the entire crust. If true, this is very important, because it is commonly assumed that lava data can be used to estimate the proportions of the primary melts that form arc crust.

4. GENESIS OF WESTERN ALEUTIAN HIGH MG# ANDESITES

4.1 Crystal Fractionation From Primitive Basalt? No.

Kelemen [1995] reviewed experimental data on partial melting of natural basalt and peridotite compositions, with and without added H_2O, at a variety of oxygen fugacities, over a range of pressures from 1 bar to 3.5 GPa. Very few

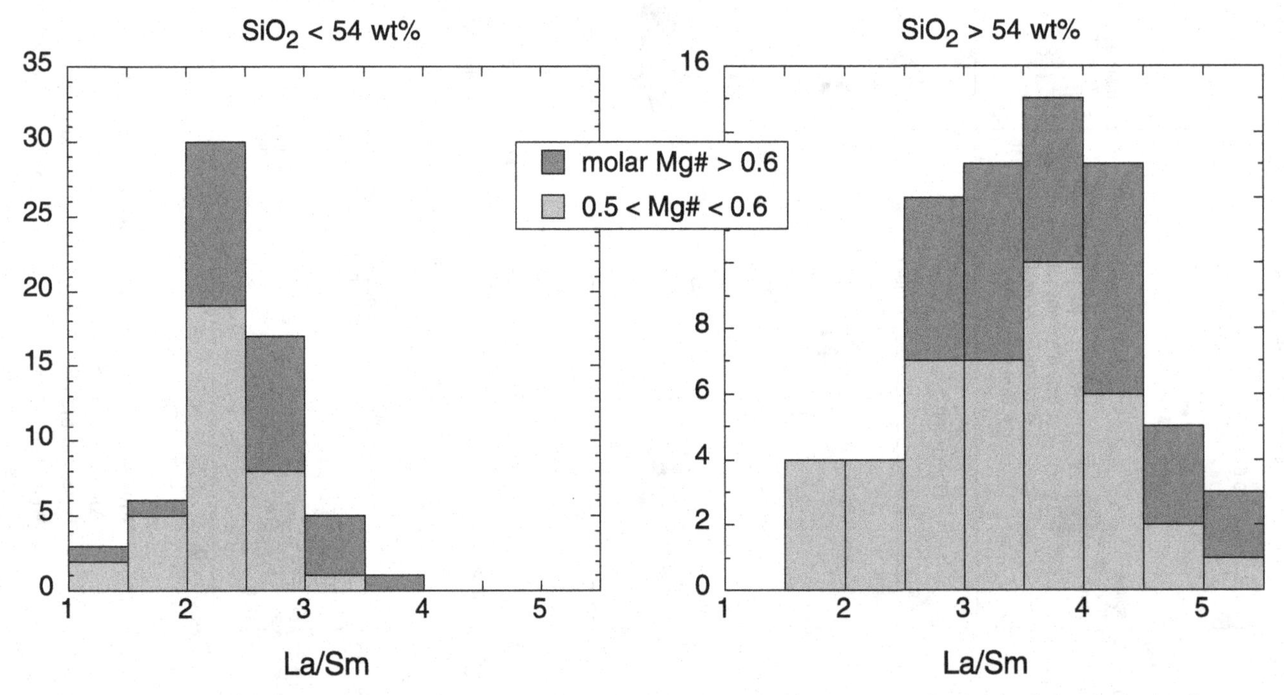

Figure 17. Histograms of La/Sm for high Mg# Aleutian lavas (molar 0.5 < Mg# < 0.6), and primitive lavas (molar Mg# > 0.6), showing a clear difference between basaltic lavas (SiO_2 < 54 wt%) and andesites (SiO_2 > 54 wt%). This difference is present in both primitive and high Mg# groups, so it suggests that primitive andesites are genetically related to more "normal" high Mg# andesites. Sources of data in caption for Figure 3.

of these experiments have produced high Mg# andesite liquid compositions, and even fewer have produced high Mg# andesites with K_2O concentrations comparable to Aleutian high Mg# andesites. *Kawamoto* [1996] melted basalts at high, unbuffered oxygen fugacity and produced a series of melt compositions that were very similar to some Aleutian high Mg# andesites. However, according to Kawamoto (pers. comm., 1997), the oxygen fugacities in those experiments were probably much higher than in most arc lavas (arc andesites record fO_2 within 2 log units of the Ni - NiO oxygen buffer [*Gill,* 1981]).

In considering the experimental data, it is also worthwhile to recall that partial melting/crystallization experiments are performed in a "closed system", in which melts continuously undergo Fe/Mg exchange with residual crystals. In fractional crystallization, which is often closely approximated in nature [e.g., *Carmichael,* 1964], residual crystals are removed so that melts evolve to lower Mg# at a given SiO_2 and K_2O content compared to closed-system crystallization. Thus, the experimental data approximate an upper bound on the SiO_2 and K_2O contents which can be achieved at a given Mg# via basalt crystallization. From this perspective, it is striking that only a few experimental melts fall into the range of Aleutian high Mg# andesite compositions, and the range of estimated continental crust compositions (two from Sisson and Grove, one in which the fO_2 buffer failed, and another at 965°C saturated in hornblende and magnetite, plus the high fO_2 experiments of Kawamoto).

Continental crust contains a significant proportion of andesitic rocks with Mg# higher than the average value, similar to primitive Aleutian andesites and dacites (summary and compilation in [*Kelemen,* 1995]). At and west of Adak, Aleutian andesites with Mg# between 0.45 and 0.6 are spatially associated with primitive andesites with Mg# > 0.6. Also, high Mg# andesites (0.5 < Mg# < 0.6) and primitive andesites (Mg# > 0.6) have similar trace element patterns, which are distinct from trace element patterns in primitive basalts (Figures 16 and 17). Therefore, it is likely that high Mg# andesites and primitive andesites are genetically related. No experimental liquids formed by partial melting of basalts approach the compositions of Aleutian primitive andesites. For these reasons, we believe that it is unlikely that western Aleutian high Mg# andesites are produced primarily via crystal fractionation from primitive basalts.

4.2 Mixing of Primitive Basalt and Evolved Dacite? Not for Enriched, Primitive Andesite

In the Aleutians, it has long been argued that both tholeiitic and calc-alkaline lava series have a common parental magma, a mantle-derived picrite with a nearly flat REE pattern [e.g., *Kay and Kay,* 1994; *Nye and Reid,* 1986; *Kay and Kay,* 1985a]. In this view, the tholeiitic series evolves mainly via crystal fractionation, perhaps at relatively low H_2O fugacity, whereas the calc-alkaline series has a more complex origin. Some light REE enriched dacite magmas are produced by crystal fractionation involving hornblende, which fractionates middle REE from light REE. Then, mixing of these evolved dacites with primitive basalt produces light REE enriched, high Mg# andesites such as those on Buldir Island in the western Aleutians.

It is well documented that Aleutian magmas, and calc-alkaline andesites in general, commonly contain oscillatory zoned clinopyroxene crystals which could be indicative of magma mixing [e.g., *Brophy,* 1987; *Kay and Kay,* 1985a; *Conrad and Kay,* 1984; *Conrad et al.,* 1983]. One possible explanation for both oscillatory zoning in clinopyroxene and high Mg# andesite lavas is mixing of primitive basalt with evolved dacite or rhyolite. Because the primitive basalt end-member has much higher Mg and Fe concentrations than the evolved end-member, mixing produces a concave downward hyperbola on plots of, e.g., Mg# versus SiO_2 or K_2O. If the evolved end-member is also light REE enriched and heavy REE depleted, perhaps as a result of crystal fractionation in the crust, then a similar trend is produced on plots of Mg# versus La/Yb (Figure 18).

Such mixing processes, or very similar processes in which primitive basalt assimilates evolved, granitic rocks, have long been proposed to explain the origin of high Mg# andesites in general, and more specifically in the Aleutians [e.g., *Brophy,* 1987; *Kay and Kay,* 1985a; *Conrad and Kay,* 1984; *Conrad et al.,* 1983, and references therein]. Mixing of low viscosity basalt and high viscosity, siliceous liquids may be physically improbable [*Campbell and Turner,* 1985]. On the other hand, there are very well documented cases of assimilation processes that produce calc-alkaline andesites [*Grove et al.,* 1988; *McBirney et al.,* 1987]. Furthermore, in considering the genesis of continental crust, tectonic juxtaposition of basaltic and evolved rocks could produce an average with the composition of high Mg# andesite without any chemical mixing.

However, it is apparent that mixing of primitive basalt and evolved dacite or rhyolite cannot explain the origin of enriched, high Mg# andesites in the Aleutians at and west of Adak. As noted in Section 2.6, among primitive Aleutian magmas, high Sr/Y, La/Sm and Dy/Yb ratios are found only in andesites and dacites, never in basalts (Figures 16 and 17). Furthermore, crystallization of plagioclase, in which Sr is compatible, lowers Sr/Y, so that all evolved Aleutian magmas, with Mg# < 0.45, have Sr/Y < 60. Thus, in the Aleutians, neither primitive basalts nor evolved magmas

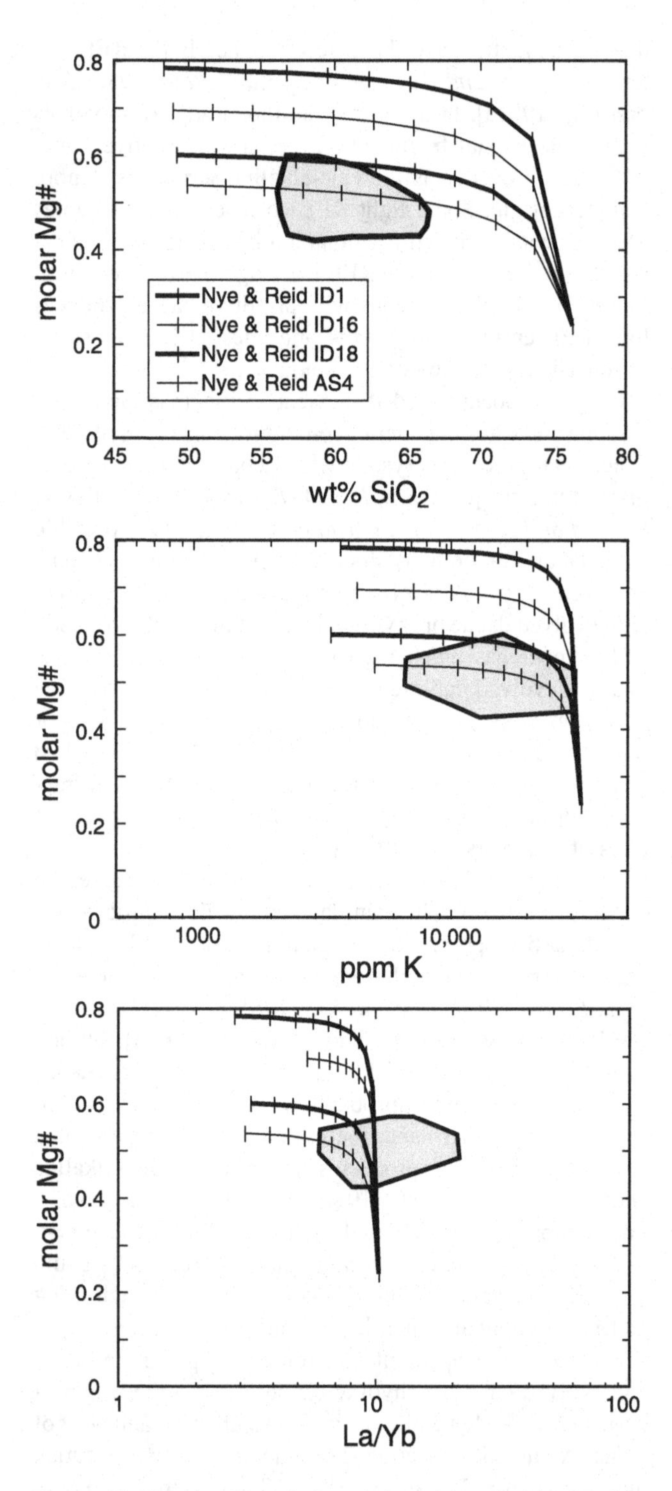

Figure 18. Calculated trends for mixing of (a) primitive basaltic lavas from Okmok volcano [*Nye and Reid,* 1986] with (b) rhyolite ia32 (also called 47ABy28) from Rechesnoi volcano [*Kay and Kay,* 1994; *Byers,* 1959]. The grey field in each panel shows estimated continental crust compositions (See Figure 4 caption for references).

have the trace element characteristics which distinguish enriched, primitive andesites.

Yogodzinski and Kelemen [1998] investigated the trace element contents of zoned clinopyroxene crystals in enriched, primitive Aleutian andesites, and compared them to trace element concentrations in clinopyroxenes from other Aleutian magma types. Primitive andesites contained clinopyroxenes with high Sr/Y, Dy/Yb and La/Yb, while primitive basalts contained clinopyroxenes with much lower ratios. Evolved lavas had relatively low Sr/Y, Dy/Yb and La/Yb, similar to the primitive basalts. In zoned clinopyroxenes in the primitive andesites, we found that the highest Sr/Y, Dy/Yb and La/Yb ratios were in the parts of crystals with the highest Mg#. Lower Mg# portions of the same crystals had lower Sr/Y, Dy/Yb and La/Yb, indicative of mixing of a primitive, highly enriched component with an evolved, less enriched component.

As Yogodzinski and Kelemen noted, trends of Aleutian lava compositions parallel those observed in the zoned clinopyroxene crystals. Primitive basalts and most evolved lavas—basaltic, andesitic, and even more silica rich—have relatively low Sr/Y, Dy/Yb and La/Yb, whereas enriched, primitive andesites have much higher Sr/Y, La/Sm and Dy/Yb (Figure 19). In fact, this bimodal distribution of trace elements may be reflected in a similar, bimodal distribution of major element contents of primitive lavas (Figures 3 and 5). Among western Aleutian lavas, a mixing trend can be seen from enriched, primitive andesites toward less enriched trace element ratios in andesites with moderate Mg#s. Thus, it is apparent that there are two primitive lava types in the western Aleutians. As previously emphasized [e.g., *Yogodzinski and Kelemen,* 1998, *Kay and Kay,* 1994; *Kay and Kay,* 1985b], magma mixing has played an important role in the genesis of many high Mg# andesites at and west of Adak. *However, the highest Mg# end-member in this mixing process was an enriched, primitive andesite, not a primitive basalt.*

It is not clear whether calc-alkaline lavas and plutonic rocks with high Mg# andesite compositions in the central Aleutians include a component derived from enriched, primitive andesites, or whether, instead, they formed via mixing of primitive basalt with enriched, evolved dacite or rhyolite. In the Adak area, hornblende-bearing, dacitic tephra include glasses with La/Yb generally between 6.5 and 7.5, and one from Kanaga has La/Yb=8.9 [*Romick et al.,* 1992]. However, there are very few dacites and rhyolites in the central and eastern Aleutians with sufficiently high La/Yb (greater than ~ 10) to explain the light REE enrichment in high Mg# Aleutian plutonic rocks (Figure 4). In our database, only 5 of ~ 900 samples from the central and eastern Aleutians have more than 60 wt% SiO_2 and La/Yb > 7. Thus, it seems plausible that magma mixing is not the general cause of light REE enrichment in Aleutian plutonic rocks. Instead, an enriched, primitive andesite component may have been an important mixing end-member throughout the arc.

Apparently, even in the Komandorsky block and on Adak, where primitive lavas include both enriched andesites and more "normal" basalts, there has been little mixing between these different, primitive magma compositions. Perhaps this is due to the different viscosities of andesitic and basaltic liquids. The mixing trend observed between enriched, primitive andesites and more evolved, but less enriched andesites may be facilitated by their similar viscosities.

4.3 An Important Role for Recycled, Continental Sediment? No.

Leaving aside the major element composition of Aleutian lavas for a moment, it might be proposed that the enrichments in Th, U, Pb, Rb, K, Sr, light REE, and so on, in western Aleutian lavas—different from other intraoceanic arcs and similar to continental crust—is simply the result of recycling of a continentally-derived sediment component. We believe that the isotopic and geologic evidence clearly indicate that this is not the case (see Sections 1.6 and 2.3). Although the concentration of some elements in primitive arc magmas worldwide may be related to the flux of these elements in subducted sediments [*Plank and Langmuir,* 1993], this is not the case in the Aleutians, at least not at and west of Adak Island.

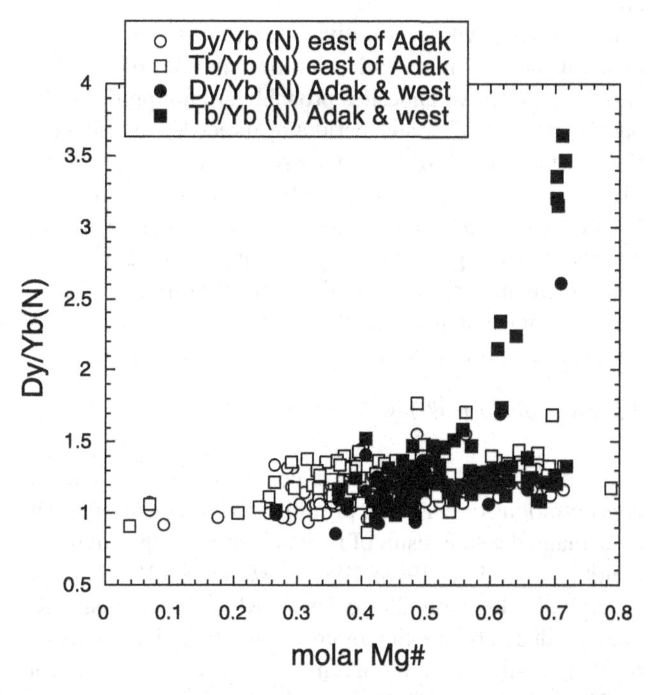

Figure 19. Dy/Yb(N) (and Tb/Yb(N)) versus molar Mg# for Aleutian lavas, illustrating that the most primitive lavas—with the highest Mg#—have the highest Dy/Yb (and Sr/Y and La/Sm, Figure 16). This is inconsistent with the mixing trajectories in Figure 18, involving primitive lavas with flat REE patterns mixing with evolved lavas with enriched REE patterns. Instead, mixing of primitive, enriched lavas with evolved lavas having flat REE patterns is consistent with the Aleutian data. Also, please see figures and discussion of this topic in [*Yogodzinski and Kelemen,* 1998]. Sources of data in caption for Figure 3.

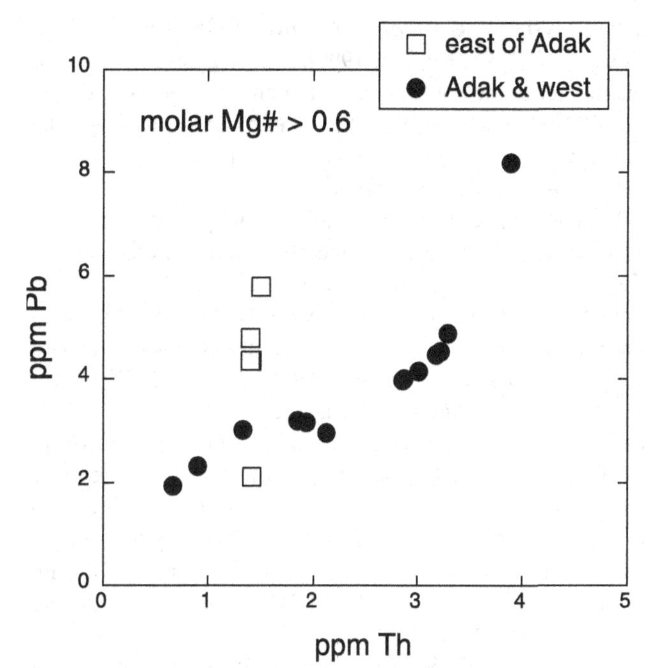

Figure 20. Concentration of Th versus Pb in primitive Aleutian lavas. Sources of data in caption for Figure 3.

4.4. Partial Melting of Spinel Peridotite Metasomatized by Aqueous Fluid? No.

As noted in Section 1.5, relatively slow convergence in the Aleutians west of Adak probably cools the mantle wedge by conduction into the subducting plate. Conversely, slow convergence produces a relatively hot subduction zone, in which most metamorphic dehydration reactions occur beneath the forearc, limiting the amount of H_2O that can flux melting beneath the arc. For these reasons, it could be that the degree of mantle melting is lower west of Adak compared to the central Aleutians. Could this explain why high Mg# andesites are abundant in the west?

High H_2O and alkali contents stabilize high SiO_2 in olivine-saturated melts [*Hirschmann et al.,* 1998; Ryerson, 1985; *Kushiro,* 1975], as demonstrated in experiments on natural systems [*Gaetani and Grove,* 1998; *Hirose,* 1997; *Baker et al.,* 1995; *Kushiro,* 1990]. Low degrees of melting of mantle peridotite could produce a melt enriched in H_2O and alkalis, and thus andesitic rather than basaltic. However, the maximum SiO_2 content is controversial [*Draper and Green,* 1999; *Baker et al.,* 1996; *Falloon et al.,* 1996].

In general, alkali-rich melts of nominally anhydrous lherzolite at ≥ 1 GPa are nepheline normative, whereas primitive andesites are quartz normative. Though H_2O tends to raise SiO_2 in mantle melts relative to NaAl and KAl, yielding quartz-normative andesite (calculated H_2O-free) at 1 to 1.5 GPa [*Gaetani and Grove,* 1998; *Hirose,* 1997; *Baker et al.,* 1995; *Kushiro,* 1990], at pressures > 1.5 or 2 GPa hydrous melts of lherzolite may be alkaline [e.g., *Kushiro,* 1969; *Kushiro,* 1968]). Partial melts of harzburgite are more SiO_2-rich [e.g., *Walter,* 1998; *Falloon et al.,* 1988; *Falloon and Green,* 1988; *Falloon and Green,* 1987]. Hydrous primitive andesites from Japan and California are in equilibrium with harzburgite at 1 to 1.5 GPa [*Baker et al.,* 1994; *Tatsumi and Ishizaka,* 1982].

In summary, primitive andesites might arise from melting of spinel peridotites. However, the trace element contents of *enriched* primitive andesites do not result from this process alone. Enrichments in Sr/Y, Dy/Yb and La/Yb must be explained via metasomatism prior to melting. If temperatures in subduction zones are too low for partial melting (see Section 4.8), transport of incompatible elements from subducted material must be in aqueous fluid. It may be that aqueous fluids carrying incompatible elements *cause* partial melting. If so, fluid composition can be estimated from H_2O in glasses combined with assumptions about the mantle source [e.g., *Grove et al.,* 2002; *Class et al.,* 2000; *Eiler et al.,* 2000; *Stolper and Newman,* 1992].

Fluid compositions inferred from this approach differ from those in experiments on aqueous fluids plus mantle peridotite or basaltic eclogite at 1 to 4 GPa [e.g., *Ayers,* 1998]. Concentrations of Th and the light REE in experimental fluids are 10 to 1000 times lower than required to explain arc enrichments [*Hawkesworth et al.,* 1993a; *Hawkesworth et al.,* 1993b]. Also, experimental data show large fractionations of soluble elements, such as Ba, Pb and Sr with fluid/rock distribution coefficients ~100, from insoluble elements such as Th and REE, with fluid/rock distribution coefficients from ~ 0.01 to 1 [*Johnson and Plank,* 1999; *Ayers,* 1998; *Stalder et al.,* 1998; *Ayers et al.,* 1997; *Kogiso et al.,* 1997; *Brenan et al.,* 1996; *Brenan et al.,* 1995a; *Brenan et al.,* 1995b; *Tatsumi et al.,* 1986]. In contrast, fluids required to metasomatize the mantle source for enriched andesites show little fractionation of Ba, Pb and Sr from light REE and Th [*Grove et al.,* 2002; *Stolper and Newman,* 1992]. Also, enriched, primitive andesites in the western Aleutians have positively correlated Pb and Th (Figure 20, and [*Kay and Kay,* 1994; *Yogodzinski et al.,* 1994; *Kay,* 1980;]). Thus, enrichment of the Aleutian mantle may be via small degree partial melts of subducted material [e.g., *Class et al.,* 2000; *Elliott et al.,* 1997; *Hawkesworth et al.,* 1997; *Plank and Langmuir,* 1993]), and not due to aqueous fluid transport.

NaCl-rich fluids in equilibrium with silicates can be "supercritical", completely miscible with H_2O-rich silicate melt at subduction zone pressures [e.g., *Keppler,* 1996]. If NaCl-rich fluids are abundant, then the distinction between aqueous fluids and melts may be artificial. However, compaction and dehydration are likely to remove much of the H_2O in subducted material beneath the fore arc. Since the solubility of NaCl in H_2O increases with temperature, aqueous fluids hotter than seawater are likely to remove most subducted NaCl at shallow depths. Thus, it seems unlikely that NaCl-rich, supercritical fluids are common beneath arcs.

In summary, while primitive andesites can be in major element equilibrium with shallow mantle peridotites, enriched, primitive andesites probably are not produced by partial melting of spinel peridotite, preceded by aqueous fluid metasomatism of the mantle source. More likely, primitive andesites form by reaction between silicate liquids—from deeper in the subduction system—and shallow mantle peridotite. During reaction at high melt/rock ratios (~1), major elements approach equilibrium with olivine + pyroxene + spinel, but incompatible trace elements are largely unchanged, and reflect processes at greater depth.

4.5 Residual Garnet? Yes.

The high Sr/Y, Dy/Yb and La/Yb ratios and low Y and Yb concentrations of enriched, primitive Aleutian andesites can be explained as the result of residual garnet in the source of melting [e.g., *Kay,* 1978; *Yogodzinski et al.,* 1995]. More specifically, high middle to heavy REE ratios must have been produced by melting or crystal fractionation involving residual garnet. In melting, if the source had chondritic Tb/Yb and Dy/Yb (Tb/Yb(N) ≈ Dy/Yb(N) ≈ 1), observed Tb/Yb(N) up to 4 and Dy/Yb(N) up to 2.5 in primitive Aleutian andesites require that the ratios of the bulk rock/liquid distribution coefficients, D(Tb/Yb) and D(Dy/Yb), must be less than 0.25 and 0.4, respectively. This is only likely where the residual assemblage includes garnet.

The main phases controlling REE abundance in melting of mantle peridotite and basaltic eclogite compositions are garnet and clinopyroxene. Over the range of melt compositions from basalt to dacite, clinopyroxene/liquid D(Tb), D(Dy) and D(Yb) are all about ~ 0.3 to 0.8, with D(Tb/Yb) and D(Dy/Yb) ≈ 1 (e.g., data compiled at http://www.earthref.org/). Measured garnet/liquid D(Tb), D(Dy) and D(Yb) are about 1 to 20, 1 to 30, and 4 to 40, respectively, with D(Tb/Yb) from 0.1 to 0.75 and D(Dy/Yb) from 0.2 to 0.7 (http://www.earthref.org/). Within this range, the highest values of D(Tb), D(Dy), D(Yb), D(Tb/Yb) and D(Dy/Yb) are for partitioning between garnet and dacite. For example, *Rapp and Shimizu* [manuscript in preparation] determined garnet/liquid D(Dy) and D(Yb) from 1 to 22 and 6 to 40, respectively, with D(Dy/Yb) ranging from 0.36 to 0.55, in the products of experimental, hydrous partial melting of basaltic eclogites [*Rapp et al.,* 1999]. Bulk rock/liquid D(Tb/Yb) and D(Dy/Yb) of ~ 0.25 and 0.4 can be attained with a garnet/clinopyroxene ratio in the source of ~ 1 or more from a variety of combinations of the garnet/melt partitioning data. Both basaltic eclogites and garnet-bearing mantle peridotites have a garnet/clinopyroxene ratio ~ 1 or more [e.g., *Rapp et al.,* 1991; *Cox et al.,* 1987; and F.R. Boyd, pers. comm. 1987].

We envision four different scenarios in which garnet plays an important role in the genesis of REE fractionation in enriched, primitive Aleutian andesites.

(1) Partial melting of unusually, hot, subducted basaltic rocks in eclogite facies gives rise to the enriched component in Aleutian primitive andesites (Figure 21B). The western Aleutian subduction zone may be unusually hot because of slow convergence rates (Section 1.5).

(2) In the unusually cold wedge in the western Aleutians, only deep-seated, garnet peridotites are hot enough to undergo fluid-fluxed partial melting (Figure 21A). The mantle wedge in the western Aleutians might be unusually cold and/or relatively H_2O-poor because slow convergence leads to increased cooling of the wedge via conduction into the subducting plate, and because a hot subducting plate may undergo most of its dehydration at shallow depths, beneath the forearc area. Low temperatures would facilitate garnet stability to depths of ~ 50 km in the wedge. For example, basaltic melt was saturated in a garnet peridotite assemblage at 1.6 GPa and temperatures less than ~ 1250°C in a system with ~ 5 wt% H_2O in the melt [*Gaetani and Grove,* 1998]. Higher H_2O contents could lead to even lower pressures for garnet stability on the mantle peridotite solidus.

(3) Subducted sediments and/or basalts melt beneath the entire arc. In the eastern and central Aleutians this component is difficult to detect because the subduction zone melt component is swamped by abundant basalts produced by mantle melting. In the west, where the mantle is colder and fluxed by a smaller amount of aqueous fluid, basalts are rare and the subduction zone melt component—with a garnet-rich, eclogite residue—is seen in its least diluted form.

(4) Cooling melts in the uppermost mantle might begin to undergo crystal fractionation while reacting with mantle peridotite (Figure 21C). Experimental data show that hydrous high Mg# andesite melts are garnet + pyroxene saturated at 1.5 GPa [*Carroll and Wyllie,* 1989]. If reaction between such melts and surrounding mantle peridotite produced a garnet pyroxenite while maintaining a high Fe/Mg ratio in derivative liquids, it could form a high Mg# light REE enriched, heavy REE depleted liquid product.

Crystal fractionation involving garnet in the lower crust cannot explain Aleutian primitive andesite compositions for several reasons. First, all of the most light- and middle-REE enriched andesites in the Aleutians are also the most primitive lavas, with Mg#'s greater than 0.6 and in some cases greater than 0.7. This seems to rule out an important role for crystal fractionation in the crust. Second, crustal thickness in the Aleutians is ~ 30 to 35 km [*Fliedner and Klemperer,* 1999; *Holbrook et al.,* 1999; *Grow,* 1973]. Recent investigations on crystallization of a hydrous, primitive andesite composition show that such melts are not garnet saturated at 1.2 GPa [*Müntener et al.,* 2001]. Garnet was stable in andesitic and basaltic bulk compositions only after large degrees of crystallization decreased the Mg# to less than 0.5. Thus, primitive melts probably are not garnet saturated at Aleutian Moho pressures.

The role of residual garnet seems clear, but it is difficult to determine from REE systematics whether residual garnet remains in subducted eclogite or in the mantle wedge. At the end of this section, we are left with four possible scenarios: (1) Partial melting of eclogite in a relatively hot subduction zone in the west. (2) Partial melting of garnet peridotite in an unusually cold mantle wedge in the west. (3) Partial melting of eclogite in the subduction zone throughout the Aleutians, obscured by extensive partial melting of relatively hot, H2O-rich mantle peridotite in the east. (4) Combined crystallization of garnet pyroxenite and reaction with peridotite in the uppermost mantle.

4.6 Abundant High Mg# Andesites Related to Slow Subduction of Young Crust

REE systematics discussed in the preceding section require a role for residual garnet in enriched, primitive andesite genesis, but do not allow us to determine whether the residual garnet resides in subducted crust or in the mantle wedge.

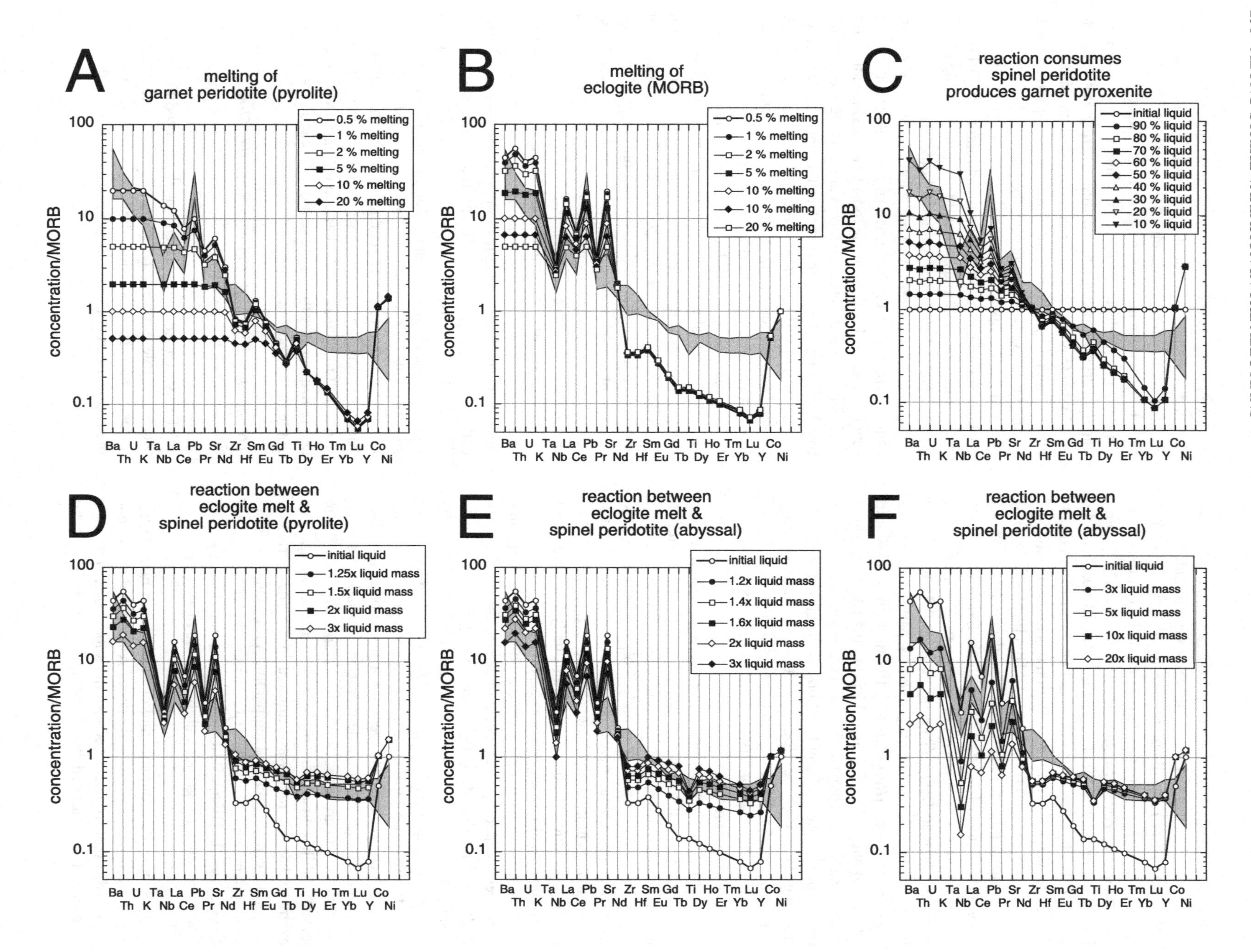
A
melting of
garnet peridotite (pyrolite)
0.5 % melting
1 % melting
2 % melting
5 % melting
10 % melting
20 % melting
B
melting of
eclogite (MORB)
0.5 % melting
1 % melting
2 % melting
5 % melting
10 % melting
10 % melting
20 % melting
C
reaction consumes
spinel peridotite
produces garnet pyroxenite
initial liquid
90 % liquid
80 % liquid
70 % liquid
60 % liquid
50 % liquid
40 % liquid
30 % liquid
20 % liquid
10 % liquid
D
reaction between
eclogite melt &
spinel peridotite (pyrolite)
initial liquid
1.25x liquid mass
1.5x liquid mass
2x liquid mass
3x liquid mass
E
reaction between
eclogite melt &
spinel peridotite (abyssal)
initial liquid
1.2x liquid mass
1.4x liquid mass
1.6x liquid mass
2x liquid mass
3x liquid mass
F
reaction between
eclogite melt &
spinel peridotite (abyssal)
initial liquid
3x liquid mass
5x liquid mass
10x liquid mass
20x liquid mass
concentration/MORB
100
10
1
0.1
Ba U Ta La Pb Sr Zr Sm Gd Ti Ho Tm Lu Co
Th K Nb Ce Pr Nd Hf Eu Tb Dy Er Yb Y Ni

It might be expected that the amount of partial melting of subducted basalt would be larger in the Aleutians west of Adak, where the trench-orthogonal convergence rate is slow and therefore subducted crust has a long time to heat conductively as it descends below the arc [*Yogodzinski et al.,* 1994; 1995]. Additionally, beneath the Komandorsky block the subducting plate may have a "torn", exposed edge, and thus be heated from one side as well as from the top [*Yogodzinski et al.,* 2001]. However, the relatively hot subduction zone in the western part of the arc is probably combined with low flux of H_2O into relatively cold mantle beneath the arc. Which of these factors is most important in producing abundant, primitive andesites?

Y and Yb concentrations in andesitic-to-dacitic lavas are low where subducted crust was younger than ~ 25 Ma, and higher where subducted crust is older [*Defant and Drummond,* 1990]. This correlation might indicate that subducted basalt in eclogite facies undergoes partial melting only when hot, young crust is subducted. Defant and Drummond [1990] incorrectly inferred that subducting crust beneath the Buldir area in the Aleutians (where < 600,000 yr enriched andesite has been dredged) is 15 Ma, whereas in fact crust subducted beneath Buldir is more than 40 Ma (Section 1.5). Despite this discrepancy, if subduction of crust younger than 40 Myr old led to formation of high Sr/Y, Dy/Yb, La/Yb magmas, while subduction of older

Figure 21. Results of trace element modeling. Grey field shows estimated bulk composition of continental crust, from Figure 6A with data sources in caption. Crystal/liquid distribution coefficients, model rock compositions, and data sources for these values are given in Table 3. Figure 21A illustrates compositions of aggregated liquids formed by fractional melting [*Shaw,* 1970; *Gast,* 1968] of garnet peridotite (pyrolite). Figure 21B illustrates compositions of aggregated liquids formed by fractional melting of eclogite (MORB). Figure 21C illustrates liquid compositions resulting from reaction of MORB with spinel peridotite (pyrolite) to produce garnet pyroxenite, using the AFC equations of [*DePaolo,* 1981], with the mass of solid reactants/mass of solid products set to 0.97. The ratio of initial liquid mass/integrated mass of peridotite reactant ranges from ~0.3 at a melt fraction of 0.9 to ~0.03 at a melt fraction of 0.1.

Partial melting of eclogite produces a trace element pattern similar to some enriched, primitive andesites in the Aleutians, but with much lower heavy REE than in continental crust. Partial melting of garnet peridotite, and reaction producing garnet pyroxenite, produce steep REE patterns in resulting liquids, but without the fractionations of Nb/La, Ce/Pb and Sr/Nd seen in Aleutian lavas and continental crust.

Figure 21D shows liquids resulting from reaction of a small degree melt of eclogite with spinel peridotite (pyrolite), with no change in solid phase proportions and the mass of solid reactants/mass of solid products, Ma/Mc, of 1.05. The melt/rock ratio (initial mass of liquid/integrated mass of peridotite reactant) in these models ranges from ~0.2 where the liquid mass is 1.25 times its initial mass to ~0.02 where the liquid is 3x initial mass. Figure 21E shows results of reaction between the same eclogite melt and depleted spinel peridotite (abyssal peridotite), with no change in solid phase proportions, Ma/Mc of 1.02, and melt/rock ratio from ~0.1 to 0.01. Figure 21F shows results of reaction between the same eclogite melt and depleted peridotite, with no change in solid phase proportions, Ma/Mc of 1.04, and melt/rock ratio from ~0.02 to 0.002, extending to very large increases in liquid mass. Crystal/liquid distribution coefficients and model rock and liquid compositions are given in Table 3.

Moderate amounts of reaction between partial melts of eclogite and spinel peridotite, under conditions of increasing liquid mass, produce liquids with trace element patterns similar to continental crust. Only Zr and Hf in continental crust are not well fit by these models. Large extents of reaction between eclogite partial melts and spinel peridotite produce liquids with nearly flat REE patterns that retain enrichments in Ba, Th, U, and K and fractionations of Nb/La, Ce/Pb and Sr/Nd seen in arc lavas worldwide.

These models are presented to show *possible* outcomes of various igneous processes. We make no claim that the these are unique explanations for the trace element contents of Aleutian lavas or continental crust. Also, the models could be incorrect if we have chosen inappropriate crystal/liquid distribution coefficients (Table 3). For the most part, small variations in distribution coefficients resulting from mineral and melt composition in basaltic systems are insignificant for our purposes in modeling. However, there is a substantial difference between distribution coefficients between minerals and basaltic melt, on the one hand, and minerals and andesitic to dacitic melt on the other hand. Whereas mineral/basalt partitioning has been extensively investigated, mineral/dacite partitioning is less well studied. In compiling mineral/dacite distribution coefficients for modeling partial melting of eclogite, we have relied mainly on studies of trace element partitioning between phenocrysts and matrix in volcanic rocks. For some elements (Th, U, Nb, Pb in clinopyroxene/dacite melt) we have had to interpolate values based on the pattern of distribution coefficients for mineral/basalt, and better known values for mineral/dacite for other elements (Ba, K, Ce, Nd in clinopyroxene/dacite melt). For highly incompatible elements, such as Ba, Th, U and K, our interpolated values are not crucial to the results of modeling. However, in considering our model results which show fractionation of Ce/Pb and Sr/Nd, the choice of distribution coefficients could be crucial. Thus, readers should be cautious in evaluating these results. With this said, we note that our models are consistent with the experimental data of [*Rapp et al.,* 1999], and the trace element models of Tatsumi ([*Tatsumi,* 2001], and pers. comm., 2000).

Also, Ti is not a trace element in processes involving residual rutile. As a result, strictly speaking, it should not be modeled here. However, we adjusted the distribution coefficients for Ti in clinopyroxene, garnet and rutile to provide results that are consistent with the experimental data of [*Rapp et al.,* 1999], which show no fractionation of Ti from REE such as Dy and Tb.

Table 3. Values used in trace element modeling.

Table 3A: peridotite melting, magma/mantle interaction

element	D ol/liq	refs	D opx/liq	refs	D cpx/liq	refs	D gar/liq	refs	D sp/liq	refs
olivine	1									
opx			1							
cpx					1					
garnet							1			
spinel									1	
rutile										
Ba	1.0E-09	estim. ≈ D(K)	0.00001	estim. ≈ D(K)	0.00068	5	0.00001	estim. ≈ D(K)	0	estim. ≈ D(K)
Th	1.0E-09	estim. ≈ D(K)	0.00001	estim. ≈ D(K)	0.01 6 0.001	9	0	estim. ≈ D(K)		
U	1.0E-09	estim. ≈ D(K)	0.00001	estim. ≈ D(K)	0.003	7	0.005	9	0	estim. ≈ D(K)
K	1.0E-09	1	0.00001	1	0.0072	1,5	0.00001	1	0 1	
Nb	0.001	1	0.0029	1	0.0077	1,5	0.013	1	0.01	1
La	0.000007	1	0.001	1	0.0536	1,5	0.001	1	0.0006	1,12
Ce	0.00001	1	0.003	1	0.0858	1,5	0.008	1	0.0006	1,12
Pb	0.00001	estim. ≈ D(Ce)	0.003	estim. ≈D(Ce)	0.072	1,5	0.0005	10	0	estim. ≈ D(K)
Pr	0.00004	interpolated	0.006	interpolated	0.13	interpolated	0.033	interpolated	0.0006	1,12
Sr	0.00001	1	0.003	1	0.1283	1,5	0.007	1	0	estim. ≈ D(K)
Nd	0.00007	1	0.009	1	0.1873	1,5	0.057	11	0.0006	1,12
Zr	0.004	1	0.04	1	0.1234	1,5	0.5	1	0.07	1
Hf	0.004	estim. ≈ D(Zr)	0.04	estim. ≈ D(Zr)	0.256	1,5	0.5	1	0.07	1
Sm	0.0007	1	0.02	1	0291	1,5	0.217	11	0.0006	1,12
Eu	0.00095	1	0.03	1	0.33	1,5	0.45	11	0.0006	1,12
Gd	0.0012	interpolated	0.04	interpolated	0.37	interpolated	0.9	interpolated	0.0006	interpolated
Tb	0.0026	interpolated	0.05	interpolated	0.41	interpolated	1.5	interpolated	0.006	interpolated
Ti	0.015	1,2	0.15	1,2	0.4	1,2,5	0.6	1	0.15	1
Dy	0.004	1	0.06	1	0.442	1,5	2	11	0.0015	1,12
Ho	0.007	interpolated	0.065	interpolated	0.43	interpolated	2.8	interpolated	0.0023	interpolated
Er	0.009	1	0.09	1	0.43	8	3.5	11	0.003	1,12
Yb	0.023	1	0.1	1	0.43	1,5	7	11	0.0045	1,12
Lu	0.03	interpolated	0.12	interpolated	0.433	1,5	9	11	0.0053	1,12
Y	0.023	estim. ≈ D(Yb)	0.1	estim. ≈ D(Yb)	0.467	1,5	7	estim. ≈ D(Yb)	0.0045	estim. ≈ D(Yb)
Co	2	bulk D estim.≈ 2	2	bulk D estim. ≈ 2	2	bulk D estim. ≈ 2	1	bulk D estim. ≈ 2	2	bulk D estim. ≈ 2
Ni	10	3	3.5	4	3	4	5	4	10	4

estim. = estimated

Table 3: Values used in trace element modeling, continued.

Table 3B: eclogite melting

element	D gar/liq	refs	D cpx/liq	refs	D rutile/liq	refs
olivine						
orthopyroxene						
clinopyroxene	1					
garnet	1					
spinel						
rutile	1					
Ba	0.02	13	0.02	13,20	0	guessed
Th	0.001	estim. ≈ peridotite	0.03	21,22	0	guessed
U	0.005	estim. ≈ peridotite	0.04	21,22	0	guessed
K	0.02	13	0.02	13	0	guessed
Nb	0.05	14	0.02	21	30	25, 30
La	0.08	15	0.04	23	0	guessed
Ce	0.2	13,15	0.08	23	0	guessed
Pb	0.005	estim. ≈ 10x perid	0.1	21	0	guessed
Pr	0.4	interpolated	0.14	interpolated	0	guessed
Sr	0.03	13,16,17	0.07	13,22,24	0	guessed
Nd	0.8	13,15	0.2	23	0	guessed
Zr	5	16	0.3	21,23	40	estim. > D(Nb)
Hf	5	15	0.3	20,23	40	estim. ≈ D(Zr)
Sm	5	13,15,16,18	0.4	23	0	guessed
Eu	7	13,15	0.45	13,21,22,23	0	guessed
Gd	10	13,15	0.6	23	0	guessed
Tb	14	13,15	0.65	interpolated	0	guessed
Ti	12	19	0.6	19	100	19, 30
Dy	16	13,18	0.7	13,23	0	guessed
Ho	18	15,18	0.7	interpolated	0	guessed
Er	20	13	0.7	23	0	guessed
Yb	25	13,15,18	0.7	13,21,22,23	0	guessed
Lu	30	13,15	0.7	13,21,22,23	0	guessed
Y	25	16	0.8	21	0	guessed
Co	2	15	1	22	0	guessed
Ni	1	bulk D estim. ≈ 1	1	bulk D estim. ≈ 1	0	guessed

crust did not, and this was *independent of subduction rate,* then we could infer that subduction zone temperature was crucial in enriched, high Mg# andesite genesis, and thus that eclogite melting is essential.

There are many localities with enriched, primitive andesites similar to those in the Aleutians: e.g., SW Japan, the Cascades, Baja California, Mexico, and the Astral Volcanic Zone in southern Chile [*Defant et al.,* 1991; *Defant et al.,* 1989; *Hughes and Taylor,* 1986; *Rogers et al.,* 1985; *Puig et al.,* 1984; *Tatsumi and Ishizaka,* 1982]. Unfortunately, in all of these areas the subducting oceanic crust is younger than 25 Ma and trench-orthogonal convergence rates are less than 40 mm/yr. Thus, it is difficult to separate the effect of subducting young crust from the effect of slow convergence. Young crust is subducting at high convergence rates along the Andean subduction zone north of the Austral Andes. With the exception of the Puna Plateau area [*Kay and Kay,* 1994], primitive andesites are not found north of the Austral volcanic zone[3]. Perhaps, subduction of young oceanic crust alone is not sufficient to ensure the presence of abundant, enriched primitive andesite lavas.

The Lesser Antilles and the South Sandwich arcs have among the slowest arc convergence rates [*DeMets et al.,* 1994; *Argus and Gordon,* 1991; *DeMets et al.,* 1990]. High Mg# andesites are found in neither arc. Oceanic crust older than 60 Ma is being subducted beneath the Lesser Antilles and the northern half of the South Sandwich arc. Crust subducting beneath the southern half of the South Sandwich arc

[3]The primitive andesites of the Puna Plateau have been interpreted as the result of delamination and partial melting of exceptionally thick lower crust [*Kay and Kay,* 1993].

Table 3: Values used in trace element modeling, continued.

Table 3C: bulk distribution coefficients | **Table 3D:** bulk compositions/MORB

element	eclogite source & melt mode (MORB)	garnet peridotite source mode (pyrolite)	garnet peridotite melt mode (pyrolite)	spinel peridotite source mode (pyrolite)	spinel peridotite source mode (abyssal peridotite)	MORB	"pyrolite" (approx MORB source)	abyssal peridotite
reference	26	27	27	27	28		Ringwood, 1966	29
olivine	54	10	46	75				
orthopyroxene	17	18	28	20				
clinopyroxene	0.495	9	30	18				
garnet	0.495	20	42					
spinel	8	5						
rutile	0.01							
Ba	0.020	6.5E-05	2.1E-04	1.3E-04	2.0E-06	1	0.1	0.001
Th	0.015	1.1E-03	3.4E-03	1.8E-03	2.0E-06	1	0.1	0.001
U	0.022	1.3E-03	3.0E-03	5.4E-04	2.0E-06	1	0.1	0.001
K	0.020	6.5E-04	2.2E-03	1.3E-03	2.0E-06	1	0.1	0.001
Nb	0.335	4.3E-03	8.4E-03	3.3E-03	1.8E-03	1	0.1	0.001
La	0.059	5.2E-03	0.017	0.010	2.4E-04	1	0.1	0.001
Ce	0.139	0.010	0.030	0.016	6.4E-04	1	0.1	0.015
Pb	0.052	7.1E-03	0.022	0.014	6.1E-04	1	0.1	0.01
Pr	0.267	0.019	0.054	0.025	1.3E-03	1	0.1	0.02
Sr	0.050	0.013	0.042	0.024	6.1E-04	1	0.1	0.02
Nd	0.495	0.030	0.082	0.036	1.9E-03	1	0.1	0.03
Zr	3.024	0.120	0.255	0.039	0.015	1	0.1	0.03
Hf	3.024	0.132	0.294	0.063	0.015	1	0.1	0.03
Sm	2.673	0.073	0.182	0.058	0.005	1	0.1	0.03
Eu	3.688	0.125	0.293	0.068	0.007	1	0.1	0.03
Gd	5.247	0.221	0.496	0.078	0.009	1	0.1	0.03
Tb	7.252	0.347	0.762	0.089	0.012	1	0.1	0.03
Ti	7.237	0.190	0.401	0.130	0.049	1	0.1	0.03
Dy	8.267	0.452	0.984	0.098	0.015	1	0.1	0.03
Ho	9.257	0.614	1.317	0.099	0.018	1	0.1	0.03
Er	10.247	0.759	1.616	0.107	0.025	1	0.1	0.03
Yb	12.722	1.468	3.089	0.116	0.037	1	0.1	0.03
Lu	15.197	1.876	3.935	0.126	0.047	1	0.1	0.03
Y	12.771	1.471	3.100	0.123	0.037	1	0.1	0.03
Co	1.485	1.800	1.580	1.960	2.000	1	2	2
Ni	0.990	7.265	4.630	6.720	8.700	1	10	10

Table 3: Values used in trace element modeling, continued. Notes

1. Kelemen et al., 1993, references cited therein, and our unpublished ion probe data
2. Kelemen et al., 1990b
3. Arndt, 1977; Hart & Davis, 1978; Kinzler et al., 1990
4. Olivine/liquid from refs. in 3, plus mineral/olivine from Bodinier et al., 1987; Kelemen et al., 1998
5. Hart & Dunn, 1993
6. Value based on Hauri et al., 1994 and LaTourrette & Burnett 1992; for our purposes, these are sufficiently close to other values suggested by Beattie, 1993; Lundstrom et al., 1994; Salters & Longhi, 1999; Wood et al., 1999
7. Value based on Beattie, 1993; for our purposes, these are sufficiently close to other values suggested by Hauri et al., 1994; LaTourrette & Burnett 1992; Lundstrom et al., 1994; Salters & Longhi, 1999; Turner et al., 2000
8. Adjusted from value in Hart & Dunn (1993) to give smooth REE pattern for garnet peridotite and eclogite melting.
9. Values based on LaTourrette & Burnett, 1992; Beattie, 1993; and Hauri et al., 1994; for our purposes these are sufficiently close to other values suggested by Salters & Longhi, 1999
10. Hauri et al., 1994; Beattie, 1993; Salters et al., 2001
11. Shimizu & Kushiro, 1975
12. Stosch, 1982
13. Philpotts & Schnetzler 1970a,b; Schnetzler & Philpotts, 1970
14. Interpolated based on garnet D pattern in ref. 1 and K, La values
15. Irving & Frey, 1978
16. Green et al., 1989
17. Jenner et al., 1993
18. Nicholls & Harris, 1980
19. Ti is a major element during partial melting with residual rutile, and its concentration is controlled by phase equilibrium rather than partitioning. However, for modeling, Ti value is interpolated based on garnet and cpx D patterns in refs. 1, 5, plus and Tb, Dy values in this Table, then adjusted to fit observation by Rapp et al., 1999, that Ti is not fractionated from REE during rutile saturated partial melting of eclogite.
20. Luhr & Carmichael, 1984
21. Larsen, 1979
22. Dostal et al., 1983
23. Fujimaki et al., 1984
24. Hart & Brooks, 1974
25. Green & Pearson, 1987
26. Rapp papers
27. Kelemen et al., 1992
28. Dick, 1989
29. Based on ~5% cpx (exsolved from high temperature opx) and cpx data from Johnson et al., 1990; Johnson & Dick, 1992
30. Foley et al., 2000

formed at the super-slow spreading American-Antarctic Ridge. Super slow spreading crust is anomalously cold, at a given age, compared to oceanic crust created at full spreading rates greater than ~ 0.01 m/yr [e.g., *Henstock et al.,* 1993; *Phipps Morgan and Chen,* 1993; *Reid and Jackson,* 1981; *Sleep,* 1975], and is sometimes composed mainly of serpentinized mantle peridotite [*Dick,* 1989; *Snow,* 1995]. However, the examples of the Lesser Antilles and the South Sandwich arc illustrate that slow convergence alone is insufficient to ensure the presence of abundant, enriched primitive andesites.

In summary, both slow convergence and young subducting crust may be necessary in order to produce abundant primitive andesite lavas.

4.7 Residual Rutile? Yes, in Partial Melting of Eclogite

Kay [1978] presented a simple model of a small degree of melting of MORB in eclogite facies, which accounted for the REE and Sr contents of the enriched, high Mg# andesites on Adak. Noting that the Mg# and Ni contents of these lavas were much higher than in a small degree melt of MORB, Kay suggested that eclogite melts reacted with the mantle during ascent. More recently, this model has been extended to include additional incompatible elements. In particular, rutile is residual during small to moderate degrees of melting of MORB-like compositions in eclogite facies [*Rapp and Watson,* 1995; *Rapp et al.,* 1991]. Elements such as Ta and Nb are compatible in rutile, and thus the depletion of Ta and Nb relative to La in enriched, high Mg# andesites—and other arc lavas—can result from eclogite melting [*Yogodzinski et al.,* 1995; *Kelemen et al.,* 1993; *Ryerson and Watson,* 1987].

Rutile might also be stable in some peridotite/melt systems [e.g., *Bodinier et al.,* 1996], perhaps under conditions where melts are hydrous, silica-rich, TiO_2-rich, and low temperature [*Schiano et al.,* 1995; *Schiano and Clocchiatti,* 1994; *Schiano et al.,* 1994]. To evaluate this, we calculated TiO2 required for rutile saturation ([*Ryerson and Watson,* 1987], using temperatures from Section 2.1) and compared it to observed TiO_2 in Aleutian primitive andesites. At 1 to 3 GPa, with 0 to 10 wt% H_2O, assuming an uncertainty of 0.5 wt%, more than 75% of 50 primitive Aleutian andesites have too little TiO_2 for rutile saturation.

Thus, residual rutile *in the mantle* does not cause Ta and Nb depletions in primitive andesites. Instead, we infer that a component in *all* primitive Aleutian andesites—regardless of the degree of light REE enrichment—is a rutile-saturated eclogite melt. Ta and Nb in primitive Aleutian lavas are weakly correlated with La concentration. Therefore, Ta and Nb depletion probably form in the same manner as light REE enrichment, due to residual rutile and garnet in an eclogite source.

4.8 Thermal Models and Partial Melting in the Aleutian Subduction Zone

Up to this point, we have reviewed a variety of possible explanations for genesis of enriched, primitive Aleutian andesites. At the end of Section 4.5, we were left with four possibilities in explaining the genesis of primitive Aleutian andesites, two of which involved partial melting of subducted eclogite. In Section 4.7, we concluded that eclogite melting is required to explain depletions of Nb and Ta relative to REE in these lavas. Therefore, we can narrow the list to two possibilities: (1) Eclogite melting in a hot subduction zone in the west, or (2) eclogite melting throughout the arc, obscured in the east by extensive melting of hot, H_2O-rich mantle.

If a sediment melt component is clearly required in the central Aleutians (Section 2.4), and a basaltic eclogite melt component is required in the western Aleutians (Sections 2.6 and 4.7), then we should choose option (2); partial melting of subducted material is ubiquitous throughout the arc. Indeed, as already mentioned in Section 2.4, Plank, Langmuir, Miller, Class and colleagues have proposed that a sediment melt component is important in the central Aleutians, based on correlation between subducted Th in sediments and Th concentration in arc lavas, correlation between Th concentration and Nd isotope ratios, and data on Th partitioning between sediments, aqueous fluids, and melts [*Class et al.,* 2000; *Plank and Langmuir,* 1998; *Miller et al.,* 1994; *Plank and Langmuir,* 1993; *Miller et al.,* 1992]. However, because this relies mainly on ideas about Th partitioning, this hypothesis may be open to question due to lingering uncertainty about aqueous fluid/rock partitioning.

To further evaluate these two options, we consider the thermal state of subduction zones and the overlying mantle wedge, both for the Aleutians and on a worldwide basis. We turn first to Aleutian data. With one exception, the 18 Aleutian high Mg# andesites with the strongest light REE enrichment (La/Yb > 9) are Miocene [*Kay et al.,* 1998; *Yogodzinski et al.,* 1995]. These samples, found on Adak Island and in the Komandorsky block, have Sr/Y > 50 and other distinctive characteristics. They have been interpreted as partial melts of subducted basalt in eclogite facies, which underwent reaction with the mantle during transport to the surface [*Yogodzinski and Kelemen,* 1998; *Yogodzinski et al.,* 1995; *Defant and Drummond,* 1990; *Kay,* 1978]. One might infer that partial melting of subducted basalt occurred during the Miocene but has since ceased.

However, the exception is a single dredged lava from a submarine cone near Buldir Island in the western Aleutians (sample 70B29, 184.7°W) has 63 wt% SiO_2, an Mg# of 0.56, 9960 ppm K, La/Yb of 9.8, and Sr/Y of 73 (or Sr/(9.6*Yb) of 104). This sample was too young to date by conventional K/Ar methods [*Scholl et al.,* 1976] and so is estimated to be less than 600,000 years old. The down-dip subduction velocity beneath this site is estimated to be ~ 0.04 m/yr and the age of the plate beneath this site is estimated to be ~ 45 to 50 Ma.

Additional evidence for a young, enriched high Mg# andesite component in the western Aleutians comes from plutonic xenoliths in Holocene lavas from Adak. These xenoliths include clinopyroxene crystals with fine-scale oscillatory zoning, and interstitial glass that was quenched against the host lava [*Conrad and Kay,* 1984; *Conrad et al.,* 1983]. The preservation of fine scale zoning (zones with distinct Fe and Mg contents as small as 25 microns) and quenched interstitial glass both indicate that the xenoliths are not much older than their host lavas. Even assuming that the interstitial glasses were hydrous and therefore relatively low temperature, one can use a lower limit diffusivity of ~ 10^{-22} m^2/s for Fe/Mg interdiffusion in clinopyroxene (e.g., summary in [*Van Orman et al.,* 2001]) to calculate that the 25 micron zones are less than ~ 200,000 years old.

Our recent ion probe analyses of clinopyroxene in these xenoliths [*Yogodzinski and Kelemen,* 2000] show that some zones have Sr/Y as high and higher than the clinopyroxene phenocrysts in previously analyzed, enriched primitive andesites [*Yogodzinski and Kelemen,* 1998], suggesting that the xenoliths crystallized, in part, from Holocene, enriched primitive andesites. The down-dip subduction velocity beneath Adak is ~ 0.06 m/yr and the age of the subducting plate beneath this site is ~ 45 to 50 Ma. Tectonically, Adak overlies a normal Benioff zone [e.g., *Boyd et al.,* 1995; *Boyd and Creager,* 1991], not a "torn slab".

Thermal models for arcs are reviewed in two separate papers in this volume [*Kelemen et al.,* this volume; *Peacock,* this volume], to which the reader is referred for references, models, and discussion. Here we give a short summary. Published thermal models up to 2002 suggest that partial melting of subducted oceanic crust more than 20 Myr old, and sediments overlying such crust, is unlikely. This is the chief objection to the geochemically-based hypothesis that there is a sediment melt component in the central Aleutians. However, this modeling result is inconsistent with evidence for an eclogite melt component in the enriched, high Mg# andesites at Adak Island. Also, these thermal models are inconsistent with petrological PT estimates for melt-mantle equilibration and metamorphic conditions in arc shallow mantle and lower crust, and do not account well for seismic data which indicate a region of anomalously low velocities (probably, a few percent melt) at 35 to 50 km depth in the shallow mantle beneath the NE Japan and Tonga arcs. Very recent work shows that incorporation of temperature-dependent viscosity and non-Newtonian mantle rheology yields thermal models consistent with fluid saturated melting of subducted sediment and/or basalt beneath the central Aleutians [*Kelemen et al.,* this volume; *van Keken et al.,* 2002; *Conder et al.,* 2002]. Some of these models are also consistent with petrological PT constraints from arc shallow mantle and lower crust [*Kelemen et al.,* this volume].

These new models do not "prove" that subducted sediment and/or basalt beneath the central Aleutians must undergo partial melting. However, they do show that well-constrained thermal models using widely accepted formulations for temperature- and stress-dependent mantle rheology, can be consistent with petrological PT constraints. Thus, the most recent results of thermal modeling show that melting of subducted material beneath the entire Aleutian arc may be likely, rather than impossible.

4.9 Melting of Subducted Material Throughout the arc, or Only in the West? Throughout.

The results of very recent thermal modeling show that subducted sediment and basalt could melt beneath both the central and western Aleutians. We find the incompatible trace element evidence for a partial melt component derived from subducted material in the central Aleutians persuasive [*Class et al.,* 2000]. Similarly, in Sections 4.5 and 4.7, we have argued that there is strong trace element evidence for partial melting of subducted eclogite in the western Aleutians, even beneath Adak where subducting crust is 45 to 50 Ma and the subduction zone geometry is "normal". Therefore, we hypothesize that partial melting of subducted material is ubiquitous throughout the arc.

If partial melting of subducted material is ubiquitous beneath the Aleutian arc, then why are the highly enriched trace element signatures indicative of an eclogite melt component most clearly observed in lavas at and west of Adak? One possible explanation is that the colder mantle wedge in the western Aleutians, convecting more slowly and fluxed by a smaller proportion of aqueous fluid, undergoes relatively little partial melting. As a result, in the western Aleutians partial melts of eclogite rise through and react with the surrounding mantle (Section 4.10), but generally do not mix with basaltic magmas derived from melting of peridotite. Further east, the hotter mantle wedge undergoes extensive melting, due to rapid convective upwelling and abundant hydrous fluids derived from the colder subducting plate. In this region, the mantle-derived basalts are far more voluminous than the component derived from partial melting of subducted material. As a result, the subduction zone melt component is difficult to detect unambiguously, except via enrichments of highly incompatible and insoluble elements such as Th.

This conclusion is somewhat at odds with the conclusions of *Miller et al.* [1992] who argued that the geochemical component derived from subducted basalt at Umnak Island in the central Aleutians was transported entirely in aqueous fluids. Miller et al. used the relationship of Ce/Pb and $^{207}Pb/^{204}Pb$ in lavas from Umnak Island in the central Aleutians to argue that Pb derived from subducted basalts is transported into the mantle wedge in a low Ce/Pb fluid rather than a high Ce/Pb melt phase. This, combined with evidence for sediment melting described in the previous paragraph (and in Section 2.4) gave rise to the aphorism, "sediments melt, basalts dehydrate". However, as we have shown in Section 2.5 and Figure 15, MORB-like Pb isotopes are observed in both high and low Ce/Pb Aleutian lavas. Because Ce/Pb is correlated with Dy/Yb and Sr/Y, thought to be indicative of an eclogite melt component in the Aleutians, we infer that the MORB-like Pb isotope signature in high Ce/Pb Aleutian lavas is derived from a partial melt of subducted basalt, not from the mantle wedge. While high Ce/Pb lavas are mainly found at and west of Adak, they are also present at Bogoslof and Amak Islands, behind the main arc trend in the central Aleutians.

Where sediments are present, their high Ba, Pb, Sr and REE contents and extreme isotope ratios (compared to MORB and the mantle) will dominate most trace element and isotopic characteristics of a partial melt derived from both sediment and basalt. Under these circumstances, it is difficult or impossible to detect a basalt melt component. Only where sediments are rare or absent is it easy to clearly detect a component derived from partial melting of basaltic eclogite.

At a given pressure, the fluid-saturated solidus should be at very similar temperatures in both basaltic and sedimentary bulk compositions, since both pelitic sediments and basalts in eclogite facies are composed of mixtures of garnet, omphacite, kyanite, coesite and phengite, plus various minor phases (Max Schmidt, pers. comm. 2000). This is borne out by experimental data [*Johnson and Plank,* 1999; *Schmidt and Poli,* 1998; *Nichols et al.,* 1994; *Carroll and Wyllie,* 1989; *Stern and Wyllie,* 1973; *Lambert and Wyllie,* 1972], in which some sediments apparently melt at higher temperatures than basalt, while others apparently melt at lower temperatures. Thus, it is difficult to envision a realistic scenario in which subducted sediments consistently undergo partial melting while subducted basalts never do. Instead, it seems likely that, for any given subduction zone geotherm, both undergo partial melting, or both do not.

4.10 Reaction Between Partial Melts of Subducted Eclogite and Overlying Mantle Peridotite

4.10.1 Phase equilibria. Small degree partial melts of eclogite are not in equilibrium with mantle olivine. Since the primitive andesite magmas which carry the trace element signature of eclogite melting show signs of having equilibrated with mantle peridotite—at least in terms of Fe/Mg and Ni exchange—we infer that they did not ascend in cracks all the way from the subduction zone to the crust. However, it is possible to envision scenarios in which melt is initially transported in cracks that terminate within the mantle, and then react with surrounding peridotite. Alternatively, eclogite melts may begin to react with the mantle wedge immediately above the subduction zone.

At constant temperature and pressure, siliceous melts of eclogite react with olivine to produce pyroxene (± garnet) and modified, lower SiO_2 melt. In simple chemical systems, and probably in some natural systems, such reactions consume more liquid than they produce, ultimately leading to complete solidification, or "thermal death" [*Rapp et al.,* 1999; *Yaxley and Green,* 1997]. This is particularly likely where reaction takes place below the solidus temperature for the peridotite, but also could occur at conditions above the peridotite solidus in systems that have a "thermal divide" between eclogite and peridotite melts [*Yaxley and Green,* 1997]. It is possible that some eclogite melts do undergo thermal death, and then hybridized, pyroxene-rich mantle peridotite circulates to some other part of the mantle wedge, later melting to give rise to an enriched magma [e.g., *Yogodzinski et al.,* 1994; *Ringwood,* 1974].

However, the observation of ^{230}Th excess in enriched, primitive andesites from Mt. Shasta in the Cascades [*Newman et al.,* 1986] and in the Austral Volcanic Zone in southernmost Chile [*Sigmarsson et al.,* 1998] provides an important constraint on the transport time of eclogite melt from source to surface. Radiogenic ^{230}Th excess is produced by small degrees of partial melting under conditions in which daughter ^{230}Th is less compatible in residual solids than parent ^{238}U. Present knowledge indicates that U is more compatible than Th in garnet [*Salters and Longhi,* 1999; *Hauri et al.,* 1994; *Beattie,* 1993; *LaTourrette and Burnett,* 1992] and—perhaps—high pressure, Na-rich clinopyroxene [*Wood et al.,* 1999]. Thus, the ^{230}Th excess in enriched, primitive andesites is consistent with derivation of the Th enriched component in such magmas by partial melting of subducted eclogite. The half-life of ^{230}Th is ~ 75,000 yrs. After a few half lives, excess ^{230}Th will decay, and $^{230}Th/^{238}U$ ratios will return to steady state, "secular equilibrium". Therefore, one can infer that the eclogite melt component in primitive andesites from Mt. Shasta and the Austral Volcanic Zone was transported from the subduction zone to the surface in less than ~ 300,000 years. This is plausible for magma transport via porous flow, hydrofracture, or, perhaps, in partially molten diapirs, but implausible for transport via solidification, solid state convection to a hotter part of the mantle wedge, and remelting.

For this reason, we prefer a model in which the eclogite melt component interacts with mantle peridotite in the wedge, and is modified but never completely solidified. As emphasized by Kelemen and co-workers [*Kelemen,* 1995; *Kelemen,* 1993; *Kelemen,* 1990c; *Kelemen et al.,* 1986], a key to considering reaction between eclogite melt and overlying peridotite is that above subduction zones, melts must heat up as they rise and decompress [*Kelemen and Hirth,* 1998]. *Grove et al.* [2002] recently made a similar point for partial melts of mantle peridotite deep within the mantle wedge. The combined effect of increasing temperature and decreasing pressure will move melt in a closed system away from its liquidus. In a reacting system in which decompressing melt is heated, these effects might counterbalance constant temperature and pressure reactions that lead to "thermal death". Thus, at temperatures below the peridotite solidus, melt mass might be nearly constant. And, if H_2O-rich melt is present at temperatures greater than the fluid-saturated peridotite solidus, melt mass will increase due to the combined effects of heating and decompression.

We also must consider the effect of alkalis and H_2O on melt/peridotite equilibria, via freezing point depression and shifting of melt composition. If reactions between subduction zone melts and peridotite consume melt, alkali and H_2O contents will rise in the remaining liquid. The aqueous fluid-saturated solidus for mantle lherzolite at 2.4 GPa is at less than 920°C [*Grove,* 2001], at least 500°C lower than the nominally anhydrous solidus [e.g., *Hirschmann,* 2000]. This is very close to experimentally constrained fluid-saturated solidii for sediment and basalt at the same pressure, ~ 650 to 750°C [*Schmidt and Poli,* 1998; *Nichols et al.,* 1994; *Lambert and Wyllie,* 1972][4]. Furthermore, because high alkali and H_2O contents stabilize olivine relative to pyroxenes [*Hirschmann et al.,* 1998; *Ryerson,* 1985; *Kushiro,* 1975], stable peridotite melt compositions at the fluid-saturated solidus will be more "felsic" than basalt; i.e., they will have relatively low normative olivine and high normative quartz or nepheline. In an open system, addition of alkalis from a subduction zone melt will heighten this effect. Thus, fluid-saturated partial melts of subducted material need not change much in temperature or composition to be stable in fluid-saturated mantle peridotite. This process is poorly constrained, but it seems likely that subduction zone melts interact with mantle peridotite in the wedge, and are modified but never completely solidified, to become an important component in Aleutian arc lavas.

4.10.2 Trace element modeling. As noted by *Gill* [1981], most arc magmas, including light REE enriched, calc-alkaline andesites, do not have REE slopes as steep as those in partial melts of eclogite, or even garnet peridotite. This can be seen by comparing trace element results for partial melting of garnet lherzolite (Figure 21A) and eclogite (Figure 21B) with trace element data for Aleutian lavas in Figure 6. Gill inferred from this that partial melting of lherzolite or eclogite with residual garnet did not play an important role in arc petrogenesis. However, *Kelemen et al.* [1993] reasoned that reaction between ascending eclogite melts and spinel peridotite in the mantle wedge might modify REE patterns, particularly if the reaction involved increasing magma mass, as discussed in Section 4.10.1. Middle to heavy REE concentrations can be modified by such reactions to values for melt in equilibrium with spinel peridotite (Figure 21D, E and F).

As a result of mass balance, light REE concentrations are modified much more slowly, and remain high. A rule of thumb is that trace element concentrations are modified by this type of reaction when the integrated melt/rock ratio (the mass of liquid to solid reactants) approaches the bulk crystal/liquid distribution coefficient for a particular element. Thus, in models in which the integrated melt/rock ratio is less than 1 but greater than 0.01, the light REE enriched, heavy REE depleted pattern characteristic of eclogite melts is transformed to produce a light REE enriched melt with a flat middle to heavy REE pattern, similar to common, light REE enriched arc magmas, and to the estimated composition of continental crust. Such a pattern is typical for lavas throughout the Aleutians, and we infer that it is possible—though by no means certain—that the light REE enrichment in Aleutian magmas may be due to the presence of a cryptic eclogite melt component, modified by extensive reaction with spinel peridotite in the mantle wedge.

A more subtle result of this type of modeling is that intermediate concentrations of elements modified by reaction are rare. Thus, for the models in Figure 21D for example, liquids with integrated melt/rock ratios greater than ~ 0.5 preserve the heavy REE depleted characteristics of an eclogite melt, though Ni and Cr (and Mg, Fe, Si) are modified to values in equilibrium for mantle peridotite. In the same models, liquids with integrated melt/rock ratios less than ~ 0.2 have heavy REE close to equilibrium with mantle peridotite. There is a very narrow range of integrated melt/rock ratios (not shown in the Figures) that yield intermediate heavy REE slopes. This is why there are no intermediate heavy REE slopes in Figures 21D, E and F. This may also explain why, in the western Aleutians, we see lavas whose heavy REE preserve a clear eclogite melting signature, and lavas with nearly flat heavy REE slopes, but very few with intermediate heavy REE slopes.

Subduction zone melts may react with mantle peridotite in ways that modify their trace element characteristics substantially. It is even possible that arc tholeiites with unexceptional trace element patterns started out as partial melts of subducted basalt and sediment in eclogite facies. In this view, reaction between subduction zone melts and overlying peridotite may lead to large increases in magma mass, modifying and diluting the original melt composition (Figure 21F). Thus, an arc tholeiite might derive more than 90% of its K_2O and light REE from a partial melt of subducted basalt, but 90% of its heavy REE and MgO from overlying, mantle peridotite.

We do not mean to imply that no aqueous fluid is derived from subducting material and incorporated into arc magmas. In the hypothetical arc tholeiite discussed in the previous paragraph, perhaps 90% of the B is derived from a fluid. In this paper, we have not concentrated on characterizing an aqueous fluid component, except to note that some proposed fluid component indicators (e.g., high Ba/La) do not appear to be separable from proposed sediment melt component indicators (e.g., high Th/La) in the Aleutians (Section 2.5). This does not mean that there are

[4]However, see [*Johnson and Plank,* 1999] for a hotter sediment solidus.

no fluids in the Aleutian arc source, but it does make it much more difficult to discern their nature and mode of transport.

For simplicity, we have used the same type of model in Figure 21F as we used for 21D and E. However, liquid mass increases by a factor of up to 20 in 21F. For such conditions, it might be more appropriate to use a fluxed melting model [e.g., *Grove et al.,* 2002; *Eiler et al.,* 2000; *Stolper and Newman,* 1992], or to approximate this effect by mixing a small degree eclogite melt with a larger degree melt of spinel peridotite, rather than using the AFC model we applied. However, it is easy to show that the trace element results of these alternative modeling approaches would be very similar to those in 21F.

5. IMPLICATIONS FOR CONTINENTAL GENESIS

5.1 Reaction Between Eclogite Melt and Mantle Peridotite

Figures 21C and D illustrate results of trace element models of reaction between a small degree partial melt of subducted eclogite and spinel peridotite under conditions of increasing melt mass. As previously proposed by *Kelemen et al.* [1993], the trace element contents of calculated liquids are strikingly similar to the estimated composition of the continental crust. Similarly, major element contents of liquids produced by experimental reaction of silicic melt with mantle peridotite closely approach those of high Mg# andesites and continental crust [*Carroll and Wyllie,* 1989], and of primitive andesites [*Rapp et al.,* 1999]. Thus, we believe that reaction of melts of subducted eclogite with the mantle wedge is a viable model for continental genesis.

In a parallel line of reasoning, many investigators have suggested that unusually orthopyroxene-rich mantle xenoliths derived from Archean cratons underwent SiO_2 enrichment as a result of reaction between highly depleted mantle residues and rising partial melts of subducted eclogite [*Rudnick et al.,* 1994; *Kelemen et al.,* 1992; *Kesson and Ringwood,* 1989]. Recently, *Kelemen et al.* [1998] showed that this hypothesis was compatible with the observed correlation of Ni in olivine with modal orthopyroxene in cratonic mantle xenoliths worldwide, and emphasized that orthopyroxene-rich mantle xenoliths cannot be simple residues of high pressure partial melting of primitive mantle peridotite. Thus, both continental crust and cratonic mantle peridotites may be related by a process of reaction between subduction zone melts and sub-arc mantle peridotites.

A point of clarification is required here. Enriched primitive andesites [e.g., *Kay,* 1978], partial melts of eclogite ([e.g., *Gill,* 1981] and our Figure 21B), and products of small amounts of reaction between eclogite melts and mantle peridotite under conditions of decreasing magma mass [*Rapp et al.,* 1999], all have middle to heavy REE slopes that are much steeper than those in typical Aleutian, calc-alkaline andesite (Figure 6), and much steeper than those estimated for continental crust (Figure 6A). Thus, in detail, we disagree with numerous authors who have proposed that direct partial melts of eclogite—without magma/mantle interaction—form an important component in the continental crust [e.g., *Defant and Kepezhinskas,* 2001; *Martin,* 1999; *Rapp and Watson,* 1995; *Rapp et al.,* 1991; *Drummond and Defant,* 1990; *Martin,* 1986]. We believe that enriched, primitive andesites have undergone relatively little reaction with surrounding mantle. More extensive reaction, with gradually increasing melt mass and melt/rock ratios ~ 0.1 to ~0.01, is required to increase heavy REE concentrations to the levels observed in most calc-alkaline andesites and in continental crust (e.g., Figures 21C and D).

At small melt/rock ratios, such as are required to explain REE systematics, compatible element concentrations and ratios (e.g., Mg#, Ni) would be completely modified to values in equilibrium with mantle peridotite, and thus the liquids parental to high Mg# andesites and continental crust are inferred to be primitive andesites with Mg# > 0.6. Thus, in addition to the reaction of eclogite melt with mantle peridotite, an additional process is required to explain the relatively low Mg# of continental crust (0.45 to 0.55) compared to primitive andesite. We believe the most likely explanation is that primitive andesites fractionate ultramafic cumulates in the uppermost mantle or at the base of the crust [*Müntener et al.,* 2001; *Miller and Christensen,* 1994; *DeBari and Coleman,* 1989], followed by "delamination" and return of these cumulates to the convecting mantle [*Jull and Kelemen,* 2001; *Kay and Kay,* 1993; *Kay and Kay,* 1991; *Arndt and Goldstein,* 1989].

5.2 Archean Versus Present-day Thermal Regime Beneath Arcs

It has often been proposed that Archean continental crust was formed by a process similar to that which forms "adakite" magmas today. *Martin* [1986] suggested that strong light REE enrichment and heavy REE depletion in Archean granitoids was the result of ubiquitous partial melting of eclogite in hotter, Archean subduction zones. Archean subduction zones may have been hotter than today's because of higher mantle potential temperatures and faster rates of mantle convection (and plate tectonics) in the past. In support of Martin's hypothesis, *Defant and Drummond* [1990] noted that modern "adakites" and arc granitoids, both with trace element contents similar to Archean granitoids, are found mainly in arcs where young oceanic crust (< 25 Ma) is being subducted. They inferred that this is due to partial melting of eclogite in unusually hot, modern subduction zones.

In this paper, we have developed the idea that partial melting of subducted sediment and/or basalt may be common in modern subduction zones. This is supported by a variety of trace element evidence from arcs worldwide. We argue that what is uncommon at present is a relative lack of basaltic magmas generated by partial melting of mantle peridotites in arcs such as the western Aleutians, the Cascades, parts of Central America, the Austral volcanic zone in Chile, and SW Japan. We suggest that the lack of basaltic melts in a few unusual localities, due to cooling of the mantle wedge and extensive shallow dehydration of the subducting plate at slow convergence rates, makes it possible to detect end-member magmas with a trace element signature reflecting a large input from partially molten eclogite.

If primitive andesites in modern arcs are the result of partial melting of subducted eclogite and a *cold* mantle wedge, why should this form magmas that are compositionally similar to those which played a large role in *hot*, Archean crustal genesis? Following *Martin* [1986] and many subsequent workers, we believe that Archean subduction zones were hotter than today, on average, due to higher mantle potential temperatures and more rapid rates of plate formation and subduction. Thus, most dehydration reactions in subducting volcanic rocks and sediments would have occurred at shallower depths than today, reducing the amount of H_2O available to flux partial melting in the mantle wedge. In addition, we believe that the Archean shallow mantle was much more depleted in basaltic components than at present. For example, there is good evidence that Archean decompression melting, beneath ridges and/or hot spots, extended to 40% melting at shallow depths, and produced residues with Mg#'s of 0.93 or more [e.g., *Bernstein et al.,* 1998]. Such refractory residues can undergo very little additional melting, even under hydrous conditions. For these reasons, there may have been very little melting in an Archean mantle wedge even though mantle potential temperatures in the Earth's interior may have been much higher than at present. Thus, partial melts of subducted volcanic rocks, reacting with overlying mantle peridotite, might have constituted a large proportion of the total magma flux in Archean arcs.

6. CONCLUSIONS

High Mg# andesites are abundant at and west of Adak at ~ 174°W. In this region, there is a systematic along-strike increase in SiO_2 from east to west. Aleutian high Mg# andesite lavas are important because they are similar in major and trace element composition to the continental crust. Such lavas are rare or absent in intraoceanic island arcs other than the Aleutians. Also, the western Aleutians show the smallest influence of a subducted sediment component of any part of the Aleutians, so that it is unlikely that enrichments in elements such as U, Th, K and light rare earths are due to recycling of subducted, continental sediments. Thus, it appears that juvenile continental crust is being produced in the western Aleutians.

As proposed by *Kay* [1978], enriched, primitive andesites in the western Aleutians are produced by partial melting of subducted oceanic crust in eclogite facies, followed by reaction of these melts with peridotite in the overlying mantle wedge. Such melts probably form beneath the central Aleutians as well as the western Aleutians. In the central Aleutians, they are obscured by coeval partial melts of subducted sediment, with higher incompatible element contents and distinctive isotope ratios, and by mixing with abundant tholeiitic basalts produced by partial melting of the mantle wedge. Enriched, primitive andesites are commonly observed in the western Aleutians because partial melts of subducted sediment and mantle peridotite are much less abundant there. Sediment flux is greatly reduced in the western Aleutians compared to the central arc. Slow convergence rates in the western part of the arc lead to heating of subducting crust and cooling of the mantle wedge. High subduction zone temperatures cause extensive dehydration of sediments and oceanic crust at shallow depths, beneath the forearc region, leaving little H_2O to flux mantle melting.

Partial melts of subducted eclogite are transported through the coldest, deepest portion of the mantle wedge in cracks or partially molten, pyroxenite diapirs. At shallower depths, they interact with mantle peridotite via reactions which increase melt mass, increase molar Mg#, Ni contents, and MgO contents of derivative liquids, and decrease SiO_2 contents of derivative liquids. The andesitic—rather than basaltic—nature of the derivative liquids in the western Aleutians probably reflects the effect of alkalis and H_2O, increasing the silica content of olivine-saturated liquids. Where the mantle wedge is hotter and/or more hydrous, similar reactions may increase magma mass still further, producing a less enriched, basaltic liquid product. In this context, partial melts of eclogite could be a ubiquitous "flux" component in many arcs, even where enriched, primitive andesites are not observed.

End-member, enriched primitive andesites in the western Aleutians provide important insights into the nature of eclogite melt components which may be present in other arcs. They have high Th concentrations, similar to proposed partial melts of sediment in the Marianas, Lesser Antilles and central Aleutian arcs. Thus, high Th, and high Th/Nb, are not good discriminants between eclogite and sediment melt components. However, Th/La correlates well with Pb isotope ratios in Aleutian lavas, so Th/La may be used to distinguish between eclogite and sediment melt components. Interestingly, Th/La also correlates with Ba/La, suggesting that Ba in Aleutian lavas may be transported in partial melts of subducted material rather than, or in addition to, aqueous fluids. Enriched, primitive andesites also have very high Na contents at a given Mg#, compared to other Aleutian lavas. It has been proposed that Na can be used as an indicator of the degree of partial melting in the mantle source of arc magmas, with high Na indicative of low degrees of melting and vice versa [*Plank and Langmuir,* 1992]. However, this can only be true where the "subduction component", added to the mantle source, contains a small and/or constant proportion of Na. In the data we have compiled, Na concentration in primitive lavas correlates with trace element ratios such as Sr/Y and Dy/Yb which are indicative of an eclogite melt component in Aleutian magmas. This suggests that a large proportion of the total Na in primitive magmas is transported in partial melts of subducted eclogite, and thus that Na contents of primitive magmas cannot be used as an indicator of the degree of melting of the mantle wedge, at least in the western Aleutians.

Moderate amounts of reaction of enriched primitive andesites with upper mantle peridotite, followed by crystal fractionation and mixing with evolved lavas, have produced high Mg# andesite lavas and plutonic rocks with compositions very similar to continental crust. We consider it likely that the processes which form high Mg# andesites in the western Aleutians are similar to those which formed continental crust. In the Archean, the mantle wedge above subduction zones was composed of highly depleted, shallow residues of ~ 40% decompression melting beneath spreading ridges or hot spots. Enriched, primitive andesite magmas were transported through the mantle wedge without extensive dilution with mantle-derived melts because of the highly refractory nature of these residual peridotites.

Acknowledgments. We gratefully used Jimm Myers' database (http://www.gg.uwyo.edu/aleutians/index.htm), and Myers provided additional aid to us. The database includes unpublished data from Jim Brophy, Bruce Marsh, Travis McElfrish, Julie Morris, Jimm Myers, Kirsten Nicolaysen, and Mike Perfit; we thank them for their generosity. Sue and Bob Kay provided samples, unpublished data, advice and encouragement. Tim Elliott shared his compilation of high quality geochemical data for island arcs worldwide. Magali Billen, Tim Elliott, Tim Grove, Peter van Keken, Simon Peacock, Bob Rapp, Vincent Salters, and Linda Elkins Tanton provided preprints of papers in press. Terry Plank and Roberta Rudnick provided up-to-date information on continental crust composition. Jurek Blusztajn made new Sr, Nd and Pb isotope analyses. Karen Hanghøj labored to convert figures and text, for this paper and two others in the same volume (*Kelemen et al., Hirth and Kohlstedt*) to JGR camera ready format. We've been fortunate to receive critical comments and suggestions from a host of colleagues, particularly Greg Hirth, Matthew Jull, Marc Parmentier, Stan Hart, Charlie Langmuir, Tim Grove, Chris Hawkesworth, Chris Nye, and Dan McKenzie. An informal review by Sue Kay, plus brief comments on a draft by Terry Plank and by Simon Turner, Rhiannon George and Chris Hawkesworth, were very helpful, as were formal reviews by Mark Defant and John Eiler. This work was supported in part by the Charles Francis Adams Chair at Woods Hole Oceanographic Institution (Kelemen), and NSF Research Grants EAR-0087706, EAR-9910899 and EAR-9814632 (Kelemen), and EAR-9419240 (Yogodzinski & Kelemen).

REFERENCES

Anders, E., and N. Grevesse, Abundances of the elements: Meteoritic and solar, *Geochim. Cosmochim. Acta, 53,* 197-214, 1989.

Argus, D.R., and R.G. Gordon, No-net-rotation model of current plate velocities incorporating plate motion model NUVEL-1, *Geophys. Res. Lett., 18,* 2039, 1991.

Arndt, N.T., Partitioning of nickel between olivine and ultrabasic and basic komatiite liquids, *Carnegie Inst. Washington Yrbk., 76,* 553-557, 1977.

Arndt, N.T., and S.L. Goldstein, An open boundary between lower continental crust and mantle: Its role in crust formation and crustal recycling, *Tectonophysics, 161,* 201-212, 1989.

Atwater, T., Plate tectonic history of the northeast Pacific and western North America, in *The Eastern Pacific Ocean and Hawaii: The Geology of North America, Vol. N,* edited by E.L. Winterer, D.M. Hussong, and R.W. Decker, pp. 21-72, Geological Society of America, Boulder, 1989.

Avé Lallemant, H.G., and J. Oldow, Active displacement partitioning and arc-parallel extension of the Aleutian volcanic arc based on Global Positioning System geodesy and kinematic analysis, *Geology, 28,* 739-742, 2000.

Ayers, J.C., Trace element modeling of aqueous fluid-peridotite interaction in the mantle wedge of subduction zones, *Contrib. Mineral. Petrol., 132,* 390-404, 1998.

Ayers, J.C., S.K. Dittmer, and G.D. Layne, Partitioning of elements between peridotite and H2O at 2.0 - 3.0 GPa and 900 - 1100°C, and application to models of subduction zone processes, *Earth Planet. Sci. Lett., 150,* 381-398, 1997.

Baker, D.R., and D.H. Eggler, Fractionation paths of Atka (Aleutians) high-alumina basalts: Constraints from phase relations, *J. Volcanol. Geotherm. Res., 18,* 387-404, 1983.

Baker, M.B., T.L. Grove, and R. Price, Primitive basalts and andesites from the Mt. Shasta region, N. California: Products of varying melt fraction and water content, *Contrib. Mineral. Petrol., 118,* 111-129, 1994.

Baker, M.B., M.M. Hirschmann, M.S. Ghiorso, and E.M. Stolper, Compositions of near-solidus peridotite melts from experiments and thermodynamic calculations, *Nature, 375,* 308-311, 1995.

Baker, M.B., M.M. Hirschmann, L.E. Wasylenki, and E.M. Stolper, Quest for low-degree mantle melts (Scientific Correspondence, Reply), *Nature, 381,* 286, 1996.

Bartels, K.S., R.J. Kinzler, and T.L. Grove, High pressure phase relations of primitive high-alumina basalts from Medicine Lake Volcano, Northern California, *Contrib. Mineral. Petrol., 108,* 253-270, 1991.

Beattie, P., Uranium-thorium disequilibria and partitioning on melting of garnet peridotite, *Nature, 363,* 63-65, 1993.

Bernstein, S., P.B. Kelemen, and C.K. Brooks, Highly depleted spinel harzburgite xenoliths in Tertiary dikes from East Greenland, *Earth Planet Sci. Lett., 154,* 221-235, 1998.

Bloomer, S.H., R.J. Stern, E. Fisk, and C.H. Geschwind, Shoshonitic volcanism in the Northern Mariana Arc, 1: Mineralogic and major and trace element characteristics, *J. Geophys. Res., 94,* 4469-4496, 1989.

Bodinier, J.-L., C. Merlet, R.M. Bedini, F. Simien, M. Remaidi, and C.J. Garrido, Distribution of niobium, tantalum, and other highly incompatible trace elements in the lithospheric mantle: The spinel paradox, *Geochim. Cosmochim. Acta, 60* (3), 545-550, 1996.

Bodinier, J.L., C. Dupuy, J. Dostal, and C. Merlet, Distribution of trace transition elements in olivine and pyroxenes from ultramafic xenoliths: Application of microprobe analysis, *Am. Mineral., 72,* 902-913, 1987.

Boyd, T.M., and K.C. Creager, The geometry of Aleutian subduction: Three dimensional seismic imaging, *J. Geophys. Res., 96,* 2267-2291, 1991.

Boyd, T.M., E. Engdahl, and W. Spencer, Seismic cycles along the Aleutian arc: Analysis of seismicity from 1957 through 1991, *J. Geophys. Res., 100,* 621-644, 1995.

Brenan, J.M., H.F. Shaw, F.J. Ryerson, and D.L. Phinney, Erratum to "Experimental determination of trace element partitioning between pargasite and a synthetic hydrous andesitic melt" [Earth Planet. Sci. Lett. 135 (1995) 1-11], *Earth and Planetary Science Letters, 140,* 287-288, 1996.

Brenan, J.M., H.F. Shaw, and R.J. Ryerson, Experimental evidence for the origin of lead enrichment in convergent margin magmas, *Nature, 378,* 54-56, 1995a.

Brenan, J.M., H.F. Shaw, R.J. Ryerson, and D.L. Phinney, Mineral - aqueous fluid partitioning of trace elements at 900°C and 2.0 GPa: Constraints on trace element chemistry of mantle and deep crustal fluids, *Geochim. Cosmochim. Acta, 59,* 3331-3350, 1995b.

Brophy, J.G., The Cold Bay Volcanic Center, Aleutian volcanic arc, I: implications for the origin of hi-alumina arc basalt, *Contrib. Mineral. Petrol., 93,* 368-380, 1986.

Brophy, J.G., The Cold Bay volcanic center, Aleutian volcanic arc, II. Implications for fractionation and mixing mechanisms in calc-alkaline andesite genesis, *Contrib. Mineral. Petrol., 97,* 378-388, 1987.

Brophy, J.G., and B.D. Marsh, On the origin of high alumina arc basalt and the mechanics of melt extraction, *J. Petrol., 27,* 763-789, 1986.

Byers, F.M.J., Geology of Umnak and Bogoslof Islands, Aleutian Islands, Alaska, *U.S. Geol. Surv. Bull., 1028-L,* 263-369, 1959.

Campbell, I.H., and J.S. Turner, Turbulent mixing between fluids with different viscosities, *Nature, 313,* 39-42, 1985.

Carmichael, I.S.E., The petrology of Thingmuli, a Tertiary volcano in eastern Iceland, *J. Petrol., 5,* 435-460, 1964.

Carroll, M.R., and P.J. Wyllie, Experimental phase relations in the system tonalite-peridotite-H_2O at 15 kb: Implications for assimilation and differentiation processes near the crust-mantle boundary, *J. Petrol., 30,* 1351-1382, 1989.

Christensen, N.I., and W.D. Mooney, Seismic velocity structure and composition of the continental crust: A global view, *J. Geophys. Res., 100,* 9761-9788, 1995.

Citron, G., The Hidden Bay pluton, Alaska: Geochemistry, Origin and Tectonic Significance of Oligocene Magmatic Activity in the Aleutian Island Arc, Ph.D. thesis, Cornell University, Ithaca, NY, 1980.

Class, C., D.L. Miller, S.L. Goldstein, and C.H. Langmuir, Distinguishing melt and fluid components in Umnak Volcanics, Aleutian Arc, *Geochemistry, Geophysics, Geosystems (G-cubed), 1,* 2000.

Conder, J.A., D.A. Weins, and J. Morris, On the decompression melting structure at volcanic arcs and back-arc spreading centers, *Geophys. Res. Lett., 29,* 15, 4 pp., 2002.

Conrad, W.K., and R.W. Kay, Ultramafic and mafic inclusions from Adak Island: Crystallization history, and implications for the nature of primary magmas and crustal evolution in the Aleutian Arc, *J. Petrol., 25,* 88-125, 1984.

Conrad, W.K., S.M. Kay, and R.W. Kay, Magma mixing in the Aleutian arc: Evidence from cognate inclusions and composite xenoliths, *J. Volcanol. Geotherm. Res., 18,* 279-295, 1983.

Cox, K.G., M.R. Smith, and S. Beswtherick, Textural studies of garnet lherzolites: Evidence of exsolution origin from high-temperature harzburgites, in *Mantle Xenoliths,* edited by P.H. Nixon, pp. 537-550, John Wiley & Sons, Chichester, UK, 1987.

DeBari, S.M., and R.G. Coleman, Examination of the deep levels of an island arc: Evidence from the Tonsina ultramafic-mafic assemblage, Tonsina, Alaska, *J. Geophys. Res., 94* (B4), 4373-4391, 1989.

Defant, M.J., L.F. Clark, R.H. Stewart, M.S. Drummond, J.Z. deBoer, R.C. Maury, H. Bellon, T.E. Jackson, and J.F. Restrepo, Andesite and dacite genesis via contrasting processes: The geology and geochemistry of El Valle Volcano, Panama, *Contrib. Mineral. Petrol., 106,* 309-324, 1991.

Defant, M.J., and M.S. Drummond, Derivation of some modern arc magmas by melting of young subducted lithosphere, *Nature, 347,* 662-665, 1990.

Defant, M.J., D. Jacques, R.C. Maury, J. DeBoer, and J.-L. Joron, Geochemistry and tectonic setting of the Luzon arc, Philippines, *Geol. Soc. Amer. Bull., 101,* 663-672, 1989.

Defant, M.J., and P. Kepezhinskas, Evidence suggests slab melting in arc magmas, *EOS, 82,* 65-69, 2001.

DeMets, C., R.G. Gordon, D.F. Argus, and S. Stein, Current plate motions, *Geophys. J. Int., 101,* 425-478, 1990.

DeMets, C., R.G. Gordon, D.F. Argus, and S. Stein, Effect of recent revisions to the geomagnetic reversal timescale on estimates of current plate motions, *Geophys. Res. Lett., 21,* 2191-2194, 1994.

DePaolo, D., J., Trace element and isotopic effects of combined wallrock assimilation and fractional crystallization, *Earth Planet. Sci. Lett., 53,* 189-202, 1981.

Dick, H.J.B., Abyssal peridotites, very slow spreading ridges and ocean ridge magmatism, Geol. Soc. Spec. Pub. *(Magmatism in the Ocean Basins), 42,* 71-105, 1989.

Dixon, T.H., and R. Batiza, Petrology and chemistry of recent lavas in the northern Marianas: implications for the origin of island arc basalts, *Contrib. Mineral. Petrol., 70,* 167-181, 1979.

Dostal, J., C. Dupuy, J.P. Carron, M. Le Guen de Kerneizon, and R.C. Maury, Partition coefficients of trace elements: Application to volcanic rocks of St. Vincent, West Indies, *Geochim. Cosmochim. Acta, 47,* 525-533, 1983.

Draper, D.S., and T.H. Green, P-T phase relations of silicic, alkaline, aluminous liquids: New results and applications to mantle melting and metasomatism, *Earth Planet. Sci. Lett., 170,* 255-268, 1999.

Drewes, H., G.D. Fraser, G.L. Snyder, and H.F.J. Barnett, Geology of Unalaska Island and adjacent insular shelf, Aleutian Islands, Alaska, *U.S. Geol. Surv. Bull., 1028-S,* 583-676, 1961.

Drummond, M.S., and M.J. Defant, A model for trondjhemite-tonalite-dacite genesis and crustal growth via slab melting, *J. Geophys. Res., 95,* 21,503-21,521, 1990.

Eiler, J., A. Crawford, T. Elliott, K.A. Farley, J.V. Valley, and E.M. Stolper, Oxygen isotope geochemistry of oceanic arc lavas, *J. Petrol., 41,* 229-256, 2000.

Elliott, T., T. Plank, A. Zindler, W. White, and B. Bourdon, Element transport from slab to volcanic front at the Mariana Arc, *J. Geophys. Res., 102,* 14,991-15,019, 1997.

Elliott, T., Slab Geochemical Tracers, this volume.

Engebretson, D.C., A. Cox, and R.G. Gordon, *Relative motion between oceanic and continental plates in the Pacific Basin: Geological Society of America Special Paper 206,* 59 pp., Geological Society of America, Boulder, 1985.

Falloon, T.J., and D.H. Green, Anhydrous partial melting of MORB pyrolite and other peridotite compositions at 10 kbar: Implications for the origin of primitive MORB glasses, *Mineral. Petrol., 37,* 181-219, 1987.

Falloon, T.J., and D.H. Green, Anhydrous partial melting of peridotite from 8 to 35 kb and the petrogenesis of MORB, *J. Petrol.,* Special Lithosphere Issue, 379-414, 1988.

Falloon, T.J., D.H. Green, C.J. Hatton, and K.L. Harris, Anhydrous partial melting of a fertile and depleted peridotite from 2 to 30 kb and application to basalt petrogenesis, *J. Petrol., 29,* 1257-1282, 1988.

Falloon, T.J., D.H. Green, and H.S.C. O'Neill, Quest for low-degree mantle melts (Scientific Correspondence, Comment), *Nature, 381,* 285, 1996.

Fliedner, M., and S.L. Klemperer, Structure of an island arc: Wide-angle seismic studies in the eastern Aleutian Islands, Alaska, *J. Geophys. Res., 104,* 10,667–10,694, 1999.

Foley, S.F., M.G. Barth, and G.A. Jenner, Rutile/melt partition coefficients for trace elements and an assessment of the influence of rutile on the trace element characteristics of subduction zone magmas, *Geochim. Cosmochim. Acta, 64,* 933-938, 2000.

Fournelle, J.H., B.D. Marsh, and J.D. Myers, Age, character, and significance of Aleutian arc volcanism, in The Geology of Alaska: *The Geology of North America, Vol. G-1,* edited by G. Plafker, and H.C. Berg, pp. 723-758, Geological Society of America, Boulder, 1994.

Fraser, G.D., and H.F. Barrett, Geology of the Delarof and westernmost Andreanof Islands, Alaska, *U.S. Geol. Surv. Bull., 1028-I,* 211-245, 1959.

Fraser, G.D., and G.L. Snyder, Geology of Southern Adak and Kagalaska Island, Alaska, *US Geol. Surv. Bull., 1028-M,* 371-408, 1956.

Fujimaki, H., M. Tatsumoto, and K. Aoki, Partition Coefficients of Hf, Zr, and REE between phenocrysts and groundmasses, *J. Geophys. Res., 89,* 662-672, 1984.

Gaetani, G.A., and T.L. Grove, The influence of water on melting of mantle peridotite, *Contrib. Mineral. Petrol., 131,* 323-346, 1998.

Gast, P.W., Trace element fractionation and the origin of tholeiitic and alkaline magma types, *Geochim. Cosmochim. Acta, 32,* 1057-1086, 1968.

Gates, O., H.A. Powers, and R.E. Wilcox, Geology of the Near Islands, Alaska, *US Geol. Survey Bull., 1028-U,* 709-822, 1971.

Geist, E.L., J.R. Childs, and D.W. Scholl, The origin of summit basins of the Aleutian Ridge: implications for block rotation of an arc massif, *Tectonics, 7,* 327-341, 1988.

Geist, E.L., and D.W. Scholl, Application of continuum models to deformation of the Aleutian island arc, *J. Geophys. Res., 97,* 4953-4967, 1992.

Geist, E.L., and D.W. Scholl, Large-scale deformation related to the collision of the Aleutian Arc with Kamchatka, *Tectonics, 13,* 538-560, 1994.

Gill, J., *Orogenic Andesites and Plate Tectonics,* 390 pp., Springer-Verlag, Berlin, 1981.

Goldstein, S.L., Isotopic studies of continental and marine sediments, and igneous rocks of the Aleutian island arc, PhD thesis, Columbia University, New York, 1986.

Green, T.H., and N.J. Pearson, An experimental study of Nb and Ta partitioning between Ti-rich minerals and silicate liquids at high pressure and temperature, *Geochim. Cosmochim. Acta, 51,* 55-62, 1987.

Green, T.H., S.H. Sie, C.G. Ryan, and D.R. Cousens, Proton microprobe-determined partitioning of Nb, Ta, Zr, Sr and Y between garnet, clinopyroxene and basaltic magma at high pressure and temperature, *Chem. Geol., 74,* 201-216, 1989.

Grove, T.L., Vapor-saturated melting of fertile peridotite revisited: A new experimental approach and re-evaluation of the hydrous peridotite solidus, *EOS, 82,* F1173-1174, 2001.

Grove, T.L., R.J. Kinzler, M.B. Baker, J.M. Donnelly Nolan, and C.E. Lesher, Assimilation of granite by basaltic magma at Burnt Lava flow, Medicine Lake volcano, California: Decouplilng of heat and mass transfer, *Contrib. Mineral. Petrol., 99,* 320-343, 1988.

Grove, T.L., S.W. Parman, S.A. Bowring, R.C. Price, and M.B. Baker, The role of H2O-rich fluids in the generation of primitive basaltic andesites and andesites from the Mt. Shasta region, N. California, *Contrib. Mineral. Petrol., 142,* 375-396, 2002.

Grow, J., Crustal and upper mantle structure of the central Aleutian arc, *GSA Bull., 84,* 2169-2192, 1973.

Hart, S.R., and C. Brooks, Clinopyroxene-matrix partitioning of K, Rb, Cs, Sr, and Ba, *Carnegie Inst. Washington Yrbk., 73,* 949-954, 1974.

Hart, S.R., and K.E. Davis, Nickel partitioning between olivine and silicate melt, , *40,* 203-219, 1978.

Hart, S.R., and T. Dunn, Experimental cpx/melt partitioning of 24 trace elements, *Contrib. Mineral. Petrol., 113,* 1-8, 1993.

Hauri, E.H., Major-element variability in the Hawaiian mantle plume, *Nature, 382,* 415-419, 1996.

Hauri, E.H., and S.R. Hart, Re-Os isotope systematics of HIMU and EMII oceanic island basalts from the south Pacific Ocean, *Earth Planet. Sci. Lett., 114,* 353-371, 1993.

Hauri, E.H., T.P. Wagner, and T.L. Grove, Experimental and natural partitioning of Th, U, Pb and other trace elements between garnet, clinopyroxene and basaltic melts, *Chem. Geol., 117,* 149-166, 1994.

Hawkesworth, C.J., K. Gallagher, J.M. Hergt, and F. McDermott, Mantle and slab contributions in arc magmas, *Annu. Rev. Earth Planet. Sci., 21,* 175-204, 1993a.

Hawkesworth, C.J., K. Gallagher, J.M. Hergt, and F. McDermott, Trace element fractionation processes in the generation of island arc basalts, *Phil. Trans. R. Soc. Lond. A, 342,* 179-191, 1993b.

Hawkesworth, C.J., S.P. Turner, F. McDermott, D.W. Peate, and P. van Calsteren, U-Th isotopes in arc magmas: Implications for element transfer from the subducted crust, *Science, 276,* 551-555, 1997.

Henstock, T.J., A.W. Woods, and R.S. White, The accretion of oceanic crust by episodic sill intrusion, *J. Geophys. Res., 98,* 4143-4161, 1993.

Hirose, K., Melting experiments on lherzolite KLB-1 under hydrous conditions and generation of high-magnesian andesitic melts, *Geology, 25,* 42-44, 1997.

Hirschmann, M.M., Mantle solidus: Experimental constraints and the effects of peridotite composition, *G-cubed, 1,* 2000.

Hirschmann, M.M., M.B. Baker, and E.M. Stolper, The effect of alkalis on the silica content of mantle-derived melts, *Geochim. Cosmochim. Acta, 62,* 883-902, 1998.

Hochstaedter, A.G., J.G. Ryan, J.F. Luhr, and T. Hasenaka, On B/Be ratios in the Mexican volcanic belt, *Geochim. Cosmochim. Acta, 60,* 613-628, 1996.

Hofmann, A.W., Chemical differentiation of the Earth: The relationship between mantle, continental crust, and oceanic crust, *Earth Planet. Sci. Lett., 90,* 297-314, 1988.

Holbrook, S.W., D.Lizarralde, S.McGeary, N. Bangs, and J. Diebold, Structure and composition of the Aleutian island arc and implications for continental crustal growth, *Geology, 27,* 31-34, 1999.

Hughes, S.S., and E.M. Taylor, Geochemistry, petrogenesis, and tectonic implications of central High Cascade mafic platform lavas, *Geol. Soc. Amer. Bull., 97,* 1024-1036, 1986.

Irvine, T.N., and W.R. Baragar, A guide to the chemical classification of the common volcanic rocks, *Can. J. Earth Sci., 8,* 523-548, 1971.

Irving, A.J., and F.A. Frey, Distribution of trace elements between garnet megacrysts and host volcanic liquids of Kimberlitic to Rhyolitic composition, *Geochim. Cosmochim. Acta, 42,* 771-787, 1978.

Jenner, G.A., Foley, S. F., S.E. Jackson, T.H. Green, B.J. Fryer, and H.P. Longerich, Determination of partition coefficients for trace elements in high pressure-temperature experimental run products by laser ablation microprobe-inductively coupled plasma-mass spectrometry (LAM-ICP-MS), *Geochim. Cosmochim. Acta, 57,* 23-24, 1993.

Johnson, K.T.M., and H.J.B. Dick, Open system melting and temporal and spatial variation of peridotite and basalt at the Atlantis II Fracture Zone, *J. Geophys. Res., 97,* 9219-9241, 1992.

Johnson, K.T.M., H.J.B. Dick, and N. Shimizu, Melting in the oceanic upper mantle: An ion microprobe study of diopsides in abyssal peridotites, *J. Geophys. Res., 95,* 2661-2678, 1990.

Johnson, M.C., and T. Plank, Dehydration and melting experiments constrain the fate of subducted sediments, *Geochemistry, Geophysics, Geosystems (G-cubed), 1,* 1999.

Jull, M., and P.B. Kelemen, On the conditions for lower crustal convective instability, *J. Geophys. Res., 106,* 6423-6446, 2001.

Kawamoto, T., Experimental constraints on differentiation and H2O abundance of calc-alkaline magmas, *Earth Planet. Sci. Lett., 144,* 577-589, 1996.

Kawate, S., and M. Arima, Petrogenesis of the Tanzawa plutonic complex, central Japan: Exposed felsic middle crust of the Izu-Bonin-Mariana arc, *The Island Arc, 7,* 342-358, 1998.

Kay, R.W., Geochemical constraints on the origin of Aleutian magmas, in *Island Arcs, Deep Sea Trenches and Back Arc Basins: AGU Monograph 1,* edited by M. Talwani, and W. Pitman III, Amer. Geophy. Union, Washington DC, 1977.

Kay, R.W., Aleutian magnesian andesites: Melts from subducted Pacific ocean crust, *J. Volc. Geotherm. Res., 4,* 117-132, 1978.

Kay, R.W., Volcanic arc magmas: Implications of a melting-mixing model for element recycling in the crust-upper mantle system, *J. Geol., 88,* 497-522, 1980.

Kay, R.W., and S.M. Kay, Creation and destruction of lower continental crust, *Geol. Rundsch., 80,* 259-278, 1991.

Kay, R.W., and S.M. Kay, Delamination and delamination magmatism, *Tectonophysics, 219,* 177-189, 1993.

Kay, R.W., S.M. Kay, and P. Layer, Original Aleutian Adakite dated at 11.8 Ma: Slab melt thermal dilemma resolved?, *EOS, 79,* F396, 1998.

Kay, R.W., J.L. Rubenstone, and S.M. Kay, Aleutian terranes from Nd isotopes, *Nature, 322,* 605-609, 1986.

Kay, R.W., S.-S. Sun, and C.-N. Lee-Hu, Pb and Sr isotopes in volcanic rocks from the Aleutian Islands and Pribilof Islands, Alaska, *Geochim. Cosmochim. Acta, 42,* 263-273, 1978a.

Kay, R.W., S.S. Sun, and C.N. Lee-Hu, Pb and Sr isotopes in volcanic rocks from the Aleutian Islands and Pribilof Islands, Alaska, *Geochimica et Cosmochimica Acta, 42,* 263-273, 1978b.

Kay, S.M., and R.W. Kay, Aleutian tholeiitic and calc-alkaline magma series I: The mafic phenocrysts, *Contrib. Mineral. Petrol., 90,* 276-290, 1985a.

Kay, S.M., and R.W. Kay, Role of crystal cumulates and the oceanic crust in the formation of the lower crust of the Aleutian Arc, *Geology, 13,* 461-464, 1985b.

Kay, S.M., and R.W. Kay, Aleutian magmas in space and time, in The Geology of Alaska: *The Geology of North America, Vol. G-1,* edited by G. Plafker, and H.C. Berg, pp. 687-722, Geological Society of America, Boulder, 1994.

Kay, S.M., R.W. Kay, H.K. Brueckner, and J.L. Rubenstone, Tholeiitic Aleutian arc plutonism: The Finger Bay Pluton, Adak, Alaska, *Contrib. Mineral. Petrol., 82,* 99-116, 1983.

Kay, S.M., R.W. Kay, and G.P. Citron, Tectonic controls on tholeiitic and calc-alkaline magmatism in the Aleutian arc, *J. Geophys. Res., 87,* 4051-4072, 1982.

Kay, S.M., R.W. Kay, and M.R. Perfit, Calc-alkaline plutonism in the intraoceanic Aleutian arc, Alaska, in Plutonism from Antarctica to *Alaska: Geological Society of America Special Paper 241,* edited by S.M. Kay, and C.W. Rapela, pp. 233-255, Boulder, 1990.

Kelemen, P.B., Assimilation of ultramafic rock in subduction-related magmatic arcs, *J. Geol., 94,* 829-843, 1986.

Kelemen, P.B., Genesis of high Mg# andesites and the continental crust, *Contrib. Mineral. Petrol., 120,* 1-19, 1995.

Kelemen, P.B., H.J.B. Dick, and J.E. Quick, Formation of harzburgite by pervasive melt/rock reaction in the upper mantle, *Nature, 358,* 635-641, 1992.

Kelemen, P.B., S.R. Hart, and S. Bernstein, Silica enrichment in the continental upper mantle lithosphere via melt/rock reaction, *Earth Planet. Sci. Lett., 164,* 387-406, 1998.

Kelemen, P.B., and G. Hirth, What happens to melts of subducted sediment and meta-basalt?, *EOS, 79,* F1002, 1998.

Kelemen, P.B., K.T.M. Johnson, R.J. Kinzler, and A.J. Irving, High-field-strength element depletions in arc basalts due to mantle-magma interaction, *Nature, 345,* 521-524, 1990a.

Kelemen, P.B., D.B. Joyce, J.D. Webster, and J.R. Holloway, Reaction between ultramafic rock and fractionating basaltic magma II. Experimental investigation of reaction between olivine tholeiite and harzburgite at 1150-1050°C and 5 kb, *J. Petrol., 31,* 99-134, 1990b.

Kelemen, P.B., Reaction between ultramafic rock and fractionating basaltic magma I. Phase relations, the origin of calc-alkaline magma series, and the formation of discordant dunite, *J. Petrol., 31,* 51-98, 1990c.

Kelemen, P.B., J.L. Rilling, E.M. Parmentier, L. Mehl, and B.R. Hacker, Thermal structure due to solid state flow in the mantle wedge beneath arcs, this volume.

Kelemen, P.B., N. Shimizu, and T. Dunn, Relative depletion of niobium in some arc magmas and the continental crust: Partitioning of K, Nb, La, and Ce during melt/rock reaction in the upper mantle, *Earth Planet. Sci. Lett., 120,* 111-134, 1993.

Keppler, H., Constraints from partitioning experiments on the composition of subduction-zone fluids, *Nature, 380,* 237-240, 1996.

Kesson, S.E., and A.E. Ringwood, Slab-mantle interactions 2. The formation of diamonds, *Chem. Geol., 78,* 97-118, 1989.

Kincaid, C., and I.S. Sacks, Thermal and dynamical evolution of the upper mantle in subduction zones, *J. Geophys. Res., 102,* 12,295-12,315, 1997.

Kinzler, R.J., T.L. Grove, and S.I. Recca, An experimental study on the effect of temperature and melt composition on the partitioning of nickel between olivine and silicate melt, *Geochim. Cosmochim. Acta, 54,* 1255-1265, 1990.

Kogiso, T., Y. Tatsumi, and S. Nakano, Trace element transport during dehydration processes in the subducted oceanic crust, I: Experiments and implications for the origin of ocean island basalts, *Earth Planet. Sci. Lett., 148,* 193-205, 1997.

Kushiro, I., Compositions of magmas formed by partial zone melting of the Earth's upper mantle, *J. Geophys. Res., 73,* 619-634, 1968.

Kushiro, I., The system forsterite-diopside-silica with and without water at high pressures, *Am. J. Sci., 267-A,* 269-294, 1969.

Kushiro, I., On the nature of silicate melt and its significance in magma genesis: Regularities in the shift of the liquidus boundaries involving olivine, pyroxene, and silica minerals, *Am. J. Sci., 275,* 411-431, 1975.

Kushiro, I., Partial melting of mantle wedge and evolution of island arc crust, *J. Geophys. Res., 95,* 15,929-15,939, 1990.

Lambert, I., B., and P. Wyllie, J., Melting of gabbro (quartz eclogite) with excess water to 35 kilobars, with geological applications, *J. Geol., 80,* 693-708, 1972.

Larsen, L.M., Distribution of REE and other trace elements between phenocrysts and peralkaline undersaturated magmas, exemplified by rocks from the Gardar igneous province, South Greenland, *Lithos, 12,* 303-315, 1979.

LaTourrette, T.Z., and D.S. Burnett, Experimental determination of U and Th partitioning between clinopyroxene and natural and synthetic basaltic liquid, *Earth Planet. Sci. Lett., 110,* 227-244, 1992.

Lonsdale, P., Paleogene history of the Kula plate: Offshore evidence and onshore implications, *Geol. Soc. Amer. Bulletin, 100,* 733-754, 1988.

Luhr, J., and I.S.E. Carmichael, Volatiles and trace element partitioning in the El Chichon trachyandesite, *EOS, 65,* 299, 1984.

Luhr, J.F., J.F. Allan, I.S.E. Carmichael, S.A. Nelson, and T. Hasenaka, Primitive calc-alkaline and alkaline rock types from the Western Mexican Volcanic Belt, *J. Geophys. Res., 94,* 4515-4530, 1989.

Lundstrom, C.C., H.F. Shaw, F.J. Ryerson, D.L. Phinney, J.B. Gill, and Q. Williams, Compositional controls on the partitioning of U, Th, Ba, Pb, Sr and Zr between clinopyroxene and haplobasaltic melts; implications for uranium series disequilibria in basalts, *Earth Planet. Sci. Lett., 128,* 407-423, 1994.

Marsh, B.D., Some Aleutian andesites: Their nature and source, *J. Geol., 84,* 27-45, 1976.

Marsh, B.D., The Aleutians, in *Andesites: Orogenic Andesites and Related Rocks,* edited by R.S. Thorpe, pp. 99-115, John Wiley, New York, 1982a.

Marsh, B.D., unpublished data available at http://www.gg.uwyo.edu/aleutians/index.htm, , 1982b.

Martin, H., Effect of steeper Archean geothermal gradient on geochemistry of subduction-zone magmas, *Geology, 14,* 753-756, 1986.

Martin, H., Adakitic magmas: Modern analogues of Archaean granitoids, *Lithos, 46,* 411-429, 1999.

McBirney, A., R., H. Taylor, P., and R. Armstrong, L., Paricutin reexamined: A classic example of crustal assimilation in calc-alkaline magma, *Contrib. Mineral. Petrol., 95,* 4-20, 1987.

McCarthy, J., and D.W. Scholl, Mechanisms of subduction accretion along the central Aleutian Trench, *Geol. Soc. Amer. Bull., 96,* 691-701, 1985.

McCulloch, M.T., and M.R. Perfit, 143Nd/144Nd, 87Sr/86Sr and trace element constraints on the petrogenesis of Aleutian island arc magmas, *Earth Planet. Sci. Lett., 56,* 167-179, 1981.

McLennan, S.M., and S.R. Taylor, *The Continental Crust: Its Composition and Evolution: An Examination of the Geochemical Record Preserved in Sedimentary Rocks,* Oxford: Blackwell Scientific, Palo Alto, Calif, 1985.

Miller, D.J., and N.I. Christensen, Seismic signature and geochemistry of an island arc; a multidisciplinary study of the Kohistan accreted terrane, northern Pakistan, *J. Geophys. Res., 99,* 11,623-11,642, 1994.

Miller, D.M., S.L. Goldstein, and C.H. Langmuir, Cerium/lead and lead isotope ratios in arc magmas and the enrichment of lead in the continents, *Nature, 368,* 514-520, 1994.

Miller, D.M., C.H. Langmuir, S.L. Goldstein, and A.L. Franks, The importance of parental magma composition to calc-alkaline and tholeiitic evolution: Evidence from Umnak Island in the Aleutians, *J. Geophys. Res., 97* (B1), 321-343, 1992.

Miyashiro, A., Volcanic rock series in island arcs and active continental margins, *Am. J. Sci., 274,* 321-355, 1974.

Morris, J.D., and S.R. Hart, Isotopic and incompatible element constraints on the genesis of island arc volcanics from Cold Bay and Amak Islands, Aleutians and implications for mantle structure, *Geochim. Cosmochim. Acta, 47,* 2015-2030, 1983.

Müntener, O., P.B. Kelemen, and T.L. Grove, The role of H2O and composition on the genesis of igneous pyroxenites: An experimental study, *Contrib. Mineral. Petrol., 141,* 643-658, 2001.

Myers, J.D., Marsh, B.D., and Sinha, A.K., Strontium isotopic and selected trace element variations between two Aleutian volcanic centers (Adak and Atka): implications for the development of arc volcanic plumbing systems, *Contrib. Mineral. Petrol., 91,* 221-234, 1985.

Myers, J.D., Possible petrogenetic relations between low- and high-MgO Aleutian basalts, *Geol. Soc. Am. Bulletin, 100,* 1040-1053, 1988.

Myers, J.D., C.D. Frost, and C.L. Angevine, A test of a quartz eclogite source for parental Aleutian magmas: A mass balance approach, *J. Geol., 94,* 811-828, 1986a.

Myers, J.D., and B.D. Marsh, Aleutian lead isotopic data: Additional evidence for the evolution of lithospheric plumbing systems, *Geochim. Cosmochim. Acta, 51,* 1833-1842, 1987.

Myers, J.D., B.D. Marsh, and A.K. Sinha, Geochemical and strontium isotopic characteristics of parental Aleutian Arc magmas: Evidence from the basaltic lavas of Atka, *Contrib. Mineral. Petrol., 94,* 1-11, 1986b.

Mysen, B.O., I. Kushiro, I.A. Nicholls, and A.E. Ringwood, A possible mantle origin for andesitic magmas: Discussion of a paper by Nicholls and Ringwood, *Earth Planet. Sci. Lett., 21,* 221-229, 1974.

Newman, S., J.D. MacDougall, and R.C. Finkel, Petrogenesis and 230th - 238U disequilibrium at Mt. Shasta, California and in the Cascades, *Contrib. Mineral. Petrol., 93,* 195-206, 1986.

Nicholls, I.A., and K.L. Harris, Experimental rare earth element partition coefficients for garnet, clinopyroxene and amphibole coexisting with andesitic and basaltic liquids, *Geochim. Cosmochim. Acta, 44,* 287-308, 1980.

Nichols, G.T., P.J. Wyllie, and C.R. Stern, Subduction zone melting of pelagic sediments constrained by melting experiments, *Nature, 371,* 785-788, 1994.

Nye, C.J., and M.R. Reid, Geochemistry of primary and least fractionated lavas from Okmok Volcano, Central Aleutians: implications for arc magmagenesis, *J. Geophys. Res., 91,* 10271-10287, 1986.

Peacock, S.M., Thermal structure and metamorphic evolution of subducting slabs, this volume.

Perfit, M.R., The petrochemistry of igneous rocks from the Cayman Trough and Captains Bay Pluton, Unalaska Island: Their relation to tectonic processes, PhD thesis, Columbia University, New York, 1977.

Perfit, M.R., unpublished data available at http://www.gg.uwyo.edu/aleutians/index.htm, , 1983.

Perfit, M.R., H. Brueckner, J.R. Lawrence, and R.W. Kay, Trace element and isotopic variations in a zoned pluton and associated volcanic rocks, Unalaska Island, Alaska: a model for fractionation in the Aleutian calc-alkaline suite, *Contrib. Mineral. Petrol., 73,* 69-87., 1980a.

Perfit, M.R., H. Brueckner, J.R. Lawrence, and R.W. Kay, Trace element and isotopic variations in a zoned pluton and associated volcanic rocks, Unalaska Island, Alaska: a model for fractionation in the Aleutian calcalkaline suite, *Contrib. Mineral. Petrol., 73,* 69-87, 1980b.

Peucker-Ehrenbrink, B., A.W. Hofmann, and S.R. Hart, Hydrothermal lead transfer from mantle to continental crust; the role of metalliferous sediments, *Earth Planet. Sci. Lett., 125,* 129-142, 1994.

Philpotts, J.A., and C.C. Schnetzler, Phenocryst-matrix partition coefficients for K, Rb, Sr and Ba, with applications to anorthosite and basalt genesis, *Geochim. Cosmochim. Acta, 34,* 307-322, 1970a.

Philpotts, J.A., and C.C. Schnetzler, Potassium, rubidium, strontium, barium, and rare earth concentrations in lunar rocks and separated phases, *Science, 167,* 493-495, 1970b.

Phipps Morgan, J., and Y.J. Chen, The genesis of oceanic crust: Magma injection, hydrothermal circulation, and crustal flow, *J. Geophys. Res., 98,* 6283-6297, 1993.

Plank, T., and C.H. Langmuir, Effects of the melting regime on the composition of the oceanic crust, *J. Geophys. Res., 97,* 19,749-19,770, 1992.

Plank, T., and C.H. Langmuir, Tracing trace elements from sediment input to volcanic output at subduction zones, *Nature, 362,* 739-743, 1993.

Plank, T., and C.H. Langmuir, The chemical composition of subducting sediment and its consequences for the crust and mantle, *Chem. Geol., 145,* 325-394, 1998.

Puig, A., M. Herve, M. Suarez, and A.D. Saunders, Calc-alkaline and alkaline Miocene and calc-alkaline Recent volcanism in the southernmost Patagonian cordillera, Chile, *J. Volc. Geotherm. Res., 20,* 149-163, 1984.

Rapp, R.P., N. Shimizu, M.D. Norman, and G.S. Applegate, Reaction between slab-derived melts and peridotite in the mantle wedge: Experimental constraints at 3.8 GPa, *Chem. Geol., 160,* 335-356, 1999.

Rapp, R.P., and E.B. Watson, Dehydration melting of metabasalt at 8-32 kbar: Implications for continental growth and crust—mantle recycling, *J. Petrol., 36* (4), 891-931, 1995.

Rapp, R.P., E.B. Watson, and C.F. Miller, Partial melting of amphibolite/eclogite and the origin of Archean trondhjemites and tonalites, *Precambrian Research, 51,* 1-25, 1991.

Reid, I., and H.R. Jackson, Oceanic spreading rate and crustal thickness, *Marine Geophysical Researches, 5,* 165-172, 1981.

Ringwood, A.E., The petrological evolution of island arc systems, *J. Geol. Soc. Lond., 130,* 183-204, 1974.

Roeder, P.L., and R.F. Emslie, Olivine-liquid equilibrium, *Contrib. Mineral. Petrol., 29,* 275-289, 1970.

Rogers, G., A.D. Saunders, D.J. Terrell, S.P. Verma, and G.F. Marriner, Geochemistry of Holocene volcanic rocks associated with ridge subduction in Baja California, Mexico, *Nature, 315,* 389-392, 1985.

Romick, J.D., The igneous petrology and geochemistry of northern Akutan Island, Alaska, MSc. thesis, Univ. of Alaska, Fairbanks, 1982.

Romick, J.D., S.M. Kay, and R.W. Kay, The influence of amphibole fractionation on the evolution of calc-alkaline andesite and dacite tephra from the central Aleutians, Alaska, *Contrib. Mineral. Petrol., 112,* 101-118, 1992.

Romick, J.D., M.R. Perfit, S.R. Swanson, and R.D. Schuster, Magmatism in the eastern Aleutian arc: Temporal characteristics of igneous activity on Akutan Island, *Contrib. Mineral. Petrol., 104,* 700-721, 1990.

Rostovtseva, Y.V., and M.N. Shapiro, Provenance of the Palaeocene-Eocene rocks of the Komandorsky Islands, *Sedimentology, 45,* 201-216, 1998.

Rubenstone, J.L., Geology and geochemistry of early Tertiary submarine volcanic rocks of the Aleutian Islands, and their bearing on the development of the Aleutian island arc, PhD thesis, Cornell University, Ithaca, NY, 1984.

Rudnick, R.L., Making continental crust, Nature, 378, 571-577, 1995.

Rudnick, R.L., and D.M. Fountain, Nature and composition of the continental crust: A lower crustal perspective, *Reviews of Geophysics, 33* (3), 267-309, 1995.

Rudnick, R.L., W.F. McDonough, and A. Orpin, Northern Tanzanian peridotite xenoliths: A comparison with Kapvaal peridotites and inferences on metasomatic interactions, in *Kimberlites, Related Rocks and Mantle Xenoliths, Vol. 1, Proceedings 5th Int. Kimberlite Conf.,* edited by H.O.A. Meyer, and O. Leonardos, CRPM, Brasilia, 1994.

Ryan, H.F., and D.W. Scholl, The evolution of forearc structures along an oblique convergent margin, central Aleutian Arc, *Tectonics, 8,* 497-516, 1989.

Ryerson, F.J., Oxide solution mechanisms in silicate melts: systematic variations in the activity coefficient of SiO_2, *Geochim. Cosmochim. Acta, 49,* 637-649, 1985.

Ryerson, F.J., and E.B. Watson, Rutile saturation in magmas: Implications for Ti-Nb-Ta depletion in island arc basalts, *Earth Planet. Sci. Lett., 86,* 225-239, 1987.

Salters, V.J.M., and J. Longhi, Trace element partitioning during the initial stages of melting beneath mid-ocean ridges, *Earth Planet. Sci. Lett., 166,* 15-30, 1999.

Salters, V.J.M., J.E. Longhi, and M. Bizimis, Near mantle solidus trace element partitioning at pressures up to 3.4 GPa, *Geochemistry, Geophysics, Geosystems (G-cubed), 3,* 7(148), 15 pp. 2002.

Schiano, P., J.-L. Birck, and C.J. Allegre, Osmium-strontium-neodymium-lead isotopic covariations in mid-ocean ridge basalt glasses and the heterogeneity of the upper mantle, *Earth Planet. Sci. Lett., 150,* 363-379, 1997.

Schiano, P., and R. Clocchiatti, Worldwide occurrence of silica-rich melts in sub-continental and sub-oceanic mantle minerals, *Nature, 368,* 621-623, 1994.

Schiano, P., R. Clocchiatti, N. Shimizu, R.C. Maury, K.P. Jochum, and A.W. Hofmann, Hydrous, silica-rich melts in the sub-arc mantle and their relationship with erupted arc lavas, *Nature, 377,* 595-600, 1995.

Schiano, P., R. Clocchiatti, N. Shimizu, D. Weis, and N. Mattielli, Cogenetic silica-rich and carbonate-rich melts trapped in mantle minerals in Kerguelen ultramafic xenoliths: Implications for metasomatism in the oceanic upper mantle, *Earth Planet. Sci. Lett., 123,* 167-178, 1994.

Schmidt, M.W., and S. Poli, Experimentally based water budgets for dehydrating slabs and consequences for arc magma generation, *Earth and Planetary Science Letters, 163* (1-4), 361-379, 1998.

Schnetzler, C.C., and J.A. Philpotts, Partition coefficients of rare earth elements between igneous matrix material and rock-forming mineral phenocrysts, II, *Geochim. Cosmochim. Acta, 34,* 331-340, 1970.

Scholl, D., W, A.J. Stevenson, M.A. Noble, and D.K. Rea, The Meiji drift body of the northwestern Pacific—modern and paleoceanographic implications, in *From Greenhouse to Icehouse: The Marine Eocene-Oligocene Transition: Geological Society of America Special Paper,* edited by D. Prothero, pp. in press, Geol. Soc. Am., Boulder CO, 2001.

Scholl, D.W., and R.V. Huene, New geophysical and geological studies support higher, but comparable, rates of both arc growth and crustal recycling at subduction zones, *Geol. Soc. Amer. Abstracts with Programs, 30* (7), A-209, 1998.

Scholl, D.W., M.S. Marlow, N.S. MacLeod, and E.C. Buffington, Episodic Aleutian Ridge igneous activity: Implications of

Miocene and younger submarine volcanism west of Buldir Island, *Geol. Soc. Am. Bull., 87,* 547-554, 1976.

Scholl, D.W., T.L. Vallier, and A.J. Stevenson, Geologic evolution and petroleum geology of the Aleutian ridge, in *Geology and Resource Potential of the Continental Margin of Western North America and Adjacent Ocean Basins - Beaufort Sea to Baja California,* edited by D.W. Scholl, A. Grantz, and J.G. Vedder, pp. 103-122, Circum-Pacific Council on Energy and Mineral Resources, Houston, 1987.

Shaw, D.M., Trace element fractionation during anatexis, *Geochim. Cosmochim. Acta, 34,* 237-243, 1970.

Shimizu, N., and I. Kushiro, The partitioning of rare earth elements between garnet and liquid at high pressures: Preliminary experiments, *Geophys. Res. Lett., 2,* 414-416, 1975.

Sigmarsson, O., H. Martin, and J. Knowles, Melting of a subducting oceanic crust from U-Th disequilibria in austral Andean lavas, *Nature, 394,* 566-569, 1998.

Singer, B.S., and J.D. Myers, Intra-arc extension and magmatic evolution in the central Aleutian arc, Alaska, *Geology, 18,* 1050-1053, 1990.

Singer, B.S., J.D. Myers, and C.D. Frost, Mid-Pleistocene basalts from the Seguam Volcanic Center, Central Aleutian arc, Alaska: local lithospheric structures and source variability in the Aleutian arc, *J. Geophys. Res., 97,* 4561-4578, 1992a.

Singer, B.S., J.D. Myers, and C.D. Frost, Mid-Pleistocene lavas from the Seguam Island volcanic center, central Aleutian arc: Closed-system fractional crystallization of a basalt to rhyodacite eruptive suite, *Contrib. Mineral. Petrol., 110,* 87-112, 1992b.

Singer, B.S., J.R. O'Neil, and J.G. Brophy, Oxygen isotope constraints on the petrogenesis of Aleutian arc magmas, *Geology, 20,* 367-370, 1992c.

Sisson, T.W., and T.L. Grove, Temperatures and H2O contents of low MgO high-alumina basalts, *Contrib. Mineral. Petrol., 113,* 167-184, 1993.

Sleep, N.H., Formation of oceanic crust: some thermal constraints, *J. Geophys. Res., 80,* 4037-4042, 1975.

Snow, J.E., Ultramafic ocean crust; implications for mantle-seawater interaction, *EOS, 76,* 275-276, 1995.

Stalder, R., S.F. Foley, G.P. Brey, and I. Horn, Mineral - aqueous fluid partitioning of trace elements at 900 - 1200°C and 3.0 - 5.7 GPa: New experimental data for garnet, clinopyroxene and rutile and implications for mantle metasomatism, *Geochim. Cosmochim. Acta, 62,* 1781-1801, 1998.

Stein, M., and A.W. Hofmann, Mantle plumes and episodic crustal growth, *Nature, 372,* 63-68, 1994.

Stern, C.R., and P.J. Wyllie, Melting relations of basalt-andesite-rhyolite H2O and a pelagic red clay at 30 kb, *Contrib. Mineral. Petrol., 42,* 313-323, 1973.

Stern, R.J., On the origin of andesite in the northern Mariana Island Arc: Implications from Agrigan, *Contrib. Mineral. Petrol., 68,* 207-219, 1979.

Stern, R.J., and L.D. Bibee, Esmeralda Bank: Geochemistry of an active submarine volcano in the Mariana Island Arc, *Contrib. Mineral. Petrol., 86,* 159-169, 1984.

Stolper, E., and S. Newman, The role of water in the petrogenesis of Mariana Trough magmas, *Earth Planet. Sci. Lett., 121,* 293-325, 1992.

Stosch, H.-G., Rare earth element partitioning between minerals from anhydrous spinel peridotite xenoliths, *Geochim. Cosmochim. Acta, 46,* 793-811, 1982.

Sun, S.S., Lead isotopic study of young volcanic rocks from mid-ocean ridges, ocean islands, and island arcs, *Phil. Trans. Roy. Soc. London, A279,* 409-445, 1980.

Tatsumi, Y., Continental crust formation by delamination in subduction zones and complementary accumulation of the enriched mantle I component in the mantle, *Geochemistry, Geophysics, Geosystems (G-cubed), 1,* 1-17, 2000.

Tatsumi, Y., Geochemical modeling of partial melting of subducting sediments and subsequent melt-mantle interaction: Generation of high-Mg andesites in the Setouchi volcanic belt, Southwest Japan, *Geology, 29* (4), 323-326, 2001.

Tatsumi, Y., D.L. Hamilton, and R.W. Nesbitt, Chemical characteristics of fluid phase released from a subducted lithosphere and origin of arc magmas: evidence from high-pressure experiments and natural rocks, *J. Volc. Geotherm. Res., 29,* 293-309, 1986.

Tatsumi, Y., and K. Ishizaka, Origin of high-magnesian andesites in the Setouchi volcanic belt, southwest Japan, I. Petrographical and chemical characteristics, *Earth Planet. Sci. Lett., 60,* 293-304, 1982.

Taylor, S.R., Island arc models and the composition of the continental crust, in *Island Arcs, Deep Sea Trenches, and Back-Arc Basins: Geophysical Monograph 1,* edited by M. Talwani, and W.C. Pitman, pp. 325-335, American Gephysical Union, Washington, D.C., 1977.

Todt, W., R.A. Clift, A. Hanser, and A.W. Hofmann, Evaluation of a ^{202}Pb-^{205}Pb double spike for high precision lead analysis, in *Earth processes: Reading the Isotopic Code: Geophysical Monograph 95,* edited by A. Basu, and S.R. Hart, pp. 429-437, Am. Geophys. Union, Washington, DC, 1996.

Tsvetkov, A.A., Magmatism of the westernmost (Komandorsky) segment of the Aleutian Island Arc, *Tectonophysics, 199,* 289-317, 1991.

Turner, S.P., J. Blundy, B. Wood, M Hole, Large ^{230}Th-excesses in basalts produced by partial melting of spinel lherzolite, Chemical *Geology, 162,* 127-136, 2000.

Turner, S., C. Hawkesworth, P. van Calsteren, E. Heath, R. Macdonald, and S. Black, U-series isotopes and destructive plate margin magma genesis in the Lesser Antilles, *Earth Planet. Sci. Lett., 142,* 191-207, 1996.

Vallier, T.L., D.W. Scholl, M.A. Fisher, R. von Huene, T.R. Bruns, and A.J. Stevenson, Geologic Framework of the Aleutian Arc, in *The Geology of Alaska: The Geology of North America, Vol. G-1,* edited by G. Plafker, and D.L. Jones, pp. 367-388, Geological Society of America, Boulder, 1994.

van Keken, P.E., B. Kiefer, and S.M. Peacock, High resolution models of subduction zones: Implications for mineral dehydration reactions and the transport of water into the deep mantle, *G-cubed, 3,* 10, 20 pp, 2002.

Van Orman, J.A., T.L. Grove, and N. Shimizu, Rare earth element diffusion in diopside: Influence of temperature, pressure and ionic radius, and an elastic model for diffusion in silicates, *Contrib. Mineral. Petrol., 141*, 6, 687-703, 2001.

von Drach, V., B.D. Marsh, and G.J. Wasserburg, Nd and Sr isotopes in the Aleutians: multicomponent parenthood of island arc magmas, *Contrib. Mineral. Petrol., 92,* 13-34, 1986.

von Huene, R., *Seismic Images of Modern Convergent Margin Tectonic Structure, American Association of Petroleum Geologists Studies in Geology No. 26,* 60 pp., Amer. Assoc. Petrol. Geol., 1986.

von Huene, R., and D.W. Scholl, Observations at convergent margins concerning sediment subduction, subduction erosion, and the growth of continental crust, *Reviews of Geophysics, 29,* 279-316, 1991.

Walter, M.J., Melting of garnet peridotite and the origin of komatiite and depleted lithosphere, *J. Petrol., 39,* 29-60, 1998.

Wood, B.J., J.D. Blundy, and J.A.C. Robinson, The role of clinopyroxene in generating U-series disequilibrium during mantle melting, *Geochim. Cosmochim. Acta, 63,* 1613-1620, 1999.

Wood, D.A., N.G. Marsh, J. Tarney, J.-L. Joron, P. Fryer, and M. Treuil, Geochemistry of igneous rocks recovered from a transect across the Mariana Trough, arc, forearc and trench, sites 453 through 461, Deep Sea Drilling Project Leg 60, *Initial Reports of the Deep Sea Drilling Project, Leg 60,* 611-645, 1981.

Woodhead, J.D., Geochemistry of the Mariana arc (western Pacific): Source composition and processes, *Chemical Geology, 76,* 1-24, 1989.

Yaxley, G.M., and D.H. Green, Reactions between eclogite and peridotite: Mantle refertilisation by subduction of oceanic crust, *Schweiz. Min. Petrog. Mitt., 78,* 243-255, 1997.

Yogodzinski, G.M., R.W. Kay, O.N. Volynets, A.V. Koloskov, and S.M. Kay, Magnesian andesite in the western Aleutian Komandorsky region: Implications for slab melting and processes in the mantle wedge, *Geol. Soc. Amer. Bull., 107* (5), 505-519, 1995.

Yogodzinski, G.M., and P.B. Kelemen, Slab melting in the Aleutians: Implications of an ion probe study of clinopyroxene in primitive adakite and basalt, *Earth Planet. Sci. Lett., 158,* 53-65, 1998.

Yogodzinski, G.M., and P.B. Kelemen, Geochemical diversity in primitive Aleutian magmas: Evidence from an ion probe study of clinopyroxene in mafic and ultramafic xenoliths, *EOS, 81, Fall Meeting Supplement,* 2000.

Yogodzinski, G.M., J.M. Lees, T.G. Churikova, F. Dorendorf, G. Woeerner, and O.N. Volynets, Geochemical evidence for the melting of subducting oceanic lithosphere at plate edges, *Nature, 409,* 500-504, 2001.

Yogodzinski, G.M., J.L. Rubenstone, S.M. Kay, and R.W. Kay, Magmatic and tectonic development of the western Aleutians: An oceanic arc in a strike-slip setting, *J. Geophys. Res., 98,* 11,807-11,834, 1993.

Yogodzinski, G.M., O.N. Volynets, A.V. Koloskov, N.I. Seliverstov, and V.V. Matvenkov, Magnesian andesites and the subduction component in a strongly calc-alkaline series at Piip Volcano, Far Western Aleutians, *J. Petrol., 35* (1), 163-204, 1994.

Peter B Kelemen, Dept. of Geology and Geophysics, Woods Hole Oceanographic Institution, Woods Hole, MA 02543. (peterk@whoi.edu)

Gene M. Yogodzinski, Dept. of Geological Sciences, University of South Carolina, Columbia, SC 29208. (gyogodzin@geol.sc.edu)

David W. Scholl, Dept. of Geophysics, Stanford University, Stanford, CA 94035.

Some Constraints on Arc Magma Genesis

Yoshiyuki Tatsumi

Institute for Frontier Research on Earth Evolution (IFREE)
Japan Marine Science and Technology Center (JAMSTEC)
Yokosuka, Japan

Improved understanding of the complex processes associated with subduction zone magmatism requires identifying and explaining tectonic, petrological, geochemical, and geophysical characteristics common to most subduction zones. These characteristics include: (1) the presence of dual volcanic chains within a single volcanic arc, (2) the constant depths to the surface of the subducting lithosphere beneath trench- and backarc-side volcanic chains, ~110 km and ~170 km, respectively, (3) greater volume of magma production beneath the trench-side volcanic chain (4) selective enrichment of particular incompatible elements, (5) systematic across-arc variations in incompatible element concentrations and isotopic/element ratios, (6) location of a high-temperature (>1350°C) region within the mantle wedge, (7) occurrence of localized high-temperatures (~1300°C) immediately beneath the crust/mantle boundary, (8) location of very low-velocity regions within the mantle wedge. It is suggested here that dehydration processes and associated selective element transport, which take place in both the subducting lithosphere and the downdragged hydrated peridotite layer at the base of the mantle wedge, and secondary convection within the mantle wedge induced by plate subduction are largely responsible for the observed characteristics of subduction zone magmatism.

INTRODUCTION

Subduction zones are regions of the Earth where cold and dense lithospheric plates sink into the mantle. Although the subducting lithosphere is expected to cool the surrounding mantle [e.g., *Davies and Stevenson,* 1992], subduction zones are sites of extensive magmatism and high heat flow (>2-3 HFU; *Watanabe et al.,* 1977; *Lewis et al.,* 1988; *Henry and Pollack,* 1988). This apparent contradiction is one of the greatest dilemmas facing those who are interested in theories of subduction processes. It has been generally accepted that H_2O, released *via.* dehydration of a subducting oceanic plate, plays a key role in this regard by lowering the solidus temperature and inducing partial melting of the mantle wedge. However, the interaction of physical and chemical processes occurring in subduction zones must be much more complex because temperatures of arc magmas are not systematically lower than those of magmas in other tectonic settings [e.g., *Tatsumi et al.,* 1983; *Takahashi and Kushiro,* 1983; *Falloon and Green,* 1988; *Draper and Johnson,* 1992]

Volcanoes in subduction zones are not distributed evenly within a volcanic arc but tend to be concentrated in certain areas or volcanic zones. For example, the trenchward limit of volcanoes tends to form a well-defined line, called the volcanic front [*Sugimura, 1960*]. Further, not all but some subduction zones, such as NE Japan, Kurile and Sangihe arcs, are characterized by the occurrence of dual volcanic chains within a single arc [*Marsh,* 1979; *Tatsumi and Eggins,* 1995]. Such spatial patterns of volcano distribution must be a surface manifestation of physical and chemical processes occurring in the region of magma generation within the mantle wedge.

The distinct chemical characteristics of subduction zone magmas compared with other tectonic settings have been well established [e.g., *Sun,* 1980; *Pearce,* 1983]. Moreover, none of the potential sources present in subduction zones such as subducting sediments/oceanic crusts and original mantle wedge match these characteristics closely as the

Inside the Subduction Factory
Geophysical Monograph 138

10.1029/138GM12

source of their origin. Therefore it has been generally concluded that these signatures are attributed to the selective metasomatic addition of particular elements from the subducting oceanic lithosphere to the mantle-wedge source of arc magmas. The mechanisms responsible for such metasomatic reactions have, however, not been fully understood.

Faced with the above mentioned complexity in arc magma generation, and further to understand comprehensively the physical and chemical consequences of the entire subduction processes, one of the practical approaches that is required at the present stage may be to recognize the petrological, geochemical, and geophysical constraints on generation of subduction zone magmas. Needless to say, our understanding will develop through the cycle of model formulation, prediction, followed by comparison of prediction and observation, and model revision. The formulation of a model should not be our final goal, but rather the beginning of the process of further understanding of complex subduction zone processes. Although several authors suggested constraints on arc magmatism and proposed a possible model for magma generation in subduction zones [e.g., *Gill,* 1981*; Wyllie,* 1988; *Tatsumi and Eggins,* 1995], key observations and theories that were obtained recently and may provide fundamental insights into the processes operating in subduction zones will be reexamined in this review.

SPATIAL DISTRIBUTION OF ARC VOLCANOES

Depth to the Subducting Lithosphere

Marsh [1979] emphasized that the width of a volcanic arc is inversely proportional to the angle of plate subduction, with a wider volcanic arc above a shallower-dipping subducting lithosphere. This is clearly documented in the Izu-Bonin-Mariana arc, where the volcanic arc width decreases southwards as the subduction angle varies from ~30° beneath the northern tip of this arc to nearly vertical in the southern part (Figure 2.20 of *Tatsumi and Eggins*, 1995) and is observed widely in modern arc-trench systems, with a concurrent decrease in the width of the forearc (Figure 1). One of the clear exceptions from this rule is the Tonga-Kermadec arc, which constitutes only a single volcanic chain and has near zero width above a rather shallow (~50°) dipping seismic zone. It may be possible to speculate, however, that the close spatial and temporal occurrence of backarc spreading behind this arc may be the cause of this exceptional distribution of volcanoes. The backarc opening is necessarily associated with input of asthenospheric materials into the mantle wedge, which may disarrange the arc magma source, the fluid flow, and the thermal structure, and hence may disturb the production of arc magmas in the mantle wedge. The volcanic hiatus during the backarc spreading that may be related to the above disarrangement has been documented for other arc-backarc systems such as the NE Japan arc—Japan Sea system [*Tatsumi et al.,* 1989; 2000].

The negative correlation between the volcanic arc width and the subduction angle provides a compelling reason to assume that pressure controls the magma production in subduction zones. This pressure dependency can be clearly demonstrated by the constant depths to the surface of the subducting plate beneath a volcanic arc (*Gill,* 1981; Figure 2.22 of *Tatsumi and Eggins*, 1995). The volcanic fronts are 108±18 km (1σ) above the slab surface and the volcanic/backarc boundary form 173±12 km above the top of the subducting plate [*Tatsumi and Eggins*, 1995]. These values are obtained by compilation of seismic data beneath the central part of a volcanic arc. It should be thus stressed that the "law of constant depth" is not necessarily applicable for the edges of volcanic arcs. An example of such an exceptional area is central Japan (Figure 2). Seismic data [*Ishida*, 1995] demonstrate that the Philippine Sea plate is lying within the mantle wedge formed by subduction of the Pacific plate beneath the region. Location of the volcanic front at exceptionally large height above the top of Pacific plate in Central Japan may be attributed to rather low temperatures within the mantle wedge that is cooled by the Philippine Sea plate. Furthermore, the subducting Pacific plate is located at depths of ~130 km immediately beneath the volcanic front at the junction of two arcs, NE Japan and Kurile (Figure 2).

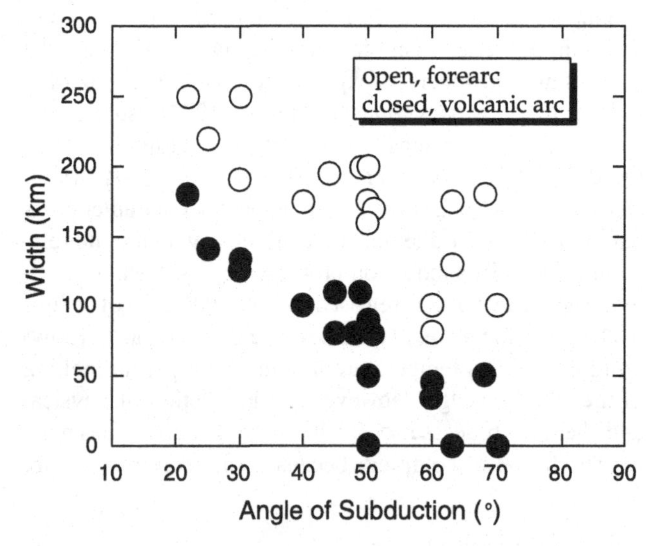

Figure 1. Negative correlation between the angle of subduction and the widths of the volcanic arc and forearc (after *Tatsumi and Eggins*, 1995), suggesting that pressure-dependent reactions control the magma production in subduction zones.

Exceptions are also found in arcs with a high angle of subduction (e.g., Vanuatu and Solomon arcs) or with a slow rate of subduction (Eolian arc). Thus, although there certainly exist arc-trench systems that do not follow the 'law of constant depth' they appear to be special cases and we still have to explain the 'magic numbers' of 110 and 170 km.

It should be stressed that the depth to the slab surface beneath the volcanic arc must be examined based on seismic data obtained by dense seismic observations. Even for NE Japan arc, one of the best seismologically surveyed arcs on the Earth, the distance between the surface of the subducting Pacific plate and volcanoes has been revised from

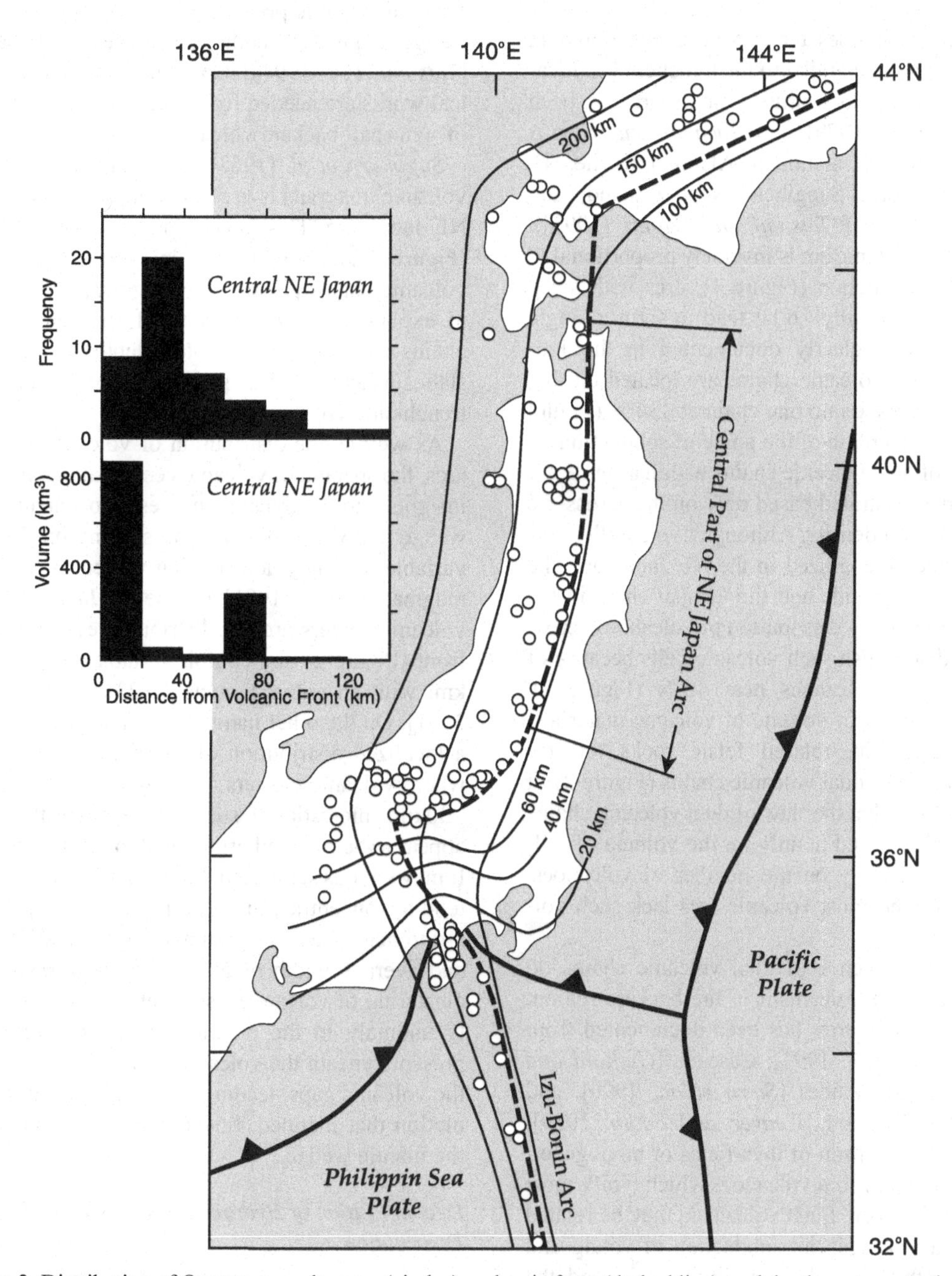

Figure 2. Distribution of Quaternary volcanoes (circles), volcanic front (dashed line), and depth contours to the surface of the dipping seismic zones within Pacific and Philippine Sea plate in NE Japan and Izu-Bonin arcs. Volume of volcanic materials, not the number of volcanoes, in central part of the NE Japan arc show the location of dual volcanic chains within a single arc.

80 km in the early 1980's to 105 km at present (Figure 2). The reason for this revision is simple; the number of seismic observatories on the NE Japan arc has increased to more than 100 during the last decade.

Dual Volcanic Chains

Subduction zone volcanoes tend to be concentrated in volcanic chains, which lie parallel to the trenches. The location of two volcanic chains within a single arc has been emphasized by *Marsh* [1979] and *Tatsumi and Eggins* [1995], examples of which include the Aleutian, Kurile, NE Japan, Bataan, Mindoro, Sangihem New Zealand, and Scotia arcs (Figure 2.7 of *Tatsumi and Eggins* [1995]). Since the width of a volcanic arc is inversely proportional to the angle of plate subduction (Figure 1), arcs with steep angles of subduction (usually >60°) tend to form a single volcanic chain. This is clearly documented in the Izu-Bonin-marian arc; two volcanic chains are located on that arc at >34°N, and coalesce into one chain at <34°N (Figure 2), with a southward increase of the angle of subduction.

The occurrence of dual volcanic chains within a single arc may not be well demonstrated based only on the across-arc variation in the volcano density. Although two parallel volcanic belts have been recognized in the NE Japan arc, the Nasu chain in the trench-side and the Chokai chain in the backarc-side, the frequency distribution of volcanoes across this arc does not distinguish such volcanic belts because of a clustering of small volcanoes near 40°N (Figure 2). However, the distribution of volume of volcanic materials, excluding caldera-forming-related felsic rocks, clearly reveals the occurrence of dual volcanic chains (Figure 2). It should be thus stressed that the 'law of dual volcanic chains within a single arc' is based mainly on the volume of volcanic materials, not simply on the number of volcanoes. Unfortunately, however, most volcanic arcs lack such volume data.

Exceptions to the presence of dual volcanic chains do exist. The occurrence of volcanism in the backarc region, probably not on volcanic arcs, has been documented from Kamchatka [*Fedotov, et al,* 1991], Cascade [*Gufftani and Weaver*, 1988], Chilean Andes [*Stern et al.,* 1990], and North Island, New Zealand [*Weaver and Smith*, 1989]. These examples take the form of lava fields or monogenetic cones rather than the central volcanoes which typify most subdcution zone volcanism. Such volcanism may be related to unusual tectonic settings: the subduction of young and hot lithosphere in the Cascade and Chile cases, and the occurrence of a transform-convergent plate boundary in the Kamkchatka case. For example, thermal modeling indicates that the temperature at the side-edge of the subducting Pacific plate that contacts with a rather high temperature mantle could be ~200°C higher than that of the layer of some tens of kilometers inside from the edge beneath the Kamchatka arc [*Tatsumi et al.,* 1994a]. It may be thus speculated that atypically high temperatures are responsible for partial melting of the subducting plate and/or the down-dragged hydrous peridotite layer at the base of the mantle wedge at greater depths, i.e., beneath the backarc region [*Tatsumi et al.,* 1994a]. Combined geochemical and theoretical works are needed for better understanding of the origin of 'unusual' backarc volcanism.

Sugimura et al. [1963] demonstrated that the volume of volcanic materials is greatest along the volcanic front in the NE Japan arc. This is certainly true for the NE Japan arc (Figure 2) and may also be the case in other arcs with dual volcanic chains [*Tatsumi and Eggins*, 1995]. We thus have to explain simultaneously the formation of two volcanic chains at different and constant depths above the subducting slab surface and the greater volume erupted along the trench-side volcanic chain.

As well as the distribution of volcanoes across volcanic arcs, the spacing of volcanic centers along arcs may provide insights into the dynamic processes occurring in the mantle wedge. However, the volcano spacing on the arc is highly variable and may add nothing to genetic questions on arc magmatism [*Gill,* 1981]. On the NE Japan arc, for example, volcanic centers are not distributed evenly on the volcanic front (Figure 2) and exhibit variable spacing from 5 to 75 km, with an average value of 23±16 km [*Tamura et al.,* 2001]. On the other hand, *Tamura et al.* [2001] emphasized a localized distribution of volcanoes in the NE Japan arc where volcanic clusters are separated by along-arc gaps of >30km. Interestingly, such spaces along the volcanic front appear to be oriented orthogonal to the arc-strike (Figure 2). It may be thus suggested that volcanoes on the NE Japan arc tend to concentrate in volcanic regions perpendicular to the arc-trench system with an average separation of 79±16 km and average width of 50 km [*Tamura et al.,* 2001]. Such clustering of volcanoes is closely related to the low-velocity anomaly in the mantle wedge; low-velocity zone are present beneath the volcanic regions but are absent beneath the volcanic gaps, leading *Tamura et al.* [2001] to the speculation that inclined, finger-like hot regions are present in the mantle wedge.

Decomposition of Hydrous Phases and Origin of Volcano Distribution

It has been widely accepted that the subducting oceanic lithosphere plays a key role in the generation of arc magmas. Since first emphasized by *Nicholls and Ringwood*

[1973], many earth scientists have favored a mechanism involving melting of the subducting slab. An alternative mechanism is the partial melting of mantle wedge peridotites by fluxing of aqueous fluids formed by dehydration reactions within the subducting slab [e.g., *Coats*, 1962]. In subduction zones where the subduction of a young and hot plate takes place, slab melting may be the case [*Drummond and Defant*, 1990; *Peacock*, 1990; *Furukawa and Tatsumi*, 1999]. Further, unusual arc magmas such as high-Mg andesites may be produced by partial melting of subducting sediments and/or oceanic crusts and subsequent melt-mantle interaction [*Kay*, 1978; *Yogodzinski et al.*, 1994; *Kelemen*, 1995; *Shimoda et al.*, 1998; *Tatsumi*, 2001]. A significant majority of researchers, however, believe that the subducting lithosphere does not melt under normal circumstances. If so, the above mentioned characteristic volcano distribution in subduction zones is implied to be controlled by stabilities of hydrous phases in both the subducting lithosphere and the mantle wedge.

Figure 3 demonstrates stability limits of hydrous phases in both hydrous basalt and peridotite compositions based on current experimental and theoretical results [*Pawley and Holloway*, 1993; *Tatsumi and Eggins*, 1995; *Poli and Schmidt*, 1995; *Schmidt and Poli*, 1998] together with a possible geothermal gradient at the slab-mantle boundary [*Furukawa*, 1993] and H_2O contents bound in hydrous phases [*Tatsumi and Kogiso*, 1997; *Schmidt and Poli*, 1998]. The major dehydration reactions in subducting basaltic oceanic crusts, which liberates >1wt% H_2O, is the decomposition of amphibole and zoisite at a pressure of ~2.5 GPa or at a depth of ~80 km. Phases carrying H_2O to deeper levels are lawsonite and phengite, which may be stable to depths greater than 200 km. These observations suggest that the subducting oceanic crust experiences *not complete but major* dehydration processes beneath the forearc region, i.e. at depths shallower than those beneath the volcanic arc. Nevertheless, it is clear that continuous dehydration reactions of lawsonite take place in the downgoing oceanic crusts if it were cool enough [*Poli and Schmidt*, 1995; *Schmidt and Poli*, 1998]. The role of lawsonite breakdown will be further examined in the following discussion.

Interestingly, even in normal subduction zones, dehydration melting of the oceanic crust may take place *via.* decomposition of amphibole and/or zoisite, because the temperature of the slab-mantle interface may be higher than the wet solidus of basalt (Figure 3). Rising hydrous slab-melts, which are not saturayted with an olivine component and possess silicic compositions, will dissolve peridotite and crystallize pyroxene, amphibole, and phlogopite in the mantle wedge. This process was investigated experimentally by *Sekine and Wyllie* [1982] and *Carroll and Wyllie* [1989] and theoretically by Kelemen [1995] and Tatsumi [2001]. Provided that the slab melts should not be consumed through melt-mantle interaction immediately above the downgoing lithosphere, they will be heated under inverted

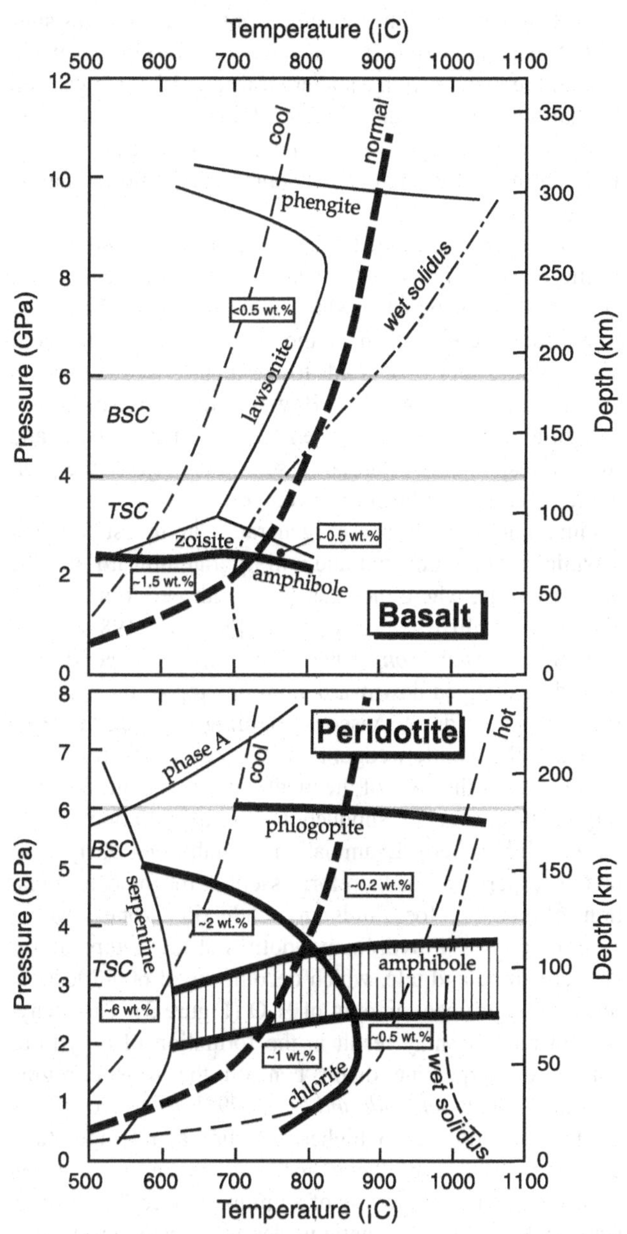

Figure 3. Stabilities of hydrous phases in basalt and peridotite systems [*Tatsumi and Eggins*, 1995; *Schimidt and Poli*, 1998] and possible temperature distribution along the plate/mantle wedge boundary (modification after *Furukawa*, 1993). Numbers in Boxes show maximum H_2O contents bound in hydrous phases [*Tatsumi and Kogiso*, 1997; *Schmidt and Poli, 1998*]. BSC and TSC indicate the depths to the top of the subducting lithosphere beneath backarc-side and trench-side volcanic chains, respectively.

geothermal gradients, thereby increasing in mass. It is thus possible that such slab-melting-induced magmatism would take place even in 'normal' subduction zones. However, this seems to be unlikely, because the depth of slab melting (~80km) is much shallower than that of the slab surface immediately beneath the volcanic front. It may be thus suggested that the slab-melts are trapped in peridotites at the base of the mantle wedge and are transported downward on the slab surface, In conclusion, breakdown of hydrous phases in the down dragged peridotite layer plays a key role in addition of H_2O to the magma source region in the mantle wedge.

If we accept the extensive hydration of a peridotite layer of the oceanic lithosphere, then serpentine present in the inner part of a subducting slab, where temperatures can be considerably cooler than at the slab surface, could contribute to the supply of fluids to the mantle wedge beneath the volcanic arc (Figure 3). However rather fresh abyssal peridotites have been recovered from the ridge walls, and such widespread distribution of serpentine within the peridotite layer has not been observed yet.

Fluid and/or melt phases generated in the subducting oceanic crust are buoyant and will migrate upward into the mantle wedge, where they can react with peridotite minerals and crystallize hydrous phases [*Tatsumi*, 1986, 1989; *Davies and Stevenson*, 1992]. Such hydrous peridotites should be dragged downward along the top of the subducting lithosphere due to viscous coupling between the rigid plate and the mantle wedge. Formation of the hydrous peridotite layer at the base of the mantle wedge is thus a necessary consequence of subduction and dehydration of the oceanic lithosphere. Examination of stabilities of hydrous phases in peridotite is therefore essential for the determination of sources for fluids in subduction zones. Major hydrous phases in hydrous peridotites at low temperatures are serpentine and talc, which decompose at 600-700°C to release a significant amount of H_2O (Figure 3). This dehydration reaction may result in the formation of a hydrous curtain or serpentine diapir beneath the forearc region [*Tatsumi*, 1989; *Schimidt and Poli*, 1998]. As hydrous peridotites are dragged to higher pressure and temperature, amphibole is one of the major hydrous phases. *Tatsumi* [1986] emphasized the role of amphibole decomposition in supplying H_2O to the mantle wedge and for the production of trench-side arc magmas, because the maximum depth of amphibole stability in peridotites is close to the depth to the surface of the subducting lithosphere immediately beneath the volcanic front (~ 110 km). On the other hand, *Schmidt and Poli* [1998] concluded that amphibole decomposes at shallower levels than 110 km, based on the results of Schreinemaker analysis and compilation of experimental data. Although the stability maximum of amphibole depends on peridotite compositions, it is safe to assume, at the present stage, that amphibole breakdown and subsequent fluid addition to the overlying mantle wedge may not directly trigger partial melting of mantle wedge peridotites (Figure 3). Alternatively, if the lateral migration of amphibole-derived fluid is the case, the above inconsistency could be overcome [*Davies and Bickle*, 1991; *Davies and Stevenson*, 1992]. The fluids may traverse laterally by a combination of vertical movement as a fluid, and downward movement fixed in secondarily-crystallized amphibiole carried down by induced mantle flow. However, such a large scale of downward flow would not be the case, because the downdragged flow may be limited to a narrow zone at the base of the mantle wedge with low viscosity.

Instead of amphibole, chlorite may play an important role as the source of subduction fluids. The reasons for this are twofold. Firstly, chlorite-bearing peridotites can contain much greater amounts of H_2O (~ 2 wt.%, *Tatsumi and Kogiso*, 1997; *Schmidt and Poli*, 1998) than amphibole peridotite (Figure 3). Consequently, fluid released by amphibole breakdown may be stored in downdragged peridotites by crystallizing chlorite:

Amphibole → clinopyroxene + garnet + H_2O

Garnet + olivine + H_2O

→ chlorite + orthopyroxene

Secondly, the stability limit of chlorite in a peridotite system possesses a rather gentle slope [*Goto and Tatsumi*, 1990], indicating that chlorite in downdragged peridotites decomposes *via.* a rather pressure-dependent reaction at depths of 100-130 km. Such figures are consistent with the distance to the top of the subducting lithosphere beneath the trench-side volcanic chain (Figure 3). In subduction zones where a very cool slab is located, however, chlorite decomposition takes place at deeper levels, resulting in the formation of the volcanic front above the deeper slab than the normal subduction zones (Figure 3). It should be stressed that, although rather pressure-dependent reaction governs the stability of chlorite, a gentle slope of such decomposition may be enough to cause variations in depths to the slab surface beneath the volcanic front due to variations in temperatures at the slab/mantle wedge interface.

The hydrous minerals in downdragged hydrous peridotites may include the presence of K-bearing phases such as phlogopite, because addition of such elements may occur *via.* aqueous fluids migrating from the subducting oceanic lithosphere [*Tatsumi et al.*, 1986; *Brenan et al.*, 1995; *Ayers et al.*, 1997; *Kogiso et al.*, 1997; *Tatsumi and Kogiso*, 1997; *Aizawa et al.*, 1999; *Johnson and Plank*, 1999]. For example, phlogopite may form through decomposition of K-bearing pargasitic amphibole in the hydrous peridotite layer:

K-bearing pargasite + orthopyroxene

→ phlogpite + clinopyroxene + olivine + garnet + H_2O

This reaction may also contribute to crystallization of chlorite in that layer through reactions between olivine, garnet, and H_2O. It should be mentioned that the decomposition of phlogopite in a peridotite system takes place at ~ 6 GPa [*Sudo and Tatsumi*, 1990], a pressure which in most volcanic arcs corresponds approximately with the depth to the surface of the subducting lithosphere beneath the backarc-side volcanic chain (Figure 3). K-amphibole with compositions close to $K(K,Na)CaMg_5Si_8O_{22}(OH)_2$ can form as high-P breakdown products of phlogopite in peridotite bulk compositions [*Trønnes et al.*, 1988; *Sudo and Tatsumi,* 1990; *Luth,* 1997; *Inoue et al.*, 1998; *Konzett and Fei,* 2000]. The hydrous potassium phases such as K-amphibole and and phase X are stable at pressures up to 113-15 and 20-23 GPa, respectively [*Konzett and Fei,* 2000] and may transport H_2O to such great depths.

As mentioned earlier, under unusually low temperature conditions in the subducting lithosphere, H_2O may be released continuously from the slab by lawsonite breakdown. Such a small amount of H_2O could be trapped in hydrous phases (e.g., amphible, chlorite, phlogopite, and Phase A) crystallized in the downdragged peridotite layer at the base of the mantle wedge. Alternatively, slab-derived H_2O liberated through continuous dehydration reactions could exude from the hydrous peridotite layer and migrate further upward into the mantle wedge. If this is the case, then such fluid fluxing, as well as that from the hydrous peridotite layer, may cause partial melting of the mantle wedge. Further experimental works are needed for understanding the fluid migration in the slab-mantle interface region.

One of the characteristic features of arc volcano distribution is the greater density of volcanoes and volume of volcanism at the trench-side chain as demonstrated in the previous section. It is plausible that such observations may be attributed to a greater rate of melt production beneath the trench-side chain. This is related to the difference in a fluid flux into the magma source region [*Tatsumi and Eggins,* 1995; *Tatsumi and Kogiso*, 1997]; a greater amount of H_2O from the breakdown of chlorite (± amphibole) beneath the trench-side chain compared with the decomposition of phlogopite beneath the backarc chain.

LAVA CHEMISTRY

It is widely accepted that subduction zone lavas are noted for their distinct chemistry compared with those in other tectonic settings. Figure 4 demonstrates that subduction zone magmas are relatively enriched in incompatible elements with larger ionic radii compared to smaller elements of the same charge. This systematic geochemical variation occurring within arc basalts is broadly consistent with transport of elements by aqueous fluids from both subducting sediments/oceanic crusts and downdragged hydrous peridotites [*Tatsumi et al.*, 1986; *Brenan et al.*, 1995; *Ayers et al.,* 1997; *Kogiso et al.*, 1997; *Tatsumi and Kogiso,* 1997; *Aizawa et al.*, 1999; *Johnson and Plank,* 1999]. The origin of such overall geochemical signatures in arc magmas is examined by Tim Elliott in this monograph. Therefore I shall here focus on the geochemical variation within a single volcanic arc, especially on the across-arc variation in lava chemistry. Before doing so, however, I have to emphasize that further experimental investigation for the trace element compositions of fluids or melts derived from both subducting lithosphere and downdragged hydrous peridotites is needed, especially for those formed by continuous dehydration reactions.

Across-Arc Variation in Incompatible Element Concentrations

A remarkable characteristic of arcs with dual volcanic chains is the systematic variation in concentrations of incompatible elements that occurs in the across-arc direction, known as the *K-h* relationship [*Dickinson,* 1975]. A compilation of K_{51}-values (K_2O wt% at 51 wt% SiO_2 calculated from linear regression of SiO_2 and K_2O; *Tatsumi and Eggins*, 1995) confirms this relationship (Figure 5a).

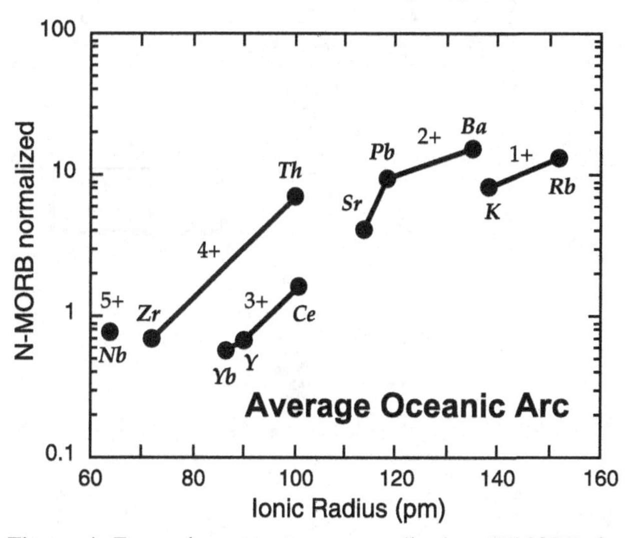

Figure 4. Trace element patterns normalized to N-MORB for average oceanic arc basalts [*Tatsumi and Eggins*, 1995].

Although at least two exceptions exist, the Vanuatu arc [*Barsdell et al.,* 1982] and the Miocene NE Japan arc [*Togashi*, 1978; *Shuto*, 1988], such clear across-arc variations in incompatible element concentrations should be explained. I also mention that an across-arc variation is also associated with across-arc change in bulk compositions of magmas, decreasing normative quartz towards the backarc side (Figure 5b).

Variation in concentrations of incompatible elements is readily explained by varying degrees of partial melting of a single source material, with decreasing partial melting towards the backarc-side of a volcanic arc being the cause of the across-arc variation trends [e.g., *Sakuyama and Nesbitt*, 1986]. Decreasing degree of partial melting at constant pressure could also cause a decrease of normative quartz in the partial melt [e.g., *Falloon and Green*, 1988; *Hirose and Kushiro*, 1993]. If this were the case, however, the temperature of magma would need to decrease systematically away from the volcanic front, which is not supported by experimental results [*Tatsumi et al.*, 1983]. On the other hand, results of melting experiments on Mg-rich basalts from the NE Japan arc [*Tatsumi et al.*, 1983] suggest that the observed across-arc variation in incompatible element concentrations and normative compositions accompanies deeper magma separation and smaller degrees of partial melting toward the backarc. Such deeper separation of magmas from the mantle may be caused by the distribution of thicker lithosphere beneath the backarc side of a volcanic arc [*Tatsumi et al.*, 1983], which is clearly demonstrated by both seismic tomography data [e.g., *Zhao et al.*, 1997] and numerical simulation of the mantle wedge [e.g., *Furukawa*, 1993] as discussed later.

H_2O is not fractionated from silicate melts by anhydrous phases such as olivine, pyroxenes, and plagioclase, which crystallize in the early stage of basaltic magma differentiation at shallow depths, and may behave as an incompatible

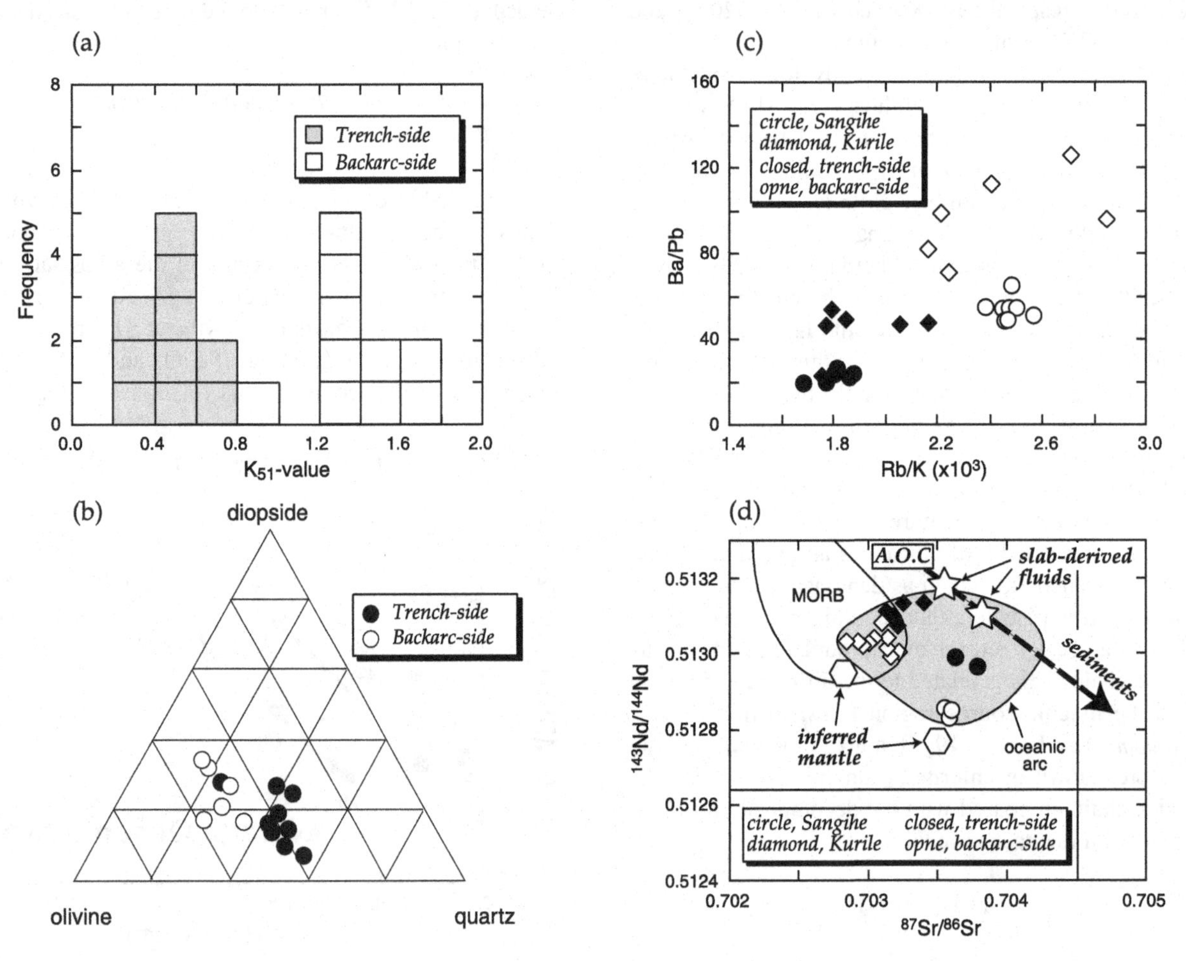

Figure 5. Across-arc variations in K_{51}-values, Nd-Sr isotopic compositions, and Ba/Pb and Rb/K ratios [*Tatsumi and Eggins*, 1995]. A.O.C., altered oceanic crust.

'element' such as potassium. *Sakuyama* [1979] examined across-arc variation in phenocryst assemblages such as hornblende, biotite, and quartz, in order to assess the lateral variation of H_2O contents in arc magmas, and suggested that H_2O contents are greater in the backarc side magmas. It should be stressed that such across-arc variation of H_2O in magmas does not necessarily mean across-arc variation of H_2O added to the magma source region, because the degree of melting is smaller in backarc-side than trench-side magmas. As discussed above, a greater fluid supply to the magma source region from the downdragged hydrous peridotite layer is expected beneath the trench-side chain than beneath the backarc-side chain.

The across-arc variation of lava chemistry is emphasized here in the context of the formation of dual volcanic chain. However, rather gradual and continuous changes in lava compositions have been clearly documented in some arcs such as the Izu-Bonin [*Ishikawa and Nakamura*, 1994] and Kurile-Kamchatka arc [e.g., *Ryan et al.*, 1995]. For example, enrichment factors for boron and cesium decrease across the arc in Kurile arc with increasing the distance between the slab surface and the volcano, leading *Ryan et al.* [1995] to the conclusion that these observations may be related to gradual changes in the compositions of fluids. Such gradual across-arc variations should be further examined with respect to the phase-control of compositions of fluid phases during dehydration processes.

Across-Arc Variation in Element and Isotope Ratios

Ratios of incompatible elements and isotopes are observed to vary systematically in the across-arc direction within individual volcanic arcs. Because these ratios are unchanged during partial melting and differentiation processes, the observed across-arc variation requires changes in across-arc source chemistry. It is reasonable to assume that the original mantle wedge before subduction possesses homogeneous chemistry. Therefore, analyses of across-arc variation in those ratios should provide insights into the processes supplying fluids and elements to the original mantle wedge. Before doing so, we have to keep in mind that such analyses must be done for oceanic island arcs with no backarc basin. The reasons for emphasizing this are twofold. Firstly, variably metasomatized, hence chemically heterogeneous subcontinental upper mantle materials are distributed beneath continental arcs and may strongly affect the lava chemistry, especially in incompatible element ratios and isotopic compositions. For continental arcs, therefore, it may be difficult to detect to what degree differences in fluid chemistry contribute to the across-arc variation observed in arc lavas. Secondly, backarc basin formation should be associated with large-scale transportation of material within the upper mantle beneath both the backarc and volcanic arc regions. This is clearly documented by temporal variation of isotopic and trace element compositions in mafic lavas that were formed in the NE Japan arc and Sikhote Alin during the opening of the Japan Sea backarc basin [*Nohda et al.*, 1988, *Tatsumi et al.*, 2000].

It has been repeatedly demonstrated for arc lavas that the ratios such as Rb/K, Rb/Sr, La/Ba, Nb/Zr, Th/U, Ba/Sr, and U/Pb *usually* increase at equivalent SiO_2 contents away from the volcanic front [e.g., *Gill*, 1981]. In order to assess how certain phases in the magma source may control the magma chemistry, the partition coefficients between these phases and the coexisting melts or fluids need to be known. In the absence of well-constrained partition coefficients, however, the identification of residual phases can be examined qualitatively using ratios of *elements having different ionic radii but equivalent charge* based on the crystal-structure-control for element mineral-melt, and possibly mineral-fluid partitioning [*Matsui et al.,* 1977].

Examples of across-arc variations in Ba/Pb and Rb/K ratios are shown in Figure 5c for lavas from Sangihe [*Tatsumi et al.,* 1991] and Kurile [*Bailey et al.,* 1989] arcs, where the above mentioned simple tectonic situation probably is satisfied. The observed across-arc variations led *Tatsumi et al.* [1991] to the conclusion that aqueous fluids produced in the presence of phlogopite, which may correspond to dehydration reactions occurring in downdragged hydrous peridotites beneath the volcanic front, may possess Ba/Pb and Rb/K ratios different from those produced in the presence of K-amphibole that is crystallized by the breakdown of phlogopite beneath the backarc-side volcanic chain.

Although there is no general rule for across-arc variation of isotopic ratios in subduction zone magmas [e.g., *Tatsumi and Eggins,* 1995], the Sangihe and Kurile arcs, probably having ideal settings mentioned above, show across-arc variations in Sr-Nd isotopic ratios, lower Sr- and Nd-ratios toward the backarc side (Figure 5d). The observed Sr-Nd isotopic trend can be explained by the simple mixing of two components, the original mantle wedge and the added subduction components. The compositions of the slab-derived component are controlled by relative contribution of altered oceanic crust and subducting sediments. How can the across-arc variation in Sr-Nd isotopic ratios be explained? One of the possible explanations for this is higher concentrations both of Sr and Nd in adding fluids beneath the volcanic front compared to those beneath the backarc volcanic chain, because both elements would be more partitioned into K-amphibole and phlogopte, existing as residual hydrous phases beneath backarc- and trench-side chains, respectively [*Tatsumi et al.,* 1991; *Tatsumi and Eggins,* 1995].

THERMAL STRUCTURE

Do dynamic processes occurring in subduction zones contribute to the resolution of the paradox of arc magmatism—that is, the occurring of melting in a setting where cold oceanic lithosphere sinks into and cools the mantle? One of the most important constraints that should be taken into account for solving the above paradox is the temperature of subduction zone magmas. It may be intuitively accepted that a simple model, involving fluid supply from the downgoing slab to the mantle wedge and resultant melt generation by lowering the peridotite solidus under hydrous condition, is not likely to explain arc magmatism. The reason for this is based on the observation that arc basalts, especially Mg-rich primitive basalts possess major element compositions almost identical to those in other tectonic settings such as MORBs. A simple explanation for this may be that arc magmas are produced under P-T conditions similar to those for MORB magmas. This suggestion will be further reviewed below based on experimental data.

Temperature of Magmas

Application of the pyroxene geothermometry for Mg-rich, but still differentiated basalts in the Quaternary NE Japan arc provides minimum estimates on magma temperatures of ~1200°C [e.g., *Wada,* 1981]. In order to estimate temperatures of magma generation more quantitatively, melting experiments on primitive basalt compositions are required, together with evaluation of the effect of H_2O on the magma temperature. Since it is difficult to control the H_2O content in melt during direct partial melting experiments on peridotite compositions, the multiple-saturation experiments have been applied for that purpose. The P-T conditions of points of multiple-saturation with essential peridotitic phases under dry conditions are shown in Figure 6. A rather old experimental work by *Tatsumi et al.* [1983] further examined the effect of H_2O on the shift of such multiple-saturation points and found that backarc-side magmas with 1-3 wt.% H_2O are in equilibrium with mantle phases at depths greater (60-80 km) than those for trench-side magmas (~40 km). Also, the residual mantle phases are different in those magmas: lherzolite and harzburgite for backarc- and trench-side chains, respectively. It is thus suggested that the backarc-side magmas are produced at greater depths and by lower degrees of partial melting compared with the trench-side magmas, which is consistent with the explanation for the across-arc variation in incompatible element concentrations.

More importantly, it is suggested based on experimental results on multiple saturation of peridotite minerals that arc primary basalt magmas, which are separated from the mantle wedge, possess temperatures of ~1300°C [*Takahashi,* 1980; *Tatsumi et al.,* 1983; *Kushiro,* 1987; *Draper and Johnson,* 1992; *Tatsumi et al.,* 1994c] (Figure 6). Given adiabatic ascent of partially molten mantle materials, the temperature at deeper levels of melting must be higher than the shallower levels, due to the heat of fusion required for melt generation. If we adopt a heat of fusion of 200-400 $Jkg^{-1}K^{-1}$ [*McKenzie,* 1984; *Fukuyama,* 1985], temperatures around 1350°C or more are necessary somewhere in the mantle wedge to produce arc basalt magmas. Melting experiments by *Tatsumi et al.* [1983] assumed mantle olivine composition of Fo_{89} that equilibrates with arc primary magmas with MgO contents of 10-13 wt.%. In some cases, however, more magnesian primary magmas up to 16 wt.% MgO have been recognized in other subduction zones [e.g., *Nye and Reid,* 1986; *Ramsay et al.,* 1984; *Eggins,* 1993]. These magnesian primary basalt magmas may require hotter mantle wedge temperatures.

As suggested by melting experiments, primary volcanic front (trench-side) magmas are separated from the mantle

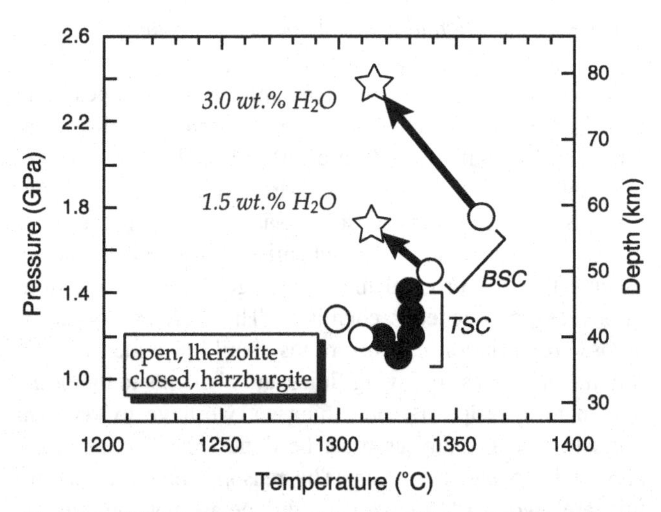

Figure 6. P-T conditions of points of multiple saturation of peridotitic minerals obtained by melting experiments of Mg-rich basalts (after *Tatsumi and Eggins,* 1995). Taking the H_2O contents in backarc side magmas into account, it is suggested that arc magmas are produced at temperatures of ~1300°C and at depths of 30~80 km (at greater depths beneath the backarc-side chain).BSC, backarc-side basalts; TSC, trench-side basalts.

wedge at ~1300°C and ~1.0 GPa (Figure 6). These temperatures at depths immediately beneath the crust/mantle boundary are ~100°C and ~250°C higher than the solidus temperatures of dry peridotite and basalt, respectively. If such high temperatures were present constantly at such depths, therefore, large amounts of partial melts or magmas should be expected to be present in the lowermost crust and uppermost mantle. This is inconsistent with geophysical observations that partially molten regions are not pervasively distributed at such depths beneath the entire volcanic arc [e.g., *Zhao et al.,* 1990]. Instead, high temperature regions may be localized and be attained locally beneath each volcano.

This is one of several lines of evidence for diapiric ascent of mantle materials from deeper and hotter parts of the mantle wedge [*Sakuyama*, 1983; *Tatsumi et al.*, 1983]. One of the counterarguments against diapirism in the mantle wedge is the cooling which a mantle diapir will undergo as it ascends into the overlying cooler mantle. Simple calculations based on Stokes' law and thermal diffusion indicate significant cooling of the diapir. However, this calculation does not take into account the reduced viscosity of the surrounding mantle due to heat loss from the diapir, which has been known to as hot Stokes [*Marsh*, 1978; *Marsh and Kantha*, 1978]. If this is the case, then a diapir rising from the high-temperature (e.g., 1300°C) core of the mantle wedge at 70 km may reach a depth of 30 km without appreciable cooling in period of 10^5 years [*Tatsumi and Eggins*, 1995].

The transport of magmas *via.* diapirism in the mantle wedge in ~ 10^5 years is inconsistent with the recent suggestion based on disequilibria between the short-lived U-series isotopes [e.g., *Williams and Gill*, 1989; *McKenzie*, 2000]. *Turner et al.* [2001] demonstrated that arc lavas possess ^{226}Ra excesses that extend to higher values than those observed in mid-oceanic ridge basalts, and proposed that such excesses are likely to have been introduced into the base of the mantle melting region by fluids released from the subducting lithosphere. If so, then preservation of this signal requires ultrafast melt transport to the surface in only a few hundreds of years. It may be thus suggested that mantle melt velocities are too fast for transport to occur by grain-scale percolation or mantle diapirisim. Instead, melt segregation and channel formation can occur rapidly in the mantle wedge.

Geophysical Constraints

Seismic tomography imaging of the mantle wedge is one of the most powerful tools to reveal the structure of the mantle wedge. NE Japan arc is a subduction zone where such tomographic studies have been done intensively with a dense network of seismic observations. Figure 7 shows one of the results [*Zhao et al.*, 1992], which documents clearly that the low- velocity zone located immediately beneath the volcanic front dips to the west in the mantle wedge and extends to a depth of >150 km. A similar low-velocity structure has also been documented in other subduction zones [e.g., *Zhao et al.*, 1997]. Such a low-velocity zone may be explained if partially molten or high-temperature materials are present in that region.

Subduction of a rigid oceanic plate into the viscous upper mantle will induce convection in the mantle wedge. Since such induced convection should influence the heat advection and the thermal structure in the mantle wedge, many models have been developed to investigate the flow pattern associated with induced convection. Most of such models failed to attain temperatures of >1350°C in the mantle wedge [e.g., *Peacock*, 1990; *Davies and Stevenson*, 1992; *Iwamori*, 1998], which is required for production of arc magmas as mentioned above. The highest temperature in the mantle wedge immediately below the volcanic front simulated in the thermal modeling by *Davies and Stevenson* [1992] is ~1000°C. Such temperatures are too low for producing basalt magmas by melting of peridotites as discussed before. In order to overcome this problem, *Honda* [1985] incorporated relatively high shear stress on the slab/wedge interface and placed a high-temperature region in the mantle wedge beneath the backarc. These rather unrealistic and assumptions were criticized by *Furukawa and Uyeda* [1989] and *Davies and Stevenson* [1992]. The problem in low-temperature models, on the other hand, may be that they do not take into account the temperature dependence of the viscosity of mantle wedge materials. Actually, *Furukawa* [1993] indicated that the high-temperature mantle is preferentially transported toward the shallower mantle wedge corner by subduction-induced convection, due to its low viscosity and ease of flow (Figure 7).

The location of the high-temperature region found by numerical simulation does, however, not match that of the low-velocity zone obtained by seismic tomography (Figure 7). One explanation of this apparent disagreement is that the low-velocity zone is not related to the location of the high-temperature zone in that region. This is further supported by the observation that the depth of magma segregation, estimated by multiple-saturation experiments on basalt magmas, or the depth of the base of the lithosphere coincides largely with that of the top of the very low-velocity region in the mantle wedge [*Tatsumi et al.,* 1983; *Zhao et al.,* 1992]. Such a very low-velocity region may thus be

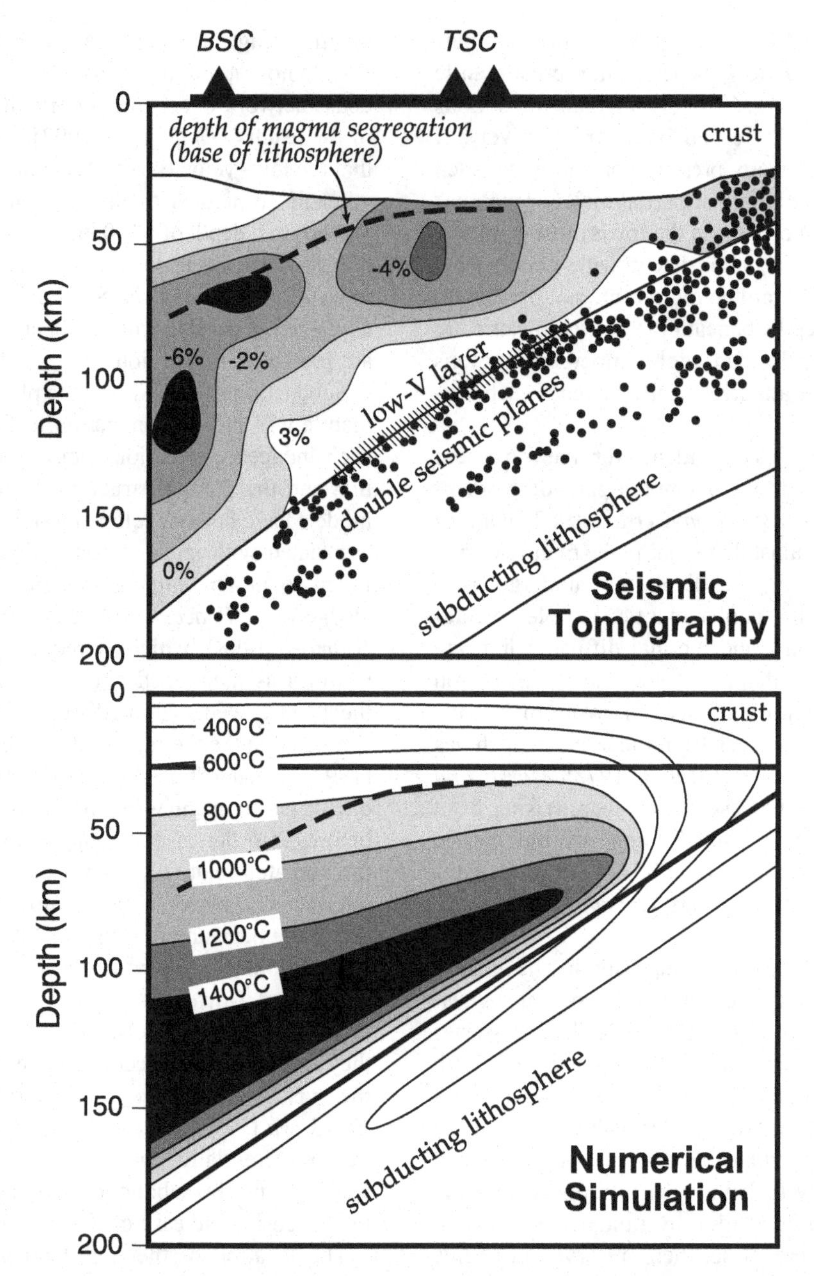

Figure 7. The mantle wedge structure inferred from seismic tomography data (upper; *Zhao et al.*, 1990) and numerical simulation (lower, *Furukawa,* 1993). Dashed lines indicate the depth of magma segregation from the mantle based on melting experiments on primary basalt magma compositions [*Tatsumi et al.*, 1983], which may correspond to the lithosphere/asthenosphere boundary. TSC, trench-side chain; BSC, backarc-side chain.

indicative of the location of partially molten mantle materials or the mantle diapir at that depth.

The efficient conversion between the resemblance of ScS and ScSp(converted from ScS to P at the boundary) phases led *Okada* [1979] to suggest the presence of a low-velocity zone in the vicinity of the mantle wedge/slab interface. Matsuzawa et al. [1986] further demonstrated that the low-velocity layer has a thickness less than 10 km near the top of the subducting lithosphere at depths between 60 and 150 km (Figure 7). *Tatsumi et al.* [1994b] and *Ito and Tatsumi* [1995] speculated that such a low-velocity layer may correspond with the downdragged hydrous peridotite layer formed at the base of the mantle wedge. Since the downgoing oceanic crust is not perfectly anhydrous at such depths (Figure 3), the low-velocity layer may be located within the subducting oceanic crust. Seismic velocity measurements on slab and mantle materials at high pressures and temperatures are needed to address this question.

CONCLUSIONS

The general characteristics of subduction zone magmatism, which are common to most convergent plate margins, may be explained largely by dehydration processes in both the subducting plate and the base of the mantle wedge and by the transport of deep-seated and hot mantle materials to the shallow mantle wedge due to induced convection. However, second-order, but still important observations require further explanation. Also, it is now clear that subduction zones need to be considered not only in their shallow upper mantle context but on the scale of the whole mantle, as residual slabs, which experienced complex chemical processes during subduction, have been stored somewhere in the deep mantle and may have evolved as possible geochemical mantle reservoirs. Lastly, the origin of continental crust with, on average, andesitic compositions may be one of the greatest unsolved problems in subduction zone magmatism. Further studies are required, which aim at a more comprehensive understanding of the role of the subduction factory in the evolution of the solid Earth.

Acknowledgments. Discussions with Yoshitsugu Furukawa and Yoshihiko Tamura are gratefully acknowledged. I am very thankful to Udo Fehn, John Eiler, Tim Grove and an anonymous reviewer for constructive comments and Miki Fukuda, Emi Takiguchi for preparation of a manuscript and figures. This work was partly supported by a Special Coordination Fund for Promoting Science and Technology (Superplume) from the Science and Technology Agency of Japan.

REFERENCES

Aizawa, Y., Y. Tatsumi, and Y. Yamada, Element transport during dehydration of subducting sediments: implication for arc and ocean island magmatism, *The Island Arc*, *8*, 38-46, 1999.

Ayers, J. C., S.K. Dittmer, and G.D. Layne, Partitioning of elements between peridotite and H_2O at 2.0-3.0 GPa and 900-1100 degrees C, and application to models of subduction zone processes, *Earth Planet. Sci. Lett., 150,* 381-398, 1997.

Bailey, J.C., T.I. Frolova, and L.A. Burikova, Mineralogy, geochemistry and petrogenesis of Kurile islamd-arc basalts, *Contrib. Mineral. Petrol., 102*, 265-280, 1989.

Barsdell, M., I.E.M. Smith, and K.B. Sporli, The origin of reversed geochemical zoning in the northern New Hebrides volcanic arc, *Coontrib Mineral. Petrol., 81*, 148-155, 1982.

Brenan, J.M., H.F. Shaw, and F.J. Ryerson, Experimental evidence for the origin of lead enrichment in convergent-margin magmas, *Nature*, *378*, 54-56, 1995.

Carroll, M.R., and P.J. Wyllie, Experimental phase relations in the system tonalite-peridotite-H_2O at 15 kb; implications for assimilation and differentiation processes near the crust-mantle boundary, *J. Petrol, 30*, 1351-1382, 1989.

Coats, R.R., Magma type and crustal structure in the Aleutian arc, in *Crust of the Pacific Basin*, pp. 92-109, Geophys. Monogr., Am. Geophys. Union, 1962.

Davies, J.H., and M.J. Bickle, A Physical model for the volume and composition of melt produced by hydrous fluxing above subduction zones, *Phil. Trans. R. Soc. London 335*, 355-364, 1991.

Davies, J.H., and D.J. Stevenson, Physical model of source region of subduction zone volcanics, *J. Geophys. Res.*, *97*, 2037-2070, 1992.

Dickinson, W.R., Potassium-depth (K-h) relations in continental margin and intraoceanic magmatic arcs, *Geology*, *3*, 53-56, 1975.

Draper, D.S., and A.D. Johnston, Anhydrous PT phase relations of an Aleutian high-Mgo basalt : an investigation of the role of olivine-liquid reaction in the generation of arc high-alumina basalts, *Contrib. Mineral. Petrol.*, *112*, 501-519, 1992.

Drummond, M.S., and M.J. Defant, A model for trondhjemite-tonalite-dacite genesis and crustal growth via slab melting: Archean to modern comparisons, *J. Geophys. Res.*, *95*, 21503-21521, 1990.

Eggins, S.M., Origin and differentiation of picritic arc magmas, Ambae (Aoba), Vanuatu, *Contrib. Mineral. Petrol.*, *1993*, 114, 1993.

Fallon, T.J., and D.H. Green, Anhydrous partial melting of peridotite from 8 to 35 kb and the petrogenesis of MORB, *J. petrol.*, 379-414, 1988.

Fukuyama, H., Heat of fusion of basaltic magma, *Earth Planet. Sci. Lett.*, *73*, 407-414, 1985.

Fedotov, S.A., Y.P. Masurenkov, and A.E. Svyatlovsky, On Quaternary and modern volcanism of Kamchatka. in *Active Volcanoes of Kamchatka, vol. 1*, edited by S. A. Fedotov, Y.P. Masurnkov, and A.E. Svayatlovsky, 12-14, Nauka, Moscow, 1991.

Furukawa, Y., Magmatic processes under arcs and formation of the volcanic front, *J. Geophys. Res.*, *98*, 8309-8319, 1993.

Furukawa, Y., and S. Uyeda, Thermal state under the Tohoku arc with consideration of crustal heat generation, *Tectonics, 164,* 175-187, 1989.

Furukawa, Y., and Y. Tatsumi, Melting of a subducting slab and production of high-Mg andesite magmas: unusual magmatism in SW Japan at 13-15 Ma, *Geophys. Res. Lett.*, *26*, 2271-2274, 1999.

Gill, J.B., Orogenic andesite and plate tectonics, New York: Springer-Verlag, 1981.

Goto, A., and Y. Tatsumi, Stability of chlorite in the upper mantle, *Am. Mineral.*, *75*, 105-108, 1990.

Guffanti, M., and C.S. Weaver, Distribution of late Cenozoic volcanic vents in the Cascade Range: volcanic arc segmentation and regional tectonic considerations, *J. Geophys. Res., 93,* 6513-6529, 1988.

Henry, S.G., and H.N. Pollack, Terrestrial heat flow above the Andean subduction zone in Bolivia and Peru, *J. Geophys. Res.*, *93,* 15153-15162, 1988.

Hirose, K., and I. Kushiro, Partial melting of dry peridotites at high pressures: determination of compositions of melts segregated from peridotite using aggregates of diamond. *Earth Planet. Sci. Lett.*, *114,* 477-489, 1993.

Honda, S., Thermal structure beneath Tohoku, northeast Japan-a case study for understanding the detailed thermal structure of the subduction zone, *Tectonophysics, 112,* 69-102, 1985.

Inoue, T., T. Irifune, H. Yurimoto, and I. Miyagi, Decomposition of K-amphibole at high pressures and implications for subduction zone volcanism, *Phys. Earth Planet. Interiors, 107,* 221-231, 1998

Ishida, M., The seismically quiescent boundary between the Philippine Sea Plate and the Eurasian Plate in central Japan, *Tectonophysics*, *243*, 241-253, 1995.

Ishikawa, T., and E. Nakamura, Origin of the slab components in arc lavas from across-arc variation of B and Pb isotopes, *Nature*, *370*, 205-208, 1994.

Ito, K., and Y. Tatsumi, Measurement of elastic wave velocities in granulite and amphibolite having identical H_2O-free bulk compositions up to 850°C at 1 GPa, *Earth Planet. Sci. Lett.*, *133*, 255-264, 1995.

Iwamori, H., Transportation of H_2O and melting in subduction zones, *Earth Planet. Sci. Lett.*, *160*, 65-80, 1998.

Johnson, M.C., and T. Plank, Dehydration and melting experiments constrain the fate of subducted sediments, *Geochem. Geophys. Geosys.*, 1, 1999gc000014, December, 1999.

Kay, R.W., Aleutian magnesian andesites: melts from subducted pacific ocean crust, *J. Volcanol. Geothem. Res.*, *4*, 117-182, 1978.

Kelemen, P.B., Genesis of high Mgndesites and continental crust, *Contrib Mineral. Petrol.*, *120,* 1-19, 1995.

Kogiso, T., Y. Tatsumi, and S. Nakano, Trace element transport during dehydration processes in the subducted oceanic crust: 1. experiments and implications for the origin of ocean island basalts, *Earth Planet. Sci. Lett.*, *148*, 193-206, 1997.

Konzett, J., and Y. Fei, Transport and storage of potassium in the Earth's upper mantle and transition zone; an experimental study to 23 GPa in simplified and natural bulk compositions, *J. Petrol, 41,* 583-603, 2000.

Kushiro, I., A petrological model of the mantle wedge and lower crust in the Japanese island arcs. In: B.O. Mysen, ed. Geochem, *Soc. Spec. Publ.*, *1,* 165-181, 1987.

Lewis, T. J., W.H. Bentkowski, E.E. Davis, R.D. Hyndman, J.G. Souther, and J.A. Wright, Subduction of the Juan de Fuca Plate: Thermal consequences, *J. Geophys. Res.*, *93, 15,* 207-15, 225, 1988.

Luth, R.W., Experimental study of the system phlogopite-diopside from 3.5 to 17 Gpa, *Am. Mineral.*, *82,* 1198-1209, 1997.

Marsh, B.D., On the cooling of ascending andesitic magma, *Phil. Trans. R. Soc. London A 288*, 611-625, 1978.

Marsh, B.D., Island arc development: some observations, experiments, and speculations, *J. Geol.*, *87*, 687-713, 1979.

Marsh, B.D., and L.H. Kantha, On the heat and mass transfer from an ascending magma, *Earth Planet. Sci. Lett.*, *39,* 435-443, 1978.

Matsui, Y., N. Onuma, H. Nagasawa, H. Higuchi, S. Banno, and Crystal structure control in trace element partition between crystal and magma, *Bill. Soc. Fr. Mineral. Cristallogr.*, *100*, 315-324, 1977.

Matsuzawa, T., N. Umino, A. Hasegawa, and A. Takagi, Upper mantle velocity structure estimated from PS-converted wave beneath the north-eastern Japan Arc, *Geophys. J. R. Astron. Soc.*, *86*, 767-787, 1986.

McKenzie, D., The generation and compaction of partially molten, *J. Petrol.*, *25*, 713-765, 1984.

McKenzie, D., Constraints on melt generation and transport from U-series activity ratios, *Chem. Geol.*, *162*, 81-94, 2000.

Nicholls, I.A., and A.E. Ringwood, Effect of water on olivine stability in tholeiites and production of silica-saturated magmas in the island arc environment, *J. Geol.*, *81*, 285-300, 1973.

Nohda, S., Y. Tatsumi, Y. Otofuji, T. Matsuda, and K. Ishizaka, Asthenospheric injection and back-arc opening: isotopic evidence form NE Japan, *Chem. Geol.*, *68*, 317-327, 1988.

Nye, C.J., and M.R. Reid, Geochemistry of primary and least fractionated lavas from Okmok Volcano, Central Aleutians: implications for arc magmagenesis, *J. Geophys. Res.*, *91*, 10271-10287, 1986.

Okada, H., New evidences of the discontinuous structure of the descending lithosphere as revealed by ScSp phase, *J. Phys. Earth*, *27*, S53-S63, 1979.

Pawley, A. R., and J. R, Holloway, Water sources for subduction zone volcanism, new experimental constraints, *Science, 260,* 664-667, 30, 1993.

Peacock, S.M., Numerical simulation of metamorphic pressure-temperature-time paths and fluid production in subducting slabs, *Tectonics*, *9*, 1197-1211, 1990.

Pearce, J. A., Role of the sub-continental lithosphere in magma genesis at active continental margin, in *Continental Basalts and Mantle, Xenoliths,* edited by. C. J. Hawkesworth, and M. J. Norry, pp.230-249, Shiva, Nantwich, 1983.

Poli, S., and M.W. Schmidt, H_2O transport and release in subduction zones: experimental constraints on basaltic and andesitic systems, *J. Geophys. Res.*, *100*, 22299-22314, 1995.

Ramsay, W.R.H., A.J. Crawford, and J.D. Foden, Field setting, mineralogy, chemistry and genesis of arc picrites, New Georgia, Solomon Islands, *Contrib. Mineral. Petrol.*, *88*, 386-402, 1984.

Ryan, J.G., J. Morris, F. Tera, W.P. Leeman, and A. Rsvetkov, Cross-arc geochemical variations in the Kurile arc as a function of slab depth, *Science*, *270*, 625-627, 1995.

Sakuyama, M., Lateral variation of H_2O contents in Quaternary magmas of northeastern Japan, *Earth Planet. Sci. Lett.*, *43,* 103-111, 1979.

Sakuyama, M., Petrology of arc volcanic rocks and their origin by mantle diaper, *J. Geophys. Res.*, *88*, 5815-5825, 1983.

Sakuyama, M., and R.W. Nesbitt, Geochemistry of the Quaternary volcanic rocks of the Northeast Japan arc, *J. Volcanol. Geothem. Res.*, *29*, 413-450, 1986.

Schmidt, M.W., and S. Poli, Experimentally based water budgets for dehydrating slabs and consequences for arc magma generation, *Earth Planet. Sci. Lett.*, *163*, 361-379, 1998.

Sekine, T., and P.J. Wyllie, Phase relationships in the system $KalSio_4$-Mg_2SiO_4-SiO_2-H_2O as a model for hybridization between hydrous siliceous melts and peridotite, *Contrib. Mineral. Petrol.*, *79*, 368-374, 1982.

Shimoda, G., Y. Tatsumi, S. Nohda, K. Ishizaka, and B.M. Jahn, Setouchi high-Mg andesites revisited: geochemical evidence for melting of subducting sediments, *Earth Planet. Sci. Lett.*, *160*, 479-492, 1998.

Shuto, K., Miocene abyssal tholeiite-type basalts from Tobishima Island, eastern margin of the Japan Sea, *J. Mineral. Petrol. Econ. Geol.*, *83*, 257-272, 1988.

Stern, C.R., F.A. Frey, K. Futa, R.E. Zartman, Z. Peng, and T.K. Kyser, Trace-element and Sr, Nd, Pb, and O isotopic composition of Pliocene and Quaternary alkali basalts of the Patagonian Plateau lavas of southernmost South America, *Contrib. Mineral. Petrol.*, *104*, 294-308, 1990.

Sudo, A., and Y. Tatsumi, Phlogopite and K-amphibole in the upper mantle: implication for magma genesis in subduction zones, *Geophys. Res. Lett.*, *17*, 29-32, 1990.

Sugimura, A., Zonal arrangement of some geophysical and petrological features in Japan and its environs, *J. Fac. Sci. Univ. Tokyo*, *12* (Sect II), 133-153, 1960.

Sugimura, A., T. Matsuda, K. Chinzei, and K. Nakamura, Quantitative distribution of late Cenozoic volcanic materials in Japan, *Bull. Volcanol.*, *26* (2), 125-140, 1963.

Sun, S.S., Lead isotopic study of young volcanic rocks from mid-ocean ridges, ocean islands arcs, *Philos. Trans. R. Soc London Ser. A* 297, 409-445, 1980.

Takahashi, E., Melting relations of an alkali-olivine basalt to 30kbar and their bearing on the origin of alkali basaltic magmas, *Yearb. Carnegie Inst. Wash.*, *79*, 271-276, 1980.

Takahashi, E., and I. Kushiro, Melting of a dry peridotite at high pressures and basalt magma genesis, *Am. Mineral.*,*68*, 859-879, 1983.

Tamura, Y., Y. Tatsumi, D. Zhao, Y. Kido, H. Shukuno, Distribution of Quaternary volcanos in the Northeast Japan arc: geologic and geophysical evidence of hot fingers in the mantle wedge, *Proc. Japan Acad.*, *77*,*135-139*, 2001.

Tatsumi, Y., Formation of the volcanic front in subduction zones, *Geophys. Res. Lett.*, *13*, 717-720, 1986.

Tatsumi, Y., Migration of fluid phases and genesis of basalt magmas in subduction zones, *J. Geophys. Res.*, *94*, 4697-4707, 1989.

Tatsumi, Y., Slab melting: its role in continental crust formation and mantle evolution, *Geophys. Res. Lett.*, 27, 3941-3944, December, 2000.

Tatsumi, Y., Geochemical modeling of partial melting of subducting sediments and subsequent melt-mantle interaction; generation of high-Mg andesites in the Setouchi volcanic belt, Southwest Japan, *Geology*, *29*, 323-326, 2001.

Tatsumi, Y., and S. Eggins, *Subduction zone magmatism*, 211 pp., Blackwell, Boston, 1995.

Tatsumi, Y., Y. Furukawa, T. Kogiso, K. Yamanaka, T. Yokoyama, S.A. Fedotov, Unusual three volcanic chainins in the Kamchatka, *Geophys. Res. Lett.*, *21*, 537-540,1944a.

Tatsumi, Y., Furukawa,Y., and Yamashita, S., Thermal and geochemical evolution of the mantle wedge in the NE Japan arc: I. Contribution from experimental potrogy, *J. Geophys. Res.*, *99*, 22275-22283, 1994b.

Tatsumi, Y., D.L. Hamilton, and R.W. Nesbitt, Chemical characteristics of fluid phase released from a subducted lithosphere and origin of arc magmas: evidence from high-pressure experiments and natural rocks, *J. Volcanol. Geotherm. Res.*, *29*, 293-309, 1986.

Tatsumi, Y., K. Ito, and A. Goto, Elastic wave velocities in isochemical granulite and amphibolite: origin of a low-velocity layer at the slab/mantle wedge interface, *Geophys. Res. Lett.*, *21*, 17-20, 1994b.

Tatsumi, Y., and T. Kogiso, Trace element transport during dehydration processes in the subducted oceanic crust: 2. origin of chemical characteristics in arc magmas, *Earth Planet. Sci. Lett.*, *148*, 207-222, 1997.

Tatsumi, Y., M. Sakuyama, H. Fukuyama, and I. Kushiro, Generation of arc basalt magmas and thermal structure of the mantle wedge in subduction zones, *J. Geophys. Res.*, *88*, 5815-5825, 1983.

Tatsumi, Y., M. Murasaki, E.M. Arsadi, and S. Nohda, Geochemistry of Quaternary lavas from NE Sulawesi: transfer of subduction components into the mantle wedge, *Contrib. Mineral. Petrol.*, *107*, 137-149, 1991.

Tatsumi, Y., K. Sato, T. Sano, R. Arai, and V.S. Prikhodko, Transition from arc to intraplate magmatism associated with backarc-rifting: evidence from the Sikhote Alin volcanism, *Geophys. Res. Lett.*, *27*, 1587-1590, 2000.

Togashi, S., Petrology of Miocene calc-alkaline rocks of northeastern Honshu, Japan, *Sci. Rep. Tohoku Univ. ser. III 14*, 1-52, 1978.

Trønnes, R.G., E. Takahashi, and C.M. Scarfe, Stability of Krichterite and phlogopite to 14 Gpa, EOS *Transactions, Am. Geophys. Union*, *69*, 1510-1511, 1988.

Turnner, S., P. Evans, and C. Hawkesworth, Ultrafast Source-to-Surface Movement of Melt at Island Arcs from ^{226}Ra-^{230}Th Systematics, *Science*, *292*, 1363-1366, 2001.

Wada, K., Contrasted petrological relations between tholeiitic and calc-alkalic series from Funagata volcano, Northeast Japan, *J. Japan. Asso. Mineral. Petrol. Econ. Geol.*, *76*, 215-232, 1981.

Watanabe, T., M.G. Langseth, and R.N. Anderson, Heat flow in back-arc basins of the western Pacific. In: Talwani , M., I.W.C.Pitman, eds. Ialand arcs, deep sea trenches, and back-arc basins, Maurice Ewing Ser. I.Washington, DC: *Am. Geophys. Union*, 137-167, 1997.

Weaver, S.D., and I.E.M. Smith, New Zealand intraplate volcanism. In: R.W. Johnson, ed. Intraplate volcanism in eastern Australia and New Zealand, *R. Soc. Austr.*, 157-188,1989.

Williams, R., and J.B. Gill, Effects of partial melting on the uranium decay series, *Geochim. Cosmochim. Acta, 53,* 1607-1619, 1989.

Wyllie, P.J., Magma genesis, plate tectonics, and chemical differentiation of the earth, Rev. *Geophys, 26,* 370-404, 1988.

Yogodzinski, G.M., O.N. Volynets, A.V. Koloskov, N.I. Seliverstov, and V.V. Matvenkov, Magnesian andesites and the subduction component in a strongly calc-alkaline series at Piip volcano, far Western Aleutian, *J. Petrol.*, *35*, 163-204, 1994.

Zhao, D., S. Horiuchi, and A. Hasegawa, A., Seismic velocity structure of the crust beneath the Japan Islands, *Tectonophysics, 212,* 289-301, 1992.

Zhao, D.P., Y.B. Xu, D.A. Wiens, L. Dorman, J. Hildebrand, and S. Webb, Depth extent of the Lau back-arc spreading center and its relation to subduction processes, *Science*, *278*, 254-257, 1997.

Zhao, T., S. Horiuchi, and A. Hasegawa, 3-D seismic velocity structure of the crust and the uppermost mantle in the Northeastern Japan Arc, *Tectonophysics*, *181*,135-149,1990.

Thermal Structure due to Solid-State Flow in the Mantle Wedge Beneath Arcs

Peter B. Kelemen

Dept. of Geology & Geophysics, Woods Hole Oceanographic Institution, Woods Hole, Massachusetts

Jennifer L. Rilling and E. M. Parmentier

Dept. of Geological Sciences, Brown University, Providence, Rhode Island

Luc Mehl and Bradley R. Hacker

Dept. of Geological Sciences & Institute for Crustal Studies, University of California, Santa Barbara, California

We summarize petrological and seismic constraints on the temperature of arc lower crust and shallow mantle, and show that published thermal models are inconsistent with these constraints. We then present thermal models incorporating temperature-dependent viscosity, using widely accepted values for activation energy and asthenospheric viscosity. These produce thin thermal boundary layers in the wedge corner, and an overall thermal structure that is consistent with other temperature constraints. Some of these models predict partial melting of subducted sediment and/or basalt, even though we did not incorporate the effect of shear heating We obtain these results for subduction of 50 Myr old oceanic crust at 60 km/Myr, and even for subduction of 80 Myr old crust at 80 km/Myr, suggesting that melting of subducted crust may not be not restricted to slow subduction of young oceanic crust.

1. INTRODUCTION

Modern thermal models suggest that subducted sediments, and subducted oceanic crust that is more than 20 million years old, are unlikely to partially melt beneath arcs undergoing near-steady state subduction. These calculations are essentially of three types: (1) analytical approximations including various assumptions about coupling between the subducted crust and the overlying mantle and about convection in the mantle wedge [e.g., *Davies*, 1999; *Molnar and England*, 1995; *Molnar and England*, 1990], (2) purely plate-driven models with uniform viscosity, in which the thermal regime is calculated numerically using analytical expressions for corner flow in the mantle wedge, with model results depending on various input parameters including the thickness of the arc "lithosphere" and the depth of coupling between subducting crust and overlying mantle [e.g., *Peacock*, 2002; *Peacock and Hyndman*, 1999; *Peacock and Wang*, 1999; *Iwamori*, 1997; *Peacock*, 1996; *Ponko and Peacock*, 1995; *Peacock et al.*, 1994; *Pearce et al.*, 1992; *Peacock*, 1991; *Peacock*, 1990a; *Peacock*, 1990b], and (3) dynamic models in which the mantle flow field as well as the thermal regime are calculated numerically, with model results depending on parameters such as thermal buoyancy, chemical buoyancy and mantle viscosity [*van Keken et al.*, 2002; *Furukawa and Tatsumi*, 1999; *Kincaid and Sacks*, 1997; *Furukawa*, 1993a; *Furukawa*, 1993b; *Davies and Stevenson*, 1992]. These models differ in many respects, but most agree that subduction of oceanic crust that is more than 20 million years old at down-dip rates greater than 20 km/Myr will not produce temperatures at the top of the subducting plate that are high enough to allow fluid-saturated melting of sediment or basalt. The sole exceptions are recent models by *Conder et al.* [2002] which, like ours, incorporate only temperature-dependent viscosity, and a model by *van Keken et al.*, [2002] which incorporates both temperature- and stress-dependent viscosity.

In contrast, a variety of geochemical and petrological inferences suggest that partial melting of subducted sediment and/or basalt is common in many arcs. These inferences may be divided into two main lines of reasoning.

Inside the Subduction Factory
Geophysical Monograph 138

10.1029/138GM13

(1) Partial melting of subducted metabasalt (or metagabbro) at eclogite facies conditions is inferred based on the similarity of trace element patterns in rare, primitive andesites with partial melts of eclogite [e.g., *Defant and Kepezhinskas*, 2001; *Grove et al.*, 2001; *Tatsumi et al.*, 2001; *Yogodzinski et al.*, 2001; *Rapp et al.*, 1999; *Yogodzinski and Kelemen*, 1998; *Yogodzinski et al.*, 1995; *Yogodzinski et al.*, 1994; *Defant and Drummond*, 1990; *Drummond and Defant*, 1990; *Kay*, 1978]. Such lavas are primarily observed where subduction rates are slow, and/or the subducting oceanic crust is young (< 20 million years old), and/or the subducting plate terminates along strike allowing heating from the side as well as from the top [*Yogodzinski et al.*, 2001; *Defant and Drummond*, 1990]. Thus, it is commonly inferred that this evidence applies mainly or exclusively to unusual thermal conditions, and does not require modification of 2D, steady-state thermal models for arcs.

(2) Efficient recycling of elements such as Th and Be from subducted sediments into arc magmas in most arcs worldwide is inferred from correlation of, e.g., Th/La with Pb and Nd isotope ratios in arc lavas [*Class et al.*, 2000; *Elliott et al.*, 1997; *Hawkesworth et al.*, 1997], and from correlation of Th enrichment in arc lavas with the flux of subducted, sedimentary Th [*Plank and Langmuir*, 1998; *Plank and Langmuir*, 1993]. Because Th and Be are relatively insoluble in aqueous fluids, even under simulated subduction zone conditions, but are highly mobile incompatible elements during partial melting of both metasediments and metabasalt, it is argued that their efficient recycling requires melting of subducting sediment [e.g., *Johnson and Plank*, 1999; *Brenan et al.*, 1995a; *Brenan et al.*, 1995b]. Unlike inference (1), this second line of reasoning applies to lavas formed via subduction of relatively old oceanic crust (> 20 million years old). For example, a sediment melt component is apparently important in the Marianas arc, where subducting crust is more than 140 Ma [e.g., *Elliott et al.*, 1997].

These two types of evidence are often treated separately, and indeed inference (2) is often summarized with the aphorism "sediments melt, basalts dehydrate". However, the solidus temperatures for fluid-saturated metasediment and metabasalt are very similar [e.g., *Johnson and Plank*, 1999; *Schmidt and Poli*, 1998; *Nichols et al.*, 1994, *Schmidt and Poli,* in press.]. Thus, geochemical data support the inference that partial melting of subducted sediment and/or basalt is common in present-day subduction zones.

The main objection to the conclusion that partial melting of subducted sediment and/or basalt is common comes from the thermal models cited in the first paragraph of this section. This has led to a long-standing disagreement, with one group confident that thermal models rule out melting in most subduction zones, and another group maintaining that "slab melting" occurs in the same subduction zones. Debate has largely centered on geochemical arguments, particularly ongoing research into the possibility that super-critical, Na-rich aqueous fluids have transport properties for elements such as Th which are very similar to the transport properties of melts [e.g., *Keppler*, 1996; *Plank*, 1996]. However, incorporating temperature-dependent viscosity into thermal models may explain "slab melting" [e.g., *van Keken et al.*, 2002], Appendix B, and this paper). In general, thermal models for arcs deserve more intensive investigation.

In the course of the debate over "slab melting", some useful constraints may have been temporarily overlooked. First, thermal models predict temperatures in the uppermost mantle and at the base of arc crust that are hundreds of degrees lower than petrological estimates of temperature at these depths (Plate 1). Magmas and metamorphic rocks formed at these shallow depths reach the surface with relatively little modification, permitting more robust pressure and temperature estimates than those inferred for partial melting deep in subduction zones. (For references, please see caption to Plate 1).

Because arcs are magmatically active, and crystallizing melt lenses do not have to lie on a steady-state geotherm, it could be that the PT estimates for magmas and metamorphosed igneous rocks need not coincide with the steady-state thermal structure predicted in models (Figure 1). However, our second point is that thermal models do not account well for the shape of low velocity anomalies in the mantle wedge beneath arcs. Observation of a 6% low P wave velocity anomaly at the base of arc crust for long distances along the strike of the NE Japan and Tonga arcs [e.g., *Zhao et al.*, 1997; *Zhao et al.*, 1992a] suggests to us that melt is present in the mantle below a "permeability barrier" at the base of arc crust. Because this is observed over long distances along strike, reasoning that "almost everywhere" is indicative of "nearly all the time", we infer that melt is present near the base of the crust at steady state.

Third, a key constraint used in all thermal models is that below a specified depth (40 to 100 km) the subducting plate and the overlying mantle wedge are mechanically coupled. However, this assumption is apparently inconsistent with arc topography and gravity data, which are better fit with a weak coupling between the subducting plate and the overlying mantle, implying a low viscosity mantle wedge [*Billen and Gurnis*, 2001; *Zhong and Gurnis*, 1992].

Fourth, the difference between predicted, steady state temperatures in subduction zones and the fluid-saturated solidii for metasediment and metabasalt is small; it may be

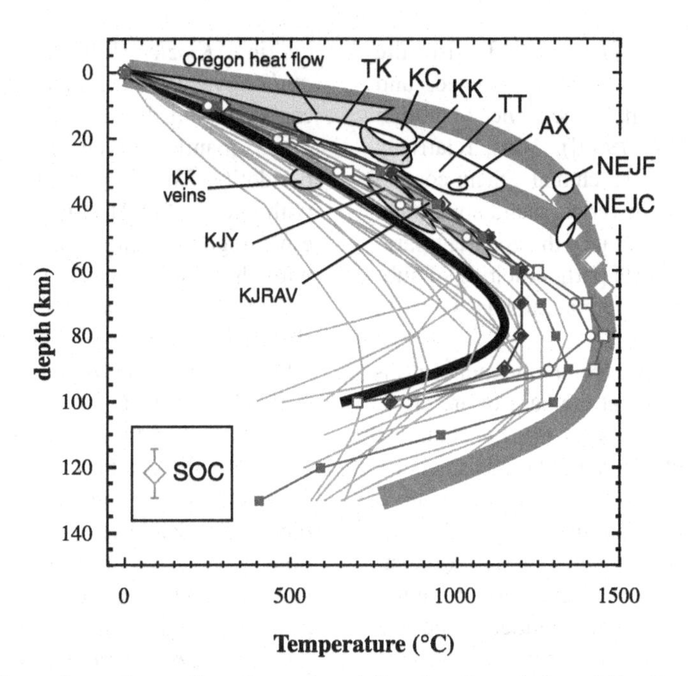

Plate 1. Predicted geotherms beneath arcs from thermal modeling (small symbols and fine lines), compared to petrological estimates of PT conditions in the uppermost mantle and lowermost crust in arcs (large symbols and thick lines). Most petrological estimates are several hundred degrees hotter than the highest temperature thermal models at a given depth. Wide grey lines illustrate a plausible thermal structure consistent with the petrological estimates. Such a thermal structure requires adiabatic mantle convection beneath the arc to a depth of ~ 50 km, instead of minimum depths of ~ 80 km or more in most thermal models.

Petrological estimates for arc crust:

Kohistan arc section, northern Pakistan: Chilas complex, lower crust (KC, [*Jan*, 1988; *Jan and Howie*, 1980] Kamila "amphibolites", lower crust (KK [*Anczkiewicz and Vance*, 2000; *Yoshino et al.*, 1998; *Jan and Karim*, 1995; *Yamamoto*, 1993; *Jan*, 1988]. In addition, high P/T metamorphism is recorded by the Jijal complex, along the Main Mantle thrust at the base of the section (KJRAV [*Anczkiewicz and Vance*, 2000; *Ringuette et al.*, 1999]; KJY [*Yamamoto*, 1993]) and by late veins in the Kamila amphibolites (KK veins [*Jan and Karim*, 1995]). This late, high P/T is probably related to continental collision and exhumation of high P rocks along the Indian-Asian suture zone [e.g., *Gough et al.*, 2001; *Treloar et al.*, 2001; *Treloar*, 1995]. **Oceanic Aleutian arc, Kanaga volcano xenoliths, base of crust?** (AX [*DeBari et al.*, 1987]). **Talkeetna arc section, south-central Alaska:** Tonsina area, garnet gabbros a few hundred meters above the Moho (TT [*DeBari and Coleman*, 1989] and this paper, Figure 2 and Table 1); and Klanelneechina klippe, lower crust (TK, this paper, Figure 2 and Table 1).

Oregon Cascades:

Triangular grey field encloses geotherms inferred from heat flow data in the Oregon Cascades arc [*Blackwell et al.*, 1982]; although interpretation of heat flow data is often controversial, the inferred geotherm is broadly consistent with metamorphic PT estimates for arc crust.

Petrological estimates of mantle-melt equilibration conditions:

NEJF, NE Japan arc front, and NEJC, NE Japan, central arc [*Tatsumi et al.*, 1983]; SOC, southern Oregon Cascades [*Elkins Tanton et al.*, 2001].

Thermal models with symbols:

Open squares and circles, NE Japan and Cascadia, non-linear rheology [*van Keken et al.*, 2002]; open diamonds and closed circles, 40 km and 70 km decoupling models, NE Japan [*Furukawa*, 1993a]; closed squares, fast subduction of thin plate, assuming mantle potential temperature of 1400°C [*Kincaid and Sacks*, 1997], heavy solid line with no symbols, isoviscous corner flow, NE Japan [*van Keken et al.*, 2002].

Other thermal models:

NE and SW Japan [*Peacock and Wang*, 1999]; Izu-Bonin [*Peacock*, 2002]; Aleutians [*Peacock and Hyndman*, 1999]; 100 km decoupling model, NE Japan [*Furukawa*, 1993a]; slow subduction of thin plate, old plate and young plate, plus fast subduction of old plate and young plate, all assuming potential temperature of 1400°C [*Kincaid and Sacks*, 1997]; Alaska Range hot and cold models [*Ponko and Peacock*, 1995]; fast and slow subduction, with and without shear heating [*Peacock*, 1996]; general models [*Peacock*, 1990a].

as narrow as 50 °C [e.g., *Nichols et al.*, 1994]. Also, the fluid-saturated solidii are not precisely determined, and have poorly quantified compositional dependence (e.g., discussion in [*Johnson and Plank*, 1999]). Thus, relatively small changes in thermal models, which make them consistent with PT constraints from the lower crust and uppermost mantle, might also lead to a resolution of the apparent paradox in which geochemistry suggests partial melting of subducted sediment and basalt, while thermal models appear to rule this out.

This paper reviews PT estimates for mantle/melt equilibration and for metamorphic rocks at the base of arc crust, including new thermobarometric data for the Talkeetna arc section in Alaska, and then presents thermal models that are consistent with these PT estimates. We show that, with a temperature-dependent rheology, cooling of material in the mantle wedge causes this material to adhere to the subducting plate, advectively removing part of the thermal boundary layer in the wedge corner, provided the thermal activation energy for mantle deformation is sufficiently large ($Q/(RT_\infty) \geq 10$, where T_∞ is the asthenospheric temperature). Also, when the asthenospheric viscosity is sufficiently low (~ 10^{17} Pa s), a density current carries relatively cold, dense material downward along the slab, further thinning the thermal boundary layer. Entrainment of cold, viscous material with the subducting plate, and convective flow, both drive enhanced return flow of hot material diagonally upward into the wedge corner. Thinning of thermal boundary layers and enhanced return flow raise temperatures at the top of the subducting plate above the fluid-saturated solidus for metabasalt and metasediment. Even though our models do not include any shear heating along the subduction zone, thermal models that yield subduction zone temperatures higher than the fluid saturated solidus arise for widely accepted values for mantle viscosity and thermal activation energy are used, and for subduction of oceanic lithosphere as old as 80 Myr at a rate as high as 80 km/yr, suggesting that melting of fluid-saturated subducted sediment or basalt in normal, steady-state subduction may be inevitable, rather than impossible.

2. PRESSURE AND TEMPERATURE IN THE SHALLOW MANTLE AND LOWER CRUST BENEATH ARCS

2a. Magma/Mantle Equilibration at High Temperature and low Pressure

Over the past decade, methods have been developed for estimating the composition of partial melts of mantle peridotite as a function of the depth and temperature of melting [*Kinzler*, 1997; *Kinzler and Grove*, 1993; *Kinzler and Grove*, 1992b; *Langmuir et al.*, 1992]. These methods can constrain the final depth and temperature at which a given, mantle-derived melt, might have equilibrated with peridotite. This is particularly straightforward for basalts with known, low H_2O-content. For such lavas in the southern Oregon Cascades and in Indonesia, last equilibration with the mantle was ~ 1290 to 1450°C and 1 to 2 GPa [*Elkins Tanton et al.*, 2001; *Sisson and Bronto*, 1998]. As shown in Plate 1, Kinzler & Grove [*Kinzler and Grove*, 1993; *Kinzler and Grove*, 1992a] estimate uncertainties of 0.25 GPa and 25°C for such estimates. Similarly, *Tatsumi et al.* [1983] conducted phase equilibrium experiments to determine the conditions at which primitive lavas from NE Japan could have equilibrated with mantle peridotite, over a range of possible H_2O-contents. They also inferred conditions of ~ 1325 to 1360°C and 1 to 2 GPa. Similar conclusions have also been reached in phase equilibrium studies of primitive, high aluminum basalts from arcs [*Draper and Johnston*, 1992; *Bartels et al.*, 1991].

No published thermal models for subduction zones predict such high temperatures in the uppermost mantle, near the base of the arc crust (Plate 1). Instead, all but one study predict temperatures less than 900°C at depths of 45 km (~ 1.5 GPa) beneath an arc. The exception is the recent paper by *van Keken et al.*, [2002], who used a stress and tempera-

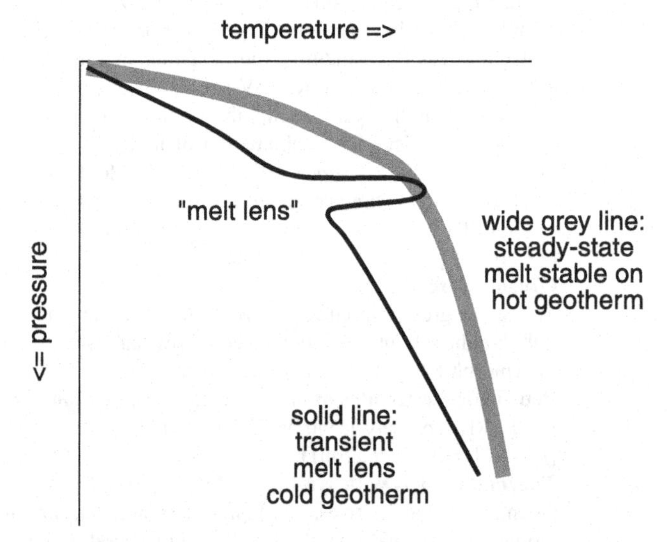

Figure 1. Schematic illustration of why petrological estimates of pressure and temperature for meta-plutonic rocks and mantle-melt equilibration do not necessarily lie along a steady state geotherm, and instead could represent temperature depth conditions for transient magma chambers and/or contact metamorphism. However, as argued in the text, seismic data suggest that high temperatures are present in the uppermost mantle and lowermost crust beneath arcs at steady state.

ture-dependent viscosity in the mantle wedge, and predicted temperatures of ~ 950 to 1000°C at 45 km beneath NE Japan and the Cascades. It is instructive to extend this comparison by asking, at what depth beneath the arc do thermal models predict temperatures of 1300°C or more? The shallowest 1300°C isotherms are at ~ 65 km in the NE Japan and Cascades models of *van Keken et al.*, [2002].

Many thermal models do not predict temperatures as high as 1300°C for the mantle wedge beneath the arc at any depth, and thus appear to be inconsistent with petrologic constraints. One caveat is that trajectories of melt migration beneath an arc might not be vertical. Porous flow might lead to melt migration vectors that ascend diagonally toward the wedge corner either due to pressure gradients in the mantle flow [*Spiegelman and McKenzie*, 1987; *Phipps Morgan*, 1987] or due to gravity driven porous flow beneath a sloping permeability barrier [*Sparks and Parmentier*, 1991]. Melt transport in fractures might follow a similar trajectory [e.g., *Davies*, 1999]. However, most thermal models predict a minimum depth of ~ 90 km for the 1300°C isotherm in the wedge. *van Keken et al.* [2002] predict 65 km beneath the Cascades and NE Japan. Thus, petrologic constraints are, indeed, inconsistent with previous thermal models of the mantle wedge.

One possible explanation for the disagreement between petrologic constraints and thermal models might be that temperatures of magmatic equilibration with mantle peridotite need not lie along a steady-state geotherm. In this case, the PT estimates from mantle/melt equilibration and from metamorphic closure conditions would not be representative of the pressure and temperature typical in the shallow mantle and lower crust most of the time. For example, melt in a transient magma chamber could equilibrate with wall rock that was heated to temperatures well above steady-state values at a given depth (Figure 1). *Elkins-Tanton et al.* [2001] noted that previous studies predict substantial partial melting of the lower crust if mantle temperatures at 36 km are as hot as 1290°C [*Mareschal and Bergantz*, 1990]. Because seismic data suggest that arc lower crust is mostly solid, Elkins-Tanton et al. suggested that relatively high, lower crustal temperatures must be localized and transient. However, models of lower crustal melting depend critically on composition; lower crust composed of refractory, plutonic rocks formed by partial crystallization of primitive magmas at high temperature could be entirely solid at the liquidus of mafic basalt.

Finally, many primitive arc magmas do not have low H_2O contents, and some could have equilibrated with the mantle at temperatures lower than 1300°C. Nonetheless, if current estimates of H_2O contents in primitive arc basalts (~ 2 to 4 wt%, e.g., [*Sisson and Layne*, 1993]) are correct, the ubiquity of primitive basalts and basaltic andesites in arcs worldwide suggests that magmatic temperatures of 1200°C or more must be common (see results for the Aleutian arc in *Kelemen et al.*, this volume). Together with seismic data (Section 3), which suggest that melt is present in much of the shallow mantle beneath arcs, this suggests that the shallow mantle must commonly be at 1200°C or more.

2b. Metamorphic Closure Temperatures at the Base of arc Crust

Two tectonically exposed arc sections, the Kohistan section in Pakistan [e.g., *Tahirkheli et al.*, 1979] and the Talkeetna section in south central Alaska [e.g., *DeBari and Coleman*, 1989; *Plafker et al.*, 1989; *Burns*, 1985], include exposures of the base of the crust and the uppermost mantle. In the Kohistan section, metamorphic assemblages in the lower crustal Chilas and Kamila complexes record 700 to 850°C at 0.5 to 0.8 GPa. Along the Indian-Asian suture zone (the Main Mantle thrust), meta-plutonic mineral temperatures record higher P/T. However, the higher pressure end of this spectrum may record partial subduction and then exhumation during continental collision.

For the Talkeetna section, DeBari & Coleman estimated metamorphic conditions during garnet formation in gabbros at the base of the crust at ~ 1 GPa, 825-900°C [*DeBari and Coleman*, 1989]. In Figure 2 and Table 1, we present seven new PT determinations for hornblende-bearing garnet gabbros and garnet gabbronorites from the Talkeetna section made with THERMOCALC v3.1 ([*Holland and Powell*, 1988], May 2001 database). Calculations made with TWQ v1.02 [*Berman*, 1991; *Lieberman and Petrakakis*, 1991; *Berman*, 1990; *Berman*, 1988] confirm these results. Ellipses shown in Figure 2 are 1σ uncertainties determined by the intersection of multivariant equilibria. The equilibria with steepest PT slopes (thermometers) are exclusively Fe–Mg exchange among garnet, orthopyroxene, clinopyroxene, and amphibole, whereas reactions with gentle PT slopes (barometers) are generally net-transfer reactions involving garnet, pyroxene, amphibole, and plagioclase, plus quartz in quartz-bearing samples. Garnets show broad core-to-rim zoning toward higher Fe/Mg, whereas pyroxenes display only near-rim, steep decreases in Fe/Mg and Al_2O_3; plagioclase crystals are unzoned. We interpret this to be the result of early cooling (during or after garnet growth), which produced garnet zoning and likely homogenization of pyroxene, followed by later, more-rapid cooling that produced the steep zoning in pyroxene rims. Because of the inferred pyroxene homogenization, temperatures calculated from mineral cores are minima.

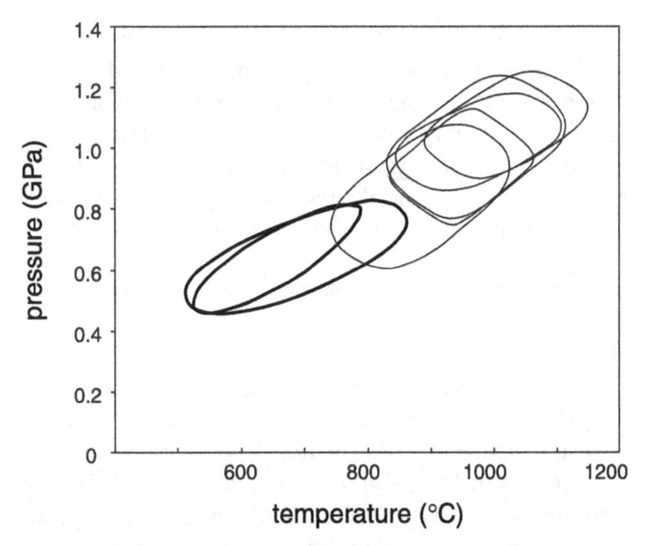

Figure 2. Peak metamorphic PT estimates for the lower crust of the Talkeetna arc section, south central Alaska (based on mineral core compositions in Table 1, with equilibria calculated using Thermocalc v. 3.1 and the May 2001 thermodynamic database). Ellipses with heavy lines are quartz-bearing samples from newly discovered garnet gabbronorites in the Klanelneechina klippe, about 50 km SW of the Tonsina area. Ellipses with thin lines are for Tonsina area samples, from a thin layer of garnet gabbro immediately above the petrological Moho at the base of the arc section in the Tonsina area [*DeBari and Coleman*, 1989]; these lack quartz, so the PT estimates are based on a variety of equilibria involving amphibole as well as pyroxenes, plagioclase and garnet.

The mineral core compositions suggest that two garnet gabbronorites at the base of the arc section near Tonsina crystallized at ~980–1025°C, 1.0–1.1 GPa, and three garnet gabbros exposed as mafic pods or thin garnet-bearing bands within garnet-absent, layered gabbros indicate conditions of ~890–980°C, 0.8–1.0 GPa. Two samples from higher in the arc section, in the Klanelneechina klippe, about 50 km SW of the Tonsina area, indicate significantly lower core temperatures and pressures of ~665–690°C, 0.6 GPa.

In addition to the data from exposed arc sections, *DeBari et al.* [1987] estimated temperatures of 980-1030°C in deformed, metaplutonic xenoliths from Adagdak volcano in the Aleutian arc. The xenoliths are presumed to be derived from the base of the arc crust, which is about 35 km thick [*Holbrook et al.*, 1999].

The metamorphic temperatures described in the previous paragraphs are almost certainly closure temperatures, rather than peak temperatures; that is, the plutonic protoliths of the rocks crystallized at higher, igneous temperatures. *Sisson and Grove* [1993] and *Yang et al.* [1996] developed methods for estimating the temperature, pressure and water content of crystallization for basaltic melts saturated in olivine, clinopyroxene and plagioclase. Using these methods, one can calculate that crystal fractionation of primitive lavas beneath the southern Oregon Cascades occurs at the base of the crust, at conditions of 1100–1300 °C and 0.8 to 1.0 GPa [*Elkins Tanton et al.*, 2001]. Similarly, *DeBari & Coleman* [1989] estimated that gabbronorites at the base of the Talkeetna arc crust originally crystallized at temperatures of 1100°C or more, and then cooled to the metamorphic conditions at which garnet formed. Thus, it is likely that the metamorphic assemblages in these rocks equilibrated along an arc geotherm, at temperatures greater than or equal to the closure temperatures they now record.

As for the magma-mantle equilibration temperatures described in Section 2a, it is conceivable that all the PT estimates for metaplutonic rocks near the base of the crust in the Kohistan and Talkeetna arc sections are "contact metamorphic", reflecting high temperatures associated with transient melt lenses or magma chambers. In this latter view, they do not necessarily represent constraints on a steady-state arc geotherm. However, in the next section we argue that consistent observation of low seismic velocities in the uppermost mantle beneath arc crust indicates that melt is commonly present at Moho depths.

3. Seismic Constraints on Melt Distribution in the Uppermost Mantle

Teleseismic tomography reveals that a –4 to –6% P wave velocity anomaly and a –6 to –10% S-wave velocity

Table 1.

Sample	Minerals	T (°C)	P (GPa)	Cor
0713B07C	gar hb cpx opx plg	982 ± 108	1.01 ± 0.13	0.46
0713B07F	gar hb cpx plg	954 ± 92	0.93 ± 0.15	0.26
0731L01	gar hb cpx opx plg	1025 ± 104	1.07 ± 0.14	0.44
1709L03A	gar hb cpx plg	978 ± 113	1.00 ± 0.19	0.41
1709L04	gar hb cpx plg	889 ± 117	0.83 ± 0.19	0.43
1712P03	gar opx cpx plg qz	663 ± 110	0.63 ± 0.14	0.84
1719P02	gar opx qz	692 ± 141	0.63 ± 0.15	0.70

gar = garnet, hb = amphibole, cpx = clinopyroxene, opx = orthopyroxene, plg = plagioclase, qz = quartz, Cor = correlation coefficient.

anomaly in the uppermost mantle underlies at least 50% of the Moho beneath the NE Japan arc over ~ 500 km along strike [*Zhao and Hasegawa*, 1994; *Zhao and Hasegawa*, 1993; *Zhao et al.*, 1992a; *Zhao et al.*, 1992b]. This low velocity anomaly—relative to a reference model with mantle P- and S-wave velocities from 7.7 and 4.34 km/s at 40 km depth to 7.9 and 4.45 km/s at 90 km—extends from the base of the crust at 35 km to ~ 50 km depth beneath the arc, and has a width of ~ 50-75 km in sections perpendicular to the arc. Similar results have been obtained for the Tonga arc [*Zhao et al.*, 1997]. While the results for both NE Japan and Tonga could be artifacts due to undetected seismic anisotropy, preliminary analysis suggests that the actual effect of anisotropy in these areas is small [*Zhao*, pers. comm. 2000]. Similar seismic anomalies in the MELT region beneath the East Pacific Rise have been taken to indicate the presence of melt in mantle peridotite at temperatures greater than 1300°C [*Forsyth et al.*, 1998a; *Forsyth et al.*, 1998b]. A similar inference may be warranted for the mantle beneath NE Japan, though temperatures could be lower because H_2O will stabilize mantle melts at lower temperatures. Making another analogy to mid-ocean ridges, we think melt fractions are high near the base of arc crust due to the presence of a "permeability barrier", formed by crystallization of cooling melt in intergranular porosity [e.g., *Kelemen and Aharonov*, 1998; *Sparks and Parmentier*, 1991]. The regionally extensive nature of the seismic anomaly suggests that the presence of melt is a steady-state feature of the uppermost mantle beneath the NE Japan and Tonga arcs.

Tamura et al. [*Tamura et al.*, 2002] have proposed that groups of volcanoes in NE Japan are underlain by low velocity anomalies as described in the preceding paragraph, with an along strike extent of 50–100 km, separated by regions with higher seismic velocity in the uppermost mantle. They suggest that this is a result of three-dimensional mantle convection in the wedge, perhaps due to transient diapirs. We agree that advective heat transport via three-dimensional mantle convection might give rise to substantial lateral temperature variation along strike on a scale of 50–100 km. However, this provides little insight into whether melt is present along much of the Moho at steady-state, or not.

Obviously, subduction zones and arcs evolve with time, so that the idea of a "steady state" is an over-simplification. Also, seismic data can only tell us about the present, so that one could ask if the regionally extensive thermal anomalies in the shallow mantle, observed beneath the NE Japan and Tonga arcs, are simply snapshots of an ephemeral though widespread phenomenon. Perhaps, most of the time, most of the shallow mantle does not have anomalously slow seismic velocities. However, we consider this alternative to be unlikely, and seek thermal models that are consistent with petrological PT constraints at steady-state.

4. RAPID ADVECTION IN THE MANTLE WEDGE

Data in the preceding sections, taken together, suggest that temperatures beneath arcs are higher than 1000°C at ~ 30 km, and approach 1300°C or more at 45 km. This is inconsistent with all published thermal models for arcs, as seen in Plate 1. One potential resolution of this discrepancy is to consider the effects of rapid advection of mantle peridotite in the shallowest part of the mantle wedge. In this section, we present results of numerical models that incorporate temperature-dependent viscosity to model solid-state flow by thermally activated creep, with a range of activation energies and "asthenospheric" viscosities.

4a. Experimentally Constrained Mantle Viscosities

As discussed by Hirth & Kohlstedt [*Hirth and Kohlstedt*, this volume] viscosities in the range of 10^{17} to 10^{21} Pa s are within the range experimentally measured for olivine at upper mantle temperatures and pressures. Viscosities in the low end of this range probably require both dissolved hydrogen [*Mei and Kohlstedt*, 2000; *Hirth and Kohlstedt*, 1996; *Kohlstedt et al.*, 1996], as might be expected in the relatively H_2O-rich mantle above a subduction zone, and melt fractions greater than ~ 3 % [*Kelemen et al.*, 1997; *Hirth and Kohlstedt*, 1995a; *Hirth and Kohlstedt*, 1995b]. In particular, a combination of partial melt and dissolved H_2O could produce mantle viscosities less than 10^{18} Pa s at pressures of 1 to 6 GPa. However, to date no thermal model of a subduction zone has incorporated such low viscosities.

4b. Activation Energy in Solid State Creep of the Mantle

The temperature dependence of viscosity during diffusion creep of olivine aggregates can be written

$$\mu/\mu_\infty = exp[Q/(RT_\infty)*(T_\infty/T-1)]$$

with an experimentally determined activation energy Q of 315 kJ/mol [*Hirth and Kohlstedt*, 1995b], equivalent to a value of $Q/(RT_\infty)$ of ~ 24 to 26 at asthenospheric temperatures. μ_∞ and T_∞ are the asthenospheric viscosity and temperature, respectively. Stress-dependence might lower the "effective" activation energy [*Christensen*, 1984]. An activation energy of 250 kJ/mol, yields $Q/(RT_\infty)$ ~ 18 to 21, though higher values of Q might be appropriate for cases involving dislocation creep in which stress is maintained at a high value. In our modeling, we have used values of $Q/(RT_{\infty})$ ranging from 0 to 40. Most of the models presented by Kincaid & Sacks [*Kincaid and Sacks*, 1997] used $Q/(RT_{\infty})$ of 30.

4c. Description of Modeling Technique

Our models of solid state flow and temperature in the mantle wedge above a subducting slab are formulated using a hybrid finite-element/finite difference method. Viscous flow with large viscosity variations is calculated using a standard penalty function finite element formulation with linear interpolation functions on quadrilateral elements [cf. *Reddy*, 1993]. The resulting algebraic equations are solved by a Gaussian elimination method, which is stable and reasonably accurate for the large viscosity variations expected with a strongly temperature-dependent viscosity. The conservation equation describing the advection and diffusion of heat is solved using a finite difference or finite volume approximation on a non-uniform rectangular grid with volumes centered on finite element node points. A second order accurate solution is obtained using centered spatial differences for the diffusion terms and a diffusion-corrected upwind method [*Smolarkiewicz*, 1983] for the advection terms. We think that this advection method is preferable to the first order accurate streamline upwind method usually employed in finite element discretizations of the temperature field which can produce significant numerical diffusion. Time-dependent steady-state solutions are obtained by a simple forward time step of the advection-diffusion equation with a stability limited time step. These solutions represent deviations from an adiabatic temperature gradient which can be simply added to the solutions. We do not account for the heat of melting or mineral reactions or the contribution of melt migration to heat transport.

The models that we present in this paper use a uniformly spaced mesh in a 900 km wide by 600 km deep domain (see Plate 2). In the horizontal direction nodes are spaced every ~10km, and in the vertical direction every ~7 km. This is comparable to the resolution in the recent models of [*Kincaid and Sacks*, 1997]. The top of the domain is a no-slip boundary on which the temperature is set to zero. On the bottom, we have experimented with both a free-slip, flow-through boundary on which the viscous stresses vanish and a free-slip, closed boundary on which the normal velocity component vanishes. The results presented here use the flow-through boundary, but this choice makes little difference within about 300 km of the wedge corner. The right vertical boundary is a flow-through boundary. In areas of inward flow, temperature is set to the prescribed mantle potential temperature, 1350°C for the cases presented here. This corresponds physically to placing a back-arc spreading center at this boundary. The choice of a mantle potential temperature of 1350°C is by analogy to mid-ocean ridges. Experimental and theoretical constraints on melt productivity during decompression melting have been combined with geochemical studies of the trace element contents of mantle peridotites and mid-ocean ridge basalts (MORB), and with seismic determinations of oceanic crustal thickness, to yield estimates ranging from 1280 to 1400°C for potential temperature in the upwelling mantle beneath ridges [e.g., *Braun et al.*, 2000; *Langmuir et al.*, 1992; *White et al.*, 1992; *McKenzie and Bickle*, 1988; *Allègre et al.*, 1973].

Velocities along the left vertical boundary and in a triangular region in the lower left corner of the domain are prescribed to be the subducting plate velocity. The temperature distribution on this boundary is an error function that would result from vertical conductive cooling of the upper mantle. We examine mantle thermal ages of 25, 50, 80 and 100 Myr, and convergence rates of 20, 40, 60, 80, and 100 km/Myr with a slab dip of 45°. This is a reasonable value based on the wide range of slab dips that are observed [cf. *Jarrard*, 1986].

A very viscous region develops near the cold top boundary. In this region significant flow should not occur by thermally activated creep. A maximum viscosity that is a factor of 10^4 greater than that of the deeper mantle was prescribed. To avoid concerns that this very viscous material would "stick to" the plate and be carried downward with it, a layer of very weak nodes was introduced along the top of the slab at depths shallower than 50 km (see Plate 2). *Kincaid & Sacks* [1997] also used this method of introducing a "fault" consisting of a dipping layer of weak nodes. Shear heating along the fault would increase temperatures in the subducting plate and mantle wedge compared to our model results. In contrast to the models of [*van Keken et al.*, 2002], we introduce no heating on the fault. However, later in this paper we estimate the temperature increase that could be associated with this effect.

We consider μ_{∞} values ranging from 10^{17} to 10^{21} Pa s. Each case is run to a time-dependent steady state in which the temperature and velocity fields are relatively constant and the overall thickness of the thermal boundary layer no longer changes (Figure 3). At the lower values of μ_{∞} a time-independent steady state is not attained because the cold, dense thermal boundary layer beneath the overriding plate "drips" into the mantle wedge (see Plate 4 and Figure 3). This behavior has only a small effect on temperatures within a few hundred km of the wedge corner. The time to steady state in the models is ~ 100 Myr. However, temperatures in the wedge corner approach a steady state in ~ 10 Myr (Figure 3).

4d. Model Results

As also seen in earlier thermal models incorporating temperature-dependent viscosity [*Kincaid and Sacks*, 1997;

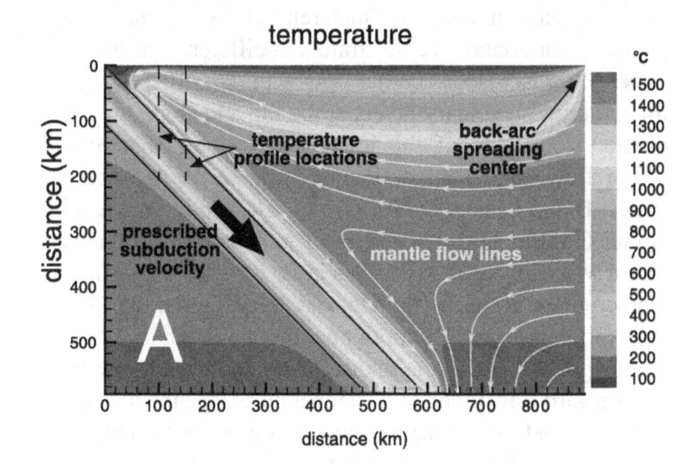

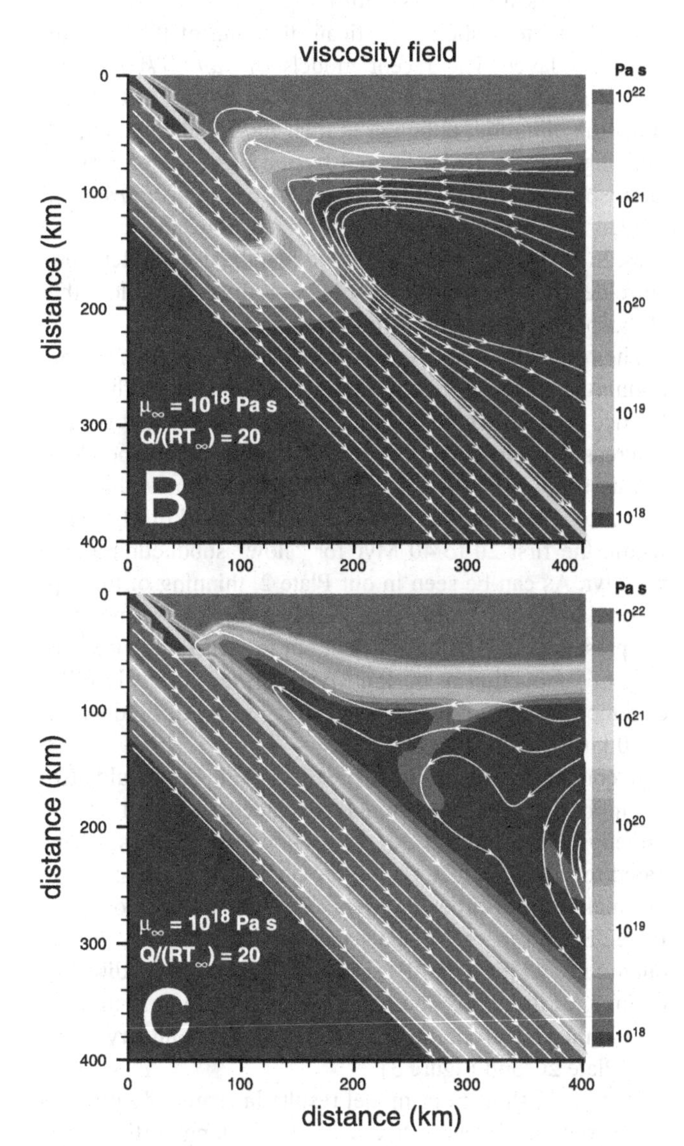

Plate 2. A. Illustration of the entire model domain with "steady state" isotherms and flow lines for model with μ_∞ of 10^{21} Pa s and $Q/(RT_\infty)$ of 20. Prescribed subduction angle in all models is 45°. Subduction velocity is also prescribed. Dashed lines indicate position of temperature/depth sections in Plate 3. **B, C.** Viscosity contours and flow lines (white) for the upper righthand corner of a model with μ_∞ of 10^{18} Pa s and $Q/(RT_\infty)$ of 20. Panel B shows an early time (2.6 Myr) with the thermal boundary layer in the overthrust plate increasing in thickness from ~ 50 km on the right to ~ 75 km adjacent to the subduction zone, and subduction initiating with a ~ 75 km thick plate. A dipping zone of weak nodes in the upper 50 km at the top of the subducting plate is used to model the presence of a fault zone (see text for discussion). Panel C illustrates the "steady state" model (developed by about 15 Myr) after mechanical erosion of wedge corner due to entrainment of viscous mantle with the subducting plate. Time dependence arises from buoyantly generated cold drips that form in the thermal boundary layer beneath the overthrust plate at a depth where the viscosity increases by about one order of magnitude (light blue) from the uniform temperature asthenosphere (dark blue).

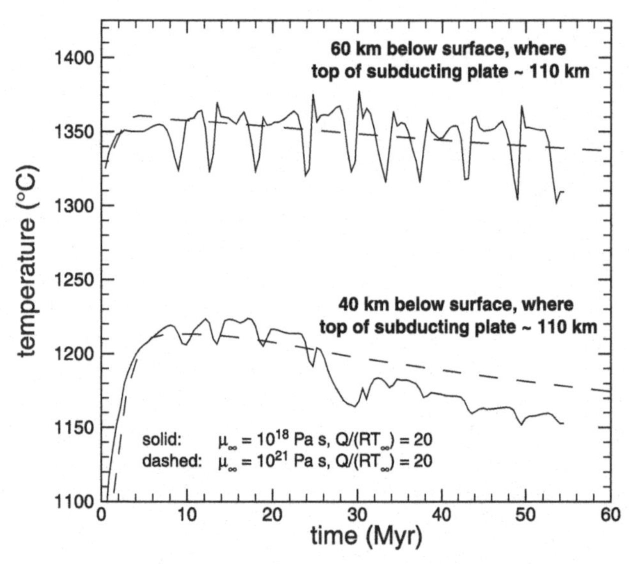

Figure 3. Temperature as a function of time at depths of 40 and 60 km, at x=120 km where the subduction zone is 110 km below the surface, for models with 50 Myr old plate subducting at 60 km/Myr, and μ_∞ of 10^{18} Pa s (solid lines) and 10^{21} Pa s (dashed lines), both with $Q/(RT_\infty)$ of 20. Temperatures increase dramatically in the first ~ 10 Myr of subduction, as the thermal boundary layer is thinned by viscous entrainment with the subducting plate, and then approach time-dependent, near steady state values after several tens of Myr. In the model with μ_∞ of 10^{18} Pa s, quasiperiodic density instabilities drive downwelling of relatively cold material from the base of the thermal boundary layer, reflected in the temperature fluctuations shown. Dripping of cold material from the base of the thermal boundary layer leads to generally higher shallow mantle temperatures in the model with μ_∞ of 10^{18} Pa s, compared to that with μ_∞ of 10^{21} Pa s.

Furukawa, 1993a; *Furukawa*, 1993b], cooling of the top and subduction-side of the mantle wedge forms an increasing volume of high viscosity material that is incorporated in thermal boundary layers. Paradoxically, this leads to thinning of the thermal boundary layers, especially near the wedge corner, because of the enhanced entrainment of cold, viscous mantle with the subducting plate. This advective thinning of the thermal boundary layer in the wedge drives rapid return flow of hot mantle into the wedge corner. Also, in cases with low μ_∞, buoyancy forces drive convection (a cold density current running down the top of the slab and "drips" from the bottom of the overriding plate), so that mantle convection velocities may actually exceed the subduction velocity.

Perhaps the most obvious manifestation of this effect is that flow of hot material into the wedge follows a diagonal upward trajectory toward the wedge corner, accelerating at shallow depths, as seen in previous studies. This contrasts with isoviscous models, in which return flow into the wedge is nearly horizontal. We illustrate this difference in Plate 3. Panel A, for isoviscous corner flow with $\mu_\infty = 10^{21}$ Pa s, has thick thermal boundary layers at the top and along the subduction zone side of the mantle wedge, and horizontal return flow of the upper mantle into the wedge corner. Panels B and C have $Q/(RT_\infty) = 20$. Though the models in B and C differ by three orders of magnitude in asthenospheric viscosity ($\mu_\infty = 10^{21}$ Pa s in B, $\mu_\infty = 10^{18}$ Pa s in C), temperatures in the two models are very similar, and both very different from A, with thinner thermal boundary layers and diagonally ascending return flow into the wedge corner. The upward flow induced by temperature dependent viscosity could lead to decompression melting of the mantle.

Our results for models with temperature-dependent viscosity in Plate 3 show significant thinning of the thermal boundary layer. The recent models of *van Keken et al.* [2002] (see their Figure 3) and *Conder et al.* [2002] also show this thinning, although in their models the amount of thinning appears to be limited by the presence of a rigid "lithospheric" layer of prescribed thickness. *Furukawa* [1993a, 1993b] also prescribed a fixed lithospheric thickness. In contrast, we have assumed that the use of a temperature-dependent viscosity should be sufficient to allow the "lithosphere" to find its own thickness.

The results of *Kincaid and Sacks* [1997] do not show the thinning we observe in our models. One reason for this difference could be that our models are run to "steady state", whereas Kincaid & Sacks concentrated on time-dependent results soon after the initiation of subduction, within the first 2 to 4 Myr for "fast" subduction at 100 km/Myr, and within the first 20 to 40 Myr for "slow" subduction at 13 km/Myr. As can be seen in our Plate 2, thinning of the top thermal boundary layer in the wedge corner takes time; it is not present at the initial stages of our models (e.g., after 2.6 Myr of subduction at 60 km/Myr in panel B) but is fully developed at "steady state" (e.g., after 15 Myr of subduction at 60 km/Myr in panel C).

In comparing our results to previous work, we should also re-emphasize that our models extended to much lower asthenospheric viscosities (μ_∞ as low as 10^{17} Pa s, compared to $\mu_\infty \geq 10^{20}$ Pa s in all previous models). This does not make a first-order difference in our models (compare Plate 3B with $\mu_\infty = 10^{21}$ Pa s and 6C with $\mu_\infty = 10^{18}$ Pa s), but models with μ_∞ of 10^{18} and 10^{17} Pa s do show additional thinning of the 'lithosphere' due to the dripping of relatively cold material from the base of the thermal boundary layer (see Plate 2C and Figure 3).

Plate 4 illustrates our model results in terms of temperature variation at steady state, with depth along vertical sec-

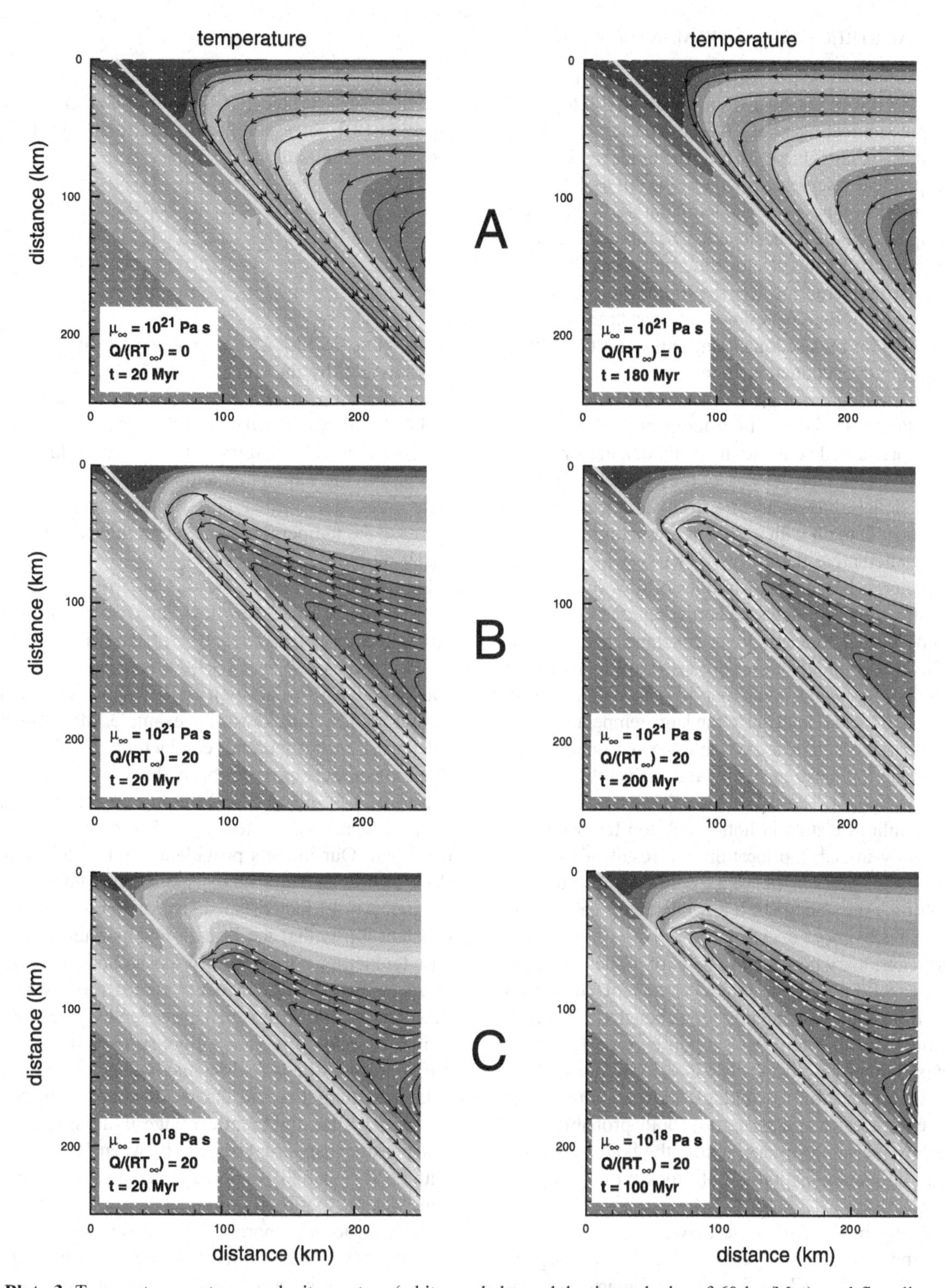

Plate 3. Temperature contours, velocity vectors (white, scaled to subduction velocity of 60 km/Myr), and flow lines (black) for the upper lefthand corner of the model domain in Plate 2A after 20 Myr of subduction (lefthand column) and 100 Myr or more (righthand column). Temperatures are nearly steady state so that temperature distributions are almost identical in the two different columns. Top of subducting plate is indicated by the straight line to which streamlines of wedge flow converge. **(A):** $\mu_\infty = 10^{21}$ Pa s and $Q/(RT_\infty) = 0$ (isoviscous model) **(B):** $\mu_\infty = 10^{21}$ Pa s and $Q/(RT_\infty) = 20$; **(C):** μ_∞ of 10^{18} Pa s and $Q/(RT_\infty) = 20$. Note horizontal return flow into wedge for the isoviscous model (top), compared to diagonally ascending flow for temperature-dependent viscosity.

tions where the top of the over-riding plate is 100 and 150 km above the subducting plate. We emphasize two major features of these results. First, *most* of the models with temperature-dependent viscosity provide a close fit to the petrological constraints on pressure and temperature in the lower crust and upper mantle beneath arcs. Second, all models with μ_∞ of 10^{17}, 10^{18} and 10^{21}, together with $Q/(RT_\infty)$ of 10 and 20, predict temperatures near the top of the subducting plate that are higher than experimentally constrained fluid-saturated solidii for metasediment and metabasalt at 100 to 150 km depth in the subduction zone.

Identifying the location of the slab top in our models introduces uncertainty in interpreting temperatures along it. According to prescribed velocities in the model, the top of the region that moves with the velocity of the subducting plate is defined by a straight line emanating from the upper left corner of the model domain with a slope representing the dip of the slab. An alternative definition of the slab top, the one which we prefer, is based on streamlines of flow in the mantle wedge. Streamlines through the corner region near the top of the mantle wedge converge to an envelope that is essentially parallel to the initially prescribed slab but slightly above it (see Plate 2 or 3) by an amount that depends on the numerical resolution of the model. Due to the large temperature gradients normal to the slab near the slab top, even small differences in the slab top location significantly affect the temperatures inferred along it. The shallower slab top defined on the basis of streamlines results in hotter slab top temperatures. This uncertainty in slab top location is a result of the very strong velocity gradients in the wedge corner at the bottom of the fault. Shifting the slab top by a few kilometers can change the temperatures by more than 50°C.

With this cautionary note, comparison of model results to experimental solidii suggests that a variety of conditions can cause fluid-saturated melting of subducted sediment, while melting of subducted basalt might be rare. However, since metasediment and metabasalt in eclogite facies both have the same mineral assemblage (garnet, omphacitic clinopyroxene, coesite and phengite), both probably have approximately the same solidus, within the bounds of variation due to bulk composition in both rock types [*Schmidt*, pers. comm., 2000].

Shear heating in the fault zone between the subducting and overriding plates will increase temperatures along the top of the slab. Based on heat flow in the forearc [*Peacock and Wang*, 1999], *van Keken et al.* [2002] adopt a shear heating rate of about 30 mW/m^2 along this fault for NE Japan. A simple estimate of the temperature increase along the fault can be obtained from the solution for transient heat conduction into a halfspace with a prescribed heat flux at the halfspace surface [*Carslaw and Jaeger*, 1959, p. 75, equation (8)] corresponding to the fault. The temperature increase on the fault is

$$\Delta T = \frac{\sigma V}{\rho c_p}\sqrt{\frac{t}{\pi \kappa}}$$

where time in this, t, corresponds to distance along the fault divided by the slip velocity V, and σ is the effective shear stress on the fault. For a plate thickness of 50 km with a 45° dipping fault, $V = 60$ km/Myr, and a fault heating rate $\sigma V = 30$ mW/m^2, $\Delta T = 40$°C. Adding this temperature to our calculated temperature along the top of the slab gives the dashed curve shown in Plate 4B. This increase in temperature enhances the possibility of wet basalt melting.

Figure 4 illustrates our results in terms of predicted heat flow, compared to heat flow data from northeast Japan (compiled by [*Furukawa*, 1993a]) and from southwestern Oregon [*Blackwell et al.*, 1982]. It is clear that the isoviscous model fails to account for heat flow in arcs, as previously noted by Furukawa, for example. Model heat flow is calculated using thermal conductivity of 3.0 W/(m K). Previous models have not been able to explain heat flow in excess of 100 mW/m^2, and the interpretation of such high values in terms of a "steady state" geotherm has been controversial. Our models provide a good fit to the high heat flow in arcs and the large heat flow gradient toward the forearc. An artifact of specifying a subducting plate with constant dip is that the heat flow maximum in our models with temperature-dependent viscosity appears to occur very close to the "trench". Models with a more realistic plate geometry with low dip at shallow depths, steepening below ~ 50 km, should produce a more realistic distance between the trench and the heat flow maximum.

In presenting these results, we do not mean to suggest that we "know" that the mantle wedge in all subduction zones has an asthenospheric viscosity between 10^{17} and 10^{21} Pa s, with $Q/(RT_\infty)$ of 20. As noted in the previous paragraph, we have not yet explored the effects of varying slab dip. Our models do not incorporate shear heating and stress-dependent viscosity (unlike the models of *van Keken et al.* [2002]), nor have we incorporated pressure-dependent viscosity. Further consideration could also be given to specifying a suitable fault geometry in the upper 50 km of the subduction zone. Perhaps most importantly, our models are two-dimensional, whereas there is abundant evidence that mantle flow in the wedge is three-dimensional.

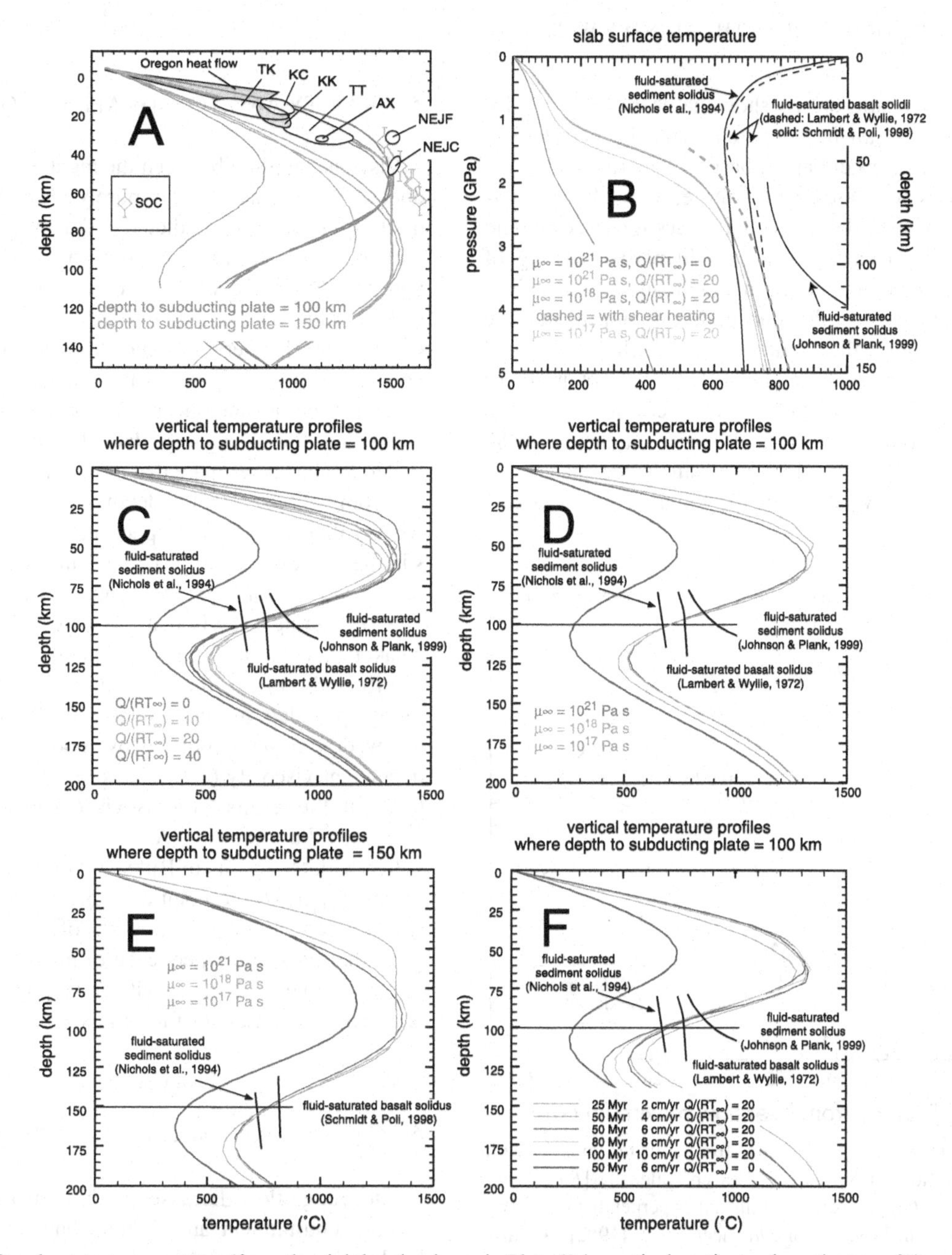

Plate 4. Steady state temperatures (from the righthand column in Plate 3) in vertical sections where the top of the subducting plate is at a depths of 100 or 150 km (Panels A and C-F), and along the top of the subducting plate (Panel B). Panels A-E are for a subducting plate thermal thickness of 75 km (50 Myr) and a convergence rate of 60 km/Myr. Panel A shows selected results compared to constraints on arc temperature discussed in the text (100 km sections in red, 150 km in blue). Thin lines are for isoviscous models with μ_∞ = 10^{21} Pa s. Bold lines are for μ_∞ = 10^{21}, 10^{18}, and 10^{17} Pa s, all with $Q/(RT_\infty)$ = 20. Clearly, a large variety of thermal models with temperature-dependent viscosity satisfy the constraints on arc lower crust and shallow mantle temperature. Panels B-F also show experimental, fluid-saturated solidii for sediment [*Johnson and Plank*, 1999; *Nichols et al.*, 1994] and basalt [*Lambert and Wyllie*, 1972; *Schmidt and Poli*, 1998]. Panels C-E compare an isoviscous model with μ_∞ = 10^{21} Pa s to models with different values of μ_∞ and $Q/(RT_\infty)$. Panel C emphasizes variation in activation energy in a section where the subducting plate is at 100 km depth, while panels D and E emphasize variation in "asthenospheric" viscosity in sections with subducting plate depth of 100 and 150 km, respectively. Panel *F* shows the effect of varying subducting plate age and subduction velocity in a section with subducting plate depth of 100 km. In Panel F, all cases have μ_∞ =10^{21} Pa s, and all but the isoviscous case have $Q/(RT_\infty)$ = 20.

However, our results clearly show that thermal models for subduction zones *can* be consistent with the petrological estimates for temperature at the base of the crust and in the upper mantle beneath arcs. Furthermore, some of these thermal models, using widely accepted values for viscosity and thermal activation energy, can explain partial melting of subducted sediment and basalt in subduction zones. This is the case despite the fact that our models do not incorporate shear heating along the "fault zone", which would yield higher subduction zone temperatures. We obtain these results for subduction of 50 Myr old oceanic crust at 60 km/Myr, and even for 80 Myr old crust subducting at 80 km/Myr, suggesting that melting of subducted material is not restricted to slow subduction of young oceanic crust.

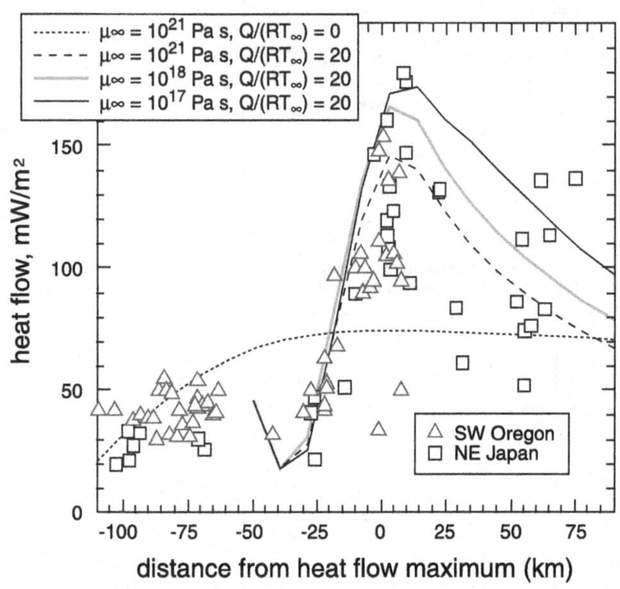

Figure 4. Comparison of heat flow from selected models with observed heat flow in NE Japan (data compiled by [*Furukawa*, 1993a]) and SW Oregon (digitized distances perpendicular to 100 mW/m^2 contour in Figure 2 of *Blackwell et al.* [1982]. All distances are in km from the maximum observed or calculated heat flow. Models with temperature-dependent viscosity account for the high heat flow values observed at the arc and for the large variation in heat flow from the forearc to the maximum in the arc. The isoviscous model clearly does not explain these heat flow characteristics. Our models do not incorporate a realistic shallow plate geometry. Therefore the location of the trench and the dimensions of the forearc region are not well described in the models. For example, the distance from the "trench" to the heat flow maximum is only ~ 50 km, whereas in most actual subduction zones, e.g. Oregon and Japan, it is much greater. The shallow plate geometry should not greatly affect the temperature distribution and flow pattern in the mantle wedge, which account for the high arc heat flow shown here.

5. GRAVITY AND TOPOGRAPHY IN OCEANIC ARCS

Viscous coupling between the mantle wedge and the subducting plate is assumed in purely plate-driven, isoviscous models for arcs, and predicted in dynamical models for arcs with mantle viscosities $\geq 10^{20}$ Pa s. However, coupling between the mantle wedge and the dense, subducting plate should produce a Bouguer gravity low over arcs [*Zhong and Gurnis*, 1992], whereas no such lows are observed. This discrepancy can be resolved if the mantle wedge has a lower viscosity, permitting mechanical decoupling between the wedge and the subducting plate [*Billen and Gurnis*, 2001]. Billen & Gurnis estimate that a region in the wedge extending from ~ 20 to $\geq$ 100 km depth, with a viscosity $\geq$ 10 times weaker than the ambient upper mantle viscosity (~ 10^{20} Pa s in their parameterization), yields the best fit to topographic and gravity data for the Tonga-Kermadec and Aleutian arcs. Lowering the viscosity to values 100 to 1000 times weaker than the ambient upper mantle viscosity does not significantly affect their results. Thus, the gravity and topography data are consistent with the low viscosities in the wedge corner produced by models with temperature-dependent viscosity (e.g., Plates 6B and 6C), and inconsistent with the results for an isoviscous wedge with a viscosity of 10^{21} Pa s in Plate 3A.

In detail, the results of Billen & Gurnis imply that the low viscosity part of the mantle wedge should be restricted to shallow depths, within ~ 200 km of the surface. This suggests that pressure dependence of mantle viscosity may be important in explaining gravity and topography in arcs. This is an important area for future investigation.

6. CAVEATS

6a. Compositional Convection in the Wedge?

The wedge flow discussed in this paper is driven only by viscous entrainment and thermal buoyancy, and not from the buoyancy of retained melt and/or reduced Fe/Mg in the residues of melting. Such compositional convection has been predicted in numerical studies of subduction zone magmatism [*Iwamori*, 1997; *Davies and Stevenson*, 1992]. Buoyancy forces arising from the presence of low density melt within its mantle source are very sensitive to assumptions regarding the permeability of intergranular melt networks in mantle peridotite. High permeabilities will yield efficient melt transport and little melt buoyancy. Geophysical and geochemical studies, even at very fast-spreading mid-ocean ridges suggest that very little melt is retained in the mantle [e.g., *Forsyth et al.*, 1998a]. As a result, we suspect that permeabilities are high within partially molten mantle peridotite, at least within those regions with an adiabatic or hotter geothermal gradient, in which ascending melt can dissolve minerals and increase permeability downstream [e.g., *Kelemen et al.*, 1995]. This implies that convection driven by buoyancy due to the presence of melt within partially molten aggregates in the mantle wedge is unlikely.

A variety of evidence favor the hypothesis that adiabatic decompression of the mantle is important in the genesis of some arc lavas [e.g., *Elkins Tanton et al.*, 2001; *Sisson and Bronto*, 1998; *Plank and Langmuir*, 1988; and Langmuir & Spiegelman, pers. comm. 2002], and this is sometimes taken as evidence for vertical diapirism due to melt buoyancy [e.g., *Tatsumi et al.*, 1983]. However, models for arcs that incorporate temperature-dependent viscosity [e.g., *van Keken et al.*, 2002; *Furukawa*, 1993a; *Furukawa*, 1993b] predict an upward component of mantle flow behind and beneath the arc, in the absence of melt-driven convection. Thus, adiabatic decompression melting in arcs does not require vertical diapirism due to melt buoyancy.

6b. Analytical Corner Flow Models, ARC Position, and ARC Magma Composition

As mentioned in Section 1, analytical, isoviscous corner flow calculations have long been a popular and simple way to estimate flow and temperature in the mantle wedge above subduction zones. The results of these physically simple calculations are strikingly similar to those of more complicated numerical calculations, which accounts for the lasting appeal of the analytical approach. In this paper, however, we argue that analytical, isoviscous corner flow calculations, together with their more complicated numerical siblings, systematically fail to predict the first-order temperature distribution in the shallow mantle and lowermost crust beneath arcs.

Recent work has demonstrated that simple corner flow solutions provide insight into subduction-related magmatism. England [*England*, 2001] found that analytical estimates of temperature in the mantle wedge are correlated with the depth to the subducting plate beneath arcs and with the major element composition of arc basalts. A similar correlation is expected in models like ours that incorporate realistic mantle viscosity variation. The wedge thermal structure, while necessarily different than in isoviscous models, is expected to have a qualitatively similar dependence on parameters such as age of subducting plate and subduction rate. From correlations like those observed, we would infer that the temperature distribution in the mantle wedge is, indeed, a strong control on the location of arcs and the composition of lavas. This should be a fertile area for future research.

7. CONCLUSION

In sections 2 and 3, we show that published thermal models fail to account for a variety of important constraints on the temperature and viscosity structure of the mantle wedge beneath arcs. A clear alternative is that thermal boundary layers at the base of arc crust and along the subduction zone are thinner than has been assumed or calculated in models thus far. This would satisfy constraints from petrology on the temperature in the wedge at the base of arc crust and at ~ 45 km depth within the mantle wedge. Also, temperatures in subducted metabasalt and metasediment could rise above the fluid-saturated solidus, permitting partial melting if fluid is available.

In section 4, we show that thermal models incorporating temperature-dependent viscosity, using widely accepted values for activation energy and asthenospheric viscosity, produce a thin thermal boundary layer beneath the arc, and an overall thermal structure that is consistent with petrological constraints on temperature in arc lower crust and shallow mantle. Some of these models are hot enough to allow for partial melting of subducted sediment and/or basalt, even though we have not incorporated the effect of shear heating. We obtain these results for subduction of 50 Myr old oceanic crust at 60 km/Myr, and even for faster subduction of older crust, suggesting that melting of subducted crust may not be restricted to slow subduction of very young oceanic crust.

Acknowledgments. We are very grateful to Greg Hirth, Magali Billen and Jack Whitehead for extensive advice on mantle rheology, dynamical modeling, and arc magmatism. Deiping Zhao and Max Schmidt shared unpublished results and thoughts on wedge temperature and partial melting of subducted sediment and basalt. Peter van Keken generously provided a preprint of a paper using a non-linear rheology to model wedge convection. We thank Karen Hanghøj for help in compiling the results from thermal models and arc thermobarometry in Plate 1. Sue DeBari, Laurel Burns, and many others guided us in field work in the lower crustal exposures of the Talkeetna arc section. Kelemen, Mehl and Hacker were supported by NSF EAR-9910899. Kelemen was also supported in part by NSF EAR-0087706 and EAR-9814632.

REFERENCES

Allègre, C.J., R. Montigny, and Y. Bottinga, Cortège ophiolitique et cortège océanique, géochimie comparée et mode de genèse, *Bull. Soc. géol. France, 15*, 461-477, 1973.

Anczkiewicz, R., and D. Vance, Isotopic constraints on the evolution of metamorphic conditions in the Jijal-Patan complex and the Kamila Belt of the Kohistan arc, Pakistan Himalaya, *Geol. Soc. Spec. Pub., 170*, 321-331, 2000.

Bartels, K.S., R.J. Kinzler, and T.L. Grove, High pressure phase relations of primitive high-alumina basalts from Medicine Lake Volcano, Northern California, *Contrib. Mineral. Petrol.*, *108*, 253-270, 1991.

Berman, R.G., Internally consistent thermodynamic data for stoichiometric minerals in the system Na_2O-K_2O-CaO-MgO-FeO-Fe_2O_3-Al_2O_3-SiO_2-TiO_2-H_2O-CO_2, *J. Petrol.*, *29*, 445-522, 1988.

Berman, R.G., Mixing properties of Ca-Mg-Fe-Mn garnets, *Am. Min.*, *75*, 328-344, 1990.

Berman, R.G., Thermobarometry using multiequilibrium calculations: A new technique with petrologic applications, *Can. Min.*, *29*, 833-855, 1991.

Billen, M.I., and M. Gurnis, A low viscosity wedge in subduction zones, *Earth Planet. Sci. Lett.*, *193*, 227-236, 2001.

Blackwell, D.D., R.G. Bowen, D.A. Hull, J. Riccio, and J.L. Steele, Heat flow, arc volcanism, and subduction in northern Oregon, *J. Geophys. Res.*, *87*, 8735-8754, 1982.

Braun, M.G., G. Hirth, and E.M. Parmentier, The effects of deep damp melting on mantle flow and melt generation beneath mid-ocean ridges, *Earth Planet. Sci. Lett.*, *176*, 339-356, 2000.

Brenan, J.M., H.F. Shaw, and R.J. Ryerson, Experimental evidence for the origin of lead enrichment in convergent margin magmas, *Nature*, *378*, 54-56, 1995a.

Brenan, J.M., H.F. Shaw, R.J. Ryerson, and D.L. Phinney, Mineral - aqueous fluid partitioning of trace elements at 900°C and 2.0 GPa: Constraints on trace element chemistry of mantle and deep crustal fluids, *Geochim. Cosmochim. Acta*, *59*, 3331-3350, 1995b.

Burns, L.E., The Border Ranges ultramafic and mafic complex, south-central Alaska: Cumulate fractionates of island arc volcanics, *Can. J. Earth Sci.*, *22*, 1020-1038, 1985.

Carslaw, H.S., and J.C. Jaeger, *Conduction of Heat in Solids*, Oxford University Press, 510p., 1959.

Christensen, U., Convection with pressure- and temperature-dependent non-Newtonian rheology, *Geophys. J. R. Astron. Soc.*, *77*, 343-384, 1984.

Class, C., D.L. Miller, S.L. Goldstein, and C.H. Langmuir, Distinguishing melt and fluid components in Umnak Volcanics, Aleutian Arc, *Geochemistry, Geophysics, Geosystems (G-cubed)*, *1*, 2000.

Conder, J.A., D.A. Weins, and J. Morris, On the decompression melting structure at volcanic arcs and back-arc spreading centers, *Geophys. Res. Lett.*, *29*, 15, 4 pp., 2002.

Davies, J.H., The role of hydraulic fractures and intermediate-depth earthquakes in generating subduction-zone magmatism, *Nature*, *398*, 142-145, 1999.

Davies, J.H., and D.J. Stevenson, Physical model of source region of subduction zone volcanics, *J. Geophys. Res.*, *97*, 2037-2070, 1992.

DeBari, S., S.M. Kay, and R.W. Kay, Ultramafic xenoliths from Adagdak volcano, Adak, Aleutian Islands, Alaska: deformed igneous cumulates from the Moho of an island arc, *J. Geol.*, *95*, 329-341, 1987.

DeBari, S.M., and R.G. Coleman, Examination of the deep levels of an island arc: Evidence from the Tonsina ultramafic-mafic assemblage, Tonsina, Alaska, *J. Geophys. Res.*, *94* (B4), 4373-4391, 1989.

Defant, M.J., and M.S. Drummond, Derivation of some modern arc magmas by melting of young subducted lithosphere, *Nature*, *347*, 662-665, 1990.

Defant, M.J., and P. Kepezhinskas, Evidence suggests slab melting in arc magmas, *EOS*, *82*, 65-69, 2001.

Draper, D.S., and A.D. Johnston, Anhydrous P-T phase relations of an Aleutian high-MgO basalt: An investigation of the role of olivine-liquid reaction in the generation of arc high-alumina basalts, *Contrib. Mineral. Petrol.*, *112*, 501-519, 1992.

Drummond, M.S., and M.J. Defant, A model for trondjhemite-tonalite-dacite genesis and crustal growth via slab melting, *J. Geophys. Res.*, *95*, 21,503-21,521, 1990.

Elkins Tanton, L.T., G. T. L, and J. Donnelly Nolan, Hot shallow mantle melting under the Cascades volcanic arc, *Geology*, *submitted*, 2001.

Elliott, T., T. Plank, A. Zindler, W. White, and B. Bourdon, Element transport from slab to volcanic front at the Mariana Arc, *J. Geophys. Res.*, *102*, 14,991-15,019, 1997.

England, P.C., Abstract T22D-11, Why are the arc volcanoes where they are?, *EOS*, *82*, F1156, 2001.

Forsyth, D.W., D.S. Scheirer, S.C. Webb, L.M. Dorman, J.A. Orcutt, A.J. Harding, D.K. Blackman, J. Phipps Morgan, R.S. Detrick, Y. Shen, C.J. Wolfe, J.P. Canales, D.R. Toomey, A.F. Sheehan, S.C. Solomon, and W.S.D. Wilcock, Imaging the deep seismic structure beneath a mid-ocean ridge; the MELT experiment, *Science*, *280* (5367), 1215-1218, 1998a.

Forsyth, D.W., S.C. Webb, L.M. Dorman, and Y. Shen, Phase velocities of Rayleigh waves in the MELT experiment on the East Pacific Rise, *Science*, *280* (5367), 1235-1238, 1998b.

Furukawa, Y., Depth of the decoupling plate interface and thermal structure under arcs, *J. Geophys. Res.*, *98*, 20,005-20,013, 1993a.

Furukawa, Y., Magmatic processes under arcs and formation of the volcanic front, *J. Geophys. Res.*, *98*, 8309-8319, 1993b.

Furukawa, Y., and Y. Tatsumi, Melting of a subducting slab and production of high-Mg andesite magmas: Unusual magmatism in SW Japan at 13 approximately 15 Ma, *Geophys. Res. Lett.*, *26* (15), 2271-2274, 1999.

Gough, S.J., M.P. Searle, D.J. Waters, and M.A. Khan, Igneous crystallization, high pressure metamorphism, and subsequent tectonic exhumation of the Jijal and Kamila complexes, Kohistan, *Abstracts: 16th Himalaya-Karakorum-Tibet Workshop, Austria, in J. Asian Earth Sciences*, *19*, 23-24, 2001.

Grove, T.L., S.W. Parman, S.A. Bowring, R.C. Price, and M.B. Baker, The role of H_2O-rich fluids in the generation of primitive basaltic andesites and andesites from the Mt. Shasta region, N. California, *Contrib. Mineral. Petrol.*, *in press*, 2001.

Hawkesworth, C.J., S.P. Turner, F. McDermott, D.W. Peate, and P. van Calsteren, U-Th isotopes in arc magmas: Implications for element transfer from the subducted crust, *Science*, *276*, 551-555, 1997.

Hirth, G., and D. Kohlstedt, Rheology of the mantle wedge, in *AGU Monograph Series*, this volume.

Hirth, G., and D.L. Kohlstedt, Experimental constraints on the dynamics of the partially molten upper mantle 2. Deformation in the dislocation creep regime, *J. Geophys. Res.*, *100*, 15,441-15,449, 1995a.

Hirth, G., and D.L. Kohlstedt, Experimental constraints on the dynamics of the partially molten upper mantle: Deformation in the diffusion creep regime, *J. Geophys. Res.*, *100*, 1981-2001, 1995b.

Hirth, G., and D.L. Kohlstedt, Water in the oceanic upper mantle: Implications for rheology, melt extraction and the evolution of the lithosphere, *Earth Planet. Sci. Lett.*, *144*, 93-108, 1996.

Holbrook, S.W., D.Lizarralde, S.McGeary, N. Bangs, and J. Diebold, Structure and composition of the Aleutian island arc and implications for continental crustal growth, *Geology*, *27*, 31-34, 1999.

Holland, T.J.B., and R. Powell, An internally consistent thermodynamic data set for phases of petrological interest, *J. metamorphic Geol.*, *16*, 309-343, 1988.

Iwamori, H., Heat sources and melting in subduction zones, *J. Geophys. Res.*, *102*, 14,803-14,820, 1997.

Jan, M.Q., Geochemistry of amphibolites from the southern part of the Kohistan arc, N. Pakistan, *Min. Mag.*, *52*, 147-159, 1988.

Jan, M.Q., and R.A. Howie, Ortho- and clinopyroxenes from the pyroxene granulites of Swat Kohistan, northern Pakistan, *Min. Mag.*, *43*, 715-726, 1980.

Jan, M.Q., and A. Karim, Coronas and high-P veins in metagabbros of the Kohistan island arc, northern Pakistan; Evidence for crustal thickening during cooling, *J. metamorphic Geol.*, *13*, 357-366, 1995.

Jarrard, R.D., Relations among subduction parameters, *Rev. Geophys.*, *24*, 217-284, 1986.

Johnson, M.C., and T. Plank, Dehydration and melting experiments constrain the fate of subducted sediments, *Geochemistry, Geophysics, Geosystems (G-cubed)*, *1*, 1999.

Kay, R.W., Aleutian magnesian andesites: Melts from subducted Pacific ocean crust, *J. Volc. Geotherm. Res.*, *4*, 117-132, 1978.

Kelemen, Peter B., Gene M. Yogodzinski, and David W. Scholl, Along-strike variation in the Aleutian Island Arc: Genesis of high Mg# andesite and implications for continental crust, this volume.

Kelemen, P.B., and E. Aharonov, Periodic formation of magma fractures and generation of layered gabbros in the lower crust beneath oceanic spreading ridges, in *Faulting and Magmatism at Mid-Ocean Ridges, Geophysical Monograph 106*, edited by W.R. Buck, P.T. Delaney, J.A. Karson, and Y. Lagabrielle, pp. 267-289, Am. Geophys. Union, Washington DC, 1998.

Kelemen, P.B., G. Hirth, N. Shimizu, M. Spiegelman, and H.J.B. Dick, A review of melt migration processes in the asthenospheric mantle beneath oceanic spreading centers, *Phil. Trans. Roy. Soc. London*, *A355*, 283-318, 1997.

Kelemen, P.B., N. Shimizu, and V.J.M. Salters, Extraction of mid-ocean-ridge basalt from the upwelling mantle by focused flow of melt in dunite channels, *Nature*, *375*, 747-753, 1995.

Keppler, H., Constraints from partitioning experiments on the composition of subduction-zone fluids, *Nature*, *380*, 237240, 1996.

Kincaid, C., and I.S. Sacks, Thermal and dynamical evolution of the upper mantle in subduction zones, *J. Geophys. Res.*, *102*, 12,295-12,315, 1997.

Kinzler, R.J., Melting of mantle peridotite at pressures approaching the spinel to garnet transition: Application to mid-ocean ridge basalt petrogenesis, *J. Geophys. Res.*, *102*, 853-874, 1997.

Kinzler, R.J., and T.L. Grove, Primary magmas of mid-ocean ridge basalts 1. Experiments and methods, *J. Geophys. Res.*, *97*, 6885-6906, 1992a.

Kinzler, R.J., and T.L. Grove, Primary magmas of mid-ocean ridge basalts 2. Applications, *J. Geophys. Res.*, *97*, 6907-6926, 1992b.

Kinzler, R.J., and T.L. Grove, Corrections and further discussion of the primary magmas of mid-ocean ridge basalts, 1 and 2, *Journal of Geophysical Research*, *98*, 22,339-22,347, 1993.

Kohlstedt, D.L., H. Keppler, and D.C. Rubie, Solubility of water in alpha, beta, and gamma phases of $(Mg,Fe)_2SiO_4$, *Contrib. Mineral. Petrol.*, *123*, 345-357, 1996.

Lambert, I., B., and P. Wyllie, J., Melting of gabbro (quartz eclogite) with excess water to 35 kilobars, with geological applications, *J. Geol.*, *80*, 693-708, 1972.

Langmuir, C.L., E.M. Klein, and T. Plank, Petrological systematics of mid-ocean ridge basalts: Constraints on melt generation beneath ocean ridges, in *Mantle flow and melt generation at mid-ocean ridges, AGU Monograph 71*, edited by J. Phipps Morgan, D.K. Blackman, and J.M. Sinton, pp. 183-280, American Geophysical Union, Washington DC, 1992.

Lieberman, J., and K. Petrakakis, TWEEQU thermobarometry: Analysis of uncertainties and applications to granulites from western Alaska and Austria, *Can. Min.*, *29*, 857-887, 1991.

Mareschal, J.-C., and G. Bergantz, Constraints on thermal models of the Basin and Range province, *Tectonophysics*, *174*, 137-146, 1990.

McKenzie, D., and M.J. Bickle, The volume and composition of melt generated by extension of the lithosphere, *J. Petrol.*, *29*, 625-679, 1988.

Mei, S., and D.L. Kohlstedt, Influence of water on plastic deformation of olivine aggregates 2. Dislocation creep regime, *J. Geophys. Res.*, *105*, 21471-21481, 2000.

Molnar, P., and P. England, Temperatures, heat flux, and frictional stress near Major thrust taults, *J. Geophys. Res.*, *95*, 4833-4856, 1990.

Molnar, P., and P. England, Temperatures in zones of steady state underthrusting of young oceanic lithosphere, *Earth and Planetary Science Letters*, *131*, 57-70, 1995.

Nichols, G.T., P.J. Wyllie, and C.R. Stern, Subduction zone melting of pelagic sediments constrained by melting experiments, *Nature*, *371*, 785-788, 1994.

Peacock, S.M., Fluid processes in subduction zones, *Science*, *248* (4953), 329-337, 1990a.

Peacock, S.M., Numerical simulation of metamorphic pressure-temperature-time paths and fluid production in subducting slabs, *Tectonics*, *9* (5), 1990b.

Peacock, S.M., Numerical simulation of subduction zone pressure-temperature-time paths: Constraints on fluid production and arc magmatism, in *Phil. Trans. Roy. Soc. London A 335: The Behaviour and Influence of Fluids in Subduction Zones*, edited by J. Tarney, K.T. Pickering, R.J. Knipe, and J.F. Dewey, pp. 341-353, London, United Kingdom, 1991.

Peacock, S.M., Thermal and petrologic structure of subduction zones, in *Subduction Zones, Top to Bottom: Geophysical Monograph 96*, edited by G.E. Bebout, D.W. Scholl, S.H. Kirby, and J.P. Platt, pp. 119-133, Am. Geophys. Union, Washington DC, 1996.

Peacock, S.M., Thermal structure and metamorphic evolution of subducting slabs, in press.

Peacock, S.M., and R.D. Hyndman, Hydrous minerals in the mantle wedge and the maximum depth of subduction thrust earthquakes, *Geophys. Res. Lett.*, *26*, 2517-2520, 1999.

Peacock, S.M., T. Rushmer, and A.B. Thompson, Partial melting of subducting oceanic crust, *Earth Planet. Sci. Lett.*, *121*, 227-244, 1994.

Peacock, S.M., and K. Wang, Seismic consequences of warm versus cool subduction zone metamorphism: Examples from northeast and southwest Japan, *Science*, *286*, 937-939, 1999.

Pearce, J.A., S.R. van der Laan, R.J. Arculus, B.J. Murton, T. Ishii, D.W. Peate, and I.J. Parkinson, Boninite and harzburgite from Leg 125 (Bonin-Mariana Forearc): A case study of magmagenesis during the initial stages of subduction, *Proc. ODP, Scientific Results*, *125*, 623-659, 1992.

Phipps Morgan, J., Melt migration beneath mid-ocean spreading centers, Geophys. Res. Lett., 14, 1238-1241, 1987.

Plafker, G., W.J. Nokleberg, and J.S. Lull, Bedrock geology and tectonic evolution of the Wrangellia, Peninsular, and Chugach terranes along the trans-Alaska crustal transect in the Chugach Mountains and Southern Copper River Basin, Alaska, *J. Geophys. Res.*, *94* (B4), 4255-4295, 1989.

Plank, T., Geochemistry; the brine of the Earth, *Nature*, *380*, 202-203, 1996.

Plank, T., and C.H. Langmuir, An evaulation of the global variations in the major element chemistry of arc basalts, *Earth Planet. Sci. Lett.*, *90*, 349-370, 1988.

Plank, T., and C.H. Langmuir, Tracing trace elements from sediment input to volcanic output at subduction zones, *Nature*, *362*, 739-743, 1993.

Plank, T., and C.H. Langmuir, The chemical composition of subducting sediment and its consequences for the crust and mantle, *Chem. Geol.*, *145*, 325-394, 1998.

Ponko, S.C., and S.M. Peacock, Thermal modeling of the southern Alaska subduction zone: Insight into the petrology of the subducting slab and overlying mantle wedge, *J. Geophys. Res.*, *100*, 22,117-22,128, 1995.

Rapp, R.P., N. Shimizu, M.D. Norman, and G.S. Applegate, Reaction between slab-derived melts and peridotite in the mantle wedge: Experimental constraints at 3.8 GPa, *Chem. Geol.*, *160*, 335-356, 1999.

Reddy, J.N., *An Introduction to the Finite Element Method*, 684 pp., McGraw-Hill, New York, 1993.

Ringuette, L., J. Martignole, and B.F. Windley, Magmatic crystallization, isobaric cooling, and decompression of the garnet-bearing assemblages of the Jijal sequence (Kohistan terrane, western Himalayas), *Geology*, *27*, 139-142, 1999.

Schmidt, M.W., and S. Poli, Experimentally based water budgets for dehydrating slabs and consequences for arc magma generation, *Earth and Planetary Science Letters*, *163* (1-4), 361-379, 1998.

Sisson, T.W., and S. Bronto, Evidence for pressure-release melting beneath magmatic arcs from basalt at Galunggung, Indonesia, *Nature*, *391*, 883-836, 1998.

Sisson, T.W., and T.L. Grove, Temperatures and H_2O contents of low MgO high-alumina basalts, *Contrib. Mineral. Petrol.*, *113*, 167-184, 1993.

Sisson, T.W., and G.D. Layne, H_2O in basalt and basaltic andesite glass inclusions from four subduction-related volcanoes, *Earth and Planetary Science Letters*, *117*, 619-635, 1993.

Smolarkiewicz, P.K., A fully multidimensional positive definite advection transport algorithm with small implicit diffusion, *J. Comp. Physics*, *54*, 325-362, 1983.

Sparks, D.W., and E.M. Parmentier, Melt extraction from the mantle beneath mid-ocean ridges, *Earth Planet. Sci. Lett.*, *105*, 368-377, 1991.

Spiegelman, M., and D. McKenzie, Simple 2-D models for melt extraction at mid-ocean ridges and island arcs, *Earth Planet. Sci. Let.*, *83*, 137-152, 1987.

Tahirkheli, R.A.K., M. Mattauer, F. Proust, and P. Tapponnier, The India-Eurasia suture zone in northern Pakistan; synthesis and interpretation of recent data at plate scale, in *Geodynamics of Pakistan*, edited by A. Farah, and K.A. DeJong, pp. 125-130, Geological Survey of Pakistan, Quetta, Pakistan, 1979.

Tamura, Y., Y. Tatsumi, D. Zhao, Y. Kido, and H. Shukuno, Abstract T31F-07, Hot fingers in the mantle wedge: New insights into magma genesis in subduction zone, *Earth and Planetary Science Letters,* 197, 1-2, 105-116, 2002.

Tatsumi, Y., N. Ishikawa, K. Anno, K. Ishizaka, and T. Itaya, Tectonic setting of high-Mg andesite magmatism in the SW Japan Arc: K-Ar chronology of the Setouchi volcanic belt, *Geophys. J. International*, *144* (3), 625-631, 2001.

Tatsumi, Y., M. Sakuyama, H. Fukuyama, and I. Kushiro, Generation of arc basalt magmas and thermal structure of the mantle wedge in subduction zones, *J. Geophys. Res.*, *88*, 5815-5825, 1983.

Treloar, P.J., Pressure-temperature-time paths and the relationship between collision, deformation and metamorphism in the northwest Himalaya, *Geol. J.*, *30*, 333-348, 1995.

Treloar, P.J., P.J. O'Brien, and M.A. Khan, Exhumation of early Tertiary, coesite-bearing eclogites from the Kaghan valley, Pakistan Himalaya, *Abstracts: 16th Himalaya-Karakorum-Tibet Workshop, Austria, in J. Asian Earth Sciences*, *19*, 68-69, 2001.

van Keken, P.E., B. Kiefer, and S.M. Peacock, High resolution models of subduction zones: Implications for mineral dehydration reactions and the transport of water into the deep mantle, *G-cubed*, *3*, 10, 20 pp. 2002.

White, R.S., D. McKenzie, and R.K. O'Nions, Oceanic crustal thickness from seismic measurements and rare earth element inversions, *Journal of Geophysical Research, 97*, 19,683-19,715, 1992.

Yamamoto, H., Contrasting metamorphic P-T-time paths of the Kohistan granulites and tectonics of the western Himalayas, *J. Geol. Soc. London, 150*, 843-856, 1993.

Yang, H.-J., R.J. Kinzler, and T.L. Grove, Experiments and models of anhydrous, basaltic olivine-plagioclase-augite saturated melts from 0.001 to 10 kbar, *Contrib. Mineral. Petrol., 124*, 1-18, 1996.

Yogodzinski, G.M., R.W. Kay, O.N. Volynets, A.V. Koloskov, and S.M. Kay, Magnesian andesite in the western Aleutian Komandorsky region: Implications for slab melting and processes in the mantle wedge, *Geol. Soc. Amer. Bull., 107* (5), 505-519, 1995.

Yogodzinski, G.M., and P.B. Kelemen, Slab melting in the Aleutians: implications of an ion probe study of clinopyroxene in primitive adakite and basalt, *Earth Planet. Sci. Lett., 158*, 53-65, 1998.

Yogodzinski, G.M., J.M. Lees, T.G. Churikova, F. Dorendorf, G. Woeerner, and O.N. Volynets, Geochemical evidence for the melting of subducting oceanic lithosphere at plate edges, *Nature, 409*, 500-504, 2001.

Yogodzinski, G.M., O.N. Volynets, A.V. Koloskov, N.I. Seliverstov, and V.V. Matvenkov, Magnesian andesites and the subduction component in a strongly calc-alkaline series at Piip Volcano, Far Western Aleutians, *J. Petrol., 35* (1), 163-204, 1994.

Yoshino, T., H. Yamamoto, T. Okudaira, and M. Toriumi, Crustal thickening of the lower crust of the Kohistan arc (N. Pakistan) deduced from Al-zoning in clinopyroxene and plagioclase, *J. metamorphic Geol., 16*, 729-748, 1998.

Zhao, D., and A. Hasegawa, P wave tomographic imaging of the crust and upper mantle beneath the Japan Islands, *J. Geophys. Res., 98* (3), 4333-4353, 1993.

Zhao, D., and A. Hasegawa, Teleseismic evidence for lateral heterogeneities in the northeastern Japan arc, *Tectonophysics, 237* (3-4), 189-199, 1994.

Zhao, D., A. Hasegawa, and S. Horiuchi, Tomographic imaging of P and S wave velocity structure beneath northeastern Japan, *J. Geophys. Res., 97* (13), 19,909-19,928, 1992a.

Zhao, D., S. Horiuchi, and A. Hasegawa, Seismic velocity structure of the crust beneath the Japan Islands, *Tectonophysics, 212* (3-4), 289-301, 1992b.

Zhao, D., X. Yingbiao, D.A. Weins, L. Dorman, J. Hildebrand, and S. Webb, Depth extent of the Lau back-arc spreading center and its relation to subduction processes, *Science*, 254-257, 1997.

Zhong, S., and M. Gurnis, Viscous flow model of a subduction zone with a faulted lithosphere: Long and short wavelength topography, gravity and geoid, *Geophys. Res. Lett., 19*, 1891-1894, 1992.

Peter B. Kelemen, Dept. of Geology and Geophysics, Woods Hole Oceanographic Institution, Woods Hole, MA 02543. (peterk@whoi.edu)

Jennifer L. Rilling and E.M. Parmentier, Dept. of Geological Sciences, Brown University, Providence, RI 02912.

Luc Mehl and Bradley R. Hacker, Dept. of Geological Sciences & Institute for Crustal Studies, University of California, Santa Barbara, CA 93106-9630.